Handbook of
Polymer Degradation

Environmental Science and Pollution Control Series

1. Toxic Metal Chemistry in Marine Environments, *Muhammad Sadiq*
2. Handbook of Polymer Degradation, *edited by S. Halim Hamid, Mohamed B. Amin, and Ali G. Maadhah*
3. Unit Processes in Drinking Water Treatment, *Willy J. Masschelein*
4. Groundwater Contamination and Analysis at Hazardous Waste Sites, *edited by Suzanne Lesage and Richard E. Jackson*
5. Plastics Waste Management: Disposal, Recycling, and Reuse, *edited by Nabil Mustafa*
6. Hazardous Waste Site Soil Remediation: Theory and Application of Innovative Technologies, *edited by David J. Wilson and Ann N. Clarke*
7. Process Engineering for Pollution Control and Waste Minimization, *edited by Donald L. Wise and Debra J. Trantolo*
8. Remediation of Hazardous Waste Contaminated Soils, *edited by Donald L. Wise and Debra J. Trantolo*
9. Water Contamination and Health: Integration of Exposure Assessment, Toxicology, and Risk Assessment, *edited by Rhoda G. M. Wang*
10. Pollution Control in Fertilizer Production, *edited by Charles A. Hodge and Neculai N. Popovici*
11. Groundwater Contamination and Control, *edited by Uri Zoller*
12. Toxic Properties of Pesticides, *Nicholas P. Cheremisinoff and John A. King*
13. Combustion and Incineration Processes: Applications in Environmental Engineering, Second Edition, Revised and Expanded, *Walter R. Niessen*
14. Hazardous Chemicals in the Polymer Industry, *Nicholas P. Cheremisinoff*
15. Handbook of Highly Toxic Materials Handling and Management, *edited by Stanley S. Grossel and Daniel A. Crow*
16. Separation Processes in Waste Minimization, *Robert B. Long*
17. Handbook of Pollution and Hazardous Materials Compliance: A Sourcebook for Environmental Managers, *Nicholas P. Cheremisinoff and Nadelyn Graffia*
18. Biosolids Treatment and Management, *Mark J. Girovich*
19. Biological Wastewater Treatment: Second Edition, Revised and Expanded, *C. P. Leslie Grady, Jr., Glen T. Daigger, and Henry C. Lim*
20. Separation Methods for Waste and Environmental Applications, *Jack S. Watson*
21. Handbook of Polymer Degradation, Second Edition, Revised and Expanded, *S. Halim Hamid*
22. Bioremediation of Contaminated Soils, *edited by Donald L. Wise, Debra J. Trantolo, Edward J. Cichon, Hilary I. Inyang, and Ulrich Stottmeister*

Additional Volumes in Preparation

Handbook of
Polymer Degradation
Second Edition, Revised and Expanded

edited by

S. Halim Hamid
King Fahd Institute of Petroleum & Minerals
Dhahran, Saudi Arabia

MARCEL DEKKER, INC.   NEW YORK · BASEL

Library of Congress Cataloging-in-Publication Data

Handbook of polymer degradation / edited by S. Halim Hamid. — 2nd ed., rev. and expanded.
  p. cm. — (Environmental science and pollution control ; 21)
  Includes index.
  ISBN 0-8247-0324-3 (alk.paper)
  1. Polymers—Deterioration. I. Hamid, S. Halim. II. Series.

QD381.9.D47 H36 2000
620.1'920422—dc21   00-024049

This book is printed on acid-free paper.

Headquarters
Marcel Dekker, Inc.
270 Madison Avenue, New York, NY 10016
tel: 212-696-9000; fax: 212-685-4540

Eastern Hemisphere Distribution
Marcel Dekker AG
Hutgasse 4, Postfach 812, CH-4001 Basel, Switzerland
tel: 41-61-261-8482; fax: 41-61-261-8896

World Wide Web
http://www.dekker.com

The publisher offers discounts on this book when ordered in bulk quantities. For more information, write to Special Sales/Professional Marketing at the headquarters address above.

Copyright © 2000 by Marcel Dekker, Inc.   All Rights Reserved.

Neither this book nor any part may be reproduced or transmitted in any form or by any means, electronic or mechanical, including photocopying, microfilming, and recording, or by any information storage and retrieval system, without permission in writing from the publisher.

Current printing (last digit):
10 9 8 7 6 5 4 3 2 1

PRINTED IN THE UNITED STATES OF AMERICA

To H. E. Dr. Bakr Abdullah Bakr
my mentor, who always inspires me to take challenges

Preface to the Second Edition

Since the publication of the first edition of the *Handbook of Polymer Degradation* in 1992, a multitude of developments in the manufacturing, processing, application, and use of polymers have occurred. Keeping this in view and encouraged by the success of the first edition, it was decided to bring out the revised and expanded second edition, covering a broad spectrum of topics in this ever-developing and broadening field of polymer degradation and stabilization. Focusing on the basics of photo- and biodegradability, as well as on environmental issues engendered by increased use of polymers in many industries, this edition exhaustively examines the life cycles of polymers from the most current theoretical and practical perspectives.

The present volume is composed of 20 contributions from over 30 experts of international repute. It contains a wealth of up-to-date information on polymer degradation and stabilization for environmental, health, and materials scientists; polymer, plastics, and chemical engineers; and upper-level undergraduate and graduate students in these disciplines. The broad spectrum of authorship from 11 countries represents the global nature and concerns of the subject matter. Each chapter attempts to achieve balance between theory and applications. The reader may find some overlap in theoretical aspects between chapters, but this is useful and necessary from the point of view of completeness.

One of the significant features of this volume is the authorship of several chapters by world authorities who have provided the essence of lifelong investigations of their subjects. For example, François Gugumus' chapter on polyolefin and agricultural greenhouse film stabilization presents an excellent blend of theoretical background and real-life application of polymeric film stabilization for extended lifetime. Peter Klemchuk's chapter on photodegradable polymers is the culmination of long years of theoretical and practical experience accumulated during his association with Ciba-Geigy. The list goes on, but I would also like to mention Jan Pospíšil, Guennadi Efremovich Zaikov, Ann-Christine Albertsson, Norma D. Searle, and Jacques Verdu.

For the sake of clarity, the contributions have been organized into four broad parts. The first part, "Additives and Stabilizers in Polymers," contains six chapters covering this important topic. The following part, "Mechanisms of Polymer Degradation and Stabilization," contains two chapters—one by George and Celina, and the other by Zaikov and coworkers. Six special topics have been covered in the chapters presented in Part III, "Controlled Degradation and Stabilization." The last part, "Wavelength Sensitivity of Polymers" was introduced in this edition to respond to the emerging questions of radiation effects, particularly in the light of ozone depletion and increased UV radiation with shifting wavelength.

The aim of this book is to present an up-to-date account of polymer degradation and stabilization. This volume is obviously an important reference monograph for researchers studying the stability of polymers for outdoor use, producers, end-users, quality controllers, environmentalists, and students who want to apply their knowledge in practical conditions.

I acknowledge the support of the King Fahd University of Petroleum & Minerals (KFUPM) for this project. I wish to express my gratitude to the Rector of KFUPM, H. E. Dr. Abdulaziz A. Al-Dukhayyil, for his encouragement. I would particularly like to thank Nihal Ahmad for his many suggestions and invaluable assistance, which helped me immensely in the successful completion of this project. I wish to express my appreciation to all the staff of Marcel Dekker, Inc., who did their jobs with utmost professionalism. Finally, to my wife and children, I give my deepest thanks for their understanding and forbearance during the development and completion of this book.

S. Halim Hamid

Preface to the First Edition

The subject of polymer degradation and stabilization is becoming increasingly complex because of its dedication to specialized polymeric applications. Rapid advances of polymer applications into new fields inevitably leave a significant gap between demand and inherent polymer limitations. We have made an attempt to include a chapter on each application detailing the behavior of polymers in a specific environment. This handbook contains 19 chapters on varied topics by internationally known authorities who have shared their experiences in the application of polymers to specialized fields.

Chapters were chosen primarily to cover the complete spectrum of polymer degradation and stabilization from the point of view of polymer producers, processors, end users, additives producers, designers, and environmentalists. In each chapter, wherever possible, theory is outlined to provide the user with a broad perspective on a particular topic. It is anticipated that the major users of this handbook will be researchers, but students and professors will also find it a unique and helpful reference book.

This handbook has been divided into three parts—general topics, degradability, and specialized topics. In the first part, we begin with a chapter on degradation and stabilization of polypropylene that discusses the structure and morphology of polypropylene and how these are influenced by radiation-induced degradation. Next, we turn to the susceptibility of polymers to UV radiation, which is well known. We focus on stratospheric ozone depletion, which generally has detrimental effects because of increased UV radiation. This chapter introduces new data and clarifies various aspects of the effects of increased UV radiation on polymers and includes measures to reduce these effects. In Chapter 3, chemical degradation of polymers is discussed in detail with emphasis on various modes of chemical degradation and its impact on applications. Polymer blends are gaining popularity because of economical and technical advantages. Chapter 4 includes various aspects of degradation of polymer blends. Next, artificially accelerated weathering and its relationship to natural weather are authoritatively explained by Musa Kamal and Bing Huang. Following this is a rather detailed treatment of the general concepts of photodegradation of polymers, which is vital to the understanding of weathering and its effects. Weather cannot be changed but polymers can be made compatible with weather. This topic is covered at length with experimental results in Chapter 7. Following this, various aspects of polymer stabilization are very well explained for different polymers by A. J. Chirinos-Padrón and N. S. Allen. Synergism and antagonism of stabilizers are included in Chapter 9.

In Part II, special attention has been accorded to the subject of environmental aspects and waste management of polymers because this is always a difficult concept to explain. Chapter 10 presents a critical review of photo- and bio-degradable plastics. A.-C. Albertsson's authoritative viewpoint on biodegradation of polymers is presented in this section. The material covered in Chapter 12 demonstrates various aspects of starch-based degradable plastics. The increased significance of environmental toxicology of plastics has required a chapter on this subject and includes legislative issues.

The chapters in Part III, on specialized topics, introduce new topics that are not currently available from other single literature sources. F. Henninger of Ciba-Geigy has contributed an excellent chapter on aspects of greenhouse film formulations. The durabililty of geomembranes and geotextiles is covered in Chapter 15. The chapter on application of polymers as a coating material, by George Mills, includes a rather complete discussion of polymer durability in various industrial applications. Medical and biomedical fields are incorporating polymers in a number of applications. We have included two chapters—one by experts from industry (Boehringer Mannheim) and the other from academia (Case Western Reserve University)—to provide a general view of polymer degradation in medical application and biodegradability of biomedical polymers. Polymeric insulators are replacing conventional ceramic insulation; however, the lifetime of polymeric insulators is dependent on weather-induced as well as electrical stresses. Chapter 18 covers different aspects of polymer degradation in insulation of high-voltage transmission.

We owe special thanks to the contributors for their enthusiastic support. This effort would not have been possible without their willingness to share valuable knowledge and experiences. As editors who have reason to know and appreciate the cost of their contributions in time and effort, we join with others in the profession in acknowledging our indebtedness to them and to the universities, companies, and organizations they serve.

We would like to take this opportunity to express our sincere thanks and appreciation to Dr. Bakr A. Bakr, Rector of KFUPM, for his encouragement and personal interest in this project. Our thanks are also due to Dr. Abdallah Dabbagh, Director of the Research Institute, for his constant support. With sincere appreciation we acknowledge the assistance of the administrative affairs department of our institute, particularly its past and present managers. Mr. Mahmoud Sourani and Mr. Hamza Garatly. Abdullah Aitani's critical observations helped in improving the final product, and we thank him for his comments. Our special thanks and appreciation go to Nihal Ahmad, without whose persistent help and long hours this book could not have taken final shape.

S. Halim Hamid
Mohamed B. Amin
Ali G. Maadhah

Contents

IV. Wavelength Sensitivity of Polymers

Contributors

Laurence Achimsky, Ph.D. Materials Department, École Nationale Supérieure d'Arts et Métiers (ENSAM), Paris, France

Naim Akmal, Ph.D. Business Analytical Specialist, Department of Hydrocarbons, Union Carbide Technical Center, South Charleston, West Virginia

Ann-Christine Albertsson, Ph.D. Professor, Department of Polymer Technology, Royal Institute of Technology, Stockholm, Sweden

Anthony L. Andrady, Ph.D. Senior Research Scientist, Chemistry and Life Sciences, Research Triangle Institute, Research Triangle Park, North Carolina

Ludmila Audouin, Ph.D., Docteur d'État Associate Professor, Materials Department, École Nationale Supérieure d'Arts et Métiers (ENSAM), Paris, France

Santosh Kumar Awasthi, Ph.D.* General Manager, Research Center, Indian Petrochemicals Corporation Limited, Vadodara, Gujarat, India

Ingmar Bauer, Ph.D. Staff Chemist, Institute of Organic Chemistry, Dresden University of Technology, Dresden, Germany

Mathew Celina, Ph.D. Staff, Research and Development, Department of Aging and Reliability, Bulk Materials, Sandia National Laboratories, Albuquerque, New Mexico

Evguenii T. Denisov, D.Sc. Professor, Department of Kinetics and Catalysts, Institute of Problems of Chemical Physics, Russian Academy of Sciences, Chernogolovka, Russia

Richard M. Fischer, Ph.D. Division Scientist, Weathering Resource Center, 3M Corporation, St. Paul, Minnesota

Jean-Luc Gardette, Sc.D. Professor, Laboratoire de Photochimie, Université Blaise Pascal (Clermont-Ferrand), Aubière, France

---

* Deceased.

Graeme A. George, Ph.D. Professor, Faculty of Science, Queensland University of Technology, Brisbane, Australia

François Louis Gugumus, Ph.D.* Consultant, Ciba Specialty Chemicals Limited, Basel, Switzerland

Ludmila Nikolaevna Guseva, Ph.D. Senior Scientist, Department of Chemical and Biological Kinetics, N. M. Emanuel Institute of Biochemical Physics, Russian Academy of Sciences, Moscow, Russia

Wolf D. Habicher, Sc.D. Lecturer, Institute of Organic Chemistry, Dresden University of Technology, Dresden, Germany

S. Halim Hamid, Ph.D. Associate Professor, Department of Chemical Engineering, and Manager, Petroleum Refining and Petrochemicals Research Institute, King Fahd University of Petroleum & Minerals, Dhahran, Saudi Arabia

Ikram Hussain, M.S. Research Engineer and Project Manager, Center for Refining and Petrochemicals, King Fahd University of Petroleum & Minerals, Dhahran, Saudi Arabia

Warren D. Ketola, B.S. Division Scientist, Traffic Control Materials Division, 3M Corporation, St. Paul, Minnesota

Peter P. Klemchuk, Ph.D. Polymer Group, Institute of Materials Science, University of Connecticut, Storrs, Connecticut

Anand Kumar Kulshreshtha, Ph.D. Senior Manager, Research Center, Indian Petrochemicals Corporation Limited, Vadodara, Gujarat, India

Yurii Arsenovich Mikheev, Ph.D., D.Sc. Head, Chain Reactions in Polymers Group, Department of Chemical and Biological Kinetics, N. M. Emanuel Institute of Biochemical Physics, Russian Academy of Sciences, Moscow, Russia

Stanislav Nešpůrek, Ph.D., D.Sc. Professor, Chemical Faculty, Technical University of Brno, Brno, and Head, Department of Electronic Phenomena, Institute of Macromolecular Chemistry, Academy of Sciences of the Czech Republic, Prague, Czech Republic

James E. Pickett, Ph.D. Staff Chemist, Corporate Research and Development, General Electric Company, Schenectady, New York

Alexander Yakovlevich Polishchuk, Ph.D. Head, Transport Phenomena in Polymers Group, Department of Chemical and Biological Kinetics, N. M. Emanuel Institute of Biochemical Physics, Russian Academy of Sciences, Moscow, Russia

---

* Retired.

Jan Pospíšil, Ph.D., D.Sc. Chief Research Fellow, Department of Electronic Phenomena, Institute of Macromolecular Chemistry, Academy of Sciences of the Czech Republic, Prague, Czech Republic

Norma D. Searle, Ph.D. Consultant, Plastics and Chemicals, Deerfield Beach, Florida

Ayako Torikai, Ph.D. Assistant Professor, Department of Applied Chemistry, Graduate School of Engineering, Nagoya University, Nagoya, Japan

Arthur M. Usmani, Ph.D. Chief Scientific Officer, Biomedical Research Department, ALTEC USA, Indianapolis, Indiana

Jacques Verdu, Docteur d'État Professor, École Nationale Supérieure d'Arts et Métiers (ENSAM), Paris, France

Guennadi Efremovich Zaikov, Ph.D., D.Sc. Professor and Deputy Director, Department of Chemical and Biological Kinetics, N. M. Emanuel Institute of Biochemical Physics, Russian Academy of Sciences, Moscow, Russia

Handbook of
Polymer Degradation

I
ADDITIVES AND STABILIZERS IN POLYMERS

1

Polyolefin Stabilization: From Single Stabilizers to Complex Systems

FRANÇOIS LOUIS GUGUMUS

Ciba Specialty Chemicals Limited, Basel, Switzerland

I. INTRODUCTION

Polypropylene (PP) needs protection in every stage of its life cycle. It starts with storage, immediately after manufacturing. Then, a small amount of a phenolic antioxidant usually gives sufficient protection. It continues with processing, during which an adequate stabilization is a prerequisite for minimizing degradation of PP in the molten state at temperatures between 200° and 300°C. It ends with a suitable stabilization for the application foreseen. This is a stabilization against thermal oxidation only, if PP is not to be exposed to light. It will be a stabilization against photothermal oxidation if ultraviolet (UV) light exposure is involved. The requirements for polyethylene (PE) are similar to those for PP, but the stabilizer levels are usually lower because PE is inherently less sensitive to oxidative attack than PP.

Processing stabilization is usually achieved with combinations of high molecular mass phenolic antioxidants with phosphites or phosphonites. These aspects will not be developed here. For details the reader is referred to other publications (1–4). Further developments in this respect have been reported more recently (5,6).

Long-term heat stabilization traditionally involved high molecular mass phenolic antioxidants. More recently, it was recognized that high molecular mass or polymeric hindered amine light stabilizers (HALS) are even more efficient for thermo-oxidative stabilization of polyolefins in the solid state. In the following, the evolution of UV stabilization of polyolefins from the beginning to the actual trends will be documented. Then, some aspects of thermal stabilization with HAS will be developed.

The polymers used were either unstabilized commercial resins stored in a refrigerated room to minimize oxidation, or polymers already containing a small

1

amount of a phenolic antioxidant for storage stabilization. The most important characteristics of the polymers as well as the basic stabilization are indicated in the tables and figures. The antioxidants used for the basic stabilization were commercial products, as identified in the Appendix. The antacid used was usually Ca stearate (Merck). The light stabilizers were also commercial products, the structures are given in the Appendix. The concentrations of the additives are expressed in weight percent unless otherwise stated.

The accelerated-exposure devices used in the experiments were equipped exclusively with xenon lamps. The backing of the tapes and multifilaments in the accelerated-exposure devices was white cardboard.

Natural weathering was performed in Florida, the samples facing south with an angle of 45°C, in direct sunlight, without glass filter. The energy received is about 140 kLy/yr (mean value). The samples were mounted without tension on transparent poly(methyl methacrylate) plates, unless otherwise stated.

Long-term heat aging (LTHA) was performed in forced draft air ovens (HORO, Model 080 V/EL).

Mechanical testing (i.e., determination of elongation, tensile strength, impact strength, and tensile impact strength) was in agreement with the procedures outlined in American Society for Testing and Materials (ASTM) D 638, D 882, and D 1822.

For LTHA, embrittlement times were assessed, in part, visually, by periodic control of the samples in the ovens. The results are expressed in time to embrittlement (in days). For other samples, the test criterion was brittleness on bending, and the corresponding results are reported as time to brittleness.

II. STABILIZATION OF THIN SECTIONS

Light stability of polyolefins has been improved successively in the past decades as more and more powerful light stabilizers became available. At first mainly UV absorbers (UVAs) were available. The most important classes, the benzophenone and the benzotriazole type UV absorbers are still used today. The stability imparted by UV absorbers was improved when nickel (Ni) stabilizers such as Ni-1 were developed. With the development of the benzoate (B_z)-type light stabilizers such as Bz-1 another improvement was achieved, especially for PP. However, before the benzoates were used commercially on a large scale, a new class of light stabilizers was introduced more than 20 years ago, the hindered amine light stabilizers (HALS) (7–9). Since these early times of HALS, much work has been devoted to the elucidation of the performance and mechanisms (3,4,10–29, and references cited therein).

At the beginning, only low molecular mass HALS were available. HALS-1 is the most typical and is still the most used representative of this class of compounds. To circumvent some disadvantages associated with low molecular mass stabilizers (e.g., high migration rate and moderate resistance to extraction), polymeric HALS were developed. HALS-2 and HALS-3 were the first polymeric HALS commercially available. An additional advantage of high molecular mass HALS, becoming increasingly important in the last years, is their pronounced contribution to long-term thermal stability (LTTS) of polyolefins.

In the last years, many new hindered amine-type compounds were developed. Some of these compounds show advantages in very specific applications. However, the UV protection conferred by these new compounds is usually of the same order

Table 1 Influence of UV Stabilizer Type on Light Stability of PP Tapes[a]

	E_{50} (kLy) Florida[b]	
UV stabilization	PP-1 (2nd-generation)	PP-2 (3rd-generation)
Control	30	56
0.5% UVA-1	65	90
0.5% Bz-1	140	165
0.5% Ni-1	65	90
0.5% UVA-4	70	80
0.05% HALS-1	205	370
0.10% HALS-3	205	420

[a] PP homopolymer + 0.1% AO-3 + 0.1% Ca stearate.
[b] Florida exposure started July 1986 for PP-1 and August 1990 for PP-2.
E_{50}: energy (kLy) to 50% retained tensile strength.

as that of the compounds that have already been available for many years. Notably, no fundamentally new class of compounds showing an efficiency comparable with that of the hindered amines has yet been commercialized. This readily explains why the most important developments in UV stabilization of polyolefins are based on combinations of UV stabilizers.

A. PP Tapes

The data of Table 1 illustrate the performance in PP tapes of the various aforementioned light stabilizer classes. Thus, on natural-weathering in Florida, the benzotriazole UV absorber UVA-1, the benzophenone UV absorber UVA-4, and the nickel stabilizer Ni-1 show comparable efficiency. The benzoate Bz-1 is significantly better than the foregoing stabilizers, confirming that a marked improvement of UV stability had been achieved with the development of this stabilizer class. However, the low molecular mass HALS, HALS-1, yields an even better light stability than Bz-1, although it is used at a concentration ten times lower (0.05 vs. 0.5%). The superiority of polymeric HALS, such as HALS-3, is less pronounced than that of the low molecular mass HALS-1. Nevertheless, it is still considerable. The data obtained with third-generation PP confirm those obtained with second-generation polymer. The main difference between both polymer types consists in the significantly higher stability level reached with third-generation PP.

This result is confirmed by an additional series of data presented in Table 2. The corresponding data generated on artificial exposure of the same tape series are also shown in Table 2. It can be seen from this table that, on artificial exposure, there is no superiority of the third-generation polymer. This confirms previous results in which the superiority of third-generation PP over second-generation PP was also restricted to natural weathering (3). The comparison of the data from natural and accelerated weathering shows once more that extrapolation from artificial weathering to natural weathering must be done with due care, even if the artificial weathering device is equipped with filtered xenon lamps (13).

Table 2 Influence of UV Stabilizer and PP Type on Light Stability of PP Tapes[a,b]

| UV stabilization | T$_{50}$ (h) Xenotest 1200[c] | | E$_{50}$ (kLy) Florida[d] | |
	PP-A (2nd-generation)	PP-B (3rd-generation)	PP-A (2nd-generation)	PP-B (3rd-generation)
Control	600	530	24	56
0.5% UVA-1	1120	1020	40	95
0.5% Bz-1	2900	1970	130	160
0.5% UVA-4	1920	1080	50	75
0.05% HALS-1	3020	2100	135	400
0.10% HALS-3	3340	4080	90	440

[a] PP homopolymer + 0.1% AO-3 + 0.1% Ca stearate.
[b] Tapes 50-μm thick, stretch ratio 1:6.
[c] Xenotest 1200, black panel temperature ~53°C, no water spraying.
[d] Florida exposure started April 1990 for PP-A and August 1991 for PP-B.
T$_{50}$, time (h) to 50% retained tensile strength; E$_{50}$, energy (kLy) to 50% retained tensile strength.

Table 3 Contribution of Various UV Stabilizers to the Performance of HALS-1 in PP Tapes[a,b]

UV stabilization	T$_{50}$ (h)[c]
Control	530
0.1% HALS-1	4900
0.2% HALS-1	7800
0.1% HALS-1 + 0.1% UVA-2	4000
0.1% HALS-1 + 0.1% UVA-4	4500
0.1% HALS-1 + 0.1% Bz-1	5200

[a] PP homopolymer PP-6 + 0.1% Ca stearate + 0.1% AO-3.
[b] Tapes: 50-μm thick, stretch ratio 1:5.25.
[c] Weather-ometer Ci 65: black panel temperature 63 ± 2°C, no water spraying.
T$_{50}$, time (h) to 50% retained tensile strength.

The performance of the combinations of various UV stabilizers with HALS-1 on artificial exposure is presented in Table 3. The data show quite clearly that the benzotriazole-type UV absorber UVA-2 and the benzophenone-type UV absorber UVA-4 do not contribute to the UV stability already conferred by HALS-1. On the contrary, the results even point to a slight antagonistic effect. It is only the benzoate Bz-1 that gives an improvement of UV stability when added on top of HALS-1. However, this contribution of Bz-1 is significantly lower than that of the same amount of HALS-1. These results are in line with previous results generated with second-generation PP (3,29).

Thus, the combination of two UV stabilizers from two distinct classes of UV stabilizers does not yield any synergistic effect. The combination of two UV stabilizers from which no synergistic effect could be expected, a priori, notably two HALS, gives the most astonishing results. The use of combinations of low mol-

Table 4 Performance of a Combination Low–High Molecular Mass HALS in PP Tapes[a,b]

HALS (0.2%)	E_{50} (kLy) Florida[c]	Days of embrittlement at 120°C
None	30	46
HALS-1	460	41
HALS-5	370	115
HALS-3	190	163

[a] PP homopolymer PP-7 + 0.1% AO-3 + 0.1% Ca stearate.
[b] Tapes 50-μm thick, stretch ratio 1:6.
[c] Florida exposure started in March 1982.
E_{50}, energy (kLy) to 50% retained tensile strength.

Table 5 Performance of a Combination Low–High Molecular Mass HALS in PP Tapes[a,b]

HALS (0.1%)	E_{50} (kLy) Florida[c]	Days to embrittlement at 120°C
None	60	49
HALS-1	565	57
HALS-5	520	101
HALS-3	360	130

[a] PP homopolymer PP-2 (3rd-generation) + 0.1% AO-3 + 0.1% Ca stearate.
[b] Tapes 50-μm thick, stretch ratio 1:6.
[c] Florida exposure started in November 1990.
E_{50}, energy (kLy) to 50% retained tensile strength.

ecular mass HALS with polymeric or nonpolymeric high molecular mass HALS led to a significant improvement in polyolefin stabilization.

Data obtained with the 1:1 combination of the low molecular mass HALS, HALS-1 with the polymeric HALS, HALS-3, on natural weathering are presented in Table 4. One can see that the combination, HALS-5, not only shows significantly more than additive effects for light stability, but it also comes very close to the low molecular mass HALS used alone at the same concentration. An additional advantage of the combination HALS-5 resides in its contribution to long-term thermal stability of the PP tapes. Hence, HALS-5 gives the possibility to combine the best of the low molecular mass HALS (i.e., superior contribution to light stability) with the best of the polymeric HALS (i.e., superior contribution to long-term thermal stability).

These data were obtained with second-generation PP. The results of natural exposure of a third-generation PP are presented in Table 5. These data confirm those obtained previously with second-generation PP, but on a higher level. The data in Table 6 show that, from a concentration of 0.2% on, the combination HALS-5 is already equivalent to the low molecular mass HALS-1 used alone.

Table 6 Concentration Effect with a Combination
Low–High Molecular Mass HALS in PP Tapes[a,b]

	E_{50} (kLy) Florida[c]	
HALS	0.1% HALS	0.2% HALS
None	60	60
HALS-1	565	800
HALS-5	520	800
HALS-3	360	590

[a] PP homopolymer PP-5 (3rd-generation) + 0.1% AO-3 + 0.1% Ca
stearate.
[b] Tapes 50-μm thick, stretch ratio 1 : 6.
[c] Florida exposure started in November 1990.
E_{50}, energy (kLy) to 50% retained tensile strength.

The linear increase of HALS performance with their concentration, found
initially for second-generation PP (20,24), had already been extended to
third-generation PP on artificial weathering (3,29). This is confirmed once more with
the plot of Fig. 1 relative to Xenotest 1200 exposure. Furthermore, the preliminary
results of natural weathering in Florida reveal the same tendency. This is shown
in Fig. 2. Therefore, the expression of the lifetime of PP tapes deduced from previous
studies seems to be generally valid. It is given by the following relation:

$$T_{50} = t_i = (t_i)_0 + b \, (\text{HALS})_0$$

where $(t_i)_0$ is the lifetime in the absence of HALS (i.e., that of the base stabilization),
$(\text{HALS})_0$ the concentration of HALS and b is a proportionality factor.

The results obtained with HALS-5 are especially interesting because the combination of two secondary hindered amines is involved. Thus, the main difference
between the two components resides in the backbone structure and in the molecular
mass. Similar effects can be obtained combining the low molecular mass HALS,
HALS-1, with the polymeric HALS, HALS-2. With the last combination, in addition
to the difference in molecular mass, the substitution of the tetramethylpiperidine
nitrogen atom is also different.

The HALS combinations envisaged so far involve a low molecular mass HALS
and a polymeric HALS. It is even more unexpected that the combination of two
polymeric HALS, such as HALS-2 and HALS-3, also yields synergistic effects in
PP tapes. Thus, the data in Table 7 show the superiority of their 1 : 1 combination,
HALS-6, over both compounds at the same concentration.

Again, the combination also has a remarkable contribution to LTTS of the PP
tapes. In the example shown in Table 7, the contribution to thermal stability is not as
outstanding as the contribution to UV stability. However, the effect found is significantly more than additive. The effect of HALS concentration on performance
is shown in Table 8. It can be seen that the superiority of HALS-6 over its components holds in the whole concentration range. Preliminary results with
third-generation PP also yield significant superiority for HALS-6 over the com-

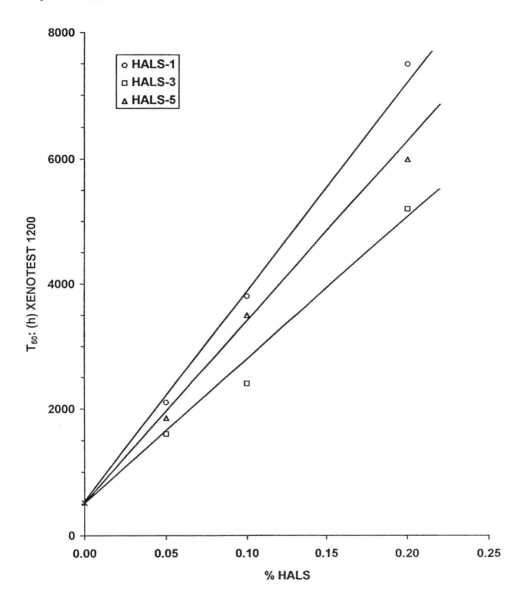

Figure 1 Effect of HALS type and concentration on UV stability of PP tapes on artificial exposure: PP-2 (3rd-generation) + 0.1% AO-3 + 0.1% Ca stearate.

ponents HALS-2 and HALS-3 (Table 9). Hence, the synergistic effect seems to be quite general with PP tapes.

B. PP Multifilaments

The light stabilizers used for PP multifilaments are overall the same as those used for PP tapes. Thus, HALS are state of the art in PP multifilaments as much as in PP

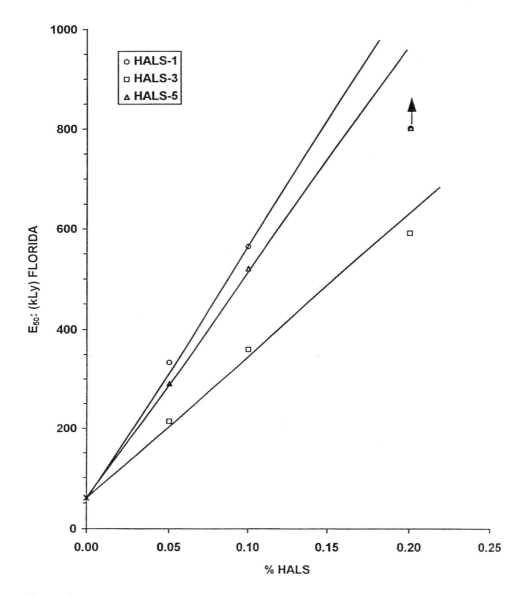

Figure 2 Effect of HALS type and concentration on UV stability of PP tapes on natural weathering: Florida exposure started in November 1990; PP-2 (3rd-generation) + 0.1% AO-3 + 0.1% Ca stearate.

tapes. However, with PP multifilaments, the choice of HALS is usually more restricted than with PP tapes. This restriction is because some applications of PP multifilaments involve somewhat drastic thermal treatments. These treatments can lead to more or less pronounced loss of low molecular mass HALS by migration or extraction. Data illustrating this behavior of low molecular mass HALS have been published many times (3,18,25). The same holds for the performance of HALS in PP

Table 7 Performance of Polymeric HALS in PP Tapes[a,b]

HALS (0.2%)	E_{50} (kLy) Florida[c]	Days to embrittlement at 120°C
None	36	45
HALS-2	155	120
HALS-3	235	276
HALS-6	260	216

[a] PP homopolymer PP-8 + 0.1% AO-3 + 0.1% Ca stearate.
[b] Tapes 50-μm thick, stretch ratio 1:6.
[c] Florida exposure started in August 1988.
E_{50}, energy (kLy) to 50% retained tensile strength.

Table 8 Performance of Polymeric HALS in PP Tapes[a,b]

HALS	E_{50} (kLy) Florida[c]			
	0.1% HALS	0.2% HALS	0.4% HALS	0.8% HALS
None	25	25	25	25
HALS-2	105	145	210	365
HALS-6	156	286	535	770
HALS-3	172	270	460	640

[a] PP homopolymer PP-8 + 0.1% AO-3 + 0.1% Ca stearate.
[b] Tapes 50-μm thick, stretch ratio 1:6.
[c] Florida exposure started in July 1988.
E_{50}, energy (kLy) to 50% retained tensile strength.

Table 9 Performance of Polymeric HALS in 3rd-Generation PP Tapes[a,b]

UV stabilization	T_{50} (h) Weather-ometer Ci 65[c]	E_{50} (kLy) Florida[d]
Control	565	65
0.05% HALS-2	1560	125
0.05% HALS-6	2120	280
0.05% HALS-3	2040	280
0.1% HALS-2	2640	225
0.1% HALS-6	3600	>400
0.1% HALS-3	3120	>400

[a] PP homopolymer PP-6 (3rd-generation) + 0.1% AO-3 + 0.1% Ca stearate.
[b] Tapes 50-μm thick, stretch ratio 1:6.
[c] Weather-ometer Ci 65, black panel temperature ∼63°C, no water spraying.
[d] Florida exposure started in August 1994.
T_{50}, time (h) to 50% retained tensile strength; E_{50}, energy (kLy) to 50% retained tensile strength.

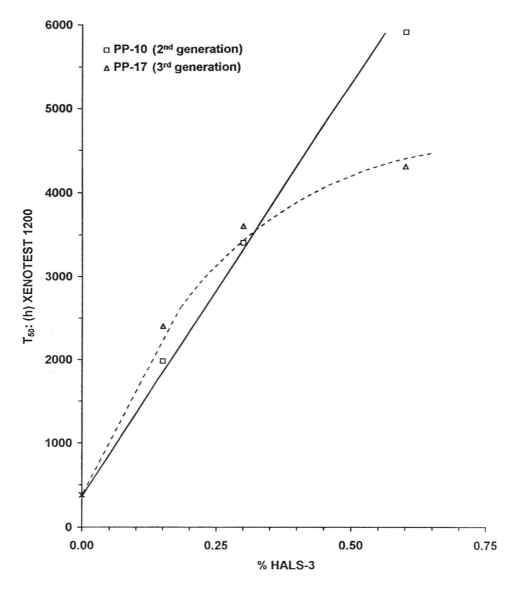

Figure 3 Effect of HALS-3 concentration and PP type on the UV stability of PP multifilaments: PP fiber grade $+0.2\%$ AO-6$+0.1\%$ Ca stearate$+0.25\%$ TiO$_2$ (rutile); multifilaments exposed without thermal treatment.

multifilaments on artificial exposure. It increases, often linearly, with HALS concentration (3,18,20,24,25). It is the same dependency as that found with PP tapes, as discussed in the foregoing. However, some data with PP multifilaments show linear increase of the performance with the square root of the HALS concentration (20,24). This difference in behavior was attributed to variations in polymer quality combined with the fact that the stabilizer concentrations used with PP multifilaments

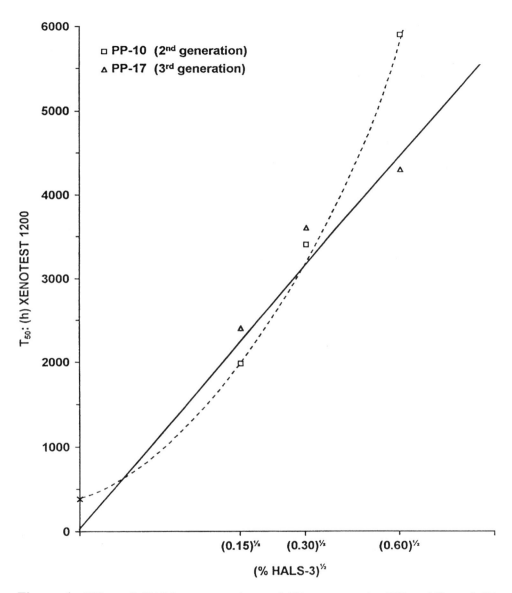

Figure 4 Effect of HALS concentration and PP type on the UV stability of PP multifilaments: PP fiber grade + 0.2% AO-6 + 0.1% Ca Stearate + 0.25% TiO_2 (rutile); multifilaments exposed without thermal treatment.

are usually significantly higher than the concentrations in PP tapes. Although there are not yet enough data, it appears as if the low amount of catalyst residues in third-generation PP would lead not only to a higher level of UV stability, but also to a linear increase of the UV stability with the square root, rather than directly with the HALS concentration. The corresponding behavior is illustrated in Figs. 3 and 4.

Table 10   Combination of Polymeric HALS in PP Multifilaments[a,b]

UV stabilization	T$_{50}$ (h) Xenotest 1200[c]	Days to embrittlement at 110°C
Control	450	10
0.3% HALS-2	3200	37
0.3% HALS-3	3600	103
0.3% HALS-6	4500	65

[a] PP fiber grade PP-9 + 0.1% AO-5 + 0.1% Ca stearate + 0.25% titanium dioxide (coated rutile).
[b] Multifilaments 130/37 exposed after thermal treatment (20 min at 120°C).
[c] Xenotest 1200, black panel temperature ∼53°C, no water spraying.
T$_{50}$, time (h) to 50% retained tensile strength.

Table 11   Combination of High Molecular Mass HALS in PP Multifilaments[a,b]

UV stabilization	T$_{50}$ (h) Xenotest 1200[c]
Control	250
0.6% HALS-2	3800
0.6% HALS-7	4850
0.6% HALS-4	4300

[a] PP fiber grade PP-10 + 0.1% AO-5 + 0.1% Ca stearate + 0.25% titanium dioxide (coated rutile).
[b] Multifilaments 130/37 exposed without thermal treatment.
[c] Xenotest 1200, black panel temperature ∼53°C, no water spraying.
T$_{50}$, time (h) to 50% retained tensile strength.
Source: Ref. 29.

As already shown for PP tapes, combinations of HALS can also be advantageous in PP multifilaments. However, because low molecular mass HALS may be lost on various thermal treatments of the PP multifilaments, mainly polymeric or other higher molecular mass HALS are used for the stabilization of such multifilaments. Hence, the use of combinations of low molecular mass HALS with polymeric HALS is not generally recommended. However, under special conditions, the performance of such combinations can be outstanding (3). Nevertheless, the combinations of polymeric HALS, such as HALS-2 and HALS-3, are more generally applicable. As with PP tapes, synergistic effects can be observed (Table 10).

Not only combinations of the tertiary HALS, HALS-2, with the secondary HALS, HALS-3, are interesting (3); the combination of two tertiary HALS such as HALS-2 and HALS-4 may also yield improved performance (Table 11). An additional advantage of such a combination (HALS-7) consists in reduced discoloration under special conditions ("gas fading"). With another experimental series, one can see (Table 12), that the synergism is also observed on natural weathering.

Table 12 Combination of High Molecular Mass
HALS in PP Multifilaments[a,b]

UV stabilization	E_{50} (kLy) Florida[c]
Control	7
0.6% HALS-2	125
0.6% HALS-7	190
0.6% HALS-4	185

[a] PP fiber grade $9+0.1\%$ AO-5 $+0.1\%$ Ca stearate $+0.25\%$
titanium dioxide (coated rutile).
[b] Multifilaments 130/37 exposed after thermal treatment
(20 min at 120°C).
[c] Florida exposure started in July 1991.
E_{50}, energy (kLy) to 50% retained tensile strength.

Table 13 Performance of Combinations HALS/UV Absorber in PE-HD Tapes[a,b]

UV stabilization	T_{50} (h)[c]	E_{50} (kLy) Florida[d]
Control	1170	105
0.05% HALS-1	7200	235
0.05% HALS-1 + 0.05% UVA-4	5800	190
0.10% HALS-1	9600	255(290[e])
0.10% HALS-1 + 0.10% UVA-4	7900	255
0.20% HALS-1	15950	370

[a] PE-HD-1 (Ti catalyst, $d=0.950$) $+0.05\%$ AO-1 $+0.1\%$ Ca stearate.
[b] Tapes 50-μm thick, stretch ratio 1:8.5.
[c] Weather-ometer WRC 600, black panel temperature ~ 53°C, no water spraying.
[d] Florida exposure started in October 1981.
[e] Figure deduced by interpolation.
T_{50}, time (h) to 50% retained tensile strength;
E_{50}, energy (kLy) to 50% retained tensile strength.
Source: Ref. 29.

C. PE-HD Tapes

The superiority of HALS over the other light stabilizer classes is as pronounced in
high-density PE (PE-HD) tapes as in PP tapes or PP multifilaments (3,18,25). It
was shown in the foregoing (see Table 3) that the addition of a benzotriazole type
UV absorber UVA-1 to HALS-1, contributed nothing to the light stability of
PP tapes. The data in Table 13 clearly show a significant detrimental effect of
the benzophenone-type UV absorber UVA-4 on the light stability conferred by
HALS-1, both on artificial and natural weathering. The combination of the low
molecular mass HALS-1 with the polymeric HALS-3 does not show such
detrimental effects. However, in a first series of experiments, summarized in Table
14, there was no performance advantage for HALS-5 over its constituents on natural
weathering. This must be attributed to the excessively high stretch ratio of the

Table 14 Performance of a Combination Low–High Molecular Mass HALS in PE-HD Tapes[a,b]

UV stabilization	T_{50} (h) Xenotest 1200[c]	E_{50} (kLy) Florida[d]
Control	1560	140
0.10% HALS-1	12400	550
0.10% HALS-5	12300	550
0.10% HALS-3	17500	600
0.20% HALS-1	17750	760
0.20% HALS-5	30000	700
0.20% HALS-3	27800	700

[a] PE-HD-2 (Ti catalyst, d = 0.945) + 0.05% AO-1 + 0.1% Ca stearate.
[b] Tapes 50-μm thick, stretch ratio 1 : 8.5.
[c] Xenotest 1200, black panel temperature ~53°C, no water spraying.
[d] Florida exposure started in March 1982.
T_{50}, time (h) to 50% retained tensile strength; E_{50}, energy (kLy) to 50% retained tensile strength.

Table 15 Performance of a Combination Low–High Molecular Mass HALS in PE-HD Tapes[a,b]

UV stabilization	E_{85} (kLy) Florida[c]
Control	25
0.05% HALS-1	205
0.05% HALS-5	255
0.05% HALS-3	200
0.10% HALS-1	260
0.10% HALS-5	300
0.10% HALS-3	200

[a] PE-HD-3 (Ti catalyst, d = 0.950) + 0.15% AO-4 + 0.1% Ca stearate; unpigmented.
[b] Tapes 50-μm thick, stretch ratio 1 : 6.
[c] Florida exposure started in August 1994.
E_{85}, energy (kLy) to 85% retained tensile strength.

PE-HD tapes used for the experiments (i.e., 1 : 8.5 instead of the more usual 1 : 5 to 1 : 6. In fact, the preliminary results of natural weathering of a new exposure series with tapes stretched to a more normal ratio, show a significant advantage of the combination HALS-5 over the components HALS-1 and HALS-3 used alone (Table 15).

Combinations of the polymeric light stabilizers HALS-2 and HALS-3 showed additive effects only on artificial exposure of highly stretched PE-HD tapes (3). The preliminary results of a new series of experiments show a distinct advantage for HALS-6 over both components on natural weathering (Table 16). The results also show that HALS-6 yields performance comparable with that obtained with HALS-5. As a preliminary conclusion, it can be stated that combinations of HALS are as advantageous for UV stabilization of PE-HD tapes as for PP tapes.

Table 16 Performance of Polymeric HALS in PE-HD Tapes[a,b]

UV stabilization	E_{85} (kLy) Florida[c]
Control	25
0.05% HALS-2	130
0.05% HALS-6	240
0.05% HALS-3	200
0.10% HALS-2	160
0.10% HALS-6	340
0.10% HALS-3	200

[a] PE-HD-3 (Ti catalyst, d = 0.950) + 0.15% AO-4 + 0.1% Ca stearate; unpigmented.
[b] Tapes 50-μm thick, stretch ratio 1 : 6.
[c] Florida exposure started in August 1994.
E_{85}, energy (kLy) to 85% retained tensile strength.

The performance of HALS in polyethylene usually increases proportionally to the square root of the concentration. This is illustrated for PE-HD tapes in Fig. 5.

The lifetime reached with HALS in PE-HD tapes can be expressed by the following relation:

$$T_{50} = t_i = b(\text{HALS})_0^{1/2}$$

where $(\text{HALS})_0$ stands for the initial HALS concentration and b is a proportionality factor. This dependence has been found many times with PE-HD tapes, both on artificial and natural exposure (20,24). It is quite different from that found for PP tapes when linear increase with HALS concentration seems mostly valid. The difference of behavior of HALS in PP and in PE-HD is explained by fundamentally different photooxidation, especially photoinitiation mechanisms (20, 24).

D. PE-LD Films

Combinations of UV stabilizers have long been used for the stabilization of low-density PE (PE-LD) films used for agricultural purposes (e.g., in greenhouse covers). The combination of the nickel stabilizer Ni-1 with the benzophenone-type UV absorber UVA-4 is the most typical example of this. Hence, it was straightforward, as soon as the polymeric HALS had been commercialized and used for stabilization of PE-LD, to combine them with benzophenone-type UV absorbers. It can be seen in Table 17 that, when used alone in PE-LD, HALS perform better than nickel stabilizers, which are more efficient than UV absorbers.

The performance of combinations with the benzophenone-type UV absorber UVA-4 is shown in Table 18 as a function of film thickness. Again, HALS perform better than the nickel quencher, and the effect of the UV absorber in the combination

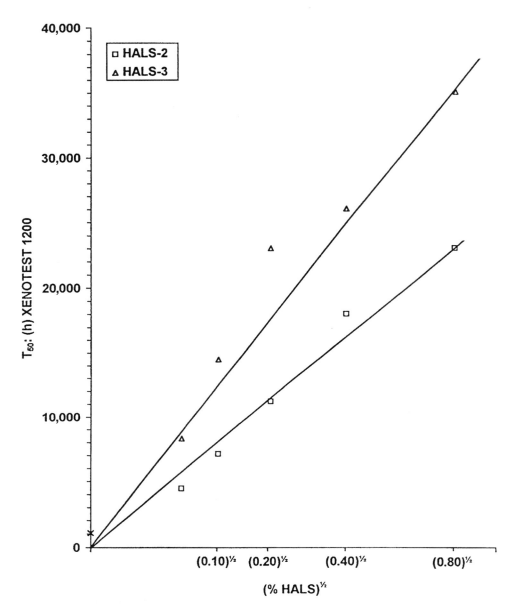

Figure 5 Effect of HALS structure and concentration on the UV stability of PE-HD tapes on artificial exposure: PE-HD-4 (Ziegler-type, d = 0.945) + 0.05% AO-1 + 0.1% Ca Stearate.

increases considerably with film thickness. More data with PE-LD films and the corresponding discussion can be found in Chapter 2 on greenhouse film stabilization.

The combination of polymeric HALS, HALS-6, also shows synergistic effects in PE-LD films. This is also shown in Chapter 2 for kaolin-containing PE-LD blown films. For concentrations above 0.3%, HALS-6 has practically the same performance as the best component under the conditions of the testing.

Table 17 Influence of Film Thickness and UV Stabilizer Type on Light Stability of PE-LD Blown Films[a]

Light stabilization	E_{50} (kLy) Florida[b]		
	50 μm	100 μm	200 μm
Control	27	31	34
0.15% UVA-4	40	55	90
0.15% Ni-1	70	115	140
0.15% HALS-2	105	120	170
0.15% HALS-3	195	260	450

[a] PE-LD homopolymer, film grade, MF(190°C/2.16 kg) = 0.3 + 0.03% AO-1.
[b] Florida exposure started in May 1980, without backing.
E_{50}, energy (kLy) to 50% retained elongation.

Table 18 Influence of Film Thickness and UV Stabilization System on Light Stability of PE-LD Blown Films[a]

Light stabilization	E_{50} (kLy) Florida[b]		
	50 μm	100 μm	200 μm
Control	27	31	34
0.15% Ni-1 + 0.15% UVA-4	115	155	210
0.15% HALS-2 + 0.15% UVA-4	105	190	295
0.15% HALS-3 + 0.15% UVA-4	200	295	535

[a] PE-LD homopolymer, film grade, MF(190°C/2.16 kg) = 0.3 + 0.03% AO-1.
[b] Florida exposure started in May 1980, without backing.
E_{50}, energy (kLy) to 50% retained elongation.

III. STABILIZATION OF THICK SECTIONS

Because absorption of UV radiation according to the Beer–Lambert law is one of the main mechanisms of stabilization by UV absorbers, it can be expected that UV absorbers will contribute increasingly to light stability when sample thickness is increased. However, this contribution of UV absorbers manifests itself mainly by a better protection of the deep layers of the polymer. Hence, it will result in a better protection of the polymer properties, mostly depending on these deep layers (i.e., some mechanical properties). It is obvious, that sample thickness has no direct effect on protection of the superficial layers and of the polymer characteristics depending on these superficial layers, such as gloss, chalking, surface roughness, and so on. Sample thickness can have a positive effect on surface protection in an indirect way if the stabilizers, protected from UV light in the deep-lying layers, are able to migrate to the surface layers. The protection mechanisms of UV absorbers that are not related to UV absorption (e.g., quenching of excited states) are also efficient in the superficial layers of the polymer samples.

Table 19 Performance of Combinations of HALS-3 with UV Absorbers in PP Plaques (2 mm)[a]

UV stabilization	E_{15} (kLy)[b]	$E_{chalking}$ (kLy)[b]
Control	14	40
0.10% HALS-3	135	130
0.10% HALS-3 + 0.10% UVA-2	205	190
0.10% HALS-3 + 0.10% UVA-4	190	190
0.20% HALS-3	215	310

[a] PP homopolymer PP-11 + 0.1% AO-3 + 0.1% Ca stearate.
[b] Florida exposure started in July 1980.
E_{15}, energy (kLy) to 15 kJ/m^2 retained impact strength; $E_{chalking}$, energy (kLy) to appearance of chalking.
Source: Ref. 29.

Table 20 Performance of Combinations of HALS-1 with UV Absorbers in PP Plaques (2 mm)[a]

UV stabilization	E_{15} (kLy)[b]	$E_{chalking}$ (kLy)[b]
Control	14	40
0.10% HALS-1	260	220
0.10% HALS-1 + 0.10% UVA-1	>450[c]	250
0.10% HALS-1 + 0.10% UVA-4	430	310
0.20% HALS-1	435	>450[c]

[a] PP homopolymer 10 + 0.1% AO-3 + 0.1% Ca stearate.
[b] Florida exposure started in July 1980.
[c] No samples left.
E_{15}, energy (kLy) to 15 kJ/m^2 retained impact strength; $E_{chalking}$, energy (kLy) to appearance of chalking.

A. PP Thick Sections

The general considerations discussed in the foregoing are illustrated in Table 19 summarizing data obtained in 2-mm–thick injection-molded PP plaques. The UV stability was evaluated according to two different criteria. They involve, on the one hand, impact strength (E_{15}) as representative for mechanical properties. On the other hand, they involve chalking ($E_{chalking}$) as an indication for the state of the surface of the polymer samples. The data in Table 19 are limited to the polymeric HALS-3 and its combinations with the benzotriazole- and benzophenone-type UV absorbers UVA-2 and UVA-4.

Table 19 shows that both UV absorbers contribute significantly to the preservation of the mechanical properties. This contribution is of the same magnitude as the contribution of an additional, equal quantity of HALS-3. The contribution of the UV absorbers to protection of the surface layers, (i.e., to inhibition of chalking) is less pronounced. In the example shown, the contribution of the UV absorbers is smaller than that obtained with the same additional amount of HALS-3.

The low molecular mass HALS, HALS-1, and the polymeric HALS, HALS-2, show overall behavior similar to that of HALS-3. This is shown in Tables 20

Table 21 Performance of Combinations of HALS-2 with UV Absorbers in PP Plaques (2 mm)[a]

UV stabilization	E_{15} (kLy)[b]	$E_{chalking}$ (kLy)[b]
Control	14	40
0.10% HALS-2	60	100
0.10% HALS-2 + 0.10% UVA-2	240	190
0.10% HALS-2 + 0.10% UVA-4	260	160
0.20% HALS-2	75	160

[a] PP homopolymer 10 + 0.1% AO-3 + 0.1% Ca stearate.
[b] Florida exposure started in July 1980.
E_{15}, energy (kLy) to 15 kJ/m^2 retained impact strength; $E_{chalking}$, energy (kLy) to appearance of chalking.

Table 22 Performance of Combinations of HALS-2 with UV Absorbers in PP Plaques (2 mm):[a] Influence of Pigments on Impact Strength

UV stabilization	E_{15} (kLy) Florida[b]		
	Unpigmented	0.5% titanium dioxide	0.5% phthalocyanine blue
Control	14	16	26
0.10% HALS-2	60	230	265
0.10% HALS-2 + 0.10% UVA-2	240	185	315
0.10% HALS-2 + 0.10% UVA-4	260	185	315
0.20% HALS-2	75	470	390

[a] PP homopolymer PP-11 + 0.1% AO-3 + 0.1% Ca stearate.
[b] Florida exposure started in July 1980.
E_{15}, energy (kLy) to 15 kJ/m^2 retained impact strength.
Source: Ref. 29.

and 21, respectively. By comparing the results one can see that the stability level reached depends on the HALS; in this connection, the protection conferred by HALS-2 to the mechanical properties is rather limited in unpigmented samples. Consequently, it is recommended that it be used in combination with a UV absorber in the absence of adequate pigments.

Table 22 shows the influence of pigments on retention of mechanical properties conferred by HALS-2 and its combinations with UV absorbers. In the absence of any pigment, the UV absorbers UVA-2 and UVA-4 contribute significantly to the preservation of the mechanical properties. However, in the presence of the pigments titanium dioxide or phthalocyanine blue, the contribution of the UV absorbers to retention of mechanical properties can be practically neglected. Anyway, this contribution is always inferior, often significantly inferior, to that of a same amount of HALS-2.

The comparison of the results obtained with HALS-2 as the only UV stabilizer, in the samples containing titanium dioxide, with the results obtained with the combinations HALS-2/UV absorbers in the unpigmented plaques is especially

Table 23 Performance of a Combination Low–High Molecular Mass HALS in PP Plaques (2 mm)[a]

UV stabilization	E_{50} (kLy)[b]	$E_{chalking}$ (kLy)[b]
Control	28	80
0.1% HALS-3	175	160
0.1% HALS-5	280	250
0.1% HALS-1	300	250

[a] PP homopolymer PP-12 + 0.1% AO-3 + 0.1% Ca stearate.
[b] Florida exposure started in June 1985.
E_{50}, energy (kLy) to 50% retained impact strength; $E_{chalking}$, energy (kLy) to appearance of chalking.
Source: Ref. 29.

Table 24 Performance of a Combination Low–High Molecular Mass HALS in PP Plaques (2 mm)[a]

UV stabilization	E_{50} (kLy) Florida[b]			
	No HALS	0.05% HALS	0.1% HALS	0.2% HALS
Control	28	—	—	—
0.1% HALS-1	—	170	300	500
0.1% HALS-5	—	175	280	570
0.1% HALS-3	—	110	175	280

[a] PP homopolymer 12 + 0.1% AO-3 + 0.1% Ca stearate.
[b] Florida exposure started in June 1985.
E_{50}, energy (kLy) to 50% retained impact strength.

instructive. It shows that the role of the UV absorber in the unpigmented plaques has been taken over quantitatively by the pigment, such that addition of a UV absorber becomes of no use. The effect of phthalocyanine blue is even more pronounced than that of titanium dioxide, especially with low stabilizer concentrations. These observations are valid only for retention of mechanical properties. If chalking is the test criterion chosen, it is with phthalocyanine blue that the results are the worst. The data discussed so far have been generated with PP homopolymer. The combination of UV absorbers with HALS yields analogous results with PP copolymers. Nevertheless, sometimes significant differences can be observed. However, this goes beyond the scope of this chapter.

Combinations of HALS with UV absorbers are used preferentially in thick sections. They present significant advantages in unpigmented samples. However, some combinations of low molecular mass HALS with polymeric HALS also yield synergistic effects in thick polymer samples. This is shown in Table 23 for HALS-5, the 1:1 combination of HALS-1 with HALS-3. It is practically equivalent to the low molecular mass HALS, HALS-1, used alone. This is true not only if retention of impact strength is taken into account, it is also valid for protection from chalking. Table 24 shows that for conservation of the mechanical properties, HALS-5 is significantly better than HALS-1 for a concentration as low as 0.2%. The performance

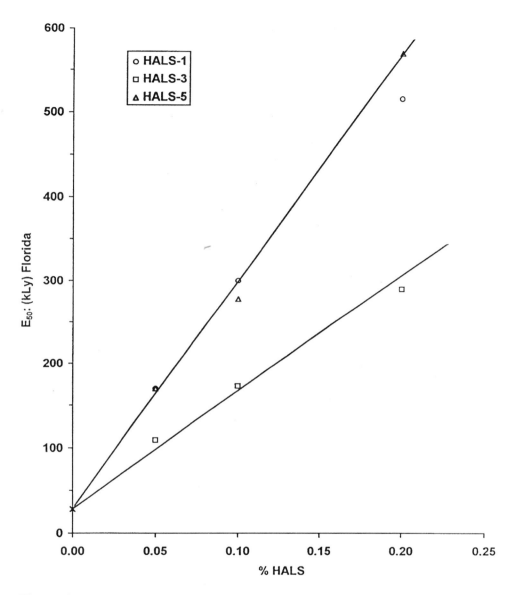

Figure 6   Effect of HALS concentration on the UV stability of 2-mm PP plaques: PP homopolymer PP-12 + 0.10% AO-3 + 0.1% Ca Stearate; Florida exposure started in June 1985.

of HALS in PP plaques usually increases proportionally to the concentration as long as the last is not too high. This is the same behavior as that found with PP tapes. It is shown in Fig. 6 for the low molecular mass HALS-1, the polymeric HALS-3 and their combination (i.e., HALS-5).

HALS combinations are also useful with filled PP. Table 25 shows the performance of HALS-5 in talc-filled PP in comparison with HALS-1 used alone at

Table 25   Light Stability of Talc-Filled PP Plaques (2 mm)[a,b]

UV stabilization	Incident energy (kLy)[c] to	
	Loss of gloss	Chalking
Control	20	20
0.4% HALS-1	150	175
0.4% HALS-5	175	200

[a] PP homopolymer PP-13 + 0.1% AO-2 + 0.3% S-1 + 0.1% Ca stearate.
[b] 40% talc, additives in weight percent relative to PP.
[c] Florida exposure started in April 1984.
Source: Ref. 29.

Table 26   Stabilization of PP Block Copolymer Injection-Molded Plaques (2 mm)[a]

Light stabilization	$T_{chalking}$ (h)	$E_{chalking}$ (kLy) Florida[b]
Control	505	80
0.1% HALS-2	2065	110
0.1% HALS-3	2235	270
0.1% HALS-6	2235	270

[a] PP copolymer PP-14 (3rd-generation) + 0.1% Ca stearate + 0.15% AO-4.
[b] Florida exposure started in January 1995.
[c] Accelerated exposure in a Weather-ometer Ci 65, without water-spraying, black panel temperature: $63 \pm 2°C$.
$T_{chalking}$, time (h) to appearance of chalking; $E_{chalking}$, energy (kLy) to appearance of chalking.

the same concentration. The combination is more efficient than the low molecular mass HALS-1 although the test criteria envisaged pertain essentially to surface properties.

Preliminary results also show synergistic effects in light stabilization of PP thick sections for HALS-6, the combination of the polymeric HALS-2 and HALS-3. Table 26 shows that HALS-6 not only yields significantly more than additive effects in comparison with its components HALS-2 and HALS-3, it is also equivalent to the best of the components. The data of artificial exposure in a Weather-ometer Ci 65 in Table 27 show the superiority of HALS-6 over HALS-3 increasing with the concentration. It can be seen in Tables 28 and 29 that the results are similar for phthalocyanine blue and titanium dioxide pigmented plaques, respectively.

B.   PE-HD Thick Sections

For UV stabilization of PE-HD thick sections, the same stabilizers are used, in principle, as for PP thick sections. They are essentially HALS, low molecular mass HALS and polymeric HALS, and UV absorbers. Data obtained with the polymeric

Table 27 Stabilization of PP Block Copolymer Injection-Molded Plaques (2 mm)[a]

Light stabilization	$T_{1.0}$ (h)[b]	$T_{chalking}$ (h)[b]
Control	230	505
0.1% HALS-3	1960	2235
0.2% HALS-3	3000	3340
0.4% HALS-3	3600	5225
0.1% HALS-6	2170	2235
0.2% HALS-6	4160	4250
0.4% HALS-6	6480	6650

[a] PP copolymer PP-14 (3rd-generation) + 0.1% Ca stearate + 0.15% AO-4.
[b] Accelerated exposure in a Weather-ometer Ci 65, without water-spraying, black panel temperature: $63 \pm 2°C$.
$T_{chalking}$, time (h) to appearance of chalking; $T_{1.0}$, time (h) to 1.0 carbonyl absorbance.

Table 28 Stabilization of Blue-Pigmented PP Block Copolymer Injection-Molded Plaques (2 mm)[a]

Light stabilization	$T_{chalking}$ (h)[b]	$E_{chalking}$ (kLy) Florida[c]
Control	710	60
0.1% HALS-3	2990	270
0.2% HALS-3	3480	>300
0.4% HALS-3	4450	>300
0.1% HALS-6	2990	270
0.2% HALS-6	3810	>300
0.4% HALS-6	5340	>300

[a] PP copolymer PP-14 (3rd-generation) + 0.1% Ca stearate + 0.15% AO-4.
[b] Accelerated exposure in a Weather-ometer Ci 65, without water-spraying, black panel temperature: $63 \pm 2°C$.
[c] Florida exposure started in January 1995.
$T_{chalking}$, time (h) to appearance of chalking; $E_{chalking}$, energy (kLy) to appearance of chalking.

HALS-2 and HALS-3 in Ziegler-type PE-HD are presented in Table 30. In this test series, the polymeric HALS-2 shows only fair performance if used alone in unpigmented samples; however, if HALS-2 is used in combination with the benzotriazole-type UV absorber UVA-2, it becomes much more efficient and comes closer to HALS-3 than in the absence of the UV absorber. In titanium dioxide-pigmented samples, the contribution of the UV absorber to light stability is small for both HALS-2 and HALS-3. Furthermore, the performance of HALS-2 is comparable with that of HALS-3 in the presence and in the absence of the UV absorber. In unpigmented and titanium dioxide-pigmented samples, the performance of the low molecular mass HALS-1 is close to that of the polymeric HALS-3 (data for HALS-1 not shown in Table 30).

The data presented in Table 30 cannot be considered as representative for all PE-HD thick sections. As a matter of fact, data obtained with another PE-HD batch

Table 29 Stabilization of White-pigmented PP Block Copolymer Injection-Molded Plaques (2 mm)[a,b]

Light stabilization	$T_{0.5}$ (h)	$E_{chalking}$ (kLy) Florida[c]
Control	475	110
0.1% HALS-3	6560	>300
0.2% HALS-3	8080	>300
0.4% HALS-3	13360	>300
0.1% HALS-6	5800	>300
0.2% HALS-6	10080	>300
0.4% HALS-6	16650	>300

[a] PP copolymer PP-14 (3rd-generation) + 0.1% Ca stearate + 0.15% AO-4.
[b] Accelerated exposure in a Weather-ometer Ci 65, without water-spraying, black panel temperature: $63 \pm 2°C$.
[c] Florida exposure started in January 1995.
$T_{0.5}$, time (h) to 0.5 carbonyl absorbance; $E_{chalking}$, energy (kLy) to appearance of chalking.

Table 30 Light Stability of PE-HD Plaques (2 mm)[a,b]

	E_{50} (kLy) Florida[c]	
UV stabilization	Unpigmented	0.5% titanium dioxide
Control	40	60
0.05% HALS-2	70	550
0.05% HALS-2 + 0.05% UVA-2	330	580
0.10% HALS-2	135	550
0.05% HALS-3	320	600
0.05% HALS-3 + 0.05% UVA-2	470	670
0.10% HALS-3	320	680

[a] PE-HD-7 (Ti catalyst, d = 0.960) + 0.03% AO-1 + 0.2% Ca stearate.
[b] Pigment: coated rutile.
[c] Florida exposure started in March 1981.
E_{50}, energy (kLy) to 50% retained tensile impact strength.
Source: Ref. 29.

(Table 31) show comparable performance for HALS-2 and HALS-3 on artificial exposure. However, the combination of both polymeric HALS (i.e., HALS-6) is better than the best of the components. In another PE-HD batch, the superiority of HALS-6 over HALS-3 observed on artificial exposure is already confirmed on Florida exposure (Table 32).

The data in Table 33 show clearly that, both the absolute and relative performance of HALS-3 and HALS-6 are heavily dependent on the high-density polyethylene used for sample preparation.

It is not yet possible to determine the characteristics of the PE-HD responsible for these differences, although the catalyst residues and their more or less careful deactivation may at least partly account for them. Tentatively, it is proposed to systematically envisage the combination HALS-6 for stabilization of PE-HD thick

Table 31 Performance of a Combination of Polymeric HALS in PE-HD Plaques (2 mm)[a,b]

Light stabilization	$T_{0.5}$ (h)		
	0.05% HALS	0.1% HALS	0.2% HALS
Control	315	315	315
HALS-2	4240	7200	12240
HALS-6	8300	14030	>21600
HALS-3	5920	7120	10160

[a] PE-HD-9 (Ti catalyst, d = 0.960) + 0.1% Ca stearate + 0.03% AO-1, unpigmented.
[b] Accelerated exposure in a Weather-ometer Ci 65, without water-spraying, black panel temperature: $63 \pm 2°C$.
$T_{0.5}$, time (h) to 0.5 carbonyl absorbance.

Table 32 Performance of a Combination of Polymeric HALS in PE-HD Plaques (2 mm)[a,b]

Light stabilization	$T_{1.0}$ (h)	E_{50} (kLy) Florida[c]
Control	590	72
0.05% HALS-3	5850	760
0.10% HALS-3	8250	870
0.05% HALS-6	10700	1150
0.10% HALS-6	17700	>1300 (61%)

[a] PE-HD-8 (Ti catalyst, d = 0.960) + 0.1% Ca stearate + 0.03% AO-1, unpigmented.
[b] Accelerated exposure in a Weather-ometer Ci 65, without water-spraying, black panel temperature: $63 \pm 2°C$.
[c] Florida exposure started in June 1988.
$T_{1.0}$, time (h) to 1.0 carbonyl absorbance; E_{50}, energy (kLy) to 50% retained tensile impact strength.

Table 33 Performance of Polymeric HALS in PE-HD Plaques (2 mm)[a,b]

Light stabilization	$T_{0.5}$ (h)				
	PE-HD-D	PE-HD-E	PE-HD-F	PE-HD-G	PE-HD-H
Control	220	260	415	295	360
0.05% HALS-6	520	11760	6440	13440	9040
0.10% HALS-6	760	~21600	18950	>21650	16090
0.20% HALS-6	1120	>216500	>21650	>21650	~21600
0.05% HALS-3	680	6200	5760	6480	5640
0.10% HALS-3	1040	9120	6840	11910	6500
0.20% HALS-3	1660	10455	10080	13160	8640

[a] PE-HD + 0.1% Ca stearate + 0.1% AO-4, unpigmented.
[b] Artificial exposure in a Weather-ometer Ci 65, black panel temperature ~63°C, no water spraying.
$T_{0.5}$, time (h) to 0.5 carbonyl absorbance.

Table 34 Light Stability of PE-LD Plaques (2 mm)[a,b]

	E_{50} (kLy) Florida[c]	
UV stabilization	Unpigmented	0.5% titanium dioxide
Control	95	360
0.1% UVA-4	265	330
0.2% UVA-4	335	320
0.1% HALS-2	700	>1600[d]
0.1% HALS-2 + 0.1% UVA-4	970	>1600[d]
0.2% HALS-2	835	~1900

[a] PE-LD homopolymer + 0.03% AO-1.
[b] Pigment: coated rutile.
[c] Florida exposure started in November 1981.
[d] No samples left.
E_{50}, energy (kLy) to 50% retained elongation.
Source: Ref. 29.

sections because, according to the results now available it should always show excellent performance. This is being checked.

C. PE-LD and PE-LLD Thick Sections

For UV stabilization of PE-LD and linear low-density PE (PE-LLD) thick sections only polymeric or high molecular mass HALS and some UV absorbers can be used. Numerous stabilizers used in PP and PE-HD cannot be used in PE-LD and PE-LLD because they are not compatible enough. This is especially so for the low molecular mass HALS-1 and similar compounds. However, the polymeric HALS-2 and HALS-3 efficiently protect PE-LD and PE-LLD thick sections from the effects of photooxidation. Benzotriazole- and benzophenone-type UV absorbers contribute to UV protection in the absence of adequate pigments. In Table 34 HALS-2 shows excellent performance even in unpigmented PE-LD 2-mm plaques in the absence of any UV absorber. The benzophenone-type absorber UVA-4 also shows good performance in unpigmented PE-LD plaques. However, in the presence of titanium dioxide, UVA-4 no longer contributes to UV stability (see Table 34). HALS-2 shows pronounced synergism with both UVA-4 and titanium dioxide.

Combinations of polymeric or high molecular mass HALS can be as efficient in PE-LD and PE-LLD as in PP and PE-HD. They may also be useful for stabilization of goods manufactured, among others, by rotational molding. The data reported in Table 35 show comparable performance for HALS-2, HALS-3 and HALS-6 on natural weathering of 5-mm–thick rotomolded plaques (JR Pauquet, unpublished results).

IV. HINDERED AMINE STABILIZERS AS THERMAL STABILIZERS

The contribution of HAS to long-term thermal stability of PP and PE is already well-documented in the literature (1–4, 16, 17, 23, 28). This is shown in Tables 4, 5, and 7 for PP tapes, and in Table 10 for PP multifilaments.

Table 35 Performance of a Combination of Polymeric HALS in PE-LLD Rotomolded Plaques (5 mm)[a,b]

Light stabilization	Time (energy) to 50% retained elongation[c]	
	(mo)	(kLy)
0.3% HALS-2	35	410
0.3% HALS-6	33	390
0.3% HALS-3	32	370

[a] PE-LLD-2 (UNIPOL, butene comonomer) + 0.05% Zn stearate + 0.15% AO-7, unpigmented.
[b] Rotomolding performed at 305°C.
[c] Florida exposure started in October 1992.
Source: JR Pauquet, Ciba Specialty Chemicals, unpublished data.

There is usually no contribution of the standard HAS to processing stability of polyolefins. The effect of HAS seems to be restricted to stabilization of polyolefins in the solid state. However, in the solid state, performance usually increases markedly with decreasing temperature. The temperature gradient observed with HAS seems to be much more pronounced than with typical phenolic antioxidants. As a consequence, the superiority of HAS over phenolic antioxidants increases with decreasing aging temperature.

There are primarily two instances for which HAS look definitely inferior to phenolic antioxidants. One of these instances concerns processing stabilization (i.e., stabilization of the polymer in the molten state). The other instance is related to the test methods used to assess the contribution of HAS to LTTS of polyolefins. The major methods yielding a false evaluation of the potential performance of HAS are those based on the determination of oxidation induction times (OIT) by techniques such as DTA/DSC used at high temperature (i.e., again applied, to the polymer in the molten state). Therefore, these methods necessarily show poor performance for HAS in comparison with phenolic antioxidants. However, because these methods usually yield results very rapidly, they are common practice and still are included in many standard test methods proposed for the assessment of the thermooxidative stability of polymers in the solid state. Nevertheless, because HAS do not protect molten polyolefins from thermal oxidation, it is not astonishing that they perform poorly in such tests. This poses once more the problem of the relevance of such test methods discussed in detail as long as 10 years ago (13).

It is noteworthy that with HAS-stabilized PP samples, the usual failure signs of thermally oxidizing PP (i.e., disintegration to powder) may not be observed, or observed only after a prolonged time period. The samples may become brittle without exterior signs of deterioration, not counting discoloration. However, they already break on slight bending; hence, in the presence of HAS it is mandatory to use an additional test to complement visual assessment of failure. This test can be mechanical (e.g., bending the samples to test for brittleness). It can also be chemical (e.g., determining oxidation products by infrared [IR] spectroscopy). These methods are illustrated by the following examples.

Table 36   Contribution of Polymeric HALS to LTTS of PP Films[a,b]

Stabilization	Days to 0.5 carbonyl absorbance at		
	135°C	120°C	100°C
Control	11[c]	18[c]	23[a]
0.2% HALS-2	76	385	900
0.2% HALS-3	112	365	1000
0.2% HALS-4	150	540	>1550

[a] PP homopolymer PP-8 + 0.1% Ca stearate + 0.1% AO-8, unpigmented.
[b] Compression-molded 0.5-mm films.
[c] Mean values because of pronounced scattering.

Table 37   Contribution of Polymeric HALS to LTTS of PP Copolymer[a,b]

Stabilization	Days to brittleness on bending at	
	149°C	135°C
Control (base)	21	85
0.2% HALS-2	54	189
0.2% HALS-3	57	189
0.2% HALS-4	93	296

[a] PP block copolymer PP-15 + 0.15% AO − 4 + 0.1% Ca stearate, unpigmented.
[b] Injection-molded 2-mm plaques.

Table 36 shows the performance of some high molecular mass HAS in PP films with carbonyl formation measured by IR spectroscopy as test criterion. The data show, at the same time, that the HAS tested perform by themselves, even in the absence of any phenolic antioxidant. As a matter of fact, AO-8 used as a processing stabilizer is so volatile that it is rapidly lost at the aging temperatures involved. Brittleness on bending is the test criterion chosen for the experiments summarized in Table 37. The HAS contribute significantly to LTTS, although this time the efficient-processing stabilization system contains a high molecular mass phenolic antioxidant yielding an excellent LTTS.

The data discussed so far involved high molecular mass HAS used alone. However, the combination of some high molecular mass HAS can also yield synergistic effects. Table 38 shows the influence of the nature of the HAS and testing temperature on LTTS of PP tapes. Again, the temperature coefficient is much superior to that of a typical phenol used in the processing stabilizing system. Also, the *synergistic effect* of HALS-6—defined as the performance above the additive effect expected from the two components—increases with decreasing temperature. At 80°C, the performance of HALS already comes close to that of the component yielding the highest value in Table 38 (i.e., HALS-3). It is assumed that, at temperatures lower than 80°C, the contribution of HALS-6 to LTTS of PP tapes will become equal and, finally, superior to that of HALS-3. This superiority is already observed at

Table 38   Contribution of Polymeric HALS to LTTS of PP Tapes[a,b]

Stabilization	Days to embrittlement at		
	120°C	100°C	80°C
Control (base)	34	61	392
0.1% HALS-2	61	154	875
0.1% HALS-3	167	771	2246
0.1% HALS-6	108	481	1620

[a] PP homopolymer PP-8 + 0.1% Ca stearate + 0.1% AO-3, unpigmented.
[b] Tapes 50-μm thick, stretch ratio 1 : 6.

Table 39   Contribution of Polymeric HALS to LTTS of PP Films[a,b]

Stabilization	Days to brittleness at		
	120°C	100°C	80°C
Control (base)	71	137	185
0.1% HALS-2	129	166	337
0.1% HALS-3	108	236	564
0.1% HALS-6	116	197	648

[a] PP homopolymer PP-D + 0.1% Ca stearate + 0.15% AO-4, unpigmented.
[b] 120-μm–thick cast films.

Table 40   Contribution of Polymeric HALS to LTTS of PP Films at 80°C[a,b]

Stabilization	Days to brittleness at 80°C		
	0.1% HALS	0.2% HALS	0.4% HALS
Control (base)	185	185	185
HALS-2	337	773	1463
HALS-3	564	1165	2305
HALS-6	648	1426	2435

[a] PP homopolymer PP-D + 0.1% Ca stearate + 0.15% AO-4, unpigmented.
[b] 120-μm–thick cast films.

80°C with PP films. Table 39 shows that, although HALS-6 is slightly inferior to HALS-2 or HALS-3 at 120° and 100°C, it becomes superior to both components at 80°C. Hence, HALS-6 not only shows synergistic effects for UV stability, but also for thermooxidative stability as soon as the temperature comes close to the temperatures more relevant for practice than the high temperatures used mostly for testing. Table 40 shows that this synergism holds in the concentration range tested.

The Arrhenius-plot of LTTS data (Fig. 7) visualizes the respective effect of temperature on the performance of a stabilization based on a phenolic antioxidant

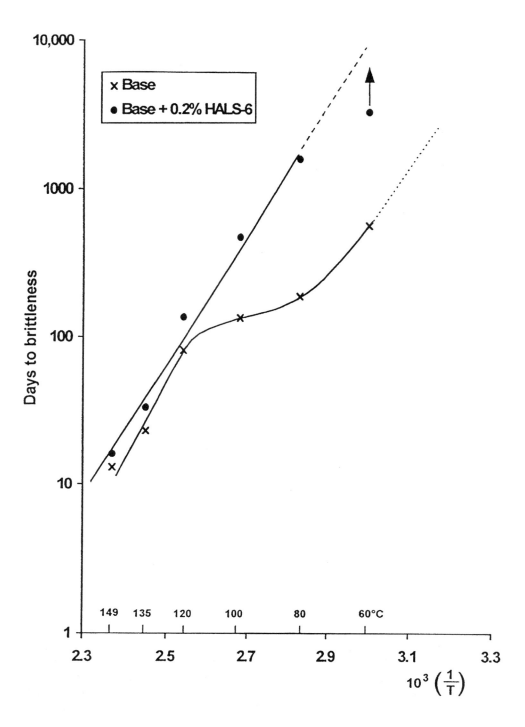

Figure 7 Arrhenius-plot of LTHA data generated with 120-μm–thick cast PP films: base stabilization: 0.1% Ca stearate + 0.1% AO-3.

Table 41   Contribution of Polymeric HALS to LTTS of PE-HD Tapes[a,b]

Stabilization	Retained tensile strength after 570 days[c] at 100°C
Control (base)	70%
0.05% HALS-2	90%
0.05% HALS-3	90%

[a] PE-HD- (Ti catalyst, d = 0.950) + 0.1% Ca stearate + 0.05% AO-1, unpigmented.
[b] Tapes 50-μm thick, stretch ratio 1 : 8.5.
[c] No samples left.
Source: Ref. 3.

Table 42   Contribution of Polymeric HALS to LTTS of Two Types of PE-HD[a,b,c]

	Days to brittleness on bending at 120°C	
Stabilization	Phillips-type PE-HD (PE-HD-5)	Ziegler-type PE-HD (PE-HD-6)
Control (base)	220	192
0.1% HALS-2	291	267
0.2% HALS-2	407	402
0.1% HALS-4	475	413
0.2% HALS-4	576	639

[a] PE-HD 5 (Cr catalyst, d = 0.960) + 0.1% AO-3, unpigmented.
[b] PE-HD 5 (Ti catalyst, d = 0.950) + 0.1% AO-3 + 0.1% Ca stearate, unpigmented.
[c] Compression-molded 0.5-mm films.

and of a stabilization containing, in addition, a polymeric HAS. It also gives an idea of the temperatures at which the superiority of HAS over the phenolic antioxidant becomes more and more pronounced.

The results obtained with polyethylene are overall similar to those obtained in PP; high molecular mass HAS can considerably enhance the thermooxidative stability of PE-HD, PE-LD, and PE-LLD. This is shown in Table 41 for PE-HD tapes. Here, the phenolic antioxidant already confers an impressive thermooxidative stability at 100°C. This is improved significantly by the addition of small amounts of the polymeric HALS-2 and HALS-3. The HAS are as effective in PE-HD films. This is shown in Table 42 for both Phillips-type and Ziegler-type high-density polyethylene. Table 42 shows that the results obtained with the two polyethylene types are comparable for the base-stabilized samples as well as for the samples containing in addition HALS-2 or HALS-4. Notably, with unstabilized samples, Phillips-type PE is much more prone to thermal oxidation than Ziegler-type PE (31). Data with PE-LD and PE-LLD films can be found in Chapter 2.

V.   CONCLUSIONS

Combinations of stabilizers very often yield optimum UV protection of polyolefins. Thus, combinations of HALS with UV absorbers give excellent results, mainly in thick cross sections. However, even in relatively thin PE-LD films (200 μm), there can be pronounced synergism between HALS and UV absorbers. In the presence of pigments, such as titanium dioxide or phthalocyanine blue, the contribution of the UV absorbers is usually too small to be of any use. The effect can even become antagonistic. With some organic pigments, especially the yellow and red ones, it is recommended that a UV absorber be added, to protect not only the polymer, but even more so, the pigment.

Specific combinations of low molecular mass HALS with high molecular mass or polymeric HALS also yield pronounced synergistic effects. More importantly, this effect does not depend on sample thickness and can be observed with thin as well as with thick sections. The synergistic effect observed by combining two polymeric HALS, such as HALS-2 and HALS-3, is even more astonishing than that observed between low molecular mass and high molecular mass HALS. By using such combinations, it is possible to achieve simultaneously good UV stability and indirect food contact approval.

The contribution of HAS to long-term heat aging of, for example, polyolefins was not given due attention for a long time. One reason for this neglection is that priority had been given to the UV-stabilizing effects of HALS. Another reason is that the very nature of some tests partly involved in the standards developed for evaluation of plastics materials were an obstacle to correct assessment of these additional possibilities of HAS. The main point here concerns the methods involving determination of oxygen induction time by DTA/DSC with the molten polymer to predict performance in the solid polymer. Another important point is the considerable difference in the failure mode of phenols and HAS in PP: "catastrophic failure" for phenols and progressive failure for HAS. Nevertheless, the views are changing and the contribution of HAS to LTTS of polyolefins is used more and more.

The choice of high molecular mass phenolic antioxidant or high molecular mass HAS to ensure LTTS of polyolefins will mainly be dictated by the testing temperature and the test criterion involved.

REFERENCES

1. F Gugumus. Oxidation inhibition in plastics. In: P Klemchuk, J Pospisil, eds. Inhibition of Oxidation Processes in Organic Materials. Vol 1. Boca Raton, FL: CRC Press, 1989, pp 61–172.
2. F Gugumus. Antioxidants. In: R Gaechter, H Mueller, eds. Plastics Additives, 4th ed. Munich: Hanser, 1993, pp 1–104.
3. F Gugumus. New trends in the stabilization of polyolefin fibers. Polymer Degrad Stabil 44:273–297, 1994.
4. F Gugumus. Mechanisms of thermooxidative stabilization with HAS. Polymer Degrad Stabil 44:299–322, 1994.
5. C Kroehnke. A major breakthrough in polymer stabilization—high performance melt processing stabilizer for polyolefins. Proceedings of the SPE Conference Polyolefins X, RETEC, Houston, TX, Feb 23–26, 1997, pp 555–569.
6. P Nesvadba, C. Kroehnke. A new class of highly active phosphorous free processing stabilizers for polymers. Proceedings of the 6th International Conference "Additives '97," New Orleans, LA. Feb 3–5, 1997.
7. F Gugumus. Progrès dans la protection des matières plastiques contre le rayonnement UV. 4e Conférence Européenne des Plastiques et des Caoutchoucs, Paris, 1974.

8. F Gugumus. Progrès dans la protection des matières plastiques contre le rayonnement UV. Kunstst Plast 22:11–19, 1975.
9. F Gugumus. Progrès dans la protection des matières plastiques contre le rayonnement UV. Caout Plast 558:67–71, 1976.
10. F Gugumus. Developments in the UV stabilisation of polymers. In: G Scott, ed. Developments in Polymer Stabilization—1. London: Applied Science, 1979, pp 261–308.
11. F Gugumus. Lichtschutzmittel. Kunststoffe 73:620–623, 1984.
12. T Kurumada, H Ohsawa, O Oda, T Fujita, T Toda, T Yoshioka. Photostabilizing activity of tertiary hindered amines. J Polym Sci Polym Chem Ed 23:1477–1491, 1985.
13. F Gugumus. The use of accelerated tests in the evaluation of antioxidants and light stabilizers. In: G Scott, ed. Developments in Polymer Stabilisation—8. Barking, UK: Elsevier Applied Science 1987, pp 239–289.
14. F Gugumus. Lichtschutzmittel. Kunststoffe 77:1065–1069, 1987.
15. F Gugumus. Photooxidation and stabilization of polyethylene. Tenth International Conference on Advances in the Stabilization and Controlled Degradation of Polymers, Lucerne, Switzerland, May 25–27, 1988.
16. F Gugumus. Advances in the stabilization of polyolefins. SPE Tech Pap 34:1447–1450, 1988.
17. F Gugumus. Advances in the stabilization of polyolefins. Polymer Degrad stabil 24:289–301, 1989.
18. F Gugumus. Photooxidation of polymers and its inhibition. In: P Klemchuk, J Pospisil, eds. Inhibition of Oxidation Processes in Organic Materials. Vol 2. Boca Raton, FL: CRC Press, 1989, pp 29–162.
19. DKC Hodgeman. Hindered amine light stabilizers—their role in the prevention of photooxidation of polypropylene. In: N Grassie, ed. Developments in Polymer Degradation—4. London: Applied Science, 1982, pp 189–234.
20. F Gugumus. Mechanisms and kinetics of photostabilization of polyolefins with HALS. Angew Makromol Chem 176/177: 241–289, 1990.
21. F Gugumus. Mechanisms of photostabilization of polyolefins with N-alkyl(methyl) HALS. Angew Makromol Chem 190:111–136, 1991.
22. P Gijsman, J Hennekens, D Tummers. The mechanism of action of hindered amine light stabilizers. Polym Degrad Stabil 39:225–233, 1993.
23. F Gugumus. Re-evaluation of the stabilization mechanisms of various light stabilizer classes. Polym Degrad Stabil 39:117–135, 1993.
24. F Gugumus. Current trends in mode of action of hindered amine light stabilizers. Polym Degrad Stabil 40:167–215, 1993.
25. F Gugumus. Light stabilizers. In: R Gaechter H Mueller, eds. Plastics Additives. 4th ed. Munich: Hanser, 1993, pp 129–270.
26. P Gijsman, J Hennekens, K Janssen. Comparison of UV degradation chemistry in accelerated (xenon) aging tests and outdoor tests (II). Polym Degrad Stabil 46:63–74, 1994.
27. K Kikkawa. New developments in polymer photostabilization. Polym Degrad Stabil 49:135–143, 1995.
28. F Gugumus. Optimized stabilizer systems for polyolefins. Proceedings of AddCon'95, Basel, Switzerland, April 5–6, 1995.
29. F Gugumus. The performance of light stabilizers in accelerated and natural weathering. Polym Degrad Stabil 50:101–116, 1995.
30. F Gugumus. Influence of stabilization mechanisms and environment on optimization of UV stability. Proceedings of the 3rd International Symposium on Weatherability. Tokyo, May 14–16, 1997, pp 94–120.
31. F Gugumus. Thermooxidative degradation of polyolefins in the solid state: part 5. Kinetics of functional group formation in PE-HD and PE-LLD. Polym Degrad Stabil 55:21–43, 1997.

APPENDIX

Abbreviation	Structure	Trade name

HALS-1 — Tinuvin 770

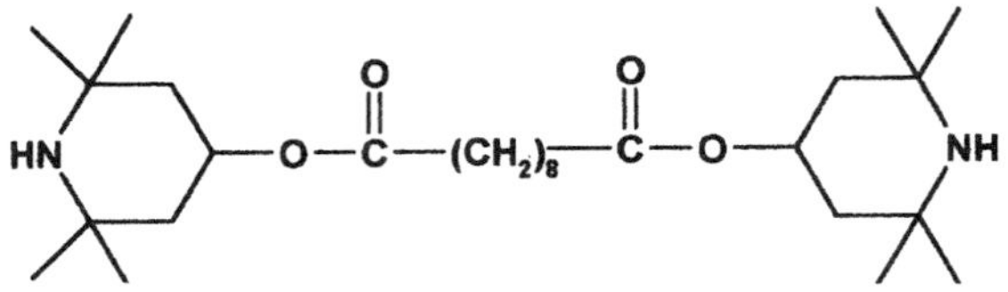

HALS-2 — Tinuvin 622

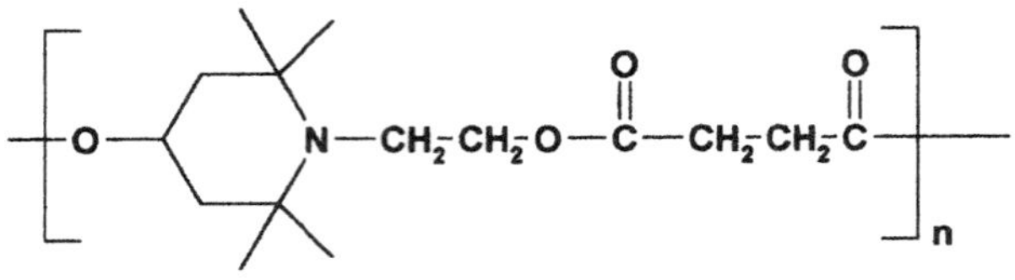

HALS-3 — Chimassorb 944

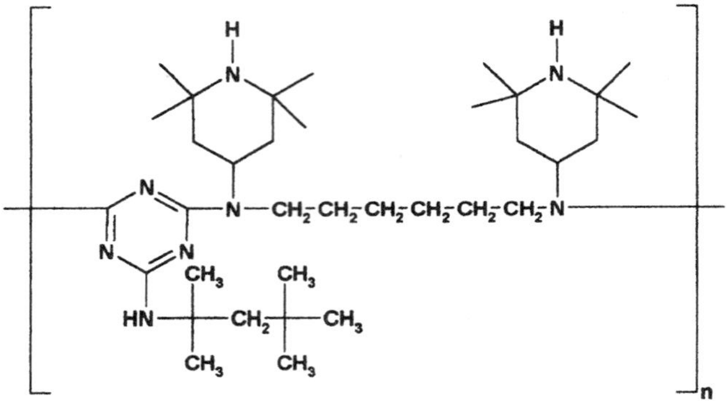

HALS-4 — Chimassorb 119

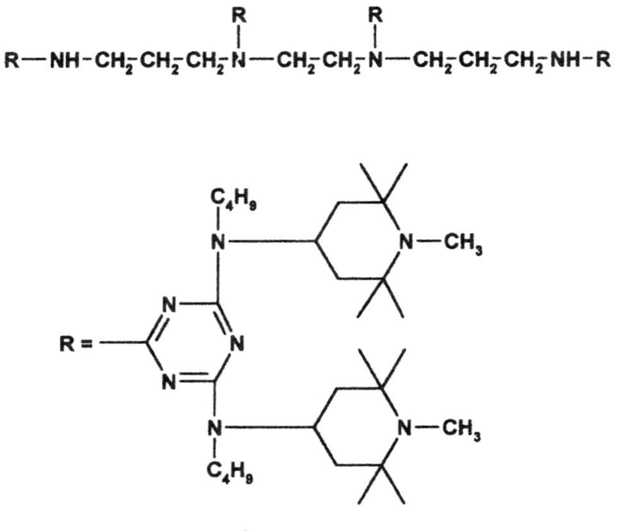

APPENDIX *Continued*

Abbreviation	Structure	Trade name
HALS-5	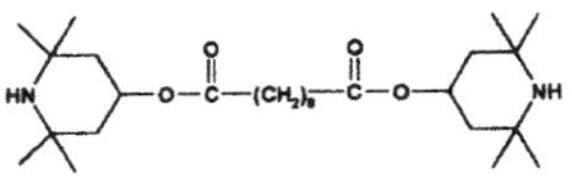 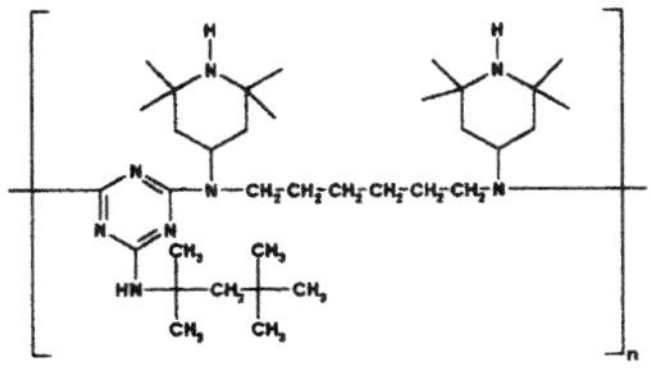 1 : 1	Tinuvin 791
HALS-6	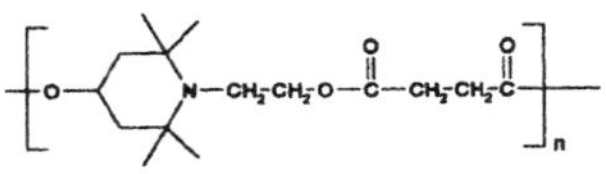 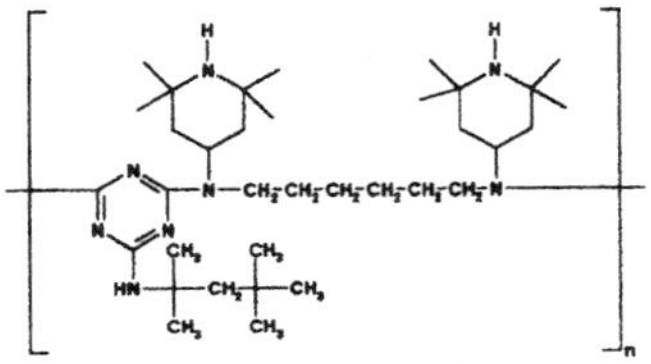 1 : 1	Tinuvin 783
HALS-7	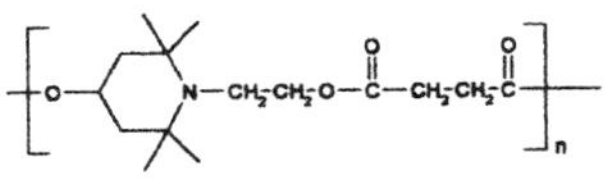 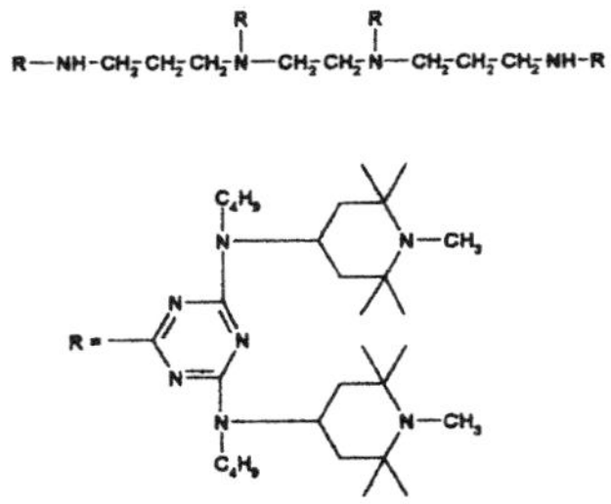 1 : 1	Tinuvin 11
Ni-1	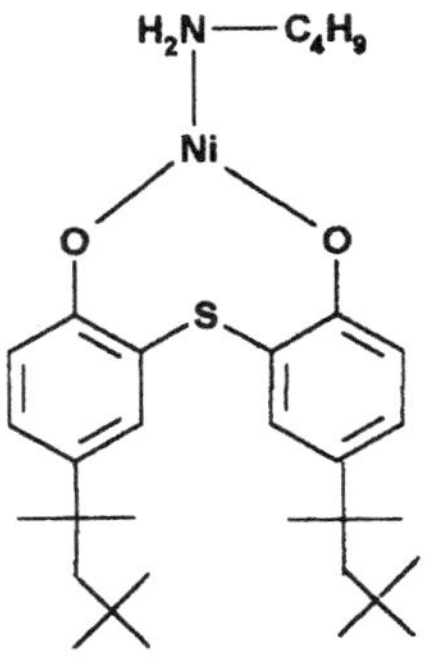	Cyasorb UV 1084

APPENDIX *Continued*

Abbreviation	Structure	Trade name
Bz-1	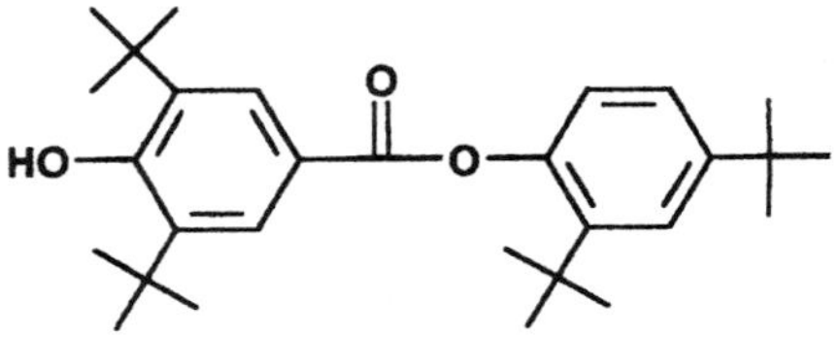	Ferro AM 340 Tinuvin 120
UVA-1	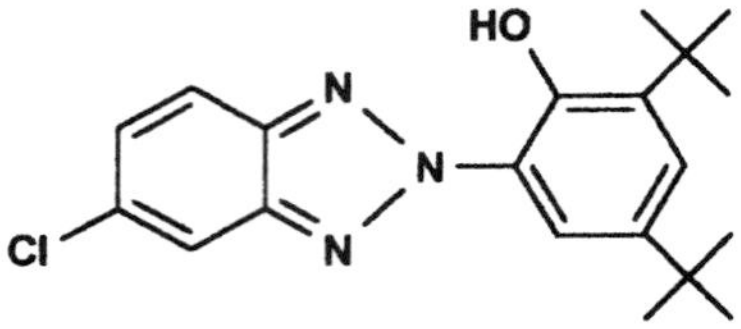	Tinuvin 327
UVA-2	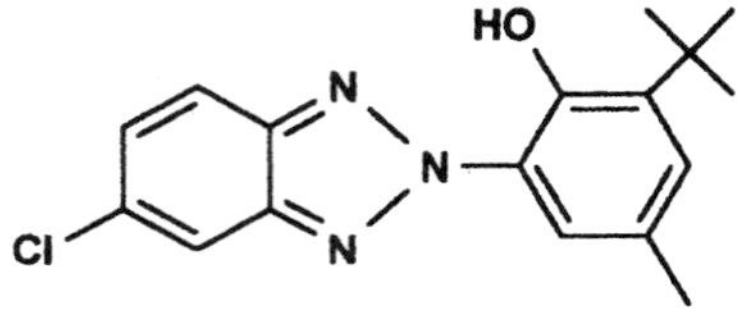	Tinuvin 326
UVA-3	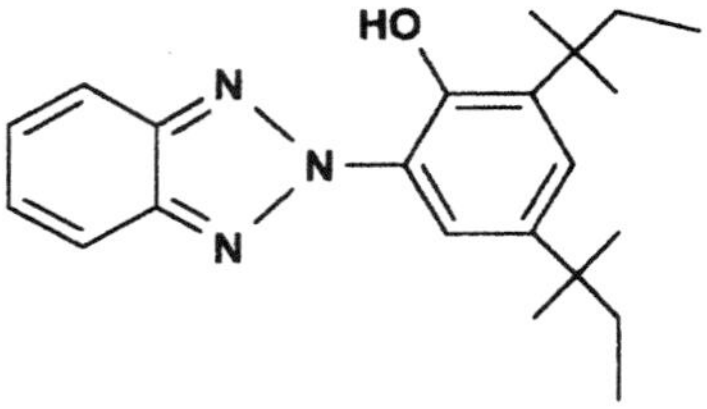	Tinuvin 328
UVA-4	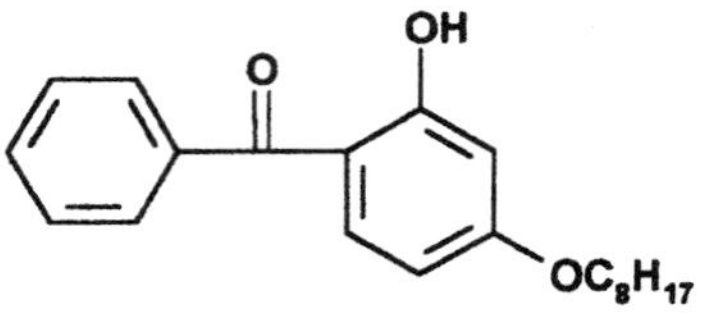	Chimassorb 81
AO-1	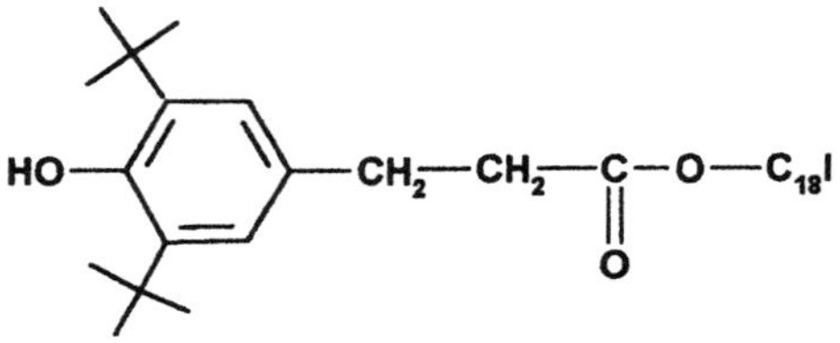	Irganox 1076

APPENDIX *Continued*

Abbreviation	Structure	Trade name
AO-2	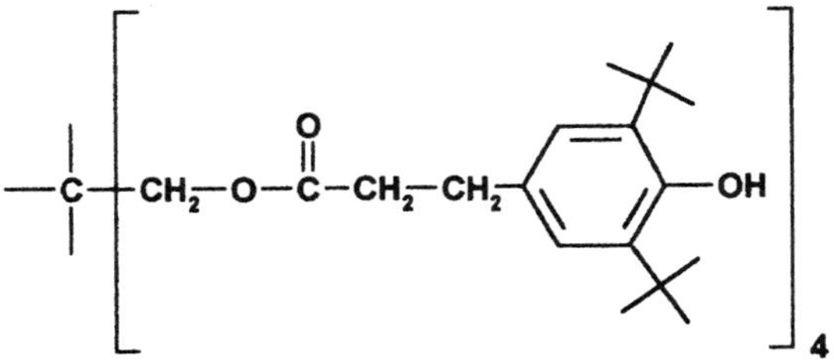	Irganox 1010
AO-3	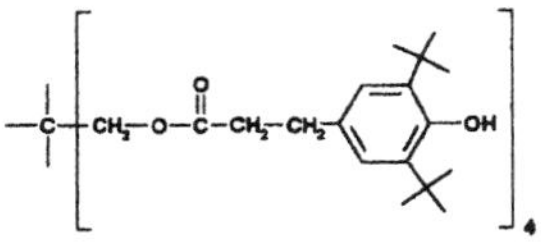 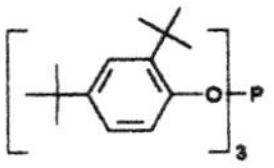 1 : 1	Irganox B 225
AO-4	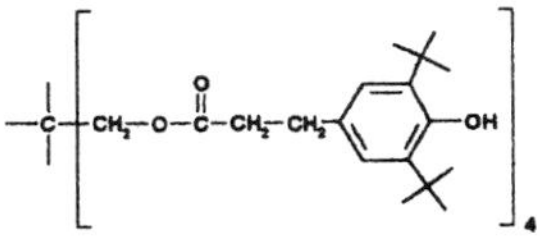 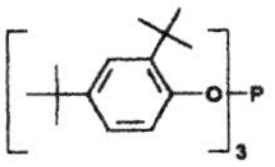 1 : 2	Irganox B 215
AO-5	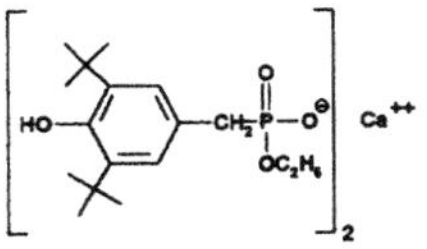 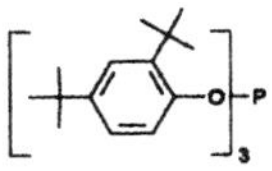 1 : 1	Irganox B 936
AO-6	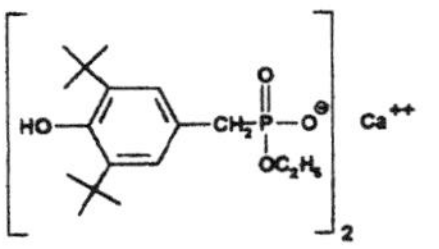 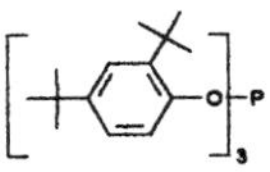 1 : 2	Irganox B B501W
AO-7	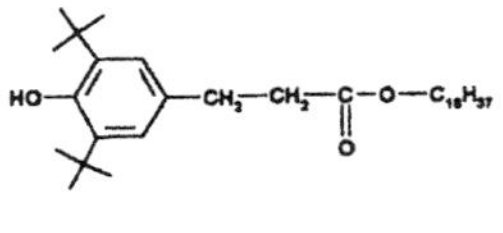 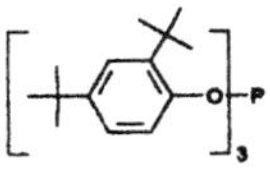 1 : 2	Irganox B 921

APPENDIX *Continued*

Abbreviation	Structure	Trade name
AO-8	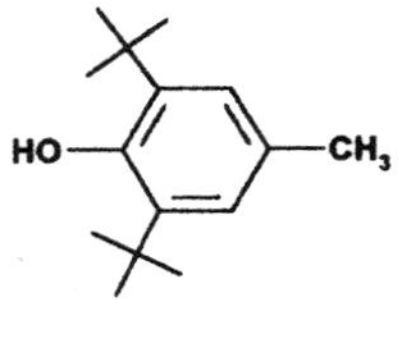	BHT, Ionol
S-1	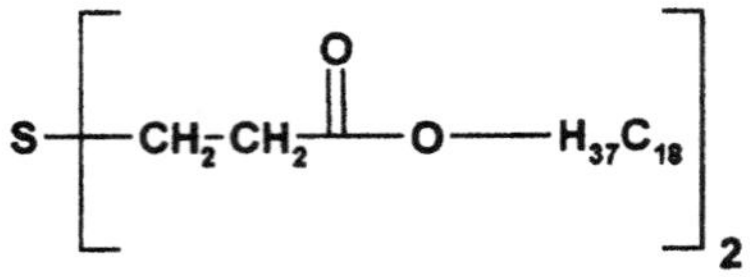	Irganox PS 802

2

Greenhouse Film Stabilization

FRANÇOIS LOUIS GUGUMUS

Ciba Specialty Chemicals Limited, Basel, Switzerland

I. INTRODUCTION

Low-density polyethylene homopolymer (PE-LD) and ethylene–vinyl acetate copolymers (EVA) are the most common plastics materials used for greenhouse cover films. There is only one major exception. In Japan, plasticized poly(vinyl chloride) (PVC) is used to a large extent. The growing importance of plastics films for greenhouse covers is best illustrated by a few figures. In 1976 the surface covered with greenhouse films worldwide was approximately 60,000 ha (1). In 1980 the corresponding surface surpassed 80,000 ha (1) and reached 220,000 ha in the 1990s (2). The impressive development of the use of greenhouses is best visualized with a country from southern Europe, such as Spain. In fact, Spain went through a tremendous evolution for the agricultural surface covered with greenhouses: from 343 ha in 1967, to 5,800 ha in 1976, to 15,000 ha in 1983 (3), and an area of 28,000 ha in the 1990s (2). Although there is some leveling off in the total area covered with greenhouses in southern Europe, the development is likely to continue in other parts of the world, in particular in South America (4).

The lifetimes required for the greenhouse cover films depend heavily on the country, especially the tradition in the country. They vary from one season (i.e., 6–9 months) to several years. However, PE-LD and EVA greenhouse films will last only a few months without the addition of light stabilizers. This clearly shows that the considerable development of greenhouses would not have been possible without the development of adequate and efficient light stabilizer systems.

Until about 1978, nickel stabilizers, such as Ni-1, combined with ultraviolet light absorbers (UV absorbers; UVAs) such as UVA-1 were state of the art for greenhouse films, In the late 1970s, polymeric hindered amine light stabilizers (HALS), such as HALS-2 and HALS-3, were introduced into this market.

Table 1   Main Parameters Influencing the Lifetime of a Greenhouse Film

1. Film-inherent parameters	
Polymer type	PE-LD, PE-LLD, EVA copolymer
Film type	Monolayer, coextruded multilayer
Film thickness	
Stabilization	Stabilizer system and concentration
Fillers	Kaolin, chalk
Pigments	
Other additives	Antiblocking, antifogging agents, etc.
Film manufacturing	
2. General environmental parameters	
Greenhouse related	
Support/frame material	Wood, aluminum, galvanized iron, etc.
Protection of contact surface	Paint, PE film layers
Design	High, low, ventilated or not
Film Fixation	
Climate related:	
Solar irradiation	UV intensity, global energy
Temperature	Mean, maximum
Wind, rain, snow	
3. Special environmental parameters	
Crop type	Tall, short
Agrochemicals	Type, frequency, and mode of application

This chapter presents some results obtained with the aforementioned light stabilizers and light stabilizer systems. The most important parameters affecting the useful lifetime of polyethylene greenhouse films will be discussed in detail. These parameters can be grouped into three classes: 1) film-inherent parameters, 2) general environmental parameters, 3) special environmental parameters. They are summarized in Table 1.

The additives mentioned in this chapter are mainly commercial products. The structures and designations of the commercial compounds used in the experiments are shown in the Appendix. The corresponding abbreviations used in the text and the tables can also be found in the Appendix.

II.   FILM-INHERENT PARAMETERS

A.   Polymer Type

Low-density PE (PE-LD) and EVA are the main polymers used for greenhouse film applications. So far, linear low-density polyethylene (PE-LLD) is mainly used blended with PE-LD or for coextruded films. In the following, the effect of the base resin on the useful lifetime of greenhouse films will be discussed.

Table 2 Comparison of PE-LD and PE-LLD Films[a]

	E_{50} (kLy) Florida[b]	
Polymer type	Without UV stabilizers	0.2% HALS-2 + 0.2% UVA-1
PE-LD[c]	60	260
PE-LLD (butene)[d]	50	260
PE-LLD (octene)[e]	60	250

[a] Blown films 200-μm thick, without backing.
[b] Florida exposure started in September 1983.
[c] PE-LD homopolymer, MF(190°C/2.16 kg) = 0.3, no base stabilization.
[d] PE-LLD (butene), gas-phase process, butene comonomer, MF(190°C/2.16 kg) = 1 + 0.1% AO-3 + 0.05% Ca stearate.
[e] PE-LLD (octene), solution process, octene comonomer, MF (190°C/2.16 kg) = 1 + 0.1% AO-1 + 0.05% Ca stearate.
E_{50}, energy (kLy) to 50% retained elongation.

1. PE-LLD and PE-LD Films

Linear low-density PE is appreciated for its superior mechanical properties compared with normal low-density polyethylene PE-LD. Among others, the elongation at break is significantly higher for PE-LLD than for PE-LD. Usually, on artificial as well as on natural weathering, a film is considered to have failed when its elongation is decreased to half that before exposure. With this failure criterion, PE-LLD and PE-LD films of the same thickness show comparable light stability. This is exemplified for outdoor exposure in Table 2.

2. PE-LLD/PE-LD Blends

As already mentioned, PE-LLD shows improved performance compared with PE-LD in many applications. However, film manufacturing with PE-LLD requires modifications of most of the equipment used previously because it is optimized for PE-LD processing. Through the use of PE-LLD–PE-LD blends, it is possible to combine the superior mechanical properties of PE-LLD with the superior optical properties of PE-LD without modifying the equipment. The data in Table 3 show the results of natural weathering of such blends as a function of the film backing. Within experimental error, the UV stability of the blends is independent of the ratio of PE-LLD/PE-LD. Furthermore, it is also the same as that of the pure polymers (5). Hence, the early fears that the light stability of the blends may be poorer than that of the components are not justified.

3. EVA and PE-LD Films

The EVA copolymers are transparent to visible light and allow all the wavelengths essential for photosynthesis to pass through. In addition, in contrast with PE-LD, EVA has absorption bands in the long-wavelength infrared (IR), corresponding to the spectral emission of a black body at 20°C. The last is comparable with that

Table 3   Light Stability of UV-Stabilized[a,b] Films[c] based on PE-LLD/PE-LD[d,e] Blends

Polymer composition	E_{50} (kLy) Florida[f]		
	PMMA backing	Without backing	Aluminum backing
100% PE-LLD	145	125	110
75% PE-LLD	140	130	105
50% PE-LLD	110	145	80
25% PE-LLD	110	125	110
0% PE-LLD	125	120	95

[a] Base stabilization: 0.075% Ca stearate + 0.1% AO-4.
[b] UV stabilization: 0.15% HALS-2 + 0.12% UVA-1.
[c] Blown films 100-μm thick.
[d] PE-LD homopolymer, tubular process, MF($190°$C/2.16 kg) = 0.3, no base stabilization.
[e] PE-LLD, gas phase process, butene comonomer, MF($190°$C/2.16 kg) = 1.
[f] Florida exposure started in May 1984.
E_{50}, energy (kLy) to 50%-retained elongation.

Table 4   Comparison of PE-LD[a] and EVA[b] Films[c]

Light stabilization	E_{50} (kLy) Florida[d]			
	Without backing		Aluminum backing	
	PE-LD	EVA	PE-LD	EVA
Control	40	50	30	40
0.3% Ni-1 + 0.3% UVA-1	290	260	180	210
0.3% HALS-2 + 0.3% UVA-1	370	385	230	315
0.15% HALS-3 + 0.15% UVA-1	400	580	300	580

[a] PE-LD homopolymer, film grade + 0.03% AO-1.
[b] EVA copolymer, 14% VA, film grade + 0.03% AO-1.
[c] Blown films 200-μm thick.
[d] Florida exposure started in March 1981.
E_{50}, energy (kLy) to 50%-retained elongation.

of the soil during the night. This spectral emission, with a wavelength between 7 and 14 μm, constitutes an important part of the energy losses from the soil and the plants.

The partial prevention of the dissipation of this thermal energy from the greenhouse is called *greenhouse effect* (see also Sec. II.D). Consequently, during the cool night hours, the temperature decrease inside the greenhouse is less pronounced. This results in earlier and possibly more abundant crops.

Films that display such a greenhouse effect are called *infrared (IR) barrier films* or *thermic films*. EVA (up to ~14% VA) is widely used for this application (see also Sect. II.D). The thermal characteristics of an EVA film are proportional to the vinyl acetate content (6). The data of Table 4 show that EVA-based films have comparable or higher UV stability than the corresponding PE-LD films. One can see from Table 4 that the polymeric HALS, HALS-2, and HALS-3, are significantly more efficient

Table 5 Light Stability[a] of Coextruded Films[b]

First layer (toward the sun)	Second layer (toward the soil)	Months to 50% retained elongation[c]
PE-LD[d]	—	18
EVA[e]	—	>21 (65%)
PE-LD	EVA	21

[a] Stabilization: 0.6% HALS-2 + 0.3% UVA-1.
[b] Films 75-μm thick, for coextruded film, each layer 37.5-μm thick.
[c] Exposure in Brazil (Sao Paulo) started in May 1989.
[d] PE-LD homopolymer, film grade.
[e] EVA copolymer, 10% VA, film grade.

than the nickel-based stabilizer Ni-1. This is especially remarkable with HALS-3 used at only half the concentration of the other two stabilizers. More data on UV stability of the polymers used for agricultural films can be found in the literature (5,7–21).

The data discussed, although generated on natural weathering, represent the performance of the films under nearly ideal conditions. They cannot be used directly to evaluate the expected lifetime of films in actual practice (i.e., films used on greenhouses).

B. Film Type: Monolayer Versus Multilayer

Coextruded films, especially PE-LD EVA and PE-LLD EVA, or combinations thereof, represent a new development in the greenhouse film market (22,23). The main purpose is to benefit from the special advantages that each polymer offers. Thus, for example, the partial opacity to long-range IR wavelengths of EVA copolymers can be combined with the stiffness of PE-LD and particularly of PE-LLD and the reduced dust accumulation on PE-LD and PE-LLD compared with EVA copolymers.

The data now available show comparable performance for monolayer and multilayer films with the same thickness (24). Table 5 shows that the lifetime of the coextruded PE-LD/EVA film seems to be intermediate between the lifetimes of the corresponding monolayer films with the same overall thickness. In fact, there is no reason for significant differences, as long as the overall thickness and the stabilization system are identical. In Table 6, the effect of the stabilization system is as pronounced with three-layer films as with monolayer films. Furthermore, poor stabilization of the upper layers cannot be compensated by an excellent stabilization system in the inner layer. As a matter of fact, as shown by the data in the last row in Table 6, the HALS stabilization in the inner layer cannot further improve the stability conferred by the nickel-based stabilization in the two exterior layers. The importance of an efficient stabilization in all the layers is also illustrated by the data in Table 7. Table 7 shows that the overall stability can be significantly reduced if the stabilization is concentrated in the top layer, although the mean stabilization is the same in both films. The difference in stability can essentially be attributed to the absence of the polymeric HALS in the two inner layers, because the

Table 6 Light Stability of Coextruded Films[a,b]

First layer (toward the sun)	Second layer (middle)	Third layer (toward the ground)	E_{50} (kLy) Florida[c]
PE-LD (A)[d]	PE-LD (A)	PE-LD (A)	205
PE-LD (B)	PE-LD (B)	PE-LD (B)	710
PE-LD (B)	PE-LD (B)	EVA (B)[e]	760[f]
PE-LD (B)	PE-LD (B)	EVA (A)	695
PE-LD (A)	PE-LD (A)	EVA (B)	190

[a] Coextruded films 150-μm thick, upper layer 40 μm, middle layer 60 μm, and ground layer 50 μm.
[b] Light stabilization: (A): 0.2% Ni-1 + 0.2% UVA-1. (B): 0.2% HALS-3 + 0.2% UVA-1.
[c] Exposure in Florida on PMMA plates started in April 1992.
[d] PE-LD homopolymer, film grade.
[e] EVA copolymer, 19% VA, film grade.
[f] Extrapolated.
E_{50}, energy (kLy) to 50%-retained elongation.

Table 7 Light Stability of Coextruded Films[a,b] with the Same Overall Stabilization

Stabilization type[c]	E_{50} (kLy) Florida[d]
0.2% HALS-2 + 0.2% UVA-1 (same in all three layers)	400
0.6% HALS-2 + 0.6% UVA-1 (in top layer only)	330

[a] PE-LD homopolymer, film grade.
[b] Three layers coextruded films, every layer 50-μm thick.
[c] Mean stabilization for both films: 0.2% HALS-2 + 0.2% UVA-1.
[d] Exposure in Florida on PMMA plates started in March 1992.
E_{50}, energy (kLy) to 50%-retained elongation.

benzophenone UV absorber UVA-1 migrates easily from the top layer to the middle layer and bottom layer such that a somewhat uniform distribution results. Hence, a stabilization system distributed as uniformly as possible in all the layers seems to be preferable to a high loading concentrated in the top layer.

C. Stabilization and Film Thickness

The effect of different light stabilizer systems on the UV stability of PE-LD films as well as the mechanisms involved have been the object of numerous publications in the past 25 years (5,7,21,25–33). This does not mean that all the problems have been definitively solved.

There are essentially three main classes of light stabilizers used in greenhouse films:

1. UV absorbers of the benzophenone and benzotriazole type, such as UVA-1 and UVA-2, respectively.

Table 8 Light Stability of PE-LD Blown Films[a,b]

UV stabilization	E_{50} (kLy) Florida[c]	
	Without backing	On aluminum
Control	32	20
0.15% Ni-1	125	90
0.15% Ni-2	62	34
0.15% Ni-3	62	33
0.15% Ni-4	23	15
0.15% UVA-1	90	52

[a] PE-LD homopolymer + 0.03% AO-1.
[b] Blown films 200-μm thick.
[c] Florida exposure started in July 1984.
E_{50}, energy (kLy) to 50%-retained elongation.

Table 9 Light Stability of PE-LD Blown Films[a,b]

UV stabilization	E_{50} (kLy) Florida[c]	
	Without backing	On aluminum
Control	32	20
0.15% Ni-4	23	15
0.30% Ni-4	48	27
0.60% Ni-4	54	39
0.15% Ni-1	125	90
0.15% UVA-1	90	52
0.15% Ni-4 + 0.15% UVA-1	70	45
0.15% Ni-1 + 0.15% UVA-1	150	105

[a] PE-LD homopolymer + 0.03% AO-1.
[b] Blown films 200-μm thick.
[c] Florida exposure started in July 1984.
E_{50}, energy (kLy) to 50%-retained elongation.

2. Nickel derivatives, often called nickel quenchers, especially Ni-1.
3. Polymeric or high molecular mass HALS, such as HALS-2, HALS-3, and HALS-4.

The corresponding stabilization mechanisms are discussed at length in the literature (8,9,16,19,25–32) and will be referred to only if necessary to understand the data.

Nickel derivatives, such as Ni-1, were among the first compounds to be used for stabilization of greenhouse films. Research for improved nickel compounds was continued well into the 1980s, in spite of the availability of HALS. However, it can be seen from Table 8 that none of the compounds commercialized after Ni-1 showed better performance. On the contrary, Table 9 indicates that some nickel compounds cannot even compete with Ni-1 when they are used at much higher concentrations or in combination with the benzophenone UV absorber UVA-1. This is partly

Table 10 Influence of Film Thickness and UV Stabilizer Type on Light
Stability of PE-LD Blown Films[a]

Light stabilization	E_{50} (kLy) Florida[b]		
	50 μm	100 μm	200 μm
Control	21	25	32
0.15% UVA-1	32	44	54
0.15% Ni-1	45	68	85
0.15% HALS-2	50	60	150
0.15% HALS-3	125	175	280

[a] PE-LD homopolymer, film grade, MF (190°C/2.16 kg) = 0.3 + 0.03% AO-1.
[b] Florida exposure started in May 1980, on aluminum.
E_{50}, energy (kLy) to 50%-retained elongation.

Table 11 Influence of Film Thickness and UV Stabilization on Light
Stability of PE-LD Blown Films[a]

Light stabilization	E_{50} (kLy) Florida[b]		
	50 μm	100 μm	200 μm
Control	27	31	34
0.15% HALS-2	105	120	170
0.15% HALS-2 + 0.15% UVA-1	105	190	295
0.3% HALS-2	155	165	215
0.3% HALS-2 + 0.3% UVA-1	140	255	420
0.6% HALS-2	180	210	330

[a] PE-LD homopolymer, film grade, MF(190°C/2.16 kg) = 0.3 + 0.03% AO-1.
[b] Florida exposure started in May 1980, without backing.
E_{50}, energy (kLy) to 50%-retained elongation.
Source: Ref. 21.

because the performance of new compounds is often assessed with accelerated
exposure devices that show poor correlation with natural weathering (13).

The UV stability of PE-LD films is largely determined by the stabilization
system and the film thickness. Because the different stabilizers respond differently
to variations of film thickness, the two parameters will be treated simultaneously.

The data in Table 10 give an idea of the response of representatives of the
different aforementioned UV stabilizer types, to an increase of film thickness.
As a rule, the lifetime increases with film thickness. This holds even for the control
sample, although the increase is rather limited. This increase is much more pronounced with the UV-stabilized films. It can also be seen in Table 10 that the performance of the UV absorber UVA-1 becomes significant for film thickness
above 100 μm. The other UV stabilizers in Table 10, nickel derivative and HALS,
are already effective for a film thickness of 50 μm. This is particularly pronounced
with HALS-3. Table 11 shows the influence of film thickness on the performance
of the polymeric HALS-2 and some of its combinations with the benzophenone-type
UV absorber UVA-1. It can be seen in Table 11 that the UV absorber does not

Table 12 Influence of Film Thickness and UV Stabilization on Light Stability of PE-LD Blown Films[a]

Light stabilization	E_{50} (kLy) Florida[b]		
	50 μm	100 μm	200 μm
Control	27	31	34
0.15% HALS-3	195	260	450
0.15% HALS-3 + 0.15% UVA-1	200	320	535
0.3% HALS-3	280	445	550

[a] PE-LD homopolymer, film grade, MF(190°C/2.16 kg) = 0.3 + 0.03% AO-1.
[b] Florida exposure started in May 1980, without backing.
E_{50}, energy (kLy) to 50%-retained elongation.

Table 13 Light Stability of PE-LD Blown Films[a,b]

UV stabilization	E_{50} (kLy) Florida[c]	
	Without backing	On aluminum
0.15% HALS-3	400	310
0.30% HALS-3	500	450
0.15% HALS-3 + 0.15% UVA-1	465	365
0.15% HALS-3 + 0.15% UVA-2	540	475
0.30% HALS-3 + 0.30% UVA-1	625	475
0.30% HALS-3 + 0.30% UVA-2	725	625

[a] PE-LD homopolymer + 0.03% AO-1.
[b] Blown films 200-μm thick.
[c] Florida exposure started in May 1981.
E_{50}, energy (kLy) to 50%-retained elongation.
Source: Ref. 21.

contribute to the performance of HALS-2 at all if the PE-LD films are only 50-μm thick. It is only from a film thickness of 100 μm on that the UV absorber contributes significantly to the light stability of the films. The contribution is higher than that of a same additional amount of HALS-2. The contribution of the UV absorber is even more pronounced with 200-μm–thick films. The results observed with combinations HALS-3–UVA-1 are quite different. Table 12 shows that the contribution of UVA-1 to the performance of the combination remains small, even for a film thickness of 200 μm. Moreover, this contribution is always inferior to that of the same amount of HALS-3. This is no more true if HALS-3 is used in combination with a benzotriazole-type UV absorber, such as UVA-2. The data in Table 13 show that the contribution of UVA-2 is not only superior to that of the same amount of UVA-1, but also to that of the same amount of additional HALS-3.

The unexpected superiority of UVA-2 over UVA-1 has been attributed to the fact that UVA-2 is protecting the polymer according to mechanisms that are complementary to the protection mechanisms of HALS (i.e., absorption of UV radiation and deactivation of the excited states of carbonyl groups; 5,28).

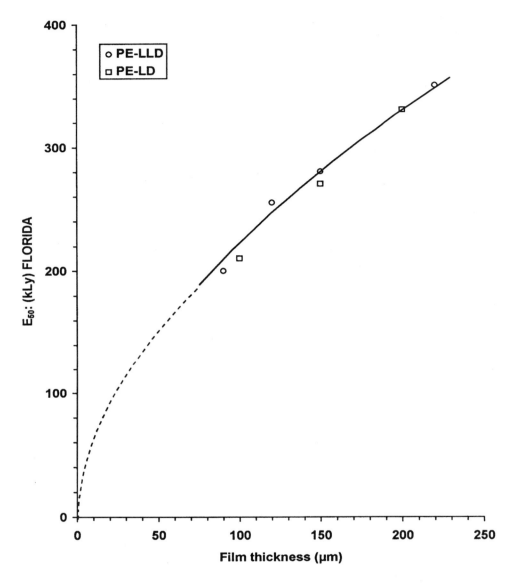

Figure 1 Effect of film thickness and polymer on performance on natural weathering: PE-LD homopolymer + 0.03% AO-1, blown films; PE-LLD, solution process, octene comonomer + 0.015% AO-2 + 0.15% Ca stearate, blown films. Light stabilization: 0.2% HALS-2 + 0.2% UVA-1; Florida exposure started in August 1984; E_{50}, energy (kLy) to 50%-retained elongation.

The effect of film thickness on performance seems to be the same for PE-LLD as for PE-LD. In Fig. 1 it can be seen that this is certainly true for the combination HALS-2–UVA-1 for film thickness between 90 and 220 μm. The exact law of variation with film thickness has not yet been established. It seems to depend on the stabilization system. However, for a rough estimation, for film thickness between

Table 14 Influence of Film Thickness on Light Stability of Coextruded Films[a,b,c,d]

Film thickness (μm)	Months to 50%-retained elongation[e]
75	21
100	>24 (65%)
150	>24 (90%)

[a] PE-LD homopolymer, film grade.
[b] EVA copolymer, 10% VA, film grade.
[c] PE-LD/EVA coextruded films, each layer half of the total thickness.
[d] Stabilization: 0.6% HALS-2 + 0.3% UVA-1.
[e] Exposure in Brazil (Sao Paulo) started in May 1989.
Source: Ref. 24.

Table 15 Light Stability of EVA Films[a,b,]

UV Stabilization	E_{50} (kLy) Florida[c]	
	Without backing	On aluminum
Control	50	40
0.30% HALS-2	220	190
0.30% HALS-2 + 0.30% UVA-1	385	215
0.60% HALS-2	360	375
0.60% HALS-2 + 0.60% UVA-1	525	475
1.20% HALS-2	460	450

[a] EVA copolymer (14% VA) + 0.03% AO-1.
[b] Blown films 200 μm thick.
[c] Florida exposure started in May 1981.
E_{50}, energy (kly) to 50%-retained elongation.

50 and 250 μm, the performance can be considered to increase proportionally to the square root of the thickness. In other terms, if film thickness passes from 50 to 200 μm, the film lifetime with the same stabilization system is doubled. Nevertheless, it must always be remembered that this rule of thumb is an approximation and cannot yield exact forecasts for many different stabilization systems.

It is expected that EVA films show a similar increase of lifetime with film thickness. First results with coextruded films PE-LD–EVA also show a significant increase of lifetime with film thickness (Table 14).

The combinations of HALS with UV absorbers are also highly effective in EVA copolymers. This is true not only for the combinations of the benzophenone-type UV absorber with HALS-2 (Table 15), but also for the corresponding combinations with HALS-3 (Table 16). Table 16 shows that the combination HALS-3–UVA-1 is also synergistic, whereas in PE-LD homopolymer it was not (see Tables 12 and 13).

In summary, under ideal conditions, the traditional formulations based mainly on combinations of the nickel stabilizer Ni-1 with the benzophenone UV absorber UVA-1 can be advantageously replaced by high molecular mass HALS used alone or in combination with benzophenone- or benzotriazole-type UV absorbers. Besides

Table 16 Light Stability of EVA Films[a,b]

UV stabilization	E_{50} (kLy) Florida[c]	
	Without backing	On aluminum
Control	50	40
0.15% HALS-3	380	350
0.15% HALS-3 + 0.15% UVA-1	580	580
0.30% HALS-3	460	450
0.30% HALS-3 + 0.30% UVA-1	670	580
0.60% HALS-3	560	560

[a] EVA copolymer (14% VA) + 0.03% AO-1.
[b] Blown films 200-μm thick.
[c] Florida exposure started in May 1981.
E_{50}, energy (kLy) to 50%-retained elongation.
Source: Ref. 21.

superior light stability and contribution to long-term heat aging, HALS offer two main additional advantages over nickel-based stabilizer formulations. First, HALS do not confer the greenish-yellow color observed with films containing nickel quenchers such as Ni-1. The films are not only colorless, but are also more transparent to UV and visible light. This can lead to accelerated growth of the plants and higher yield (34,35). Second, HALS do not present the environmental problems associated with nickel-containing stabilizers, especially on combustion or dumping of the films.

That the light stability of films is a function of the thickness should be kept in mind. It is especially important if PE-LD is replaced by PE-LLD, giving comparable mechanical properties with thinner films. Then, it is mandatory to increase the light stabilizer concentration if the same UV stability is to be maintained. Notably, films of 100–150 μm can be used for a duration of 1 year or less. However, for lifetimes of 2 or more years, the recommended thickness is 180–200 μm.

D. Fillers

Mineral fillers, such as kaolin (china clay), are used in greenhouse films to obtain an IR barrier effect (36–38). By other means, the purpose is the same as that of an EVA copolymer film (i.e., reduce the dissipation of thermal energy during night and thereby, increase the mean temperature in the greenhouse). However, the presence of kaolin can strongly affect the light stability of the films. Table 17 shows that the detrimental effect is observed in the absence as well as in the presence of a UV stabilizer system. The strong negative influence of the china clay employed in this test series is likely related to the presence of iron impurities. The aluminum silicate also included in the series originates from kaolin containing fewer impurities and in addition is calcined. As expected, the negative effect on light stability is less pronounced with the purer filler. The chalk used for the tests hardly influences light stability.

With PE-LD films containing china clay, the behavior of the benzo-phenone-type UV absorber in combination with HALS is again different from that

Table 17 Influence of Fillers on Light Stability of PE-LD Films[a,b]

Filler	E_{50} (kLy) Florida[c]	
	Without UV stabilizer	0.15% HALS-2 + 0.15% UVA-1
Without	50	255
1% Kaolin	34	140
3% Kaolin	25	90
5% Kaolin	23	85
3% Aluminum silicate	31	135
1% Chalk (coated)	48	270
3% Chalk (coated)	46	280
5% Chalk (coated)	46	300

[a] PE-LD homopolymer, MF(190°C/2.16 kg) = 0.3, 0.03% AO-1.
[b] Blown films 200-μm thick.
[c] Florida exposure started in January 1982, without backing.
E_{50}, energy (kLy) to 50%-retained elongation.
Source: Ref. 12.

Table 18 Light Stability of PE-LD Thermic Films[a,b]

UV stabilization	E_{50} (kLy) Florida[c]	
	Without backing	On aluminum
Control	22	19
0.15% HALS-3	150	135
0.15% HALS-3 + 0.15% UVA-1	210	145
0.15% HALS-3 + 0.15% UVA-2	275	235
0.30% HALS-3	270	235
0.30% HALS-3 + 0.30% UVA-1	245	180
0.30% HALS-3 + 0.30% UVA-2	500	400
0.60% HALS-3	500	400

[a] PE-LD homopolymer + 5% china clay + 0.03% AO-1.
[b] Blown films 200-μm thick.
[c] Florida exposure started in July 1986.
E_{50}, energy (kLy) to 50%-retained elongation.
Source: Ref. 17.

observed in the foregoing examples. Thus, combining UVA-1 with HALS-2 yields an increase in UV stability, but this increase is less pronounced than that caused by an additional quantity of HALS-2 (17). In combination with HALS-3, the effect of UVA-4 is negligible or even negative (Table 18). However, the combination of HALS-3 with a benzotriazole-type UV absorber in Table 18 shows interesting performance, similar to that observed in the absence of china clay (see Table 13). Even more astonishingly, the combination of the polymeric HALS-2 with the polymeric HALS-3 also shows synergistic effects with kaolin-containing PE-LD films. This is shown in Table 19.

Table 19   Combination of Polymeric HALS in PE-LD Thermic Films[a,b]

Light stabilization	E_{50} (kLy) Florida[c]			
	No HALS	0.3% HALS	0.6% HALS	1.2% HALS
Control 27	—	—	—	
HALS-2	—	95	150	230
HALS-3	—	240	400	610
HALS-5	—	175	370	600

[a] PE-LD homopolymer + 5% china clay + 0.03% AO-1.
[b] Blown films 200-µm thick.
[c] Florida exposure started in July 1986, aluminum backing.
E_{50}, energy (kLy) to 50%-retained elongation.
Source: Ref. 21.

E. Pigments

Although more commonly used in mulch films and silage films, pigments may also be used in greenhouse films. Their purpose in the last application may be to differentiate the film from competitors or to protect the culture from excessive sunshine. Another application becoming more and more important is to influence plant growth by light filtration through use of photoselective films (39–43).

Addition of pigments to plastics in general, and polyolefins in particular, must be done with due care. Pigments can have a detrimental influence on the light stability of the plastic material (16,19). This holds also with polyethylene films.

Table 20 shows results obtained with white-pigmented PE-LD films on natural weathering. It can be seen that, depending on the special pigment used, the UV stability can be higher or considerably lower than that of a corresponding nonpigmented film. The best results are obtained with coated, stabilized, rutile-type titanium dioxide.

The influence of some organic and inorganic pigments on light stability is illustrated in Table 21. In this instance also, the light stability can be considerably higher or lower than that observed with the nonpigmented film. Hence, it is always advisable to check the influence of pigments on light stability.

F. Other Additives

It has been claimed that the addition of coordination compounds of rare-earth metals, such as Europium, leads to a reduction of the UV light accompanied by an increase in the red light, with a peak at 615 nm (44). As a result, it is claimed that the activity of the phytochromes and of the overall photosynthetic process is increased and leads to significantly higher yields. The effects on the light stability of the films are unknown.

The presence of other additives, such as antiblock, slip, antistatic, and antifogging agents, can affect the light stability of polyethylene films. The negative influence can be related to low purity (e.g., iron in silica) or poor compatibility with the substrate. The last can lead to loss of stabilizers if the additional additive drags them to the film surface. In addition, poor chemical stability of the coadditive

Table 20   Influence of Titanium Dioxide on Light Stability of PE-LD Films[a,b]

Pigments (2%)[c,d,e]	E_{50} (kLy) Florida[f]
Without	220
A: Rutile, coated, stabilized (sulfate process)	380
B: Rutile, coated, stabilized (sulfate process)	>420[g]
C: Rutile, coated, stabilized (chloride process)	>420[g]
D: Rutile, not coated, unstabilized	70
E: Rutile, not coated, unstabilized	70
F: Rutile, coated, unstabilized	60
G: Anatase, coated, stabilized	150

[a] PE-LD homopolymer + 0.2% HALS-3.
[b] Blown films 150-μm thick.
[c] Commercial pigment master batches (50% in PE-LD).
[d] Pigment stabilizer: inorganic modifier, usually Al_2O_3/SiO_2.
[e] Pigment coating: usually silicones.
[f] Florida exposure started in June 1983, PMMA backing.
[g] No samples left.
E_{50}, energy (kLy) to 50%-retained elongation.

Table 21   Influence of Color Pigments on Light Stability of PE-LD Films[a,b]

Pigment[c]	E_{50} (kLy) Florida[d]		
	Without UV stabilizer	0.2% HALS-3	0.2% HALS-3 + 0.2% UVA-1
Without pigment	30	220	300
2% lead chromate (pigment yellow 34)	35	>420[e]	>420
2% lead molybdate (pigment red 104)	25	>420[e]	>420
1% diarylide yellow (pigment yellow 17)	20	120[f]	130[f]
0.05% α-phthalocyanine blue	—	300	—
0.08% ultramarine blue	—	>420[e]	—

[a] PE-LD homopolymer, film grade.
[b] Blown films 150-μm thick.
[c] Commercial pigment master-batches (50% in PE-LD).
[d] Florida exposure started in June 1983, PMMA backing.
[e] No samples left.
[f] Color faded.
E_{50}, energy (kLy) to 50%-retained elongation.

can have a prodegradant effect on the film (e.g., unsaturated groups in fatty acid amines).

G. Film Manufacturing

It is obvious for those working in the field, that some basic rules concerning film manufacturing should always be observed:

1. The additives must be distributed homogeneously in the film. This can be easily achieved either by using resins (compounds) already containing these additives or by adding them in the form of a master batch. Blending of resins in pellet form with additives in powder form should be avoided. The same is true for the addition of pure additives directly through the hopper.
2. Variations of film thickness should be kept to a minimum.

It is quite obvious that the films should be produced according to the state of the art, following the recommendations given by the polymer manufacturer.

III. GENERAL ENVIRONMENTAL PARAMETERS

A. Framework

Many different materials are used for manufacturing greenhouse structures. Aluminum, galvanized iron, and wood are most frequently used for this purpose. The detrimental influence of some materials on light stability of PE-LD films has long been known. It has been reported that the lifetime in the absence of any backing or with a neutral backing, such as (poly(methyl methacrylate) (PMMA) plates, is 1.45 times the lifetime of PE-LD films on aluminum (8).

Additional comparison data on aluminum with neutral backing have already been shown in this chapter. The influence of some additional materials on the lifetime of a film is illustrated in Table 22. It can be seen that the effects of painted iron and galvanized iron are comparable with the effect of aluminum. However, pinewood appears especially detrimental to the light stability of the films. This has been investigated in detail. A high amount of acidic species migrates from the wood into the polyethylene film; abietic acid is a major component of pine wood resins. Direct compounding into polyethylene simultaneously with HALS-3 has been used to check its effect. The results, summarized in Table 23, do not show any detrimental effect of abietic acid. Hence, the sensitizing effect from the migrating wood components, also documented in Table 23, must originate from other substances.

Usually the films break at the point of contact with the framework. This can be for several reasons (e.g., heat accumulation at the support, light reflection, and contamination of the film with prodegradants from the support). In addition to contamination by components from wooden supports, corrosion by iron from an iron structure, or defects in painted or galvanized iron, accelerates the aging of the film. In practice, excellent results have been obtained by painting the metallic frame or the film at the point of contact with a white vinyl acrylic-based paint. This coating prevents contamination of the film by the support material and, owing to the white color, it does not accumulate too much heat. Frequently, instead of painting the film, the framework is covered at the point of contact with several layers

Table 22 Influence of Backing on Light Stability of PE-LD Films[a,b]

| Light stabilization | E_{50} (kLy) Bologna[c] | | | |
	Pine wood[d]	Galvanized iron	Painted iron[e]	Aluminum
0.3% HALS-3	150	210	>220	200
0.6% HALS-3	250	350	315	345
0.3% HALS-3 + 0.3% UVA-1	245	235	280	285

[a] PE-LD homopolymer, film grade + 0.03% AO-1.
[b] Blown films 200-μm thick.
[c] Bologna, Italy, exposure started in March 1980.
[d] Untreated.
[e] White acrylic paint.
E_{50}, energy (kLy) to 50%-retained elongation.

Table 23 Effect of Pine Wood on Light Stability of PE-LD Films[a]

| Treatment/additive | $T_{0.1}$ (h) | |
	Film 1(200 μm)[b]	Film 2(100 μm)[c]
No treatment	9700	7800
800 (h) between pine[d] wood at 60°C	8200	—
0.3% abietic acid added	—	8600

[a] PE-LD homopolymer, film grade, MF(190°C/2.16 kg) = 0.3 + 0.03% AO-1.
[b] Film 1, blown film 200-μm thick.
[c] Film 2, compression molded film, 100-μm thick.
[d] Accelerated exposure in a Weather-ometer WRC 600, black panel temperature ~55°C, no water spraying.
$T_{0.1}$, time (h) to 0.1 carbonyl absorbance.

of polyethylene film. The greenhouse structure has to be sufficiently rigid and without rough surfaces.

B. Design

The importance of different factors in the design of greenhouses has been discussed in detail (45,46). It is much less known that the design of the greenhouse may influence the lifetime of the films, especially when pesticides are applied. As a matter of fact, the closer the plants are to the greenhouse film, the more pesticides can come into contact with the film. The design, open or closed, high or low, and such, and the ventilation, play an important role in this context.

C. Film Fixation

There are three important considerations deserving attention:

1. The film must be taut, otherwise the wind can cause mechanical damage.
2. The film must be properly fixed (e.g., with a two-piece connector), to avoid tearing stresses at the puncture points (no corrosive materials).
3. Rusted wires should be replaced because they often cause premature failure.

D. Climate and Solar Irradiation

The solar radiation energy is measured with pyranometers and expressed as joule per square centimeter (J/cm^2) in the SI system. The langley (abbreviation: Ly) is often more convenient and is used in this chapter. Both units are easily converted to one another by the following equivalency:

$$1 \text{ Ly} = 1 \text{ cal}/cm^2 = 4.184 \text{ J}/cm^2$$

The annual average incident energy in different parts of the world is documented in the literature (47,48). For a few exposure sites used in the work discussed here and for an angle of 45° south it is:

Florida $\sim$ 140 kLy/yr
Almeria (southern Spain) $\sim$ 150 kLy/yr
Pontecchio Marconi (near Bologna, northern Italy) $\sim$ 110 kLy/yr
Sicily (Italy) $\sim$ 160 kLy/yr
Bogota (Colombia) $\sim$ 140 kLy/yr
Sao Paulo (Brazil) $\sim$ 150 kLy/yr
Basel (Switzerland) $\sim$ 100 kLy/yr (mean value in the 10-year period 1987–1996:98.9 kLy/yr).

In some other sites around the world it is:

Arizona $\sim$ 190 kLy/yr
Bandol (southern France) $\sim$ 130 kLy/yr
Dahran (Saudi Arabia) 180–220 kLy/yr (49)
Singapore $\sim$ 140 kLy/yr (48)
Guangzhou (China) $\sim$ 100 kLy/yr
Melbourne (Australia) $\sim$ 140 kLy/yr (48)
Darwin (Australia) $\sim$ 180 kLy/yr (48)
Nairobi (Kenia) $\sim$ 160 kLy/yr (48)
Spartan (South Africa) $\sim$ 170 kLy/yr
Upington (South Africa) $\sim$ 240 kLy/yr (50)

The solar energy (kLy) to which a sample is subjected is just one parameter influencing its aging. The severity of the climate of a given region is also influenced by such factors as local temperature, wind, rain, humidity, and pollution. By exposing identical samples of a PE-LD film at different sites, it has been shown (8) that 1 year of exposure in Florida ($\sim$ 140 kLy) corresponds to 2.5–3 years of exposure in Basel (225–270 kLy). This clearly shows that a given amount of energy in a tropical climate is much more detrimental to greenhouse films than the same amount in a temperate climate.

It is noteworthy that under more or less ideal conditions, determination of UV stability in tropical climates shows quite good reproducibility. This is shown in Table 24 for natural weathering in Florida.

E. Temperature

Sample temperature plays an essential role in polymer degradation during weathering. The greenhouse film is exposed to not only ambient temperature, but also to heat stress at the point of contact with the framework. Ni-1 contributes much less to thermooxidative aging of PE-LD films at 100° and 80°C than a typical phenolic antioxidant such as AO-1 (31). On the other hand, HALS-2 was much more effective than the same antioxidant (31). This is confirmed once more by the data presented in Table 25 in which the test criterion is formation of some amount of carbonyl groups. The results obtained with PE-LLD films also show a pronounced contribution of HALS to long-term heat aging (Table 26). Notably the test criterion considered involves mechanical properties (50%-retained elongation), whereas in Table 25 the test criterion is chemical (i.e., carbonyl absorbance).

It can be concluded that the polymeric HALS-2 and HALS-3 contribute to thermooxidative stability as well as to photooxidative stability.

IV. SPECIAL ENVIRONMENTAL PARAMETERS: AGROCHEMICALS

The nature of the crop will indirectly determine the special environment of a greenhouse film. On the one hand, the crop will influence the type of pesticides to be applied. The chemical nature of the pesticides is an important factor influencing the useful lifetime of greenhouse films. On the other hand, the influence of the crop is related to the design of the greenhouse (see Sec. III.B), in the sense that a large distance between the plants and the film will reduce the risk of direct spraying of pesticides onto the film.

Pesticides used inside greenhouse are mostly sulfur- or halogen-based compounds. Hence, it is not astonishing that they can affect the light stability of greenhouse films, in general, and of HALS-stabilized greenhouse films, in particular. In fact, it was found very soon after HALS became available, that sulfur-containing additives, such as distearyl thiodipropionate (DSTDP), can have a negative effect on the light-stabilizing performance of HALS (5,16,19). Furthermore, HALS showed poor performance in halogenated polymers and in polyolefins in the presence of halogenated flame-retardants (16,19,29).

The extent to which agrochemicals affect the useful lifetime of PE films, depends on many factors. Among these factors, the composition of the chemicals, the frequency of application, the method of application, and the design of the greenhouse seem to be the most important. It is beyond the scope of this chapter to discuss the use and the interchangeability of pesticide and nematicide systems in practice. In the following, the first results will be reported as well as further developments leading to better understanding of the phenomena and possible solutions.

A. Early Results with Agrochemicals in Spain

The first greenhouse cover film failures attributed specifically to agrochemicals were reported at the beginning of the 1980s in Spain. This spurred a broad investigation of

Table 24 Data Reproducibility on Natural Weathering of PE-LD Blown Films (200 μm)[a]

			E_{50} (kLy) Florida on exposure[c] started in					
UV stabilization[b]	November 1977	March 1979	September 1979	May 1980	March 1981	July 1984	April 1989
Control	70	42	70	35	42	34	33
0.6% HALS-2	350	355	—	—	375**	290***	375****
		290*		330*			
0.3% HALS-2 + 0.3% UVA-1	—	360	320	335*	350*	310***	—

[a] Results marked * originate from a different batch of film from those marked ** or *** or **** or not marked.
[b] Base stabilization: 0.03% AO-1.
[c] Exposure without backing.
E_{50}, energy (kLy) to 50% retained elongation.
Source: Ref. 21.

Table 25 Influence of UV Stabilizers on Long-term Heat Aging of PE-LD Films[a,b]

UV stabilization	$T_{0.2}$ (days) at $100°C$[c]
Without	43
0.15% HALS-2	275
0.30% HALS-2	330
0.60% HALS-2	450
0.15% HALS-3	290
0.30% HALS-3	500
0.60% HALS-3	900

[a] PE-LD homopolymer, film grade, + 0.03% AO-1.
[b] Blown films 200-μm thick.
[c] Aging in draft air oven at $100°C$, started in February 1992.
$T_{0.2}$, time (days) to 0.2 carbonyl absorbance.

Table 26 Influence of UV Stabilizers on Long-term Heat Aging of PE-LLD Films[a,b,c]

Stabilization	T_{50} (days)
Base stabilization	21
0.03% AO-1	225
0.03% AO-2	400
0.1% AO-4	290
0.05% AO-4 + 0.05% HALS-2	635
0.05% AO-4 + 0.05% HALS-3	600

[a] PE-LLD solution process, octene comonomer, film grade, +0.15% Ca stearate +0.005% AO-2.
[b] Blown films 100-μm thick.
[c] Aging in draft air oven at $100°C$.
T_{50}, time (days) to 50%-retained elongation.
Source: Ref. 19.

the phenomena involved. The first results of field trials became available in 1985 (51). The sulfur-containing nematicide sodium methyldithiocarbamate (also called metamsodium), used as a soil disinfectant, can have a marked detrimental effect on the lifetime of greenhouse films. The compound releases methylisothiocyanate and may affect not only the performance of the light stabilizers in the films, it can also directly induce polymer degradation (52,53).

This early study has the merit of pinpointing some sulfur-based agrochemicals as the most detrimental to the lifetime of greenhouse films stabilized with HALS. Thus, a single treatment of the soil with metam sodium was sufficient to destroy the film in 1 year. However, with the same treatment, if the soil is covered with a sheet during the application of the nematicide and for 3–4 weeks afterward, the damage to the film is considerably reduced. Moreover, it lasts as long as the films treated with the other chemicals used in the trials. This certainly is an example of use of agrochemicals without excessive damage to the greenhouse film (51). There

Table 27 Stability of PE-LD Greenhouse Films
(Sicily)[a,b,c]

Light stabilization	E_{50} (kLy)
0.2% HALS-3	120[d]
0.1% HALS-3 + 0.1% UVA-1	110
0.2% HALS-2 + 0.2% UVA-1	110

[a] PE-LD homopolymer, film grade.
[b] Blown films 150-μm thick.
[c] Exposure in Sicily on greenhouses, pesticides, pine wood
support, started in August 1985.
[d] Film analysis after 14 months exposure ($\sim$180 kLy):
440 ppm sulfur + 70 ppm chlorine.
E_{50}, energy (kLy) to 50%-retained elongation.

is also another possibility to reduce film deterioration by pesticides, that is, during
pesticide application due care is taken to minimize direct spraying onto the
greenhouse cover.

B. Broad Investigations in the Field and the Laboratory

1. Field Trials

The first film failures in Spain triggered a broad investigation of the origins and
possible remedies to the unexpected behavior observed. It showed first the diversity
and complexity of the phenomena involved.

In southern Italy and Sicily, the lifetime expected for greenhouse covers was
usually only one season (i.e., 9–10 months) excluding the summer months. In Table
27 it can be seen that these requirements can be met by relatively low concentrations
of HALS used alone or in combination with an UV absorber. This is valid in spite of
an extensive use of the pesticides shown in Table 28. As an example, HALS-3, when
used at 0.2%, conferred sufficient light stability to 150-μm–thick PE-LD films
for one season under the conditions encountered.

In another area, in Colombia, near Bogota, variously stabilized films covering
a carnation plantation were tested for light stability. The results of the experiments
are presented in Table 29. The chemical nature of the pesticides used and the frequency of application are shown in Table 30. In Table 29, it can be seen that both
nickel- and HALS-based formulations confer good UV stability under the conditions of the experiments. With the usual failure criterion of 50%-retained
elongation, the nickel stabilizer-containing films lasted 1½–2 years, whereas the
corresponding HALS-stabilized films lasted 1 year longer. Hence, under the conditions of the carnation greenhouse, there is a distinct advantage for HALS.
However, the results in Spain had shown that this is not always so. There, the chemical metam sodium was especially detrimental to HALS-stabilized greenhouse films.
This sulfur-containing pesticide–nematicide is often used in combination with
the chlorine-containing pesticide permethrin. There is another widely used pesticide
combination. It is based on the sulfur-containing compound mancozeb and the

Table 28 Pesticides Used in the Experiments in Table 27

Common name	Chemical formula	Application	Frequency per season
Thionazine	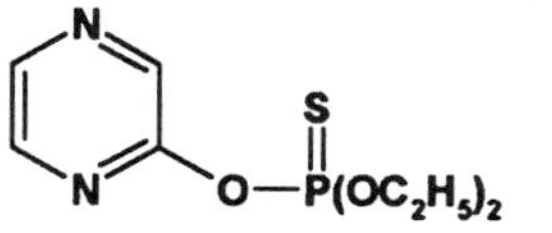	Underground (before covering with film)	1
Pirimicarb	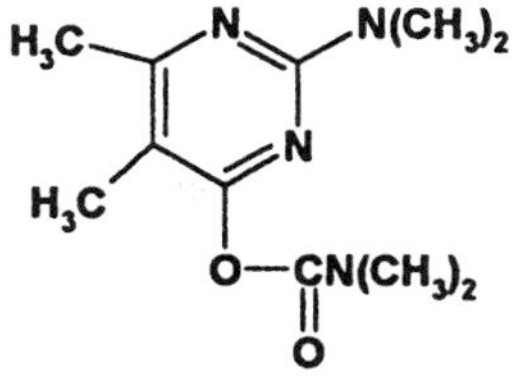	Sprayed	10
Tetradifon	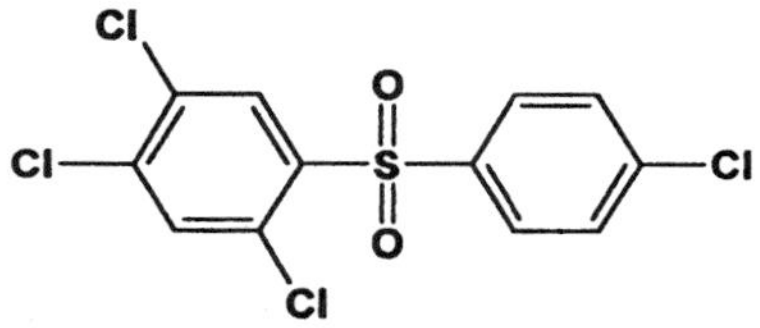	Sprayed	3–4
Dicofol	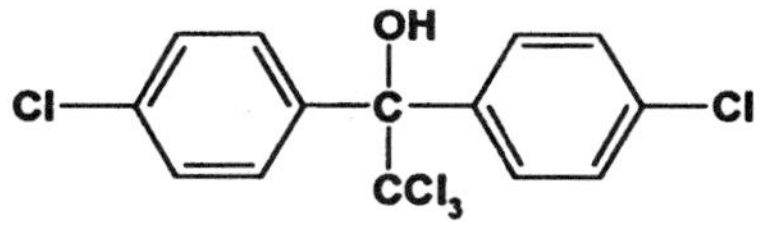	Sprayed	3–4
Parathion-methyl	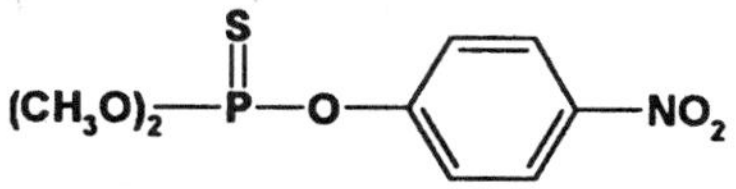	Sprayed	3–4
Fenarimol	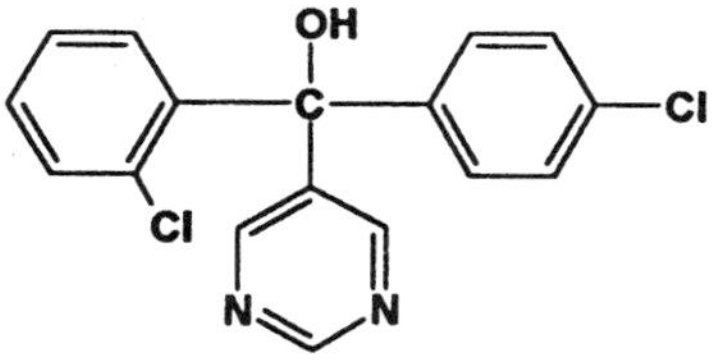	Sprayed	5

chlorine-containing compound captan. The respective effects of these two combinations on films were tested with two distinct greenhouses on the same exposure site at the same time. The results are shown in Table 31. The chemical structure of the pesticides is given in Table 32. It is noteworthy that all the pesticides involved in the trial were sprayed, although metham sodium and mancozeb are usually used in the soil for disinfecting (i.e., as a nematicide). Therefore, the tests were especially severe. It can be seen in Table 31 that the films covering greenhouse I (metam sodium plus permethrin) are significantly more affected than the films covering greenhouse II

Table 29 Stability of PE-LD Greenhouse Films
(Colombia)[a,b,c]

Light stabilization	T_{50} (mo)[d]
0.3% Ni-1 + 0.3% UVA-1	20
0.4% Ni-1 + 0.4% UVA-1	22[e]
0.3% HALS-2 + 0.3% UVA-1	32
0.4% HALS-2 + 0.4% UVA-1	34[f]
0.3% HALS-3 + 0.3% UVA-1	30
0.4% HALS-3 + 0.4% UVA-1	39

a PE-LD homopolymer, film grade.
[b] Blown films 180-μm thick.
[c] Exposure close to Bogota (Colombia), on greenhouses,
pesticides sprayed every week, eucalyptus wood support,
started in August 1986.
[d] Film analysis after 24-months exposure.
[e] 235 ppm sulfur + 55 ppm chlorine.
[f] 115 ppm sulfur + 70 ppm chlorine.
T_{50}, time to 50%-retained elongation.

(mancozeb plus captan). In fact, all the films on greenhouse II were in good condition
in every respect, even after 24-months exposure. All the films on greenhouse I were
intact for 21 months, but one HALS-stabilized film (0.75% HALS-2 + 0.375%
UVA-1) failed during the following summer. The film containing 0.75% HALS-3
was still intact after 24 months exposure on greenhouse I, but very close to breakage
with only 10%-retained elongation. As already shown by the results in Spain (see Sec.
IV.A), the difference in behavior of the films on the two greenhouses can very likely
be attributed to the use of the sulfur-containing metam sodium in greenhouse I.
This is confirmed by the very high concentrations of sulfur found in the films on
greenhouse I. From Table 31, one can see that these concentrations are higher than
the sulfur concentrations in the corresponding greenhouse II films by more than
an order of magnitude. The chlorine contents found in the films on both greenhouses
are comparable.

2. Laboratory Experiments and Interpretation

For some time it has been known that sulfur- and chlorine-containing chemicals
exert a negative influence on the light stability of HALS-stabilized polyolefins.
The explanations for this antagonistic effect have been as diverse as the explanations
for the stabilization by HALS (see Refs. 16 and 19 and discussions therein). They will
not be exposed in detail here, the discussion will be restricted to the explanation seen
as the most plausible with our current state of knowledge.

  Many halogen-containing chemicals are easily photolyzed by the UV part of
sunlight to yield halogen free radicals. These radicals will be more or less reactive,
depending on the particular halogen. The chlorine atoms formed on photolysis
of chlorine-containing chemicals are very reactive and will easily abstract a hydrogen
atom from the polymer to yield hydrochloric acid and a macroalkyl radical. These

Table 30 Pesticides Used in the Experiments in Table 29[a]

Common name	Chemical formula
Diazion	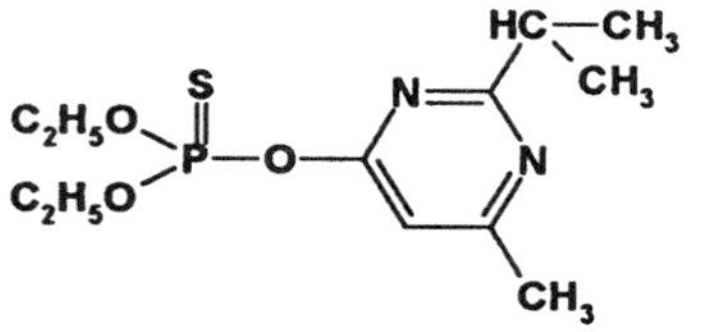 C_2H_5O, C_2H_5O—P(=S)—O—(pyrimidine ring): ring bearing $HC(CH_3)CH_3$, two N, and CH_3
Phosphamidon	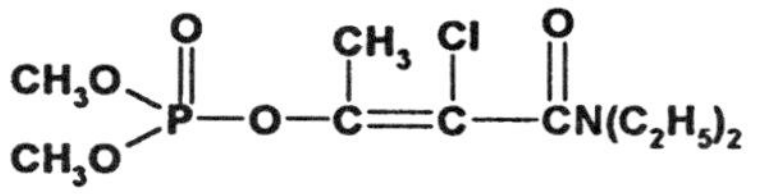 CH_3O, CH_3O—P(=O)—O—C(CH_3)=C(Cl)—C(=O)N$(C_2H_5)_2$
Methomyl	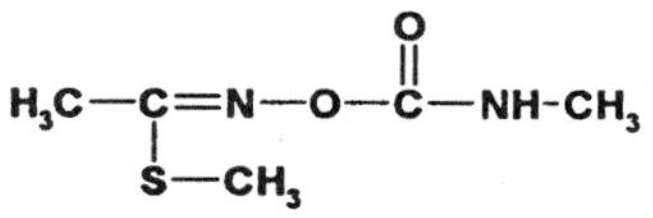 H_3C—C(S—CH_3)=N—O—C(=O)—NH—CH_3
Oxydemeton-methyl	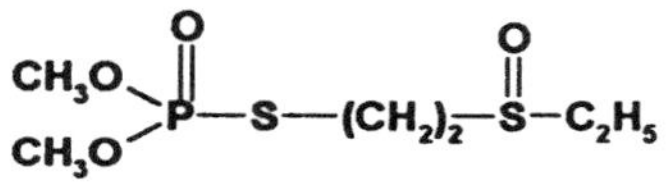 CH_3O, CH_3O—P(=O)—S—$(CH_2)_2$—S(=O)—C_2H_5
Pirimicarb	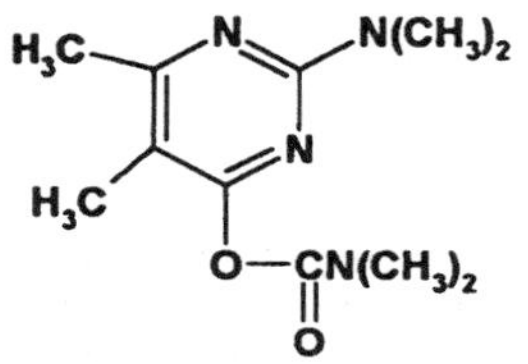 H_3C, H_3C pyrimidine ring with N(CH$_3$)$_2$ and O—C(=O)N$(CH_3)_2$
Propineb	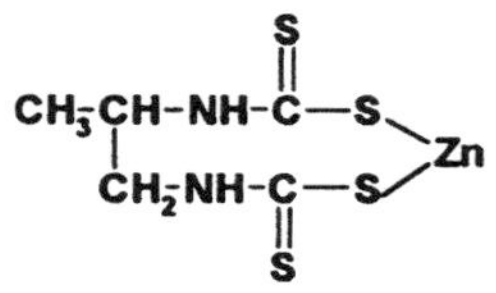 CH_3CH—NH—C(=S)—S, CH_2NH—C(=S)—S \ Zn
Benomyl	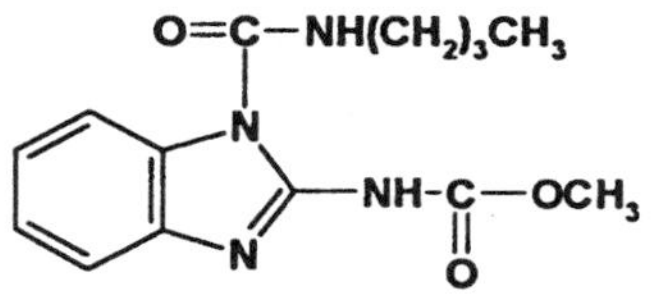 O=C—NH$(CH_2)_3CH_3$ on benzimidazole N; ring bearing —NH—C(=O)—OCH$_3$
Chlorothalonil	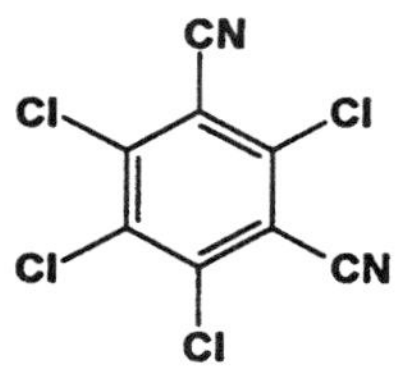 benzene ring with CN, CN, and four Cl substituents
Mancozeb	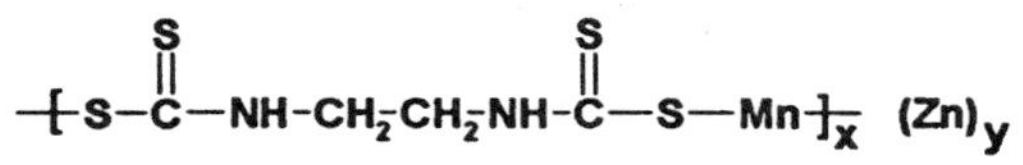 —[S—C(=S)—NH—CH_2CH_2NH—C(=S)—S—Mn]$_x$ $(Zn)_y$

[a] Application: all the chemicals sprayed. Frequency: above every 8 days.

Table 31 Influence of Pesticide Treatment on Light Stability of PE-LD[a,b] Greenhouse Films[c,f]

	T_{50}	
Light stabilization	Greenhouse I[d] (mo)	Greenhouse II[e] (mo)
0.75% Ni-1 + 0.5% UVA-1	25	26
0.75% HALS-2 + 0.375% UVA-1	<21[g]	27[h]
0.5% HALS-2 + 0.5% Ni-1	28	>34(62%)
0.5% HALS-3 + 0.5% Ni-1	25	>34(73%)
0.75% HALS-3	22[i]	>34(64%)[j]

[a] PE-LD homopolymer, film grade + 0.03% AO-1.
[b] Blown films 200-μm thick.
[c] Pontecchio Marconi (Italy) exposure on greenhouses, galvanized iron support, started in October 1984.
[d] Greenhouse I: metamsodium every 6 months, permethrin every month.
[e] Greenhouse II: mancozeb every 6 months, captan every month.
[f] Analysis performed after 24-mo exposure.
[g] 1200 ppm sulfur + 28 ppm chlorine.
[h] 92 ppm sulfur + 26 ppm chlorine.
[i] 1700 ppm sulfur + 36 ppm chorine.
[j] 135 ppm sulfur + 34 ppm chlorine.
T_{50}, time (mo) to 50%-retained elongation.

Table 32 Pesticides Used in the Experiments in Table 31[a]

Common name	Chemical formula
Metham	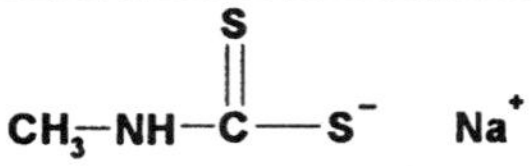
Permethrin	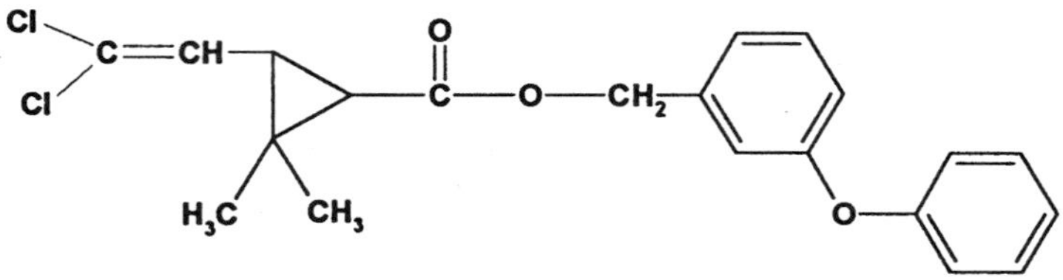
Mancozeb	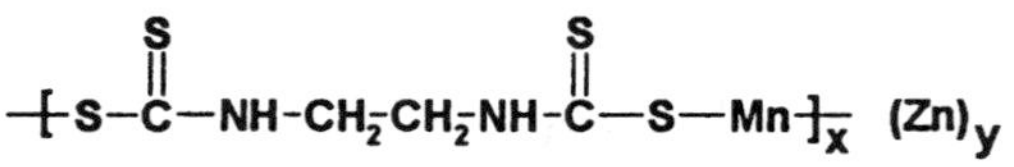
Captan	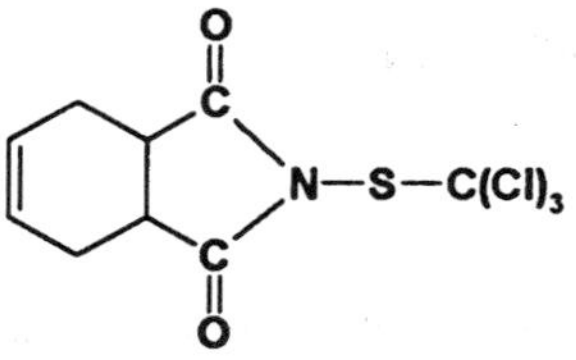

[a] Application: all the chemicals sprayed.

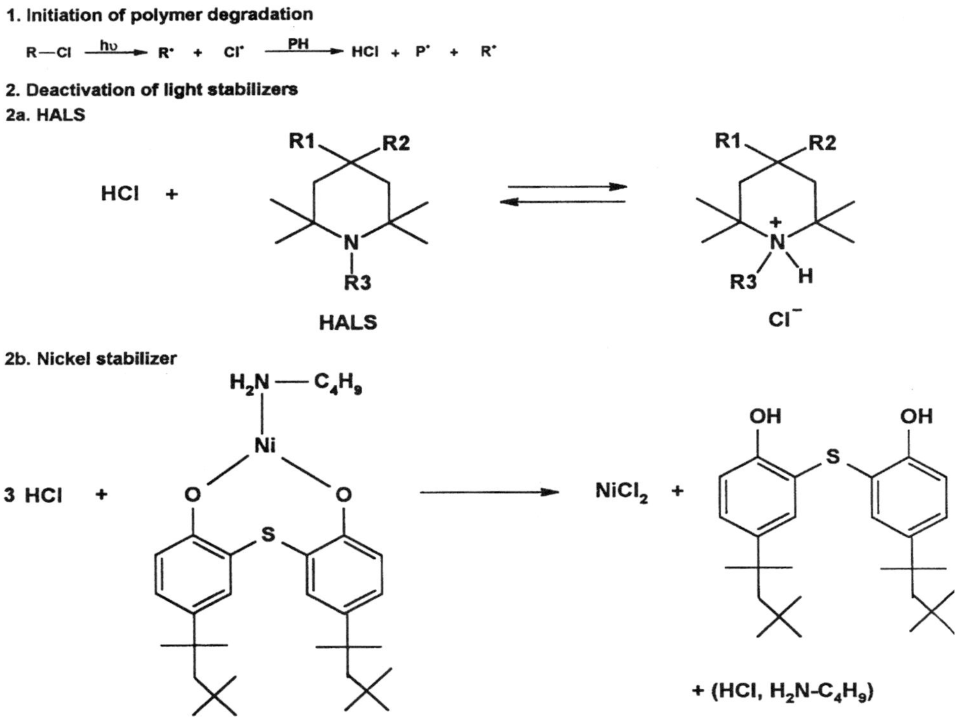

1. Initiation of polymer degradation

$$R—Cl \xrightarrow{h\upsilon} R^{\cdot} + Cl^{\cdot} \xrightarrow{PH} HCl + P^{\cdot} + R^{\cdot}$$

2. Deactivation of light stabilizers

2a. HALS

HCl +

HALS

2b. Nickel stabilizer

3 HCl +

$\longrightarrow$ NiCl$_2$ +

+ (HCl, H$_2$N-C$_4$H$_9$)

Scheme 1 Possible explanation for the detrimental effect of chlorine-based chemicals on light stability.

reactions are shown in Scheme 1. In this reaction, besides initiation of an oxidation chain and polymer degradation, there is formation of hydrochloric acid.

The experimental results plotted in Fig. 2 show the dramatic effect of this strong mineral acid on the performance of various light stabilizers in PE-LD films. The experiments consisted in overnight exposure of the films to hydrochloric acid vapor, followed by artificial weathering. In Fig. 2 one can see that the performance of the different light stabilizers is completely reversed in comparison with that observed in the absence of any treatment by hydrochloric gas (see, e.g., Table 10). In fact, at least in the early stages of exposure, the benzophenone-type UV absorber UVA-1 performs much better than the nickel-derivative Ni-1 which, in turn, is better than the polymeric HALS—HALS-2 and HALS-3. Even the control sample containing no light stabilizer, but also treated with hydrochloric gas, is superior to the HALS-stabilized samples in the early stages of exposure. However, after an exposure of several hundred hours, the rates of carbonyl increase with the HALS-stabilized films are considerably reduced. There is no such reduction of the rate of carbonyl formation with the Ni-1- or UVA-1-stabilized samples. These results are best explained by the reactions given in Scheme 1. As a matter of fact, the reaction of the acid with the basic amine yields the corresponding salt, which remains in chemical equilibrium with its constituents. The ammonium salt is not an active UV stabilizer and does not protect the polymer. This corresponds to

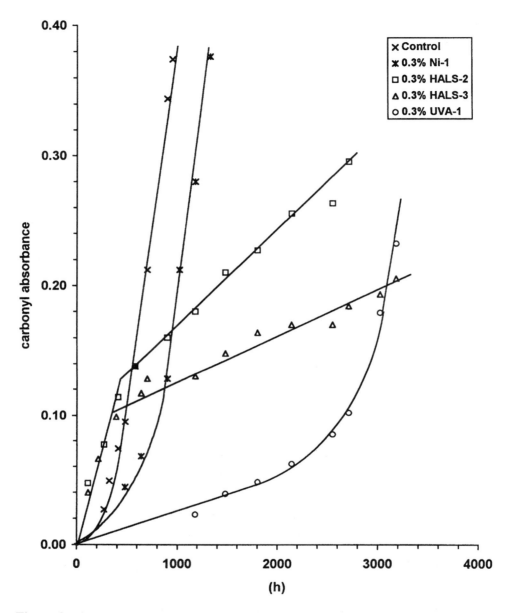

Figure 2   Light stability of PE-LD films (200 μm) after treatment with HCl gas for 16 h: exposure in Weather-ometer 600 WRC, BPT 55°C.

the early stages of the experiments in Fig. 2. With the concentration of free hydrochloric acid in the film decreasing with increasing exposure time through evaporation, the equilibrium in Scheme 1 is displaced toward the left. Therefore, increasing amounts of salt decompose with liberation of the amine, which can again act as an UV stabilizer. This corresponds to the second stage of the experiments shown in Fig. 2 with the HALS-stabilized films. The Ni-1-stabilized films do not

Table 33 Effect of Hydrochloric Acid on Light Stability of PE Films[a]

	$T_{0.1}$ (h)[b]	
UV stabilization	PE-LD homopolymer[c]	EVA copolymer[d]
Control	400	290
0.3% UVA-1	1250	830
0.3% Ni-1	470	210
0.3% HALS-2	210	340
0.3% HALS-3	250	310

[a] Blown films 200-μm thick, treated every week with hydrochloric acid vapor for 16 h.
[b] Accelerated exposure in a Weather-ometer WRC 600, with water spraying.
[c] PE-LD homopolymer, film grade, MF (190°C/2.16 kg) $= 0.3 + 0.03\%$ AO-1.
[d] EVA copolymer (14% VA), film grade, MF(230°C/2.16 kg) $= 0.6 + 0.03\%$ AO-1.
$T_{0.1}$, time (h) to 0.1 carbonyl absorbance (PE-LD homopolymer) or 0.1 hydroxyl absorbance (EVA copolymer).

show such a recovery of the stabilizing effect. This can be attributed to an irreversible destruction of the nickel stabilizer by hydrochloric acid, as shown in Scheme 1.

It can be concluded that the results of exposure in the presence of a strong and volatile mineral acid can be quite easily interpreted. Notably, the recovery of the stabilizing effect of HALS is possible only if the supply of hydrochloric acid is stopped. If there is continuous supply or repeated supply of high amounts of hydrochloric gas, there can be no recovery. The results from weekly treatments with HCl are shown in Table 33. These effects are quite dramatic, both on nickel- and HALS-stabilized films. There is an interesting effect to be deduced from the data in Table 33. In PE-LD homopolymer, HCl affects Ni-1 somewhat less than the HALS included in the test. Also, under the same conditions, in EVA copolymer, the HALS perform better than Ni-1 (see Table 33). The foregoing experiments also help us understand the results with sulfur-based chemicals in general, sulfur-based pesticides in particular. In fact, photooxidation of these chemicals yields various acidic species derived from sulfur, such as sulfurous acid SO_3H_2, sulfonic acids $R\text{-}SO_3H$, and sulfuric acid SO_4H_2. It is quite obvious that these acids will also react with the amine-based light stabilizers to yield the corresponding salts. Because these salts are not active as UV stabilizers, the reaction easily explains the reduced UV stability observed in the presence of sulfur-containing chemicals. Sulfonic acids and sulfuric acid are considerably less volatile than hydrochloric acid. Therefore, it is clear that the chemical equilibrium favors salt formation much more than with hydrochloric acid. This is shown in Scheme 2. With this mechanism, it is easily understood that, once the amines are neutralized by many oxyacids from sulfur, it is not possible to recoup part or all of the initial-stabilizing efficiency. This interpretation is in agreement with the high amounts of sulfur found by analysis in exposed films that have failed or are close to failure. It is also in agreement with the results on the quantitative effect of added sulfur-containing chemicals on the performance of HALS in polyolefins: the antagonistic effect is not a catalytic effect, but rather, a much more stoichiometric effect. It becomes really detrimental if the amount of sulfur compound becomes comparable with the amount of HALS present.

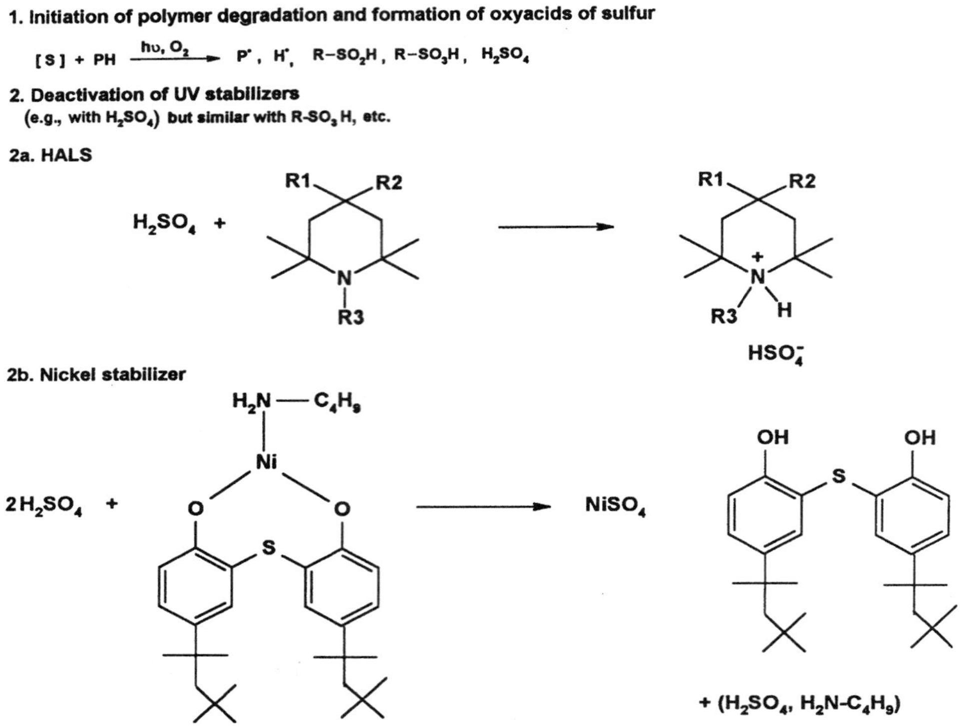

1. Initiation of polymer degradation and formation of oxyacids of sulfur

$$[S] + PH \xrightarrow{h\upsilon, O_2} P^\cdot, H^\cdot, R\text{-}SO_2H, R\text{-}SO_3H, H_2SO_4$$

2. Deactivation of UV stabilizers
  (e.g., with H_2SO_4) but similar with $R\text{-}SO_3H$, etc.

2a. HALS

HSO_4^-

2b. Nickel stabilizer

$NiSO_4$

$+ (H_2SO_4, H_2N\text{-}C_4H_9)$

Scheme 2   Interpretation of the detrimental effect of sulfur-based chemicals on polymer light stability.

   In some greenhouses, for plant protection, sulfur is burnt to give sulfur dioxide. This is also detrimental to the UV stability for various reasons. In other greenhouses, especially in rose greenhouses, the protecting properties of elemental sulfur against fungal diseases, such as "powdery mildew," are used extensively. This is achieved through heating elemental sulfur in the greenhouse at a temperature suitable for the sublimation of the required amount of sulfur.

C. Current Solution

The experimental results and discussions in the preceding section have stressed the role of acidic species from chlorine- and sulfur-containing pesticides in reducing the lifetime of greenhouse covers, especially of HALS-stabilized greenhouse films. Accordingly, it seems straightforward to add basic compounds to the stabilizer formulations to intercept preferentially the acidic species formed. However, this approach required considerable effort before an efficient and commercially valid solution could be proposed. It is based on the combination of an efficient HALS with a metal oxide and a stearate. HALS-7 and HALS-8 are the representatives of this class of compounds. The metal oxide used is acting as an acid scavenger and thus reduces the amount of acidic species that react directly with the HALS. From the experimental results discussed previously, it is clear that this should yield

Table 34 Stability of PE-LD Films[a,b] on an Experimental Greenhouse

	E_{50} (kLy) Pontecchio Marconi[c]		
Light stabilization	On pine wood	On galvanized iron	Window (no backing)
0.4% HALS-3 + 0.2% UVA-1	185	>190(60%)	175
0.4% HALS-6 + 0.2% UVA-1	165	170	165
0.7% HALS-7	>190(75%)	>190(80%)	>190(60%)

[a] PE-LD homopolymer, film grade.
[b] Blown films 150-μm thick.
[c] Pontecchio Marconi (Italy) exposure started in September 1992.
E_{50}, energy (kLy) to 50%-retained elongation.
Permethrin, chlorine-containing pesticide, applied monthly.
Metamsodium, sulfur-containing pesticide, applied every semester.
Source: Ref. 54.

Table 35 Stability of PE-LD Films[a,b] on Experimental Greenhouse

	E_{50} (mo) Pontecchio Marconi[c]		
Light stabilization	On pine wood	On galvanized iron	Window (no backing)
0.4% Ni-1 + 0.2% UVA-1	~14(interpolated)	~19(interpolated)	>22(75%)[d]
0.7% HALS-7	>24(65%)	>24(75%)	>22(60%)

[a] PE-LD homopolymer, film grade.
[b] Blown films 150-μm thick.
[c] Pontecchio Marconi, Italy, exposure started in September 1992.
[d] Film embrittled on support.
Permethrin, chlorine-containing pesticide, applied monthly.
Metamsodium, sulfur-containing pesticide, applied every semester.
Source: Ref. 55.

improved light stability. In addition, the metal oxide acts as an UV absorber and, as a consequence, directly protects the greenhouse cover film from photodegradation. The stearate, a necessary coadditive, seems to have a synergistic effect on the performance of the metal oxide (54,55).

Table 34 shows the behavior of a HALS-7-stabilized film on an experimental greenhouse with extensive pesticide spraying. In this test series HALS-7 is compared with two conventional HALS. The last, HALS-3 and HALS-6 show comparable performance, but are significantly less efficient than HALS-7. This is valid on pine wood and galvanized iron as well as in the absence of any backing. The comparison of HALS-7 with the combination Ni-1–UVA-1 in Table 35 shows similar superiority for HALS-7. This superiority is especially pronounced for the film on pine wood or galvanized iron backing. The good behavior of the Ni-1–UVA-1 combination in the film part on window (absence of backing) is noteworthy, although it does not prevent overall film failure resulting from failure on the backing. This behavior can be attributed to relative insensitivity toward sulfur-containing pesticides.

Table 36   Stability of PE-LD Films[a,b] on Experimental Greenhhouse

Light stabilization	E_{50} (kLy) Pontecchio Marconi[c]		
	On pine wood	On galvanized iron	Window (no backing)
0.9% Ni-1 + 0.5% UVA-1	250	280	>300(80%)[d]
0.9% HALS-3 + 0.5% UVA-1	>330(65%)	>300(75%)	285
0.9% HALS-2 + 0.5% UVA-1	230	295	275
1.5% HALS-8	>330(80%)	>330(90%)	>330(60%)

[a] PE-LD homopolymer, film grade.
[b] Blown films 200-μm thick.
[c] Pontecchio Marconi (Italy) exposure started in October 1993.
[d] Film embrittled on support.
E_{50}, energy (kLy) to 50%-retained elongation.
Permethrin, chlorine-containing pesticide, applied monthly.
Metamsodium, sulfur-containing pesticide, applied every semester.
Source: Ref. 54.

Table 37   Stability of PE-LD Films[a,b] Exposed in Almeria (Spain)[c] on FIAPA Experimental Greenhouse

Light stabilization	% retained elongation after ~300 (kLy)	
	On support[d]	Window (no backing)
0.75% Ni-1 + 0.5% UVA-1	25	48
0.8% HALS-3 + 0.4% UVA-1	Embrittled	Embrittled
0.8% HALS-2 + 0.3% UVA-1	Embrittled	Embrittled
1.2% HALS-7	85	80

[a] PE-LD homopolymer, film grade.
[b] Blown films 200 μm thick.
[c] Almeria exposure started in January 1995.
[d] Support consists of iron wire.
Permethrin, chlorine-containing pesticide, applied monthly.
Metamsodium, sulfur-containing pesticide, applied every semester.
Source: Ref. 55.

For especially long film life, use of HALS-8 is recommended, rather than HALS-7. This is to avoid excessive amounts of coadditives at the high concentrations required. Table 36 shows the results of a comparison of a high loading of HALS-8 with corresponding loadings of combinations of classic HALS or Ni-1 with the benzophenone-type UV absorber UVA-1. Overall, the data are similar to those obtained with HALS-7, but on a higher level. Again, the nickel-based stabilization shows good performance on window, but fails on the frame. Only HALS-8 shows good performance in the film, after 3 years of exposure, on window as well as on the backing.

The results discussed so far pertain to outdoor exposure in northern Italy. They are confirmed by preliminary results from southern Spain (Almeria). It can be seen in Table 37 that, the films stabilized with the classic HALS have already failed and the

Table 38 Light Stability of PE-LD Films[a,b] Before and After "Acid Rain" Treatment[c]

	$T_{0.1}$ (h) Weather-ometer[d]	
UV stabilization	Without treatment	After treatment
Control	285	180
0.1% HALS-3	4120	1640
0.1% HALS-4	4320	3680
0.1% HALS-9	9120	9320

[a] PE-LD (d = 0.918) + 0.03% AO-1.
[b] Compression-molded 200-μm–thick films.
[c] 24 h in 0.1 N SO_3H_2 before exposure.
[d] Weather-ometer Ci 65: black panel temperature $63 \pm 2°C$.
$T_{0.1}$, time (h) to 0.1 carbonyl absorbance.
Source: Ref. 59.

film stabilized by Ni-1–UVA-1 is close to failure after slightly more than 2-years exposure. The HALS-7-stabilized film still shows remarkable mechanical properties under the same conditions.

D. New Developments

The coadditive approach yields long film life under most of the conditions encountered in practice. However, it has still to be clarified if this approach can satisfy very special conditions, such as those encountered in rose greenhouses. As a matter of fact, the fungicidal properties of elemental sulfur are used quite extensively in rose greenhouses (56,57). The sulfur concentration in the corresponding film covers can reach very high levels, up to 3000 ppm and more.

An alternative to the coadditive approach consists in the development of a new class of HALS. It is clear that, besides low reactivity with acids, they must also protect the films according to mechanisms that are still operative under the drastic conditions encountered. Some hydroxylamine ethers derived from sterically hindered amines correspond to these expectations (32,58). The performance of such a hydroxylamine ether, HALS-9, in PE-LD films subjected to a treatment that is thought to simulate acid rain, before artificial exposure, is shown in Table 38 (59). The efficiency of the hydroxylamine ether is the same before and after the treatment. The classic HALS included in the test series, HALS-3 and HALS-4, are significantly less efficient after treatment than in the absence of any treatment. The data generated with PE-LLD films led to similar conclusions (60). Table 39 shows that HALS-9 outperforms the classic HALS included in the test, even if it is used at half the concentration. It looks as if HALS-9 was also efficiently protecting films in rose greenhouses (i.e., under exposure to elemental surfur; 57).

From the results available so far, it can be concluded that the hydroxylamine ether approach is certainly promising. One of the main problems with HALS-9 has to do with its physical state. It is a liquid, and its incorporation into solid resins presents some difficulties. This is not so with a high molecular mass hydroxylamine ether such as HALS-10. In Table 40 one can see that, with intensive pesticide

Table 39   Light Stability of PE-LLD Films[a,b] After "Acid Rain" Treatment[c]

UV stabilization	$T_{0.1}$ (h) Weather-ometer[d]
Control	205
0.3% HALS-3	4040
0.3% HALS-4	5000
0.3% HALS-6	3280
0.15% HALS-9	13440
0.3% HALS-9	20000

[a] PE-LLD-1 (butene comonomer, UNIPOL) + 0.1% Ca stearate + 0.03% AO-2.
[b] Compression molded 200 μm films.
[c] 24 hours in 0.1 N SO_3H_2 before exposure.
[d] Weather-ometer Ci 65: black panel temperature $63 \pm 2°C$.
$T_{0.1}$, time to 0.1 carbonyl absorbance.
Source: Ref. 60.

Table 40   Stability of PE-LD Films[a,b] on Experimental Greenhouse[c]

UV stabilization	Months to film failure	Failure area
0.4% Ni-1 + 0.2% UVA-1	21	Frames
0.4% HALS-3	20	Window
0.8% HALS-4	22	Window
0.4% HALS-10	29	Window

[a] PE-LD homopolymer, film grade.
[b] Blown films 200-μm thick.
[c] Pontecchio Marconi, Italy, exposure started in September.
E_{50}, energy (kLy) to 50%-retained elongation.
Permethrin, chlorine-containing pesticide, applied monthly.
Metamsodium, sulfur-containing pesticide, applied every semester.
Source: Ref. 55.

treatment, HALS-10 outperforms the conventional HALS as well as the combination of the nickel stabilizer with the benzophenone-type UV absorber. Hence, the hydroxylamine ethers derived from HALS seem to represent a new breakthrough for stabilization under very special conditions (e.g., if heavy pesticide treatments are involved).

V. CONCLUSIONS

Greenhouse film stabilization shows many facets. The considerable superiority of high molecular mass HALS over nickel-based stabilizers observed on outdoor weathering under more or less ideal conditions is not always found with similar films on actual greenhouses. This can hardly be attributed to film-inherent parameters, such as polymer, film thickness, or the presence of fillers or other additives. As a matter of fact, the superiority of HALS-based stabilization systems over nickel-based stabilization systems always is pronounced under these varying conditions. The same is true for the effects of external environmental parameters; for greenhouse-related factors, such as the support; as well as for climate-related factors, such as mean irradiation and temperature.

The special environmental conditions caused by more or less pronounced use of various pesticides or fungicides are mainly responsible for the complexity of the behavior observed with actual greenhouses. The pesticide treatment very often involves a combination of sulfur-containing and chlorine-containing chemicals. It is thought that the acidic species derived from these chemicals (i.e., hydrochloric acid and various oxyacids derived from sulfur) are essentially responsible for HALS deactivation to various extents. The coadditive approach is one way to take the formation of acidic species into account, by binding the acids to metal oxides, rather than to the HALS. This type of solution, in the form of HALS-7 and HALS-8, gives excellent results under various conditions, with pronounced superiority over nickel stabilizers. The other way to reduce or eliminate the reaction with acids is to use very specific HALS. The best solution here involves hydroxylamine ether derivatives of HALS (e.g., HALS-9 and HALS-10). They show excellent performance even with heavy use of chlorine- and sulfur-containing pesticides.

At long last, it seems possible to take advantage of the superior light and heat stability of greenhouse films stabilized with HALS in comparison with films stabilized with nickel stabilizers, even under severe pesticide treatments. Under these conditions, the additional advantages of HALS-stabilized films over nickel-stabilized films will lead to substantial improvements. HALS-stabilized films are colorless in comparison with nickel-stabilized films. This often results in faster plant growth and higher yields. Furthermore, HALS-stabilized films do not present the environmental problems associated with nickel-stabilized films, especially on film disposal, such as burning.

ACKNOWLEDGMENTS

The author expresses sincere thanks to M. Bonora and N. Lelli for data and discussions. The help of T. Franz and U. Moser is gratefully acknowledged.

REFERENCES

1. J-C Garnaud. Plastics in world agriculture 1980. Plasticulture 49:37–52, 1981.
2. J-C Garnaud. L'état de l'art de la plasticulture. 13th International Congress of CIPA, vol 1. Verona, Italy, March 8–11, 1994.
3. FR de Pedro. Los plasticos en la agricultura española: analisis, estadisticas y nuevos desarrollos. Rev Plast Modern 332:161–170, 214, 1984.
4. The "Almeria phenomenon" as model. Plasticulture 107:48, 1995.
5. F Gugumus. Lichtschutzmittel. Kunststoffe 77:1065–1069, 1987.
6. R Dartiguepeyrou. EVA for cladding plastics structures. What to choose? Plasticulture 70:19–26, 1986.
7. M Carp, F Gugumus. New developments in light stabilisation of polyethylene film for greenhouse covers. International Agricultural Plastics Congress, Lisboa (Portugal), Oct 6–11, 1980.
8. F Gugumus. UV stabilization of low-density polyethylene. Third International Conference on Advances in the Stabilization and Controlled Degradation of Polymers, Luzern, Switzerland, June 1–3, 1981.

9. F Gugumus. Advances in UV stabilization of polyethylene. In: J Kresta, ed. Polymer Additives. New York: Plenum Press, 1984, p. 17–33.
10. F Gugumus. La stabilisation UV du polyethylene. Proceedings of the 5èmes Journees d'Etudes sur le Vieillissement des Polymeres, Bandol, France, Sept 30–Oct 1, 1982.
11. KR Stahlke, F Gugumus. Bewitterungsverhalten von EVA-Copolymerisaten im Vergleich zu LDPE-Homopolymerisaten. Kunststoffe 73:432–435, 1983.
12. F Gugumus. Lichtschutzmittel. Kunststoffe 74:620–623, 1984.
13. F Gugumus. La stabilisation lumiere des polyolefines. Une mise au point. Proceedings of the 6èmes Journées d'Etudes sur le Vieillissement des Polymères, Bandol, France, Oct 11–12, 1984.
14. F Gugumus. The use of accelerated tests in the evaluation of antioxidants and light stabilizers. In: G Scott, ed. Developments in Polymer Stabilisation—8. Barking, UK: Elsevier Applied Science, 1987, pp 239–289.
15. F Gugumus. Advances in the stabilization of polyolefins. Polym Degrad Stabil 24:289–301, 1989.
16. F Gugumus. Photooxidation of polymers and its inhibition. In: P Klemchuk, J Pospisil, eds. Inhibition of Oxidation Processes in Organic Materials vol 2. Boca Raton: CRC Press, 1989, pp 29–162.
17. F Gugumus. Le rôle des combinaisons de stabilisants dans la protection UV des polyoléfines. Journee d'Etudes sur "La Stabilisation et le Vieillissement des Matieres Plastiques," Paris, Oct. 17, 1990.
18. F Henninger, F Gugumus. Evaluation of the life-time of greenhouse cover films. Plasticulture et Qualité Industrielle, Puy en Velay, France, June 3–4, 1992.
19. F Gugumus. Light stabilizers. In: R Gaechter, H Mueller, eds. Plastics Additives. 4th ed. Munich: Hanser, 1993, pp 129–270.
20. F Gugumus. Optimized stabilizer systems for polyolefins. Proceedings of AddCon'95, Basel, Switzerland, Apr 5–6, 1995.
21. F Gugumus. The performance of light stabilizers in accelerated and natural weathering. Polym Degrad Stabil 50:101–116, 1995.
22. T Daponte. Advantages of 3-layers co-extruded films in agriculture and horticulture. Plasticulture 75:5–12, 1987.
23. GW Gilby. Greenhouse cladding film developments. Plasticulture 81:19–28, 1989.
24. F Henninger, LC Roncaglione. Evaluation of greenhouse films exposed in Brazil. XII Congreso Internacional de Plasticos en Agricultura, Granada, Spain, May 3–8, 1992.
25. F Gugumus. Progres dans la protection des matieres plastiques contre le rayonnement UV. Kunstoffe Plast 22:11–19, 1975. Caoutch Plast 558:67–71, 1976.
26. F Gugumus. Developments in the UV stabilisation of polymers. In: G Scott, ed. Developments in Polymer Stabilisation—1. London: Applied Science, 1979, pp 261–308.
27. F. Gugumus. The chemistry of UV stabilisation of polyethylene. Presented at Golden Jubilee Conference, Polyethylenes 1933–1983, Paper D 12, London, June 8–10, 1983.
28. F Gugumus. Photooxidation and stabilization of polyethylene. Tenth International Conference on Advances in the Stabilization and Controlled Degradation of Polymers, Lucerne, Switzerland, May 25–27, 1988.
29. F Gugumus. Some aspects of polyethylene photooxidation. Makromol Chem Macromol Symp 27:25–84, 1989.
30. F Gugumus. Angew Makromol Chem 176/177:241–289, 1990.
31. F Gugumus. Re-evaluation of the stabilization mechanisms of various light stabilizer classes. Polym Degrad Stabil 39:117–135, 1993.
32. F Gugumus. Current trends in mode of action of hindered amine light stabilizers. Polym Degrad Stabil 40:167–215, 1993.
33. MB Amin, HS Hamid, JH Khan. Photo-oxidative degradation of polyethylene greenhouse film in a harsh environment. J Polym Eng 14:253–267, 1995.

34. J Lagier, AK Rooze, F Moens. Comparative agronomical experiment on greenhouse films stabilised with HALS and nickel quenchers. Plasticulture 96:29–34, 1992.

35. HH Friend, DR Decoteau. Spectral transmission properties of selected row cover materials and implications in early plant development. Proceeding, 22nd National Agriculture Plastics Congress, 1990, pp 1–6.

36. G Chevallier. Films polyéthylènes pour serres agricoles. Caoutch Plast (Aug/Sept) 662:75–79, 1986.

37. G Magnani. Agronomical performance of new LDPE and EVA films with mineral fillers. Two years research. Plasticulture 76:5–18, 1987.

38. M Hancock. Mineral additives for thermal barrier plastics film. Plasticulture 79:4–14, 1988.

39. A Erez, A Kadman-Zahavi. Growth of peach plants under different filtered sunlight conditions. Physiol Plant 26:210–214, 1972.

40. G Lotti, G Magnani, M Macchia, F Navari-Izzo, L Pioli. Influenza della luce sul metabolismo di *Glycine max* (L.) Merr. Nota I. Sviluppo della pianta. Agric Ital 107:219–235, 1978.

41. S Bualek, S Navachinda, K Suchiva, K Kuha. A study of wavelength selective polyethylene films for greenhouse covering materials. XII Congreso Internacional de Plasticos en Agricultura, Granada, Spain, May 3–8, 1992.

42. R Reuveni, M Raviv, R Bar, Y Ben-Efraim, D Assenheim, M Schnitzer. Development of photoselective PE films for control of foliar pathogens in greenhouse-grown crops. Plasticulture 102:7–16, 1994.

43. I Verlodt, T Daponte, P Verschaeren. Interference pigments for greenhouse films. Plasticulture 108:13–26, 1995.

44. SI Kusnetsov, GV Lepljanin, JU I Murinov, RN Schelokov, VE Karasev, GA Tolstikov. "Polisvetan," a high performance material for cladding greenhouses. Plasticulture 83:13–20, 1989.

45. J-P Frustié. The design of plastics greenhouses. Plasticulture 76:19–26, 1987.

46. A Jaffrin, A Morisot. Role of structure, dirt and condensation on the light transmission of greenhouse covers. Plasticulture 101:33–44, 1994.

47. HE Landsberg, H Lippmann, KH Paffen, C Troll. World Maps of Climatology. 3rd ed. Berlin; Springer Verlag, 1966.

48. A Davis, D Sims. Weathering of Polymers. London: Applied Science, 1983, pp 1–19.

49. FS Qureshi, MB Amin, AG Maadhah, SH Hamid. Weather induced degradation of linear low density polyethylene (LLDPE): mechanical properties. J Polym Eng 30:67–84, 1990.

50. A Schröter. Freibewitterung in Südspanien und Südafrika. Presented at the meeting "Bewitterung von Kunststoffen," Süddeutsches Kunststoff-Zentrum, Würzburg, Germany, Mar 19, 1997.

51. F Javier Barahona, JM Gomez Vasquez. Influence of pesticides on the degradation of polyethylene film greenhouse cladding on the Andalusian coast. Plasticulture 65:3–10, 1985.

52. BA Dogadkin. The mechanism of vulcanization and the action of accelerators. J Polym Sci 30:351–361, 1958.

53. JH Shorter, CE Kolb, PM Crill, RA Kerwin, RW Talbot, ME Hines, RC Harriss. Rapid degradation of atmospheric methyl bromide in soils. Nature 377:717–719, 1995.

54. N Lelli, F Gugumus. New developments in agrofilms stabilisation. Plasticulture 111:3–16, 1996.

55. N Lelli, M Bonora. New developments in agrofilms stabilization. CIPA International Congress for Plastics in Agriculture. Tel Aviv, Israel, Mar 9–14, 1997.

56. R Schachar, R Stelman, E Shai, B Efrat, Y Ashkenazi, D Asenheim. HALS stabilized LDPE agrifilms under the influence of elemental sulfur. International Conference on Environmental Impact of Polymeric Materials, Rehovot and Tel Aviv, Israel, May 12–16, 1996.

57. R Schachar, R Stelman, B Efrat, Y Ashkenazi, N Lelli. Experiments with HALS stabilized LDPE agrifilms in rose greenhouses. CIPA International Congress for Plastics in Agriculture, Tel Aviv, Israel, Mar 9–14, 1997.

58. F Gugumus. New trends in polymer photostabilization. In: NS Allen, M Edge, IR Bellobano, E Selli, eds. Current Trends in Polymer Photochemistry. New York: Ellis Horwood, 1995, pp 255–317.

59. F Gugumus. Impact de l'environnement et des mécanismes de stabilisation sur l'optimisation de la stabilité aux UV. Presented at the 12èmes Journées d'Etudes sur le Viellissement des Polymères, Bandol, France, Sept 25–26, 1997.

60. F. Gugumus. Influence of stabilization mechanisms and environment on optimization of UV stability. Proceedings of the 3rd International Symposium on Weatherability, Tokyo, Japan, May 14–16, 1997, pp. 94–120.

APPENDIX

Abbreviation	Structure	Trade name
HALS-1	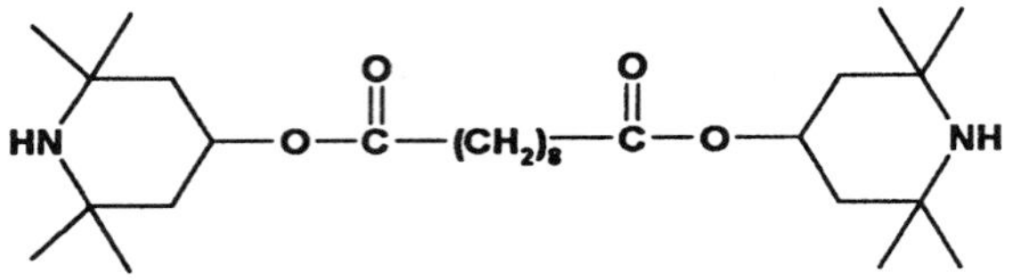	Tinuvin 770
HALS-2	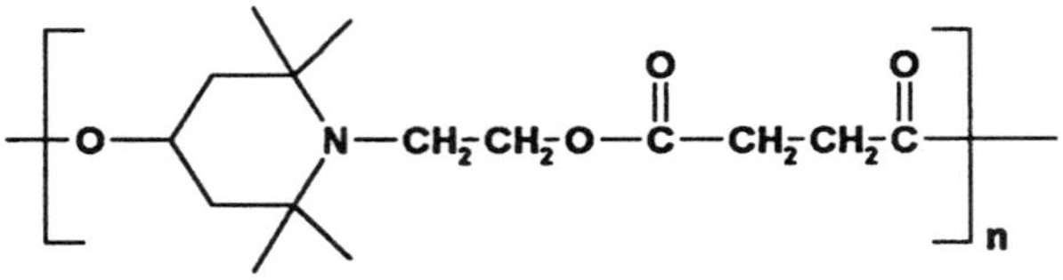	Tinuvin 622
HALS-3	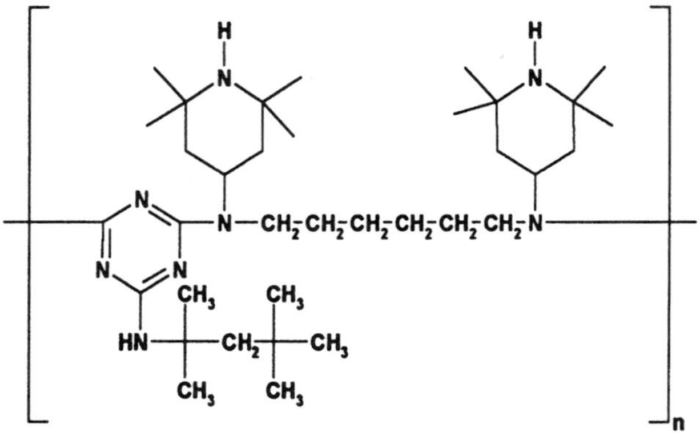	Chimassorb 944

APPENDIX *Continued*

Abbreviation	Structure	Trade name

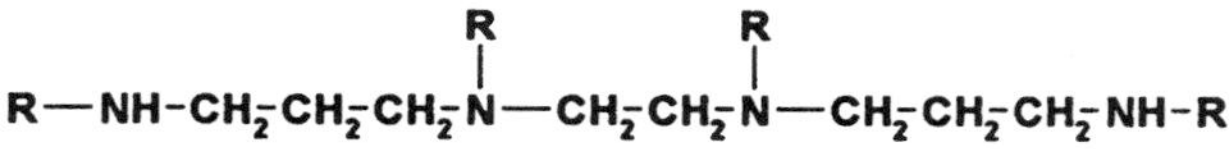

$$R-NH-CH_2CH_2CH_2-N-CH_2CH_2-N-CH_2CH_2CH_2-NH-R$$

with R substituents at the two nitrogen positions

HALS-4

R =

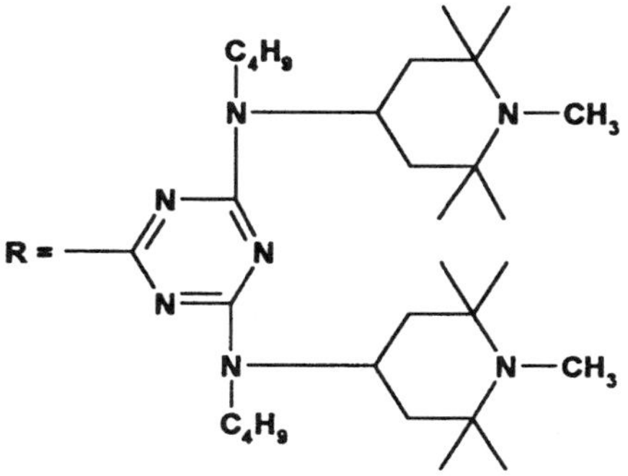

Chimassorb 119

HALS-5

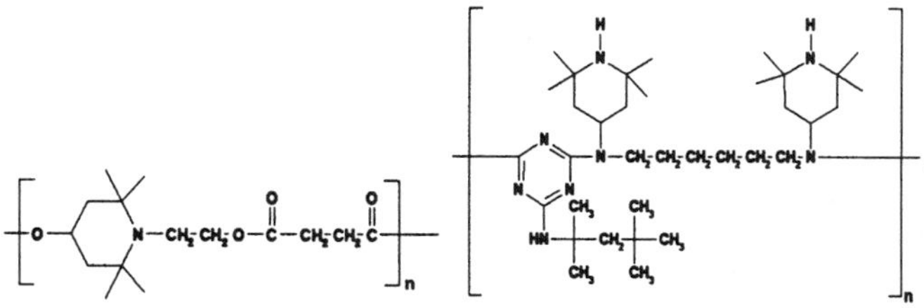

Tinuvin 783

HALS-6

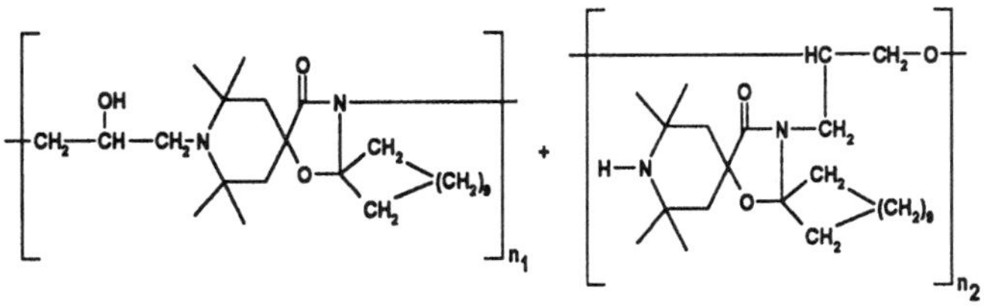

Hostavin N 30

Synergistic mixture of oxides, stearates, and

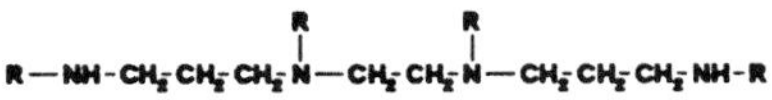

$$R-NH-CH_2CH_2CH_2-N-CH_2CH_2-N-CH_2CH_2CH_2-NH-R$$

HALS-7

R =

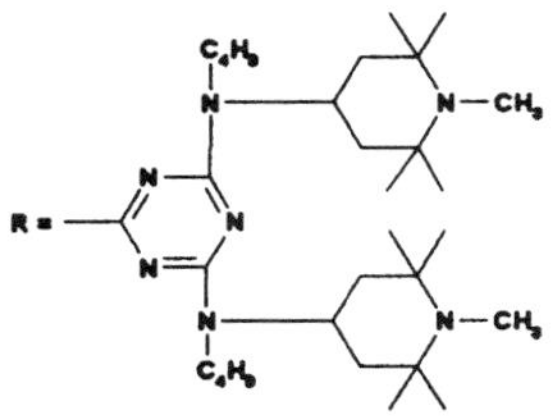

Tinuvin 492

APPENDIX *Continued*

Abbreviation	Structure	Trade name

Synergistic mixture of oxides, stearates, and

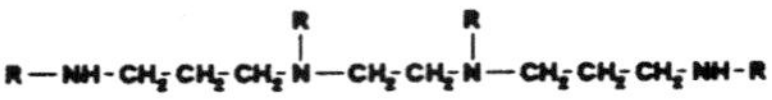

$$R-NH-CH_2CH_2CH_2-\underset{\underset{R}{|}}{N}-CH_2CH_2-\underset{\underset{R}{|}}{N}-CH_2CH_2CH_2-NH-R$$

HALS-8

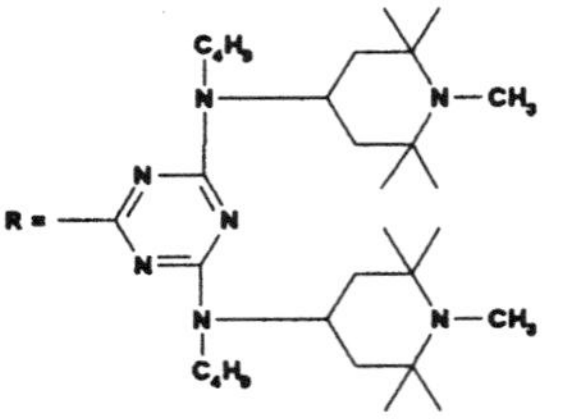

Tinuvin 494

HALS-9

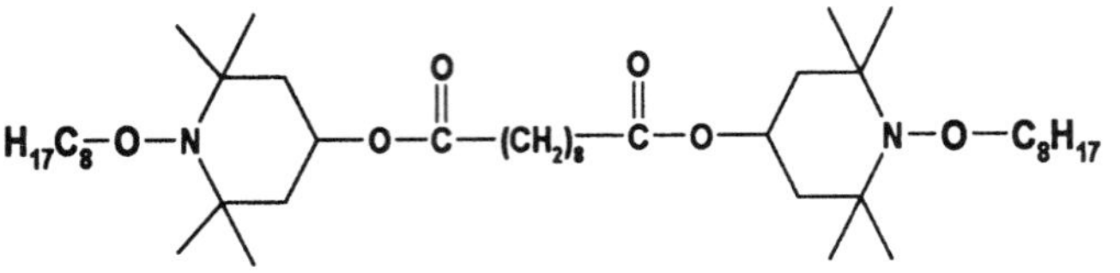

Tinuvin 123

HALS-10   Experimental hydroxylamine ether of high
           molecular mass HALS

UVA-1

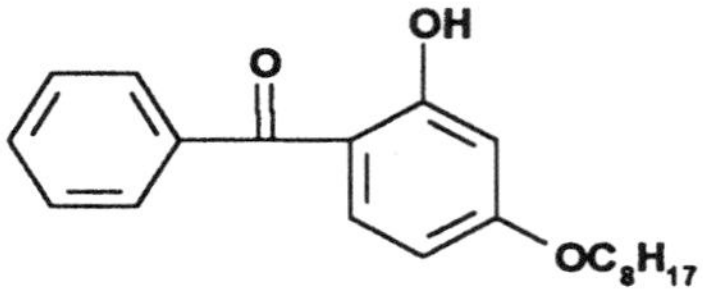

Chimassorb 81

UVA-2

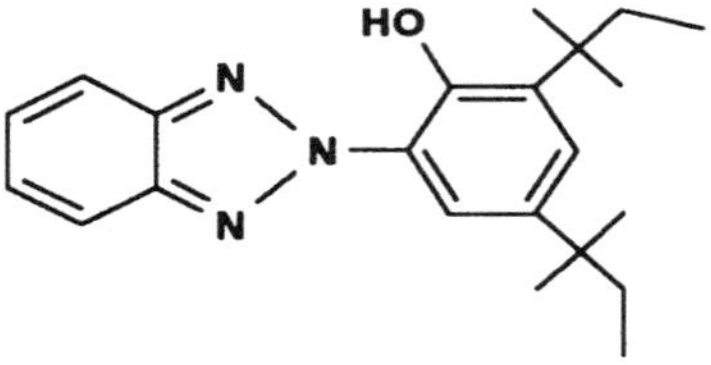

Tinuvin 328

Ni-1

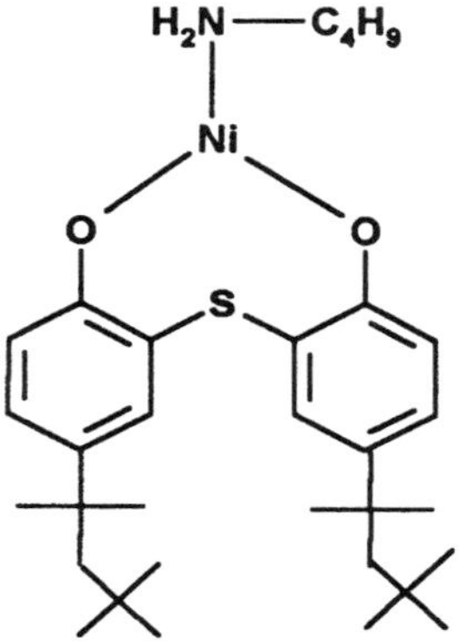

Cyasorb UV 1084

APPENDIX *Continued*

Abbreviation	Structure	Trade name
Ni-2	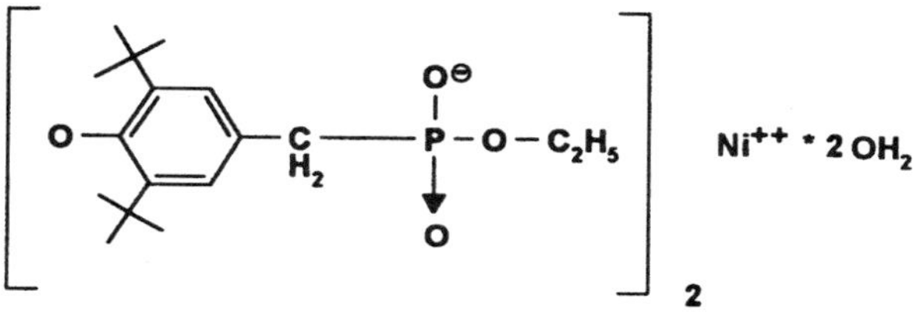	
Ni-3	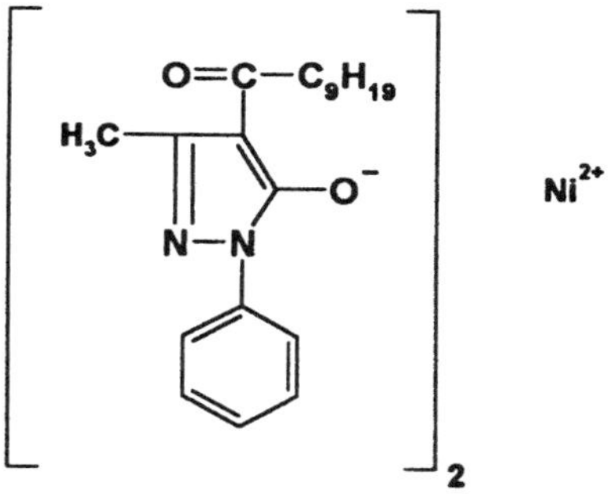	Sanduvor NPU
Ni-4	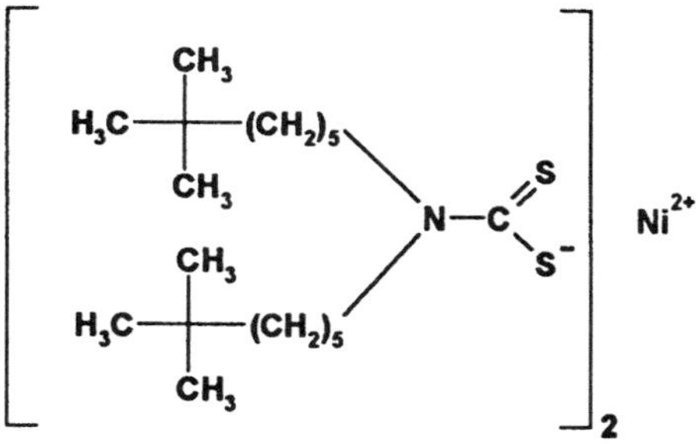	
AO-1	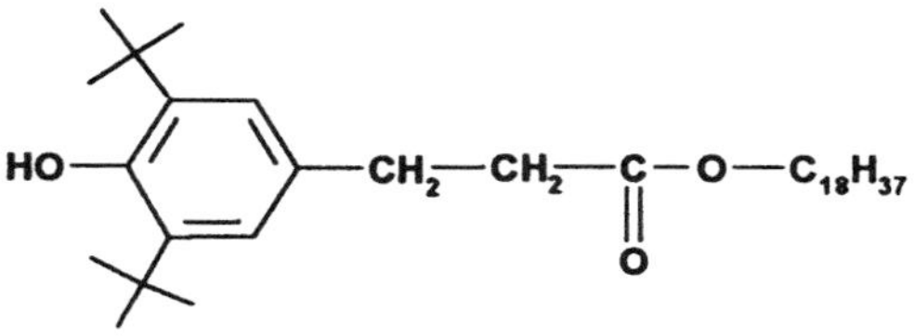	Irganox 1076
AO-2	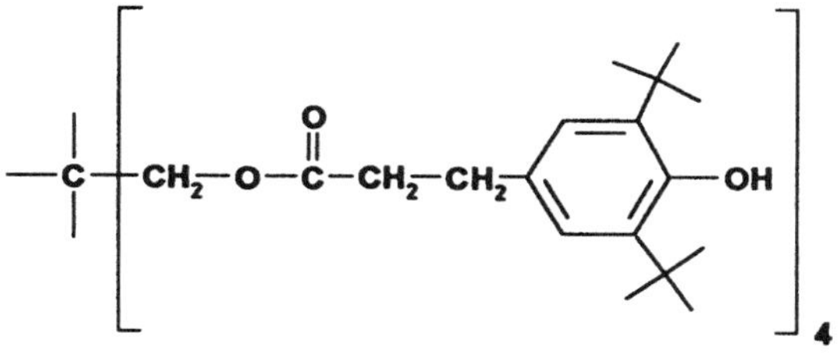	Irganox 1010

APPENDIX *Continued*

Abbreviation	Structure	Trade name

AO-3

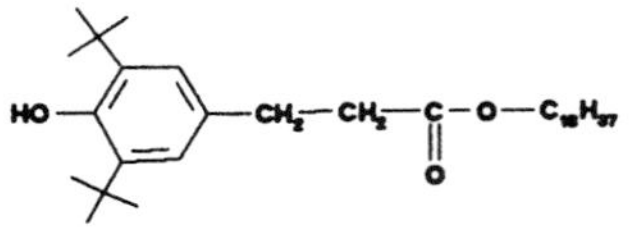

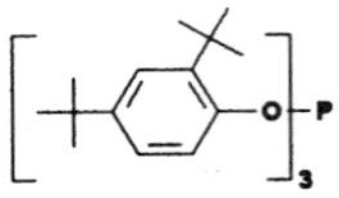

1   :   4

Irganox B 900

AO-4

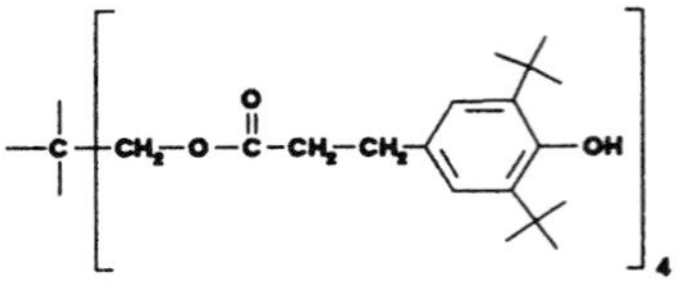

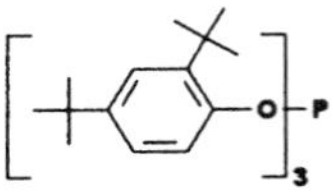

1   :   4

Irganox B 561

P-1

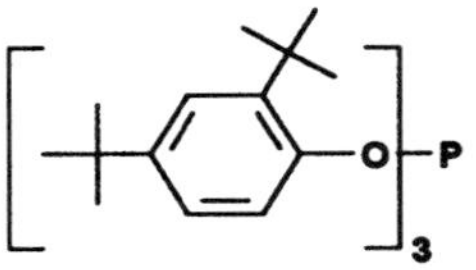

Irgafos 168

3

Phosphite Stabilizers for Polymers: Performance and Action Mechanisms

WOLF D. HABICHER and INGMAR BAUER
Dresden University of Technology, Dresden, Germany

I. INTRODUCTION

The protection of organic material against oxidative degradation processes during fabrication, processing, storage, or end use is a prerequisite to their successful application. This is particularly true for natural and synthetic polymers. Polymers are subjected to heat, mechanical stress, oxygen, sunlight, and other degradation-initiating influences of different intensities during their lifetime, which leads to changes of chemical and physical properties of the material. The extent of oxidative degradation of polymers is influenced not only by the aggressiveness of the environment and an inherent sensitivity of the polymer structure itself, but also by catalytic impurities and, especially in polymers, by the morphology of the substrate. It is evident from the enumeration of the deterioration processes that different additives, especially inhibitors of oxidation processes, must be used to protect organic materials and to guarantee a sufficient performance during their envisaged application. The steady increase in the production and use of polymers in the last decades has also been accompanied by an increasing consumption of polymer stabilizers.

The economic importance of polymer stabilizers is reflected in research and development of new highly effective compounds. Many different compounds have been reported to have antioxidant activities, by only a few chemical classes of compounds (phenols, various amines, phosphorus, and sulfur compounds) with specific structures are currently used as antioxidants.

Among a variety of additives applied to stabilize polymers effectively against oxidative deterioration, esters of phosphorous and phosphonous acid hold a special position owing to their multivalent properties. In the processing stabilization of

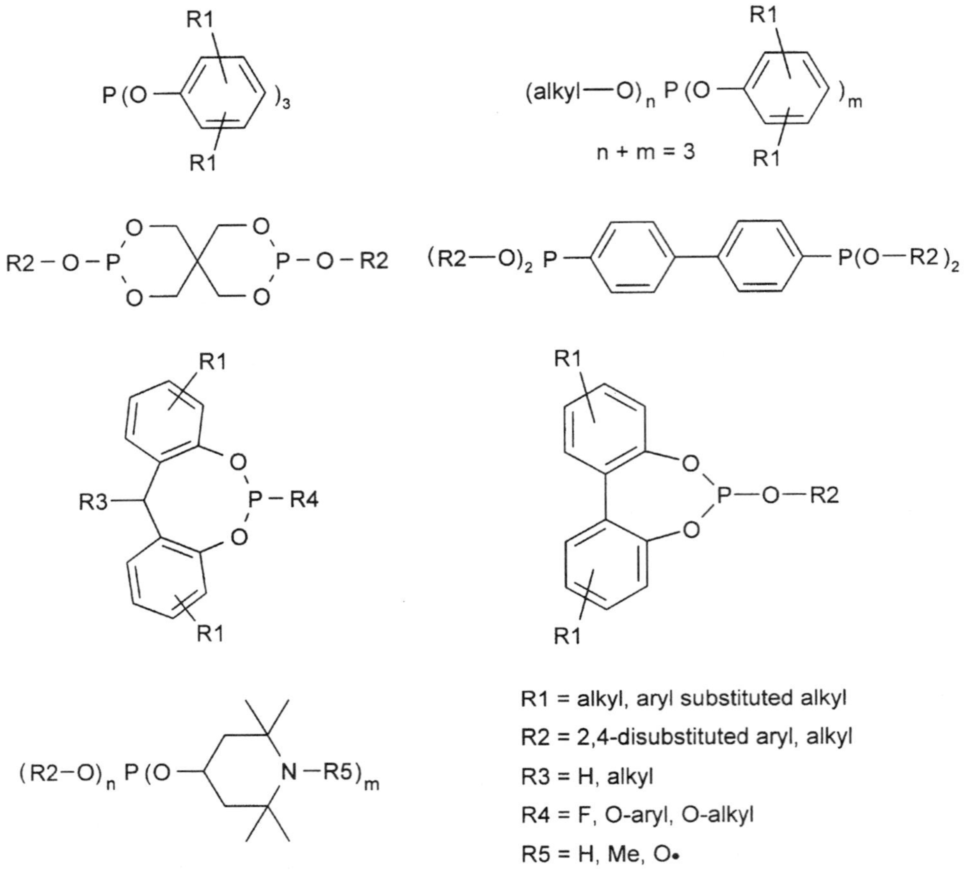

Scheme 1   Basic structure of common commercial phosph(on)ites.

polymers), (e.g., polyolefins, rubber, ABS, PVC, PET, PC, and others), phosphorus compounds are widely used in combination with high molecular weight phenolic antioxidants, and they are of crucial importance for the performance of such stabilizer formulations. Typical examples of trivalent phosphorus stabilizers commercialized as well as some derivatives which have been studied repeatedly in the literature are listed in Scheme 1.

Although phosphorus antioxidants are generally applied in combination with hindered phenols and other stabilizers, the sterically hindered aryl or aryl alkyl phosphites and phosphonites are, under some conditions, active by themselves and are able to partially substitute phenols. This is reflected in an increase of the share of phosphorus compounds in stabilizer formulations for melt stabilization in particular. Besides the high overall stabilizing effect as antioxidants, trivalent phosphorus compounds have a positive effect on the color stability of polymers and act as metal–complex-forming agents, blocking polyvalent metal ions that cause chain initiation and branching by reaction with hydroperoxides or other labile bonds in organic substrates. Moreover, most of the phosphites are quite compatible with polymers, have a low volatility and tendency to bloom out, and show a low sensitivity toward molecular oxygen at ambient temperature.

II. SYNTHESIS OF PHOSPHITES AND PHOSPHONITES

A. Phosphites

Many phosphite antioxidants are synthesized according to a well-established method by reaction of phosphorus halides with alcohols or phenols (1), Eq. (1).

$$\text{\textbackslash P---Cl} \quad + \quad \text{HOR} \quad \longrightarrow \quad \text{\textbackslash P---OR} \quad + \quad \text{HCl} \quad (1)$$

Addition of a tertiary amine as an acid scavenger favors the reaction and, for aliphatic alcohols, prevents a reaction of the alkyl phosph(on)ites formed with hydrogen halide to a phosphonate and an alkyl halide. The reaction of phosphorus acid chlorides with 2,2,6,6-tetramethyl- or 1,2,2,6,6-pentamethyl-4-piperidinol (HALS) gives the so-called HALS-phosphites (1–4). In this case the amine function of the HALS moiety competes with the acid scavenger for the HCl, thus lowering the yield of HALS-phosphites. The use of stronger amine bases than triethylamine can solve this problem (5).

Symmetric phosphites are easily obtained by reaction of phosphorous trichloride with 3 equivalents of the alcohol or phenol. Unsymmetric phosphites must be synthesized by phosphorous acid diester monochlorides or phosphorous acid monoester dichlorides. However, only with sterically hindered alcohols or phenols are these intermediates readily accessible without by-products (mainly phosphorous triesters).

Many commercial phosphite stabilizers are synthesized by transesterification of simple phosphites according to Eq. (2).

$$\text{\textbackslash P---OR} \quad + \quad \text{HOR'} \quad \longrightarrow \quad \text{\textbackslash P---OR'} \quad + \quad \text{HOR} \quad (2)$$

With different ratios of the starting material, the reaction can be controlled in such a way that one, two, or all three alcohol functions from the original phosphite, are substituted.

Recently, we have introduced the so-called phosphoramidit method, Eq. (3) (6) for the synthesis of HALS-phosphites (7). The advantage is the convenient accessibility of building blocks containing the HALS and phosphite structure, which can react with alcohols or phenols to give HALS-phosphites.

$$\text{\textbackslash P---NR}_2 \quad + \quad \text{HOR'} \quad \longrightarrow \quad \text{\textbackslash P---OR'} \quad + \quad \text{HNR}_2 \quad (3)$$

The reaction is more selective and the esterification can be carried out stepwise. Amine hydrochlorides function as catalysts.

B. Phosphonites

There are two industrially applied ways to build the P–C bond in phosphonites, such as in compound P-2 (Sandostab PEPQ). The first method is the electrophilic substitution at the aromatic ring with PCl_3 catalyzed by $AlCl_3$, Eq. (4) (8).

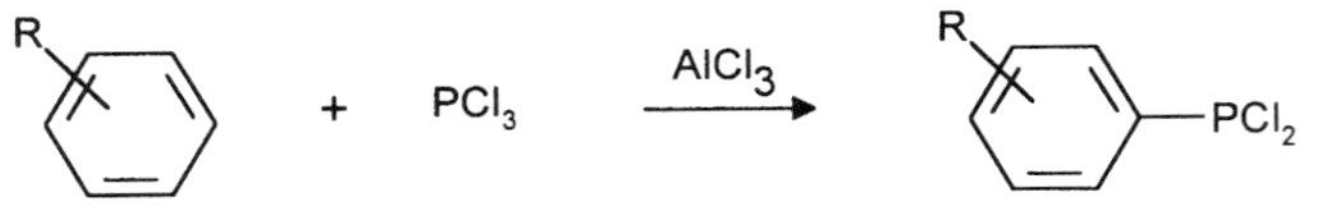

$$ \text{(4)} $$

The second approach is the reaction of Grignard reagents with PCl_3 to phosphonous acid dichlorides Eq. (5) (9).

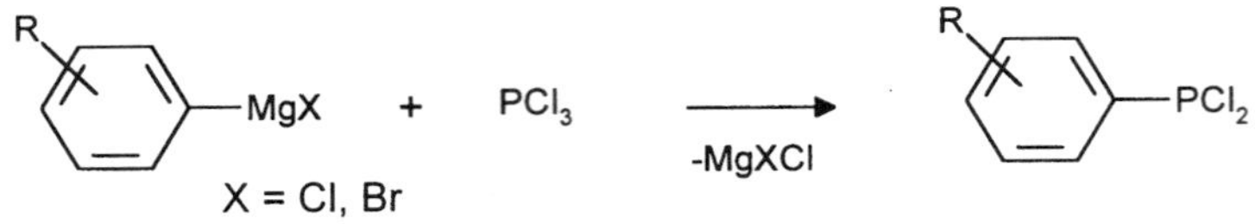

$$ \text{(5)} $$

Further esterification of these products is the same as described for the synthesis of phosphites and leads to the corresponding phosphonites.

C. Fluorophosphites

Recently Klender (10) introduced fluorophosphites as polymer antioxidants. These compounds can be synthesized starting from the corresponding chlorophosphites by treatment with KF, Eq. (6).

$$ \text{(6)} $$

III. PERFORMANCE OF PHOSPHITES AND PHOSPHONITES AS STABILIZERS FOR POLYMERS

A. Antioxidant Efficiency of Phosphites and Phosphonites During Processing of Polymers

The oxidation and stabilization of various organic materials, especially of commercial polymers, have been repeatedly reviewed (11–13). Among the stabilizers used, phosphites or phosphonites play a crucial role in the processing stabilization of a variety of polymers. They are widely used to prevent oxidative degradation and discoloration of polymers. Usually, sterically hindered aryl phosph(on)ites or alkyl aryl phosphites (see Scheme 1) are applied in combination with hindered phenols.

Polypropylene (PP) which is very sensitive toward oxidation, cannot be used without adequate stabilization. Especially for the processing of PP, usually performed at temperatures between 200° and 300°C, a phosph(on)ite costabilization is most effective. The phosphites P-1, P-2, P-3, and P-4 are commonly used in concentrations of 0.05–0.2% in combination with sterically hindered phenols such as

AO-1, AO-2, and others. Although phosphites are highly effective in processing stabilization of PP, they have no pronounced effect on long-term heat ageing of PP. For this reason thioether synergists are added.

Alkyl aryl phosphites, which are sensitive against hydrolysis, often perform better as processing and color stabilizers than more stable sterically hindered aryl phosphites. However, they should be used with caution because they may induce corrosion of the processing equipment (13,14).

As described for polypropylene, also for the processing stabilization of high-density polyethylene (HDPE) and high molecular weight polyethylene (HMWPE), an optimum of performance is obtained by using phenolic antioxidants, together with phosphites or phosphonites.

The stabilizers for linear low-density polyethylene (LLDPE) are, as a rule, the same as for HDPE. The performance of the stabilizer system (phenol–phosphite) is essentially determined by the phosphite, which is used in a small excess.

The sensitivity of elastomers such as polybutadiene, polyisoprene, styrene-butadiene rubber, ethylene-propylene-diene monomer (EPDM), butyl rubber (BR), styrene-acrylonitrile copolymers (SAN), acrylonitrile-butadiene-styrene terpolymers (ABS), styrene-butadiene-styrene (SBS), and styrene-isoprene-styrene block copolymers (SIS) to oxidation already demands a sufficient stabilization in an earlier stage of their production and during application. Sterically hindered phosphites, but also the nonhindered phosphite P-3, are used in concentrations up to 0.8% in nonstaining antioxidant packages in combination with phenols and thioethers. The improvement in polymer color by phosphites is often a desired side effect. Discoloration of polyamides during processing can also be efficiently reduced by phosphites used in concentrations of 0.2–0.4%.

The stabilizing efficiency of phosphites in polyvinyl chloride (PVC) is partly due to their antioxidant action as hydroperoxide decomposer and radical scavenger, but they are also able to take part in the Arbusov-type reaction, such as the substitution of labile, bounded chlorine. Alkyl aryl phosphites or simple trialkyl phosphites are used in concentrations of 0.05–0.15% together with metallic soaps as main stabilizers for PVC.

Processing stabilization of polycarbonate (PC) is generally achieved by the addition of phosphites or phosphonites in concentrations of 0.05–0.15%. These phosphorous compounds are superior to phenolic antioxidants in this system.

In the stabilization of polyurethanes (PUR), phosphites are also used with phenols and secondary amines. They act as hydroperoxide decomposers to stabilize polyols against autoxidation during storage and prevent discoloration (so-called scorching) in the core of the PUR material during foaming.

For the stabilization of polyterephthalates (PET) simple trialkyl and triaryl phosphites and phosphonites are still widely used. The discoloration can be reduced by combining the phenolic antioxidant with the phosphite.

The joint action of hindered phenols and phosphites has long been used in commercial applications. Moreover, it is well documented in the literature and the standard example for synergism between chain breaking and hydroperoxide decomposing antioxidants (Figs. 1 and 2) (15–17). However, such synergism is possible only under conditions for which the hydroperoxide-decomposing action of the phosphite affects the rate of oxidation. This applies for temperatures at which hydroperoxides homolytically decompose. At low temperatures, for instance, in

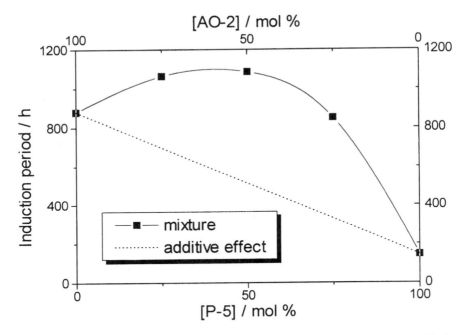

Figure 1 Induction period of carbonyl group formation during thermooxidation of PP stabilized with P-5/AO-2 mixtures ([mixture] $= 7 \times 10^{-3}$ mol/kg; 140°C). (From Ref. 17.)

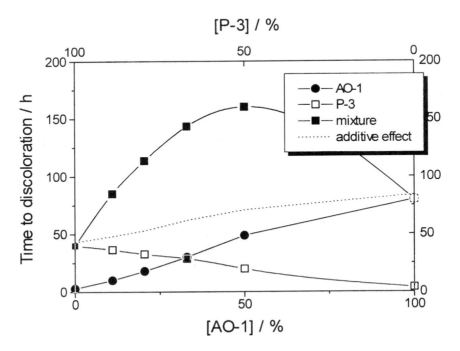

Figure 2 Results of discoloration tests of butadiene rubber (BR) containing AO-1, P-3 and AO-1/P-3, oven-aged at 100°C. (From Ref. 16.)

the oxidation of hydrocarbons and polymers below 80°C, phosphites have no effect on the antioxidative efficiency of phenols.

Color stabilization of the polymers is very important for many applications. Phenolic antioxidants, however, form colored products, by their antioxidant action, which strongly contribute to the discoloration of the polymers. The discoloration of phenol-stabilized polymers is related to the formation of quinone and quinone methide derivatives having very high extinction coefficients. These processes are obviously retarded by phosphorus compounds. The strong synergism of phosphite–phenol mixtures in their color-stabilizing performance is shown in Fig. 2 (16).

Starnes et al. (18) investigated the reaction of a quinone methide with triethyl phosphite in model experiments. They demonstrated that besides less-colored transformation products of the phenol used, the formation of the appropriate deeply colored stilbene quinone is reduced to 1–3% in contrast with about 50% in the absence of phosphites. The reaction of aromatic phosphites with quinone derivatives, however, has not yet been investigated in detail. No reaction products could be isolated. A McCormack reaction of phosphites with *o*-quinones or *o*-quinone methides to noncolored phosphoranes could also be taken into account in the color-stabilizing action of phosphites (19).

Besides phenols and phosphites, stabilizer systems often contain hindered amine light stabilizers (HALS). Discussion of the interaction of HALS and phosphites in stabilizer systems has been contradictory. The different results have been reviewed in the literature (1,4,17,19,20,21).

During thermooxidation, mixtures of HALS and phosphites mostly show synergism. We attribute the enhancement of the stabilizing efficiency of phosphites and HALS in their mixtures to a heterosynergism between a radical-trapping antioxidant, in particular the NO· radical formed from the HALS compound, and a hydroperoxide-decomposing stabilizer. Nitroxyl radicals are the most efficient thermostabilizers in this system at ambient temperatures. They show synergism with phosphites at every ratio applied (Fig. 3) (4).

Parent HALS compounds, such as HALS-1, do not show any stabilizing effect at higher temperatures. At moderate temperatures, however, they are very efficient long-term thermostabilizers. This is especially true for high molecular weight HALS compounds (22). We believe that this is because at higher temperatures the formation of nitroxyl radicals from the parent HALS proceeds slower than the autoxidation of the substrate. The polymer will be destroyed before an effective amount of NO· radicals can be formed.

In the presence of phosphites, however, the substrate will be efficiently protected at high temperatures, especially during processing. Simultaneously, the HALS compound is allowed to form enough NO· radicals to efficiently protect the polymer during long-term thermooxidation at moderate temperatures. This results in the remarkable synergism between HALS-1 and P-1 observed at 140°C (Fig. 4) (4).

Aliphatic phosphites are better synergists for HALS under these conditions than the hindered aromatic ones, indicating that the hydroperoxide-decomposing activity, which is higher for the alkyl phosphites, is crucial for the synergistic behavior (23).

In combinations of HALS with phosphites we observed antagonistic effects with aryl phosphites (Figs. 5 and 6) and synergistic effects with alkyl phosphites

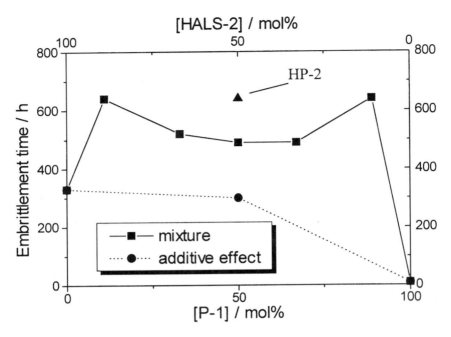

Figure 3 Embrittlement time of PP (thermooxidation at 140°C) stabilized with HALS-2/P-1 mixtures ([mixture] = 7×10^{-3} mol/kg). (From Ref. 4.)

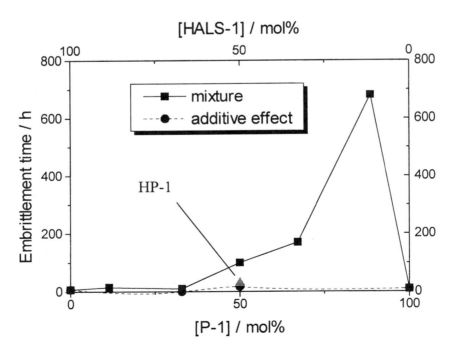

Figure 4 Embrittlement time of PP (thermooxidation at 140°C) stabilized with HALS-1/P-1 mixtures ([mixture] = 7×10^{-3} mol/kg). (From Ref. 4.)

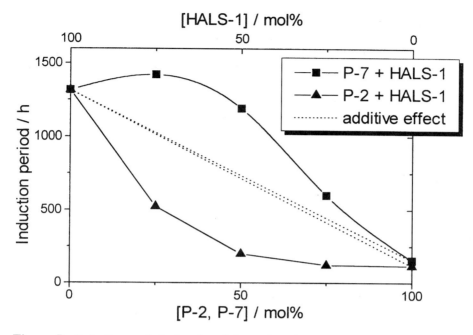

Figure 5 Induction period of carbonyl formation of PP stabilized with HALS-1/P-2 and HALS-1/P-7 mixtures during photooxidation ([mixture] = 7×10^{-3} mol/kg). (From Ref. 17.)

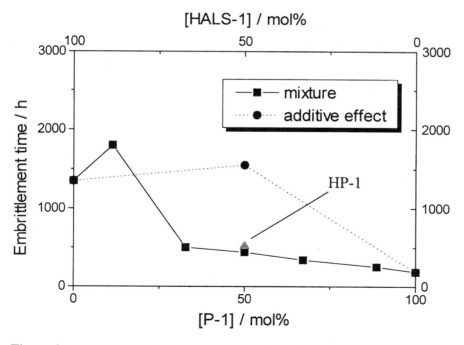

Figure 6 Embrittlement time of PP (photooxidation) stabilized with HALS-1/P-1 mixtures ([mixture] = 7×10^{-3} mol/kg). (From Ref. 19.)

(see Fig. 5) during photooxidation of polypropylene. Generally, it can also be supposed that, under photooxidative conditions, no chemical interaction between different HALS functions and P(III)-compounds takes place. The reason for the antogonistic effect of the mixture HALS–aryl phosphite is not the phosphite function itself, but the phenolic function attached to the phosphorus. This phenolic moiety can be released from the phosphite by its radical-trapping reaction, by hydrolysis or peroxidolysis. In this example, the problem is reduced to an interaction between HALS and hindered phenol derivatives, which is mainly reported to lead to antagonistic effects under photooxidative conditions (17). In our opinion the photoinitiating ability of some transformation products of hindered phenols, but not chemical interactions between both stabilizers, plays a crucial role in this system.

Usually, multifunctional HALS–phosphites containing HALS and phosphite functions in one molecule showed a better photo- and thermostabilizing performance than the corresponding mixtures of the individual compounds at the same total antioxidant concentration (see Figs. 3, 4, and 6).

IV. MECHANISMS OF ANTIOXIDANT ACTION OF PHOSPHITES AND PHOSPHONITES

The practical importance of phosphorus stabilizers has, during the last 30 years, induced numerous studies investigating the mechanistic problems of their antioxidant action without having solved all of the problems. The results of these investigations have been repeatedly reviewed (24–29). Phosphites and phosphonites can act by several different mechanisms, depending on their structure, the structure and morphology of the substrate to be stabilized, and the application conditions. All oxidative degradation processes have a common basis in the radical chain autoxidation mechanism (Scheme 2).

Also, as a consequence the efficiency of phosphorus compounds as stabilizers against oxidative degradation processes, is essentially determined by their ability to disturb the free radical chain mechanism of the autoxidation (see Scheme 2)

RH			$\longrightarrow$ R• (RO•, ROO•, HO•, HOO•)	**A**
R•	+	O_2	$\longrightarrow$ ROO•	**B**
ROO•	+	RH	$\longrightarrow$ ROOH + R•	**C**
ROOH			$\longrightarrow$ RO• + OH•	**D**
RO•	+	RH	$\longrightarrow$ ROH + R•	**E**
HO•	+	RH	$\longrightarrow$ R• + H_2O	**F**
ROO•	+	ROO•	$\longrightarrow$ O_2 + inactive products	**G**
R•	+	ROO•	$\longrightarrow$ ROOR	**H**
R•	+	R•	$\longrightarrow$ R-R	**I**

Scheme 2   Mechanism of autoxidation of hydrocarbons and polymers. (From Ref. 30.)

by removal or deactivation of the chain-propagating radical species, by nonradical decomposition or by blocking the radical precursors or initiators.

Phosphites and phosphonites can act as antioxidants by five basic mechanisms:

1. Decomposition of hydroperoxides, thus preventing the branching of the oxidation chain reaction.
2. Trapping of alkoxyl radicals by aryl phosphites, giving rise to a radical substitution reaction with release of antioxidatively acting aroxyl radicals.
3. Hydrolysis of aryl phosphites with formation of phenols and phosphonates.
4. Complexing of metal residues that are able to initiate homolytical hydroperoxide decomposition.
5. Reaction with molecular oxygen dissolved in the polymer.

The importance of these action mechanisms of phosphites in the stabilizing process is different for each individual polymer to be stabilized, and it also depends on technological parameters of their production and processing.

A. Hydroperoxide-Decomposing Action of Phosphites and Phosphonites

All phosphites and phosphonites are mainly considered as preventive antioxidants, decomposing hydroperoxides in a nonradical manner. This results in a suppression of the radical chain-branching step in the familiar radical chain mechanism of autoxidation (see Scheme 2, **D**). However, this reaction has no influence on the rate of autoxidation at the low temperatures at which hydroperoxides do not homolytically decompose. Recently, Neri et al. (30) also doubted that this process was relevant for processing stabilization by phosphites. They emphasized that, because of the rapid decomposition of hydroperoxides at higher temperatures, their ionic decomposition by phosphites is too slow to efficiently prevent this reaction. Haruna (31), however, stated that the activity of phosphites may be predicted by their reactivity with hydroperoxides.

1. Stoichiometric Reaction of Phosphites and Phosphonites with Hydroperoxides

The reduction of hydroperoxides by organic phosphites giving alcohols and the corresponding phosphates [Eq. (7)] was first investigated by Walling et al. (32). The reaction proceeds with a 1 : 1 stoichiometry by a nonradical mechanism, obeying a second-order rate law, first-order relative to both reactants. Most probably the O–O bond of the hydroperoxide is attacked in an S_N2 manner by the trivalent phosphorus atom as a nucleophile.

$$
\begin{array}{c}
RO \\
\quad\searrow \\
\qquad P{-}OR \; + \; HOOR' \longrightarrow RO{-}\overset{\overset{\displaystyle O}{\|}}{\underset{\underset{\displaystyle OR}{|}}{P}}{-}OR \; + \; HOR' \qquad (7) \\
RO \nearrow
\end{array}
$$

The reactivity of the phosphites toward hydroperoxides is mainly determined by polar and steric effects of the groups bound to phosphorus. It decreases with increas-

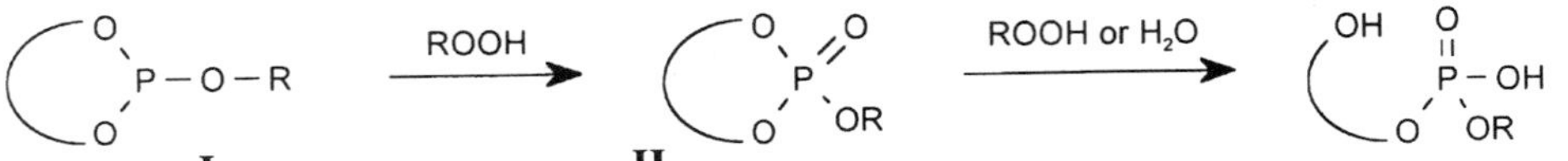

Scheme 3 Formation of acidic phosphoric acid diesters from five-membered cyclic phosphites.

ing electron-acceptor ability and bulkiness of substituent groups in the sequence (24):

Alkyl phosphites > aryl phosphites > hindered aryl phosphites.

Phosphonites react with hydroperoxides in an analogous manner, giving the corresponding phosphonates, Eq. (8). Their hydroperoxide-decomposing reactivity, however, is very much higher than that of phosphites (24).

$$\begin{array}{c} RO \\ \backslash \\ \quad P-R \\ / \\ RO \end{array} \quad + \quad HOOR' \quad \longrightarrow \quad RO-\overset{\overset{\displaystyle O}{\|}}{\underset{\underset{\displaystyle OR}{|}}{P}}-R \quad + \quad HOR' \qquad (8)$$

2. The Catalytic Decomposition of Hydroperoxides by Phosphites and Phosphonites

At higher temperatures, five-membered cyclic phosphites (1,3,2-dioxaphosphole and 1,3,2-dioxaphospholane derivatives) are able to catalyze the ionic decomposition of alkyl hydroperoxides. In this manner more than a stoichiometric amount of hydroperoxide is decomposed (33). This reaction is caused by acidic transformation products of these phosphites formed in the course of the reaction with hydroperoxides (Scheme 3) (34,35).

In a first step the phosphite **I** is oxidized by cumene hydroperoxide to the corresponding phosphate **II**. Formation of the catalytically active, acidic phosphoric acid diester **III** can occur by further reaction of **II** with cumene hydroperoxide through an unstable peroxyphosphate, which decomposes to **III** or by simple hydrolytic ring opening of phosphate **II**.

In contrast to some claims (33,36), six- and higher-membered cyclic phosphites are not capable of catalytically decomposing hydroperoxides (35,37). This parallels the well-known circumstances in the hydrolysis of cyclic phosphorus acid esters, where only the five-membered derivatives exhibit an extraordinary reactivity (38).

B. Radical Chain-Breaking Antioxidant Action of Phosphites and Phosphonites

To act as chain-breaking antioxidants, stabilizers must match two basic requirements:

1. They must be able to compete effectively with the substrate RH for chain-propagating radicals.

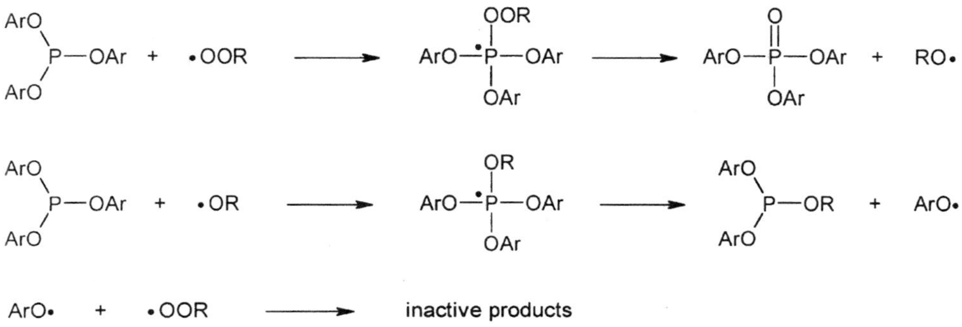

Scheme 4 Radical trapping reactions of aryl phosphites. (From Ref. 24.)

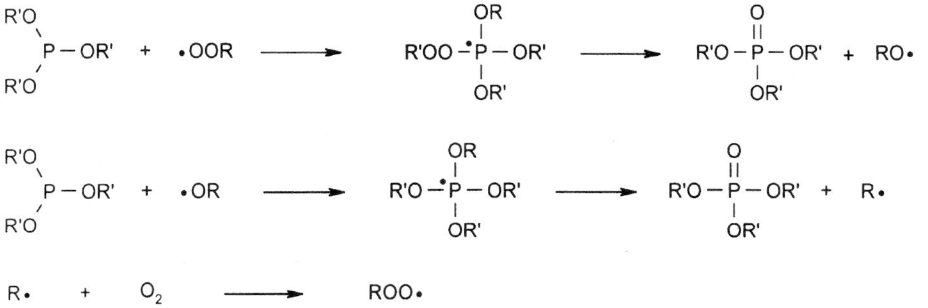

Scheme 5 Reactions of alkyl phosphites with oxyl radicals. (From Ref. 24.)

2. They must form an efficient chain-terminating agent as a result of its radical-scavenging reaction.

Aryl phosphites, particularly those based on sterically hindered phenols, show both an efficient competition with autoxidation reactions **C** and **E** (see Scheme 2) by trapping ROO• and RO• radicals and formation of relatively stable aroxyl radicals, which are not able to propagate the oxidation chain-reaction (Scheme 4) (24–27).

Alkyl phosphites, however, are not capable of functioning as chain-breaking antioxidants. Although they can trap oxyl radicals in a first step, this reaction always leads to the subsequent release of new, active radicals propagating the oxidation chain reaction (Scheme 5). Therefore, at low temperatures, these compounds even act pro-oxidatively. The same is true for both alkyl and aryl phosphonites (39).

Both aryl phosphites and alkyl phosphites are oxidized by ROO· radicals in a first-reaction step, affording the corresponding phosphate and RO· radicals. These radicals can react in a second step with the parent phosphite in different ways. This reaction is crucial for the antioxidative efficiency of phosphites. Only aryl phosphites that react with alkoxyl radicals by substitution releasing chain-terminating aroxyl radicals can act as primary antioxidants (Scheme 6). In this case the intermediate

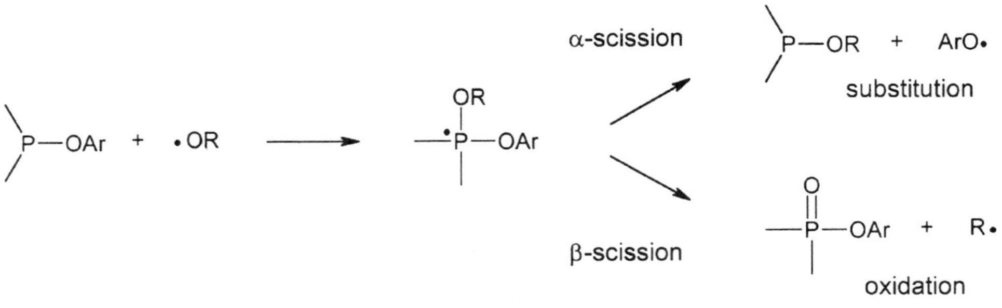

Scheme 6   Reactions of aryl phosphites with alkoxyl radicals. (From Ref. 24.)

alkoxyphosphoranyl radicals undergo α-scission, whereas β-scission leads to oxidation of the phosphite and formation of chain-propagating alkyl radicals (see Scheme 5, **K** (24).

The efficiency of aryl phosphites as chain-breaking antioxidants is much lower than those of phenols. Their stoichiometric factor, indicating the number of radicals being trapped, is less than 1 in contrast with hindered phenols, which are able to trap two peroxyl radicals per molecule. Furthermore, many phosphite molecules are destroyed in their reaction with hydroperoxides. This partial destruction gives rise to a critical antioxidant concentration below which phosphites do not inhibit the oxidation. The easier oxidizable the substrates are and the higher the temperature is, the higher is the critical concentration (24). Therefore, at moderate temperatures, hindered aryl phosphites are effective chain-breaking antioxidants only in rather high concentrations and predominantly in substrates of low oxidizability. Furthermore, strong chain-breaking action, can result from reaction products (mainly phenols) formed by hydrolysis and peroxidolysis of aryl phosphites and phosphonites in the course of oxidation under special conditions.

Because of these various action mechanisms, the antioxidant activity of phosphorous compounds is not a fixed unchangeable property, but depends strongly on the oxidation conditions and the nature of the oxidizable material.

C.   Hydrolysis of Organophosphorus Antioxidants

In the presence of small amounts of water, common phosphites undergo a fast hydrolysis. Negative and positive influences of this process on the performance of phosphites as polymer stabilizers have been discussed.

On one hand, hydrolysis has to be considered as a limiting factor for commercial applications of phosphites as polymer antioxidants. Acidic products formed during hydrolysis can cause corrosion of the processing equipment (13,14,40). Moreover, the storage and handling of these compounds may become difficult because the product mixtures tend to cake. Poor thermal stability can accompany the hydrolytic instability of phosphites and leads to the formation of "black specks" during processing of the polymer (41). It has been stated, therefore, that the development of phosphite antioxidants has to focus on the improvement in hydrolytic stability as well as performance (40).

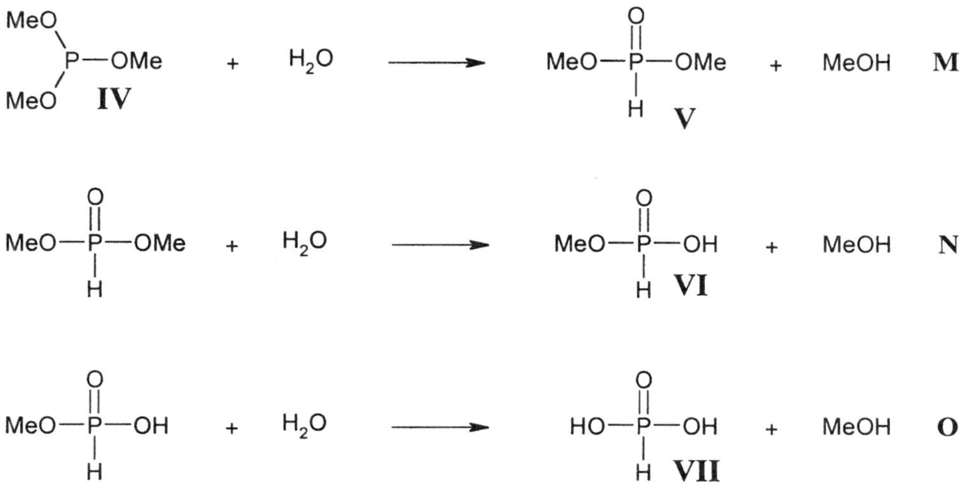

Scheme 7   Hydrolytic pathway of phosphites.

On the other hand, active-stabilizing species, such as hindered phenols, can be released by hydrolysis of phosphites. It has been suggested that these phenols are the species responsible for the antioxidant activity of phosphites at higher temperatures. They can form synergistic mixtures with the parent phosphites or phosphorus-containing hydrolysis products (42–47). Therefore, it was emphasized that a high hydrolytic stability may not always lead to maximum performance. Release of antioxidatively acting secondary compounds from the phosphites by hydrolysis in combination with the other stabilizing mechanisms can contribute to better overall performance (48).

Hydrolysis of phosphites proceeds by phosphonates (**V**) and hydrogen phosphonates (**VI**) to finally give phosphorous acid (**VII**; Scheme 7). Depending on the conditions, however, some intermediate hydrolysis products can be relatively stable.

The reaction of phosphites with water can be catalyzed either by acids or bases. Acid-catalyzed hydrolysis is the faster process and involves a protonation step at the phosphorus atom leading to a phosphonium salt (**VIII**). The following nucleophilic attack of water and the release of the equivalent amount of alcohol or phenol give phosphonates (**V**) via phosphoranes (**IX**; Scheme 8) (49).

Further hydrolysis of this compound (**V**) that finally gives phosphorous acid (see Scheme 7, **N** and **O**) is slower, as the pentavalent phosphorus atom cannot be protonated, and the protonation of the oxygen atom is less favorable. This is also why the rate of the hydrolysis of a trialkyl phosphite exceeds that of a trialkyl phosphate under acid catalysis by the factor of 10^{12} (49).

Base-catalyzed hydrolysis of phosphites leads to the formation of the anion of hydrogen phosphonates (**X**, Eq. 9).

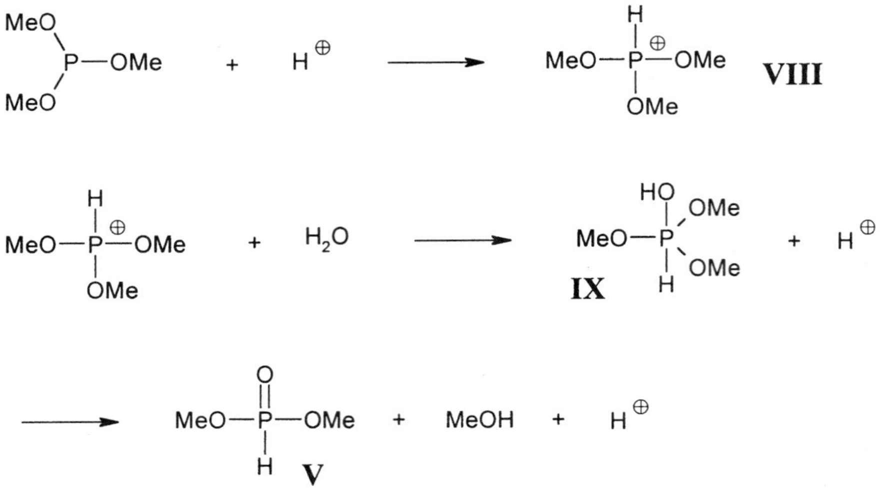

$$\underset{MeO}{\overset{MeO}{\diagdown}}P-OMe \quad + \quad H^{\oplus} \quad \longrightarrow \quad MeO-\overset{\overset{H}{|}}{\underset{\underset{OMe}{|}}{P}}\overset{\oplus}{-}OMe \quad \textbf{VIII}$$

$$MeO-\overset{\overset{H}{|}}{\underset{\underset{OMe}{|}}{P}}\overset{\oplus}{-}OMe \quad + \quad H_2O \quad \longrightarrow \quad MeO-\overset{\overset{HO}{\diagup}}{\underset{\underset{H}{|}}{P}}\overset{OMe}{\underset{OMe}{\diagdown}} \quad + \quad H^{\oplus}$$

$$\qquad\qquad\qquad \textbf{IX}$$

$$\longrightarrow \quad MeO-\overset{\overset{O}{\|}}{\underset{\underset{H}{|}}{P}}-OMe \quad + \quad MeOH \quad + \quad H^{\oplus}$$

$$\qquad\qquad\qquad\qquad\qquad \textbf{V}$$

Scheme 8 Proposed mechanism of the acid-catalyzed hydrolysis of phosphites. (From Ref. 49.)

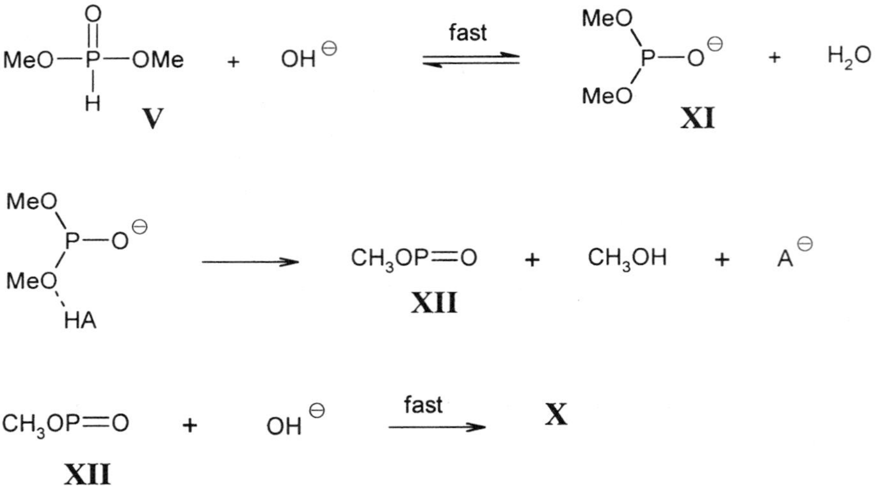

$$MeO-\overset{\overset{O}{\|}}{\underset{\underset{H}{|}}{P}}-OMe \quad + \quad OH^{\ominus} \quad \underset{\xrightarrow{\text{fast}}}{\overset{}{\rightleftharpoons}} \quad \underset{MeO}{\overset{MeO}{\diagdown}}P-O^{\ominus} \quad + \quad H_2O$$

$$\qquad\qquad \textbf{V} \qquad\qquad\qquad\qquad\qquad\qquad\qquad \textbf{XI}$$

$$\underset{MeO}{\overset{MeO}{\diagdown}}\underset{\underset{HA}{\diagdown}}{P}-O^{\ominus} \quad \longrightarrow \quad CH_3OP{=}O \quad + \quad CH_3OH \quad + \quad A^{\ominus}$$

$$\qquad\qquad\qquad\qquad\qquad \textbf{XII}$$

$$CH_3OP{=}O \quad + \quad OH^{\ominus} \quad \xrightarrow{\text{fast}} \quad \textbf{X}$$

$$\textbf{XII}$$

Scheme 9 Proposed mechanism of the base-catalyzed hydrolysis of phosphonates. (From Ref. 49.)

Under these conditions the intermediate phosphonate (**V**) hydrolyzes about 10^5 times faster than phosphorous acid triesters **IV** (49). This is because the intermediate phosphorous acid diesters (**V**) can be ionized under the influence of a base such as OH^- to a dimethyl phosphite anion (**XI**). Westheimer et al. propose a general acid-catalyzed decomposition of this anion under release of an unstable metaphosphite (**XII**), which rapidly reacts with water to give the anion of the phosphonous acid monoester (**X**) (49).

Under neutral conditions in water—acetonitril the hydrolysis of a trialkyl phosphite, such as tri-*n*-propyl phosphite, obeys a third-order time law involving simultaneously two molecules of water in the process, as reported by Aksnes et al. (50). The authors also showed that traces of a Lewis base, such as pyridine, significantly retard the rate of hydrolysis. This effect is used commercially when triethanol amine is added to easily hydrolyzable phosphites to increase their hydrolytic stability.

Investigating the hydrolysis of various phosphites in *o*-dichlorobenzene and in a mixture of *o*-dichlorobenzene with a commercial polyether alcohol in ratio 1 : 1, Hähner (51) found the following sequence of hydrolytic stability:

Sterically hindered aryl phosphites > unsubstituted aryl phosphites
   > araliphatic phosphites > aliphatic phosphites.

To increase hydrolytic stability of phosphites for application as polymer stabilizers Klender (52) proposes three possible methods:

1. Internal (a) or external (b) addition of basic components to the phosphites
2. Increase of steric hindrance around the phosphorus atom
3. Reduction of electron density at the phosphorus atom

Method 1a is realized in HALS-phosphites and triethanol amine-based compound P-6. According to method 1b, tertiary amines (e.g., triethanolamine, triisopropanolamine) are often added to hydrolytically unstable phosphites, such as P-4 and P-5. Bulky substituents (method 2) are introduced in commercial phosphites, such as P-1 and P-2. Reduction of electron density at the phosphorus atom (method 3) is achieved by a fluoro substituent (see, for instance, compound FP-1) as shown by Klender (41). This type of compound showed a unique hydrolytic stability, but also good activity as a processing and color stabilizer.

In general, however, the demands for a good hydrolytic stability are contrary to the requirements for a high activity as hydroperoxide decomposer. Therefore, a balance between hydrolytic stability and stabilizing performance must be found.

During the oxidation of hydrocarbons inhibited by aryl phosphites without basic functionalities at higher temperatures ($<100°C$) Schwetlick et al. (42) observed the formation of phosphonates (V) as well as phosphorous acid (VII) and phenols. In a polyether, which is more oxidizable, however, only a mixed alkyl aryl phosphite was transformed into hindered phenol and phosphonate, whereas the triaryl phosphites studied were completely oxidized to phosphates.

Klender (41) observed all possible products [phosphonates (V), hydrogen phosphonates (VI) and H_3PO_3 (VII)] during hydrolysis of neat P-1 being exposed to moisture from air at 50°C. Tochacek et al. (53) found the hydrolysis of P-5 during storage under the humidity of air did not proceed quantitatively. Aromatic moieties were substituted from phosphites before aliphatic groups. An equilibrium of different hydrolysis products was obtained. Partial hydrolysis even improved the performance of this compound. A good performance was found even when 48% of the phosphite was hydrolyzed.

Five-membered cyclic phosphites, but also phosphates, hydrolyze much more rapidly than larger-ring phosphites or open-chain compounds. The mechanism

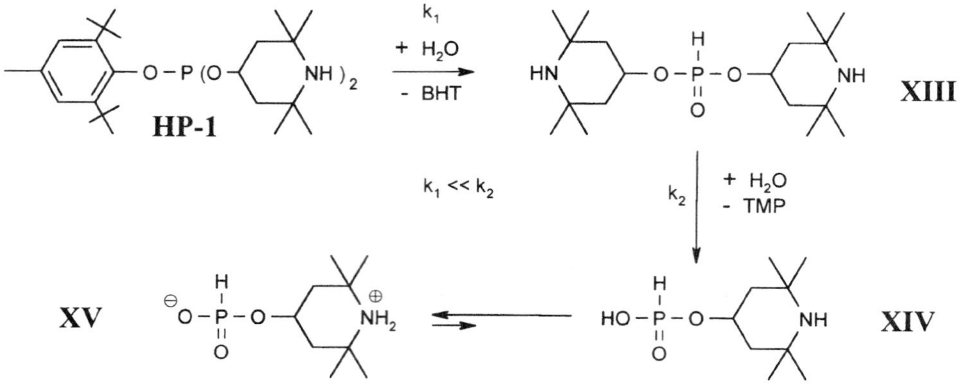

Scheme 10   Hydrolysis of HP-1 under neutral conditions.

of action of such phosphites comprises an initial oxidation and the subsequent hydrolysis of the five-membered ring phosphate formed (34; see Scheme 3).

The results of the hydrolytic behavior of the foregoing phosphites are related to systems containing no base. Under these conditions, hydrogen phosphonates (**VI**) can easily further hydrolyze to eventually form phosphorous acid (**VII**). In the presence of basic compounds, hydrogen phosphonates give hydrolytically relatively stable anions (**X**).

Pawelke (54) compared the hydrolytic behavior of the HALS-phosphite HP-1 with other phosphites in a mixture of hydrocarbons at 150°C. The HALS-phosphite turned out to be extremely stable toward hydrolysis. Under all of the conditions applied, oxidation proceeded faster than hydrolysis.

Lingner et al. (55), by monitoring the change of the pH value during hydrolysis, also found HALS-phosphites to be very stable against hydrolysis. We also investigated the course of the hydrolysis of HALS-phosphites (e.g., HP-1) in an isopropanol/water solution by ^{31}P nuclear magnetic resonance (NMR) spectroscopy and found that, in the first step, the hindered phenol is split off the molecule, leading to a phosphonate with two HALS moieties (**XIII**). In agreement with results by Westheimer et al. (49) for base-catalyzed hydrolysis, this compound reacts much faster with water than the original phosphite, leading to the hydrogen phosphonate (**XIV**) that exists predominantly as a betaine (**XV**). Because of the delocalized negative charge around the phosphorus atom, this species is very stable against water. Further hydrolysis of this compound was not observed (Scheme 10).

D.   Oxidation of Phosphites and Phosphonites with Molecular Oxygen

According to Neri et al. (30) hydroperoxide decomposition, radical scavenging, and metal complexing do not entirely explain the efficiency of phosphites as processing stabilizers. After processing, no polymer-bound phosphites or phosphates or hydrolytic products were found. The only product observed was extractable phosphate.

Although phosphites are not sensitive to oxygen at room temperature or higher, they are easily oxidized at temperatures higher than 200°C (30). As a con-

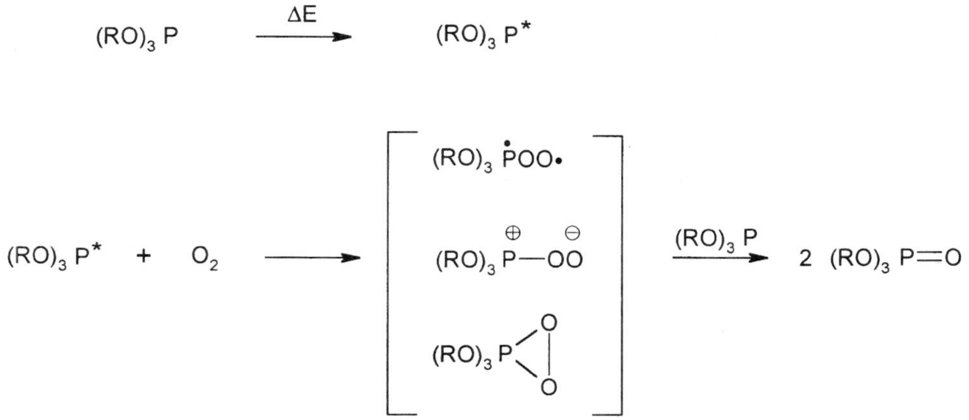

$$(RO)_3 P \quad \xrightarrow{\Delta E} \quad (RO)_3 P^*$$

$$(RO)_3 P^* \; + \; O_2 \; \longrightarrow \; \left[\begin{array}{c} (RO)_3 \overset{\bullet}{P}OO\bullet \\[2mm] (RO)_3 \overset{\oplus}{P}{-}O\overset{\ominus}{O} \\[2mm] (RO)_3 P\overset{O}{\underset{O}{<}}{\mid} \end{array} \right] \; \xrightarrow{(RO)_3 P} \; 2 \; (RO)_3 P{=}O$$

Scheme 11 Possible oxidation mechanisms of phosphites with molecular oxygen. (From Ref. 57.)

sequence, the authors attribute the efficiency of phosphites as processing stabilizers simply to their reaction with the oxygen that is dissolved in the polymer. This reaction is supposed to take place by a charge transfer mechanism from an activated phosphorus compound (Scheme 11) (30,56,57).

Reaction of phosphites with singlet oxygen ($^1O_2$) has not been observed (56).

E. Complexing of Metal Residues by Phosphites and Phosphonites

Phosphites form coordination complexes with metal ions, thereby changing the potential activity of the metal (58). In addition, some phosphite–metal complexes incorporated into polymers were active stabilizers. One particular study found that, in polyolefines containing high levels of catalyst residues, phosphites effectively improve the weathering resistance of the polymer by interacting with the metal residues (58). Pobedimskii et al. (26) reviewed the enhancement of the efficiency of organic phosphites in the presence of transition metal ions, in particular transition metal acetyl acetonates. They suggested the formation of new metal complexes with phosphite ligands, which effectively scavenge peroxyl radicals and decompose hydroperoxides. Prolonged induction periods of high- and low-density polyethylene stabilized with phosphites by factors of 2–7 were observed in the presence of various titanium and vanadium compounds.

V. CONCLUSIONS

Phosphites and phosphonites are valuable polymer additives, with a wide and increasing range of application. In combination with sterically hindered phenols they provide superior processing and color stability for a variety of different polymers.

The performance of phosphites in the stabilization process is an interaction of different chemical reactions. Those being relevant to their stabilizing efficiency are summarized in Scheme 12.

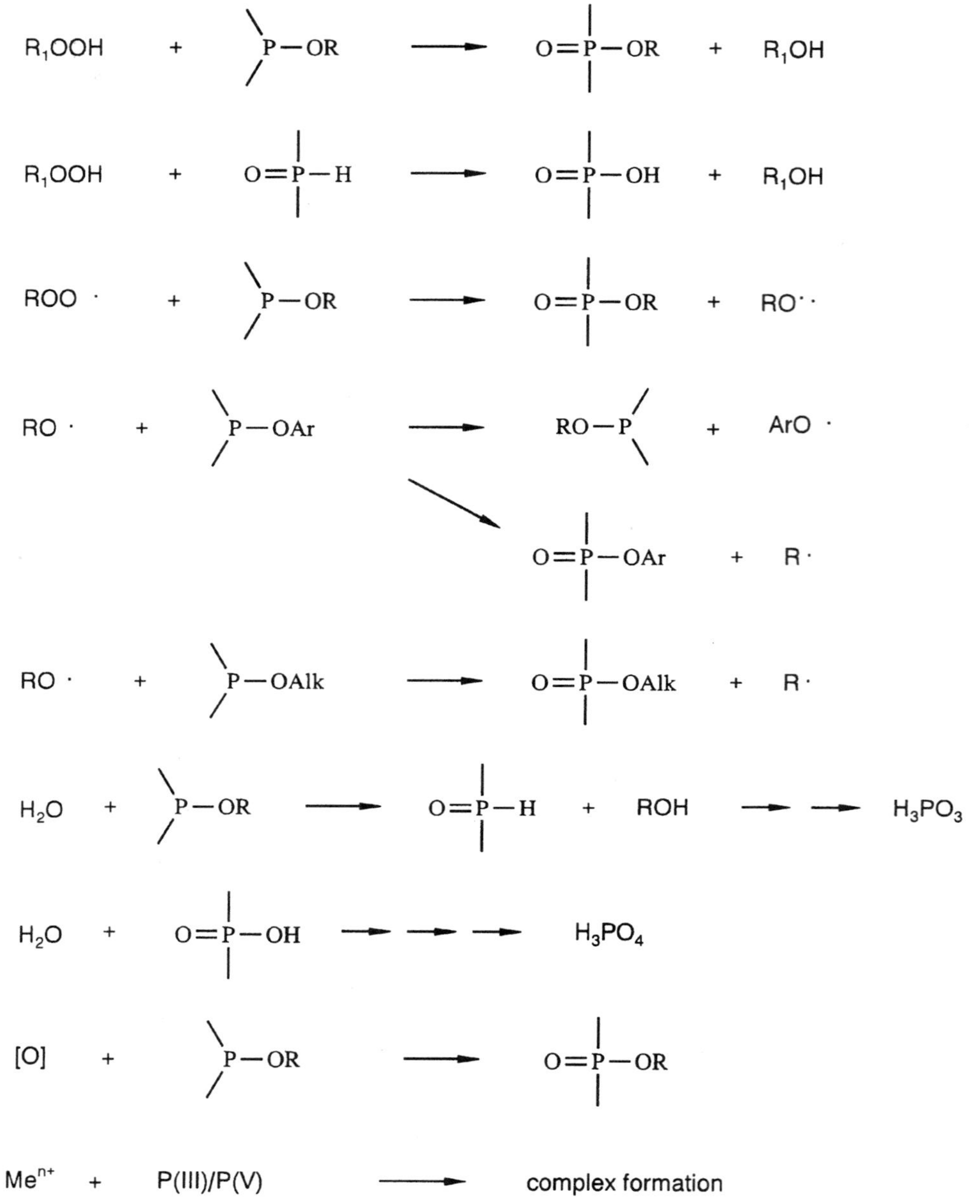

$$R_1OOH \; + \; \overset{\diagdown}{\underset{\diagup}{P}}-OR \longrightarrow O{=}\overset{|}{\underset{|}{P}}-OR \; + \; R_1OH$$

$$R_1OOH \; + \; O{=}\overset{|}{\underset{|}{P}}-H \longrightarrow O{=}\overset{|}{\underset{|}{P}}-OH \; + \; R_1OH$$

$$ROO\cdot \; + \; \overset{\diagdown}{\underset{\diagup}{P}}-OR \longrightarrow O{=}\overset{|}{\underset{|}{P}}-OR \; + \; RO\cdot\cdot$$

$$RO\cdot \; + \; \overset{\diagdown}{\underset{\diagup}{P}}-OAr \longrightarrow RO-\overset{\diagup}{\underset{\diagdown}{P}} \; + \; ArO\cdot$$

$$\searrow$$

$$O{=}\overset{|}{\underset{|}{P}}-OAr \; + \; R\cdot$$

$$RO\cdot \; + \; \overset{\diagdown}{\underset{\diagup}{P}}-OAlk \longrightarrow O{=}\overset{|}{\underset{|}{P}}-OAlk \; + \; R\cdot$$

$$H_2O \; + \; \overset{\diagdown}{\underset{\diagup}{P}}-OR \longrightarrow O{=}\overset{|}{\underset{|}{P}}-H \; + \; ROH \longrightarrow \longrightarrow H_3PO_3$$

$$H_2O \; + \; O{=}\overset{|}{\underset{|}{P}}-OH \longrightarrow \longrightarrow \longrightarrow H_3PO_4$$

$$[O] \; + \; \overset{\diagdown}{\underset{\diagup}{P}}-OR \longrightarrow O{=}\overset{|}{\underset{|}{P}}-OR$$

$$Me^{n+} \; + \; P(III)/P(V) \longrightarrow \text{complex formation}$$

Scheme 12   Reactions relevant to the performance of phosphites as polymer stabilizers.

Although the basic mechanisms of the antioxidant action of phosphites seem to be clear, some details, especially cooperative effects with other stabilizers and the contribution of the different reactions of the phosphorus compounds to the over-all-stabilizing performance in the polymer, discussions are still contradictory and need to be further investigated.

REFERENCES

1. T König, WD Habicher, U Hähner, J Pionteck, C Rüger, K Schwetlick. J Prak Chem 334:333–349, 1992.
2. U Hähner, WD Habicher, Š Chmela. Polym Degrad Stabil 41:197–203, 1993.
3. Š Chmela, WD Habicher, U Hähner, P Hrdlovic. Polym Degrad Stabil 39:367–371, 1993.
4. I Bauer, WD Habicher, C Rautenberg, S Al-Malaika. Polym Degrad Stabil 48:427–440, 1995.
5. I Bauer. Dissertation, TU Dresden, 1997.
6. EE Nifant'ev, MK Gratchev. Usp Khim 63:602–637, 1994; Russ Chem Rev 63:575–609, 1994 and literature cited therein.
7. I Bauer, WD Habicher. Phosphorus, Sulfur Silicon Related Elements 128:79–103, 1997.
8. B Bucher, LB Lockhart Jr. J Am Chem Soc 73:755–756, 1951.
9. Eur. Pat. Appl., EP 374761, 1990.
10. GJ Klender. In: RL Clough, NC Billingham, KT Gillen ed. Polymer Durability—Degradation, Stabilization and Lifetime Prediction. Advances in Chemistry Series 249. Washington, DC: ACS, 1996, pp 397–425.
11. G Scott. Mechanisms of Polymer Degradation and Stabilisation. New York: Elsevier Applied Science, 1990.
12. J Pospisil, PP Klemchuk. Oxidation Inhibition in Organic Materials. Boca Raton, FL: CRC Press, 1990.
13. F Gugumus. In: R Gächter, H Müller, eds. Taschenbuch der Kunststoffadditive. Munich: Carl Hanser Verlag, 1983, pp 1–103.
14. F Gugumus. In: J Pospisil, PP Klemchuk, eds. Oxidation Inhibition in Organic Materials, vol I. Boca Raton, FL: CRC Press, 1990, pp 61–173.
15. H Zweifel. In: RL Clough, NC Billingham, KT Gillen, eds. Polymer Durability—Degradation, Stabilization and Lifetime Prediction. Adv Chem Ser 249. Washington, DC: ACS, 1996, pp 375–396.
16. S Yachigo. In: SH Hamid, MB Amin, AG Maadhah, eds. Handbook of Polymer Degradation, New York: Marcel Dekker, 1991, pp 305–331.
17. WD Habicher, I Bauer, K Scheim, C Rautenberg, A Loßack. Macromol Symp 115:93–125, 1997.
18. WH Starnes Jr, JA Myers, JJ Lauff. J Org Chem 34:3404–3410, 1969.
19. I Bauer, WD Habicher, S Körner, S Al-Malaika. Polym Degrad Stabil 55:217–224, 1997.
20. NS Allen, A Hamidi, FF Loffelman, P McDonald, PV Susi. Plast Rubber Proc Appl 5:259–265, 1985.
21. NS Allen, A Hamidi, DAR Williams, FF Loffelman, P McDonald, PV Susi. Plast Rubber Proc Appl 6:109–114, 1986.
22. F Gugumus. Polym Degrad Stabil 24:289–301, 1989.
23. C Rautenberg. Dissertation, TU Dresden, 1991.
24. K Schwetlick. In: G Scott, ed. Mechanisms of Polymer Degradation and Stabilization. New York: Elsevier Applied Science, 1990, pp 23–61.
25. K Schwetlick, WD Habicher. In: RL Clough, NC Billingham, KT Gillen, eds. Polymer Durability—Degradation, Stabilization and Lifetime Prediction. Adv Chem Ser 249. Washington, DC: ACS, 1996, pp 349–358.
26. DG Pobedimskii, NA Mukmeneva, PA Kirpichnikov. In: G Scott, ed. Developments in Polymer Stabilization, vol 2. Barking; UK. Applied Science Publishers, 1980. pp 125–185.
27. PA Kirpichnikov, NA Mukmeneva, DG Pobedimskii. Usp Khim 52:1831–1851, 1983.
28. K Schwetlick. Pure Appl Chem 55:1629–1636, 1983.

29. K Schwetlick, WD Habicher. Angew Makromol Chem 323:239–246, 1995.
30. C Neri, S Constanzi, RM Riva, R Farris, R Colombo. Polym Degrad Stabil 49:65–69, 1995.
31. T Haruna. Proceedings of the 16th International Conference on Advances in the Stabilization and Degradation of Polymers, Luzern, Switzerland, 1994, pp 129–141.
32. C Walling, M Rabinowitz. J Am Chem Soc 81:1243–1249, 1959.
33. KJ Humphris, G Scott. J Chem Soc Perkin Trans 2:826–830, 1973.
34. K Schwetlick, C Rüger, R Noak. J Prak Chem 324:697–705, 1982.
35. C Rüger, T König, K Schwetlick. J Prak Chem 326:622–632, 1984.
36. KJ Humphris, G Scott. J Chem Soc Perkin Trans 2:831–835, 1973.
37. J Holcik, JL Koenig, JR Shelton. Polym Degrad Stabil 5:373–397, 1983.
38a. PC Haake, FH Westheimer. J Am Chem Soc 83:1102–1113, 1961.
38b. ET Kaiser, K Kudo. J Am Chem Soc 89:6725–6728, 1967.
38c. DG Gorenstein, BA Luxon, JB Findlay, R Momii. J Am Chem Soc 99:4170–4172, 1977.
39. K Schwetlick, T König, C Rüger, J Piontek. Z Chem 26:360–366, 1986.
40. M Minagawa. Polym Degrad Stabil 25:121–141, 1989.
41. GJ Klender. In: RL Clough, NC Billingham, KT Gillen, eds. Polymer Durability—Degradation, Stabilization and Lifetime Prediction. Adv Chem Ser 249. Washington, DC: ACS, 1996, pp 397–423.
42. K Schwetlick, J Piontek, A Winkler, U Hähner, H Kroschwitz, WD Habicher. Polym Degrad Stabil 31:219–228, 1991.
43. D Ryšavý, Z Sláma. Z Angew Makromol Chem 9:129–135, 1969.
44. MS Khloplyankina, ON Karpuchin, AL Buchachenko, PI Levin, Neftekhimiya 5:49, 1965.
45. SI Bass, SS Medvedev. Zh Fiz Khim 36:2537, 1962.
46. LV Novoselova, LI Zubtsova, VG Babel', VA Proskuryakov. Zh Prikl Khim 46:1329–1333, 1973.
47a. LP Zaichenko, VG Babel', VG Proskuryakov. Zh Prikl Khim 47:1168–1171, 1974.
47b. LP Zaichenko, VG Babel', VG Proskuryakov. Zh Prikl Khim 47:1354–1358, 1974.
47c. LP Zaichenko, VG Babel', VG Proskuryakov. Zh Prikl Khim 49:465–468, 1976.
48. J Tochacek, J Sedlar. Polym Degrad Stabil 41:177–184, 1993.
49. FH Westheimer, S Huang, F Covitz. J Am Chem Soc 110:181–185, 1988.
50. G Aksnes, D Aksnes. Acta Chem Scand 18:1623–1628, 1964.
51. U Hähner. Dissertation, TU Dresden, 1990.
52. GJ Klender. Proceedings of the 11th Bratislava IUPAC/FECS International Conference on Polymers, Bratislava, 1996, pp 85–86.
53. J Tochacek, J Sedlar. Polym Degrad Stabil 50:345–352, 1995.
54. B Pawelke. Dissertation. TU Dresden, 1990.
55. G Linger, P Staniek, K Stoll. Presented at the SPE Polyolefins RETEC, Houston, Texas, 1993.
56. AV Il'Yasov, AM Kibardin, VI Morozov, PI Gryaznov, AA Vafina, AN Pudovik. Zh Obshch Khim 57:1017–1019, 1987.
57. AN Pudovik, ES Batyeva, AV Il'Yasov, VZ Kondranina, VD Nesterenko, VI Morozov. Zh Obshch Khim 46:1964–1967, 1976.
58. J Scheirs, J Pospisil, MJ O'Connor, SW Bigger. In: RL Clough, NC Billingham, KT Gillen, eds. Polymer Durability—Degradation, Stabilization and Lifetime Prediction. Adv Chem Ser 249. Washington, DC: ACS, 1996, pp 359–374 and literature cited therein.

APPENDIX: STRUCTURES OF THE STABILIZERS USED

Code	Structure	Chemical and trade name
AO-1		2,6-di-*tert*-butylphenol (BHT)
AO-2		octadecyl-3-(3,5-di-*tert*-butyl-4-hydroxyphenyl)propanoate (Irganox 1076)
AO-3		tetrakis[methylene-3-(3,5-di-*tert*-butyl-4-hydroxyphenyl)-propanoate]methane (Irganox 1010)
HALS-1		bis(2,2,6,6-tetramethyl-4-piperidyl)sebacate (Tinuvin 770)
HALS-2		bis(2,2,6,6-tetramethyl-4-piperidyl-1-oxyl)sebacate
P-1		tris(2,4-di-*tert*-butylphenyl) phosphite (Irgafos 168)
P-2		tetrakis(2,4-di-*tert*-butylphenyl) 4,4'-biphenylendiphosphonite (Sandostab PEPQ)

APPENDIX *Continued*

Code	Structure	Chemical and trade name
P-3	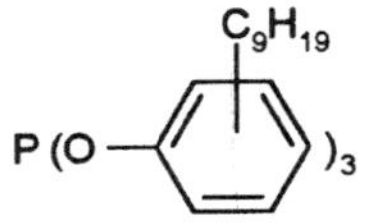	tris(4-nonylphenyl) phosphite (TNPP)
P-4	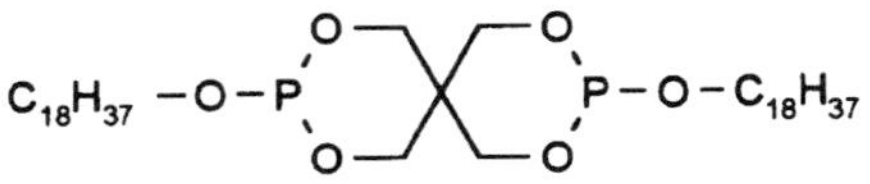	distearyl pentaerythrityl diphosphite (Ultranox 618)
P-5	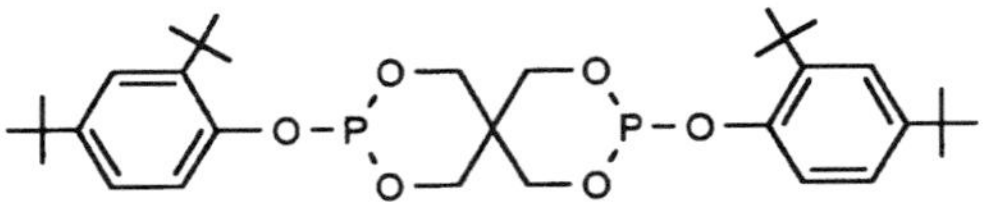	bis(2,4-di-*tert*-butylphenyl) pentaerythrityl diphosphite (Ultranox 626)
P-6	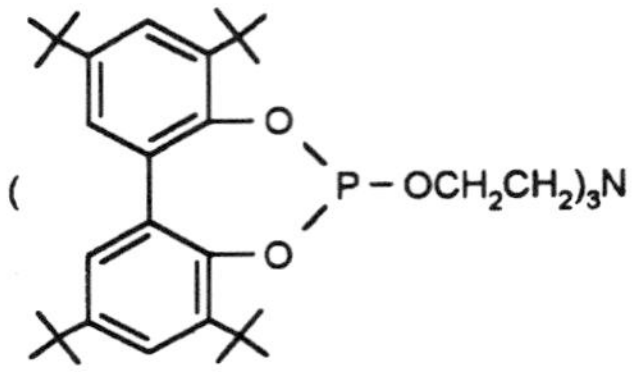	*N,N,N*-tris{2[(2,4,8,10-tetra-*tert*-butyl-12-methyl-12*H*-dibenzo [*d.g*] [1,3,2]dioxaphosphocin-6-yl) oxy]ethyl}amine (CGA 12)
P-7	P (OC$_{18}$H$_{37}$)$_3$	tristeary phosphite
FP-1	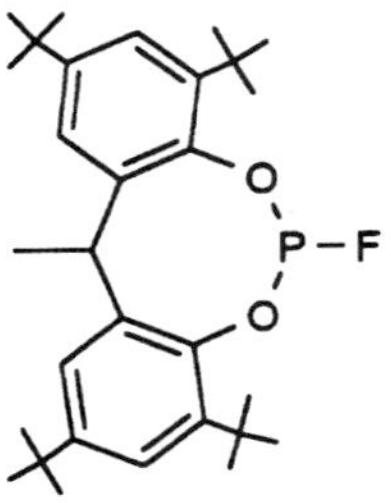	2,4,8,10-tetra-*tert*-butyl-6-fluoro-12-methyl-12*H*-dibenzo[*d,g*] [1,3,2]dioxaphos phocine (Ethanox 398)
HP-1	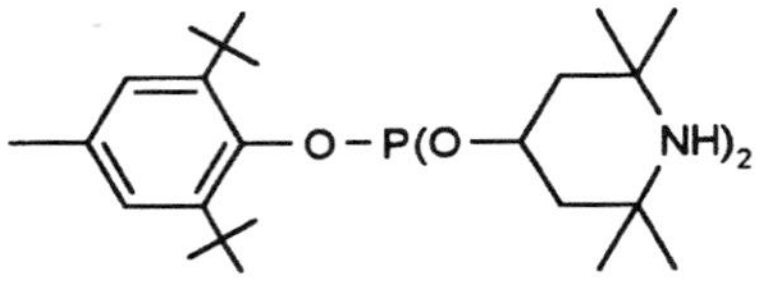	2,6-di-*tert*-butyl-4-methylphenyl bis(2,2,6,6-tetramethyl-4-piperidyl) phosphite
HP-2	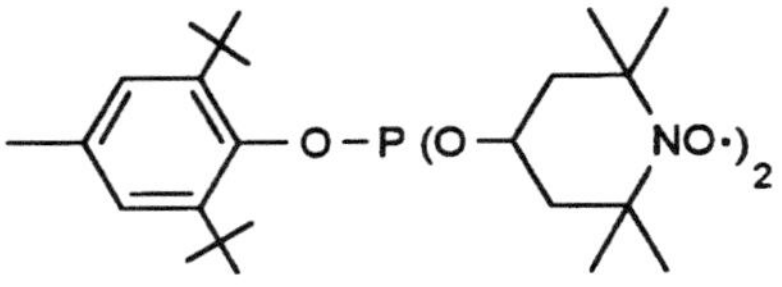	2,6-di-*tert*-butyl-4-methylphenyl bis(2,2,6,6-tetramethyl-4-piperidyl-1-oxyl) phosphite

4

Polymer-Processing Additives

ANAND KUMAR KULSHRESHTHA and SANTOSH KUMAR AWASTHI*
Indian Petrochemicals Corporation Limited, Vadodara, Gujarat, India

I. INTRODUCTION

Processing additives have been used throughout the years to improve the versatility of polymers. In fact, the following table lists all of the positive claims for processing additives. They include benefits in mixing, including mills and internal mixers; calendering, extrusion, building operations, and molding. This chapter will review the classes of additives, their chemistry, and briefly describe the various modes of action. Specific examples will be presented showing how these additives can be of benefit to compounders.

Claims for Processing Additives

Claim	Output	Quality
Improved filler dispersion		X
Reduced mixing time	X	
Reduced mixing energy	X	
Better mill handling	X	X
Faster extrusion	X	
Lower heat build up	X	X
Reduced die swell	X	X
Better calendering	X	X
Shorter injection times	X	X
Improved mold release	X	X
Easier fabrication	X	
Improved product appearance		X

---

* Deceased.

II. PROCESSING ADDITIVES

A. What Is a Processing Additive?

A processing additive is a material that, when added to a polymer compound at relatively low loading, will improve processability without adversely affecting physical properties.

It is difficult to determine who knowingly first used processing additives to modify a natural or synthetic polymer. "In 1865 Alexander Parks obtained a patent for the use of various materials, including fatty glycerides, oils, gum, and tars in lubricating nitro-cellulose; and John and Isaiah Hyatt obtained a patent on the use of camphor in nitro-cellulose in 1870. On May 11, 1858, H. L. Hall in Massachusetts was assigned Patent U.S. 220242 to Beverly Rubber Company for the use of asphalt, coal tar, resin and pitch to convert hard vulcanized rubber into a soft material which could be formed into useful articles."

III. ANTISTATIC AGENTS

1. What is an antistatic agent?
   An antistatic agent is a product that may be applied externally or incorporated into a polymer to dissipate electrostatic charges.
2. Why are antistatic agents used?
   They are used to reduce or eliminate the problems caused by static electricity, for example:

Mutual attraction of bodies of opposite charge, giving, for example, dust pick-up
Mutual repulsion of bodies of the same charge that cause problems with
    Polymer bead packing
    Film process lines
Electric shocks
Fire and explosion hazards

3. Classification of antistatic agents:
Internal antistatic agents are incorporated in the bulk of the resin.
External antistatic agents are applied over the resin surface.

A. External Antistatic Agents

In theory any molecule having surface-active or hygroscopic properties can be used as external antistat.

Products Used

    Polyols
    Fatty alcohols plus ethoxylated derivatives
    Glycerol esters
    Sorbitan esters
    Ethoxylated amine
    Quaternary ammonium compounds

Application Method

> The external antistats are dissolved in an appropriate solvent carrier (water, lower alcohols, hydrocarbon).
> The item is coated by spraying or immersion.
> The carrier is evaporated.

Example

1.	Atmer 110	2%
	Isopropanol	48%
	Water	50%
2.	Atmer 190	2%
	Water	98%

The main drawback of external antistats is the loss of antistatic protection because of surface abrasion or migration into the polymer.

The primary advantage is that the process is effective at a low percentage of relative humidity.

B. Requirements for Internal Antistatic Agents

> Be sufficiently polar to perform antistatic functionality
> Have a "balanced" affinity for the polymer
> Be compatible with the polymer matrix across the entire range of processing conditions
> Be compatible with other additives
> Be thermally stable
> Have no adverse effect on the resin properties
> Confirm to food contact legislation

1. Mode of Action of an Internal Organic Antistatic Agent

An internal antistatic agent acts by increasing the rate of charge dissipation.

1. It builds up a surface layer by a controlled migration of the antistat to the surface of the resin.

The migration rate is determined mainly by the

Relative compatibility of antistat and polymer
Polymer crystallinity
Total polymer additives formulation
Concentration of antistat used
Temperature

2. Orientation of the antistatic agent at the air–polymer interface. Surface charges are dissipated by reducing the surface resistivity.
3. The adsorption of air moisture gives a conductive surface layer. The effect is improved in the presence of electrolytes.

4. Reducing the generation of charges: Antistatic agents can act as lubricants which reduce intersurface friction.

C. Most Widely Used Internal Antistats

Nonionics

Fatty acid mono- or diglycerides (e.g., Atmer 129/122)
Ethoxylated alkanolamide
Ethoxylated fatty amines (e.g., Atmer 163)
Ethoxylated alcohols (e.g., Atmer 151)

Anionics

Alkane sulfonate (e.g., Atmer 190)

Cationics

Quaternary ammonium salts

D. Internal Antistatic Agents

Atmer 129

Glycerol monostearate with high monoester content (90% minimum)
Solid powder, easy to compound
Effective for polyolefins, especially polypropylene (PP)
Indirect food approved

Atmer 163

Ethoxylated synthetic amine, water-clear, low-viscosity liquid
Very effective and long-lasting antistatic properties in a wide range of
    polymers.
Recommended for crystal-clear materials.
Its synthetic character ensures constant quality and minimizes contaminants.
Excellent heat and color stability compared with amines based on natural raw
    materials.
Can be used as external antistat.

Atmer 170

Blend of antistat absorbed on a carrier, powder form
Effective in PP as well as in high-density polyethylene (HDPE)
Has antistatic and nucleating properties

Atmer 172

Antistat absorbed on a carrier, powder form
Has antiblock and antistat properties

E. Indirect Food Contact Approval

	Additives approved[a] according to regulations for food contact in						
	Belgium	France	Germany	Italy	Netherlands	UK	USA (FDA)
Atmer 110	*	–	–	–	*	*	*
Atmer 122	*	*	*	*	*	*	*
Atmer 129	**	*	*	*	*	*	*
Atmer 163	*	*	*	*	*	*	*
Atmer 190	*	*	*	*	*	*	*

[a] Indicates approval subject to certain detailed restrictions.
Atmer 151 is not approved for use in food contact plastics.

F. Indicative Levels of Usage

LDPE	Films	Atmer 163	0.05–0.1%
		Atmer 129	0.1–0.2
	Injection molding	Atmer 163	0.15%
HDPE	Injection molding	Atmer 163	0.15%
		Atmer 163/129 (2/1)	0.15%
PP	Films	Atmer 163	0.15%
		Atmer 129	0.2–0.4%
		Atmer 129/163 (2/1)	0.2–0.4%
	Injection molding	Atmer 163	0.3–0.6%
		Atmer 129	0.3–0.6%
		Atmer 129/163 (2/1)	0.3–0.5%
HIPS	Injection molding	Atmer 190	0.7–1.0%
		Atmer 163	Approx. 1.0%
ABS	Injection molding	Atmer 163	1.0–2.0
PVC	Flexible	Atmer 151	+1.0%
		Atmer 129	0.5–0.7%
	Rigid	Atmer 190	Approx. 1.0%

Source: Ref. 14.

IV. ANTISTATIC DETERMINATION METHOD

A. Static Charge Decay

The method determines the time required for a test specimen that has been given an electric charge at a certain voltage to lose half this charge when held under specified conditions of temperature and humidity (Fig. 1).

B. Surface Resistivity

Surface resistivity is determined through the intensity of current flowing over the surface of a plastic when a given potential is applied between two electrodes (Figs. 2 and 3).

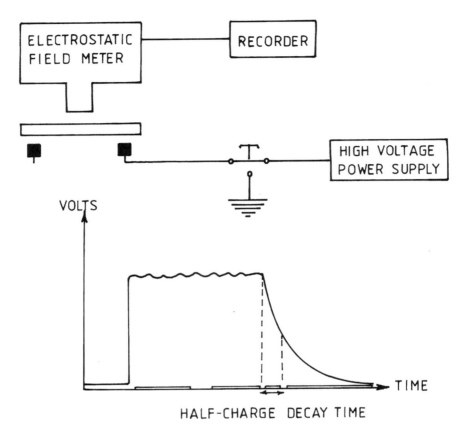

Figure 1 Determination of the efficiency of antistatic agent from the rate of static charge dissipation. (From Ref. 14.)

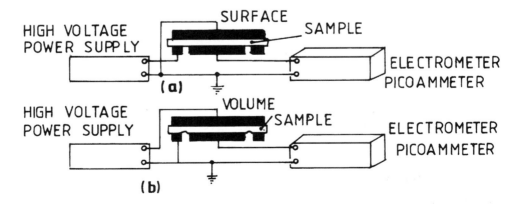

Figure 2 Standard ASTM test method for measuring antistatic agent efficiency from electric current flow on a plastic surface.

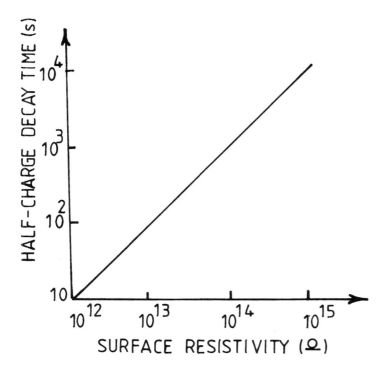

Figure 3 Correlation of surface resistivity/half-charge decay time. (From Ref. 14.)

$$\text{Surface resistivity (SR)} = \text{cst.}\,\frac{V}{\gamma}\,\text{(ohms)}$$

Measurements are made according to ASTM D 257-61.

Surface resistivity DIN 53482	Half-charge decay time (s) Honestometer	Rating
10^9	0	Excellent
10^9–10^{10}	1	Very good
10^{10}–10^{11}	2–10	Good
10^{11}–10^{12}	10–60	Moderate
10^{12}	>60	Poor

Source: Ref. 14

C. Dust Pickup Test

The subject is charged by rubbing the surface with a cloth and qualitatively assessing the attraction when the article is brought close to ashes in a tray (Fig. 4).

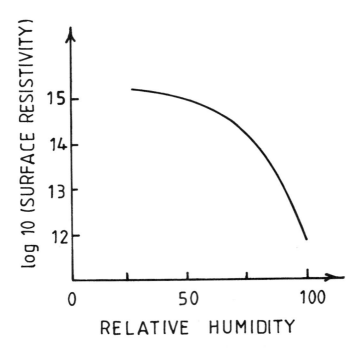

Figure 4   Conditions for the dust pickup test. (From Ref. 14.)

V. FOG FORMATION ON PLASTIC FILMS

The term *fog* is used to describe the condensation of water vapor on a plastic film surface in the form of small (typically 0.01–0.05 cm in a diameter or less), discrete water droplets. The film then becomes opaque because of reduced light transmission (Fig. 5). Reduced light transmission is a consequence of total reflection of the light hitting the droplet–air interface at an angle of incidence higher than the limit angle, 1, for water–air.

The light transmission is influenced by (Fig. 6):

Droplet density
Film surface (%) covered by water droplets
Droplet size distribution

A.   Prevention of Fog

Fog prevention can be achieved by causing the small polydispersed, high-contact–angle droplets to spread and coalesce to ideally form a uniform, thick, continuous film. We have focused our work on those internal antifog agents that modify the film surface properties.

For a liquid to spread over the solid surface, it is necessary for the surface tension of the liquid to be spread over the solid surface, such that the surface tension of the liquid is higher than the initial surface tension of the solid.

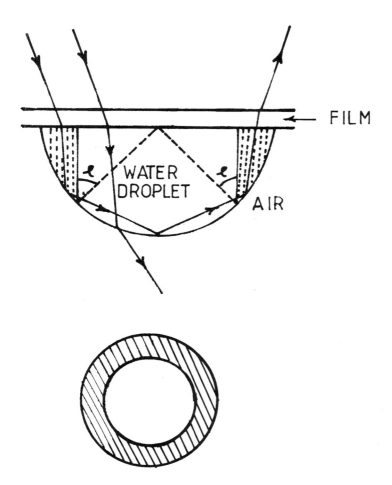

Figure 5 Opacity of plastics caused by the total internal reflection inside the water drop condensed on the film. (From Ref. 14.)

Parameters Controlling Condensation Rate

> Temperature (T) between film and the enclosed air
> The relative humidity (% RH)
> The absolute temperature

1. Mode of Action

Because of a controlled incompatibility with the resin the antifog agent slowly migrates to the air film interface. The emerging heat disperses the condensed water by lowering the film interfacial tension and, from there, spreading over the film surface. The presence of the antifog agent at the film–air interface will increase its critical surface tension.

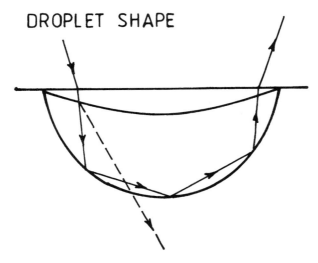

DROPLET SHAPE

Figure 6   Factors affecting the opacity of plastics film. (From Ref. 14.)

2.   Selection of Antifogging Agents

The choice of the optimum antifogging agent for a given film formulation depends mainly on

> Polymer type
> Formulation ingredients
> Film-processing conditions

B.   Application Areas for Antifog Film

1.   Food-Wrapping Films

The presence of condensed water droplets affects visual appeal, detracts from observable food quality, and causes deterioration of packaged food.

Figure 7 illustrates a food-wrapping application and a temperature profile, as a function of time, that reaches an equilibrium, with no significant washing-off effect of the antifogging agent, which maintained its antifogging properties for approximately 1 month.

The water condensation was ± 4 g condensed water per kilogram of dry air.

2.   Agricultural Films

> The occurrence of fog may result in

> 1.   Plant damage
>   Burns from "lens" effect
>   Continuous water drip
> 2.   Reduction in light transmission with consequent reduction in plant growth and crop yield
> 3.   Water condensation

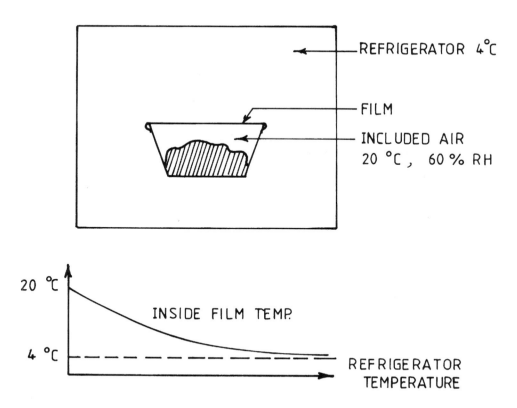

Figure 7 Food-wrapping application. (From Ref. 14.)

Figure 8 illustrates the antifogging test for food-wrapping film.

3. Greenhouse Application

Figure 9 shows a typical greenhouse situation and a temperature profile as a function of time. Figure 10 illustrates some typical applications of antifogging films.

C. Antifog Agents

1. Food-wrapping Films

ATMER 100.   Sorbitan ester, liquid, excellent for PE, EVA, and soft PVC food wrapping films.

ATMER 116.   Ethoxylated sorbitan ester liquid. 1 part Atmer 116 with 3 to 5 parts Atmer 121 provides an excellent antifogging effect for soft PVC wrapping film.

COLD FOG TEST

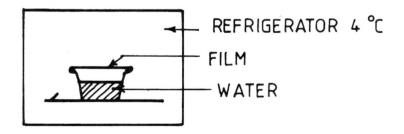

HOT FOG TEST

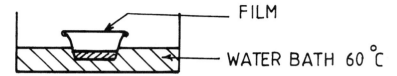

Figure 8   Antifogging test for food-wrapping film. (From Ref. 14.)

ATMER 121.   Glycerol oleate, paste is an excellent antifog for PVC food-wrapping film alone or in combination with Atmer 116. It also imparts adhesion properties (cling effect to the film).

ATMER 122–129.   GMS, solid, excellent for LDPE, imparts antistatic effect.

2.   Greenhouse

ATMER 103.   Sorbitan ester, solid; effective and long-lasting effect in LDPE, soft PVC, and EVA (low VA content) agriculture films.

ATMER 184.   Glycerol fatty acid ester, solid; highly effective in LDPE, EVA with high VA level. Offers additional antistatic effect.

	Indirect food contact approval[a]						
	Belgium	France	Germany	Italy	Netherlands	U.K.	USA (FDA)
Atmer 100	+		+	+	+	+	+
Atmer 102	+	+	+	+	+	+	+
Atmer 103	+		+	+	+	+	+
Atmer 114					+		
Atmer 116	+		+	+	+	+	+
Atmer 121	+	+	+	+	+	+	+
Atmer 122	+	+	+	+	+	+	+
Atmer 129	+	+	+	+	+	+	+
Atmer 184			+	+	+		+

[a] + indicates approval subject to certain detailed restrictions.

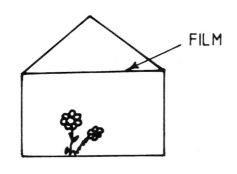

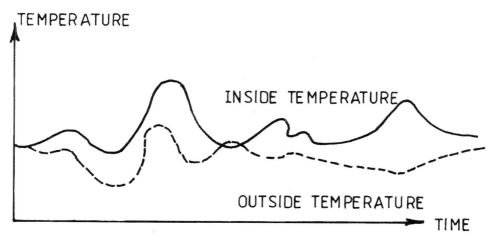

Figure 9 Greenhouse application: typical situation. (From Ref. 14.)

VI. CLING ADDITIVES

Cling additives are those incorporated in thermoplastic films to improve the adhesion of film layers to themselves or to the surface of a container.

1. Application area
   Food-packaging film
   Stretch wrapping film
2. Level usage
   Food-packaging film 0.5–1.0%
   Stretch wrapping film 1.0%

The benefits of Imperial Chemistry Industries' (ICI) specialty chemicals cling additives include

Effectiveness at low levels of addition.
Wide acceptance under regulations relating to food-packaging applications.
Improvement in film transparency.
Compatibility with other cling additives (e.g., polyisobutylene), and they can be used for fine-tuning of film adhesion and tackiness.

HEAT SCREEN INSIDE GREENHOUSE

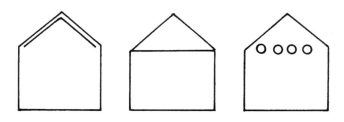

FILM REQUIREMENTS— 50μM THICK
                    — A.F. EFFECT
                    2-4 MONTHS

TEMPORARY PROTECTION FILM

                    — 50μM THICK

LARGE DOME

                    — 200μM THICK
                    — A.F. EFFECT
                    2 SEASONS

Figure 10   Typical applications for a greenhouse.

A. Systems to Obtain Cling Films

ADHESIVE TACK.   PE formulated with ± 5% low molecular weight (500–2000) polyisobutylene rubber; PE formulated with ± 5% atactic PP (molecular weight 2000–4000).

OILY TACK.   PE formulated with ± 1% surfactant

B. Cling Additive Product Table

Product	Preferential application	Suggested for use in
Atmer 100	Food packaging	LDPE, LLDPE
Atmer 121	Food packaging and stretch wrapping	LDPE, LDPE–LLDPE mixtures
Atlas G-695	Food packaging and stretch wrapping	LDPE, LDPE–LLDPE mixtures
Atlas G-2127	Food packaging and stretch wrapping	LDPE, LDPE–LLDPE mixture
Atmer 106	Stretch wrapping	LDPE
Atlas G-1086	Stretch wrapping	LDPE
Atlas G-1096	Stretch wrapping	LDPE

C. Testing Method for Cling Additives

Measure of the blocking force by using an elongation test apparatus with a load cell (Fig. 11).

1. The blocking force is measured between two pieces of film, pressed together, with an overlap of 2 in.
2. Condition the film at 22°C and 40% RH.
3. Take measurements after 1 day, 1 week, 2 weeks, and 4 weeks.
4. Test in triplicate and average the results.

VII. NUCLEATING AGENTS

It is a characteristic of many crystallizable polymers, such as PP, that a melt of the polymer can be cooled below the melting point of its crystallized form before the molten polymer begins to crystallize. The structure of the resulting crystallized product and many of its properties are dependent on the extent of the melt's super-cooling before crystallization. Nucleating agents decrease the tendency of supercool, thereby affecting crystal structure in the following ways:

Rate of crystallization: higher nucleation density
Number of crystals: a higher number of nuclei per unit volume
Size of the crystals
Total percentage crystallinity

The mechanism of action is poorly understood, but it is generally accepted that the nucleating agents behave as heterogeneous substrates, lowering the energy bar-

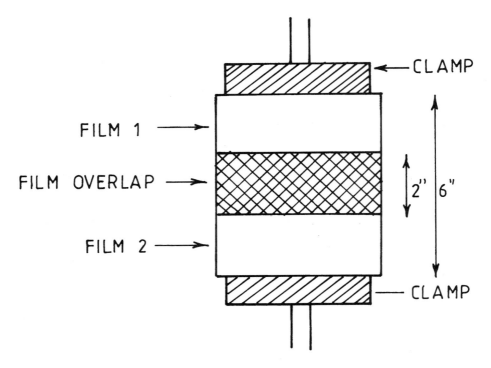

Figure 11 Testing method for cling additives. (From Ref. 14.)

rier toward crystal nucleation. The phenomenon of nucleation is a physical effect, at least in PP.

A. End Effects of the Nucleating Agent

REDUCED CYCLE TIME.   Crystallization temperature is about 10–15°C higher, which allows faster processing.

TRANSPARENCY.   There is a reduction of large crystallites or sperulites that reflect or scatter the light.

SURFACE GLOSS.   The reflection of incident light is improved by increased crystallinity and homogeneous distributed crystals.

MECHANICAL PROPERTIES.   Tensile strength and flexual modulus are improved, whereas impact strength is reduced. All of which increase the percentage of crystallinity.

APPLICATION AREA.   Any area for which optimum transparency is required and it is desirable for the content to be visible (e.g., packaging film and thin-walled containers)

B. Nucleating Agents Used by the Industry

Talc

Effect on
Clarity: none

Stiffness: increased
Crystallization temperature: moderate
Easy to incorporate
Low cost
Toxicological savings

Sodium Benzoate

Effect on
Clarity: mediocre
Stiffness: excellent
Dispersion is a problem: They use a specially ground product (particle size
±2 μm).

Al-PTEB

Effect on clarity: good
Potential carcinogen

DBS

Effect on
Clarity: excellent
Crystallization temperature: increase of 10°C
Tensile strength and flexural modulus: moderate improvement
Impact strength and heat deflection: unchanged
Surface gloss: reflection of incident light improved by increased crystallinity
and homogeneously distributed crystals
Mechanical properties
Improved: tensile strength and flexural modulus, yielding an increased
percentage crystallinity
Reduced: impact strength

C. Application Area

Packaging film: desirable for the content
Thin-walled containers: to be visible

D. Nucleating Agents Used by the Industry

Talc

Effect on clarity: none
Stiffness: increased
Crystallization temperature: moderate
Easy to incorporate
Low cost
Toxicological savings

Sodium Benzoate

Effect on
    Clarity: mediocre
    Stiffness: excellent
Dispersion is a problem: they use a specially ground product (particle size $\pm 2\ \mu m$).

Al-PTBB

Effect on clarity: good
Potentially carcinogenic

DBS

Effect on clarity: excellent
Crystallization temperature: increase of 10°C
Tensile strength and flexural modulus: moderate improvement
Impact strength and heat deflection: unchanged
Surface gloss: 10% improvement

E. Nucleating Agents Specific

1. Atmer 165: dibenzylidene sorbitol, white powder
2. Atmer 166

Advantage
    Better dispersion of active product especially in homo-PP.
    Reduction of migration; less blooming
Level of usage: 0.1–0.25%
Indirect food approval: FDA 0.25% PP/homocopolymer; France 0.25 PP homo/copolymer
Usage conditions
    Polymer should be processed at temperature higher than 220°C, and lower than 270°C.
    Formulation should contain sufficient antiacid
    Effectivity especially in PP random copolymer may be used in LLDPE

3. Atmer 170: blend of talc and antistatic agents

VIII. WHAT IS AN ANTIOXIDANT?

An *antioxidant* is a molecule capable of reducing the rate of degradation of a polymer during manufacture, processing, and service.

A. Why Does a Polymer Need Protection?

Loses mechanical properties (e.g., embrittlement, tensile properties)
Changes in melt index (viscosity)

Discoloration (e.g., yellowing)
Loses electrical insulation properties
Develops odor and taste

B.   Oxidation Is Made More Severe by

High temperatures
Contact with metals, such as iron or copper
Exposure to UV light (e.g., outdoor application)

C.   When Does a Polymer Oxidize?

1. During manufacture
    Vacuum stripping of residual monomer (impact polystyrene)
    Latex drying (ABS latex).
2. During the processing and forming
    Compounding with pigments, fillers, plasticizers, and such
    All operations, such as extrusion, molding, film blowing, fiber spinning,
    and so on.
3. During the life of the polymer
    Temperatures ranging from ambient to 100°C or hither may be
    encountered (e.g., automotive, cables)

D.   Polymer Properties and Evaluation

Key Properties

Tensile strength
Embrittlement
Color
Melt viscosity

Evaluation of Polymer

1. Life stability
    Accelerated oven aging for color and loss of physical properties.
2. Process stability
    Multiple extrusion for color and loss of melt viscosity
    Mill stability
    Brabender stability
3. UV stability
    Weatherometer aging for color and loss of physical properties
4. Electrical properties
    Resistivities
5. Differential scanning calorimetry (DSC)
6. Thermal gravimetric analysis (TGA)
7. Extraction testing

E. Methods Used for Measurement of Polymer Degradation

Measurements of static oxidation: samples in oven for certain time at certain
   temperature
Measurements of oxidation during processing: two-roll mill; extruder with different dies (cast sheets/blown films)

Characteristics	Apparatus used
1. Loss of mechanical properties	Tensile and impact testers
2. Changes in melt index (viscosity)	Melt indexer
3. Discoloration (e.g. yellowing)	Human assessment
4. Loss of electrical insulation properties	Surface and volume resistivity tester
5. Development of odor and taste	Human assessment

F. Types of Antioxidants

The different types of antioxidants are summarized by the following structures.

1. **Phenolic Antioxidants**

MONOPHENOLIC (e.g., BHT)

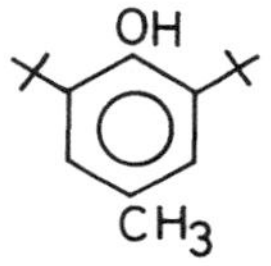

Available from many suppliers
Monophenols are not covered by patents and are sold as commodities

2. **High Molecular Weight Antioxidants**

For example:
Topanol CA (A Trisphenol)

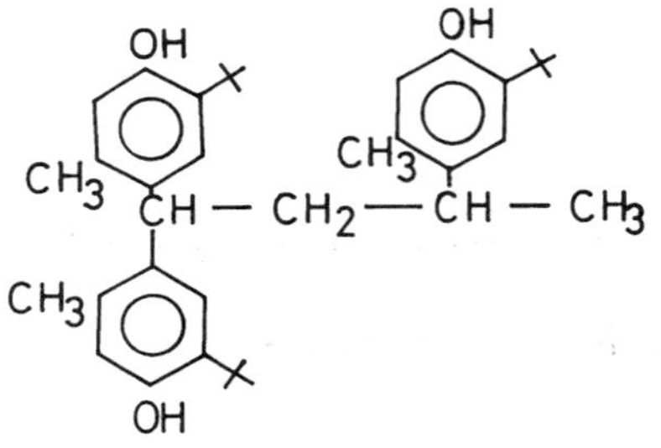

Irganox 1076 (Ciba-Geigy)

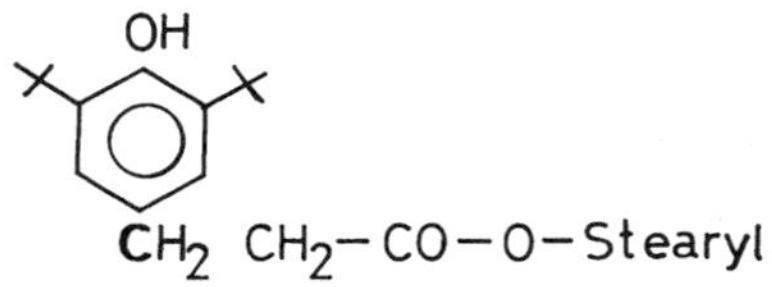

Irganox 1010 (Ciba-Geigy)

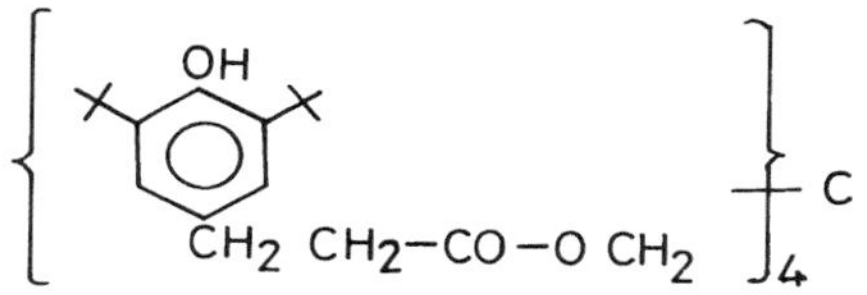

Goodrite 3114 (B.F. Goodrich)

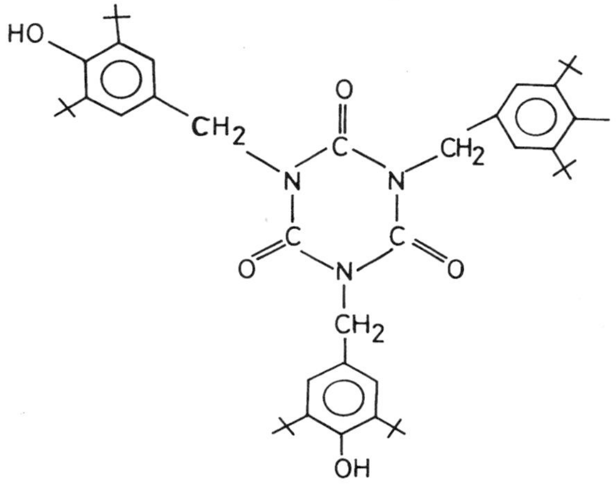

Ethyl 330 (Irganox 1330)

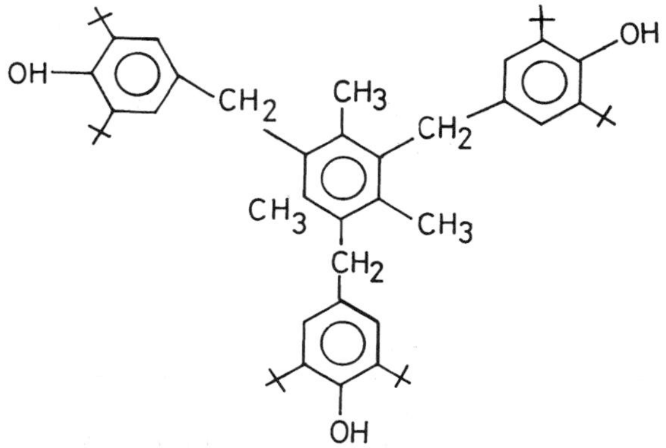

All high molecular weight antioxidants are securely patented and supplied by a single manufacturer.

3. Synergists

    a. *Thioester Synergist*

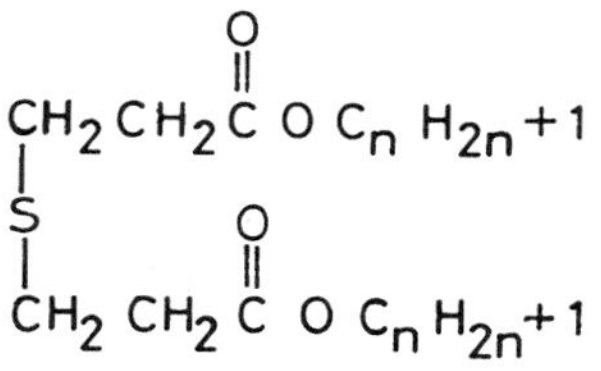

DLTDP: $n = 12$
DSTDP: $n = 18$

b.   Phosphite Synergists

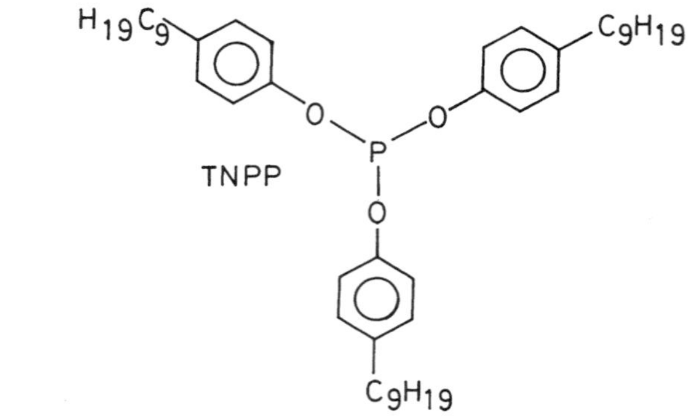

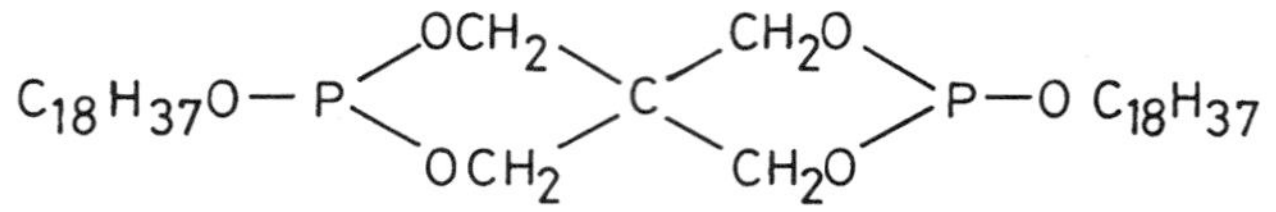

TNPP

Weston 618

IX.   SYNERGISM

Chemical *synergism* may be defined as the result of the combined components is greater than the sum of the individual components.

Topanol CA is capable of producing highly synergistic systems in several important polymers. Synergism is exhibited with both thioester and phosphite secondary antioxidants (Fig. 12).

A.   Topanol CA

Features	Benefits
Efficient	Low usage levels
High molecular weight	Excellent performance with synergist
Versatile	Low volatility
Food contact clearances	Used in a broad range of polymers
Nonstaining	Broadest approvals for HMW
	phenolics; will not discolor resin
Powder form	Free-flowing, easily dispersed in polymer
Soluble in a wide range of solvents (not aliphatic)	Can be used in a wide range of physical systems

B.   Choice of Type and Level of Synergist

1.   The type of polymer may affect the selection of the synergist (e.g., polypropylene: DSTDP, HIPS/ABS-DLTDP).

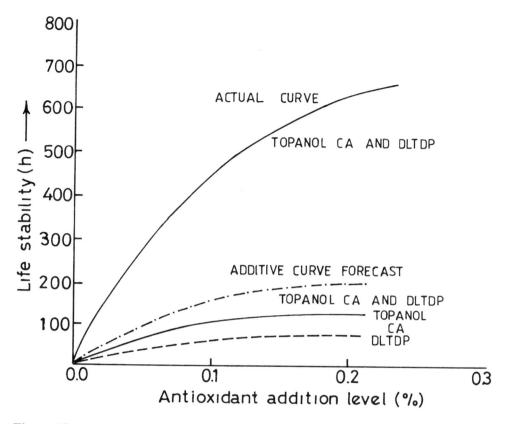

Figure 12   Schematic display of synergism between topanol CA and DLTDP. (From Ref. 14.)

2. Thioesters are used because they impart improved thermal stability to the polymer
3. Phosphites are used because they impart improved color and process stability to the polymer.
4. Cost considerations are of the utmost importance in the selection of antioxidant systems; synergist can help reduce cost.
5. The Topanol CA/synergist optimum ratio is 1 : 3.

1. Low-Density Polyethylene

Competition

BHT
Irganox 1076 ⎤
Irganox 1010 ⎬ synergistic mixtures with thioesters
Santowhite ⎦ phosphites
Santonox

Suggested Formulations

Topanol CA alone:	0.05% or more
Topanol CA/DLTDP:	0.03–0.05%/0.1%
Topanol CA/Phosphite:	0.03–0.05%/0.03–0.10%

Advantages of Topanol CA

Less volatile than BHT
More cost-efficient than other HMW phenolic systems

2. Topanol CA Performance: LDPE Wire and Cable

Topanol CA shows excellent retention of its antioxidant activity on oven-aging in the presence of copper, and it is recommended for use in LDPE for wire and cable applications:

	Degradation time (h): 1% copper dust	
Antioxidant	120°C	140°C
0.10% Topanol CA	235	45
0.10% Santowhite	155	30
0.05% Topanol CA/0.10%DLTDP	735	65
0.05% Santowhite/0.10% DLTDP	135	25

3. Polypropylene

Topanol CA works best in general-purpose, molding, and fiber grades. New applications in this resin arise from expanding use of new, low-residue catalyst systems.

Competition

Ethyl 330
Irganox 1010 } and synergistic mixtures with BHT, thioesters,
Goodrite 3114 and phosphites.

Suggested Formulations

For good processing and life stability	Topanol CA/DSTDP	0.05–0.1%/ 0.15–0.25%
For good processing and color, and fair life	Topanol CA/phosphite	0.05–0.1%/ 0.15–0.25%
For good processing, good color, and good life	Topanol CA/ phosphite/DSTDP	0.05–0.1%/ 0.05%/0.25%
For good processing and good UV stability	Topanol CA/ phosphite or DSTDP/ benzophenone	0.10%/0.25%/ 0.30%
	or: Topanol CA/ phosphite/HALS	0.10%/0.25%/ 0.15%

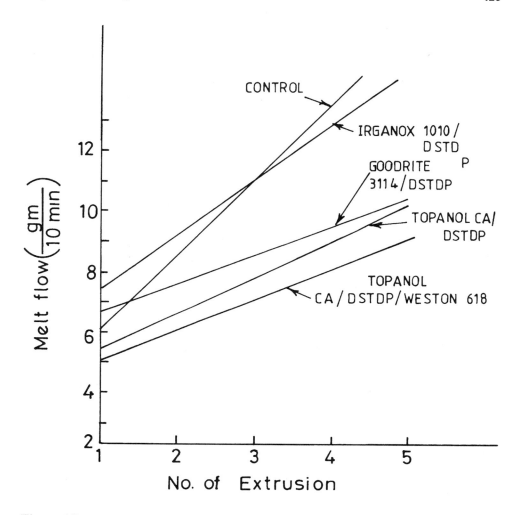

Figure 13 Process stability of polypropylene. (From Ref. 14.)

ADVANTAGES. Topanol CA/ESTDP/phosphite combinations, being highly synergistic provide more effective processing and life stabilization (Fig. 13).

4. Polystyrene

Impact polystyrene, containing 5–15% butadiene rubber, requires significant antioxidant additions. Clear (crystal) polystyrene and foam polystyrene use minimal amounts of antioxidant (usually BHT).

Suggested Formulations

Topanol CA (alone)	0.15% (approx.)
Topanol CA/TNPP	0.03–0.05%/0.1–0.15%
Topanol CA/DSTDP	0.03–0.05%/0.1–0.15%

Advantages of Topanol CA

Low volatility ensures that the resin will be protected during monomer stripping and extrusion.

Topanol CA is nonyellowing in styrenic polymers.

Topanol CA/synergist systems are more effective and more cost competitive than other HMW phenolic systems.

Topanol CA imparts good long-term stability to the polymer.

5. Topanol CA Antioxidant Addition to ABS Processes

Topanol CA antioxidant can be incorporated into ABS resin either during the polymerization process or afterward at the compounding stage (Fig. 14).

Addition of an aqueous dispersion of Topanol CA antioxidant to the ABS latex, enables the antioxidant to become intimately dispersed within the polymer during the coagulation step. Addition of the antioxidant at this stage protects the polymer during processing and drying.

Topanol CA antioxidant can also be added at the compounding stage, before extrusion. The Topanol CA antioxidant may be added at both stages.

6. ABS

The presence of acrylonitrile–butadiene–styrene (ABS)—the butadiene rubber, which may make up 25% of the polymer—is why the antioxidant is necessary.

Suggested Formulations

Topanol CA/DSTDP   0.01–0.15%/0.3–0.45%
Topanol CA/phosphite   0.1–0.15%/0.3–0.45%

Advantages of Topanol CA

Low volatility ensures that resin will be protected during monomer stripping and drying.

Nonyellowing.

Topanol CA/synergist systems are more effective than and more cost competitive with other HMW phenolic systems.

7. Additives to Reduce Molding Cycle

Foamed plastics have good insulation properties (98% of entrapped air in closed cells). Therefore, the molded articles must be cooled for a long time period (three-fourths of total molding operating time).

Energy saving is an important objective in the further development of products (reducing cooling time; increasing demolding temperature) and production process (vacuum technique).

8. Cooling Time Can Be Reduced by Additives

1. Incorporated during the impregnation stage
   Recommendation: G 3858 (Brij 58)
                     G 3998 (Brij 98)   1–3
                     G 1650 (Renex 650) 1

[1] POLYMERIZATION : S-POLYMERIZATION

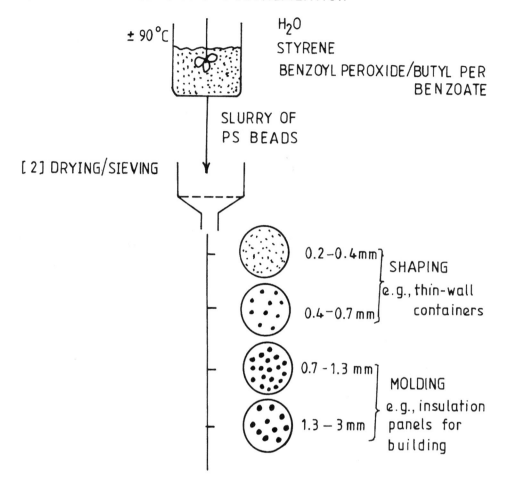

± 90°C

H₂O
STYRENE
BENZOYL PEROXIDE/BUTYL PER
BENZOATE

SLURRY OF
PS BEADS

[2] DRYING/SIEVING

0.2-0.4mm ⎫
⎬ SHAPING
0.4-0.7 mm ⎭ e.g., thin-wall
containers

0.7-1.3 mm ⎫
⎬ MOLDING
1.3-3mm ⎭ e.g., insulation
panels for
building

[3] IMPREGNATION
STAGE

± 90°C

PS BEADS
EXPANDING AGENT e.g., low boiling
hydrocarbon

E-PS BEADS

Figure 14 Production process of expanded PS articles. (From Ref. 14.)

(4) Preexpansion stage

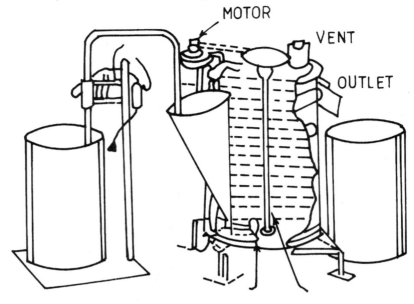

(5) Conditioning

—TO DRY EXPANDED ARTICLES : GIVE BETTER MOLD
                                               FILLING

—PRESSURE EQUILIBRATION   : RESTORE VACUUM

(6) Molding

STEAM CHEST MOLDING
steam

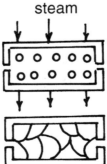

Figure 14 *Continued.*

G 1678 (Renex 678)
0.15–0.3% on PS beads
2. Coated on EX-PS beads
   Recommendation
   a. Sorbitan esters: Atmer 102–103–106
   b. Fine particle size GMS
   c. Necessary: To avoid cluster formation and to obtain uniform distribution
   d. Products: Atmer 123 (EL 1614)
      $\pm$49% mono-
      $\pm$37% di-
      $\pm$14% tri-
      >69 μm: 12%
      Atmer 128 (EL 1603)
      90% mono-
      >68 μm: 25%

X. ICI SPECIALTY SURFACTANTS FOR EXPANDABLE POLYSTYRENE

1. Antistatic agent: reduces the generation of static electricity during seiving and transportation.
   Recommendation: Atmer 163 as external coating 0.05–0.15%
2. Additives to reduce molding cycle: foamed plastics have good insulation properties so the molded article must be cooled for a long time ($\pm$ three-fourths of total operation time).
3. Incorporation during the impregnation stage
   Recommendation: G 3858, G 3998–Eo alcohol
                    G 1650, G 1678–Eo NP
                    0.15–0.25%
   Note: can also be added at the end of the polymerization.
4. Coating of the foamable resin: also reduces the cluster formation during the preexpansion.
   Recommendation: Atmer 102 ⎫
                    Atmer 103 ⎬ sorbitan esters
                    Atmer 106 ⎭

                    EL 1603 ⎱ fine particles size gl/cerol
                    EL 1604 ⎰ fatty acid partial ester
   Note: Can also be added during impregnation stage. GMS also acts as an antistatic.
5. Ross-Miles foam height test (Fig. 15): ASTM D 1173.
   a. Fill foam receiver with the solution upto 50 mL mark.
   b. Fill the pipet with the solution to the 200 mL mark.
   c. Place the pipet on the top of the receiver.
   d. Let the solution run out of the pipet.
   e. Start a stop watch: take a reading of the foam height.
   f. Take a second reading at the end of 5 min.

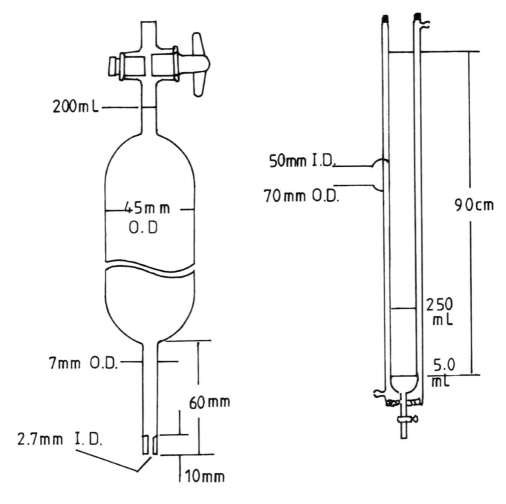

Figure 15   Testing method for antifoam additives.

XI. PLASTICIZERS AND PLASTISOLS

Plasticizers are nonvolatile solvents, the purpose of which is to make the mixture of polymer and plasticizer softer, more flexible, and more easily worked. Normally, the rigid polymer molecules are held together by van der Waals forces, hydrogen bonding, polar forces and, as described earlier, these bonds may be broken and the molecules separated by a solvent: This process may require heat. There may be crystalline regions into which the plasticizer (solvent) penetrates with great difficulty and noncrystalline regions into which it penetrates slowly without heating. Thus, rigid polyvinyl chloride, for example, may be quite insoluble at moderate temperatures in solvents or plasticizers that have the same cohesive energy density (CED) as the polymer; hence, a dispersion of the polymer in the plasticizer may be made that is essentially a stable nonaqueous dispersion. The dispersion is called a plastisol, and it may have the consistency of slush.

When this plastisol is heated for a few minutes, at about 150°C, the plasticizer penetrates the polymer and they combine into a uniform mass that is now soft and flexible, the hard, rigid, plastic polyvinyl chloride has been plasticized by the nonvolatile solvent or plasticizer to yield the flexible, vinyl materials with which we are familiar, such as vinyl-coated fabrics used for boots. The mixture can be molded and fabricated in many ways. One of the processes is called slush molding and another rotomolding. Later we will describe what the plasticizer has done, but at this stage, it will suffice to say that the plasticizer molecules are dispersed between the polymer molecules, polyvinyl chloride in this example, such that these molecules are held apart and are able to move more easily.

This is noted as a greater free volume and a lower glass transition temperature.

The use of plasticizers need not be by only the plastisol process, and indeed, the usual process consists merely of mixing steps at elevated temperatures to achieve the desired product having the flexibility and strength required. This subject will again recur from time to time, but one must recall that a compatible or good plasticizer will have a CED very close to that of the polymer, whereas other organic materials that differ markedly from the polymer as far as the CED is concerned will not act as a plasticizer, will not make a plastisol, and may separate readily or bleed from the mixture with time, forming a second phase or a surface film, respectively.

The use of volatile solvents yields what is known as an organosol, but the process is exactly the same, except that the volatile solvent leaves the polymer behind in its original rigid form, similar to a lacquer, for example. Also, a polymerizable (monomer) solvent may be used, and after the plastisol has been made uniform, this monomer is polymerized to yield a plastic, the two interpenetrating networks of polymer molecules forming a rigid sol. Sometimes one adds to plastisols, organosols, or the rigid sols, a thickening agent before use so that the mass is thixotropic, or more easily spread onto a surface, such as a fabric, in a thick layer without running. The whole subject is very broad and only the main concepts have been touched on.

Plastizers are usually present in amounts from 5 to 30 parts per 100 parts of polymer. When 30–70 parts are used, the term plasticizer is still used, but extension is often implied, more plasticizer is added than is needed to soften the polymer, and the mixture is hardened again by adding a cheap filler, such as carbon black, clay, talc, magnesia, or calcium carbonate. If still more, say 70–500 parts is added, the products are heavily loaded with fillers and are called mastics or caulks.

If the amount of plasticizer exceeds that which can be retained by the polymer or the polymer and filler, the excess will come to the surface, bleed, and can be felt, or in extreme cases, it will actually separate. Plasticizers, in addition to imparting softness and flexibility, may result in cheaper compounds for certain purposes and may, by being rich in chlorine or bromine, impart flame retardancy to the mixture. This subject will be discussed later.

Although not nearly as well known or as often used as plasticizers, there are antiplasticizers. These stiffen polymers, lower the free volume, and increase the glass transition temperatures, but similar to plasticizers, they are nonvolatile solvents. The classic examples are chlorinated aromatics in polycarbonate resins that have a large free volume. The chlorinated wax, "arochlor," fits between the chains, but holds them together so that they cannot separate or move freely, just the opposite of the traditional effect of a plasticizer, which allows movement.

To polymer fabricators, the compatibility of small organic molecules with polymer molecules, as indicated by the CED or the solubility parameter, has many uses and is very important whether it be in surface coatings, adhesives, plastisols, spinning of fibers, sealants, and caulks, or mastics. A knowledge of the approximate values for the solubilities of the polymers and the various solvent and plasticizers used in a practical system may well explain production and end-use problems not attributable to the quality of any individual ingredient in the formulation.

A. Fluorescent Whitening Agents

Fluorescent whitening agents act as fluorescent dyes, although they are colorless. They improve the initial color of plastics, which is often slightly yellowish, by imparting a brilliant white to the end-use articles. Also they increase the brilliancy of colored and black-pigmented articles (17). This function is achieved by the absorption of UV radiation (360–380 nm), converting it to longer wavelengths and reemitting it as visible blue or violet light. The chemical classes of fluorescent whitening agents are benztriazoles, benzoxazoles, pyrene, and triazine.

B. Metallic Pigments

The metallic pigments are lamellar pigments having unique optical properties owing to their plate-like shape and to the high reflectivity of pigment particles. They impart pearly luster to objects. The main nacreous pigments are basic lead carbonate, which has the highest pearl brilliance and is used in HDPE, LDPE bottles, and PP articles, and titanium dioxide coated with muscovite mica, which gives a white pearl effect in blow-molded PE containers (17). Metallic flake pigments are composed of copper, bronzes, and aluminum. They are available as pastes.

XII. ADDITIVES "ANCHORED HALS"

A class of hindered amine light stabilizer (HALS) has been launched by Clariant Hunigue SA (Hunigue, France) an entity "demerged" from Sandoz Corp. Clariant claims that its new additives have the potential to transform UV stabilization practices across a broad swath of resins and markets.

Sanduvor PR-31 uses a patented "photo reactive" chemistry to graft HALS molecules to the matrix polymer's backbone, thereby reducing volatility, extraction, and migration. The mechanism that triggers grafting in Clariant's product is a 5- to 20-day exposure of stabilized end products to UV light.

A. Benefits of Anchoring

Conventional HALS of low and high molecular weight (LMW; HMW) types have served plastics well, as attested to by their widespread use and rapid growth rate. Some industry sources, however, say they suffer deficits. Short-chain LMW types are often volatile or migratory, and their UV resistance becomes depleted. Long-chain HMW types have more permanence, but their surface stability is less.

Reactive conventional HALS can also impair other stabilizers (e.g., UV absorbers), mute certain pigment colors, or disrupt consistency, according to sources.

Proponents say "chemical anchoring" of HALS alleviates "blooming" and "plateout" in processing, reduces required UV stabilizer loading, and extends the long-term weathering life of stabilized products, suppliers say.

Clariant claims to circumvent the limits of conventional HALS with its PR-31, in which the HALS molecules, in sunlight, graft to the polymer backbone precisely at the part surface, creating an outer UV-absorbing layer. This occurs in the 300- to 340-nm light band, in which UV degradation in polyolefins mostly takes place. A reservoir of unreacted HALS, meanwhile, remains in the substrate, protected by the UV-absorbing outer layer. When drawn on gradually, it extends the part's long-term weathering life.

Data indicate that PR-31 delivers a major extension of outdoor exposure life in a 100-μm–thick PP homopolymer film versus a competing olifomeric HALS using an equal loading. In molded HDPE parts, the HALS increases impact strength, reduces color change, and elevates gloss, versus competing offerings. Clariant claims its HALS yields an excellent white base color, and reports no known interference with other additives.

Other suppliers are adopting analogous (if different) grafting techniques. Ferro Corp.'s Bedford Division (Bedford, Ohio) reports market inroads with the chemically anchored HALS it recently launched. HALS AM 866 grafts HALS molecules to a methyl methacrylate backbone and synergizes with the supplier's existing workhorse benzoate ester UV absorber.

Another grafted HALS is Elf Atochem's (Paris, France), which offers a product that bonds to the polymer backbone using peroxides that react during polymerization or processing.

Cytec Industries (Stamford, Connecticut) has a new HALS concentrate that uses long-chain entanglement to achieve an improved balance of UV stability. Cyasorb UV-3853S is a 50% concentrate in an LDPE carrier with low-blooming characteristics for ABS, filled PP, and thermoplastic elastomer.

XIII. DYNAMAR (3M) (POLYMER-PROCESSING ADDITIVE FX-5920)

A. Introduction

Dynamar polymer-processing additive FX-5920 is a free-flowing fluorocarbon-based formulation that is designed for use at very low levels to improve processing of plastics. At the very low-use levels (typically 250–1000 ppm) necessary to improve processing, it neither alters nor detracts from the good mechanical properties associated with high-strength plastics (11).

Dynamar FX-5920 can offer performance and cost advantages over comparable loadings of other processing additives. It exhibits exceptional commercial usefulness in low-melt–index film-grade linear low-density polyethylene (LLDPE). It is especially effective in polyolefin resins containing silica-based antiblocking agents, titanium dioxide-based pigments, and other inorganic additives.

Dynamar FX-5920 lowers apparent melt viscosity and permits lubricators to use high-strength resins that otherwise could not be processed on available equipment. Now with the aid of Dynamar FX-5920, fabricators can produce superior films and other articles of improved strength and quality.

Dynamar[a] Polymer-Processing Additives

Demonstrated benefits in
    LLDPE, HMW-HDPE, ULDPE, LDPE
    Polypropylene
    PE/PP copolymers
    Certain PET-EVA-ano PVC-resins
For applications, such as
    Blown and cast film
    Blow molding
    Wire and cable
    Pipe and other tubing
    General extrusion
With advantages that include
    Increased throughput
    Decreased power consumption
    Decreased processing temperatures
    Surface defect elimination
    Use of tougher, stronger resins
    Decreased die buildup
    Less down time and improved quality

[a] Available as powder pellet or master batch concentrate.

As a process aid, Dynamar FX-5920 reduces or eliminates melt fracture, can reduce extruder torque, increases output, improves clarity and gloss, and yields films with more balanced bidirectional properties.

Dynamar polymer-processing additives (PPA) are multipurpose extrusion aids that are used in a variety of extrusion processes and resins. Such applications include film, sheet, and pipe extrusion, as well as wire and cable extrusion and blow molding. Resins benefiting from Dynamar PPA include polyolefins, such as LDPE, VLDPE, LIDPE, HDPE, HMW–HDPE, EVA, and PP. The extrusion of other thermoplastics may benefit as well. Dynamar products are available in several particle sizes to accommodate different incorporation procedures, and these vary in composition to optimize end-use performance.

B. Direct Addition During Resin Manufacture

1. Introduction

Dynamar PPA products have been developed with specific particle sizes and flow characteristics for direct incorporation into formulations during resin manufacture. Often, the need for melt-compounded masterbatches can be eliminated.

2. Other Additives

Contacting Dynamar PPA at ambient conditions with other powdered additives has not had negative effects. In fact, combination with other additives may improve additive distribution in the final product. Associating the Dynamar PPA at fixed weight ratios with other additives can also reduce the need for analytical testing by linking its concentration to other known levels. At elevated temperatures, certain additives may affect the performance of the Dynamar PPA.

3. Addition Methods

Ideally, Dynamar PPA can be added individually or in combination with other powdered additives through a weight loss feeder. If additive distribution is a problem, a dry preblend can be produced with powdered resin direct from the reactor. A ribbon blender with an intensive mixer is recommended to blend the materials. This will segregate and distribute the additives, ultimately leading to more uniform and precise additive levels.

4. Equipment and Conditions

The viscosity of Dynamar PPA is such that it can be mixed easily with a twin-screw extruder and continuous mixer-pelletizing trains. No modifications are required other than addition of additive feeders or preblenders, if not present. Standard conditions used for homogenizing and pelletizing the resin should be adequate. The presence of liquid additives may require additional shear to disperse the PPA. Highly viscous resins are more difficult to homogenize, and a master batch may be required.

5. Evaluation

When properly dispersed, the PPA should have particle sizes of 1 μm or less and be evenly distributed throughout the sample (see 3M method "Optical Microscopy Methods for Dispersion Analysis in Polyolefins"). Several analytic methods have been developed by 3M for determining quantitative levels of Dynamar PPA is fully formulated resins.

XIV. FLUOROCARBON ELASTOMERS AS PROCESSING AIDS FOR POLYOLEFINS

A. Introduction

Additives generally categorized as improving polymer processability encompass a wide variety of chemical species. However, few have shown the capacity to simultaneously improve both throughput and surface defects, such as sharkskin and cyclic melt fracture (1). Of those that do, fluorocarbon elastomers have been particularly effective in polyolefin applications throughput and pressure extrusions, such as pipe or wire and cable coatings. Resins used in these processes typically contain a variety of additives that include antioxidants, fillers, slip agents, and antiblocking agents, to name a few. Although additive concentrations in the final extruded products are usually expressed in parts per million (ppm), interactions among the additives can occur and are often exacerbated by the severe processing or compounding conditions that may precede extrusion. The need to identify such interactions as they affect fluorocarbon elastomer-processing additives has stimulated research efforts. Research that could guide the processor to control or possibly capitalize on specific interactions has been limited because of the need to focus on isolated PPA–additive combinations to understand such interactions. The present work, through consideration of factors, such as temperature, concentration, and compounding conditions,

Table 1   Additive Glossary

Abbreviation	Chemical name
AB-1	Diatomaceous earth (93% SiO_2)
AO-1	Octadecyl-3-(3,5-dibutyl-4-hydroxyphenyl)-propionate
AS-1	Bis(2-hydroxyethyl)tallowamine
CaSt-N	Calcium stearate (neutral)
CaSt-B	Calcium stearate (basic)
HALS-1	N,N-Bis(2,2,6,6-tetramethyl-4-piperidinyl)-1-,6-trichloro-1,3,5-polymer with 2,4,6-trichloro-1,3,5-trizaine and 2,4,4-trimethyl 1,2-pentanamine
Ni-1	2,2-Thiobis(4-*tert*-octylphenolato)n-butylamine nickle
P-1	Tris (2,4-di-*tert*-5-butylphenyl)phosphite
P-2	Bis[2,4-di-*t*-butylphenyl) pentaerythrito]diphosphite
P-3	Trisnonylphenylphosphite
PEG(400)	Polyethylene glycol (MW-400)
PEG(1450)	Polyethylene glycol (MW-1450)
PEG(3340)	Polyethylene glycol (MW-3350)
S-1	Erucamide
TiO_2-1	Titanium dioxide, w/1.5% Al_2O_3, 0.5 organic
TiO_2-2	Titanium dioxide, w/4.5% Al_2O_3
TiO_2-3	Titanium dioxide, no coating
TiO_2-4	Titanium dioxide, w/0.5% Al_2O_3, 0.25% organic
TiO_2-5	Titanium dioxide, w/2.0% Al_2O_3, 0.33% organic
UFT	Hydrated magnesium silicate (ultrafine talc)
UVA-1	2-Hydroxy-4-n-octyloxy-benzophenone
ZnO	Zinc oxide
ZnS	Zinc sulfide
ZnSt	Zinc stearate

Source: Ref. 9.

is an attempt in that direction. A glossary of additives included in both the present and previous work is provided in Table 1.

The fluorocarbon elastomer used in this study was a copolymer of vinylidine fluoride and hexafluoropropylene. In the chemical reactivity studies it was used as an unmodified raw gum. For incorporation in polyethylene, a free-flowing powder, containing 10 wt% microtalc as a partitioning agent, was used. The latter form is known commercially as Dynamar Brand Polymer Processing Additive FX9613, hereafter referred to as PPA. Plastics additives used in addition to the PPA are described in Table 2.

B.   Concentration Effect

Although dilution to the parts per million level, without any highly concentrated preliminary contact, can mitigate interactions between additives, such PPA-containing mixtures are likely to reach a concentration ratio at which optimized processability begins to decline. Table 3 summarizes the effect of increasing the concentration of a hindered amine light stabilizer (HALS-1) on the viscosity

Table 2 Summary of Functional Interaction of
Additives on Processing Effectiveness of PPA (Capillary
Rheometry Studies)

Evidence of synergism
    AO-1
    AS-1[a]*
    P-1
    P-2
    P-3
    PEG(3350)
Little or no effect
    Cast—N*, B*
    PEG(400)
    UFT
    ZnS
    ZnSt
    UVA-1
    S-1*
Evidence of interference
    AB-1
    AS-1[a]*
    HALS-1*
    HALS-2*
    Ni-1*

*, evidence for chemical reactivity with PPA; [a], depends on
processing conditions.
Source: Ref. 9.

and melt fracture reduction typical of a LLDPE containing 500 ppm PPA. This
additive has also been previously shown to chemically react with the PPA. The table
indicates that at HALS-1/PPA ratios of greater than about 1 : 2 interference is
maximized, and there is little or no drop in viscosity or change in the appearance
of the filament. However, at a ratio of 1 : 1 (500 ppm each), both the equilibrated
viscosity and minimum shear rate for melt fracture onset approach values similar
to those of the sample containing only the PPA.

A second PPA–additive system exhibiting a pronounced concentration effect is
summarized in Table 4. Here, the titanium dioxide pigment is not believed to react
chemically with the PPA and requires a much higher concentration before
influencing the processability. Comparison of a 5% TiO_2/1,000 ppm PPA sample
with a pigment-free control shows only a slight decrease in the effect of the
PPA on apparent viscosity and melt fracture elimination. However, as the
TiO_2/PPA ratio is increased from 50 : 1, to 100 : 1, the situation deteriorates to
a point at which only marginal benefits result from the presence of the PPA. A variety
of factors (e.g., abrasion at the die wall, or adsorption of PPA particles by the
pigment) may account for the concentration ratio–interference effects seen in Table
4. Other chemically unreactive additives also influence PPA, although at different
levels. Generalization of the specific results in Table 4 to other commercial TiO_2
pigments is, however, unwise because of the role of pigment surface treatment
on PPA interaction as will be described later.

Table 3   Effect of Hindered Amine Light Stabilizer Concentration on PPA Performance[a]

HALS-1 (ppm)	PPA (ppm)	Apparent viscosity at 600 s^{-1} (PaS)[b]	Shear rate for melt fracture onset (s^{-1})
0	500	360	1600
500	500	400	1600
1000	500	510	600[c]
3000	500	530	600
5000	500	525	600
5000	0	539	600

[a] In LLDPE (1-butene) at 210°C.
[b] Equilibrated value.
[c] Filament exhibits smooth and fractured streaks.
Source: Ref. 9.

Table 4   Effect of Titanium Dioxide Concentration on PPA Performance[a]

TiO$_2^{-4}$	PPA (ppm)	TiO$_2$/PPA	Apparent viscosity at 600 s^{-1} (PaS)[b]	Shear rate for melt fracture onset (s^{-1})
0	500	0 : 7	341	1800
5	1000	50 : 1	377	1600
5	500	100 : 1	400	1200
10	500	200 : 1	497	600
5	0		653	200
0	0		650	200

[a] In LLDPE (1-hexene) at 190°C.
[b] Equilibrated value.
Source: Ref. 9.

C. Process Aid Effectiveness

A chemical reaction between two additives does not in itself predispose interference in their ability to function in the polymer melt. Processing temperature, residence time, and the effect of dilution, all can mitigate the effect of a given reaction. On the other hand, two additives showing no tendency to react chemically may be incompatible for other reasons. For the fluorocarbon elastomer process aids, an interfering additive could, for example, adversely affect dispersion or interfere with adhesion modification at the die wall. For consideration of binary interactions in a diluted polymer matrix, each additive was combined with linear low-density polyethylene that had been precompounded with 500 ppm PPA. Additive levels were set at either 2,000, 5,000, or 10,000 ppm, generally corresponding, to maximum levels in most commercial applications.

Capillary rheometry was used to provide rheological measurements and to produce filaments that could be examined for evidence of melt fracture. To ensure full equilibration, each sample was subjected to a constant shear rate of 600 s^{-1} until a viscosity change of no more than 20 poise occurred over a period of 5 min.

Table 5 Effect of PPA–Additive Interaction on LLDPE Viscosity

	Apparent viscosity of LLDPE (poise)		
Additive (ppm in LLDPE)	Without PPA[a]	With PPA[b]	(%) (−)
Control (no additive)	6,130	3,720	39
AB-1 (10,000)	5,850	4,370	25
AO-1 (2,000)	6,030	3,710	38
AS-1 (2,000)	5,990	3,950	34
CaST-N (2,000)	5,930	3,850	35
CaST-B (2,000)	6,060	3,870	36
HALS-1 (5,000)	5,910	5,350	9
HALS-2 (5,000)	6,010	5,290	12
Ni-1 (5,000)	5,930	4,000	33
P-1 (2,000)	6,150	3,380	45
P-2 (2,000)	6,000	3,450	43
P-3 (2,000)	5,860	3,560	39
PEG(400) (2,000)	5,890	4,180	29
PEG(3,350) (2,000)	5,850	3,900	33
S-1 (2,000)	5,650	3,580	37
UFT (10,000)	5,600	3,480	38
UVA-1 (5,000)	6,090	4,560	25
ZnO (2,000)	6,050	4,660	23
ZnS (10,000)	5,700	3,330	42
ZNSt (2,000)	5,760	3,770	35

[a] Base resin of 1.0 MI LLDPE (butene comonomer) with 300 ppm AO-1.
[b] Base resin plus 500 ppm FX9613 fluoroelastomer PPA.
Source: Ref. 9.

The resin containing only PPA showed a 39% drop in viscosity relative to the PPA-free control with equilibration occurring in 30 min. Each additive–PPA combination was then evaluated with the same criteria and compared with control resins containing only the relevant neat additive. Comparative viscosity decreases are summarized in Table 5. Although repeat analyses showed some variation in the time required to reach equilibration, the equilibrium viscosities were very reproducible.

Differences in the influence of additives on PPA effectiveness in suppressing melt fracture were also observed. Without process aid the capillary filaments displayed the familiary surface deformation, known as sharkskin, at shear rates of about 200 s^{-1} and persisted to at least 2,000 s^{-1}, after which the filament exhibited gross distortion and shear rate-independent viscosity. Polyolefin additives other than the PPA had little effect on melt fracture. Sharkskin was always present, and only slight changes were noted in the shear rate at which cyclic melt fracture (CMF) began, as seen in Table 6.

Fracture onset was observed for CaSt, UVA-1, and ZnO. Four additives, including HALS-1, HALS-2, Ni-1, and AB-1 had a detrimental effect on melt fracture elimination. Sharkskin was never completely removed from resins containing HALS-2, Ni-1, or AB-1, and in both HALS-1 and HALS-2 cyclic melt

Table 6 Effect of PPA–Additive Interaction on Melt Fracture Elimination: Capillary Rheometer Results

| Additive (ppm in LLDPE) | Shear rate for onset of melt fracture | | | Type |
| | Without PPA | | With PPA | |
	Sharkskin	Cyclic MF	With PPA	Type
Control (no additive)	600 s^{-1}	$1,000 \text{ s}^{-1}$	$1,400 \text{ s}^{-1}$	Continuous
AB-1 (10,000)	600	1,200	600[a]	Sharkskin
AO-1 (2,000)	600	1,100	1,900	Continuous
AS-1 (2,000)	600	1,100	1,800	Continuous
CaSt-N (2,000)	600	1,000	1,300	Continuous
CaSt-B (5,000)	600	1,100	1,300	Continuous
HALS-1 (5,000)	600	1,100	1,100	Cyclic
HALS-2 (5,000)	600	1,000	600[a]	Sharkskin
Ni-1 (5,000)	600	1,000	600[a]	Sharkskin
P-1 (2,000)	600	1,100	1,900	Continuous
P-2 (2,000)	600	1,100	1,800	Continuous
P-3 (2,000)	600	1,100	1,800	Continuous
PEG (400)	600	1,200	1,500	Continuous
PEG (3,350)	600	1,100	1,800	Continuous
s-1 (2,000)	600	1,200	1,700	Continuous
UFT (10,000)	600	1,000	1,400	Continuous
UVA-1 (5,000)	600	1,000	1,300	Continuous
ZnO (2,000)	600	900	1,200	Continuous
ZnS (10,000)	600	1,100	1,600	Continuous
ZnSt (2,000)	600	1,200	1,500	Continuous

[a] System did not equilibrate: some sharkskin was eliminated.
Source: Ref. 9.

fracture occurred at the same rate at which it did in the absence of PPA. To a large extent interference in the ability of the process aid to suppress melt fracture seemed to mirror the influences observed in lowering the viscosity. Chemical reactivity appears to be the dominant factor for three of the additives showing the greatest tendency to interfere in PPA effectiveness. With AB-1, effects other than chemical reaction are clearly at play. Combining the results of both viscosity and melt fracture measurements, a summary of the functional interaction of individual polyolefin additives with the PPA under capillary conditions is presented in Table 2.

D. Addition of PPA Synergist

We have also found, as noted in Table 2, that certain additives may act as synergists (9,10) toward fluorocarbon elastomer PPAs. In several instances these interactions have been corroborated with larger-scale blown film trials. The use of a synergist to reestablish optimum processability, otherwise reduced by an interfering additive, could be economically attractive. Polyethylene glycol, in particular, has shown great

Table 7   Effect of Polyethylene Glycol on AB-1/PPA/LLDPE Processability[a]

AB-1 (ppm)	PPA (ppm)	PEG(1450) (ppm)	Apparent viscosity at 600 s^{-1} (PaS)[b]	Shear rate for melt fracture onset (s^{-1})
0	500	0	406	1400
5000	500	0	463	600[c]
5000	500	2000	339	2000
5000	0	0	577	600

[a] In LLDPE (1-hexene) at 190°C.
[b] Equilibrated value.
[c] Filament exhibits smooth and fractured streaks.
Source: Ref. 9.

promise for this. Results illustrating this effect are presented in Table 7. A combined master batch of 20% diatomaceous earth and 2% PPA was compounded and let down to concentrations of 5,000 ppm and 500 ppm, respectively. At 600 s^{-1} the drop in viscosity owing to the PPA was considerably less than in the absence of the antiblock, and the extruded filament could not be completely rid of sharkskin. A sample was then compounded with 20% diatomaceous earth, 2% PPA, and 8% polyethylene glycol (PEG 1450) and let down to concentrations of 5,000 ppm, 500 ppm, and 2,000 ppm, respectively. Capillary extrusion of this sample showed that both viscosity reduction and melt fracture elimination were improved relative to either the PPA–additive combination or the sample containing only PPA. The dramatic benefit of using polyethylene glycol as a synergist for fluorocarbon elastomer PPAs has also been observed in our laboratory under blown film conditions, particularly, in highly filled or pigmented systems (8). The reasons for this synergy remain speculative, although there is some evidence of chemical association between the two species (5).

E. Troubleshooting Guide for Injection Molding of Rigid Vinyl

Defect: Short Shots, Possible Causes

1. Insufficient material
2. Injection pressure too low
3. Injection speed too slow
4. Cylinder temperatures too low
5. Mold temperatures too low
6. Insufficient venting
7. Sprues, runners, or gates too small
8. Melt temperature too low

Defect: Sink Marks or Excessive Shrinkage

1. Insufficient material
2. Injection pressure too low
3. Hold time too short

4. Cooling time too short
5. Melt temperature too high
6. Mold temperature too high
7. Sprues, runners, or gates too small (improper mold design)
8. Injection hold pressure too low

Defect: Weak Welds, Possible Causes

1. Mold temperature too low
2. Injection speed too slow
3. Melt temperature too low
4. Injection pressure too low
5. Insufficient mold venting
6. Improper gate locations or size
7. Cylinder temperature too low
8. Screw back-pressure too low
9. Nozzle diameter too small

Defect: Part Sticking in Cavity, Possible Causes

1. Injection pressure too high
2. Hold pressure too high
3. Hold time too long
4. Core side of mold too hot
5. Rough surface on cavity side of mold
6. Fill rate too fast
7. Shot size too large

Defect: Blush Marks Around Gates, Possible Causes

1. Mold temperature too cold
2. Injection fill speed too fast
3. Melt temperature too high or too low
4. Improper gate location
5. Sprue and nozzle diameter too small
6. Nozzle temperature too low
7. Insufficient cold slug well
8. Imperfections in gate openings
9. Moisture in the compound

Defect: Dullness on Molding Surface, Possible Causes

1. Cylinder temperatures too low (increase in small increments)
2. Screw backpressure too low
3. Injection fill speed too slow
4. Mold temperature too cold
5. Melt temperature too low
6. Moisture in the compound

Defect: Silver Streaks on Part Surface, Possible Causes

1. Melt temperature too high
2. Nozzle temperature too high
3. Injection speed too fast
4. Excessive moisture on material

Defect: Flashing Possible Causes

1. Injection pressure too high
2. Insufficient clamping pressure
3. Injection speed too fast
4. Melt temperature too high
5. Mold faces not plane and parallel
6. Improper venting (one cavity venting while another fails to fill)
7. Improper mold design

Defect: Dull Streaks, Flow Lines, Possible Causes

1. Melt temperature too low
2. Runners too small
3. Improper gate size or location
4. Mold temperature too low
5. Inadequate cold slug wells

Defect: Warpage, Possible Causes

1. Mold temperature too high (for thick wall sections)
2. Melt temperature too high
3. Insufficient hold time
4. Injection and holding pressure too high or too low
5. Injection speed too fast
6. Cycle time too short

Defect: Lamination, Possible Causes

1. Purging compound left in cylinder
2. Mold temperature too low
3. Melt temperature too low
4. Injection speed too fast
5. Gate size too small
6. Injection pressure too high

Defect: Temperatures Overriding on Front Zones, Possible Causes

1. Compression ratio of screw too high
2. Excessive backpressure
3. Insufficient air circulation on overriding zones
4. Screw RPM too high

Defect: Burn Streaks in Center of Sprue, Possible Causes

1. Front zone temperature too high
2. Screw speed too high
3. Excessive backpressure
4. Compression ratio of screw too high
5. Melt temperature too high

Defect: Burn Streaks at Gate, Possible Causes

1. Injection speed too fast
2. Injection pressure too high
3. Gates or nozzle diameter too small (improper design)
4. Shear burning owing to cold material

Defect: Discoloration or Burned Areas in Part Possible, Causes

1. Screw speed too fast
2. Backpressure too high
3. Cylinder temperatures too high
4. Faulty temperature controllers
5. Gates too small
6. Dead material hung up on screw or nozzle
7. Insufficient mold venting
8. Melt temperature too high
9. Moisture in compound

Defect: Weld Burning, Possible Causes

1. Injection speed too fast
2. Melt temperature too high or too low
3. Screw RPM too high
4. Backpressure too high
5. Nozzle diameter too small
6. Sprue, runner, or gates too small
7. Injection pressure too high
8. Insufficient mold venting
9. Excessive moisture on the material

XV. ADDITIVES FOR RECYCLED PLASTICS

A plastic waste management system is summarized , and projections for plastic recycling are outlined, up to the year 2000, in the following tables.

Before beginning with physical and chemical upgrading, sorted and cleaned streams of used plastics are usually needed. These two operations already belong to the upgrading chain. The market for applications of uncleaned and not presorted used plastics is only limited.

There are numerous possibilities of improving the quality of recycled plastics. Some methods are technical and some physicochemical. Technical methods mainly comprise, besides washing and separation, compounding and melt filtration.

Physicochemical methods include blending with various types of additives, such as compatibilizers, polymer modifiers, fillers and, not to be forgotten, stabilizers.

Polymers, as all organic materials, degrade during processing and end use. Aging deteriorates visual and mechanical properties. Resin producers and processors, therefore, rely on the use of processing and heat and light stabilizers, while manufacturing resins and producing articles of virgin polymers. Degradation processes are not confined to virgin resins only, but proceed in recycled materials, even more dramatically (13,15).

In developing new applications for used plastics material, the whole cycle must be taken into consideration. This means plastic streams that are available as material resource, and the envisaged end application. In between lies the whole technical know-how and the marketing skills of the recycling industry to offer on the market reprocessed resins that meet approximately the quality standards of virgin materials.

Integrated Plastics Waste Management Solutions

Source reduction
Reuse
Material recycling
Feedstock recycling (chemical recycling)
Incineration with energy recovery
Landfill

Source: Ref. 13.

Plastics Use and Material Recycling by the Year 2000

Use	Over 100 MM tons worldwide
Material recycling	10–15 MM tons (10–15%)
	(conservative estimate)
Upgraded material	20%? 50%? or more?
	3 MM tons 7.5 MM tons

Source: Ref. 15.

Upgrading Possibilities for Used Plastics

Technical methods	Physical/chemical methods
By washing or grinding	By blending with virgin material
By separation	By blending with polymer modifiers
By melt filtration	By addition of compatibilizers
By compounding	By addition of reinforcing material (fibers, fillers, ...)
By different construction methods	By addition of processing aids
By process optimization (coextrusion, ...)	By restabilization

Source: Ref. 15.

As a rule, polymers are stabilized only for their first-life application. Therefore, recyclers should not assume that this initial stabilization is sufficient for the second life. It often happens, that recyclates from short-term applications as packaging will generally find that use is insufficient for the new application.

Data concerning the use of additives in recycled plastics are becoming increasingly available. Inhouse scrap generated by polymer producers and processors is generally homogeneous and relatively clean. It is usually mixed with virgin material and restabilized with the original additive formulation. Used plastics or postconsumer recyclate (PCR), as it is called in the United States, usually requires a pretreatment before reformulation or restabilization, leading to material for high-end applications.

Before restabilizing PCR, recyclers should take into consideration the degree of degradation, the amount of residual stabilizers, and the type and quantity of contamination present. To be on the safe side, as far as second-life application is concerned, the recycler should analyze the material and study its long-term performance after restabilization.

In the following, examples of upgrading by restabilization of six different polymers are given: polyethylene, polypropylene, polyethylene terephthalate, polyvinyl chloride, and mixed plastics.

Restabilized Plastics Material (Examples)

Polyethylene	Bottle crates
Polypropylene	Automotive parts/bumpers
	Battery cases
Polystyrene	Yogurt cups
	EPS
PET	Bottles
PVC	Roofing sheets
	Window profiles
Mixed plastics	Bottle fraction (HDPE + PP)
	Material ex household collection

Source: Ref. 13.

A. Processing, Long-Term Heat and Light Stabilization of Recycled Plastics

The plastics industry upgrades the used polymeric material by the use of stabilizers.

1. Polyethylene

Meanwhile a classic example is the recycling and restabilization of HDPE bottle crates. The use of light stabilizer HALS-1 and UVA-1 in addition to an antioxidant package is imperative for achieving good results.

2. Polypropylene

Different factors speak for material recycling of automotive parts such as bumpers, battery cases, and others.

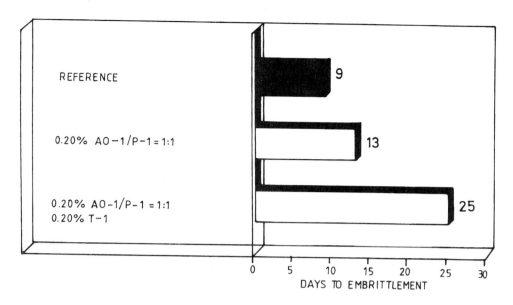

Figure 16 Long-term thermal stability of PP/EPDM scrap (100% used car bumpers). Extrusion: twin-screw, max. 260°C; test samples: 2-mm injection; molded plaques; oven aging: 150°C, circulating air oven.

Single polymers, relatively easy to dismantle
Big parts (by weight)
Expected legislation (e.g., Germany)
Image care (Europe, Japan, and United States)

A part of a bumper containing 20% recycled bumper material was restabilized with antioxidants and HALS light stabilizers. Figure 16 shows laboratory results of 100% recycled PP/EPDM bumpers in a standard oven-aging test, with and without antioxidant blend (0.1% AO-1 + 0.1% P-1). Addition of thiosynergist (T-1) is especially advantageous in this high-temperature test, resulting in oven-aging times of 25 versus 9 days, but the known influence of thiosynergists on HALS should be also taken into consideration.

In Europe and the United States there are now over 100,000 tons of battery cases recycled industrially. They are often used for low-end applications, such as flower pots, but by reformulating them and using some heat and light stabilizers, they can be upgraded to again become battery cases or even stadium seats in their second life. Figure 17 represents processing stability of used battery cases and Fig. 18 lists long-term heat-aging results.

In both instances AO-2, an antioxidant blend developed for recycled materials, shows the best performance.

3. Polystyrene

Polystyrene (PS), either in its impact (IPS; e.g., yogurt cups) or expanded (EPS: e.g., packaging) form, constitutes articles of our daily life; hence, its visibility is quite

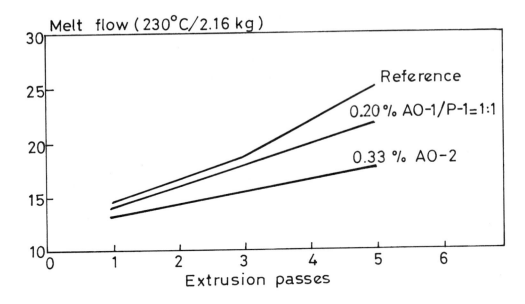

Figure 17   Processing stability of PP scrap from battery cases.

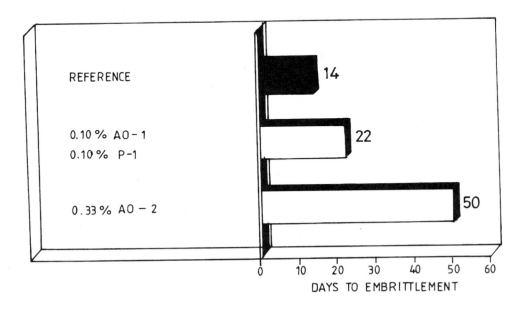

Figure 18   Long-term thermal stability of PP scrap from battery cases.

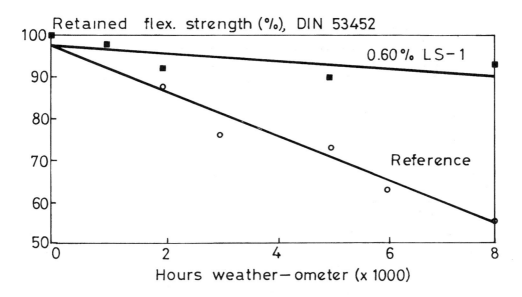

Figure 19   Light stability of mixed plastics from household collection.

high. The polystyrene-manufacturing and processing industry, therefore, started recycling schemes in different parts of the world. To achieve improved processing and long-term heat stability of recycled PS, typical antioxidants such as AO-3 can be used. Figure 19 depicts an example for restabilization of yogurt cups and Fig. 20 of foamed EPS from packaging.

4. Polyethylene Terephthalate

Recycling of polyethylene terephthalate (PET) bottles is especially popular in the United States. But also in Switzerland PET bottle recycling is strongly promoted by PRS (PET Recycling Switzerland) and had achieved over 5000 tons in 1993.

A typical second-life application would be fibers. One of the problems during PET recycling and reprocessing is strong molecular weight decrease and degradation manifesting itself by lower intrinsic viscosities. In the presence of a 4 : 1 blend of P-1 and AO-1 this degradation can be markedly reduced (Fig. 21).

B.   Mixed Plastics

1. Polyolefin Mixtures

Sometimes relatively simple mixtures of polymers can be collected in rather high amounts. One such example is represented by bottle collection (Europe, United States), which consists of 90% HDPE and 10% PP blend without restabilization and in the presence of 0.2 or 0.4% AO-2 (Fig. 22). In the presence of 0.4% AO-2 the samples last three times longer until embrittlement than without stabilizer.

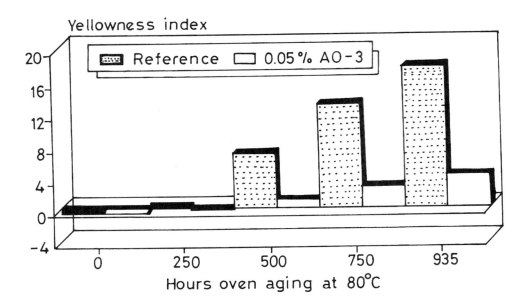

Figure 20   Long-term thermal stability of recycled PS cups.

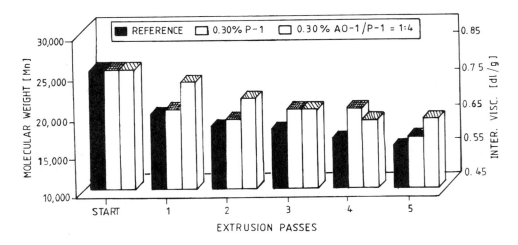

Figure 21   Processing stability of PET for bottles.

2.   Mixed Streams

Although the separation technologies for mixed plastics improved substantially in the last 3 years, there will always be a portion of collected plastics present in the mixed form (mainly for economic reasons). If this material (stemming mostly from household collection) is properly cleaned, it can be considerably upgraded by using different additives, such as modifiers, compatibilizers and stabilizers.

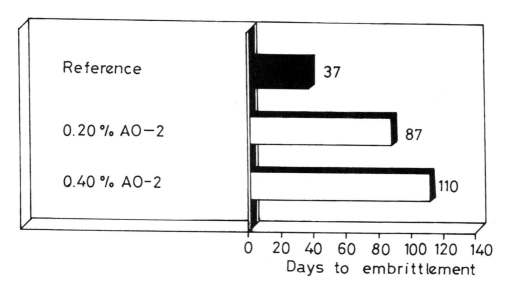

Figure 22   Long-term thermal stability of mixed plastics (PE/PP = 90 : 10).

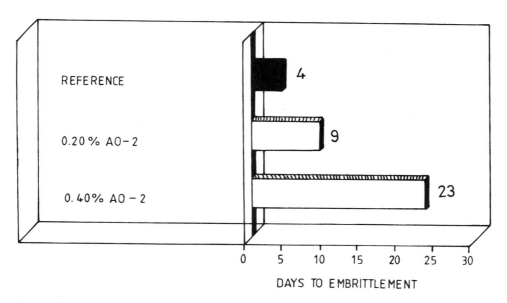

Figure 23   Long-term thermal stability of mixed plastics from household collection.

In Fig. 23 long-term heat-aging results are shown for a mixture from household collection. Thermal stability can be improved in this test by a factor of 5–6 using 0.4% AO-2. Similar improvements can be seen in light weathering using the appropriate LS blend (0.6% LS-1 in Fig. 19).

C. Conclusions

Upgrading Steps

Separate collection or careful separation
Cleaning
Reformulation, compounding, and use of proven additives

Restabilization not only upgrades recyclates for higher-end applications, but also contributes to the economic value of the end article.

Stabilizers Presented in this Chapter (Ciba-Geigy)

AO-1:	Irganox 1010
AO-2:	Recyclostab 411
AO-3:	Irganox 245
AO-4:	Irgastab CZ 2000
AO-5:	Developmental Ca/Zn stabilizer
P-1:	Irgafos 168
T-1:	Irganox PS 802
HALS-1:	Tinuvin 770
UVA-1:	Benzotriazole light stabilizer
LS-1:	Recyclostab 811

Source: Ref. 13.

XVI. DEGRADABLE PLASTICS AS ADDITIVES

Degradable plastics can be a means of solving some of the problems associated with the disposal of packaging materials as well as for litter control. An ethylene copolymer containing 1% carbon monoxide has been used to manufacture the degradable plastic carrier. Total volume now exceeds 100 million lb/yr.

Polyethylene terephthalate, PET can be made photodegradable by copolymerization with glycols and diacids containing ketone groups. Furthermore, master batches containing the ketone group can be made cheaply from recycled PET (16). A master batch can be pellet blended before extrusion to film or blow-molded to provide bottles or other containers. Degradable PET can return to the environmental cycle when it becomes litter (16).

Aluminum cans and PET containers are environment friendly. However, aluminum requires three times as much energy to manufacture and, if not recycled, a similar waste of energy. Use of aluminum cans causes a much greater drain on nonrenewable resources than do plastic containers.

APPENDIX: POLYMERIC ADDITIVES AND ADDITIVE PACKAGES

A. Polymeric Additives

1. Polybutenes in Polypropylene

Dramatic increases in impact resistance are obtained by adding polybutenes to PP (19). Impact resistance of PP improves by 400% at the 5-phr level and over 1000%

at 10-phr, compared with untreated PP. Adding polybutenes also increases MFI and elongation (19) and slightly reduces tensile and flexural strength and HDT. In PP resins containing a elastomeric impact modifier polybutenes increased impact resistance up to 250%. Polybutenes actually improve dispersion of elastomer in PP, further increasing its effectiveness as an impact modifier. Polybutenes may be suitable for partially replacing the more expensive elastomer (EPDM). Polybutenes plasticize EPDM as well as PP.

2. LLDPE Cling Film

Use of polybutene with LLDPE takes advantage of the partial incompatibility to impart cling to LLDPE film. With the aid of extruder processing, the partially incompatible polybutene becomes well dispersed into the PE; 24 h after production extensive cling development takes place (19).

3. Polypropylene

Polypropylene can be used to advantage with thermoplastic elastomers (TPEs) of the SBS type and can be blended with polystyrene. PP–SBS compounds tend to have extremely high tensile strength and higher stiffness.

Resin 18 [Poly-α-Methylstyrene]

Improves ease of processing of olefinic TPEs.
Acts as an extender and processing aid with TPUs, increasing MFI, improving physical properties, and yielding a wider molding temperature window.
Processing aid for extruded PVC pipe and injection-molded fittings. It offers several processing advantages over the acrylics.
Processing aid for floor tiles.

B. Block Copolymer Additives

1. Polyethylenes

Most useful block copolymer modifiers are thermoplastic elastomers (TPE). These include copolyester TPEs, polyurethane TPEs, and styrenic TPEs. Styrenic TPEs significantly improve the properties of molded and extruded PE and PE blown film. An impact strength improvement of 50–100% can typically be achieved in LDPE, HDPE, and LLDPE with 15% styrenic TPE. The improvement will allow manufacture of "supertough" films, or will allow films to be down-gauged. Styrenic TPEs can be continuously added in-line with PE to the blown film extruder. Processing conditions are the same as those for 100% PE.

2. Polypropylene

Styrenic TPEs significantly improve both room temperature and ($-7°$C) Izod impact strength and Gardner impact of PP homopolymer. Dramatic impact improvement occurs (19) above 10% styrenic TPE. For injection-molded part uniformity, precompounding should be done. Styrenic TPE can be used to produce food-contact grade PP, filled PP, and upgrading recycled PP, improving impact in every case.

3. Additives for PP

Additional heat stabilizers are used to improve long-term heat aging of PP. Typically, hindered amines can be used in combination with other additives to deliver service life of 5 years outdoors. This can be done without affecting the appearance of a component. PP grades of carbon black at levels about 2% can provide a service life of more than 20 years. Low levels of stabilizers are needed for PP for the protection of its components in radiation-sterilization environments (19) and in stabilization of wire-coating resins against the effects of copper. Nucleators may also affect processing performance, owing to their effect on the rate of polymer crystallization. They improve stiffness, HDT, and clarity. Monoglycerides or diglycerides are typical PP antistats that minimize dust pickup in packaged products and in PP sheets, bottles, and molded articles.

4. Additive Packages (PVC)

Weatherable Siding
(twin screw)

PVC, K = 67	100
Tin stabilizer	1.4
Calcium stearate	1.5
Paraffin wax (165)	0.9
Titanium dioxide	10–15
Processing aid (e.g., paraloid K-120N)	0.4–1.0
Impact modifier	5–7

Weatherable Window Profiles (19)
(twin screw)

PVC, K = 67	100
Tin stabilizer	1.6
Paraffin wax (165)	1.1
Titanium dioxide	7–10
Processing aid (e.g., paraloid K-120N)	1.0
Impact modifier (e.g., paraloid KM-34)	5–7

Pressure Pipe
(twin screw, tin-stabilized)

PVC K = 57–58	100
Tin stabilizer	0.25–0.45
Calcium stearate	0.9
Paraffin wax (165)	1.1
Titanium oxide	1.1
Processing aid (e.g., K-120-N)	0.4–0.8
Calcium carbonate	0–4.0

Pipe Fittings
  (tin-stabilized)

PVC, K = 57	100
Tin stabilizer	1.4–2.1
Calcium stearate	0.6
Paraffin wax (165)	0.6
Titanium dioxide	1.4
Processing aid (e.g., K-120N)	0.8–1.0
Lubricating-processing aid (e.g., paraloid K-175)	0.6
Impact modifier MBS (e.g., paraloid BTA-753)	1.2–5.2
Acrylic (e.g., paraloid KM-334)	1.2–5.2

Calendered Films
  (stabilized with tin: high-temperature process)

Suspension or bulk PVC, K = 60	100.0
Dioctyltin thioglycolate	0.9–1.4
Processing aid	1–1.5
Glycerol monooleate	0.2–0.4
Ethylene glycol	0.2–0.5
N-*N'*-Bis-stearoylethylene diamine	0.2–0.4

Extruded Films
  (stabilized with tin)

Suspension PVC, K = 60	100.0
MBS impact modifier	4–9.0
Processing aid	0.9–1.6
Dioctyltin thioglycolate	1.2–1.6
Glycerol fatty acid ester	0.9–1.1
Glycerol montanate	0.3–0.6

Bottle (19)
  (stabilized with tin)

Suspension or bulk PVC, K = 57	100.0
Dioctyltin thioglycolate	1.2–1.6
Processing aid	1.2–1.8
Ethylene glycol montanate	0.6
Polyethylene wax (MW about 9000)	0.12

Sections (High-Impact)

Chlorinated polyethylene-modified PVC	100.0
Ba-Cd stabilizer	1.4
Organic phosphite	0.6

Epoxized soybean oil	0.9–1.9
Hydroxystearic acid	0.2–0.4
Partially saponified 1,3- butanediol montanate	0.25–0.45
QZ (MW 5000)	0.1
TiO$_2$	3.1–4.1
Chalk	3.5–5.5

Mineral Water Bottles

PVC, K = 57–58	100
CdZn stabilizer (Mark 3704)	1.4
Secondary stabilizer (Rhodiastab 50)	0.25
Epoxized soybean oil	5.0
Internal lubricant (Loxiol HO B7121)	0.35
External Lubricant (Loxiol G 70S)	0.11
Processing aid (e.g., Paraloid K-120 ND)	0.8–1.8
Lubricating processing aid (e.g., Paraloid K-175)	0.4–1.2
MBS impact modifier (e.g., Paraloid BTA-733)	11–14

Plasticized PVC
(blow-molded film; cadmium–zinc stabilized)

Suspension or bulk PVC, K = 70	100.0
Dioctyl adipate	21.0
Calcium stearate	0.5–0.7
Zinc stearate	0.35–0.55
Epoxized soybean oil	4.5–5.5
Glycerol monooleate	1.3–1.9
Ethylene glycol montanate	0.7–0.9
N,N'-Bis-stearoyl ethylenediamine	0.45–0.55
Polyethylene wax (MW 9000)	0.04–0.09

Cable Sheathing
(Pb-stabilized)

Suspension PVC, K = 70	100.0
Dioctyl phthalate	65.0
Chalk	18.0
Calcined kaolin	10.0
Tribasic lead sulfate	4.0–8.0
Calcium stearate	0.2–0.6
Ethylene glycol montanate	0.5–0.7

PVC Pipe Formulation with Resin 18

PVC (0.8 I.V.)	100
Resin 18-210	2.5
Tin stabilizer	0.3
$CaCO_3$	2.4
Calcium stearate	0.9
Paraffin wax	1.1
Polyethylene wax	0.14
TiO_2	1.8

Processing equipment	Application	Suggested processing aid[a]
Twin-screw extrusion	Siding	Paraloid K-120N
	Window profiles	Paraloid K-120N
	General purpose profiles	Paraloid K-120N
	Cellular profiles sidings and sheets	Paraloid K-125
Extrusion single screw	Cellular profiles, siding	Paraloid KM-318F
Extrusion single or twin pipe	Nonpressure	Paraloid K-175
Extrusion twin screw, pipe	Pressure	Paraloid K-120N
Injection molding	Pipe fittings	Paraloid K-120N
Extrusion blow-molding	Bottles, food contact water bottles	Paraloid K-120ND
Calender	Sheets and film, clear	Paraloid K-120ND
	Sheet and film, opaque	Paraloid K-120N
Extrusion single screw	Sheet and film, clear	Paraloid K-120ND
	Sheet and Film, opaque	Paraloid K-120N

[a] Supplier: Rohm & Haas.

Manufacturer	Product name	Grade	Composition
Koreha/Rohm & Haas	Paraloid	K-120N	MMA/EA
	Paraloid	K-175	MMA/BA/ST
	Paraloid	K-120ND	MMA/EA

REFERENCES

1. J Stepek, H Daoust. Additives for Plastics. New York: Springer Verlag, 1983.
2. KS Percell, HH Tomlinson, LE Walp. Plast Eng Sept: 33–36, 1987.
3. W Brotz. Lubricants and related auxiliaries for thermoplastics materials. In: R Gachter, H Muller, eds. Plastics Additives Handbook. Munich: Hanser, 1984, pp 270–34.
4. T Riedel. Lubricants and related additives, In: R Gachter, H Muller, eds. Plastics Additives Handbook. Munich: Hanser, 1990, pp 423–480.
5. C Vasile. Additives for polyolefins: general outlook. In: C Vaisle, RB Seymour eds. Handbook of Polyolefins. New York: Marcel Dekker, 1993, pp 575–604.

6   R Wolf, BL Kaul. Plastics, additives. In: Ullmann' Encyclopedia of Industrial Chemistry A20. 1992, pp 479–482.

7. C Bluestein. Lubricants. In Encyclopedia of Polymer Science and Technology, vol 8. New York: John Wiley & Sons, 1968, pp 325–338.

8. EL White. Lubricants. In: LI Nass, ed. Encyclopedia of PVC. New York: Marcel Dekker, 1977, pp 644–710.

9. BV Johnson, TJ Blong, JM Kunde, D Duchesne. Factors affecting the interaction of polyolefin additives with fluorocarbon elastomer processing aids. Best Paper TAPPI 1988 (3M).

10. BV Johnson, JM Kunde. The influence of polyolefin additives on the performance of fluorocarbon processing aids. Best Paper ANTEC 1988 (3M).

11. Dynamar Polymer Processing Additive, 3M, 1991.

12. Hoechst Additives for Plastics. Polyethylene waxes. Hoechst Product Leaflet.

13. WO Drake, T Franz, P Hofmann, F Sitek (Ciba Geigy). The role of processing stabilizers in recycling of polymers. Recycle 1991.

14. MD Castle. Speciality Chemicals Group, ICI Europa Ltd., 1987.

15. F Sitek, H Herbst, K Hoffmann, R Pfaendner (Ciba-Geigy). Recycle 1994, Switzerland.

16. JE Guillet, HX Huber, JA Scott. In: Degradable Polymers, Recycling and Plastics Waste Management. New York: Marcel Dekker, 1995.

17. M Rusu, C Vasile, RD Deanin. In: C Vasile, RB Seymour, eds. Handbook of Polyolefins. New York: Marcel Dekker, 1993.

18. TW Taylor. Plast Bus 17(1): 1996.

19. J. Edenbaum, ed. Plastic Additives and Modifiers Handbook. New York: Chapman & Hall, 1996.

5

Permanence of UV Absorbers in Plastics and Coatings

JAMES E. PICKETT

General Electric Company, Schenectady, New York

I. INTRODUCTION

Plastics and organic coatings used outdoors are exposed to ultraviolet (UV) radiation in the range of 295–400 nm, which is often the primary cause of degradation and weathering. UV absorbers (UVAs) are commonly used in these applications to prevent UV radiation from reaching the bulk of the polymer, or from penetrating the coating and reaching a UV-sensitive substrate. Early in the development of UVAs, it was recognized that they did not perform perfectly and could be depleted during exposure. Indeed, as early as 1961 Hirt and colleagues wrote:

> The protective absorbers are not everlasting; they do photodecompose, but at a much slower rate than the materials which they are designed to protect. The photodecomposition was found to be dependent on a number of factors including the substrate in which the absorber is dispersed and the wavelength of irradiation (1).

This statement is a concise and accurate account of the permanence of UVAs. Although subsequent workers recognized that UVAs could be depleted from polyolefins, the insight of slow, photochemical UVA loss seemed gradually to disappear from common wisdom. Then around 1990, several groups began to rediscover that UVAs used in coatings could be lost through photochemical reactions, and that this loss was responsible for limiting the lifetime of the coatings. This has led to a resurgence in interest in UVA photostability and to an understanding of how rates of UVA loss affect the lifetime of coatings.

Many examples of the loss of UVAs from coatings and polyolefins will be discussed in subsequent sections, but the loss from other types of materials has also been documented. Pern and Czanderna (2) and Klemchuk et al. (3) describe the depletion of a benzophenone-type UVA from ethylene vinyl acetate encapsulants in solar cells. The UVA was lost only in regions exposed to sunlight, indicating photochemical decomposition. Bolon and Irwin (4) found that a benzotriazole-type UVA was lost from the weathered surface of polycarbonate window panels. Priddy et al. (5) have studied the loss of UV absorbers and other stabilizers from reaction with active halogen compounds (such as found in swimming pool water) as well as the photochemical loss from acrylic films. They clearly demonstrated that the UVA loss was due to photochemical reactions and not to physical loss from the films. Bonekamp and Maecker (6) examined the photochemical loss of a variety of UVAs from rubber-modified acrylic films. They described a procedure for accurately determining the absorbance of the UVA even in a medium that scatters light.

II. STRUCTURE AND PROPERTIES OF COMMERCIAL UV ABSORBERS

Ultraviolet absorbers must have at least three properties to be useful. First, they must strongly absorb UV radiation that would be harmful to the polymer or coating. Second, they must harmlessly dissipate the energy that they absorb. Finally, they must persist in the matrix for the expected lifetime of the article. Most commercially important UV absorbers are based on one of the five chromophores shown in Fig. 1. Several general reviews of UVAs' structures and properties are available (7–9). A summary of the absorption characteristics of representative absorbers is shown in Table 1.

The benzophenones (see Fig. 1, **1**) invariably have a hydroxyl group in the 2-position and either a hydroxyl group or an ether in the 4-position. A typical example is the octyl ether Cyasorb 531 (Cyasorb is a trademark of the Cytec Corporation). Other variations are sulfonate groups in the 5-position, hydroxyl groups in the 2'-position, and hydroxyl or ether substitution in the 4'-position. Simple derivatives have absorption maxima near 285 nm ($\varepsilon \sim 15,000$) and 325 nm ($\varepsilon \sim 10,000$). Derivatives with 2'-hydroxyl groups have a strong absorption near 350 nm that exhibits hypsochromic or bathochromic shift, depending on the polarity of the solvent or matrix.

Benzotriazoles (see Fig. 1, **2**) always have a hydroxyl group in the 2'-position and alkyl substitution at R_2, and there are many examples with alkyl substitution at R_1 as well. X is usually hydrogen, but there are examples where X = chlorine. A typical compound is $R_1 =$ H, $R_2 = t$-octyl, X = H, which is available as Cyasorb 5411 or Tinuvin 329 (Tinuvin is a trademark of Ciba Specialty Chemicals). Most benzotriazoles exhibit absorption maxima near 295 nm ($\varepsilon \sim 14,000$) and 345 nm ($\varepsilon \sim 16,000$).

Triazines (see Fig. 1, **3**) were investigated in the late 1960s (10), but have been commercialized only recently. They have absorption maxima near 290 nm ($\varepsilon \sim 43,000$) and 340 nm ($\varepsilon \sim 23,500$), but their higher molecular weights generally offset the higher molecular extinction coefficients. The triazine UV absorbers always have a hydroxyl group in the 2'-position and an ether group in the 4'-position. The aryl groups are most commonly 2,4-dimethylphenyl (e.g., Cyasorb 1164 and Tinuvin 400). The triazines exhibit superior photostability in many, but not all, coatings applications. A compound in which the aryl groups are phenyl (Tinuvin 1577) has outstanding photostability (11), but only limited solubility.

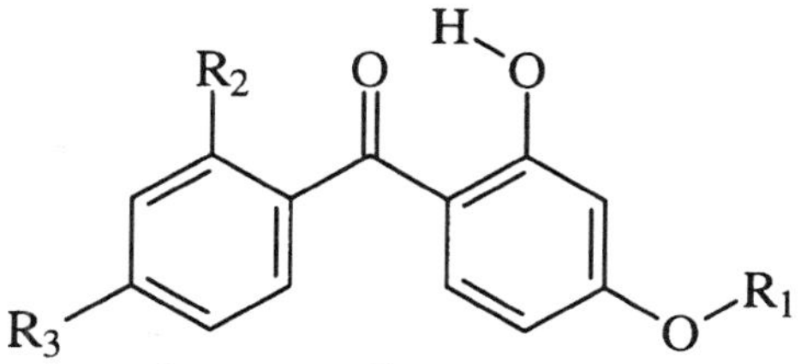

1

benzophenone

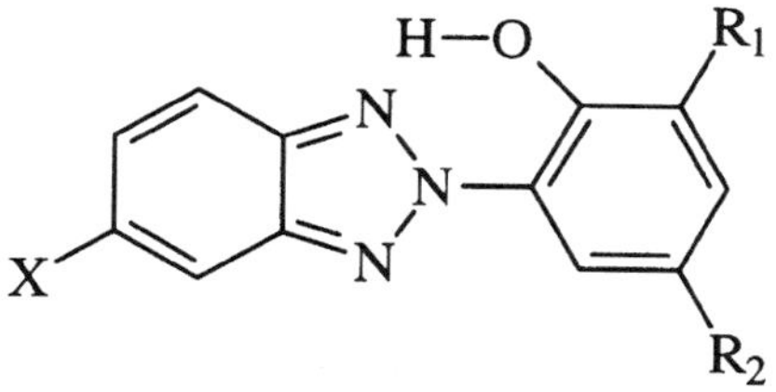

2

benzotriazole

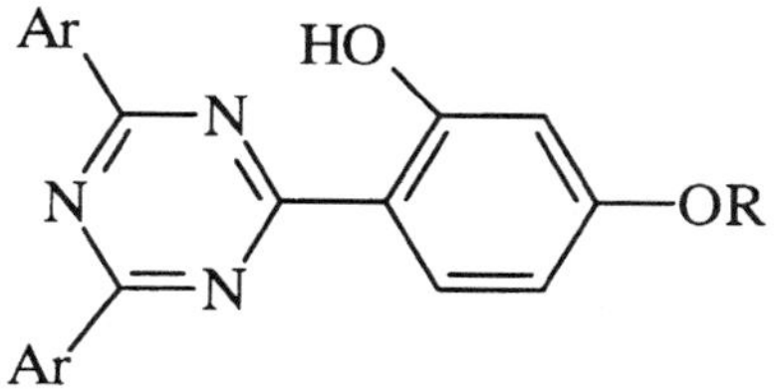

3

triazine

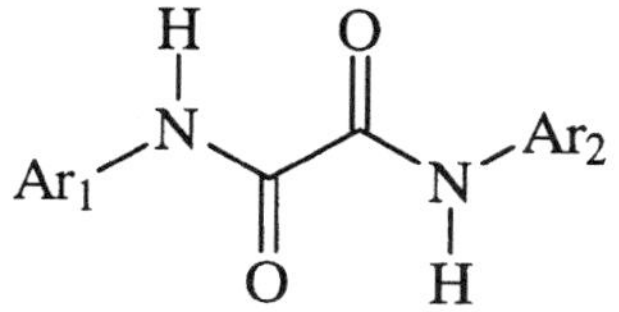

4

oxanilide

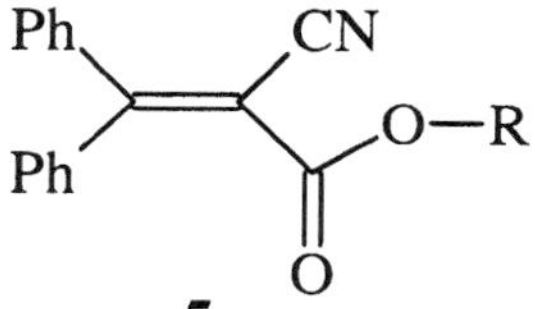

5

cyanoacrylate

Figure 1 Representative structures of the most common types of commercial UV absorbers.

Oxanilides (see Fig. 1, **4**) such as Sanduvor VSU (Sanduvor is a trademark of the Clariant Corporation) are generally asymmetrically substituted with an alkyl aryl group at one end and a 2-alkoxy aryl group at the other. This results in a broad absorption spectrum with a maximum near 300 nm ($\varepsilon \sim 15{,}000$).

The only aprotic examples among the more stable UV absorbers are the cyanoacrylates (see Fig. 1, **5**) such as Uvinul 3039 (Uvinul is a trademark of BASF). These compounds have a single absorption band with a maximum at 305 nm ($\varepsilon \sim 13{,}500$). Because the cyanoacrylates are unaffected by hydrogen bonding, they are considered to be applicable in highly polar polymers or in polymers that might be sensitive to phenolic groups.

There are three classes of commercial UV absorbers that have marginal photostability or are used in limited applications. The structures are shown in Fig. 2. The benzilidine malonates (see Fig. 2, **6**) are now available from Clariant Corporation. They have a strong absorption at 313 nm ($\varepsilon \sim 28{,}800$). These compounds are quite photoreactive, which makes them less attractive as UV absorbers. However, this photoreactivity can be exploited for other purposes. When the esters

Table 1 Absorption Characteristics of the UV Absorbers Shown in Fig. 1 in Chloroform Solution

UV absorber	λ_{max} (nm)	Extinction coefficient
Benzophenone, **1**		
$R_1 = C_8H_{17}$; R_2, $R_3 = H$	289	15,200
	327	10,800
$R_1 = CH_3$; $R_2 = OH$; $R_3 = H$	296	10,300
	350	13,900
Benzotriazole, **2**		
$R_2 = CH_3$; R_1, $X = H$	298	14,300
	340	16,600
$R_1 = t\text{-Bu}$; $R_2 = CH_3$; $X = Cl$	312	14,800
	353	16,200
1,3,5-Triazine, **3**		
$R = C_8H_{17}$; $Ar = 2,4\text{-dimethylphenyl}$	290	42,800
	338	23,500
Oxanilide, **4**		
$Ar_1 = 2\text{-ethoxyphenyl}$; $Ar_2 = 2\text{-ethylphenyl}$	280	14,000
	300	15,000
Cyanoacrylate, **5**		
$R = C_2H_5$	305	13,400

are formed from 4-hydroxy-2,2,6,6-tetramethylpiperidine, the photoreactivity of the benzilidine malonate moiety results in a graftable hindered amine light stabilizer (HALS; e.g., Sanduvor PR-31). This has been demonstrated for polyolefins (12). A second class is the formamidines (see Fig. 2, **7**) available as Givsorb UV-1 and UV-2 (Givsorb is a trademark of Givaudin). These compounds have a strong absorption at 313 nm ($\varepsilon \sim 27,700$). They are reported to have good photostability in solution (13), but seem to have poor stability in polymer films. They are recommended for use in cellulosics. Finally, Cyasorb UV-3638 is a benzoxazinone (see Fig. 2, **8**), with absorption maxima at 300, 314, 333, 348, and 365 nm ($\varepsilon \sim 23,000$, 31,800, 41,500, 46,800, and 26,000). It has relatively poor photostability and has limited applications in engineering thermoplastics, such as polyesters and polycarbonate.

III. MECHANISMS OF ENERGY DISSIPATION

The photophysics of the protic UV absorbers has been studied extensively (7,14–18). They all undergo proton-transfer reactions from their first excited state as shown in Figs. 3 and 4. On absorption of a photon, the molecule is excited to its first excited singlet state (S_1), which then undergoes an excited state intramolecular proton transfer (ESIPT) to create a proton-transferred species in its first excited singlet state (S_1'). The ESIPT process is very rapid, and fluorescence from the S_1 state usually is not observed for good UV absorbers except in highly polar media. The excited proton-transferred species (S_1') loses its energy by a nonradiative decay process, as

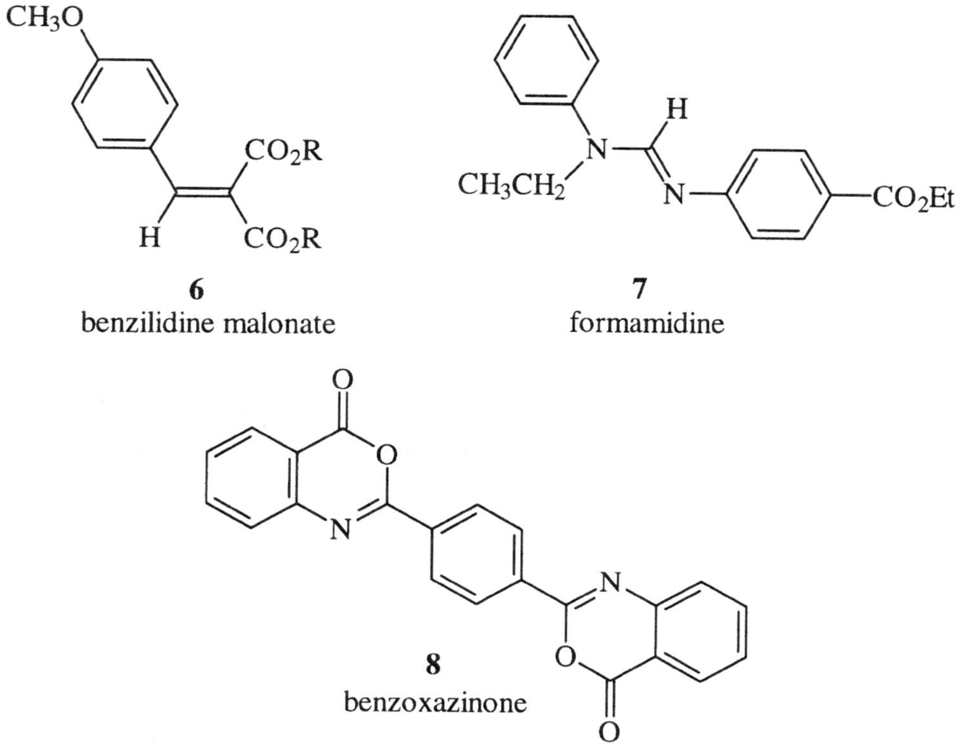

6
benzilidine malonate

7
formamidine

8
benzoxazinone

Figure 2 Other types of commercial UV absorbers.

thermal energy to the matrix, to form the ground-state proton-transferred species. (S_0'). A proton transfer then re-forms the ground-state UV absorber (S_0). This entire cycle occurs in the time period of picoseconds to nanoseconds. There is some discussion in the literature that the benzophenone class enters the triplet manifold from the proton-transferred S_1' state and then undergoes rapid nonradiative decay from T_1' to T_1 to S_0 (8,18).

The ease of the ESIPT process is key to the stability of these absorbers. The energy difference between S_1' and S_0' is thought to be sufficiently small that the excitation energy can be rapidly and efficiently dissipated by internal conversion (14); that is, by losing it as heat. The heat energy so released is a very small fraction of the total heat load on a sample (most coming from absorbed visible and infrared [IR] wavelengths) and does not contribute to thermal degradation of the material. Compounds with ethers in place of the hydroxyl groups cannot undergo ESIPT and generally show strong fluorescence, indicating that the energy dissipation mechanism has been disrupted. They are not photostable. In apolar media, good UVAs generally exhibit little or no fluorescence or phosphorescence. However, polar aprotic solvents are capable of disrupting the internal hydrogen bond of these compounds, resulting in fluorescence and photoreactivity. Any factors that affect the strength of the intramolecular hydrogen bond, such as polarity and basicity of the matrix, are likely to affect the photostability of the UVA.

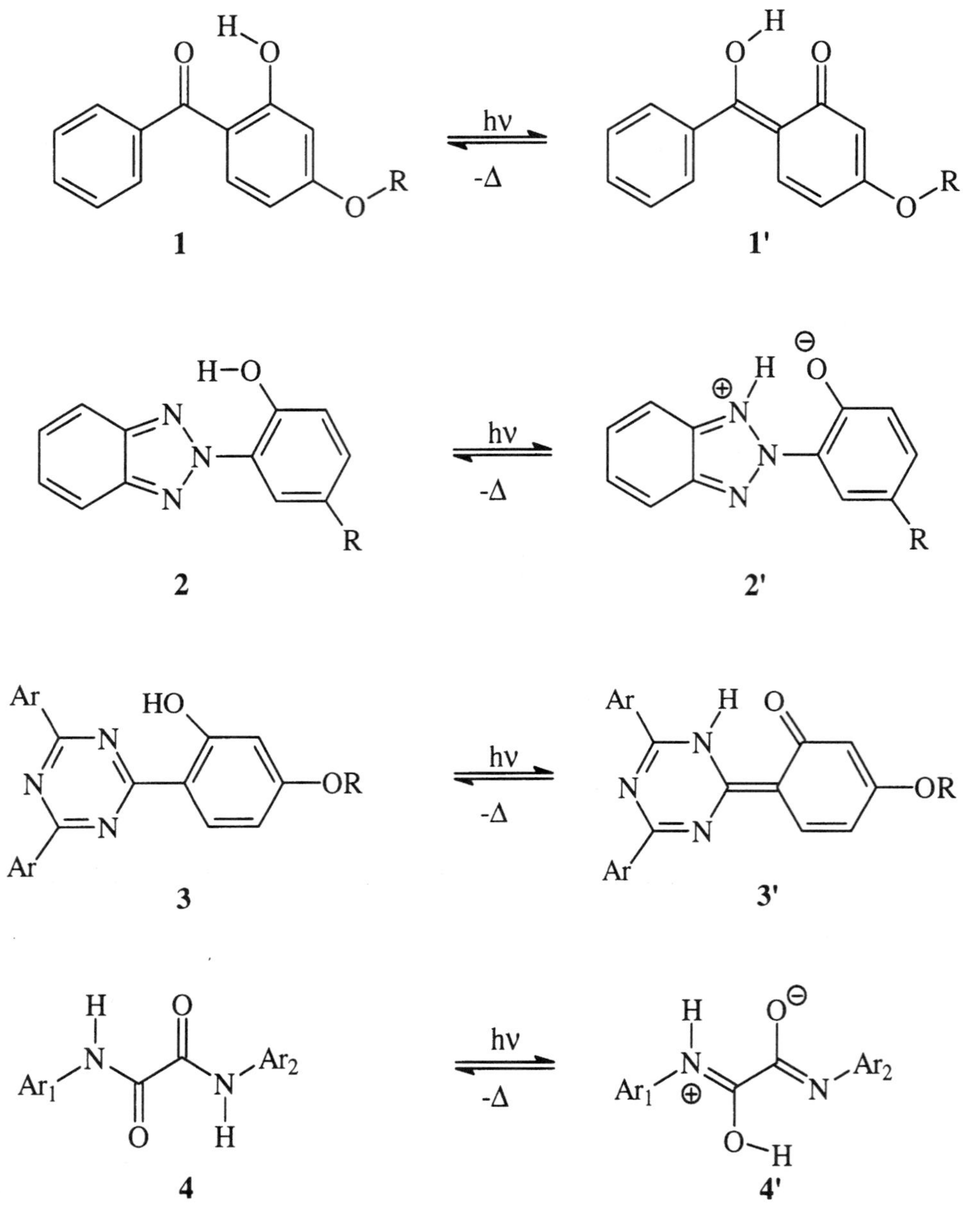

Figure 3   Energy dissipation mechanisms for protic UV absorbers. (Adapted from Ref. 26.)

We are not aware of any photophysical studies on the cyanoacrylate UVAs. They are remarkably photostable considering the lack of an ESIPT mechanism for energy dissipation. It seems likely that a charge-separated species (Fig. 5, **5'**) could be formed from the excited state. Such a structure would allow energy dissipation through rotation or increased vibration about the central bond.

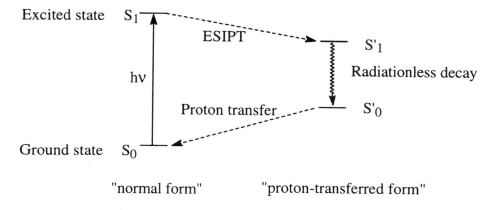

Figure 4 Cycle for energy dissipation in protic UV absorbers: first, excitation to the first excited singlet state; second, excited state intramolecular proton transfer (ESIPT); third, radiationless decay to a ground state, proton transferred species; fourth, proton transfer to reform the ground state UVA.

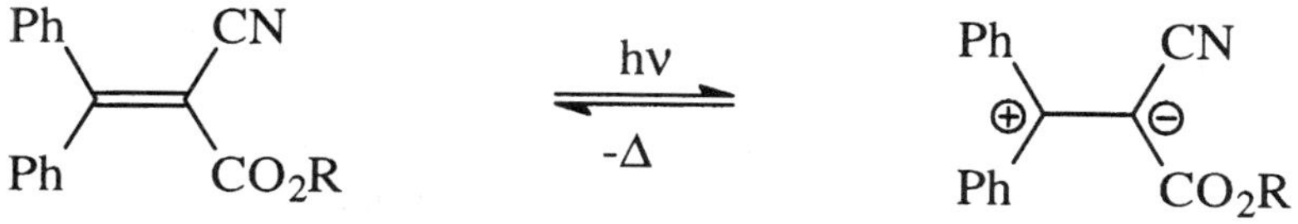

Figure 5 Possible energy dissipation mechanism for cyanoacrylate UV absorbers. (From Ref. 27.)

IV. CHEMISTRY OF UVA PHOTODEGRADATION

A. Quantum Yields for UVA Photolysis

The common UVAs are quite resistant to photodegradation, making the study of their photochemistry difficult. The quantum yields for disappearance are on the order of 10^{-7} to 10^{-6}, and are dependent on the nature of the medium (14) and on the wavelength of light. Sedlar, et al. found the quantum yield for disappearance of 2-(2-hydroxy-3,5-di-t-butylphenyl)benzotriazole (Tinuvin 327) was on the order of 10^{-7} in heptane solution under argon (19,20). Oxygen alone played little role, but the photolysis in the presence of oxygen and t-butyl hydroperoxide resulted in quantum yields on the order of 10^{-5} indicating that light plus oxygen radicals were very detrimental to the benzotriazole UVA. These authors also found very high photoreactivity in the presence of a ketone, such as acetone, with quantum yields of disappearance as high as 0.1. These results supported their conclusions that the loss of the UVA from thin films of polypropylene on photolysis was primarily due to photolysis of the UVA in the presence of the polypropylene degradation products.

The quantum yield for disappearance of 2-(2-hydroxy-5-methylphenyl-benzotriazole (Tinuvin P) in dimethyl sulfoxide (DMSO) solution was highly dependent on wavelength varying from 2.6×10^{-4} at 290 nm to 0.4×10^{-4} at 350 nm (21). The plot of quantum yield versus wavelength was coincident with the absorption spectrum calculated for the "open," nonintramolecularly hydrogen-bonded form of the UVA, suggesting that the photochemistry arose from the open form.

The wavelength dependence of the quantum yields for disappearance of a benzotriazole UVA (Cyasorb 5411) and a benzophenone UVA (Cyasorb 531) in polymethyl methacrylate (PMMA) films has been determined (22). The benzotriazole quantum yield varied from 6×10^{-7} at 300 nm to 2×10^{-7} at 350 nm, whereas the benzophenone varied from 4×10^{-6} at 300 nm to 0.2×10^{-6} at more than 365 nm. A benzotriazole UVA was reported (23) to degrade with a quantum yield of 5×10^{-7} in polycarbonate. The quantum yield for disappearance of benzotriazole UVA from a UV-curved acrylic/urethane was reported to be 4×10^{-6} (24). A much lower quantum yield reported (25) for benzotriazole loss from PMMA of 10^{-9} probably is incorrect because the authors overestimated the role of physical loss in their experiments (25). An upper limit of less than 10^{-4} has been determined for the quantum yield of disappearance for an oxanilide UVA in ethanol solution (16). The authors suggest that the actual quantum yield could be several orders of magnitude less.

To put these quantum yields into perspective, most photochemical reactions have quantum yields of at least 10^{-3} to proceed at measurable rates in the laboratory. Very long exposures are required to measure significant degradation of these UV absorbers. For example, consider a UV absorber with a quantum yield for destruction on the order of 10^{-6}. Over a pathlength that the UVA has an absorbance of 1 (i.e., absorbing 90% of the incident light at some highly absorbed wavelength), roughly one-third of the absorbance would be lost in 1 year of Florida exposure or in 1000–2000 h of xenon arc-accelerated testing (26).

B. Photochemistry of Benzophenones

Early work on the photodegradation of benzophenone UVAs indicated formation of "conjugated acids," presumably benzoic acid, and loss of the hydroxy-substituted ring (1). That same work showed that conjugated acids are not formed from 2,2'-dihydroxybenzophenones, and that the carbonyl group presumably was lost as carbon monoxide. Work by Pickett and Moore (27) also showed benzoic acid as the primary product from photolysis of a benzophenone UVA in ethyl acetate solution. Photolysis of 2,4-dihydroxybenzophenone in a PVC film reportedly gave hydroquinone as the primary product as identified by its IR spectrum and melting point (28). However, this product seems unlikely and certainly would not be photostable under these conditions. It is more likely that the product was impure benzoic acid, which would exhibit a similar IR spectrum and melting point.

Photophysical experiments have tried to find the cause of photoinstability. Work by Lamola and Sharp (29) showed that 2-hydroxy-4-methoxybenzophenone has no phosphorescence arising from a triplet state in a hydrocarbon glass at 77 K. However, in a polar aprotic glass, weak emission was observed, and in a polar protic glass, such at ether–ethanol, strong emission was observed. The authors conclude that strong intramolecular hydrogen bonding precludes formation of the triplet state

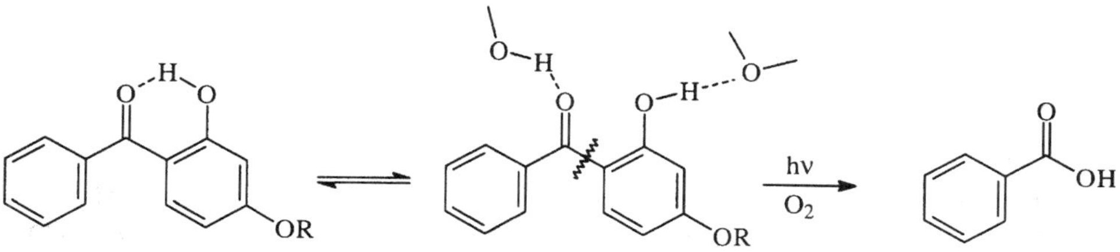

Figure 6 Photooxidation of intermolecularly hydrogen-bonded UVA to give benzoic acid.

owing to the rapid tautomerization shown in Fig. 3. In media that allow intermolecular hydrogen bonding, the triplet state can be formed, and phosphorescence and photochemistry can occur (Fig. 6). The authors also observed that a larger population of intermolecular hydrogen-bonded species was formed if the glass was rapidly cooled. Annealing the glass allowed the UVA to "relax" into the more stable intramolecular hydrogen-bonded form, and less phosphorescence was observed. In our laboratory, we have often observed rapid changes in UVA absorption during the first 100 h or so of exposure of thin glassy films or coatings, followed by a slower rate of degradation. This is presumably due to the rapid loss of a small amount of intramolecularly hydrogen-bonded UVA.

Another mode of degradation of benzophenone UVA is by reaction with free radicals. Many groups have observed the loss of benzophenone UVA from polypropylene films on irradiation (30), and much of this work has been reviewed (31). Indeed, it is suggested that the primary beneficial effect of benzophenone UVAs in polypropylene is their ability to act as a hydrogen atom donors; that is, as sacrificial antioxidants, or possibly, as excited state quenchers (32–35). Benzophenone UVAs seem to act only as UV absorbers in polyethylene, however (36). Vink (37) found destruction of a benzophenone UVA during the "induction period" of polypropylene photodegradation, and the UVA extended the time to onset of rapid oxidation. Loss of the UVA was also found (38) during exposure of a polypropylene film, and oxygen accelerated the rate of loss; however, increasing the peroxide content of the film by preoxidation did not accelerate UVA loss when the film was subsequently irradiated in a nitrogen atmosphere (38). Furthermore, the loss was not accelerated on irradiation in deoxygenated isooctane solution containing *t*-octylhydroperoxide or di-*t*-butylperoxide, although the loss seemed surprisingly fast in the deoxygenated isooctane control solution. This indicates that the benzophenone UVAs are not susceptible to oxygen radical attack on irradiation in a nonpolar medium *in the absence of oxygen*.

Chakraborty and Scott (39) found that 2-hydroxy-4-octyloxybenzophenone was only a weak thermal antioxidant: that is, the hydrogen-bonded phenolic group was resistant to radical attack. However, in the presence of light and oxygen, the phenolic group was susceptible to attack by oxygen radicals. This explained the inefficiency of this class of UVAs in abusively processed polyolefins, which would contain a high level of peroxides. It is unclear from this work whether the reactivity is due to an excited state of the UVA being more prone to hydrogen atom donation, or from more energetic radicals being formed during photooxidation. However, it is clear that radicals will accelerate the destruction of a benzophenone UVA only in the presence of light and oxygen.

C. Photochemistry of Benzotriazoles

The chemistry of benzotriazole photolysis has been recently studied by Gerlock, et al. (40,41). These authors were concerned with the loss of benzotriazole UVAs from melamine–acrylic coatings and studied photolysis in these coatings and in solution. The findings can be summarized:

1. Little or no direct photolysis of the benzotriazole UVAs in the absence of oxygen.
2. No reaction observed between the UVA and photolysis products of the coating in the absence of oxygen. In the presence of oxygen, the rate of loss of the UVA correlated with the photooxidation rate of the coating.
3. Little or no reactivity toward carbon-centered or oxygen-centered radicals in the absence of light.
4. Rapid loss of UVA in the presence of carbon or oxygen radicals on UV irradiation. Oxygen was not required for reaction, but the presence of oxygen changed the product distribution.
5. UVA loss was much faster in more polar solvents in the presence of light and a radical source.
6. Benzotriazole was the major product of UVA photolysis.
7. Benzotriazole UVAs undergo photoaddition reactions to benzophenone.

These authors suggest that benzotriazole UVAs do not act as hindered phenolic antioxidants (40). Reaction occurs not from hydrogen abstraction from the phenolic hydroxy group, but rather, from radical attack on the phenolic ring of an excited state arising from the nonplanar form of the UVA (Fig. 7). This addition results in breaking the aromaticity of the ring, and further photoreactions lead to the cleavage of the benzotriazole moiety from the phenolic ring. In the presence of oxygen, the phenolic fragment is photooxidized to small fragments and essentially "disappears," whereas benzotriazole is photostable. In coatings and polymers subject to weathering, benzotriazole is sufficiently small and has sufficient water solubility that it is extracted from the coating and is also lost.

The benzotriazole UVAs undergo rapid destruction in the presence of light and a hydroperoxide (19), and are also photoreactive with acetone. This explains the loss of UVA from polypropylene films only after the polypropylene began rapid photooxidation when high concentrations of peroxides and ketones were present. A benzotriazole UVA has reacted with t-butyl peroxy radical to give benzotriazole

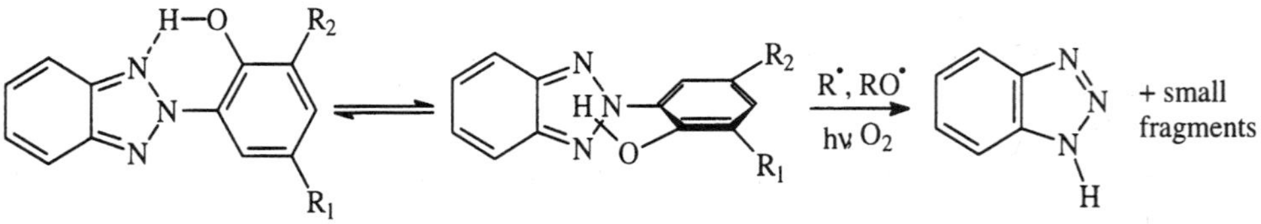

Figure 7   Photooxidation of benzotriazole UV absorbers.

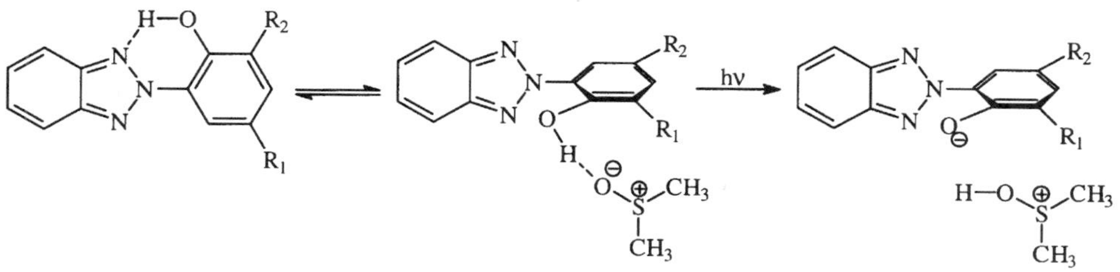

Figure 8 Proton transfer from a benzotriazole UVA to DMSO.

and peroxy addition products to the phenolic ring (42). These reactions occurred without UV light, but required the presence cobalt(II) acetylacetonate. Evidence was obtained for a cobalt–UVA complex, which may have promoted reaction with the peroxy radicals in the dark. Vink (31, 37) studied the loss of a benzotriazole UVA from polypropylene films on photooxidation (31,37). In contrast with the benzophenone UVAs that were lost during the induction period, the benzotriazole loss became fast only after the period of rapid oxygen uptake had commenced.

A study (21) of the photolysis of a benzotriazole UVA in DMSO solution suggested that the photolysis involved the triplet state of the UVA arising from excitation of the nonplanar form. However, a significant fraction of the UVA is hydrogen-bonded to the DMSO solvent (43), which suggests that the UVA transfers a proton to the DMSO on irradiation to create a phenolate anion that is highly susceptible to photooxidation, as shown in Fig. 8. It is further suggested (43) that small amounts of such intermolecularly hydrogen-bonded UVA may be responsible for the slow photolysis of the UVA in polar polymeric media.

D. Photochemistry of Other Absorbers

Oxanilides have been studied in some detail (16). The oxanilides are very photostable in solution and show no reactivity toward radicals, even under irradiation. The lifetime of the singlet state is less than 1 ps, owing to a rapid intramolecular proton transfer (ESIPT) reaction, (see Fig. 3). The photophysics of some triazine UVA has been described (17,44). These compounds possess very strong intramolecular hydrogen bonds (11) and are quite photostable. The degradation that does occur presumably results in the phenolic moiety cleaving from the triazine ring, in analogy with the benzotriazoles and benzophenones, but no product studies have been reported in the literature.

The remaining classes of absorbers have received much less attention and little is known about their photochemistry. Gupta and co-workers (45) studied the fate of a benzilidine malonate UVA (Cyasorb 1988) in poly(vinyl chloride) (PVC) and in polystyrene. They found rapid loss of the UVA in PVC, which they attributed to attack by chlorine radicals. By contrast, the UVA was quite stable in polystyrene. (We have observed very fast disappearance of this UVA from PMMA films on xenon arc irradiation in our laboratory.) Photografting reactions of benzilidine malonates onto polyolefins has been described in the foregoing (12). Pickett and Moore (26,27) speculated that the cyanoacrylates may be susceptible to nucleophilic or radical addition to the double bond, thereby disrupting the chromophore.

V. PHYSICAL LOSS OF UV ABSORBERS

Ultraviolet absorbers are most effective in the surface layers of a polymer or in relatively thin coatings. Therefore, they are often subject to loss through migration, blooming, evaporation, or leaching. Several comprehensive reviews have been written on the diffusion and loss of additives from polyolefins (46–48), and a detailed analysis is beyond the scope of this chapter. Briefly, the key variables are the solubility of the additive in the polymer, its diffusion coefficient, its vapor pressure, and its solubility in water. Typically, an additive is rapidly lost by evaporation or by being washed away once it slowly diffuses to the surface. Billingham (47,48) calculated that 90% of a typical UV absorber, such as 2-hydroxy-4-octoxybenzophenone (Cyasorb 531), would be lost in less than a year from a 100-μm–thick polyolefin film. Thus, the physical loss rate from the surface of a polyolefin is comparable with the rate of chemical loss for additives of moderate molecular weight. The problem of physical loss can be reduced by using higher molecular weight, polymeric, or grafted additives.

The existence of both physical and chemical mechanisms for additive loss in polyolefins presents a particular problem for accelerated testing. The activation energies for the physical loss mechanisms and for the photooxidative degradation of the polymer and stabilizers are unlikely to be the same (47). Accelerated tests, therefore, may unrealistically accelerate one process faster than the other. This is true for 2-hydroxy-8-octoxybenzophenone in linear low-density polyethylene (49). Under accelerated test conditions the UVA is lost mostly from photochemical mechanisms, whereas physical loss predominates under natural conditions. This makes it impossible to make lifetime predictions on the basis of stabilizer loss (49).

The problem of physical loss of stabilizers from polymers is limited to those being used or tested at temperatures higher than the glass transition temperature (T_g). Segmental mobility of the polymer is required for the relatively large stabilizer molecules to diffuse at a significant rate. The rates of diffusion drop to nearly zero when the temperature is less than the polymer's T_g. For example, 2,4-dihydroxybenzophenone had no detectable diffusion in plasticized polyvinyl chloride even a few degrees below T_g (50). Diffusion was detectable at a few degrees over T_g. Similar results are reported (51) for UVAs in polycarbonate. This means that stabilizer migration and physical loss are important mechanisms in polyolefins and other low T_g polymers, but are not likely to be important in engineering thermoplastics and other materials with T_gs higher than the testing and use temperatures.

The physical loss of stabilizers from coatings also is a possibility. If a coating has a thermoplastic matrix (such as an acrylic lacquer), then the T_g of the coating will be the T_g of the matrix polymer less any effects from plasticization by the additives. If the T_g is high, then the physical loss of the UVA should be negligible. However, many coatings are thermosets that are cross-linked by a thermal or radiation cure. To a first approximation, the T_g of a cured coating will be approximately the temperature at which it was cured. A thermally cured coating is usually cured at a temperature higher than the expected use or test temperature, so any diffusion or physical loss of the additive should occur only during the cure step. However, radiation- or UV-cured coatings can be cured at low temperature and, thereby, have relatively low T_gs. Loss of stabilizers during weathering, therefore, is a possibility, especially if the matrix degrades during exposure. Significant physical and chemical loss of UVA from a UV-cured acrylic–urethane coating has been observed (24,52). The physical loss of the UVA was attributed to loss of the polymer cross-links during weathering, allowing greater diffusion of the UVA. The rates of both chemical and physical loss were reduced when an HALS was added to the coating. By contrast, no physical loss of UVA was detected (40) from a thermally cured acrylic–melamine coating after 4000 h of accelerated testing when the coating was exposed behind a 400-nm–cutoff filter such that no photodegradation occurred.

VI. KINETICS OF UVA PHOTODEGRADATION

Many authors have described the kinetics of UVA degradation in terms of first-order kinetics, zero-order kinetics, or in plots of the fraction of remaining UVA during exposure (normalized zero-order kinetics). These kinds of analyses can be very misleading. In studies of UVA photodegradation in PMMA films during xenon arc exposure no unique rate constant could be obtained from treating the data in these ways (22). The observed rates are dependent on the initial absorption of the films or coatings. A UVA in a highly absorbing film (a thick film or at high loading) can appear to degrade faster than in a low-absorbing film when treated as zero-order kinetics. By contrast, the same UVA in the more highly absorbing film will appear to degrade more slowly than in a low-absorbing film when treated as first-order or normalized zero-order kinetics. Some of the confusion in the literature arises because it is difficult to study the loss of the absorber for more than one or two "half-lives." The data will give satisfactory straightline fits with almost any kind of kinetic treatment over this amount of conversion, especially given the scatter inherent in long-term–weathering experiments. Rate constants that are valid over an the entire absorbance range would be valuable because they would allow quantification and comparison of UVA photostability, and they are useful in making lifetime predictions for coatings, as discussed in a later section.

The counterintuitive kinetics are due to the "inner filter" effect of the UVA: the molecules in the upper surface of the film encounter the full intensity of the light, whereas molecules deeper in the film are protected. In highly absorbing films, the molecules in the "back" degrade only very slowly. The degradation is inhomogeneous across the thickness of the film or coating (26,27,52), and the system cannot obey simple first- or zero-order kinetics. The full analysis of UVA degradation kinetics has been described independently (22,53), and a similar analysis for the photolysis of methyl salicylate in PMMA films has been described (54).

Three assumptions must be made for a good kinetic analysis. First, the loss of the UV absorber must be adequately described by assuming irradiation at a single, highly absorbing wavelength. The quantum yield data described in Sec. IV.A indicate that the benzophenone and benzotriazole UVAs are more sensitive to short-wavelength light, and that the quantum yield decreases at longer, less highly absorbed wavelengths. In addition, experimental data show that the degradation can be modeled adequately assuming the importance of short-wavelength, highly absorbed light alone (27). This assumption, therefore, seems valid at least for benzophenone and benzotriazole UVAs.

Second, the products of the UVA photodegradation must not have significant absorption. Experimental data show that this is true: the UVAs seem to cleanly disappear from coatings and films on photolysis with little or no change in the absorption spectra from 280 to 400 nm, as determined in our laboratory and by others (25,40,55).

Finally, the chemistry leading to UVA degradation must arise from light absorption by the UVA. Physical loss is a separate matter. All of the degradation processes described in previous sections seem to require light, even if the reaction is with radicals generated from the polymer matrix. If the degrading matrix contributes to the UVA degradation, then these kinetics will be superimposed on UVA loss kinetics. However, if one assumes that the UVA also acts as an inner filter for the matrix degradation, then the kinetics described in the following should still be adequate. In addition, one must assume that the matrix essentially is nonabsorbing. This generally is a good assumption because if the matrix does absorb a significant amount of light, it will degrade so quickly that UVA stability will not be an issue.

Consider the example of a polymer film or coating in which the UVA has an absorbance A. The absorbance of the film is proportional to the amount of UVA in it (Beer–Lambert law), so one can simply discuss the loss of absorbance, rather than loss of UVA. Also, for the purposes of this discussion, we will assume that "time" is the same as "light dose." The loss of absorbance at any time will be proportional to the fraction of the incident light that the film absorbs [Eq. (1)], where A is absorbance, k is a rate constant (basically the product of incident light intensity and quantum yield), and T is the transmission of the film ($T = 10^{-A}$). Equation (2) is equivalent to Eq. (1).

$$dA/dt = -k(1 - T) \tag{1}$$
$$dA/dt = k(1 - 10^{-A}) \tag{2}$$

Integration of Eq. (2) and rearrangement gives Eq. (3), where T_0 and A_0 are the initial transmission and absorbance, k is a rate constant, t is time or light dose, and A_t is the absorbance at time t. Equation (4) is an equivalent form as derived by Bauer (53).

$$A_t = \log_{10}\left[(1 - T_0)10^{(A_0 - kt)} + 1)\right] \tag{3}$$
$$A_t = \log_{10}\left(10^{(A_0 - kt)} - 10^{-kt} + 1\right) \tag{4}$$

A plot of experimental data for the loss of 2-hydroxy-4-octoxybenzophenone in a PMMA film exposed to xenon arc weathering (22) is shown in Fig. 9. The solid line in Fig. 9 is calculated from Eq. (3) using a rate constant k of 0.183 A/1000 kJm^{-2}. The fit is excellent. The slight waviness in the experimental data is probably due to fluctuations in the lamp over its lifetime. It was replaced about 2000 h (4500 kJm^{-2} of exposure) into this experiment.

Equation (3) is plotted over a wide absorbance range in Figs. 10 and 11. Figure 10 is plotted on a linear scale. Note the essentially linear portion when the absorbance is greater than 1. In this range, the UVA absorbs more than 90% of

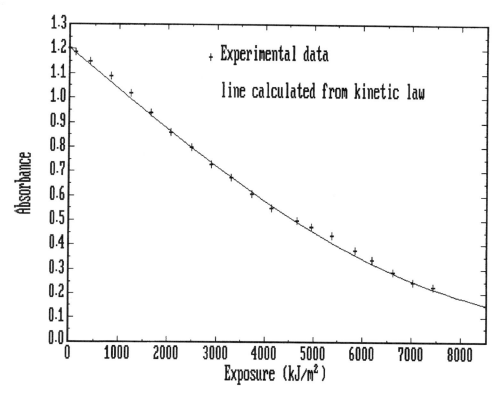

Figure 9 Experimental data for the loss of Cyasorb 531 absorbance from a PMMA film during xenon arc exposure. Line is calculated from Eq. (3) using a rate constant of 0.183 A/1000 kJm^{-1}. (Adapted from Ref. 26.)

the incident light and exhibits zero-order kinetics. That is, the rate of UVA loss is independent of the amount of UVA and depends only on a rate constant and the light dose. In this range, the system is light-limited; it makes no difference if the absorbance is 2 or 20, essentially all of the light is being absorbed, resulting in a constant maximum rate of UVA loss. This *zero-order* kinetic range is described by Eq. (5). The concept of half-life has no meaning in this high-absorbance range, for the absorbance always drops at a constant rate. The slope of the line in this highly absorbing range is the rate constant k.

$$A_t = A_0 - kt \quad for \quad A > 1 \tag{5}$$

Equation (3) is plotted on a logarithmic scale in Fig. 11, and shows a linear portion when the absorbance is less than 0.1. In this range, the coating or film absorbs less than 20% of the incident light, so the light intensity is nearly constant over the entire thickness of the film. This results in *first-order* kinetics, as described in Eq. (6). In this

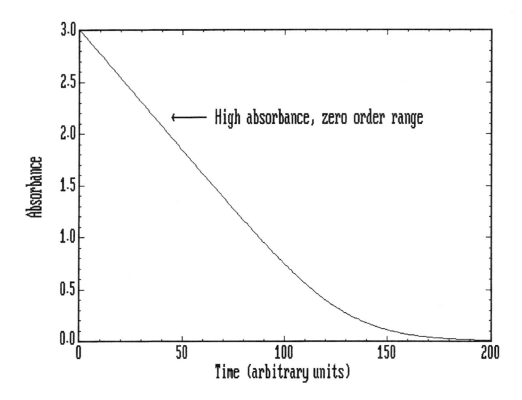

Figure 10   UVA loss calculated using Eq. (3) plotted as zero-order kinetics (linear scale). (From Ref. 22.)

range, the concept of a stabilizer half-life is valid and is shown by Eq. (7).

$$\log_{10}(A/A_0) = -kt \quad for \quad A < 0.1 \tag{6}$$

$$t_{1/2} = \log_{10}(2)/k = 0.301/k \quad for \quad A < 0.1 \tag{7}$$

A method for easily determining the rate constant k from experimental data has recently been described by Iyengar and Schellenberg (56). The absorbance of a film or coating on a transparent substrate is measured as a function of exposure time or light dose. Equation (3) can be rearranged to give Eq. (8), where A_t is the absorbance of the film at time t. Thus, plotting $\log(10^{A_t} - 1)$ as a function of time should give a straight line with slope k and an intercept related to the initial absorbance.

$$\log(10^{A_t} - 1) = -kt + A_0 + \log(1 - T_0) \tag{8}$$

The zero-order kinetic range [see Eq. (6)] is important in most practical coatings for which the UVA loading is high and the absorbance is usually much greater than 1. The first-order kinetic range [see Eq. (7)] is important in the surface of coatings and in the surface of polymers in which the absorbance is low. This has been dem-

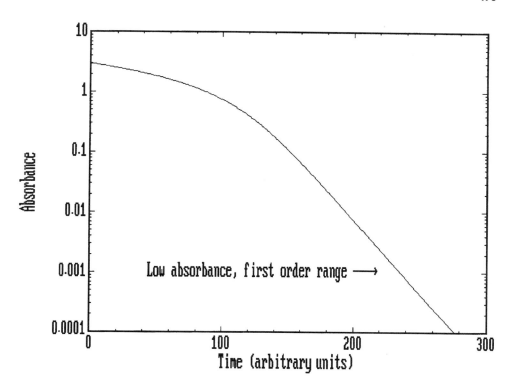

Figure 11 UVA loss calculated using Eq. (3) plotted as first-order kinetics (log scale). (From Ref. 22.)

onstrated for the loss of a benzophenone-type UVA from a silicone coating exposed to Florida weathering (22). The coating, as a whole, had an initial absorbance of 1.6, and lost absorbance with zero-order kinetics, as measured by transmission UV spectroscopy. By contrast, the UVA in the top few tenths of a micron of the surface, where the absorbance was less than 0.1, was lost by first-order kinetics, as measured by attenuated total reflectance IR. The same rate constant k was applicable to both cases.

The initial absorbance A_0 from a UVA in a coating or through some cross section of a polymer can be estimated from Eq. (9) where ε is the UVA's extinction coefficient, l is the loading of the UVA in weight percentage, d is the density of the polymer, r is the thickness of interest in microns, and M is the molecular weight of the UVA. This is only an estimate because the extinction coefficients reported in solvents may not be the same as in a polymer matrix, and scattering of light by pigments is ignored.

$$A = \frac{0.001(\varepsilon \, l \, d \, r)}{M} \qquad (9)$$

Table 2   Rates of Loss of UVAs in Various Matrices When Exposed to Xenon Arc Weathering

Absorber	Class	Rate (A/1000 kJm^{-2})		
		PMMA	Silicone hardcoat[a]	UV-cured acrylic
Cyasorb 531	Benzophenone	0.18	0.10	0.42
Cyasorb 5411	Benzotriazole	0.11	0.32	0.28
Uvinul 3039	Cyanoacrylate	0.14	0.31	0.42
Sanduvor VSU	Oxanilide	0.14	0.31	N/A
Cyasorb 1164	Triazine	0.085	0.81	0.22

[a] Absorber with a trialkoxysilyl group in place of an alkyl group, but otherwise the same chromophore.
Source: Ref. 22.

VII. FACTORS AFFECTING UVA PHOTODEGRADATION RATES

A. Matrix Effects

Several variables that could affect the photostability of UVAs have been examined and by far, the most important is the matrix in which the UVA is placed (27). This is not surprising because the nature of the matrix can affect the hydrogen bonding that is so critical to UVA function. More polar environments should be detrimental to UVA stability. Second, if the matrix is subject to photodegradation (and every organic material will photodegrade at some rate), then the radicals generated can undergo reaction with an excited state of the UVA as described in Secs. IV.B and IV.C. Unfortunately, it is very difficult to predict just how these factors will affect the degradation rate. This is illustrated in Table 2, which compares the rate of absorbance loss among the five major classes of UVAs in three different matrices. PMMA is a polar, relatively stable thermoplastic. The silicone hardcoat is highly cross-linked. Although one thinks of silicones as nonpolar, this is made of condensed methyl trimethoxysilane and colloidal silica and has some residual silanol functions that may create local polar regions. The UV-cured acrylic is less polar than PMMA because it contains a higher proportion of aliphatic units, but the residual acrylic functionality makes it susceptible to rapid photodegradation. One notes that the rates of absorbance loss [as determined using Eq. (8)] vary widely. Indeed, even the ranking of stability changes as one goes from the acrylic to the silicone matrix. This emphasizes the importance of experimentally determining the UVA stability in each particular polymer or coating of interest.

The photostability of the matrix is a very important factor in UVA loss because radicals generated from the matrix can destroy the chromophores of UVAs under irradiation conditions. Such matrix effects have also been observed by Decker (24), who found that a benzotriazole UVA was much more photostable in a fluorinated acrylate polymer (Lumiflon) than in a UV-cured acrylic urethane. The fluorinated acrylate was itself much more resistant to photodegradation than the acrylic urethane. Gerlock (57) has found that the rates of UVA loss correlate with rates of degradation of melamine coatings in which they are placed.

B. Effect of HALS

HALS have no effect on UVA stability in a photostable PMMA matrix (26,27). However, if a HALS stabilizes a coating matrix, then the rate of UVA loss is reduced (24,26,27,52,57,58). UV-cured acrylic–urethane coatings, UV-cured acrylic–silica coatings, and heat-cured melamine coatings are all susceptible to photooxidation. HALS reduce the rate of photooxidation, and thereby, they reduce the rate of free radical production. In general, any factor that improves the photostability of the matrix should improve the retention of the UVA, and any factor that accelerates the rate of matrix photooxidation should accelerate the rate of UVA loss.

C. Effect of Concentration

The degradation of benzotriazole and benzophenone UVAs in PMMA films were examined at several concentrations ranging from 1 to 20% on borosilicate-filtered xenon arc irradiation (26,27). The thickness of the films were adjusted so that the initial absorbances were all in the range of 0.8–1.4. No significant difference was observed among the UVA loss rates. Ghiggino et al. (15) reported that the fluorescence quantum yield of a benzotriazole UVA in PMMA films decreased with increasing concentration, indicating self-quenching. The independence of loss rate suggests that self-quenching is not an important mechanism for UVA stability, at least in a PMMA matrix.

D. Effect of Substituents

The effect of various substitutions on the photostability of benzophenone UVAs is shown in Table 3. The rates in this table are for the loss of absorbance owing to the UVAs in PMMA films exposed in a borosilicate-filtered xenon arc accelerated weathering device using the methodology described in Sec. VI to determine the rate constant (59). The rates, and possibly the rankings, would surely be different in a different matrix.

A hydroxyl group in the 2′-position (Table 3, entry 2) improves the photostability relative to entry 1. The NMR chemical shifts of the hydroxyl groups indicate that each hydrogen bond is weaker than the hydrogen bond (35) for entry 1, but the doubled chance of photoenolization seems to more than compensate. Oxygen substitution in the 4′-position (entries 3 and 4) leads to poorer photostability than the corresponding unsubstituted analogues. Substitution in the 6-position (entry 5) forces the phenol-bearing ring out of plane from the carbonyl group and greatly weakens the hydrogen bond, resulting in very poor photostability. Alkyl substitution in either the 3- or 2′-position (entries 6 and 7) have positive effects possibly by increasing the strength of the hydrogen bond or through sterically shielding the hydroxyl group and carbonyl from intramolecular hydrogen bonds with the PMMA matrix. 2,4-Dimethoxybenzophenone (entry 8) has no possibility of hydrogen bonding and is very rapidly photolyzed.

The photostability of several benzotriazoles with varied substitution in PMMA films (59) is shown in Table 4. As with the benzophenones, the rates and rankings are likely to be different in different matrices. There is relatively little difference among the commercially available compounds, at least in PMMA. Similar results were obtained (5,6) in acrylic films. Surprisingly, substitution *ortho* to the hydroxyl group

Table 3   Photodegradation Rates of Benzophenone Absorbers in PMMA Films Exposed to Xenon Arc

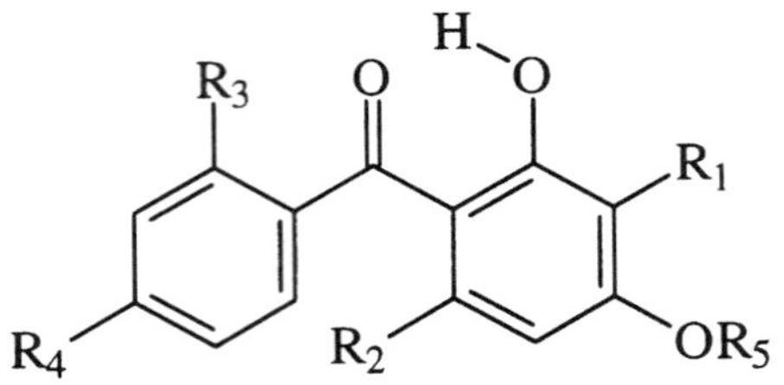

Entry	R_1	R_2	R_3	R_4	R_5	Rate (A/1000 kJm^{-2})
1	H	H	H	H	C_8H_{17}	0.16
2	H	H	OH	H	CH_3	0.06
3	H	H	OH	OCH_3	CH_3	0.13
4	H	H	H	OCH_3	CH_3	0.46
5	H	CH_3	H	H	C_3H_7	>2.5
6	CH_3	H	H	H	C_3H_7	0.11
7	H	H	CH_3	H	C_3H_7	0.09
8	2,4-Dimethoxybenzophenone					>2.5

Source: Ref. 59.

Table 4   Photodegradation Rates of Benzotriazole Absorbers in PMMA Films Exposed to Xenon Arc

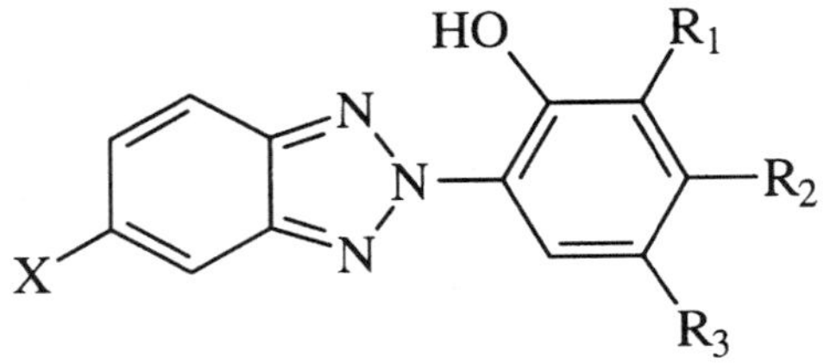

Entry	R_1	R_2	R_3	X	Rate (A/1000 kJm^{-2}
1	H	H	t-Octyl	H	0.11
2	H	H	Methyl	H	0.14
3	Cumyl	H	Cumyl	H	0.15
4	t-Butyl	H	CH_2CH_2-$COOC_8H_{17}$	H	0.19
5	t-Butyl	H	Methyl	Cl	0.14
6	t-Amyl	H	t-Amyl	H	0.22
7	CH_2 (dimer)	H	t-Octyl	H	0.15
8	H	H	CH_2CH_2OH	H	0.09
9	H	OCH_2CH_2OH	H	H	0.33

Source: Ref. 59.

(position R_1), which might be expected to improve the photostability (43), has little effect and may actually be slightly destabilizing in PMMA films. The result for entry 7 is interesting. This compound has two benzotriazole UVAs linked by a methylene unit such that they are held intimately close, but are not in conjugation with each other. Even in this apparently ideal situation for stabilization by self-quenching, no improvement in photostability results. Indeed, the rate of absorbance loss is somewhat higher than for the unlinked analogue (entry 1). Additional oxygen substitution on the phenolic ring (entry 9) is very detrimental to photostability.

There are only two major variations commercially available for triazine-type UVAs. Cyasorb 1164 (see Fig. 3, **3**, Ar = 2,4-dimethylphenyl, R = C_8H_{17}) has a loss rate of about 0.085 A/1000 kJm^{-2} in a PMMA film exposed to xenon arc irradiation. The closely-related Tinuvin 400 presumably has a similar loss rate. By contrast, Tinuvin 1577 (see Fig. 3, **3'**, Ar = phenyl, R' = C_6H_{13}) has a rate of absorbance loss of less than 0.02 A/1000 kJm^{-2} in PMMA films when exposed to borosilicate-filtered xenon arc irradiation. The latter compound is the most photostable commercially available UV absorber. Excellent photostability has also been demonstrated in polycarbonate films for Tinuvin 1577 (11). Triazine UVAs, with no substitutions on the aryl groups, exhibit no detectable low-temperature fluorescence (44), whereas compounds bearing methyl groups on the aryls (such as Cyasorb 1164) give a weak, but detectable low-temperature fluorescence. This indicates more efficient ESIPT processes for the unsubstituted triazine, although the reasons remain unclear.

VIII. UVA PHOTOSTABILITY AND LIFETIME PREDICTIONS

A. UVA as a Bulk Additive

Although a UVA is often used as a bulk additive to a polymer, only the UVA that is near the surface is actually subject to any photochemistry. The situation is very complicated for polyolefins. The UVA near the surface is subject to photodegradation that is driven mostly by the degradation of the polymer matrix, as described in Secs. IV.B and IV.C. It is also subject to physical loss through diffusion and evaporation or leaching. While this is occurring, UVA from deeper within the polymer can diffuse into the surface region, which partially restores the initial concentration. In this situation, the inherent photostability of the UVA probably has little effect on the lifetime of the polymer because rates of diffusion and rates of matrix degradation are probably controlling factors. The importance of these factors can change greatly, depending on exposure and testing conditions. The conclusions of Tidjani et al. (49) are probably correct: lifetime prediction on the basis of stabilizer loss is not likely to be possible for polyolefins.

The analysis is somewhat more straightforward for higher T_g engineering thermoplastics because diffusion of the stabilizer is negligible. Often, most of the degradation is confined to a rather narrow surface layer—generally less than about 50 µm—and failure is often due to loss of surface properties, such as loss of gloss, chalking, and discoloration. Degradation in the top 10 or even 5 µm might be sufficient to cause failure.

Application of Eq. (9) to a case in which 0.5% of a UVA with a molecular weight of 350 and $\varepsilon = 12,000$ is used in a polymer with a density of 1 gives an absorbance of 0.17 at a depth of 10 μm. This is in the absorbance range at which first-order kinetics are more or less valid and the half-life of the UVA can be calculated by using Eq. (7). Degradation rates derived from outdoor exposure data (22,26,27) show that most commercial UVAs will have a half-life of about 1 year or less in Florida when in the first-order regimen. Most engineering polymers will lose their surface properties and begin to erode within 1–2 years of outdoor Florida aging (60). The stability of the UVA is unlikely to have much effect on the lifetime of the polymer surface because the UVA contributes little to the stabilization of the surface. Heller (10) has described calculations for the efficiency of UVAs as bulk additives and found that in this low-absorbing range, the UVA can reduce the rate of polymer degradation by less than 20%. Exceptions would be if the UVA were extraordinary unstable or in polymers such as PMMA and fluorinated polymers that have much longer lifetimes. In these examples, the photostability of the UVA can play a role, especially if these polymers are used as coatings or laminates over other materials, as described in the following section. However, UVA photostability probably is not a limiting factor in the service life for normal UVAs as bulk additives in most engineering polymers.

Data have been presented recently that seem to bear this out (23). A layer of polycarbonate containing several percent of UVA was extruded over polycarbonate containing a low level of UVA. One UVA tested was a "dimeric" benzotriazole and another was a highly photostable triazine. Although the triazine was at least seven times more photostable than the benzotriazole, the difference in photoyellowing rates of the two coextruded articles was small when exposed to xenon arc testing. The observed difference in the yellowing rates could have been due as much to the higher absorbance of the triazine at short wavelengths (where the polycarbonate is particularly sensitive) as to its greater photostability.

B. UVAs in Coatings

In recent years the use of transparent coatings to protect photosensitive polymers and other articles has become more common. The coating materials often enhance properties such as hardness or gloss. Usually, UV absorbers are added to the coating to prevent harmful UV from reaching the substrate, thereby delaying discoloration and delamination. The UVA can also act as an inner filter to help stabilize the coating material itself. If the coating matrix is relatively photostable, the lifetime of the UVA can determine the service lifetime of the coated article. Bauer has described the science and mathematics of coating lifetime prediction in a recent landmark paper (53, 61). Bauer's examples are of automotive paint systems in which a UVA-bearing clearcoat protects pigment-bearing underlayers. We have applied similar methods to predicting the lifetime of abrasion resistant coatings over polycarbonate.

For a coating that is protecting a sensitive substrate, the important factor is the transmission of UV radiation through the coating that can cause degradation of the substrate. A more complete analysis would include transmission at all wavelengths, the action spectrum of the substrate, and the emission spectrum of the light source, but selection of a single wavelength at which the substrate shows maximum sensitivity in the activation spectrum—less than 340 nm for many materials (62)—provides an adequate first approximation. Figure 12 shows the situation

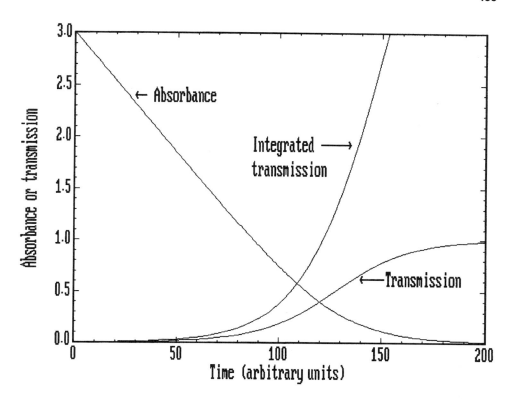

Figure 12 Decreased absorbance, increased transmission, and cumulative light dose (integrated transmission) calculated for a hypothetical coating using Eqs. (3) and (11). (From Ref. 26.)

for a hypothetical coating that has an initial absorbance of 3 at 330 nm. As the coating is exposed to weathering, the UVA degrades and the absorbance decreases following the kinetics described in Eqs. (3) and (4). As long as the absorbance is high, the transmission is low. As the absorbance decreases below 1 or so, the transmission increases more quickly. The critical variable is the total dose of UV radiation that reaches the substrate, which is the integral of the transmission. When that dose reaches some critical level, sufficient degradation will have occurred to the substrate to cause failure, such as delamination or discoloration. Bauer (53) has integrated Eq. (4) to produce Eq. (10), which is the transmitted light dose D at time t in terms of the absorbance at time t [A_t, from Eq. (4)] using the rate constant k as defined in a Sec. VI. An equivalent form in terms of the initial transmission T_0 and k is shown in Eq. (11).

$$D = t + \frac{A_t}{k} - \frac{A_0}{k} \tag{10}$$

$$D = t + \frac{1}{k}\log_{10}\left[T_0 + (1 - T_0)10^{-kt}\right] \tag{11}$$

Figure 12 shows the light dose as a function of exposure time. The critical light dose can be determined by exposing the coated system with no UVA in the coating and determining the time or amount of exposure required to cause failure. Care is required. Some substrates show strong sensitivity to the unnaturally short-wavelength light in some accelerated-testing equipment that the UVAs efficiently screen out. The samples without UVA can have a different response or correlation factor with outdoor exposure than coated samples. The rate constant of UVA degradation k, as well as the initial absorbance and transmission, A_0 and T_0, can be determined by applying the coating to a stable, transparent substrate, such as PMMA or quartz, and using the methodology described in a Sec. VI. Calculations using Eq. (10) or (11) result in a graph of transmitted dose versus time, and the predicted service life is the time required to reach the critical dose (53). Equation (11) may be solved for t to give the time required to reach the critical dose D_c, as shown in Eq. (12).

$$t_c = \frac{1}{k}\log_{10}\left[\frac{10^{kD_c} + T_0 - 1}{T_0}\right] \tag{12}$$

Because the experiments are usually performed in an accelerated-testing device, some correlation between the test and outdoor exposure for the particular coating system is also required.

The lifetime predicted by this method, in fact, represents a maximum lifetime for the coating. Many factors can cause earlier failure. One cause can be that the UVA level in the actual coating might not be as high as estimated because of losses due to volatilization or diffusion into the substrate during processing. In addition, during the long exposures required to reach failure because of UVA depletion, other mechanisms that might not be readily apparent during shorter exposures can play important roles. These might be hydrolysis at the coating–substrate interface, cracking owing to repeated thermal cycling, or degradation of the coating matrix material, leading to unacceptable loss of properties.

IX. CONCLUSIONS

The effectiveness of UVAs can be lost both by physical and chemical mechanisms. Physical loss usually is a problem only in low T_g polymers, such as polyolefins, and in coatings cured at relatively low temperatures. Chemical loss can be due to a variety of chemical mechanisms, but the overriding factor is the polymer or coating matrix itself. Polar matrices can disrupt the critical internal hydrogen bonding and lead to rapid UVA photolysis. Matrices that are subject to rapid photooxidation will usually also cause the UVAs in them to degrade rapidly. Conversely, additives that improve the stability of the matrix, such as HALS, often improve the stability of the UVA.

The kinetics of UVA loss depends on the absorbance of the coating or polymer cross section. When the absorbances is less than 0.1, such as at the surface of the polymer or coating, the loss obeys first-order kinetics. When the absorbance is greater than 1, the loss obeys zero-order kinetics. Contrary to intuition and common practice, the "fraction of initial absorbance" is never an appropriate way to treat UVA loss data. The mathematics of UVA loss is now well-understood and generally has been experimentally verified, although each type of UVA and matrix material presents its own peculiarities. The stability of the UVA seems to have a minor effect on the weathering performance of plastics when the UVA is used as a bulk additive. By contrast, UVA stability can be the determining factor in the service lifetime of coatings and films. Predictive models for coating lifetimes have been described and used with success.

ACKNOWLEDGMENTS

I express my thanks to Daniel Olson for reading the manuscript and offering helpful comments, to Mrs. Oltea Siclovan for translation of a paper, and to Joseph Webster of Clariant Corporation for useful references to papers.

REFERENCES

1. RC Hirt, NZ Searle, RG Schmitt. Ultraviolet degradation of plastics and the use of protective ultraviolet absorbers. SPE Trans 1:26–30, 1961.
2. FJ Pern, AW Czanderna. Characterization of ethylene vinyl acetate (EVA) encapsulant: effects of thermal processing and weathering on its discoloration. Solar Energy Mater Solar Cells 5:3–23, 1992.
3. P Klemchuk, M Ezrin, G Lavingne, W Holley, J Galica, S Agro. Investigation of the degradation and stabilization of EVA-based encapsulant in field-aged solar energy modules. Polym Degrad Stabil 55:347–365, 1997.
4. DA Bolon, PC Irwin. Chemical changes in engineering thermoplastics. Makromol Chem Macromol Symp 57:227–234, 1992.
5. B Bell, DE Beyer, NL Maecker, RR Papenfus, DB Priddy. Permanence of polymer stabilizers in hostile environments. J Polym Sci 54:1605–1612, 1994.
6. JE Bonekamp, NL Maecker. The permanence of UV absorbers in a rubber-modified acrylic film by UV/visible spectrophotometry. J Appl Polym Sci 54:1593–1604, 1994.
7. JF Rabek. Photostabilization of Polymers. Principles and Applications. London: Elsevier Applied Science, 1990, pp 202–278.
8. F Gugumus. Light stabilizers. In: R Gachter, H Muller, eds. Plastics Additives Handbook. 4th ed. Munich: Hanser, 1996, pp 175–194.
9. M Dexter. UV stabilizers. In: Kirk-Othmer Encyclopedia of Chemical Technology. 3rd ed, vol 23. New York: John Wiley & Sons, 1983, pp 615–627.
10. HJ Heller. Protection of polymers against light irradiation. Eur Polym J Suppl 105–132, 1969.
11. G Rytz, R Hilfiker, E Schmidt, A Schmitter. Introduction to a new class of high performance light stabilizers and the influence of light stabilizers structure on the polymer's life. Angew Macromol Chem 247:213–224, 1997.
12. L Avar. UV-light induced reaction of new radical scavengers with polymers. Second North American Research Conference on Stabilization and Degradation of Polymers, Hilton Head, NC, March 13–15, 1995.
13. JA Virgilio, E Heilweil. Synthesis and photostability of some N-alkyl-N,N'-diaryl-formamidines. Org Prep Proc Int 10:97, 1978.
14. JEA Otterstedt. Photostability and molecular structure. J Chem Phys 58:5716–5725, 1973.
15. KP Ghiggino, AD Scully, SW Bigger. Photophysics of hydroxyphenylbenzotriazole polymer stabilizers. In: E. Reichmanis, ed. Effects of Radiation on High-Technology Polymers. ACS Symp Ser 381. Washington DC: American Chemical Society, 1989, pp 57–79.

16. M Allan, T Bally, E. Haselbach, P Suppan, L Avar. Photo-physics and photo-chemistry of aromatic oxanilides used as polymer light stabilizers. Polym Degrad Stabil 15:311–325, 1986.

17. SW Bigger, KP Ghiggino, IH Leaver, AD Scully. Photophysics of 6-(2′-hydroxy-4-methoxyphenyl)-*s*-triazine photostabilizers. J Photochem Photobiol A 40:391–9, 1987.

18. W Klopffer. Intramolecular proton transfer in *ortho*-hydroxybenzophenones. J Polym Sci Symp 57:205–212, 1976.

19. J Sedlar, J Petruj, J Pac. Photochemical behavior of a benzotriazole UV stabilizer. 1st International Symposium on Photochemical Processes in Polymer Chemistry, IUPAC, Louvain, Belgium, June 1972.

20. J Petruj, J Sedlar, M Parizek, J Pac. Elliptical photochemical reactor for preparative and quantitative studies in the liquid phase. J Photochem 2:393–400, 1973/74.

21. J Catalan, JC Del Valle, F Fabero, NA Garcia. The influence of molecular conformation on the stability of ultraviolet stabilizers toward direct and dye-sensitized photo-irradiation: the case of 2-(2′-hydroxy-5′-methylphenyl)benzotriazole (Tin P). Photochem Photobiol 61:118–123, 1995.

22. JE Pickett. Review and kinetic analysis of the photodegradation of UV absorbers. Macromol Symp 115:127–141, 1997.

23. G Rytz, R Hilfiker. Influence of light stabilizers structure on the polymer's lifetime. Eighteenth Annual International Conference on Advances in the Stabilization and Degradation of Polymers, Luzern, Switzerland, June 19–21, 1996.

24. C Decker, S Biry, K Zahouily. Photostabilization of organic coatings. Polym Degrad Stabil 49:111–119, 1995.

25. A Gupta, GW Scott, D Kliger. Mechanisms of photodegradation of ultraviolet stabilizers and stabilized polymers. In: SP Pappas, FH Winslow, eds. Photodegradation and Photostabilization of Coatings. ACS Symp Ser 151. Washington, DC: American Chemical Society, 1981, pp 27–42.

26. JE Pickett, JE Moore. Photostability of UV screeners in polymers and coatings. In: RL Clough, NC Billingham, KT Gillen, eds. Polymer Durability: Degradation, Stabilization, and Lifetime Prediction. ACS Adv Chem Ser 249. Washington DC: American Chemical Society, 1995, pp 287–301.

27. JE Pickett, JE Moore. Photodegradation of UV screeners. Polym Degrad Stabil 42:231–244, 1993.

28. H Goth. Zur eigenstabilitat einiger UV Absorber. Kunststoffe Plast 3/68:96–97, 1968.

29. AA Lamola, LJ Sharp. J Phys Chem 70:2634–2638, 1966.

30. JQ Pan, WWY Lau, SF Shang, XZ Hu. Synthesis and properties of new UV-absorbers with higher MW. Polym Degrad Stabil 53:153–159, 1996.

31. P Vink. Loss of UV stabilizers from polyolefins during photo-oxidation. In: G Scott, ed. Developments in Polymer Stabilization—3. London: Applied Science, 1980, pp 117–138.

32. JP Guillory, CF Cook. Mechanism of stabilization of polypropylene by ultraviolet absorbers. J Polym Sci 9(part A1):1529–1536, 1971.

33. DJ Carlsson, T Suprunchuk, DM Wiles. Photo-oxidation of polypropylene films. VI. Possible UV-stabilization mechanisms. J Appl Polym Sci 16:615–626, 1972.

34. NS Allen, M Mudher, P Green. Photo-stabilising action of *ortho*-hydroxy aromatic compounds in polypropylene film: UV absorption versus radical scavenging. Polym Degrad Stabil 7:83–94, 1984.

35. JH Chaudet, JW Tamblyn. Some functions of 2-hydroxybenzophenones as weathering stabilizers in polymers. SPE Trans April:57–62, 1961.

36. NS Allen, M Mudher. Photo-stabilising action of orthohydroxy aromatic compounds in low density polyethylene film: UV absorption versus radical scavenging. Polym Degrad Stabil 9:145–154, 1984.

37. P Vink. Changes in concentration of some stabilizers during the photooxidation of polypropylene films. J Polym Sci Symp 40:196–173, 1973.
38. DJ Carlsson, DW Grattan, T Suprunchuk, DM Wiles. The photodegradation of polypropylene. IV. UV stabilizer decomposition. J Appl Polym Sci 22:2217–2228, 1979.
39. KB Chakraborty, G Scott. Mechanisms of antioxidant action: the role of 2-hydroxybenzophenones in photo-oxidizing media. Eur Polym J 15:35–40, 1979.
40. JL Gerlock, W Tang, MA Dearth, TJ Korniski. Reaction of benzotriazole ultraviolet light absorbers with free radicals. Polym Degrad Stabil 48:121–130, 1995.
41. MA Dearth, TJ Korniski, JL Gerlock. The LC/MS/MS characterization of photolysis products of benzotriazole-based ultraviolet absorbers. Polym Degrad Stabil 48:111–120, 1995.
42. DKC Hodgeman. Reactions of ultraviolet stabilizers: reactions of some 2′-hydroxy-2-phenylbenzotriazoles with the *tert*-butyl peroxy radical in solution. J Polym Sci Polym Lett Ed 16:161–165, 1978.
43. PF McGarry, S Jockusch, Y Fujiwara, NA Kaprinidis, NJ Turro. DMSO solvent induced photochemistry in highly photostable compounds. The role of intermolecular hydrogen bonding. J Phys Chem A 101:764–767, 1997.
44. J Keck, HEA Kramer, H Port, T Hirsch, P Pischer, G Rytz. Investigations on polymeric intramolecularly hydrogen-bridged UV absorbers of the benzotriazole and triazine class. J Phys Chem 100:14468–14475, 1996.
45. BD Gupta, L Jirackova-Audouin, J Verdu. Spectrophotometric study of the photostabilization of PVC by a benzilidine malonate. Eur Polym J 24:947–951, 1988.
46. J Luston. Physical loss of stabilizers from polymers. In: G Scott, ed. Developments in Polymer Stabilization—2. London: Applied Science, 1980, pp 185–240.
47. NC Billingham, PC Calvert. The physical chemistry of oxidation and stabilisation of polyolefins. In: G Scott, ed. Developments in Polymer Stabilization—3. London: Applied Science, 1980, pp 139–190.
48. NC Billingham. Physical phenomena in the oxidation and stabilization of polymers. In: J Pospisil, PP Klemchuk, eds. Oxidation Inhibition in Organic Materials, vol 1. Boca Raton, FL: CRC Press, 1990, pp 249–297.
49. A Tidjani, E Fanton, R Arnaud. The oxidative degradation of stabilized LLDPE under accelerated and natural conditions. Angew Makromol Chem 212:35–43, 1993.
50. M Johnson, RG Hauserman. Diffusion of stabilizers in polymers. IV. 2,4-dihydroxybenzophenone in plasticized poly(vinyl chloride). J Appl Polym Sci 21:3457–3463, 1977.
51. DR Olson, KK Webb. Thermal loss of UV stabilizers from BPA-polycarbonate. Macromolecules 23:3762, 1990.
52. C Decker, K Zahouily. Light fastness of UV Absorbers in radiation-cured acrylic coatings. Polym Mat Sci Eng 68:70–71, 1993.
53. DR Bauer. Predicting in-service weatherability of automotive coatings: a new approach. J Coatings Technol 69:85–95, 1997.
54. CL Renschler. Photodegradation kinetics of methyl salicylate in poly(methyl methacrylate). J Appl Polym Sci 29:4161–4173, 1984.
55. A Gupta, GW Scott, D Kliger, O Vogel. Photostability of ultraviolet screening transparent acrylic copolymers of [2(2-hydroxy-5-vinylphenyl)2-*H*-benzotriazole]. Polym Preprints 23(1):219–220, 1982.
56. R Iyengar, B Schellenberg. Loss rate of UV absorbers in automotive coatings. Polym Degred Stabil 61:151–159, 1998.
57. JL Gerlock, CA Smith, EM Nunez, VA Cooper, P Liscombe, DR Cummings, TC Dusibiber. In: RL Clough, NC Billingham, KT Gillen, eds. Polymer Durability: Degradation, Stabilization, and Lifetime Prediction. ACS Adv Chem Ser 249. Washington DC: American Chemical Society, 1995, pp 335–348.

58. C Decker, K Moussa, T Bendaikha. Photodegradation of UV-cured coatings II. Polyurethane–acrylic networks. J Polym Sci A Polym Chem 29:739–747, 1991.

59. JE Pickett, JE Moore. Photodegradation of UV absorbers: kinetics and structural effects. Angew Makromol Chem 232:229–238, 1995.

60. MR Kamal, B Huang. Natural and artificial weathering of polymers. In: SH Hamid, MB Amin, AG Maadhah, eds. Handbook of Polymer Degradation. New York: Marcel Dekker, 1992, p. 128.

61. DR Bauer. Application of failure models for predicting weatherability in automotive coatings. In: DR Bauer, JW Martin, eds. Service Life Prediction of Organic Coatings. ACS Symp Ser 722. Washington, DC: American Chemical Society, 1999, pp. 378–395.

62. AL Andrady. Wavelength sensitivity in polymer photodegradation. Adv Polym Sci 128:47–94, 1997.

6

Highlights in the Inherent Chemical Activity of Polymer Stabilizers

JAN POSPÍŠIL

Institute of Macromolecular Chemistry, Academy of Sciences of the Czech Republic, Prague, Czech Republic

STANISLAV NEŠPŮREK

Technical University of Brno, Brno, and Institute of Macromolecular Chemistry, Academy of Sciences of the Czech Republic, Prague, Czech Republic

I. INTRODUCTION

Plastics, elastomers, polyblends, and coatings are vulnerable to harmful effects of the environment. Deteriorative factors are of a chemical (oxygen and its active forms, humidity, metal ions, harmful anthropogenic emissions, and atmospheric pollutants such as NO_x, SO_2) and physical (heat, mechanical stress, radiation) nature. The relevant processes, melt degradation, long-term heat-aging, and weathering, are initiated by temperature, mechanical forces, chemical catalysis, and solar and high-energy radiations.

Polymers are very different as far as their inherent sensitivity to oxidation and photodegradation is concerned. Generally, differences in stability arise from a variety of chemical structures, manufacturing processes, and polymer morphology. Highly unsaturated polymers and polymers containing structure-borne chromophores are much more oxidation- and photooxidation-sensitive. Some defect structures present in trace amounts in the polymer (structural inhomogeneities, polymeric impurities), such as olefinic unsaturation C=C, or polymer-bound oxygenated groups, are formed in polymers as a consequence of adventitious oxidation during manufacture (e.g., pelletization of polyolefins), storage, or shipping, and make polymers more prone to degradation during processing and environmental exposure. The level of structural inhomogeneities that sensitize degradation gradually increases during the polymer lifetime. Polymer branching, cross-linking, chain scission and cyclization, and changes of physical properties are, together with new functional groups, degradation consequences (1). The progressive degradation is enhanced by nonpolymeric impurities, such as various metal contaminants or photoactive pigments. The appearance or aesthetics (discoloration, loss of gloss or transparency, surface cracks) occur more or less simultaneously with the loss of mechanical properties (impact and tensile strengths, elongation).

The mechanisms and theoretical background of individual degradation steps have been explained during years of elucidation. The sensitizing effects are more complex in polymer blends owing to co-oxidation phenomena between component polymers, and in recycled aged plastics reprocessed for the second-life application. In the latter case, oxygenated polymer inhomogeneities and molecular fragments formed by photolysis during the first life of component polymers, various metal oxides or ions, printing inks and paint residues, and transformation products of the first-life stabilizers are present as additional impurities. An understanding of the initial phases of the degradation, effects of deteriogens and impurities, laws of free radical chain reactions, and photophysical processes during the outdoor application of solid polymers, created a basis for development of an effective protection of polymers.

Safe processing and long-term applications of all polymers that have become commercial products are dependent on stabilizing additives. This permits the polymers to withstand chemical and physical stresses during different phases of their lifetime. The plastics as well as rubber industries seem to have recovered from the recession. Consequently, the prospects for the stabilizer industry are favorable. Virgin single and blended polymers and recycled plastics principally suffer from analogous degradation mechanisms. Therefore, they benefit from the same stabilization approaches.

According to their principal activity mechanisms, common polymer stabilizers are conventionally classified as antioxidants (see Sec. III.A), metal deactivators (see Sec. III.B), antiozonants (see Sec. III.C), photoantioxidants (see Sec. III.D), photostabilizers (see Sec. III.E), and heat stabilizers for poly(vinyl chloride) (PVC). Molecular architecture and functional moieties involved in the molecule determine the stabilization mechanism: "inherent chemical activity of the stabilizer." Active moieties characteristic of each class and chemical group of stabilizers, such as amino or hydroxy group or trivalent phosphorus in amines, phenols, or phosphites, and structural modifications optimizing reactivity of the moieties, for example, sterical hindrance, substitution preventing unwanted reactivities and, last but not least, modifications optimizing solubility and compatibility in a polymer matrix and reducing physical losses by volatilization and leaching are structural features optimizing the final effect.

The chemistry of stabilizers has been elucidated under conditions simulating chemical and physical degradation of solid polymers. Relations between structure and effectivity were determined for the principal classes of stabilizers (2). Extensive product studies explaining chemical reactions with free radical species and molecular products formed in degrading polymers were performed to explain the stabilizer activity, consumption modes during polymer protection, and the role of stabilizer transformation products (3). This approach accounts for additives meeting expectations of consumers for high effectivity, acceptable cost–performance relations, reduced unwanted side effects, and environmental acceptability.

The up-to-date knowledge of the chemistry of stabilizers protecting polymers against thermally and photochemically induced oxidative degradations and managed by the inherent chemical efficiency is interpreted in this chapter. Where relevant, review papers are reported in the manuscript.

II. SITES OF THE EFFECTIVE APPLICATION OF STABILIZERS IN DEGRADING POLYMERS

Mechanical shear forces and heat attack polymers in an oxygen-deficient atmosphere during processing. Oxygen and the ultraviolet (UV) spectrum of the terrestrial solar radiation (UV-A radiation, 320–400 nm and a part of the UV-B radiation, ranging from 290 to 320 nm), the regular components of the troposphere are, together with heat, mechanical stress, and oxidizing tropospheric pollutants (ozone, nitrogen oxides) the principal deteriogens that attack polymers during weathering. Formation of free radical and molecular polymer-bound structures characteristic of degrading polymers is briefly outlined to illustrate basic principles of the relevant stabilization.

Structural polymeric impurities characteristic of olefinic unsaturation, such as vinyl, *trans*-vinylidene, or vinylene groups (1) are responsible for light absorption by saturated carbon chain polymers, such as polyolefins or PVC. The unsaturation arises during processing in polyolefins by disproportionation of two macroalkyls (4) (Eq. 1) and by thermolysis of PVC (5) (Eq. 2). The unsaturation is effectively reduced in polyolefins by processing antioxidants (phenols, see Sec. III.A.1 and phosphites, see Sec. III.A.2) and by heat stabilizers in PVC (see Sec. III.F).

$$2 \ \sim\!\!\!\!/C^{\bullet}H_2 \ \longrightarrow \ \sim\!\!\!\!/CH_3 \ + \ \sim\!\!\!\!/\!\!\!\!/ \tag{1}$$

$$\tag{2}$$

Most oxidative mechanisms involve free radicals generated thermally, catalytically, mechanochemically, or by radiation in the presence of trace impurities. Macroalkyls $P^{\bullet}$ are formed from polymers PH by breaking C–H or C–C bonds [Eq. (3)]. In oxygen-free atmosphere, $P^{\bullet}$ transform by disproportionation, recombination, or addition reactions account for unsaturation, chain scission, branching, or cross-linking of the polymer (4). $P^{\bullet}$ fragments are oxidized with oxygen to alkylperoxyls $POO^{\bullet}$, a source of alkylhydroperoxides POOH (6) [Eq. (4)]. There is a general consensus that hydroperoxides are the most important photoinitiators in the very early stages of the (photo)degradation. The photoinitiating capacity of hydroperoxides has been attributed to the high quantum yield of their photolysis. Hydroperoxides homolyze in thermal, radiation and in metal-catalyzed reactions into alkoxyls $PO^{\bullet}$ and migrating hydroxy radicals $HO^{\bullet}$ [Eq. (5)]. After concomitant β-scission of alkoxyls, various carbonyl compounds (saturated or unsaturated polymer-bound ketones) are formed (6–10) [Eq. (6)], playing a role of principal chromophores and photosensitive moieties in the hydrocarbon backbone of polyolefins. Carbonyls have a lower photoinitiating effectiveness in comparison with hydroperoxides. After Norrish-type I photocleavage [Eq. (7)], terminal macroalkyls and acyl radicals **1** are generated

$$ PH \xrightarrow[\text{shear}]{\Delta,\ h\nu,\ M^{n+}} \quad P^{\bullet} \longrightarrow \text{Products} \qquad (3) $$

$$ P^{\bullet} + O_2 \longrightarrow POO^{\bullet} \qquad (4) $$

$$ POO^{\bullet} + PH \longrightarrow \boxed{POOH} + P^{\bullet} \qquad (5) $$

$$ POOH \xrightarrow[\gamma]{\Delta,\ h\nu,\ M^{n+}} PO^{\bullet} + HO^{\bullet} \qquad (5) $$

$$ PO^{\bullet} \xrightarrow{h\nu} \boxed{\text{Carbonyl compounds} >CO} \qquad (6) $$

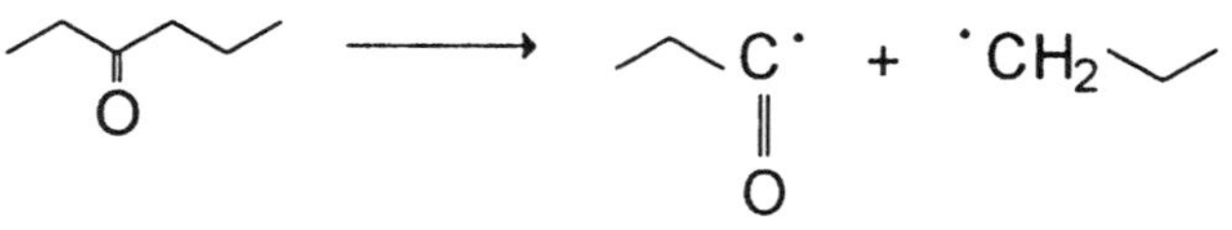

(7)

Scheme 1   **1**

The "hydroperoxide" oxidative pathway (Scheme 1) is characteristic of hydrocarbon polymers (polyolefins, styrene- and diene-based polymers) and was also adapted for polymers containing heteroatoms in their construction units (8–10). Since hydroperoxides are also formed during polymer processing, the thermal history of the polymer and the efficiency of the processing stabilization have been reflected in the outdoor performance of polymers. Polymer transformation processes and products arising from the hydroperoxide pathway and characteristic of thermal and photooxidations are basically similar.

To prevent formation of macroalkyls, hydroperoxide-decomposing antioxidants (HD AO; see Sec. III.A.2) and UV absorbers (UV abs.; see Sec. III.E.1) are used during processing and outdoor exposure, respectively (Scheme

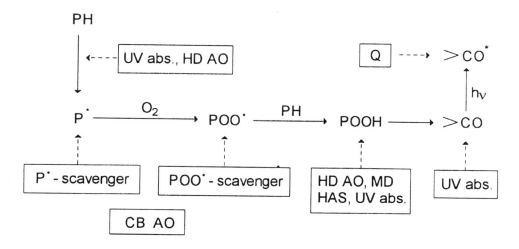

Scheme 2

2). Radical scavengers (chain-breaking antioxidants, CB AO, see Sec. III.A.1) are used to deactivate the formed macroalkyls and alkylperoxy radicals. Homolysis of hydroperoxides is prevented by HD AO (see Sec. III.A.2), photoantioxidants (hindered amine stabilizers, HAS; see Sec. III.D), UV absorbers (see Sec. III.E.1) and, in special cases, by metal deactivators (MD; see Sec. III.B) (Scheme 2).

Chromophores [Ch; carbonyl compounds formed in Eq. (6)] absorb actinic sunlight radiation and are transformed in the excited singlet (^{1}Ch*) and triplet (^{3}Ch*) states [Eq. (8)] (8,10). Excited chromophores transfer the electronic energy to the polymer PH or ground-state oxygen and sensitize formation of macroalkyls [Eq. (9a)] or singlet molecular oxygen [Eq. (9b)]. Moreover, they accelerate homolysis of hydroperoxides by an exciplex [Eq. (9b), Scheme 3].

$$Ch \xrightarrow{\text{h}\nu} {}^1Ch^* \xrightarrow{\text{h}\nu} {}^3Ch^* \qquad (8)$$

$$
{}^3Ch^* \begin{cases} \xrightarrow{PH} Ch + P^{\bullet} & (9a) \\ \xrightarrow{{}^3O_2} Ch + {}^1O_2 & (9b) \\ \xrightarrow{POOH} [Ch\ldots POOH]^* \longrightarrow Ch + PO^{\bullet} + HO^{\bullet} & (9c) \end{cases}
$$

Scheme 3

The UV absorbers (see Sec. III.E.1) prevent formation of excited chromophores. Light stabilizers with properties of quenchers (Q; see Sec. III.E.2) are able to deactivate excited chromophores and singlet oxygen (see Scheme 2) and prevent their involvement in Eqs. (9a–c).

Besides oxygenated polymeric inhomogeneities, transition metals including residues of polymerization catalysts (1) and photoactive pigments or dyes (11) may act as chromophores in outdoor degradation of polymers and enhance their breakdown. Singlet oxygen $^1O_2$, the lowest excited state of molecular oxygen, forms allylic hydroperoxides by addition on unsaturated polymer molecules [Eq. (10)] (8) or by proton transfer in saturated polymer molecules [Eq. (11)] (12).

$$\left[\!\!\!\diagdown\!\!\!\diagup\!\!\!\diagdown\!\!\!\right] \xrightarrow{\ ^1O_2\ } \left[\begin{array}{c} OOH \\ \diagup\!\!\!\diagdown\!\!\!\diagup\!\!\!\diagdown \end{array}\right] \tag{10}$$

$$\left[\!\!\!\diagdown\!\!\!\diagup\!\!\!\diagdown\!\!\!\right] \xrightarrow{\ O_2\ } \left[\begin{array}{c} O_2^{\cdot-} \\ \overset{+}{\diagup\!\!\!\diagdown\!\!\!\diagup\!\!\!\diagdown} \end{array}\right] \longrightarrow \left[\begin{array}{c} {}^{\cdot}OOH \\ \overset{\cdot}{\diagup\!\!\!\diagdown\!\!\!\diagup\!\!\!\diagdown} \end{array}\right] \longrightarrow \left[\begin{array}{c} OOH \\ \diagup\!\!\!\diagdown\!\!\!\diagup\!\!\!\diagdown \end{array}\right] \tag{11}$$

From the point of view of polymer stabilization, we also have to consider a "peracid" pathway in the integral mechanism of oxidative degradation of hydrocarbon polymers. This pathway accounts for the formation of acylperoxy radicals **2** and peroxyacids **3** from acyl radicals **1** arising according to Eq. (7) (13). Species **2** and **3** are effectively depleted by photoantioxidants (HAS; see Sec. III.D).

$$\sim CH_2C(O)OO^{\cdot} \qquad\qquad\qquad \sim CH_2C(O)OOH$$

$$\mathbf{2} \qquad\qquad\qquad\qquad\qquad\qquad \mathbf{3}$$

Tropospheric ozone is a dangerous photooxidant in the polluted urban atmosphere. It attacks unsaturated rubbers ("ozone aging") and participates, together with oxygen, in rubber flex-cracking (14,15). The molecular mechanism of ozone aging of rubbers is well understood. Ozonides **4**, reactive zwitterions **5**, cyclic peroxides **6**, polyperoxides **7**, polymeric ozonides **8**, hydroperoxides **9**, and terminal aldehydes **10** are formed. Antidegradants having properties of antiozonants (see Sec. III.C), antifatigue agents, and chain-breaking antioxidants (see Secs. III.A.1.b and III.A.1.d) stabilize rubbers (15).

Even saturated hydrocarbon polymers are not safe when stressed by ozone (16). Ozone contributes to initiation of their oxidation by generation of free radicals

$$-\overset{\displaystyle O-O}{\underset{\displaystyle O}{CH}}\overset{}{\diagdown}\overset{}{CH} \qquad\qquad -C^{+}HOO^{-} \qquad\qquad -\overset{OO}{CH}\overset{}{\diagdown}\overset{}{\underset{OO}{CH-}} \qquad\qquad \left[CH(R)OO\right]_n$$

$$\mathbf{4} \qquad\qquad\qquad \mathbf{5} \qquad\qquad\qquad \mathbf{6} \qquad\qquad\qquad \mathbf{7}$$

$$-CH(R)OCH(R)OO- \qquad\qquad -CH(R)OOH \qquad\qquad RCHO$$

$$\textbf{8} \qquad\qquad\qquad\qquad \textbf{9} \qquad\qquad\qquad \textbf{10}$$

[see Eqs. (12)–(15), Scheme 4) and by oxidation with atomic oxygen Eq. (17)], generated by ozone photolysis in the urban atmosphere in the presence of nitrogen dioxide [see Eq. (16)]. A specific antiozonant protection of saturated polymers (polyolefins) by HAS has been recently suggested (17).

$$PH + O_3 \longrightarrow POO^{\bullet} + HO^{\bullet} \tag{12}$$

$$PH + O_3 \longrightarrow PO^{\bullet} + HOO^{\bullet} \tag{13}$$

$$POOH + O_3 \longrightarrow POO^{\bullet} + HO^{\bullet} + O_2 \tag{14}$$

$$POO^{\bullet} + O_3 \longrightarrow PO^{\bullet} + O_2 \tag{15}$$

$$O_3 \xrightarrow[\; NO_2 \;]{h\nu < 430\ nm} O_2 + O \tag{16}$$

$$PH + O \longrightarrow P^{\bullet} + HO^{\bullet} \tag{17}$$

Scheme 4

Nitrogen dioxide, another oxidizing anthropogenic tropospheric pollutant, generates free radicals from hydrocarbon polymers by H-abstraction [Eq. (18)], addition to C=C unsaturation [Eq. (19)], addition to macroalkyls and subsequent photolysis of the respective nitrite [Eq. (20)], and by atomic oxygen [see Eq. (17)] arising by photolysis of nitrogen dioxide (16) [Eq. (21); Scheme 5]. Phenolic antioxidants react with nitrogen dioxide ("gas fading") and are deactivated in this way in phenol-doped polyolefins (see Sec.IV.A.1).

$$PH + NO_2 \longrightarrow P^{\bullet} + HNO_2 \tag{18}$$

$$>C=C< \; + \; NO_2 \longrightarrow \; >\overset{\displaystyle}{C}-\underset{\displaystyle NO_2}{C}< \tag{19}$$

$$P^{\cdot} + NO_2 \longrightarrow \left[PONO \right] \longrightarrow PO^{\cdot} + NO \qquad (20)$$

$$NO_2 \xrightarrow{ h\nu < 430 \text{ nm} } NO + O \qquad (21)$$

Scheme 5

Irreversible chemical changes arising from degradation are localized in the polymer matrix. Heterogeneities caused by degradation were detected at molecular and supermolecular levels (18,19). Sites with highly oxygenated structures are randomly distributed in the polymer bulk and surface layers, and they are accompanied by changes in the spherulitic structure and chemocrystallization. It follows from that that the stabilizers preferentially react in the localized oxygenated sites.

According to their inherent chemical activity, individual groups of stabilizers interfere with chemical and physical degradation mechanisms. Scheme 2 exemplifies sites of the effective application of different classes of stabilizers in protection of (photo)oxidizing polymers. According to their structure, stabilizers may involve more than a single activity. It is understandable that a combination of different stabilizers is beneficial in providing a complex protection (see Sec. V).

III. STABILIZERS FOR DEGRADING POLYMERS

Stabilizers have been conventionally classified according to their principal activity mechanisms into following groups:

> Antioxidants
> Metal deactivators
> Antiozonants
> Photoantioxidants
> Photostabilizers
> Heat stabilizers for PVC

Additives with different chemical structures have been involved within these groups. Some stabilizers contribute to the polymer protection by more than one mechanism. Mostly, the inherent chemical efficiency of both the parent stabilizer and its transformation products have been involved. Because of the complexity of deteriogens attacking polymers during their lifetime, a complex protection has to be achieved by a proper combination of two or more stabilizers differing in activity mechanism (see Sec. V). Some bifunctional stabilizers provide beneficial protection by combining intramolecularly different active moieties (see Sec. VI).

Structures of present commercial stabilizers reflect the results of the research and intensive commercial exploitation of factors influencing the inherent chemical efficiency. The market offers more than 200 additives. Many of them have identical structures and have been produced under different trade names. During the last decade, only seldom have fundamentally new structures been commercialized. Most newly introduced compounds are characterized by structural modifications improving the application spectrum of known stabilizers. They have enhanced activity at lower application levels, at increased environmental aggressivity, and with suppressed unwanted side effects. The variety of additives used depends on the product service demands. No new commodity nor engineering polymers have been introduced onto the market. The consumers expect, however, a higher performance and better appearance. New problems arise by stabilization of multiphase systems or recycled, aged polymers (20).

The growth of the stabilizer market over the last 10 years has basically reflected the consumption of plastics, coatings, and rubbers, and the increased pressure on their performance (21,22). It was reported (23) that the annual growth of plastic additives in the United States is 4% through 1996. Data estimating the consumption of principal structural types of stabilizers are indicative. Antioxidants for thermoplastics consist of 56% phenolics (this includes phenolic MD), 32% organophosphites, 9% thioethers, and 3% others (21). The consumption of photoantioxidants and photostabilizers increases. According to the chemical structures, HAS dominate, with 44% (their popularity was enhanced by novel applications as thermal stabilizers). UV absorbers that have the structures of 2-(2-hydroxyphenyl)benzotriazoles make up 29%, hydroxybenzophenones 18%, and other types of light stabilizers form only 9% (22). For thermal stabilization of PVC, the stabilizer package consists of 49% organotins, 27% calcium or zinc metallic soaps, and 18% cadmium- and lead-based compounds. The remaining 6% are formed by different costabilizers, mainly organic phosphites and epoxy compounds (23). Antioxidant and antiozonant additives for rubbers consist mostly of aromatic and heterocyclic amines: N,N'-disubstituted 1,4-phenylenediamines dominate with 50%; derivatives of 1,4-dihydroquinoline makeup 10%, substituted diphenylamines and their condensates 6%, and organophosphites and phenolics 13% each. Derivatives of urea, thiourea, and metal dithiocarbamates form 8% (24).

For an effective commercial stabilization, the high efficiency at appropriate concentration and acceptable price have been required. This involves optimum relations between inherent chemical efficiency and physical properties, such as solubility in the host polymer, and resistance against volatility and leaching; high inherent thermal, chemical, and light stability; absence of prodegradant and discoloring effects on the protected polymer; good cooperation with other additives or compounding ingredients; and handling safety and toxicity related to legislative requirements. All these aspects have to be taken into consideration.

Examples of characteristic structures of modern stabilizers are listed in this section. The structures were selected relative to the importance and mechanistic features discussed in Section IV.

A. Antioxidants

Antioxidants are added to oxidizing polymers to prolong their useful lifetime. Polymer materials react with oxygen either in purely thermal processes (usually at elevated temperatures) or with the assistance of UV solar radiation (4,10). Two groups of antioxidants have been classified (25): chain-breaking, or primary antioxidants, limit the rate of the chain propagation step by trapping oxygen- or carbon-centered free radicals; and hydroperoxide-decomposing, or secondary antioxidants, interfere with the chain initiation and transfer steps arising from hydroperoxide homolysis.

1. Chain-Breaking Antioxidants

Additives described in this part principally protect against thermal oxidation when used as sole additives. Their efficiency decreases mostly at temperatures higher than 150°C and in UV light-initiated oxidation characterized by short kinetic chains. In a combination with light stabilizers, chain-breaking antioxidants also protect polymers against atmospheric aging. Additives with specific properties of photoantioxidants are described in Section III.D.

Hindered phenols, aromatic and nonhindered heterocyclic amines, aliphatic hydroxylamines, and transformation products of HAS that have structures of *O*-alkylhydroxylamines or hydroxylamines, are effective scavengers of alkylperoxyls.

a. Phenolic Antioxidants. A substantial number of hindered phenols have been used for stabilization of polyolefins and styrene-based polymers for which their nondiscoloring properties are advantageous. Relatively small amounts are designed for application in aliphatic polyamides and rubbers. Usual concentrations in polymers are in the range of 0.025–0.25% (9).

Mononuclear **11**, various binuclear **12–14**, trinuclear **15, 16**, and tetranuclear phenols **17, 18** are effective antioxidants. Phenolic phosphonate **19** (M = Ca) and oligomeric phenols **20** rank among stabilizers with strongly enhanced physical persistence. Esters of 4-hydroxybenzoic acid **21** (R = octadecyl, 2,4-di-*tert*-butylphenyl) have been considered as phenols with enhanced inherent photo-stability.

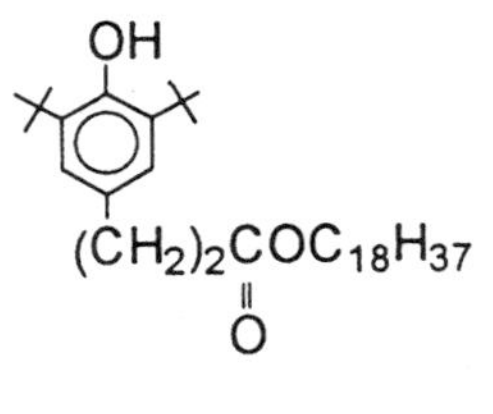

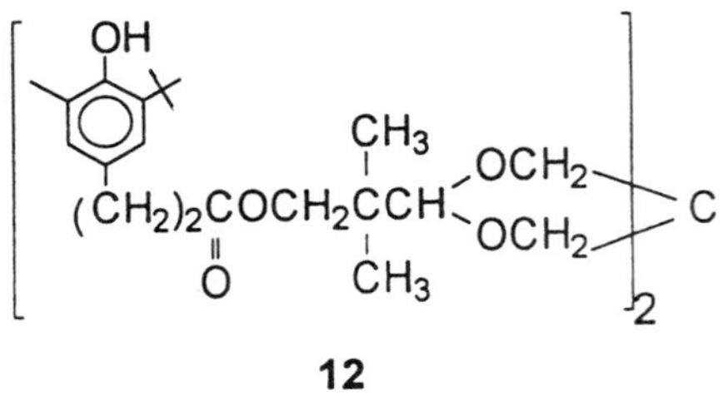

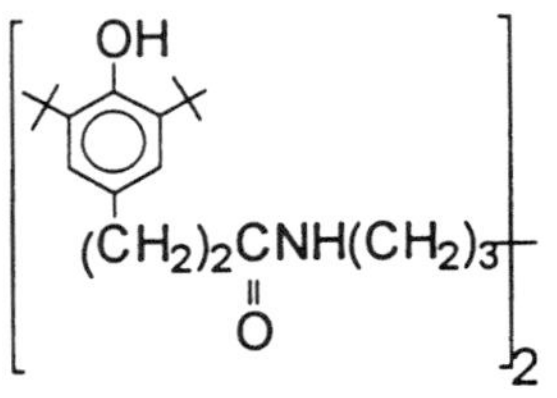

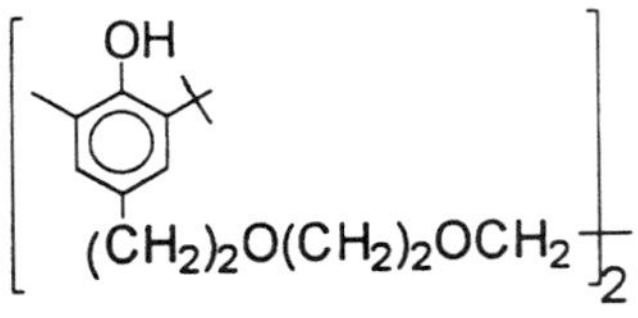

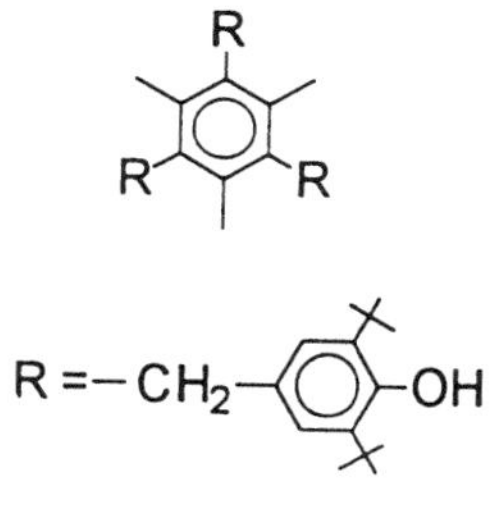

$$R = -CH_2-\!\!\raisebox{0pt}{\fbox{}}\!\!-OH$$

15

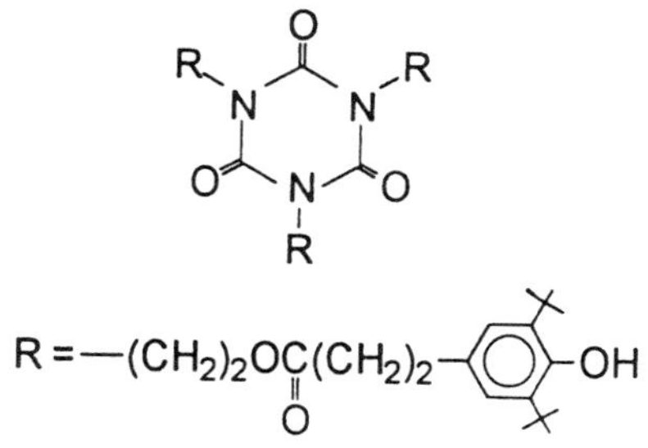

$$R = -(CH_2)_2OC(CH_2)_2-\!\!\raisebox{0pt}{\fbox{}}\!\!-OH$$

16

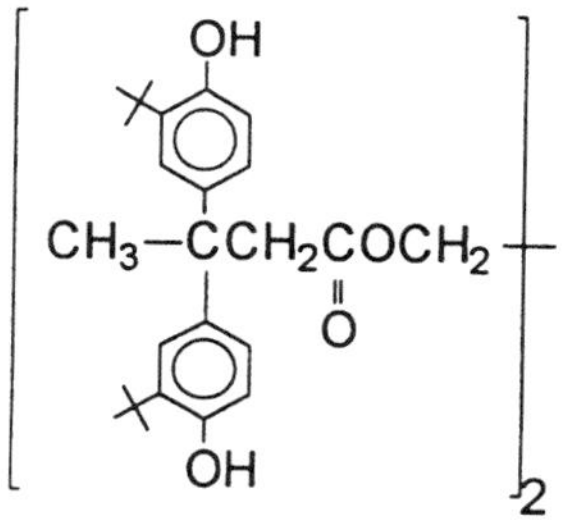

$$CH_3-CCH_2COCH_2$$

17

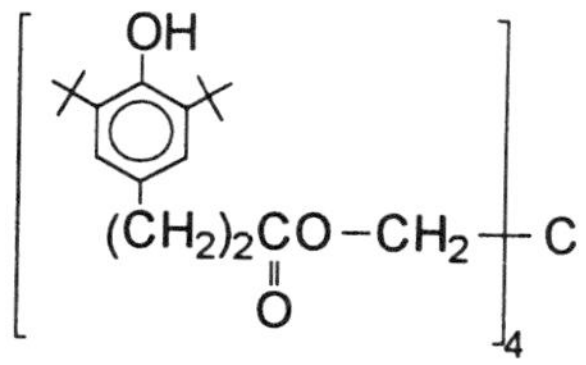

$$(CH_2)_2CO-CH_2$$

18

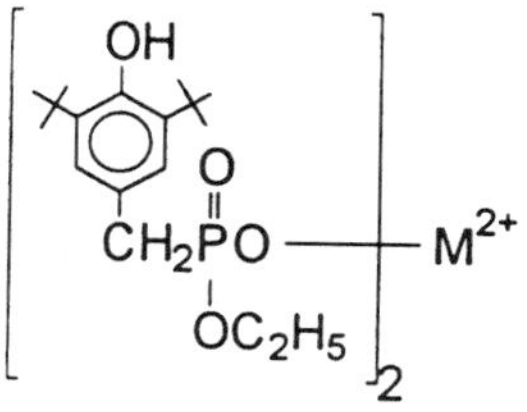

$$CH_2PO-\!\!-M^{2+}$$
$$OC_2H_5$$

19

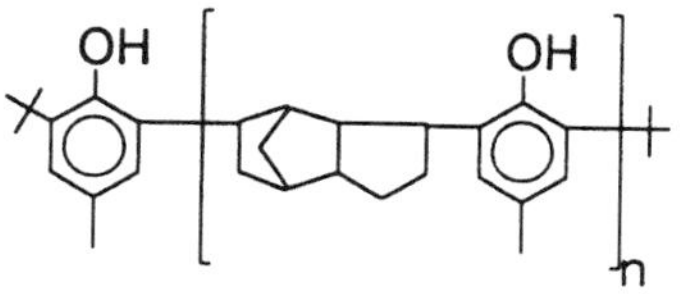

20

$$O=COR$$

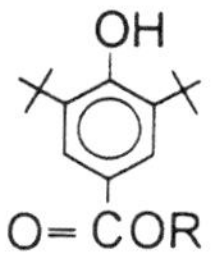

21

b.   Aromatic and Heterocyclic Amines.   Amine additives are preferentially used for the stabilization of rubbers (15). They are not usually used in plastics in which intensive discoloration, owing to formation of conjugated transformation products, is considered an undesirable property. Typical representatives are 4,4′-disubstituted diphenylamines **22a,b**, *N,N′*-disubstituted 1,4-phenylenediamines **23a–c**, 6-substituted-2,2,4-trimethyl-1,2-dihydroquinoline **24**, or 2,2-disubstituted indol-3(2*H*)-one or its phenylimino derivative **25**. Additives **26** and **27** are oligomeric condensates of aniline or diphenylamine with oxo compounds.

c.   Hydroxylamines.   Aliphatic hydroxylamines **28** and substituted dibenzyl-hydroxylamines **29** represent the developing class of chain-breaking antioxidants for polyolefins and diene-based polymers (26).

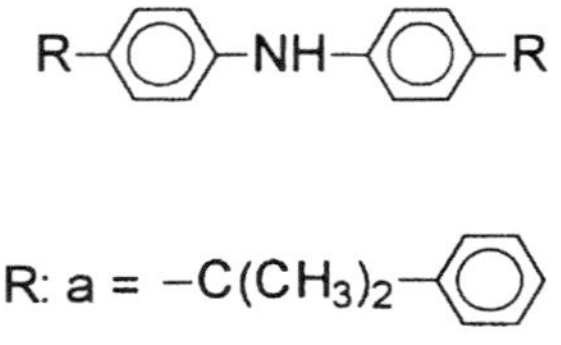

R—◯—NH—◯—R R^1NH—◯—NHR^2

R: a = $-C(CH_3)_2$—◯

b = 2,2,4,4 - tetramethylbutyl

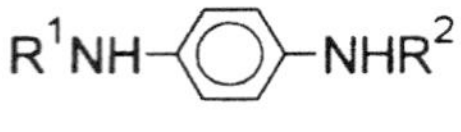

a: $R^1 = R^2$ = sec. alkyl
b: R^1 = sec. alkyl, R^2 = phenyl
c: $R^1 = R^2$ = phenyl

22   **23**

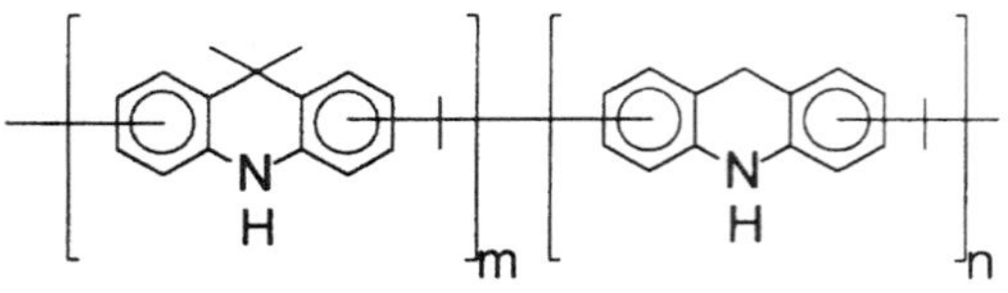

C_2H_5O—quinoline—N(H), CH_3

X = O, N—◯

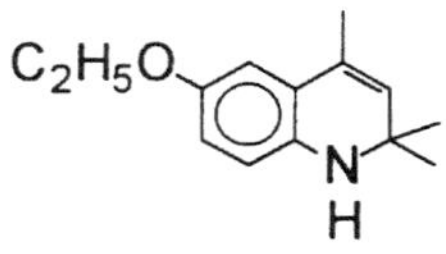

24 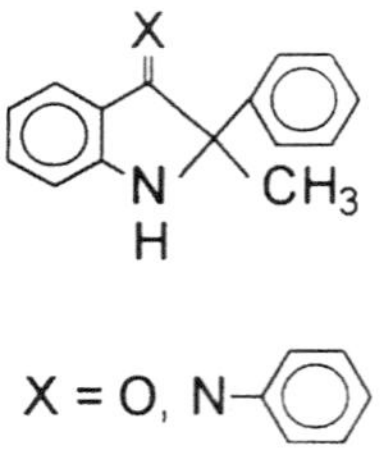   **25**

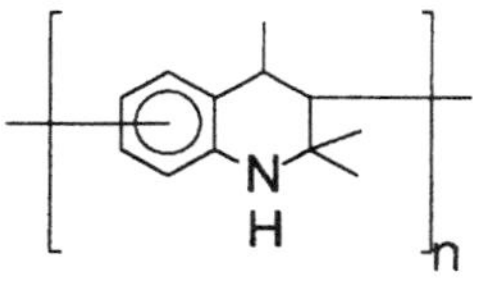

polymer **26**

polymer **27**

$(C_{12}H_{25})_2NOH$

$\left[(C_4H_9CHCH_2)_2NC\text{—}◯\text{—}CH_2 \right]_2 NOH$
  with C_2H_5 and O

28   **29**

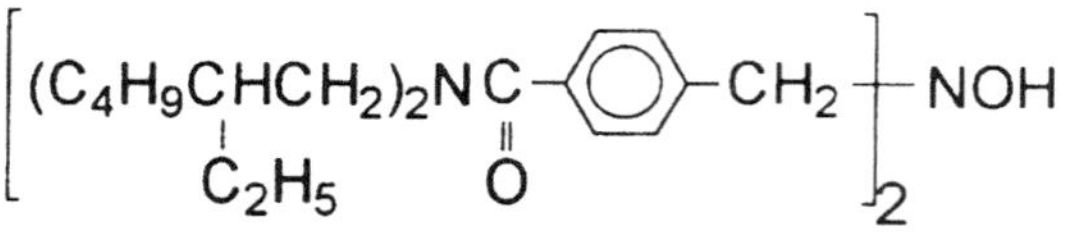

Transformation products of HAS, having structures of *O*-alkylhydro-xylamines or hydroxylamines, are very effective scavengers of alkylperoxy radicals and are mentioned in Section IV.E.3.

d. Scavengers of Carbon-Centered Radicals. Independently of the initiation mode, macroalkyls are formed from carbon-chain polymers or polymers containing segments with C–C bonds. Under oxygen-deficient conditions, the alkyls undergo self-termination reactions [see Eq. (3)] (4). In the presence of oxygen, a fast oxidation to peroxyls takes place [see Eq. (4)]. Alkyl scavengers have to compete with both reactions. It is evident that they may be effective only in oxygen-deficient environment, such as in melt processing. At present, effective nondiscoloring, nontoxic, and nonvolatile compounds that act exclusively as alkyl scavengers in polymers are rare. With the exception of the monoacrylate of 2,2′-methylenebis(4-methyl-6-*tert*-butylphenol) **30** (27) and recently announced derivative of benzo[*b*]furan-2(3*H*)-one **31** (28), none of commercially available stabilizers in their original chemical form are able to scavenge alkyls.

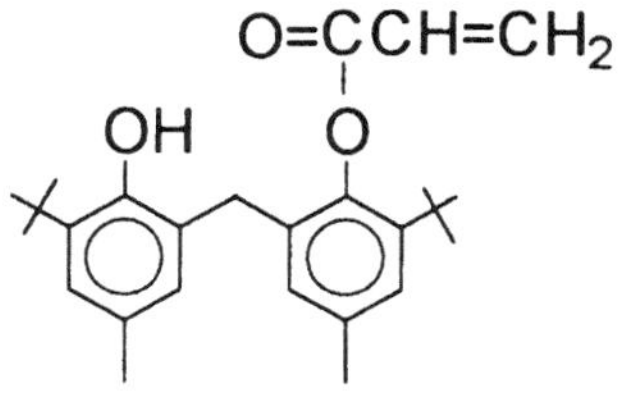

30 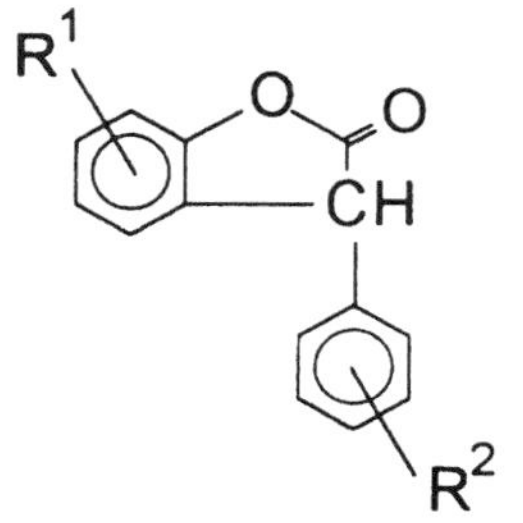   **31**

Phenol **30** is an effective processing stabilizer for unsaturated diene-based rubbers and thermoplastic elastomers (27). Benzo[*b*]furan-2(3*H*)-one **31** was declared as a nondiscoloring processing stabilizer for polyolefins (28).

Some transformation products of conventional chain-breaking antioxidants formed as a consequence of reactions with alkylperoxyls (aromatic aminyls, nitroxides, nitrones, and quinone imines; see Sec. IV.A.2) and of HAS (nitroxides; see Sec. IV.E.3) are able to trap alkyls.

2. Hydroperoxide-Decomposing Antioxidants

Hydroperoxides are the primary molecular oxidation products in thermal and light-induced oxidation and are harmful photoinitiators in polymer photodegradation. Reduction of hydroperoxides with antioxidants to alcohols or deactivation by HAS suppresses homolysis and, consequently, chain initiation and transfer (25). Two chemical classes serve as hydroperoxide-decomposing antioxidants: organic compounds of sulfur (see Sec. III.A.2.a) and compounds of trivalent phosphorus (see Sec. III.A.2.b). Both classes are conventionally used in combinations with phenolic antioxidants (29,30) (see Sec. V). A new development deals with combinations of phosphites with HAS (31).

a. Activated Compounds of Sulfur. Sulfur-containing stabilizers are used as thiosynergists with phenolic antioxidants in long-term heat stabilization of polyolefins, in particular polypropylene (9,29). Conventionally, they are used in excess to phenolics (the ratio is 1 : 3 to 1 : 5). The total concentration of both stabilizers does not surpass 0.75%. Thioethers **32** having R $= C_{12}-C_{18}$ are the most common thiosynergists. Aliphatic disulfide **33**, physically persistent thioether **34**,

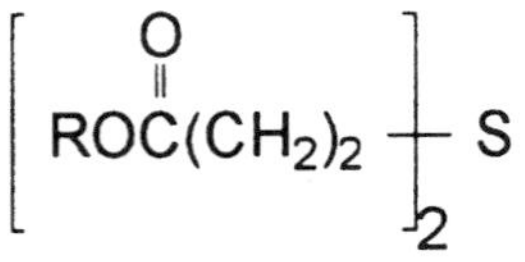

$$\left[\ \underset{\displaystyle 2}{\overset{\displaystyle O\ \ \ \ }{ROC(CH_2)_2}}\ \right] S$$

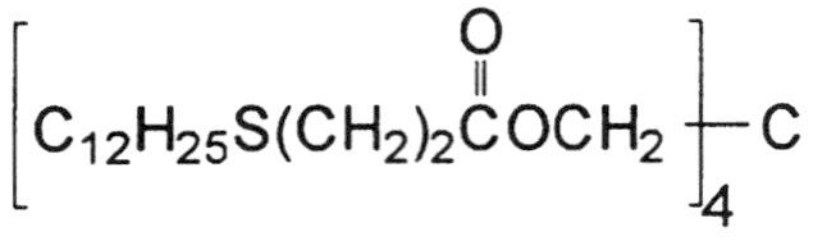

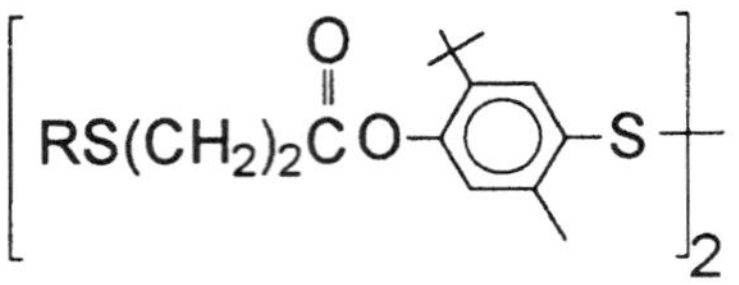

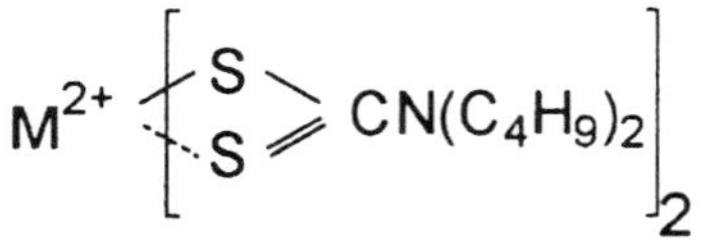

32

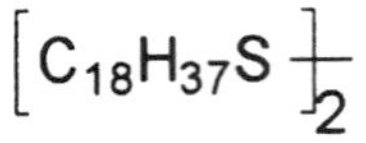

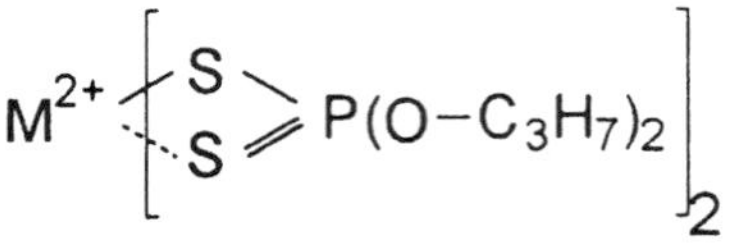

$$\left[\ C_{18}H_{37}S\ \right]_2$$

33

$$\left[\ \underset{\displaystyle 4}{\overset{\displaystyle O}{C_{12}H_{25}S(CH_2)_2COCH_2}}\ \right]C$$

34

$$\left[\ \overset{\displaystyle O}{RS(CH_2)_2C}\!-\!O\!-\!\bigcirc\!-\!S\ \right]_2$$

35

$$M^{2+}\left[\ \overset{S}{\underset{S}{>}}CN(C_4H_9)_2\ \right]_2$$

36

$$M^{2+}\left[\ \overset{S}{\underset{S}{>}}P(O\!-\!C_3H_7)_2\ \right]_2$$

37

$$M^{2+}\left[\ \overset{S}{\underset{S}{>}}COC_4H_9\ \right]_2$$

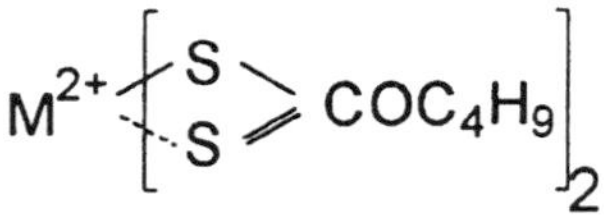

38

additive **35** combining two different sulfidic moieties, and metal thiolates (M = Zn), such as dialkyldithiocarbamate **36**, dialkyldithiophosphate **37**, or xanthate **38**, are other examples.

b. Organic Compounds of Trivalent Phosphorus. Organic phosphites and phosphonites attract attention for commercial development owing to their broad, structure-dependent activity spectrum. Aliphatic phosphites are represented by **39**. The major current use of aromatic hindered phosphites **40** (R = 2,4-di-*tert*-butylphenyl) and **44** and phosphonite **46** is in melt stabilization of polyolefins (where the phosphites are used in combinations with phenolic antioxidants; see Sec. V.A) and poly(ethylene terephthalate) (9). Phosphites minimize chain scission and color development during processing of polyolefins. Combinations of phosphites with HAS provide promising contribution to long-term thermal stability of polyolefins (31). Hindered aromatic phosphites chelate *trans*-esterification catalysts in poly(ethylene terephthalate) and prevent its discoloration. Application levels of phosphites are between 0.05 and 0.25%.

Aromatic nonhindered phosphite **43** is an effective storage stabilizer in unsaturated elastomers (9). Aliphatic **39** and mixed aliphatic–aromatic phosphites **41**, **42** are costabilizers and metal-complexing additives in PVC (32). Cyclic dioxaphosphocin **45**, containing an amine moiety, and fluorophosphonite **47** resulted from development of phosphorus stabilizers having a higher inherent resistance to hydrolysis.

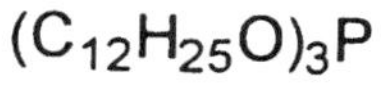

$$(C_{12}H_{25}O)_3P$$

39

$$\left[ROP \underset{OCH_2}{\overset{OCH_2}{<}} \right]_2 C$$

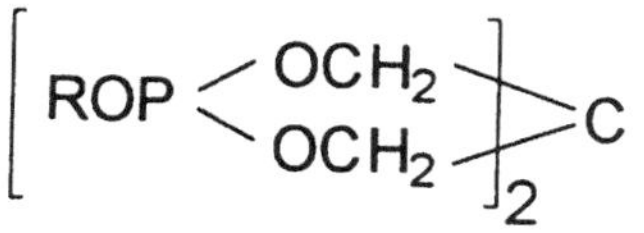

40

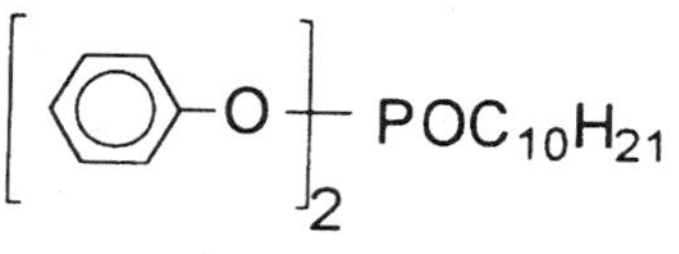

41

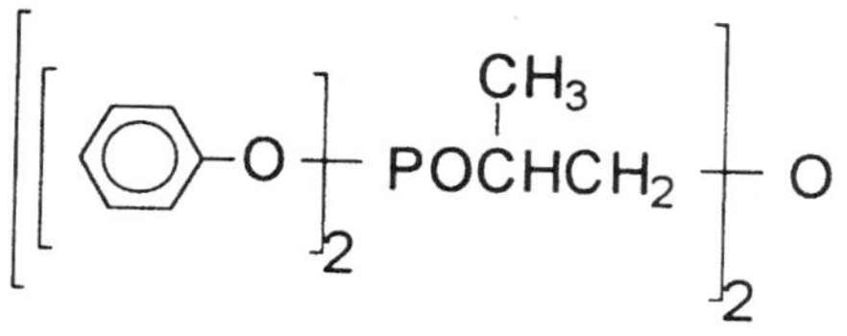

42

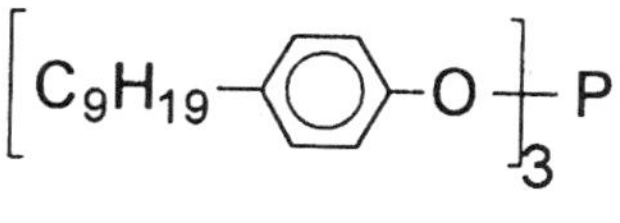

43

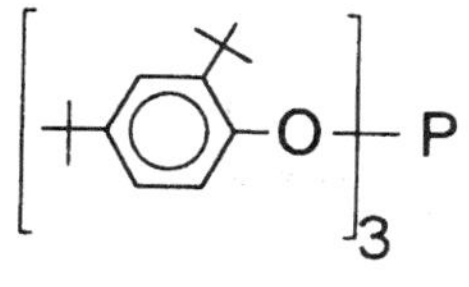

44

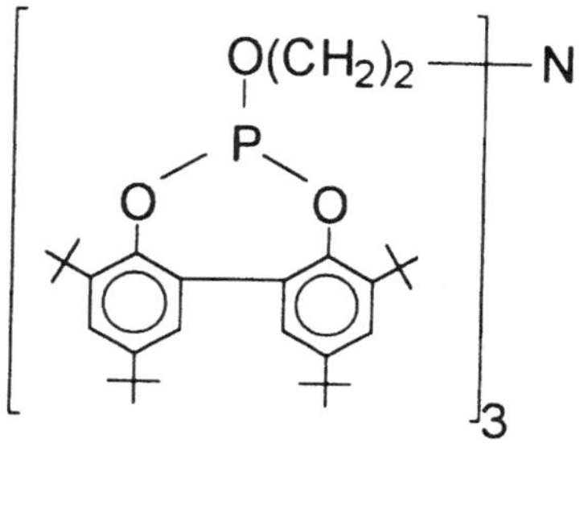

45

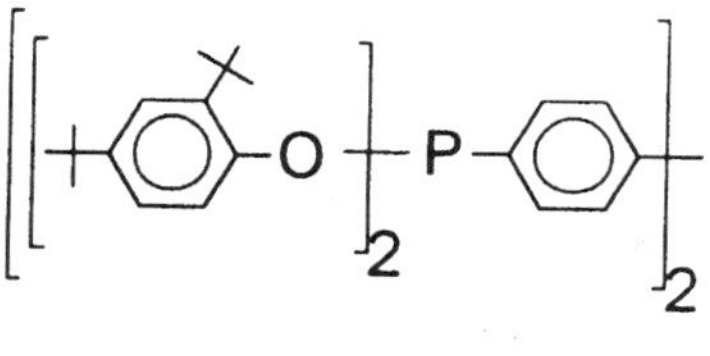

46

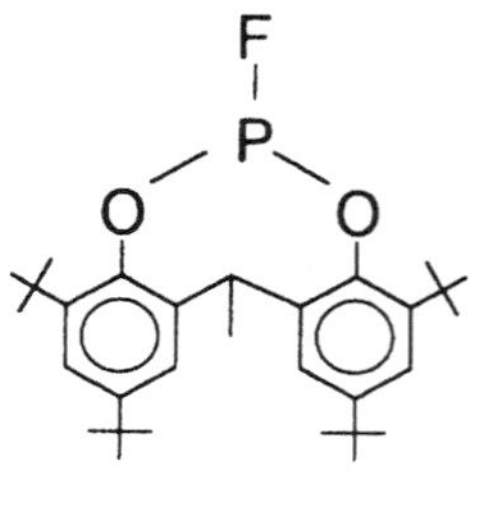

47

B. Metal Deactivators

Thermoplastic polymers very often come deliberately or inadvertently into contact with a variety of metallic compounds. This occurs during manufacture (with polyolefin polymerization catalysts), compounding with contaminated fillers and reinforcing agents, or during the use of polymers as insulating materials for wire and cables with copper conductors (33,34). The interactions between metallic prooxidants and polymers are complex and result in accelerated aging of polymers. Conventional phenolic antioxidants are ineffective in providing a long-term stability in the presence of copper. Metal deactivators reduce the deleterious effects of metal ions, copper in particular. The effective metal deactivators **48–51** contain phenolic

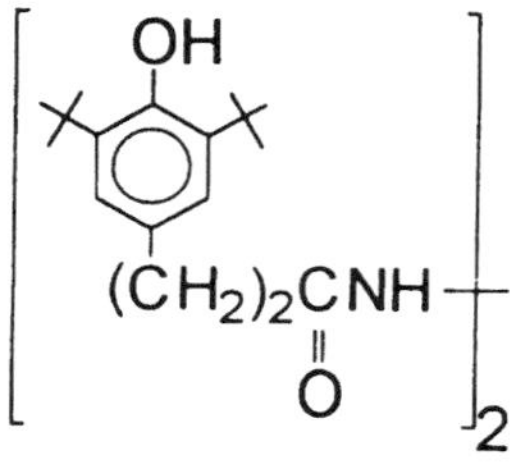

48

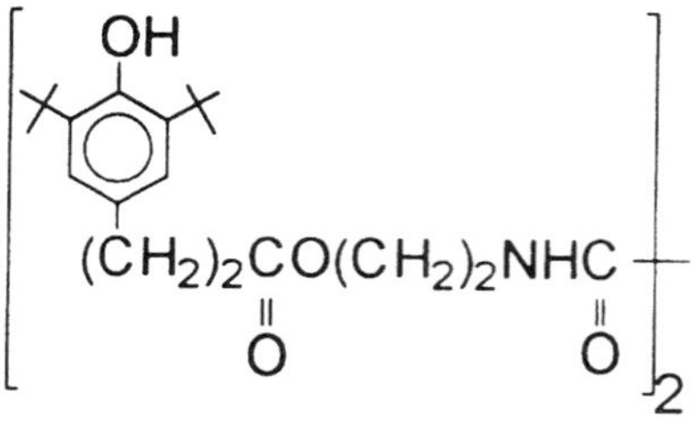

49

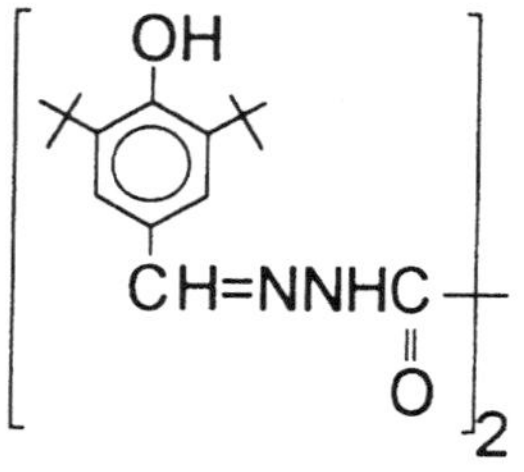

50

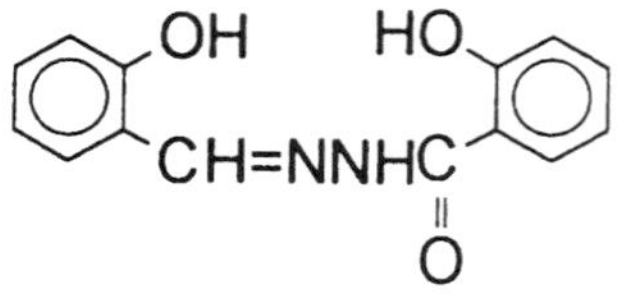

51

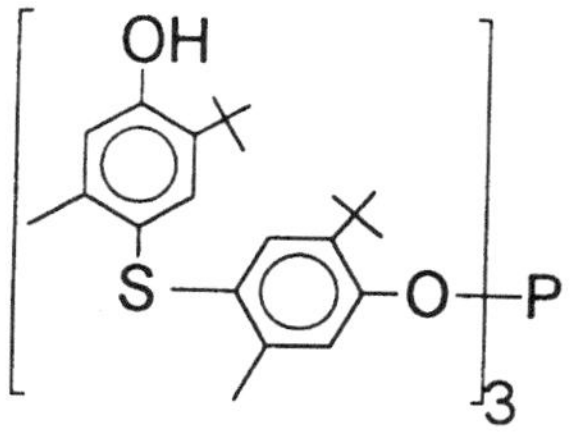

52

moiety and nitrogen functions or a combination of phenolic sulfide and phosphite moieties in **52**. Most metal deactivators should be considered as bifunctional stabilizers.

C. Antiozonants

The antiozonant group of antidegradants has been developed to chemically protect unsaturated rubbers against ozone attack (15,35). Physical protection of rubbers by waxes is not effective under dynamic ozone aging. Because of the complicated ozonation mechanism of rubbers, the selection of additives protecting them against ozone attack and reactions with rubber ozonation products are rather limited. Antiozonants are also expected to act as chain-breaking antioxidants. To provide an effective protection, antiozonants have to be used in concentrations of 2–3%.

Most antiozonants used have structures of *N,N'*-di-*sec*-alkyl- (see **23a**) or *N-sec*-alkyl-*N'*-phenyl-1,4-phenylenediamines (see **23b**). These additives provide a complex antioxidant, antiflex–crack and antiozonant protection (15,35). A weaker effect has been reached by trimethyl-1,2-dihydroquinolines (see **24** and **26**). Other chemical antiozonants include nickel(II) dibutyldithiocarbamate (see **36**; M = Ni) and substituted isothiourea **53**. Recently, a class of nondiscoloring antiozonants, based on acetals of pentaerythritol and 1,2,5,6-tetrahydrobenzaldehyde **54**, was developed (36). Only **23** (M = Ni) has an antiozonant activity comparable with diamines **23**. However, it causes green discoloration, and the presence of nickel brings about ecological problems. Isothiourea **53** accelerates cure, and acts as a scorch promoter. Acetal **54** does not have an antioxidant effect.

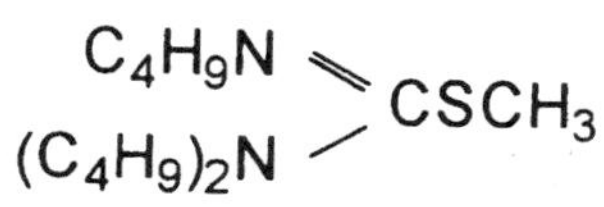

53

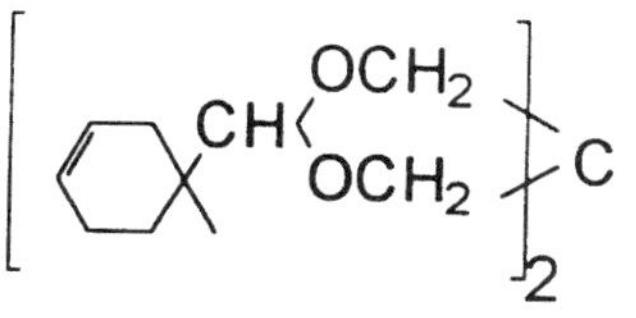

54

D. Photoantioxidants

Only stabilizers that have a high inherent photostability function effectively as deactivators of alkylperoxyls and hydroperoxides in photooxidizing polymers. The selection of suitable additives is rather limited. Conventional phenolic antioxidants generally suffer from a low photostability. Aliphatic or aromatic hydroxybenzoate **21** can be used as photostable radical scavenger (37). Metal dialkyldithiocarbamate **36** or dialkyldithiophosphate **37** (M = Ni) have been listed among photostable hydroperoxide-decomposing photoantioxidants (37,38).

Fortunately, the photoantioxidant problem was solved by invention and development of HAS. Introduction of this class of additives was really a revolution in polymer stabilization. Today, the structures result from an intensive development performed by principal companies that produce stabilizers (15,22). Recent optimization of their inherent chemical efficiency and physical persistence substantially broadened their application in plastics and coatings. Common application levels of HAS range from 0.1 to 1% (10).

Hundreds of research papers have tried to explain their intricate activity mechanism (see Sec. IV.E.3). Examples indicate diversity in structures of HAS. Most

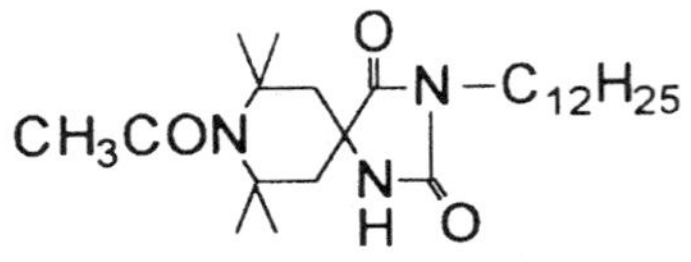

55

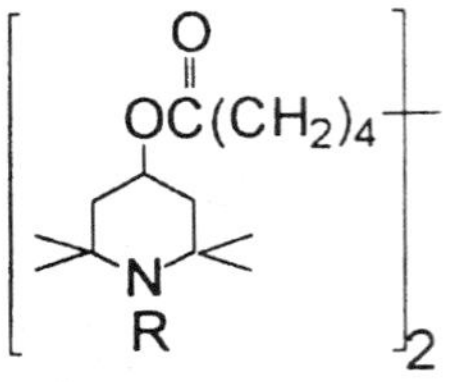

56

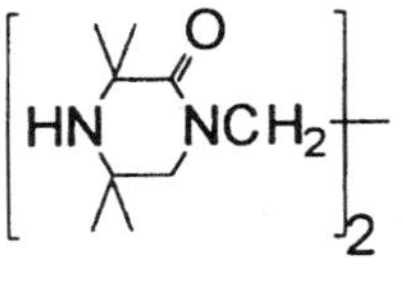

57

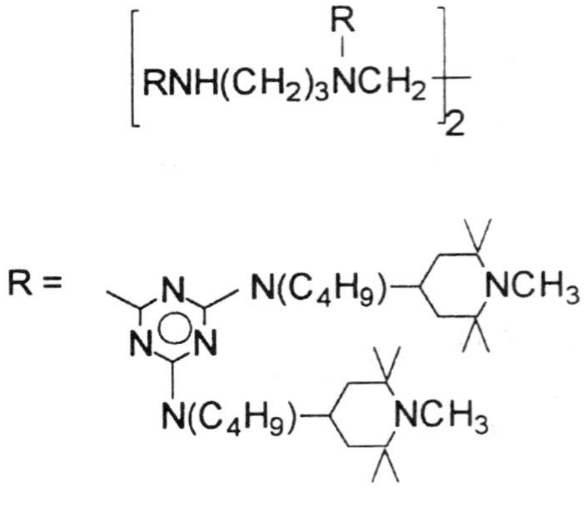

58

commercialized additives contain 2,2,6,6-tetramethylpiperidine moieties. Piperazinone **57** is an exception. The HAS differ in molecular weights, as exemplified by mononuclear **55**, binuclear **56** and **57**, and polynuclear HAS **58**. Physical persistence has been solved by introduction of different oligomeric HAS **59–63** (*tert*-octyl means 2,2,4,4-tetramethylbutyl in **59** and throughout the text). Derivatives of hindered oxazolidine, 4-imidazolidone, piperazine, hexahydro-pyrimidine, or decahydroquinoxalinone have protecting effects comparable with hindered piperidines (15).

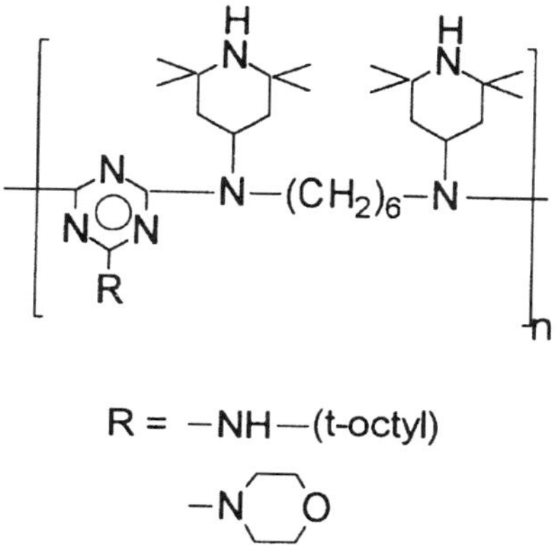

59

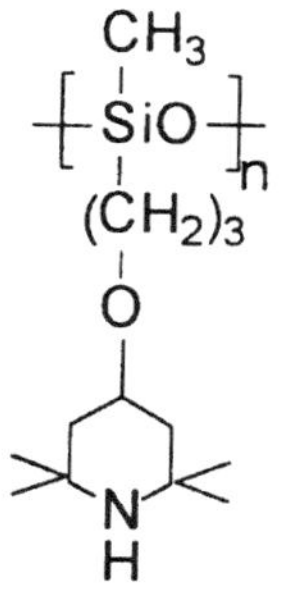

60

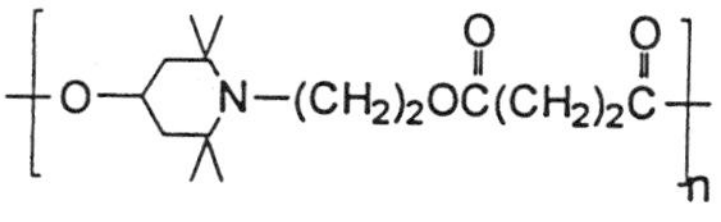

61

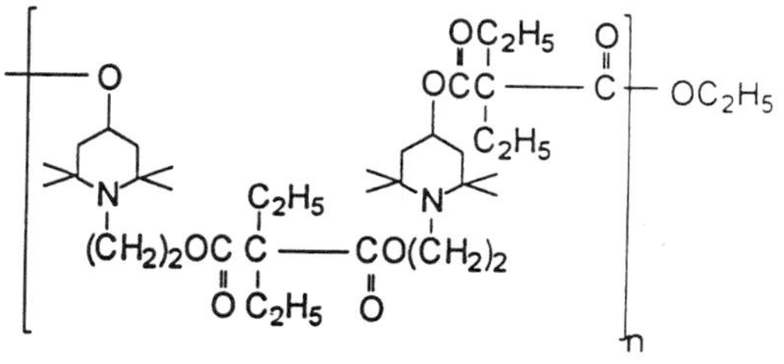

62

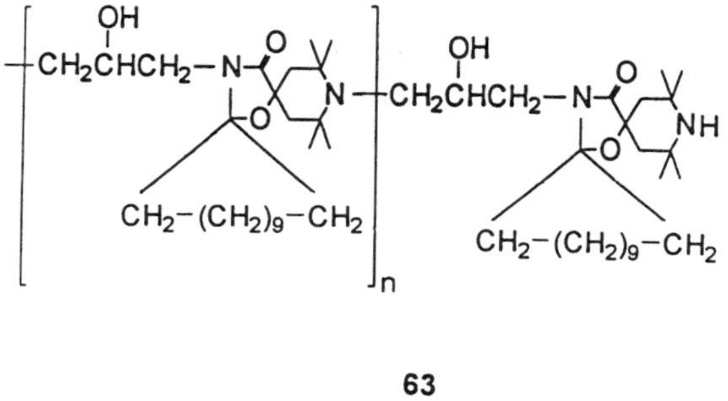

63

The success of HAS photostabilizers in numerous commodity and engineering polymers is impressive. HAS with secondary > NH and tertiary > NR moieties have been universally used. HAS having lower pK_a values (**55** or **56**; R = O-iso-C_8H_{17} ; isooctyl means 2-methylheptyl in **56** and throughout the text) may be also used in polymeric environment containing acid components or contaminants (15). The optimum effectiveness in polymer protection has been obtained using combinations of HAS with phenolic antioxidants, organic phosphites, and UV absorbers. Combinations of HAS with phosphites are expected to substitute a part of phenolic antioxidants (31).

E. Photostabilizers

Most commodity polymers do not strongly absorb the UV spectrum of the terrestrial sunlight. The presence of trace amounts of light-absorbing polymer-bound or nonpolymeric impurities is sufficient to trigger photodegradation during atmospheric aging (1,20). The concentration of chromophoric impurities increases during the long-term application of polymers.

Most polymers designed for outdoor applications need to be stabilized by additives interfering with chemical and physical processes induced by solar radiation. Structures of photostabilizers have to be adjusted to the nature of the polymer, with specific applications considered, and have to possess high inherent photostability (39). In this chapter, we concentrate on organic additives and do not consider carbon black and other photoprotecting pigments discussed recently (11,37).

Additives protect polymers by preferential absorption of the harmful solar radiation and by its dissipation as heat (UV absorbers; see Sec. III.E.1), and stabilizers capable of taking over the energy from photoexcited carbonyl chromophores and singlet oxygen and disposing of it effectively as heat or as fluorescent or phosphorescent radiations (quenchers; see Sec. III.E.2) are, besides being photoantioxidants, in particular HAS (see Sec. III.D), the principal classes of photostabilizers.

1. UV Absorbers

Commodity and engineering plastics and coatings are protected against actinic solar radiation by compounds that absorb the harmful radiation preferentially to polymers (10,22,37). Light screening, quenching of excited states, and scavenging of alkylperoxyls, to some extent, contribute to the integral activity of UV absorbers (40). Radical scavenging is characteristic of UV absorbers containing a phenolic moiety (30). An ideal UV absorbers would be expected to absorb all UV solar radiation, but no visible light (10); usually a compromise has to be made. To achieve maximum UV absorption, a slight yellow discoloration of UV absorbers may be tolerated (23). Various classes of additives fulfill the commercial expectations for a long-term performance (37) and are commercialized by many suppliers: 4-substituted-2-hydroxybenzophenone **64**, various 2-(2-hydroxyphenyl)benzotriazoles **65–68**, 2-(2-hydroxyphenyl)-1,3,5-triazine **69**, oxanilide **70**, and 2-cyanoacrylate **71** or salicylate **72** esters.

Generally, relatively high concentrations of UV absorbers (0.25–5%) have to be used in commercial stabilization (10). This may limit (because of a low solubility

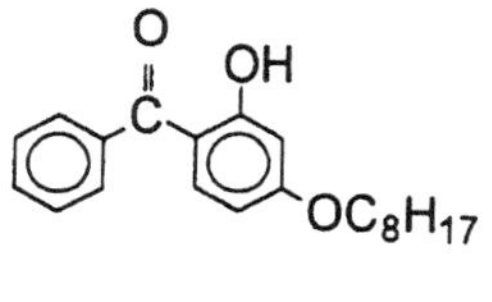

64

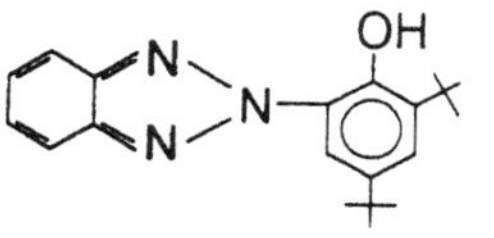

65

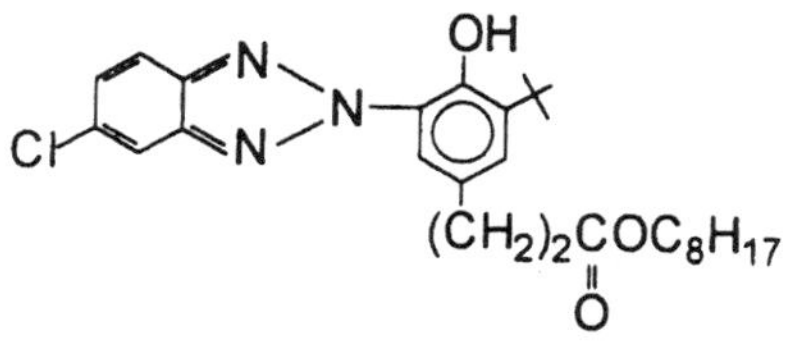

66

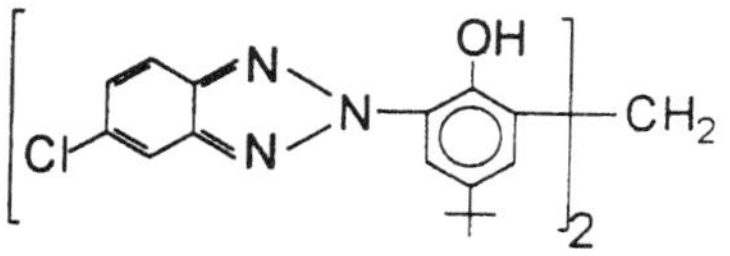

67

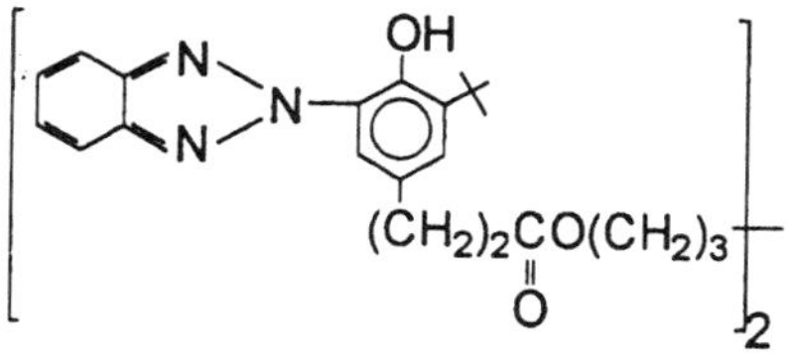

68

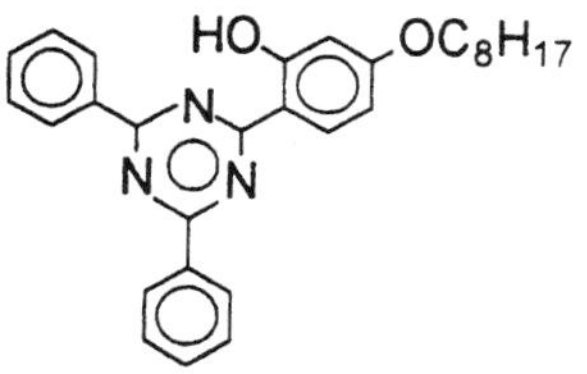

69

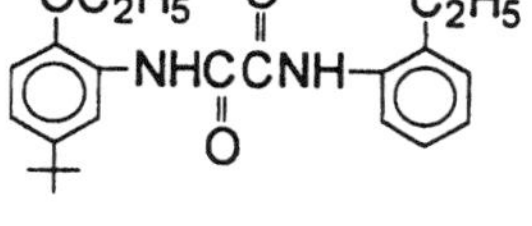

70

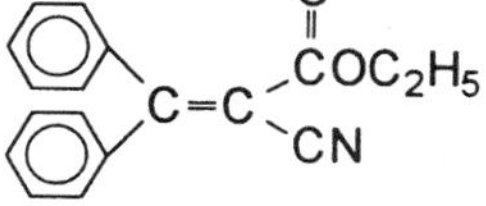

71

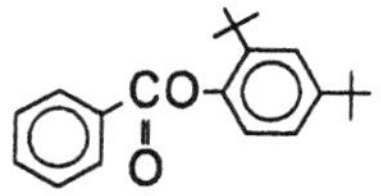

72

or poor compatibility) the effective application of UV absorbers in photosensitive polymers that have high surface/volume ratios, such as thin films or fibers, or in surface-protecting coatings. Necessary concentration levels can be achieved only with well-solubilized stabilizers; for example, with triazine **69** in coextruded polycarbonate sheets (41).

2. Quenchers

Nickel(II)-containing photostabilizers with the properties of quenchers of excited chromophores exploit the ability of the nickel atom to dissipate the excitation energy as heat through vibrations. Quenchers have been used almost exclusively in polyolefins (10,37). At common concentrations (0.2–0.5%), the nickel complexes impart green discoloration to plastic articles.

The effective quenchers are represented by nickel(II) ethyl (3,5-di-*tert*-butyl-4-hydroxybenzyl)phosphonate (see **19**; M = Ni), nickel(II) dibutyldithiocarbamate (see **36**; M = Ni), nickel(II) complex of 2,2'-thiobis(4-*tert*-octylphenol) with butylamine **73**, and nickel(II) bis(1-phenyl-3-methyl-4-decanoyl-5-pyrazolate) **74**.

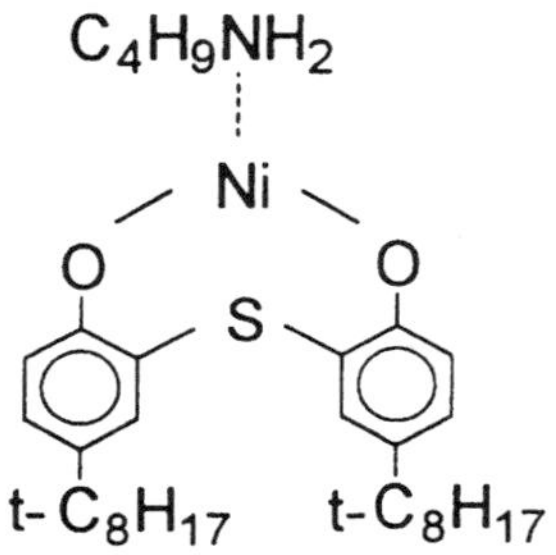

73 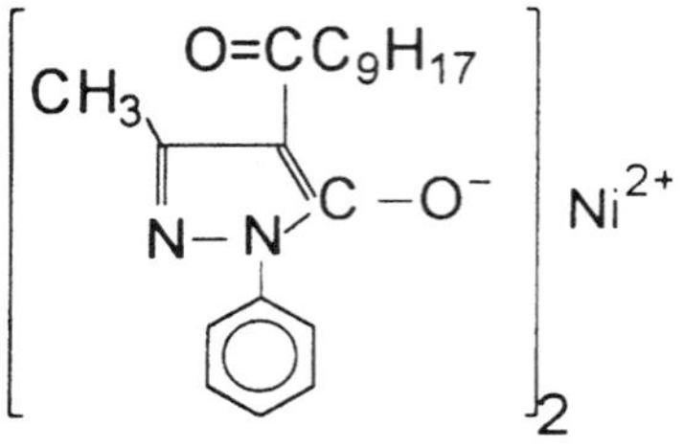   **74**

The presence of sulfur in the molecule of **73** enhances quenching potency of singlet oxygen. According to the recent analysis (40), some nickel-containing quenchers also possess properties of UV absorbers and hydroperoxide-decomposing antioxidants.

In commercial polymer stabilization, nickel-containing quenchers are used in combination with phenolics, phosphites and UV absorbers. Their application was reduced significantly after introduction of HAS. This reflects ecological expectations to reduce the environmental contamination by nickel salts.

F. Heat Stabilizers for Poly(vinyl chloride)

The heat stabilizer additives protect PVC during processing (such as molding or extrusion) and extend the useful lifetime of finished PVC products that are exposed to heat (23,32). Typical heat stabilizers consist of metal carboxylates **75** (i.e., barium, calcium, or zinc soaps of higher fatty acids C_{12}–C_{18}; mostly used as synergistic binary salts Ba/Zn or Ca/Zn) and organotin compounds **76–78** (R = alkyl).

$$(R\overset{\text{O}}{\overset{\|}{C}}O)_2M^{2+}$$

75

$$(C_4H_9)_2Sn(O\overset{\text{O}}{\overset{\|}{C}}CH=CH\overset{\text{O}}{\overset{\|}{C}}OR)_2$$

76

$$(C_8H_{17})_2Sn(SR)_2$$

77

$$(C_8H_{17})_2Sn(SCH_2\overset{\text{O}}{\overset{\|}{C}}O-i\text{-}C_8H_{17})_2$$

78

Cadmium soaps **75** should be eliminated for ecological reasons. In spite of a potential toxicity, different lead-containing salts (phosphites, carbonates, and stearates) still have a broad aplication in this field. Various costabilizers are usually used with heat stabilizers (see Sec. V.D).

IV. STABILIZER ACTIVITY AND CONSUMPTION

The present high-level performance of additives in polymer stabilization arises from the knowledge of factors governing inherent chemical efficiency of stabilizers (2,15,32–34,42–44) and physical relations in the system stabilizer–polymer matrix (45). Commercially available stabilizers guarantee the highest possible long-term performance under expected application conditions and, in a particular, polymer or polyblend. General trends include the use of more efficient stabilizers at lower concentrations to achieve a better performance and fewer side effects to meet increasing demands of end-use customers (23,24).

There is, however, no continuum in the stabilizer activity. Stabilizers are very reactive compounds in polymer degradation, and the stabilization process proceeds at the expense of the chemistry of the parent stabilizer structure. This has been categorized as *sacrificial consumption* of stabilizers, which means the sum of its chemical transformations resulting from effective protection of polymers (2). Model experiments using multiple extrusion tests revealed progressive transformation of processing stabilizers in polyolefins (4,46,47). The consumption of the original form of stabilizers continues during thermal aging and weathering of polymers. New chemical structures different from those originally added are formed from additives interfering with alkylperoxyls or hydroperoxides. The consumption is preferential at randomly distributed sites, with enhanced concentration of oxidizing species and in surface layers (18,19). The stabilizers remain unconsumed in the undamaged zones.

Besides the sacrificial consumption, the stabilizers are depleted by direct or sensitized photolysis (30,39,48–51), energetic radiation (49), oxidizing atmospheric pollutants (ozone and nitrogen dioxide) (52), residues of polymerization catalysts (53), various other metallic impurities (1), antagonistic effects of other additives (e.g., brominated flame retardants), mineral acids arising by hydrolysis of some hydroperoxide-decomposing antioxidants or their transformation products (3,15), or by physical adsorption on fillers. These processes have been categorized as *depleting* stabilizer consumption (1,2). Structures of new compounds formed from stabilizers by sacrificial and depleting consumptions are usually identical (3,15,49), and the transformation products accumulate stepwise in the aged polymer matrix.

Some particular transformation products formed from thiosynergists or HAS act as true stabilizers (their generation is a necessary condition of activity; (3,15). Some other products (e.g., conjugated cyclic dienones arising from phenolic and aminic antioxidants) only slightly contribute to the activity of the parent stabilizers (3,15,54).

Both sacrificial and depleting consumption of stabilizers ultimately account for reduction in concentration of the active form of the stabilizer, below such a level that it is no longer capable of protecting the degrading polymer that contains enhanced levels of prooxidants arising during aging (20). The residual stability and the failure time of the polymer are related to its oxidation level (55). Hence, the polymer failure is localized in initiating centers that have enhanced concentrations of oxidation products and, consequently, insufficient residual stabilizer concentration. The stabilizer consumption has also been reflected in insufficient stability of aged recycled plastics (20). Most products formed in the ultimate stage of stabilizer consumption that are accumulated in aged plastics intended for recycling are considered as nonpolymeric impurities (1). They discolor polymers, may corrode metals, or form insoluble precipitates during reprocessing. Data dealing with a potential migration of stabilizer transformation products into the extracting environment are still lacking.

A. Antioxidants Scavenging Free Radicals

Most degradation processes have a free radical character, independent of the initiation mode. This results in a continuous interest in free–radical-scavenging additives (15,21,42,49,56–58). Stabilizers that deactivate oxygen-centered free radicals (alkylperoxyls) are ever of interest. Stabilization by effective scavenging of carbon-centered free radicals (macroalkyls) remains an open problem.

1. Scavengers of Oxygen-Centered Free Radicals: Phenols

Depending on the environmental conditions during various stages of the polymer lifetime, hindered phenols act as melt (processing) stabilizers, antioxidants in thermal aging and, to some extent, as antioxidants in weathering (42,48,49). Commercially available phenolic antioxidants (see Sec. III.A.1.a) are, chemically, a most diversified group of stabilizers. Phenolic moieties also constitute an important part of molecules of metal deactivators (see Sec. III.B) and UV absorbers (see Sec. III.E.1). Scavenging of alkylperoxyls is the principal stabilization pathway for antioxidants and metal deactivators. However, this is the depleting step in UV absorbers (30).

The activity mechanism of phenolic antioxidants (AO) and properties of transformation products have been discussed from various aspects (42,48,49,54,58). The scavenging of alkylperoxyls by a hydrogen transfer from the phenolic hydroxy group is the principal pathway of the sacrificial consumption [Eq. (22a); Scheme 6]. Transition states characteristic of various degrees of charge separation, such as caged intramolecularly H-bonded species **79**, a loose π-complex **80**, and states with partial **81** or complete charge separation **82** have been envisaged (57,58). Their role has not been sufficiently appreciated in the stabilization mechanism, in particular in the light-induced reactions.

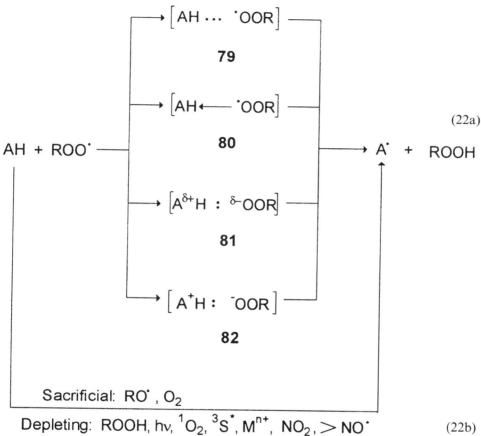

$$\text{AH} + \text{ROO}^\bullet \longrightarrow \left[\begin{array}{c} \left[\text{AH} \cdots {}^\bullet\text{OOR}\right] \\ \mathbf{79} \\ \left[\text{AH} \longleftarrow {}^\bullet\text{OOR}\right] \\ \mathbf{80} \\ \left[\text{A}^{\delta+}\text{H} : {}^{\delta-}\text{OOR}\right] \\ \mathbf{81} \\ \left[\text{A}^+\text{H} : {}^-\text{OOR}\right] \\ \mathbf{82} \end{array}\right] \longrightarrow \text{A}^\bullet + \text{ROOH} \qquad (22a)$$

Sacrificial: $RO^\bullet$, O_2

Depleting: ROOH, hv, $^1O_2$, $^3S^*$, M^{n+}, NO_2, $> NO^\bullet$ $\qquad$ (22b)

$$A^\bullet \longrightarrow \text{Products} \qquad (23)$$

Scheme 6

Reactivity with alkoxy radicals and high-temperature trapping of oxygen contribute to the sacrificial consumption of phenolics. Photoprocesses and oxidation with transition metal ions (M^{n+}), alkylhydroperoxides, nitrogen dioxide, and nitroxides developed from HAS [see Eq. (22b), Scheme 6] deplete the ability of phenols to scavenge alkylperoxyls (42,49). The reactivity of phenolics according to Scheme 6 is governed by substituent effects (42). Relatively long-living phenoxyls

A˙, primary free radical transformation products of AO, are formed independently of the generation pathways [see Eqs. (22a,b), Scheme 6] (56,58). Products formed from A˙, according to Eq. (23), play the key role in the chemistry and photochemistry of phenolic antioxidants (3,42,57,58). The significance of individual products depends on their relative concentration in the polymer matrix and on their reactivity under specific environmental conditions. This implies the influence of heterogeneity in oxidation of semicrystalline polymers on the antioxidant chemistry. At sites with high local concentrations of oxidizing agents and phenolics, bimolecular reactions are more favored. The reactivity of phenoxyls **83** or mesomeric cyclohexadienonyls **84a,b** was explained using model experiments (42,58). For

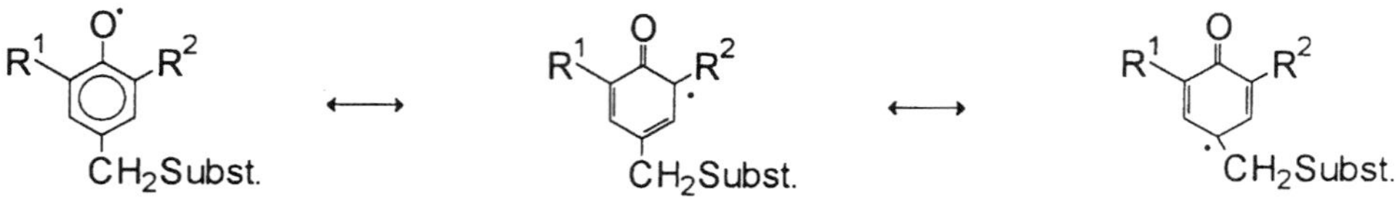

| 83 | 84a | 84b |

sterically hindered phenols ($R^1 = tert$-butyl, $R^2 =$ methyl or *tert*-butyl), "Subst." indicates a hydrocarbon group or a bridge containing various functional moieties, such as a substituted aromatic ring, ester group, sulfide, phosphonate, oxamide, hydrazide 1,3,5-triazine, 1,3,5-triazine-2,4,6-($1H,3H,5H$)-trione (e.g., in phenols, see structures **11–18**, **48–50**, Secs. III.A.1.a and III.B, or a benzotriazole group in UV absorbers see structures **64–68**, Sec. III.E.1). Major reactivities of the phenoxyl **83** include the following:

1. Disproportionation yielding the parent phenol **85** and monomeric ("primary") quinone methide **86** [Eq. (24)] (57).

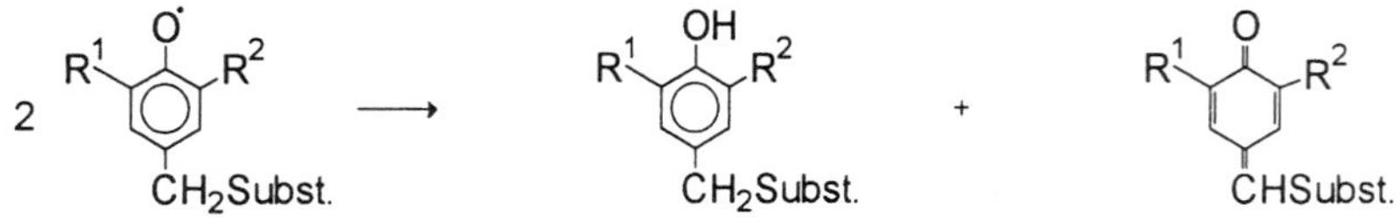

| 85 | 86 |

2. Rearrangement of **83** into the isomeric C-centered free radical **87**, dimerizing in **88** [Eq. (25)] (49,58).

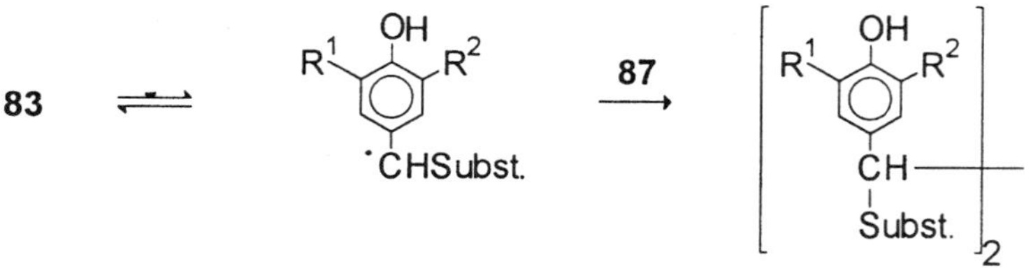

| 87 | 88 |

3. Intermolecular or intramolecular recombinations characteristic of reversible C–O coupling to **89** [Eq. (26)] and irreversible C–C coupling yielding **90** [Eq. (27)] (49).

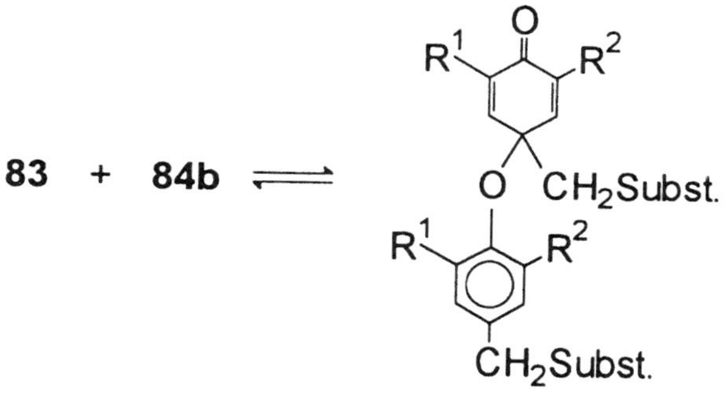

$$\mathbf{83} \ + \ \mathbf{84b} \ \rightleftharpoons \qquad \tag{26}$$

89

$$\mathbf{84b} \ + \ \mathbf{87} \ \rightleftharpoons \qquad \tag{27}$$

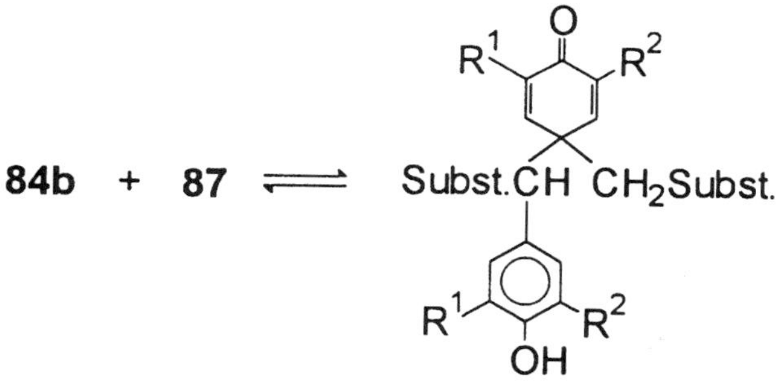

90

4. Cyclohexadienonyls **84a** recombine with polymer-derived alkylperoxyls or alkoxyls, with oxygen or free radicals (e.g., thiyls or nitroxides) arising in polymers stabilized with thiosynergists or HAS. Substituted 4-alkylperoxy-2,5-cyclohexadien-1-ones **91**, 4-alkoxy analogue **92**,

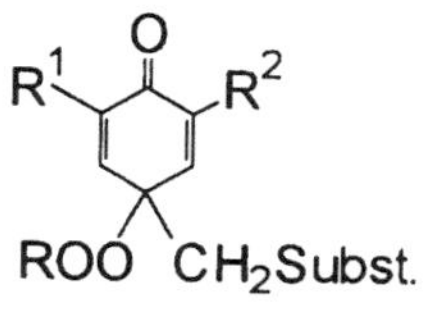

91 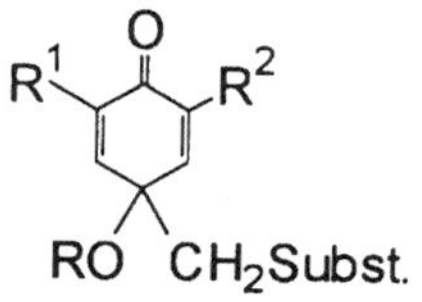   **92** 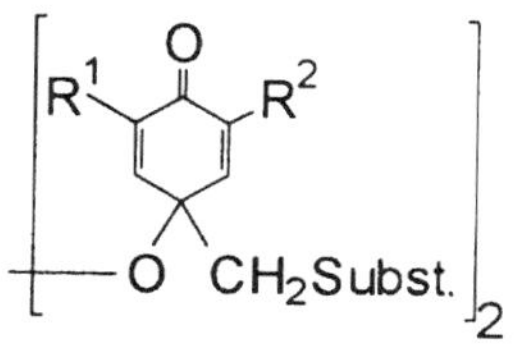   **93**

bis(cyclohexadienonyl)peroxides **93**, thio derivative **94**, or *O*-substituted hydroxylamines **95**, **96** are formed (3,15,30,42,56).

Compounds isomeric to **91–95**, such as **97**, may arise from 3,5-cyclo-hexadien-1-onyl **84b** if a lower sterical hindrance (e.g., $R^1 = tert$-butyl, $R^2 = $ methyl) enables this reactivity. In polynuclear phenols, such as **12–18**, each of the phenolic

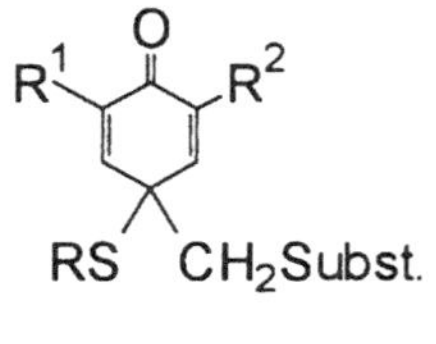

RS CH₂Subst. structure — **94**

\>NO CH₂Subst. structure — **95**

CHSubst. ON< structure — **96**

94 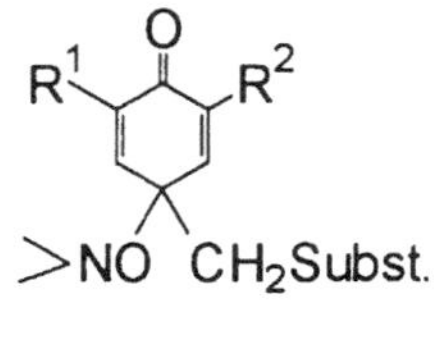   **95** 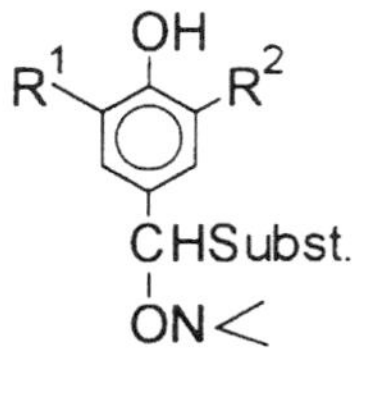   **96**

nuclei is oxidized independently and products with rather complicated structures are formed.

In photooxidized polymers, hindered phenols yield hydroperoxycyclo-hexadienones (e.g., **98**). Singlet oxygen and excited sensitizers are considered as active coreactants in their formation (42,48,58).

Some primary transformation products, such as structures **89**, **92**, **93**, or **95** are unstable and are usually formed only transiently. Other products, with general structures, **86**, **88**, **90**, **91**, **96–98**, derived from different mono- and polynuclear phenols were isolated. The C–C-coupling dimer of type **88** and quinone methide **86** rank among the most stable products (42–57,58).

Alkylperoxycyclohexadienones **91** or **97** are preferentially formed at sites of high local concentration of alkylperoxyls (3,49) in polymers oxidized at tempera-tures lower than the thermolysis of the peroxidic bond. All peroxidic derivatives of cyclohexadienone (**91**, **93**, **97**, or **98**) thermolyse or photolyse. 4-Alkylperoxy derivatives (see type **91**) are more temperature-resistant (they decompose between 105 and 125°C) than 2-isomers (see type **97**) decomposing at about 75°C (58). Homolysis of the peroxidic bond accounts for the respective oxy radicals (e.g., **99**) and results in thermo- and photoinitiation of the polymer (42,58). After

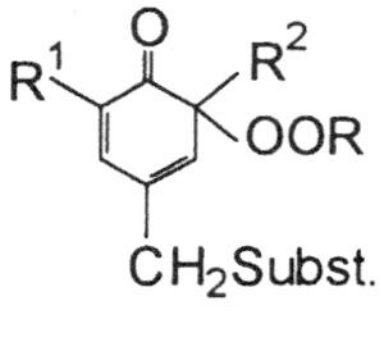

97 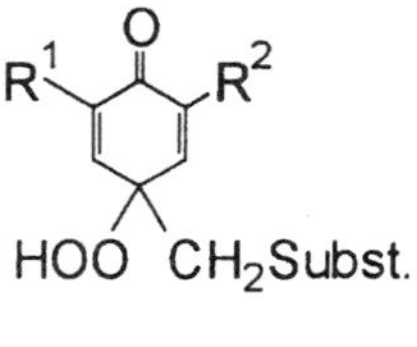   **98** 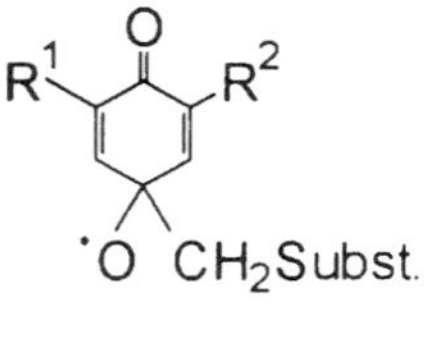   **99**

thermolysis of **91** or **93**, migrating and volatile low-molecular weight fragments, such as substituted benzoquinone **100**, diphenoquinone **101**, or exocyclic oxo com-pounds, such as 4-hydroxybenzaldehyde **102** are formed.

Compound **102** has a weak chain-breaking antioxidant effect (58). Quinones **100** and **101** do not contribute effectively to polymer stabilization (the alkyl radical-trapping ability common for unsubstituted benzoquinones (59) is too weak because of the sterical hindrance by substituents R^1, R^2). Hydroper-oxycyclohexadienone **98** thermolyses into substituted 4-hydroxycyclohexadienone **103** and causes a weak retardation of the hydrocarbon oxidation (58).

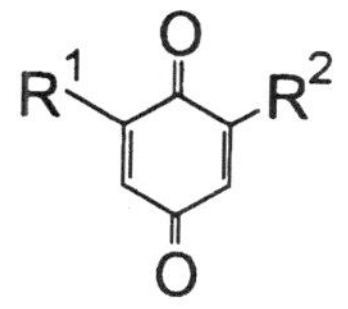

100

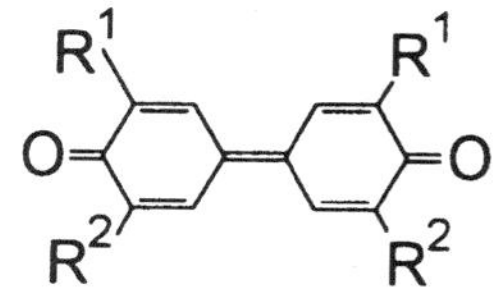

101

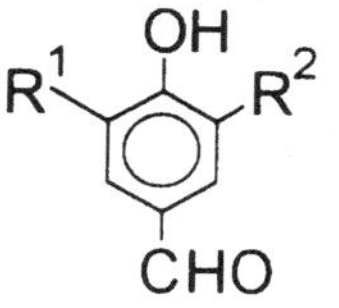

102

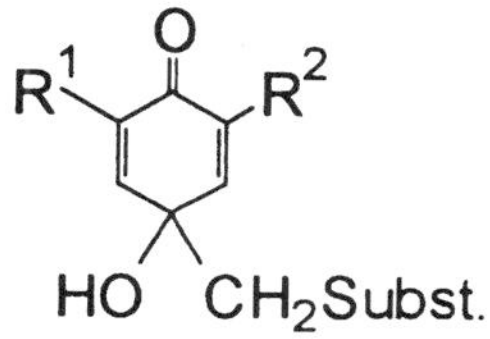

103

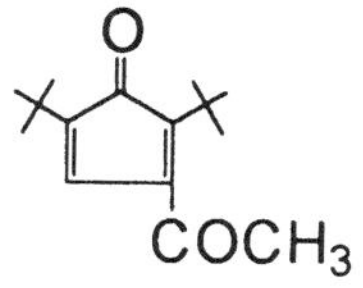

104

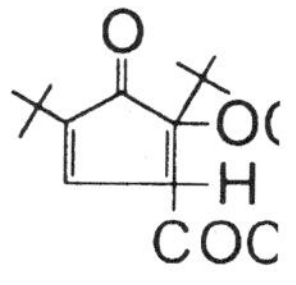

105

Peroxycyclohexadienones absorb solar UV radiation in the range of 300–380 nm (60). The photolysis of the peroxidic bond is assisted by n-π* excitation of the conjugated cyclohexadienonyl system and yields, under conditions of polymer weathering, derivatives of cyclopentadiene or cyclopentene, such as **104** or **105**. Photolysis of **91, 93, 97,** or **98** has a prooxidative effect in the initial stages of photooxidation of hydrocarbons or high-impact polystyrene. The photolytic product **104** has a pronounced light-screening effect, if present at relatively high concentration levels (about 0.1%) (49).

Carbon–carbon coupling [see Eq. (25)], yielding products of the type **88**, is a process that competes with formation of peroxycyclohexadienones. It accounts for phenolic products that have a lower volatility than the parent phenol and preserve high chain-breaking activity (3). The C–C coupling dominates at relatively low concentrations of alkylperoxyls in the matrix. The process was confirmed by isolation of 4,4′-ethylenebis(2,6-di-*tert*-butylphenol) **88** (R^1, $R^2 = $ *tert*-butyl; Subst. $= $ H). Dimers such as **88** preserve properties of phenolic antioxidants and are oxidized according to Scheme 6. As a consequence, more complicated polynuclear compounds are formed following the mechanisms mentioned for mononuclear phenols.

The C–C=coupling product, 2,6-di-*tert*-butyl-4-methyl-4-(3,5-di-*tert*-butyl-4-hydroxybenzyl)-2,5-cyclohexadienone-1-one **90** (R^1, $R^2 = $ *tert*-butyl; Subst. $= $ H), combining cyclohexadienoide and phenolic moieties, is less stable than **88** (R^1, $R^2 = $ *tert*-butyl; Subst. $= $ H), fragments under formation of free radical species and finally transforms into stable **88** (61).

Hydroxybenzyl radicals **87** effectively scavenge aromatic and heterocyclic nitroxides >NO• arising from aromatic amines or HAS (see Secs. IV.A.2 and IV.E.3, respectively) (15) and account for the corresponding hydroxybenzyl-substituted hydroxylamines **96** (R^1, $R^2 = $ *tert*-butyl; Subst. $= $ H).

The C–O-coupling products **89** (R^1, R^2 = *tert*-butyl) or analogous *ortho*-isomers are mostly unstable at ambient temperatures. A stable coupling dimer **106** was reported as an oxidation product of **21** (62).

A low steric hindrance of the phenolic hydroxy group augments the possibility of the formation of the C–O-coupling products. This was confirmed with 2-methyl-4,6-bis[(octylthiomethyl)phenol] (see Sec. VI) (63) or with 2,2'-ethylenebis(4,6-di-*tert*-butylphenol), yielding a spiroquinol ether **107** by intramolecular C–O coupling (62). Disproportionation of phenoxyl **83**, which,

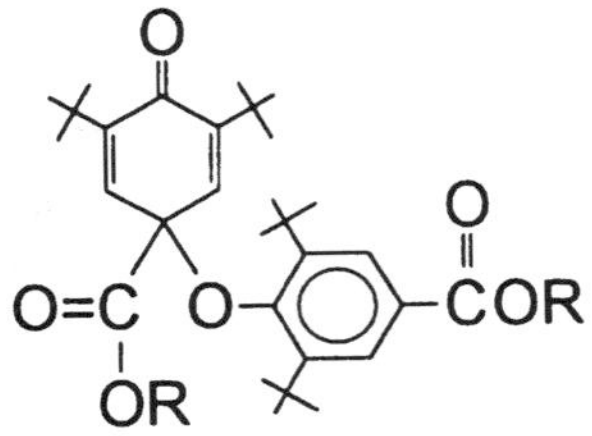

106

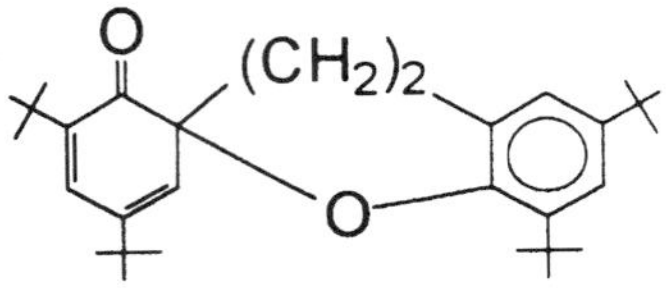

107

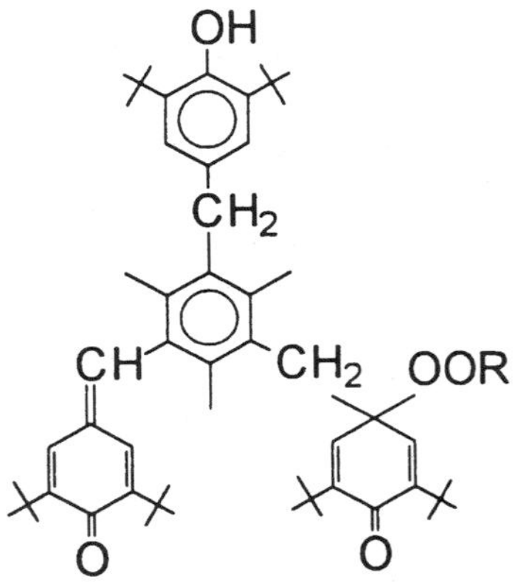

108

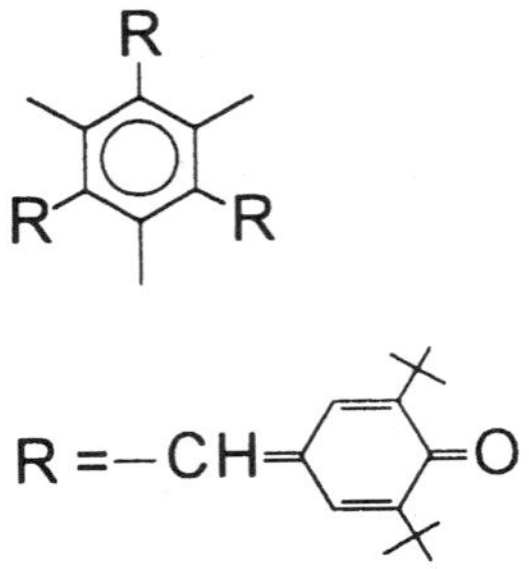

109

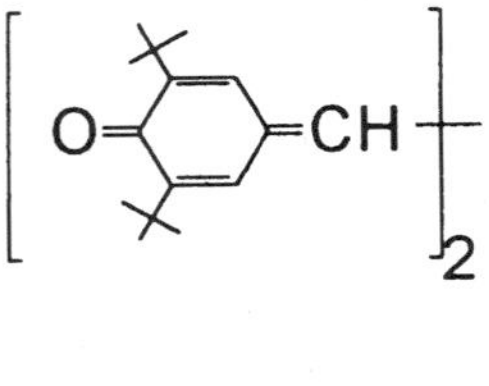

110

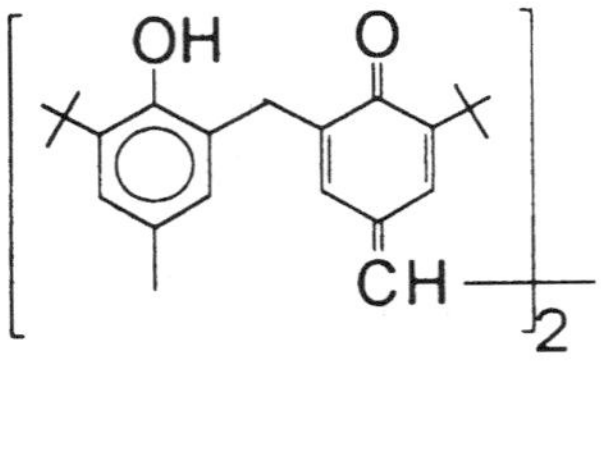

111

accounts for the colored quinone methide **86** [see Eq. (24)], is characteristic of phenolic antioxidants having a methyl or substituted methylene group, containing at least one hydrogen atom, in the para-position, such as in phenols **15** or **19** (4-hydroxylbenzyl-type phenols) or an ethylene group in phenols **11–14**, **16**, and **18** ("propionate-type" phenols) (57).

Quinone methides are important ultimate transformation products of all mentioned mono- and polynuclear phenolics **11–19** (42,58). Transformations of individual nuclei in polynuclear phenols proceed independently. This may account for a mixture of complex structures containing different intramolecularly bound transformation products. For example, compounds **108** and **109** are formed from trisphenol **15**. Structures of various complicated oxidized coupling products of polynuclear phenols detected in trace amounts in aged polymers were not deciphered.

Quinone methides arising from phenolics have attracted much attention (42,54,57,58). "Monomeric" quinone methide **86** (R^1, $R^2 = tert$-butyl; Subst. = H) is very unstable and dimerizes to a mixture containing **88** (R^1, $R^2 = tert$-butyl; Subst. = H) and 3,5,3′,5′-tetra-$tert$-butylstilbene-4,4′-quinone **110**, the oxidation product of **88**. The dimerization of quinone methides derived from the benzyl-type phenols is hindered by substitution of the exocyclic methylene group. For example, steric hindrance suppresses dimerization of the quinone methide derived from the trisnuclear phenol **15**. Its phenolic nuclei are converted stepwise into monomeric quinone methide moieties, such as in **109**. Moreover, the quinone–methinoide structure is formed preferentially at the unsubstituted methyl. This effect is exemplified with 2,2′-methylenebis(6-$tert$-butyl-4-methylphenol): its oxidation yields exclusively a reactive quinone methide on the unsubstituted methyl and not on the methylene bridge. Dimer **111** and an analogous trimer formed from the primary monomeric quinone methide were isolated (64). Analogous dimers were reported as oxidation products of **30**. Substitution of the position 4 in **30** by a tertiary alkyl group hindered formation of the quinone methide and reduced the relevant discoloration (65).

Propionate-type phenolic antioxidants **11–14**, **16** and **18** and their various analogues represent a highlight in the inherent phenolic efficiency (2) and exploitation of the chemistry of transformation products formed from phenolics. As an example, the chemistry of the principal transformation product of phenol **11**, quinone methide **112** (in all structures **112–116**, $R = C_{18}H_{37}$), is shown. The primary quinone methide **112** is prone to aromatize to a phenolic cinnamate **113**, after an intramolecular rearrangement; compound **113** is a chain-breaking antioxidant that traps alkylhydroperoxides. This results in formation of "secondary" quinone

112   **113**   **114**   **115**   **116**

methide, dimerizing into **114**. Phenolic cinnamate dimer **115** arises after rearrangement of **114**. In the ultimate phase, conjugated dimeric quinone methide **116** is formed (57). This chemistry accounts for the excellent integral antioxidant activity of the propionate-type phenolic antioxidants.

The sacrificial transformation of phenolics into quinone methides does not mean a complete loss of the stabilizing power, as revealed in model experiments. Quinone methides retard autoxidation of hydrocarbons. Their effectivity decreases with increasing temperature. The beneficial contribution of quinone methides formed in situ was shown during long-term heat aging of phenol-doped polypropylene (66). The contribution of the propionate-type phenol **18** yielding an isomerizing, polynuclear quinone methide, was more effective than that of the benzyl-type phenol **15**.

Besides isomerization, scavenging of alkyl radicals by the exocyclic methylene bond was proposed as an effect that contributes to stabilization (54). This reactivity may occur only in quinone methides without any appreciable sterical hindrance on the methylene group. Scavenging of free radicals ($Y^{\bullet} = R^{\bullet}$, $ROO^{\bullet}$, $>NO^{\bullet}$) by the free radical intermediate **117**, derived from quinone methide **112**, was postulated to yield a recombination product **118** capable of isomerizing to a derivative of 4-hydroxycinnamic acid **119**, having antioxidant properties [Eq. (28)]. The reactivity of quinone methides arising from phenolics used in combinations was HAS and phosphite processing stabilizers is mentioned in Section V.B.

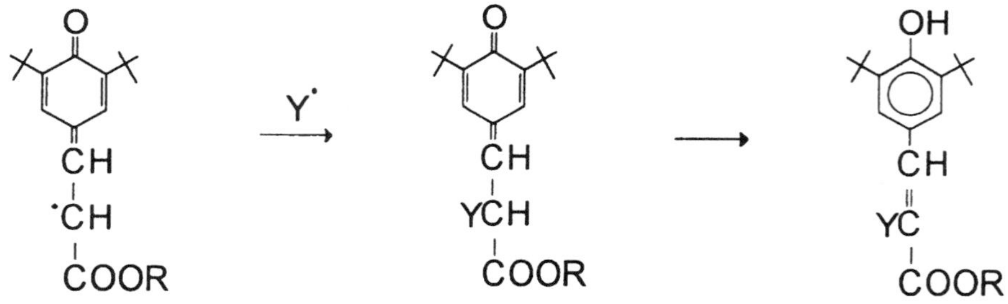

(28)

Quinone methides are very resistant against photolysis (58). They autoretard photooxidation (i.e., the depleting consumption) of the parent phenols, by quenching singlet oxygen, and slow down the photooxidative degradation of polymers, as exemplified in photooxidation of the photosensitive high-impact polystyrene. The light-screening effect most probably contributes to the stabilizing effect of quinone methides.

The presence of the conjugated system of double bonds, which accounts for a strong absorption in visible light at 420–463 nm, is a drawback of quinone methides (57). The quinone methides that gradually accumulate in aged polyolefins are, therefore, responsible for some discoloration. This may be rather intensive with trace concentrations (5 ppm) of highly conjugated quinone methides, such as **116**. This problem should be considered, particularly in deeply aged recyclates (1). (The danger of discoloration is strongly reduced using combinations of phenolics with phosphites; see Sec. V.A.)

Quinone methides are not formed from cryptophenolic antioxidants, having in positions 2 and 4 substituents bound to the phenolic nucleus by quaternary carbons. The tetranuclear phenol **17** is the only commercialized structure of this type. No disproportionation of the primary phenoxyl, according to Eq. (24), is possible. The transformation mechanism was adapted from the chemistry of 2,4-di-*tert*-butylphenol or 2-*tert*-butyl-4-(α, α-dimethylbenzyl)phenol (42,49,58). From the primary phenoxyl **120** (A = remainder of molecule **17** attached to the quaternary carbon atom), the corresponding derivative of 1,2-benzoquinone **121** arises by oxidation with alkylperoxyls. Dimer **122** is formed by C–C coupling, spiroquinol **123** after trapping of alkylperoxyl by **122** and intramolecular C–O coupling (59), and oxepinobenzofuran **124** after valence isomerization in

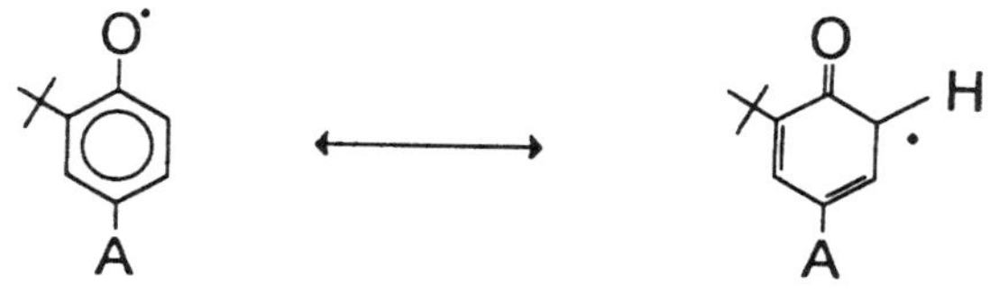

120

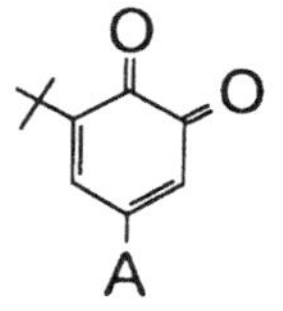

121

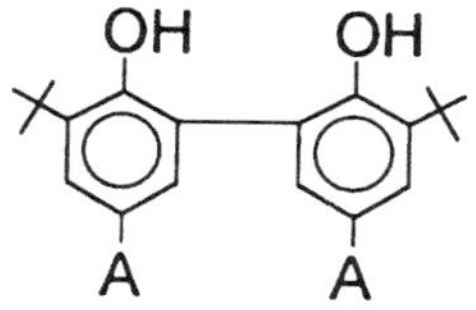

122

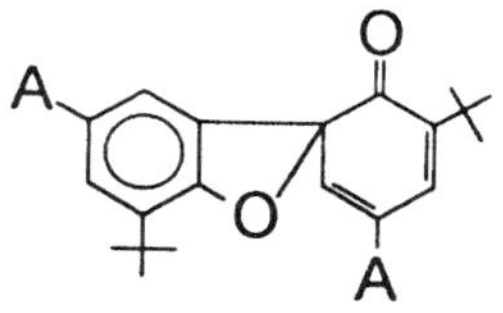

123

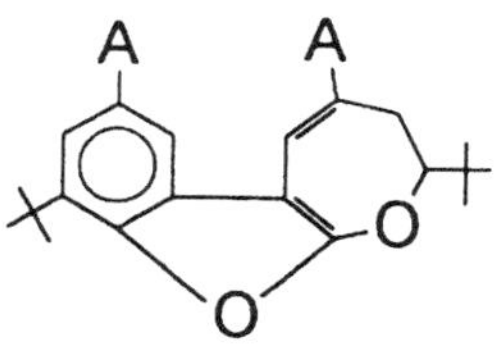

124

3,3'-di-*tert*-butyl-5-5'-disubstituted-2,2'-diphenoquinone, the oxidation product of dimer **122** (49).

a. Depleting Consumption of Phenolic Antioxidants. Attention has been aroused relative to oxidation of phenolic antioxidants by nitrogen dioxide, the oxidizing environmental pollutant. The process results in yellowing of phenol-doped polyolefins, called gas-fading. It was reported that 2,6-di-*tert*-butyl-4-methylphenol oxidizes into stilbenequinone **110** (52). Yellow nitro compounds (**125, 126**) are formed transiently. Phenol **12** is oxidized by nitrogen dioxide into a mixture of the nitro compounds **127** and **128** (A = remainder of molecule **12**) and an structural analogue of **125** (67).

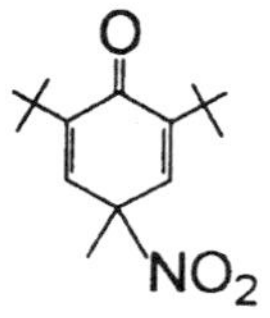

125 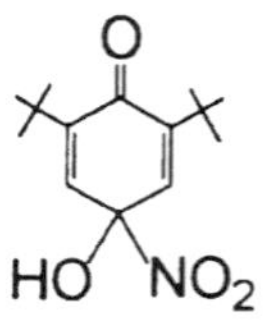   **126** 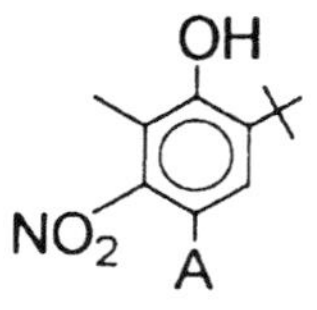   **127** 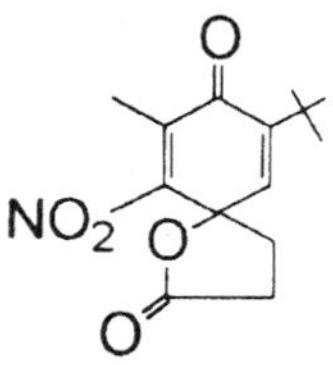   **128**

Phenols suffer from an easy oxidation [see Eq. (22b)] or sequestering with ions of transition metals that arise in traces from residues of polymerization catalysts, metal-contaminated fillers, or accidental sources (1,3,49,53,68). In this way, the chain-breaking activity is diminished or lost. Copper and iron are the most dangerous oxidants. Phenols that are chelated by residues of titanium catalysts, such as in **129** (49,68) or in analogous complexes with chromium or zirconium, and complexes **130** arising from residues of alkylaluminium cocatalysts have no more antioxidant activity. The reactivity according to Scheme 6 is impossible. Most of the complexes of **129** and **130** are discoloring and oxidation-sensitive (30,68). Stabilization of polyolefins produced by modern technologies using chromium-based catalysts characterized by a stronger prooxidative effect is a more difficult task than stabilization of the first generation of polyolefins from titanium-based technologies.

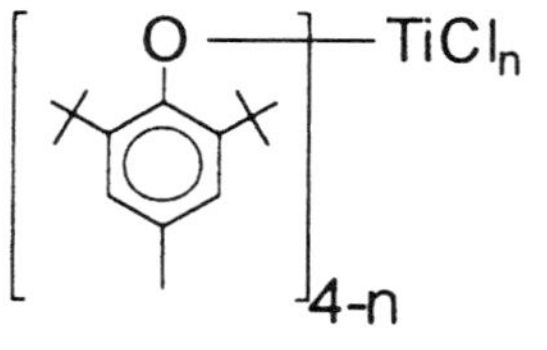

129 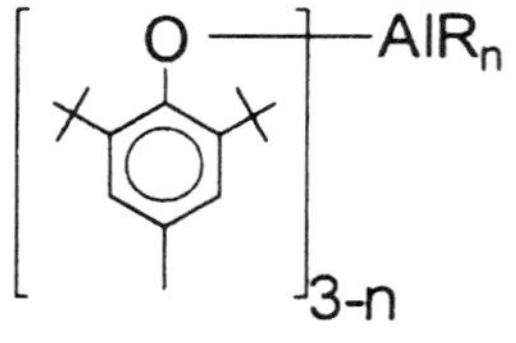   **130**

2. Scavengers of Oxygen-Centered Free Radicals: Aromatic and Heterocyclic Amines

Representative structures are given in Section III.A.1.b. Amines have a long history of application. The mechanism of activity and transformations was interpreted in detail using the up-to-date knowledge of the inherent chemical efficiency of individual structural types (2,8,15,56) and physical factors governing the final effect (69). The secondary amino group –NH– in the α-position to the aromatic nucleus is the key functionality characterized by its inherent chemical efficiency.

Aromatic and heterocyclic amines rank among the most efficient chain-breaking antioxidants. Some of them also have properties of antifatigue agents and antiozonants. Even their appreciable metal deactivation effect is reported. In spite of the multifunctional antidegradant properties and broad potential application spectrum, the technical use has been limited to carbon black-filled general-purpose rubber vulcanizates. Quinone imines, which invariably arise in the service transformations of amines (15), are a disadvantage of amine chemistry, and they are the main obstacle that limits the application in plastics and rubber-modified plastics, for which discoloration is not acceptable by consumers. It is understandable that the development of amines having a reduced tendency for discoloration is a continuous aim. The effort was successful by a proper substitution in diphenylamine series, by synthesis of 4,4'-bis(α,α-dimethylbenzyl) diphenylamine **22a** or its bis-*tert*-octyl analogue **22b**, by substitution of the position 6 in the 2,2,4-trimethyl-1,2-dihydroquinoline series, such as in **24** or by development of rather low-discoloring 2,2-disubstituted indol-3(2*H*)-ones (**25**) or 9,10-dihydroacridine (15).

As chain-breaking antioxidants, amines **22–27** scavenge alkylperoxyls according to Scheme 6, Eq. (22a; AH = amine). The charge-transfer complexes **131** and **132** have been envisaged for aromatic monoamines and diamines, respectively,

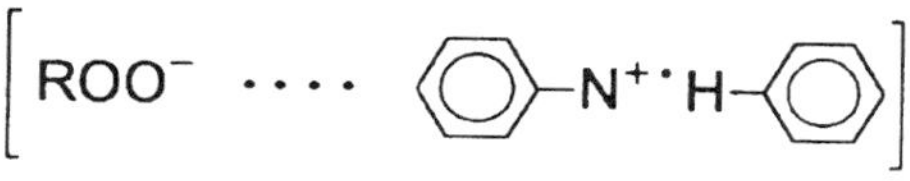

131 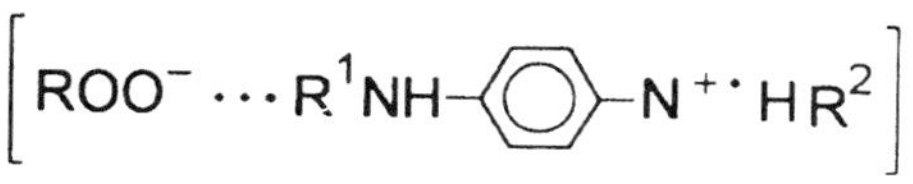   132

as steps proceeding in the formation of short-living aminyls **133a**, the primary radicals arising from amines (15). Aminyls **133a** react in mesomeric carbon-centered free-radical forms **133b,c** [see. Eq. (29)] and undergo N–N, C–N, and C–C coupling reactions.

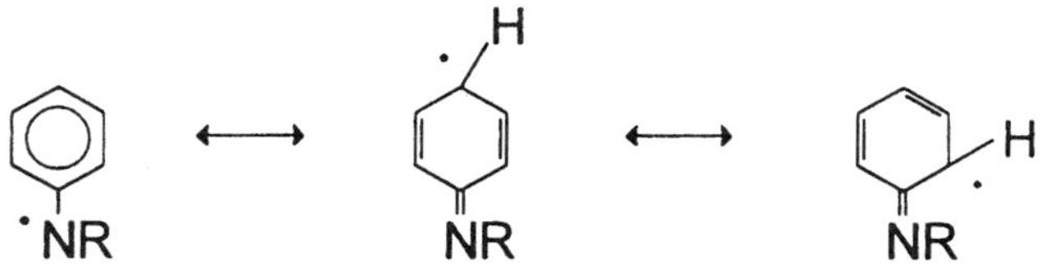

133a   133b   133c

By coupling processes, the molecular weight of the products increases (15,70). Substituted *N,N,N',N'*-tetraphenylhydrazine **134** is formed in a reversible N–N coupling from **133a** (R = phenyl). The N–N-coupling dimer **135** is reported as a product arising from phenylenediamine **23b** (14). Various dimeric products (**136–138**) have been formed irreversibly by C–N coupling from **133a** and **133c** in the diphenylamine and phenylenediamine series, respectively. The C–N-coupling dimer **139** was formed from **24**. If a secondary amino group remains in the coupling product, such as in **135**, **136**, **138**, or **139**, the chain-breaking activity is not destroyed and the dimers are able to repeatedly scavenge alkylperoxyls. New aminyls participating in coupling reactions are formed. The composition of the final mixture of products is very complex, individual "oligomeric"-coupling products are present, mostly in trace amounts.

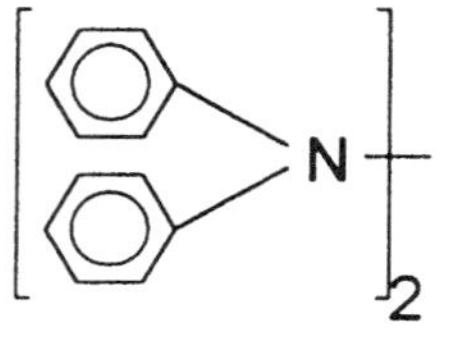

134

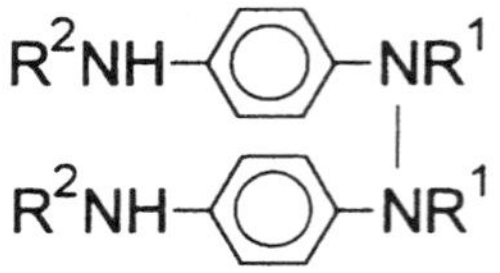

135

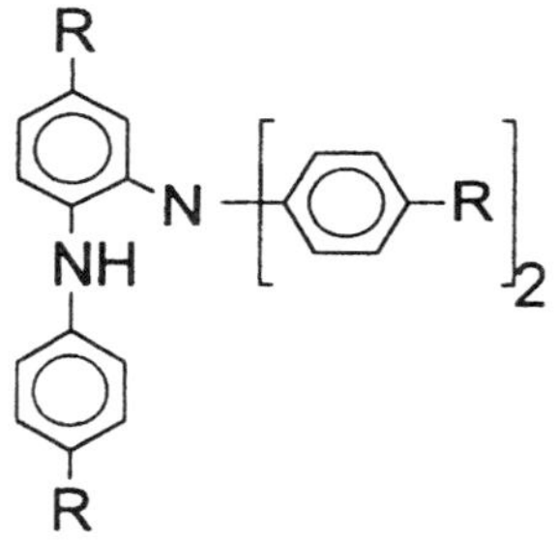

136

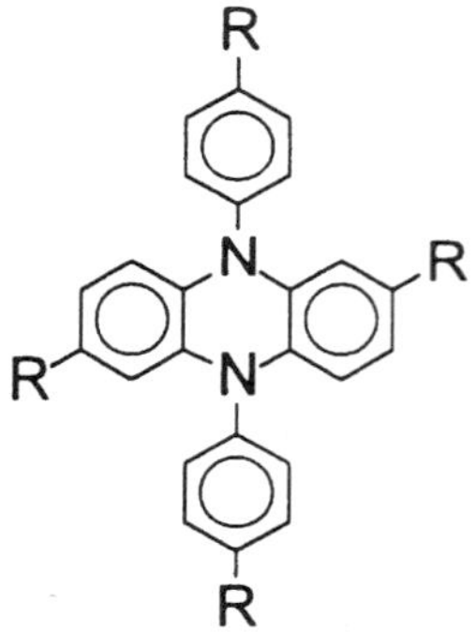

137

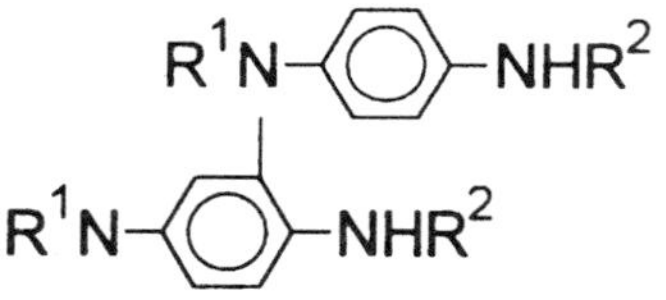

138

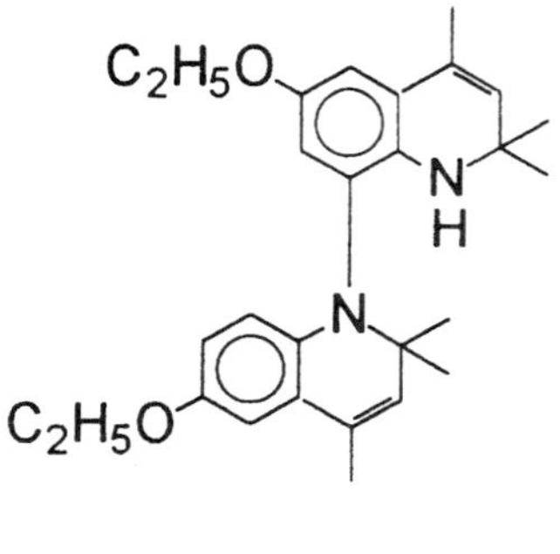

139

It is a characteristic feature of aromatic amines that the primary radical **133** has a dual free–radical-scavenging reactivity: it reacts with both alkyl- and alkylperoxyl radicals. Scavenging of alkyl radicals results in *N*- and *C*-alkylates. Formation of an *N*-alkylate **140** from **133a** has a regenerative character: the parent amine **22** is formed after thermolysis of **140** [Eq. (29)] (15).

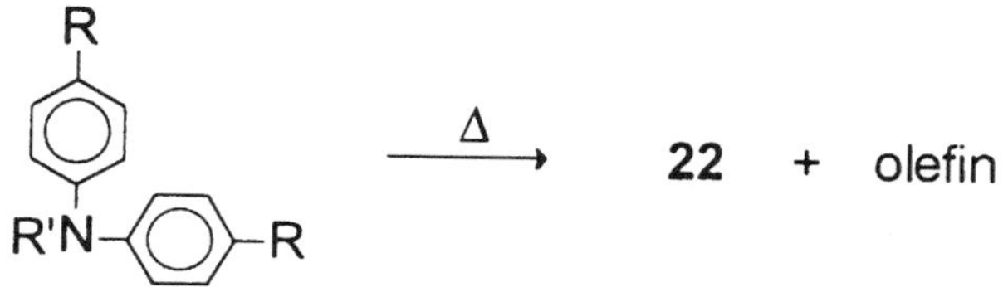

$$\xrightarrow{\Delta} \quad 22 \; + \; \text{olefin} \tag{29}$$

140

The mesomeric free radicals **133b,c** trap alkyls irreversibly. Formation of *C*-alkylates was proved in model experiments with low molecular weight alkyls (15,70). Alkylates **141** and **142** were isolated from **24** and **23b**. The alkyls are bonded in sites of the highest spin density in **133b,c**. The formation of polymer-bound species **140–141** (R is a polymeric residue) has been extrapolated from model experiments. Some structures containing amino moieties –NH– may be considered as polymer-bound stabilizers formed in situ during sacrificial transformations of amines.

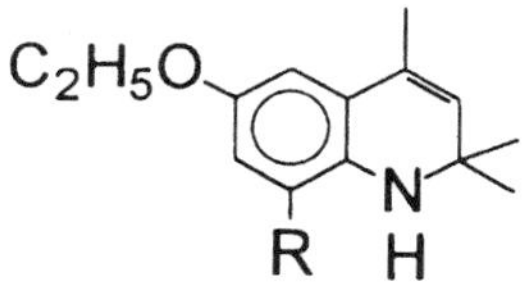

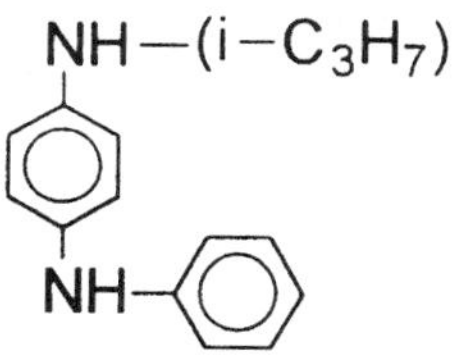

141   **142**

Scavenging of alkylperoxyls by aminyls (i.e., the second chain-breaking step in amines), derived from diphenylamine **22**, dihydroquinolines **24** or **26**, and dihydroindolidone **25**, results in oxidation and formation of the respective nitroxides **143–145** (15,56,58,70). Compound **143**, bearing a hydrogen atom in the α-position to the nitroxide function, may react in the corresponding mesomeric nitrone form **146**.

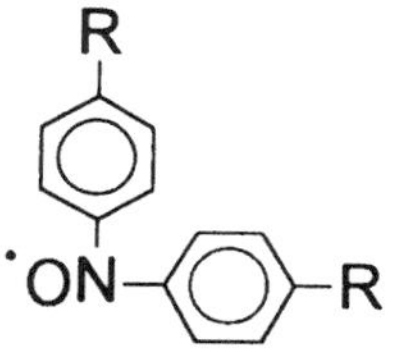

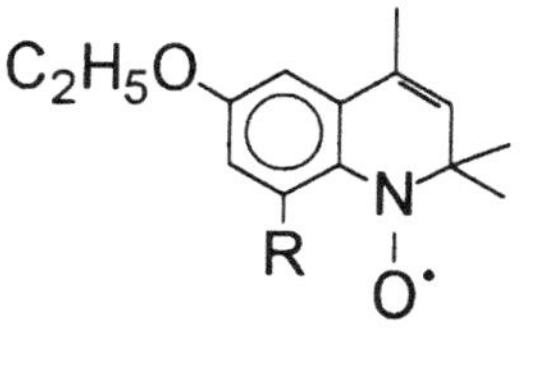

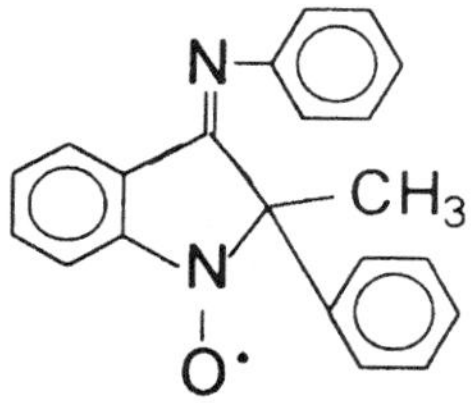

143   **144**   **145**

Monoaminyls derived from N,N'-disubstituted-1,4-phenylenediamines **23** react with alkylperoxyls, preferentially on the second amino group. Well-defined colored N,N'-disubstituted 1,4-benzoquinonediimines **147** are the major oxidation products (15,70). The relevant mono- or bis(nitroxides) were not detected by electron paramagnetic resonance (EPR) spectroscopy among the reaction products. The theoretically possible bis-nitroxide is unstable and reacts in its mesomeric nitrone form **148**.

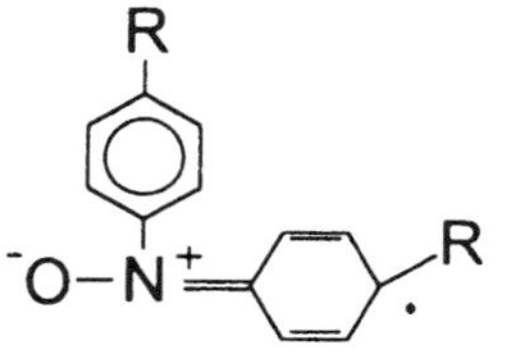

146 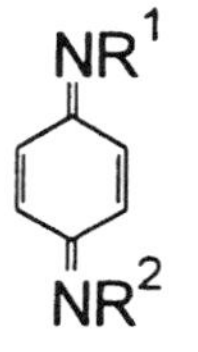   **147** 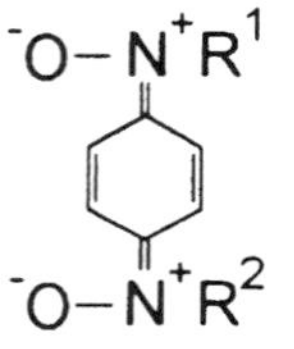   **148**

Carbon-centered radicals **133b,c** are prone to react with alkylperoxyls and yield the respective unstable alkylperoxy derivatives, such as **149**, arising from **22** (15). Compound **149** converts into colored 1,2-benzoquinoneimines **150** or **151** (R = *tert*-octyl or α,α-dimethylbenzyl) and more complex quinone imines arising from the corresponding coupling products derived from **133a–c**. Free radicals derived from dihydroquinoline **24** are oxidized into quinone imines **152** and **153**.

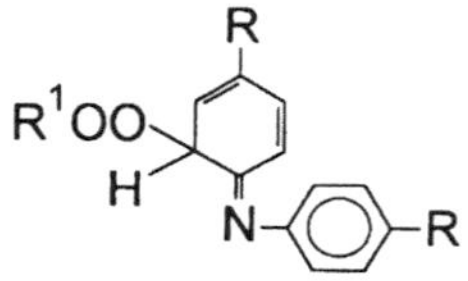

149 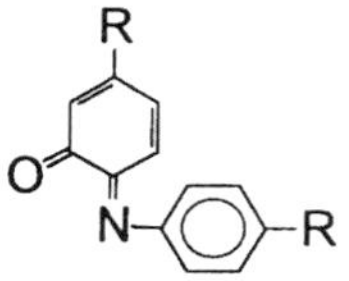   **150**

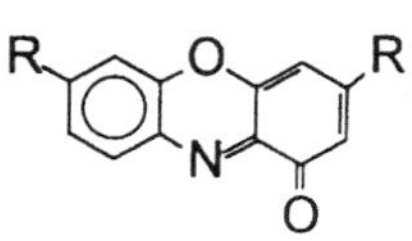

151 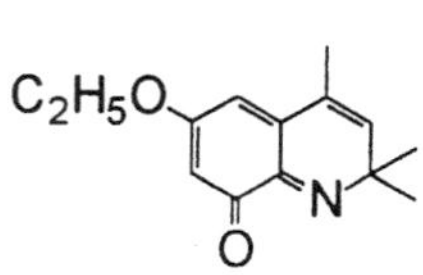   **152** 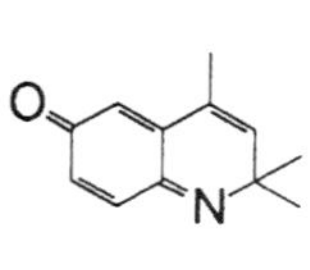   **153**

The coupling product **139** yields compound **154**. Colored quinone imine **155** arises from phenothiazine.

Nitroxides **143–145** have been considered as the key free radical intermediates in the stabilizing chemistry of monoamines **22** and **24** (15,70). The stability drops, in order, from the dihydroquinoline to dihydroindole and diphenylamine series and is influenced by the tendency of nitroxides to disproportionate.

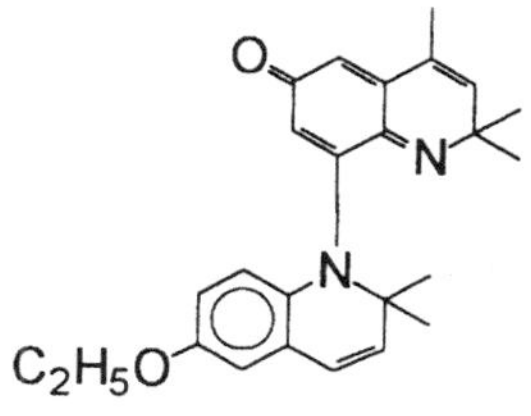

154

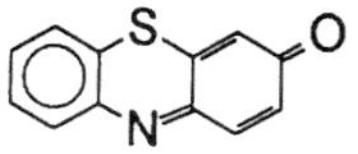

155

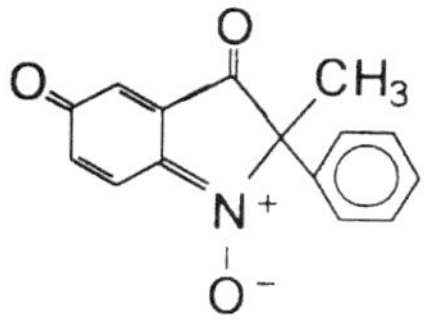

156

Nitroxides react in mesomeric forms and are prone to scavenge alkylperoxyls and alkyls (3,15,70). Trapping of alkylperoxyls by **143** (R = H), derived from unsubstituted diphenylamine **22** (R = H), results in volatile unsubstituted 1,4-benzoquinone, nitrosobenzene, and nitrobenzene. Nitroxides **143** (R = *tert*-alkyl) or **144** (R = OC$_2$H$_5$) are more stable to oxidative degradation than **143** (R = H) owing to the presence of substituents in the *para*-position to the nitroxide moiety. Nitroxide **145** is oxidized by alkylperoxyls into quinone imine *N*-oxide **156**.

The most appreciated reactivity of nitroxides or their mesomeric nitrones is their trapping of alkyl radicals. The respective aromatic *O*-alkylhydroxylamines, such as **157** and **158**, are formed from **143** and **144**, respectively (70). The bisnitrone **148** yields an alkylated "secondary" nitroxide **159**. Trapping of alkyls is very effective in an oxygen-deficient environment and accounts for the antifatigue activity observed with aromatic amines.

O-Alkylhydroxylamines > NOR **157** or **158**, formed in situ in the sacrificial transformation mechanism of amines, thermolyze above 100°C into the corresponding hydroxylamines > NOH **157** (R' = H) and **158** (R = H). Both of these

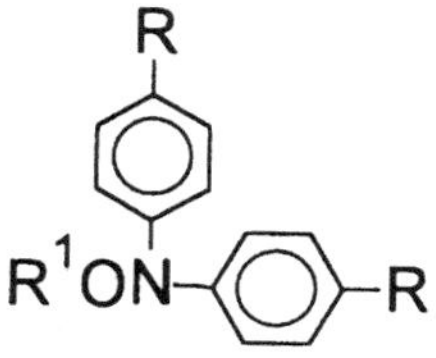

157

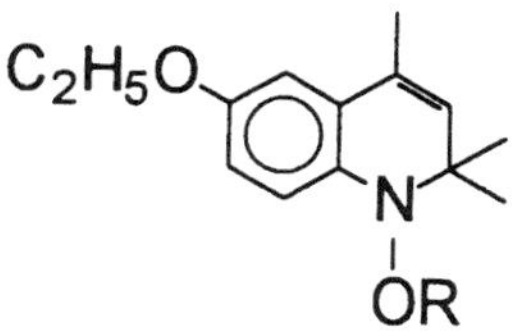

158

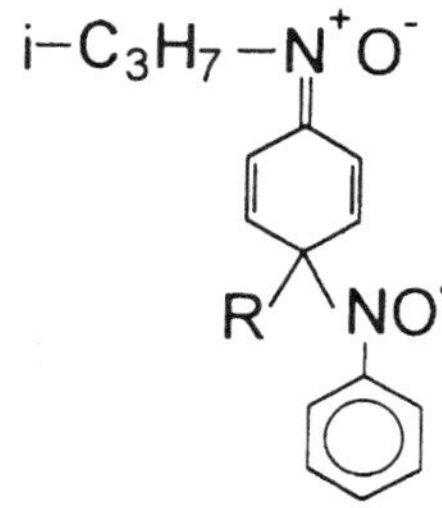

159

hydroxylamines are strong scavengers of alkylperoxyls (15). Hence, after a repeated scavenging of alkylperoxyls, the aromatic nitroxide > NO$^\bullet$ may be regenerated (Scheme 7).

Model experiments with diphenylnitroxide also disclosed a potency to scavenge other free radicals (such as thiyls or sulfinyls) formed in the aged polymer matrix (1,70).

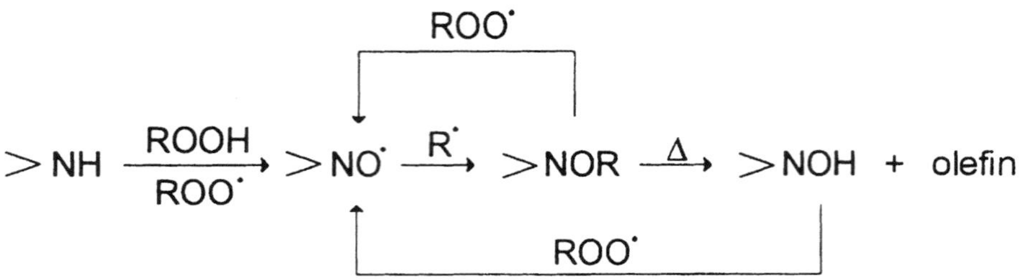

$$>\text{NH} \xrightarrow[\text{ROO}^{\bullet}]{\text{ROOH}} >\text{NO}^{\bullet} \xrightarrow{R^{\bullet}} >\text{NOR} \xrightarrow{\Delta} >\text{NOH} + \text{olefin}$$

Scheme 7

Aromatic and heterocyclic nitroxides disproportionate, as shown for nitroxide **143** and its mesomeric form **146** in Eq. (30). A coupling product **160** is formed as an intermediate. The process yields parent amine **22** and N-phenyl-1,4-benzoquinonemonoimine-N-oxide **161** (70).

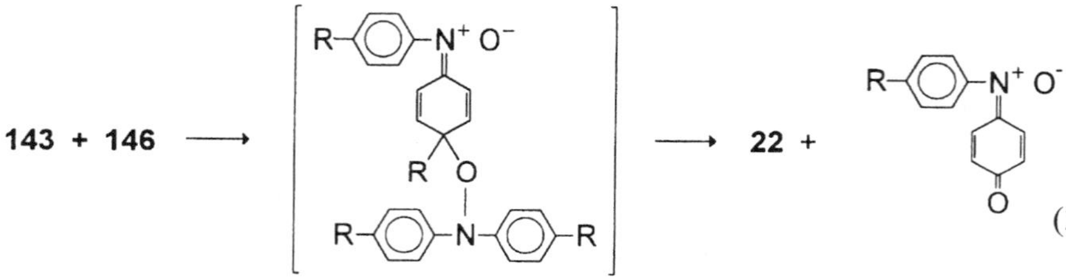

$$\mathbf{143 + 146} \longrightarrow \mathbf{22 +}$$

(30)

160   **161**

An analogous mechanism yields 2,2,4-trimethyl-6-quinolone-N-oxide **162** in the dihydroquinoline series. The tendency to disproportionate diminishes the free–radical-scavenging potency of aromatic nitroxides.

A rather complicated chemistry is characteristic of N,N'-disubstituted phenylenediamines **23**. Regardless of the character of the N,N'-disubstitution and the oxidation agent (alkylperoxyl, alkylhydroperoxide, ozone, ozonide), N,N'-disubstituted 1,4-benzoquinone diimines **147** are formed (15,56,70) that are responsible for the deep discoloration of polymers because of the strong absorption of the visible light between 440 and 480 nm. In spite of the discoloring and staining properties, benzoquinone diimines have a rather beneficial effect in aging polymers. They retard autoxidation of hydrocarbons and shorten the scorch time of rubbers (71).

The character of the substituents on nitrogens affects the reactivity of **147** (15,70). The presence of one secondary alkyl C_3–C_8 in **147** ($R^1 = sec$-alkyl; $R^2 =$ phenyl) makes the benzoquinone diimine prone to hydrolysis in the presence of acid impurities. A relatively stable N-phenylbenzoquinone monoimine **163** is formed. Quinone imines **147** and **163** are oxidation agents that form redox pairs with the relevant phenylenediamine **23** and 4-hydroxydiphenylamine **164**, respectively.

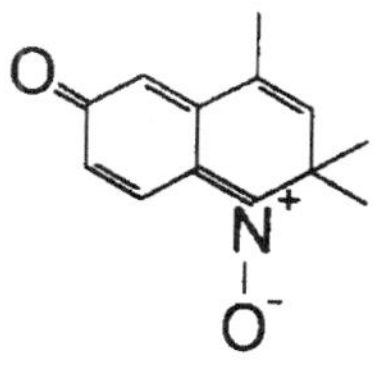

162

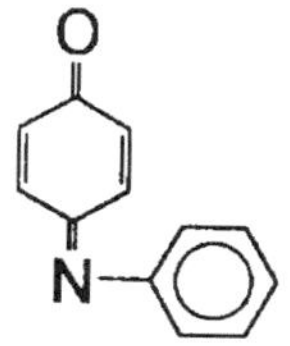

163

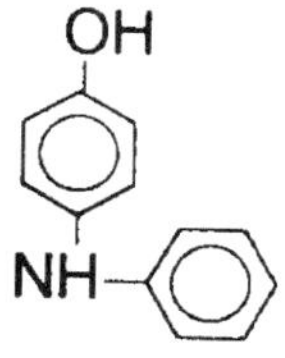

164

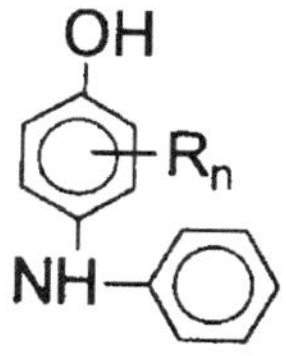

165

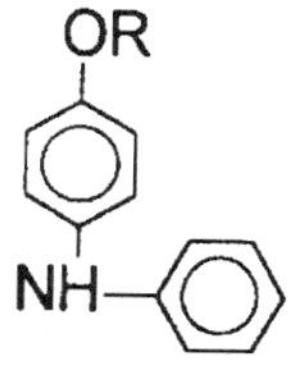

166

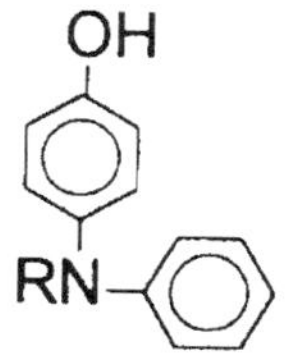

167

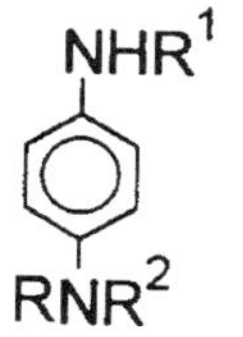

168

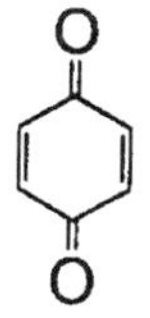

169

The redox pairs **163–164** and **147–23** are characteristic of a high alkyl radical-scavenging potency. Stable *C*-alkylates **142** or **165** (n = 1, 2), which have alkyls linked to the arylene ring, and *O*-alkylate **166** are products of trapping of alkyls. Thermolyzing *N*-alkylates **167** or **168** are generated from the quinone monoimine **163** or diimine **147**, respectively. The alkyl is split off similarly to the example in Eq. (29). An olefin is formed and the parent amines **163** or **23** are regenerated.

Benzoquinone diimines, substituted by two secondary alkyls **147** ($R^1 = R^2 = iso\text{-}C_8H_{17}$) and generated from the strong antiozonant **23a**, are hydrolyzed in the presence of weak acids to unsubstituted 1,4-benzoquinone **169**. In spite of its volatility, **169** is an efficient scavenger of alkyl radicals (59).

N,N'-Diphenyl-1,4-benzoquinonediimine **147** ($R^1 = R^2 = $ phenyl) arising from the strong antioxidant **23c** hydrolyzes only very slowly, and with low yields, into **163**. The released phenylamino group participates in the formation of Bandrowski's base **170** (azophenine) by substitution of the parent **147**. After disproportionation and cyclization of base **170**, discoloring heterocycles, such as 5-phenyl-2-(phenylamino)-3-(phenylimino)-3,10-dihydrophenazine **171** or 1,8-diphenyl-fluorindine **172**, are formed (15).

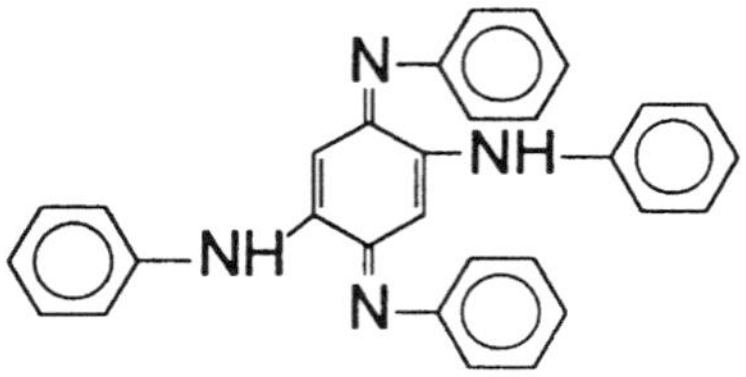

170

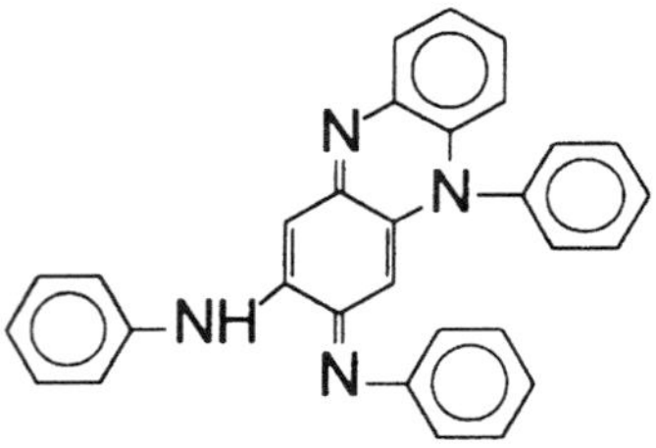

171

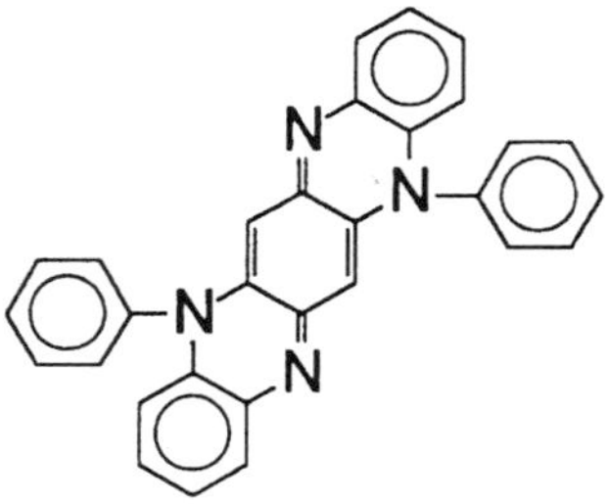

172

Regardless of the intricate chemistry of aromatic amine stabilizers, the role of the derived sacrificial transformation products, aminyls > N$^\bullet$, nitroxides > NO$^\bullet$, nitrones, and quinone imines, in the stabilization process is explained by model studies (3,15,56,70). The effective ROO$^\bullet$/R$^\bullet$ free–radical-scavenging activity, summarized in Scheme 8, from aromatic amines makes > NH high-performance stabilizers in applications for which discoloration and contact staining are not a limiting factor.

3. Scavengers of Oxygen-Centered Free Radicals: *N,N*-Dialkyl and *N,N*-Dibenzylhydroxylamines

The recently introduced additives **28** and **29** are antioxidants for polyolefins and diene-based elastomers (26). Their activity mechanism involves cyclic ROO$^\bullet$/R$^\bullet$ trapping (15). Nitroxides and nitrones, arising after trapping of alkylperoxyls and hydrogen abstraction by alkyls, form secondary nitroxides after trapping another alkyl. A repeating alkyl trapping has been postulated (Scheme 9).

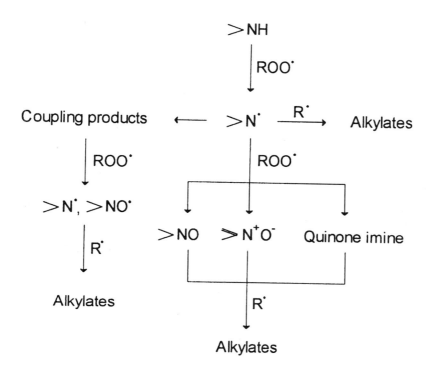

Scheme 8

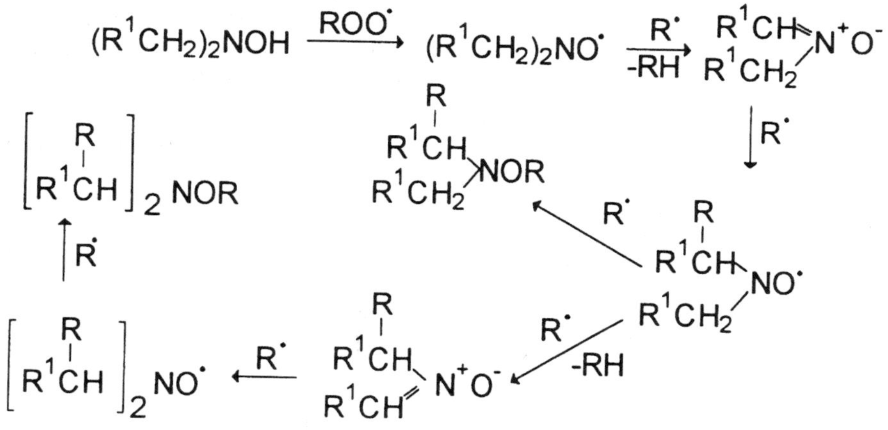

$$(R^1CH_2)_2NOH \xrightarrow{ROO^{\cdot}} (R^1CH_2)_2NO^{\cdot} \xrightarrow[-RH]{R^{\cdot}} \begin{matrix} R^1CH \\ R^1CH_2 \end{matrix} N^+O^-$$

Scheme 9

4.   Scavengers of Carbon-Centered Free Radicals

Macroalkyls are formed in polymers containing hydrocarbon chains or segments. Under oxygen-deficient conditions, self-termination reactions yielding polymer unsaturation, branching, or cross-linking take place (4,47). In the presence of oxygen, a rapid oxidation to alkylperoxyls proceeds. Hence, alkyl scavengers have to compete with both reactions.

At present, nondiscoloring and nontoxic compounds, acting in their parent structure as alkyl scavengers in polymers, are rare. Mechanisms valid for the inhibition of the free radical polymerization (59) have been exploited in the search for structures serving as alkyl scavengers. The monoacrylate of 2,2'-methylenebis(4-methyl-6-*tert*-butylphenol) 30 has been commercialized. The activity mechanism in an oxygen-deficient atmosphere reveals (27,65) alkyl scavenging by the vinyl moiety in the first reaction step. A free phenoxy radical 173 is formed after an intramolecular rearrangement of the primary product disproportionates into phenol 174 and the corresponding discoloring quinone methide 175, capable of dimerizing.

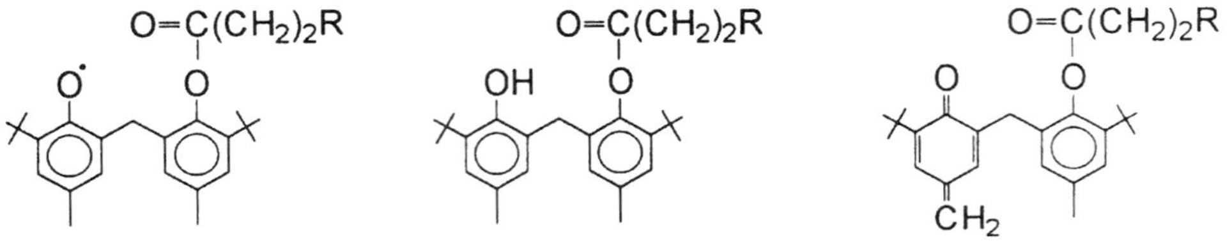

173   **174**   **175**

The monoester analogous to 30 and having a tertiary alkyl group in the position 4 does not form the discoloring quinone methide. This improvement was reached at the expense of a lower-stabilizing performance. Compound 30 is a phenol and reacts in an oxidizing atmosphere as a chain-breaking antioxidant that scavenges alkylperoxyls. The dual $R^{\bullet}/ROO^{\bullet}$ reactivity makes 30 a very efficient processing stabilizer of diene-based polymers and thermoplastic elastomers in which the presence of trace amounts of oxygen cannot be avoided (2,72).

Benzofuranone 31 was announced more recently as a nondiscoloring, processing H-donor stabilizer for hydrocarbon polymers (28). The push–pull substituent effects of R^1 and R^2 weaken the C–H bond in position 3 and make it prone to splitting-off the hydrogen by reactive alkyls or alkylperoxyls. A carbon-centered radical 176 is formed that converts into 177 after trapping another alkyl [Eq. (31)]. It was postulated that the hydrogen released from 31 saturates another alkyl.

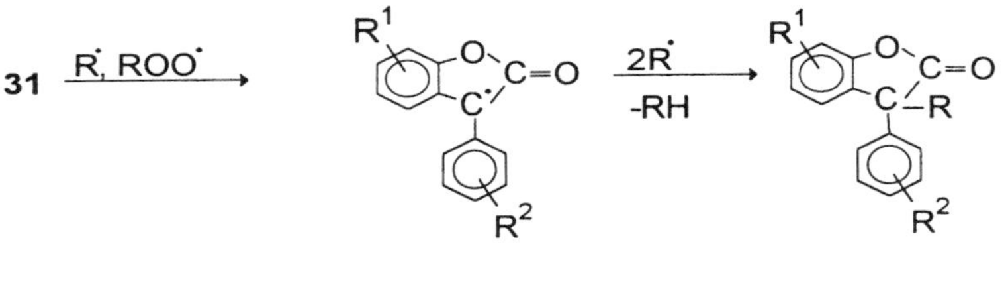

(31)

176   **177**

Heavy hydroaromatic hydrocarbons (**178** is used as a model) were recently reported as effective melt and long-term heat stabilizers for polyolefins in strictly oxygen-deficient environments (73). The H-donor mechanism is based on an easy aromatization by reaction with reactive alkyl radicals [Eq. (32)].

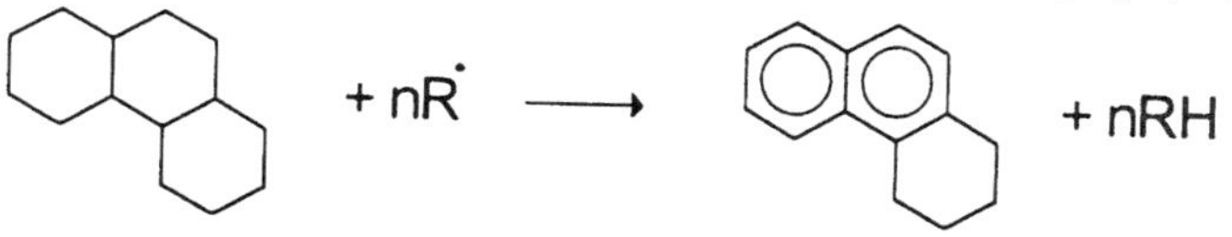

$$+ n\overset{\bullet}{R} \longrightarrow \qquad + nRH$$

(32)

178

With the exception of the hemiester **30** and its analogues, none of the commercially available phenolic and aminic antioxidants, in their original chemical form, trap alkyls (2,56). It was recognized, however, that the complementary alkyl-trapping activity of some sacrificial transformation products arising from phenols (see Sec. IV.A.1: quinone methide, yielding **118**); aromatic amines (see Sec. IV.A.2: aminyls, nitroxides, nitrones, and quinone imines, yielding *N*-alkylates **140**, **167**, **168**; *O*-alkylate **166**; and ring *C*-alkylates **141**, **142**, **165**; *O*-alkylhydroxylamines **157**, **158**; or secondary nitroxide **159**); hydroxylamines (see Sec. IV.A.3, Scheme 9); and HAS (see later in Sec. IV.E.3) plays an important role in polymer protection and results from a complex chemistry of stabilizers capable of scavenging alkylperoxyls and alkyls in a cascade process (2,15,54,56).

Trapping of alkyls by phenoxy radicals **83** or mesomeric cyclohexadienonyl radicals **84** proposed in the literature as a potential regenerative step for phenols (AH) [Eq. (33)] was reported as noneffective in the practice even with reactive macroalkyls derived from thermoplastic elastomers (72).

$$\textbf{83 (84)} + \overset{\bullet}{R} \longrightarrow AH + olefin$$

(33)

B. Hydroperoxide-Decomposing Antioxidants

The general stabilization mechanism of hydroperoxide decomposers (HD) involves displacement on the peroxidic O–O bond accounting for reduction of hydroperoxides in alcohols and for stoichiometric oxidation of the used antioxidant (Scheme 10) (25).

Some oxidation products of HD [i.e., HD(O); see Scheme 10] are unstable and, by a rather complicated mechanism, decompose to a mixture of compounds having peroxidolytic properties deactivating hydroperoxides in an overstoichiometric

$$HD \xrightarrow{\;n\,ROOH\;} HD(O)_n + n\,ROH$$
$$\Delta \downarrow$$
$$\text{Peroxidolytic products}$$

Scheme 10

manner. As a consequence, hydroperoxide-decomposing antioxidants are consumed irreversibly during the scarificial process (3,25).

The principal hydroperoxide decomposers are organic compounds of sulfur (3,25; see Sec. III.A.2.a) and organic compounds of trivalent phosphorus (25,43; see Sec. III.A.2.b). Deactivation of hydroperoxides by HAS will be discussed in Section IV.E.3.

1. Organic Compounds of Sulfur

Decomposition of hydroperoxides, according to the Scheme 10, is the principal mechanism in antioxidants **32–38** (3,38) (compounds **36–38** have also been considered as photoantioxidants; see Sec. IV.E.2). The multistep activity mechanism includes formation of effective peroxidolytic compounds (3).

Thioethers such as **32** ($R = C_{18}H_{37}$) or disulfide **33** react with alkylhydroperoxides and yield the corresponding sulfoxide **179** (n = 1) or thiosulfinate **180** (n = 1) in the first stoichiometric stabilization step (3,25,74). Both primary products **179** and **180** undergo oxidation and thermolysis and are without doubt precursors of peroxidolytic species, such as sulfenic acid **181** (n = 0) or sulfoxylic acid **182** (n = 0). At the same time, sulfurless fragments (e.g., **183**) are formed. The generation of acids **181** or **182** (n = 0) is possible in sulfides activated by the presence of an ethylene or tertiary alkyl group in the neighborhood of the sulfur atom, such as in structures $ROCO(CH_2)_2S-$, $R(CH_2)_2S-$, $C_6H_5(CH_2)_2S-$, $RC(CH_3)_2S-$, or $R(CH_2)_2SS-$. Any molecule of the strong acids **181** or **182** (n = 0) catalytically decomposes hydroperoxides and is transformed in higher-oxidized compounds; for instance, sulfinic **181** (n = 1), sulfonic **181** (n = 2), thiosulfurous **182** (n = 1), or thiosulfuric **182** (n = 2) acids.

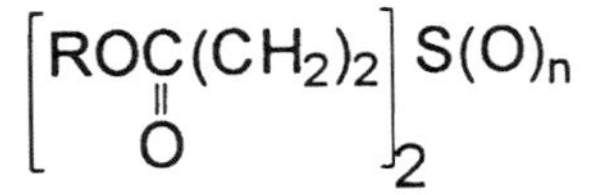

$$\left[ROC(CH_2)_2 \atop {\underset{O}{\|}} \right]_2 S(O)_n \qquad\qquad C_{18}H_{37}SS(O)_nC_{18}H_{37}$$

179  $\qquad\qquad\qquad\qquad\qquad\qquad\qquad$ **180**

$$\underset{\underset{O}{\|}}{ROC}(CH_2)_2S(O)_nOH \qquad C_{18}H_{37}SS(O)_nOH \qquad \underset{\underset{O}{\|}}{ROC}CH = CH_2$$

181 $\qquad\qquad\qquad\qquad$ **182** $\qquad\qquad\qquad$ **183**

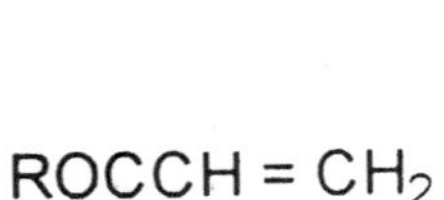

Oxidation products **184** and **185**, having two sulfur atoms, are also formed from monothioethers **32** by the "sulfoxide" pathway via **32→179** (n = 1)→**181** (n = 0). Two molecules of **181** (n = 0) yield thiosulfinate **184** that thermolyzes into the sulfoxylic acid **185** and the sulfurless fragment **183** (Scheme 11).

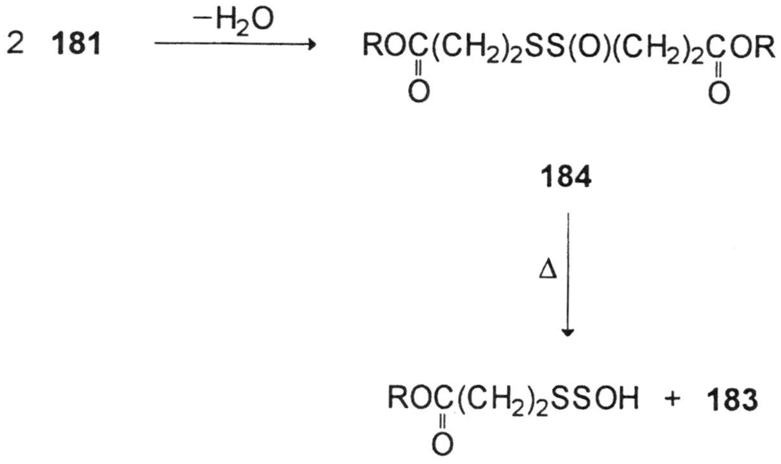

$$2 \;\; \mathbf{181} \quad \xrightarrow{\;\;-H_2O\;\;} \quad ROC(CH_2)_2SS(O)(CH_2)_2COR$$

(with carbonyl oxygens beneath, $\overset{\|}{O}$ on both carbonyls)

184

$$\Delta \downarrow$$

$$ROC(CH_2)_2SSOH \;\; + \;\; \mathbf{183}$$

(carbonyl $\overset{\|}{O}$)

185

Scheme 11

The complex reactivity of activated thioethers **32–35** and a stepwise generation of peroxidolytic products are the basic principles of their high-stabilizing effect. The peroxidolytic effect by protonic acids **181**, **182**, or **185** is reduced in the presence of fillers, such as calcium carbonate, that neutralize acids (74). The effect of acids may also be negated by formation of a complex [$RSO_nH \ldots OSR_2$] between acids **181**, **182**, or **185** and sulfoxides (e.g., **179**; $n = 1$) present in the reaction mixture in excess (75).

Thiyl $RS^{\bullet}$, sulfinyl $RS^{\bullet}(O)$, sulfonyl $RS^{\bullet}(O)_2$, and perthiyl $RSS^{\bullet}$ radicals are formed as intermediates in the very intricated thermal transformation mechanism of **32** (3). Sulfinyl radicals also arise by photolysis of sulfoxides **179** ($n = 1$) (38) and are suggestive of a photosensitization effect in polymer degradation.

In the ultimate stages of the transformation mechanism, sulfur dioxide and sulfuric acid are formed. These inorganic transformation products are also active hydroperoxide decomposers (74).

Thermolysis is a substantial process in the generation of peroxidolytic species. If the primary oxidation products **179** or **180** ($n = 1$) are not thermolyzed, they are oxidized in the second stoichiometric step to the relevant thermostable aliphatic sulfone **179** ($n = 2$) or thiosulfonate **180** ($n = 2$). None of them possess hydroperoxide decomposing activity (74).

Mechanism of deactivation of hydroperoxides by open-chain sulfites **186** ($R =$ alkyl, aryl) and cyclic sulfite **187** was studied (76). A slow redox reaction yielding the respective sulfates and alcohols takes place in the first step and is followed by a rapid catalytic decomposition of the hydroperoxide by proton acids derived from both sulfites and sulfates.

Zinc salts of thiolates **36–38** ($M = Zn$) show generally similar mechanistic patterns in their antioxidant behavior (25,38). The process is exemplified by

$(RO)_2S = O$

186   **187**

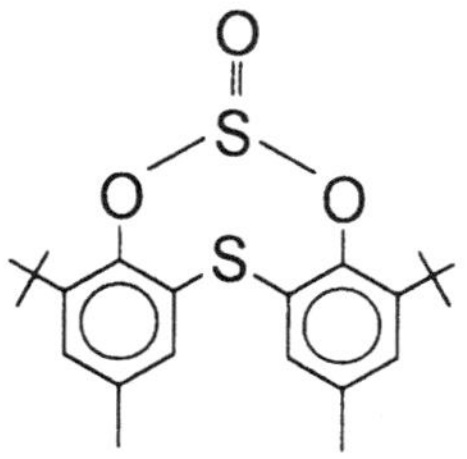

dithiophosphate **37**. Disulfide **188** arises in the first step and the metal is eliminated as zinc oxide [Eq. (34)]. Disulfide **188** oxidizes and thermolyzes in following steps into proton acid **189**, thionophosphoric acid **190**, and sulfur dioxide [Eq. (35)], all of them having peroxidolytic properties. Thiolates **36–38** were also reported as scavengers of alkylperoxyls. An electron transfer has been suggested to take place from the sulfur atom to the alkylperoxy radical, resulting in peroxide **191** and disulfide **188** [Eq. (36)].

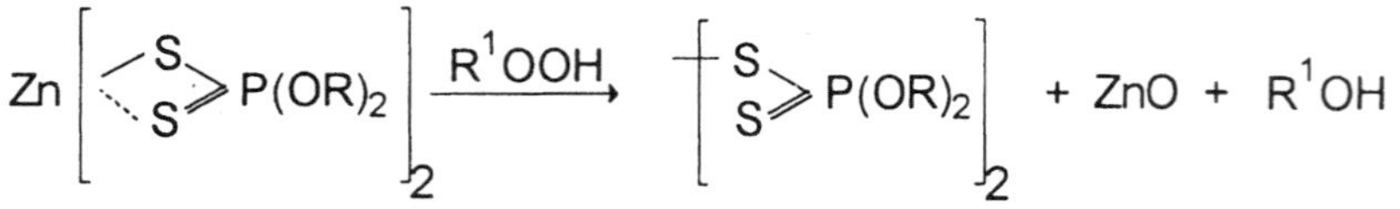

$$Zn\left[\underset{S}{\overset{S}{<}}\!\!\!\!>P(OR)_2\right]_2 \xrightarrow{R^1OOH,} \left[\underset{S}{\overset{S}{<}}\!\!\!\!>P(OR)_2\right]_2 + ZnO + R^1OH \quad (34)$$

37   **188**

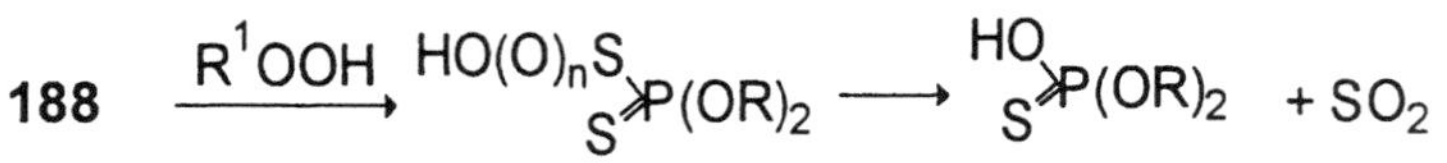

$$\textbf{188} \xrightarrow{R^1OOH,} HO(O)_n\underset{S}{\overset{S}{<}}\!\!P(OR)_2 \longrightarrow HO\underset{S}{\overset{}{<}}\!\!P(OR)_2 + SO_2 \quad (35)$$

189   **190**

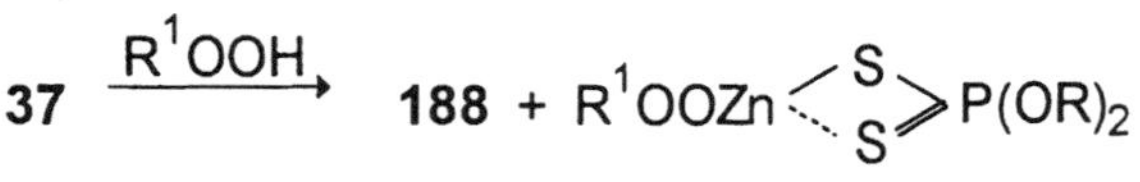

$$\textbf{37} \xrightarrow{R^1OOH,} \textbf{188} + R^1OOZn\underset{S}{\overset{S}{<}}\!\!\!\!>P(OR)_2 \quad (36)$$

191

Organosulfur compounds are used as long-term heat stabilizers in plastics and elastomers, in synergistic combinations with phenolics (see Sec. V.A) (9,29,77). They may cause undesirable organoleptic problems, owing to volatile organic acids of sulfur and sulfur dioxide. This limits their use in odor-sensitive applications.

Another problem arises from deactivation of HAS by acid transformation products of thiosynergists (see Sec.V.E).

2. Organic Compounds of Trivalent Phosphorus

The reactivity of organophosphorus stabilizers with hydroperoxides increases in the series hindered aryl phosphite **44** < nonhindered aryl phosphite **43** < alkyl phosphite **39** < aryl phosphonite **46**. The stoichiometric phase of the sacrificial transformation of phosphites **39–44** and phosphonite **46**, according to Scheme 10, results in structurally relevant compounds of pentavalent phosphorus (25,43), such as phosphate **192** arising from **44**. This transformation ceases the antioxidant activity of phosphites or phosphonites because the pentavalent phosphorus compounds are inactive as antioxidants.

The formation of **192** was confirmed experimentally by quantitative analyses during processing of polyolefins doped with **44** (4,47). Most of **44** was consumed during the first extrusion pass of high-density polyethylene. Progressively lower proportions of **44** were consumed in subsequent passes. It was reported (78) that aliphatic phosphites such as **39** or **40** (R = $C_{18}H_{37}$) are more active processing stabilizers than their aromatic analogues. However, the former are consumed faster during processing and have a lower hydrolytic stability.

Peroxidolytic activity was reported for five-membered cyclic phosphites **193** (43). The relevant phosphate **194**, formed by oxidation, hydrolyzes to the hydrogen phosphate **195** having properties of a Lewis acid able to catalytically decompose hydroperoxides and scavenge alkylperoxyls by the free phenolic hydroxy group (43). Peroxidolytic properties were also reported for the cyclic thiophosphite **196**, owing to the formation of acid transformation products **197** (n = 1–3).

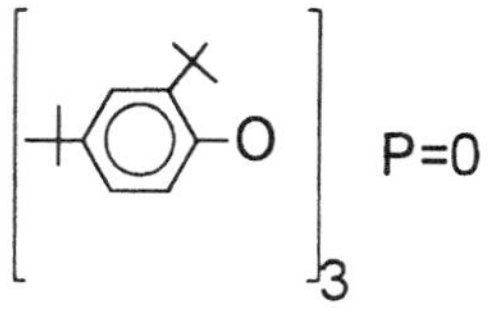

192

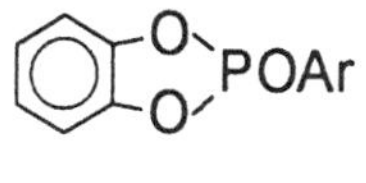

193

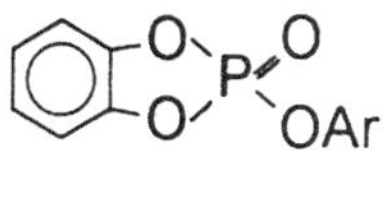

194

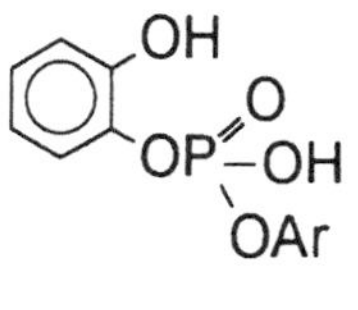

195

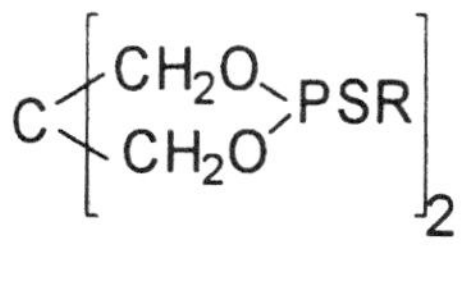

196

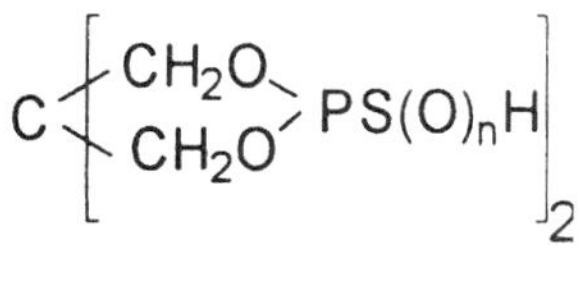

197

According to a proposal arising from liquid model experiments, the aromatic phosphite **44** also contributes to the overall stabilization by scavenging alkoxy and alkylperoxy radicals (25,43). Trapping of alkylperoxyls results in a phosphate and an alkoxyl [Eq. (37)]. The released alkoxyls are able to substitute, in repeating steps, the aromatic moieties in the form of aryloxyls ArO• [Eq. (38)] (2,4-di-*tert*-butylphenoxyl in **44**). Mixed aliphatic–aromatic phosphites result by this substitution process. The released aryloxyl acts as a scavenger of alkylperoxyls, as explained by the reactivity of **120** (A = *tert*-butyl; see Sec.IV.A.1).

$$P(OAr)_3 \ + \ ROO^• \ \longrightarrow \ O = P(OAr)_3 \ + \ RO^• \tag{37}$$

$$P(OAr)_3 \ + \ RO^• \ \longrightarrow \ ROP(OAr)_2 \ + \ ArO^• \tag{38a}$$

$$\xrightarrow{RO^•} \ (RO)_2POAr \ + \ ArO^• \tag{38b}$$

According to the recent analysis (78), the free–radical-trapping by phosphites [see Eq. (37)] seems to be questionable at high-temperature processing of polyolefins. Formation of polymer-bound species such as (polymer-*O*)P(*O*-aryl)$_2$ was not observed. This experimental proof emphasizes deactivation of alkylhydroperoxides as the principal activity pathway in the phosphite mechanism. However, the temperatures during polyolefin processing are far higher than those of hydroperoxide thermolysis. Therefore, the deactivation of hydroperoxides would contribute preferentially at temperatures below 120°C (the common decomposition temperature of hydroperoxides). We should admit that some scavenging of alkoxy radicals formed by thermolysis of hydroperoxides [see Eq. (5)] takes place during processing according to Eq. (38). Moreover, direct scavenging of molecular oxygen by phosphites competing with Eq. (4a) has been proposed (78) to take place at processing temperatures of polyolefins higher than 200°C. This prevents oxidation of alkyls to alkylperoxyls. A charge–transfer mechanism accounting for phosphate formation has been postulated for the process [Eq. (39)].

$$[(ArO)_3P \cdots O_2] \ \longrightarrow \ [(ArO)_3P^+ \ OO^-] \ + \ (ArO)_3P \ \longrightarrow \ 2\,(ArO)_3P = O \tag{39}$$

Phosphites, aliphatic **39** and nonhindered aromatic **43**, in particular, are prone to hydrolysis by humidity during storage and handling. According to model experiments (43), hydrolysis of aromatic phosphites results in hydrogen phosphates **198** and phenols ArOH (Scheme 12). In this way, hydrolysis yields a blend of a hydroperoxide-decomposing and chain-breaking antioxidant. The hydrolysis is catalyzed by acids (e.g., impurities arising from residues of polymerization catalysts, halogenated flame retardants, or PVC) (1,43,68) and accounts for phosphorous acid in the ultimate phase.

The hydrolysis was reported to proceed, particularly in slow-oxidizing liquid substrates, at 150–180°C (43) and is a common process for **43** used commercially

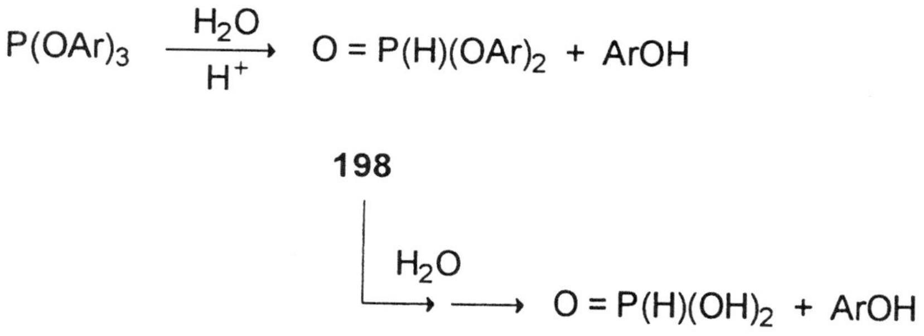

$$P(OAr)_3 \xrightarrow[H^+]{H_2O} O=P(H)(OAr)_2 + ArOH$$

198

$$\xrightarrow{H_2O} \longrightarrow O=P(H)(OH)_2 + ArOH$$

Scheme 12

in unsaturated elastomers. It may take place in solid polyolefins only with phosphites having an extremely poor hydrolytic stability. For example, no hydrolysis was evidenced with aromatic phosphite **44** in melt stabilization of polypropylene (4,46).

It was determined (79) that the partial hydrolysis of phosphites that arises during storage does not deplete their melt-stabilizing performance. The high efficiency of hydrolyzing aliphatic phosphites, accompanied by their fast consumption, was evidenced during processing of polyolefins (78,79). On the contrary, aromatic phosphites are hydrolytically resistant, and their consumption is slower at the expense of their activity. Phosphites without fully suppressed hydrolyzability, such as mixed aliphatic–aromatic phosphites $(RO)P(OAr)_2$, which release phenols after hydrolysis, may be considered as a compromise. Similarly, a combination of partially hydrolyzable and hydrolytically resistant phosphites may account for an enhanced performance. In the mentioned combination, the more effective phosphite is consumed faster.

However, the attendant corrosion of the processing equipment, formation of polymer-insoluble precipitates or clusters, such as $Ca[OP(OAr)_2]_2$, and deactivation of HAS are enhanced by the acids produced by hydrolysis of phosphites. To minimize the hydrolytic instability, basic additives such as 1% tris(2-hydroxypropyl)amine have been added to aliphatic phosphite **40** $(R=C_{18}H_{37})$, and two new classes of stabilizers have been commercialized: dibenzo[d,g][1,3,2]-dioxaphosphocin **45** (80) and fluorophosphonite **47** (81). The new stabilizers have excellent hydrolytic and thermal stabilities and synergistic cooperation with phenolics. The amine moiety in **45** also imparts some metal-deactivating properties, which is important in polyolefins produced with metallocene catalysts or used as insulators of copper wires.

In addition to catalyzed hydrolysis, phosphites such as **43** are destroyed in elastomers by sulfur or peroxide cure (35) and are converted to thiophosphate or phosphate esters, respectively.

C. Metal Deactivators

The catalytic effects of metallic impurities, introduced with fillers, pigments, or reinforcing agents, and residues of polymerization catalysts during fabrication or processing of polymers, or the deleterious effects of copper in wires and cables on polyolefin insulation markedly increase the rate of oxidation of the polymer matrix (1,33). New oxidation chains are initiated by the homolysis of hydroperoxides catalyzed by transition metals (Fe, Co, Mn, Cu, V) [Eqs. (40) and (41)[(34).

$$M^{n+} + RO\overset{.}{O}H \longrightarrow M^{(n+1)+} + RO^{.} + HO^{-} \tag{40}$$

$$M^{(n+1)+} + ROOH \longrightarrow M^{n+} + ROO^{.} + H^{+} \tag{41}$$

Chain-breaking phenolic antioxidants used as single stabilizers usually have a reduced efficiency in stabilization of polymers containing metal impurities. The local concentration of alkylperoxyls and alkoxyls is too high; consequently, phenols are consumed very quickly. The presence of chain-breaking antioxidants is, however, mandatory and the stabilization of metal-contaminated polymers is achieved either with combinations of metal deactivators, such as **51** with phenols (33), or with phenols having their structures modified with hydrazide, amide, or phosphite moieties that form polyfunctional chelating agents with ligand atoms N, O, S, or P. Additives **48–50** and **52** are examples of effective structures.

The mechanism of activity of metal deactivators is characterized by conversion of active metal species into inactive, or at least less active, hydro-peroxide-homolyzing metal complexes (33,34). The chelation capacity of a particular metal deactivator and the character of the metal ion determine the efficiency of the deactivation process. The most effective systems contain polydentate chelates (e.g., **199**; Subst. = phenolic part of the stabilizer) (34). The phenolic moiety in metal deactivators is an efficient scavenger of alkylperoxyls and transforms according to mechanisms described in Section IV.A.1.

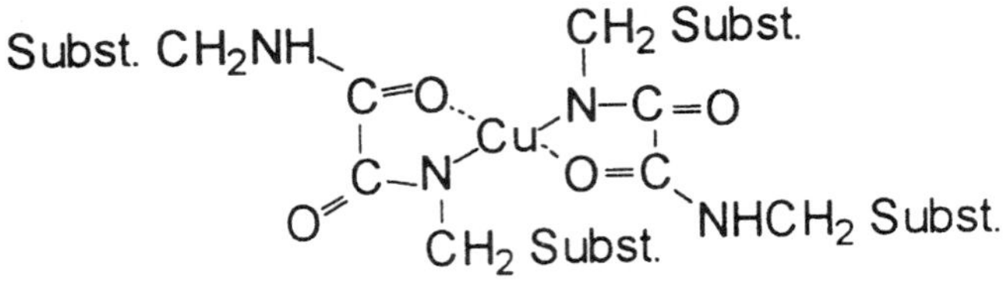

199

D.   Antiozonants

Complex permanent protection of unsaturated rubbers against thermal- and photooxidation, fatiguing, and ozone aging, taking place either as parallel or con-secutive degradation processes, is mandatory. Carbon black, used in rubber vulcanizates as reinforcing agent, acts as light screen protecting against sunlight. Structures of commercialized additives used as antiozonants are shown in Sections III.A.1.b and III.C. Most beneficial properties have additives combining antioxidant and antiozonant properties. The most common antiozonants are diamines **23a,b**, possessing also strong chain-breaking antioxidant activity (see Sec. IV.A.2) (15,35,56,70). Other commercial antiozonants, derivatives of dihydroquinoline **24**, **26**, nickel(II) dibutyl dithiocarbamate **36** (M = Ni), and *N,N*-dibutyldithiourea **53** also have antioxidant properties. Exclusively antiozonant properties have been postulated for acetal **54**.

Some theories of antiozonant action were postulated. The complexity of parallel autoxidation and ozonation mechanisms and the diversity of products formed in aging of rubbers indicate that a combination of more concerted or complementary processes participates in the mechanism of the antiozonant action:

1.   Direct scavenging of ozone by chemical antiozonants and formation of an inert surface barrier by high molecular weight transformation products of antiozonants
2.   Reaction of antiozonants or their oxidized and ozonized transformation products with ozonized rubber molecules (ozonides **4**, zwitterions **5**, hydroperoxides **9**, and aldehydes **10**), resulting in an inert protection film and a partial reparation of rubber macromolecules by relinking of severed chains and formation of polymer-bound residues of used antiozonants
3.   Reaction with oxygen and alkylperoxyls (chain-breaking antioxidant effect) and with alkyl radials (antifatigue effect)
4.   Complementary activity with physical antiozonants (waxes)

A very limited number of chemical additives can fulfill the complex requirements. Moreover, the antiozonant protection requires presence of a rather high concentration of additives (diamines **23a,b** are used in amounts up to 3%). Therefore, a high solubility of antiozonants in rubber vulcanizates is a prerequisite. Ozonation is a surface process. Hence, a high surface concentration of antiozonants is required. This suggests good migration of fresh antiozonants from the bulk to the rubber surface, to replenish the consumed additive. Waxes used as physical antiozonants may contribute to the migration. Most information about the antiozonant mechanism was obtained with *N,N'*-disubstituted 1,4-phenylenediamines **23a,b** (14,15,70).

Direct ozone scavenging is certainly the primary step preventing rubber from ozonation in the absence of any protecting layer. This reveals the importance of the high reactivity of antiozonants with ozone, enabling competition with ozonation of the rubber matrix (14,70). The initial stages of ozonation pathways of **23a,b** include formation of an ozone adduct **200** converting into an amine oxide **201** (pathway *a*, Scheme 13) or unstable nitroxide **202** (pathway *b*). The reaction products are formed by destruction of the adduct **200** through a transient **203** (pathway *c*), by coupling of the free radical intermediate **204** (pathway *d*) and by transformations of **201** and **202**.

The relative importance of pathways *a–d* depends on the character of the *N,N'*-substituents. *N,N'*-Bis(1-methylheptyl)-1,4-phenylenediamine (see **23a**) transforms mainly by pathway *a* and by oxidation of aliphatic *N*-substituents (pathway *c*) (14). Pathway *a* results in hydroxylamine **205** (R^1, R^2 = 1-methylheptyl) ozonizing to 4-nitroso- **206** (n = 1, R^1 = 1-methylheptyl) and 4-nitrophenyl(1-methylheptyl) amine **206** (n = 2, R^1 = 1-methylheptyl), and transforming into Bandrowski's base **207** (R^1, R^2 = 1-methylheptyl). The ozonation through pathway *c* (see Scheme

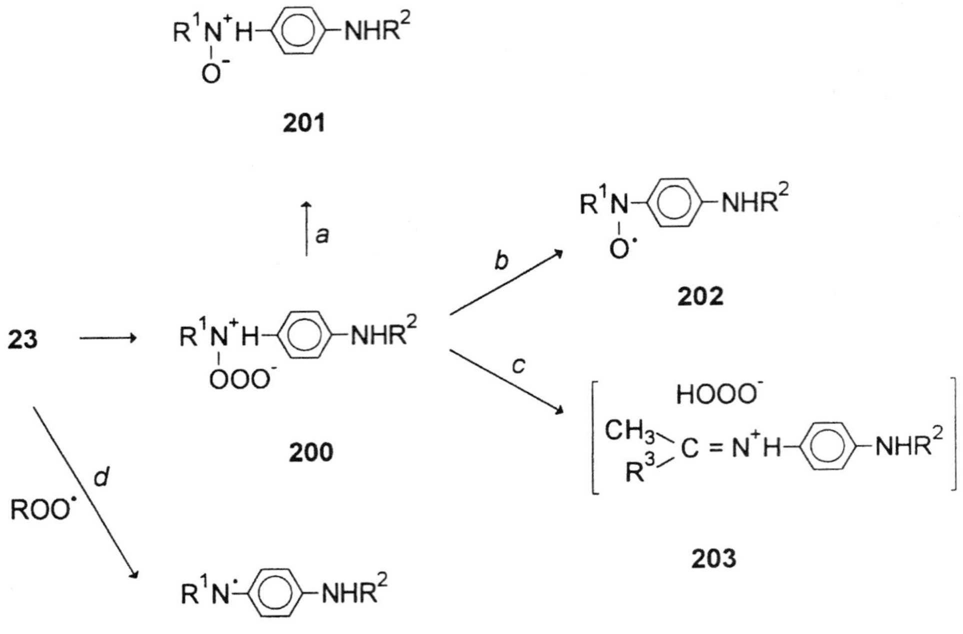

R¹N⁺H—⟨C₆H₄⟩—NHR²
    |
    O⁻

201

23 ⟶ R¹N⁺H—⟨C₆H₄⟩—NHR²
              |
              OOO⁻

200

a (upward)

b ⟶ R¹N—⟨C₆H₄⟩—NHR²
           |
           O•

202

c ⟶ [HOOO⁻ ; CH₃/R³ C = N⁺H—⟨C₆H₄⟩—NHR²]

203

d (ROO•) ⟶ R¹N•—⟨C₆H₄⟩—NHR²

204

Scheme 13

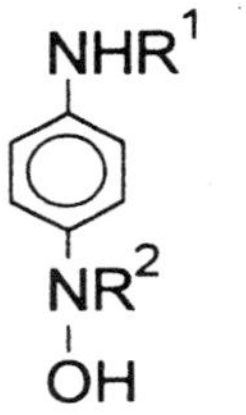

NHR¹
⟨C₆H₄⟩
NR²
|
OH

205

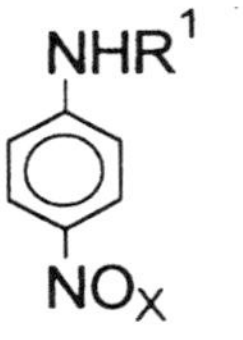

NHR¹
⟨C₆H₄⟩
NOₓ

206

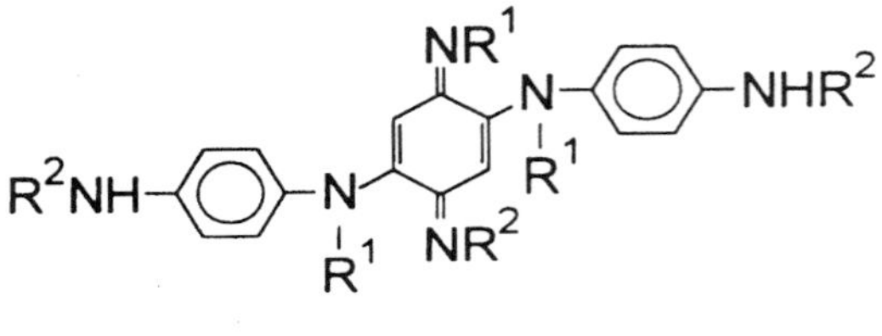

R²NH—⟨C₆H₄⟩—N(R¹)—⟨quinone: NR¹ ... NR²⟩—N(R¹)—⟨C₆H₄⟩—NHR²

207

CH₃CO(CH₂)₆CH₃

208

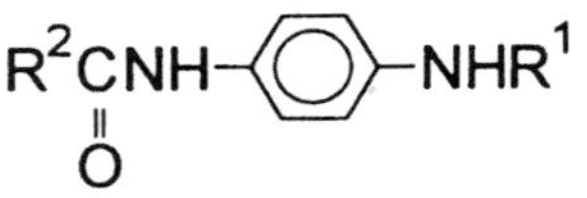

R²CNH—⟨C₆H₄⟩—NHR¹
    ‖
    O

209

N—⟨C₆H₅⟩
‖
⟨C₆H₄⟩
‖
RN⁺O⁻

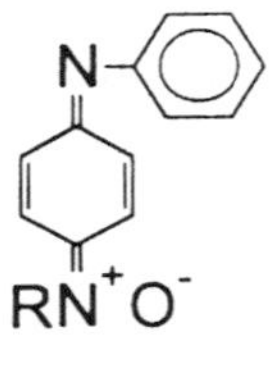

210

13) results in oxidation of secondary alkyls and yields octan-2-one **208** and amide **209** (R^1 = 1-methylheptyl; R^2 = alkyl less than C_6). Pathway *b* yields bisnitrone **148** (R^1, R^2 = 1-methylheptyl) and is not particularly important for transformations of **23a**. Pathway *d* does not takes place with **23a** at all.

Transformations of antiozonant *N*-(1,3-dimethylbutyl)-*N'*-phenyl-1,4-phenylenediamine **23b** by all four pathways are shown in Scheme 13 (15). The increased importance of the free radical process (pathway *d*) is a consequence of the substituent effect of the diphenylamine part of the molecule. Ozonation proceeds only at the side of the molecule containing the aliphatic substituent. Pathway *a* results in formation of **206** (R^1 = phenyl). The presence of Bandrowski's base was not detected. Mononitrone **210** (R = 1,3-dimethylbutyl) and amide **209** (R^1 = phenyl, R^2 = alkyl less than C_4) arise by pathways *b* and *c*, respectively. N–N- and C–N-coupling products **135** and **138** (R^1 = phenyl, R^2 = 1,3-dimethylbutyl), respectively, arise by pathway *d* (we consider formation of **135** and **138** as a result of oxidative transformations caused by scavenging of alkylperoxyls).

N,N'-Diphenyl-1,4-phenylenediamine **23c** having exclusively aromatic *N,N'*-substituents reacts by pathways *b* and *d*, i.e., in analogy to mechanisms characteristic of the antioxidant activity; (see Sec.IV.A.2). This also causes formation of the quinone imine **147** (R^1, R^2 = phenyl), arising in a concerted antioxidant–antiozonant process.

Nonmigrating and nonvolatile products of ozonation of **23** participate, together with the parent **23**, in formation of the protective film on the surface of rubber, thereby hindering contact of rubber with ozone (14,15,70). Phenylenediamines **23a,b** also react with products of the ozonized rubber (15). Diamines **23** are oxidized by ozonides **4** into the respective quinone imines **147**. Zwitterion **5** generates mono- and bishydroperoxide **211**. Aldehydes **10** arising by ozonolysis of the rubber carbon chain are potential reactants, accounting for bound-in phenylenediamines **212** (R = polymeric residue), relinked rubbers **213** (R = polymeric residue) and rather complex cross-links.

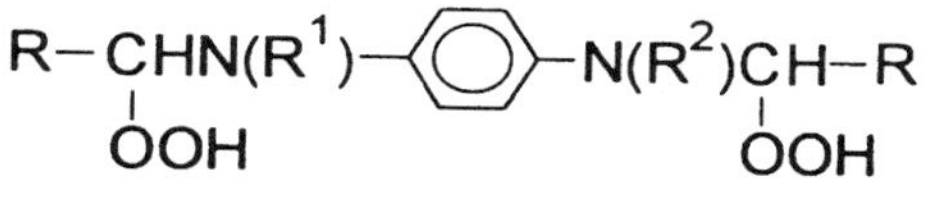

211

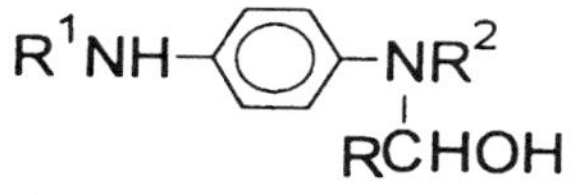

212

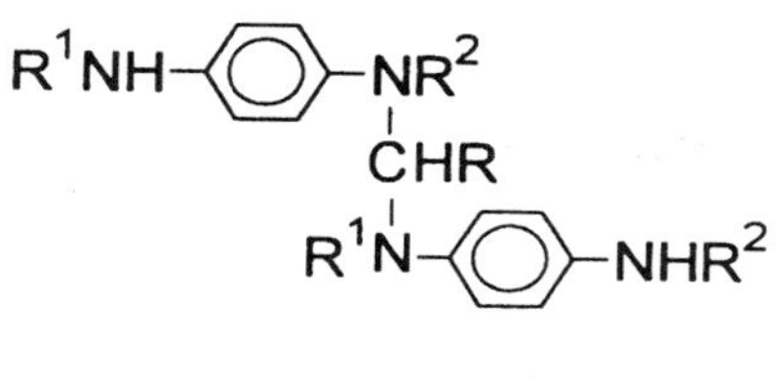

213

Products arising from N,N'-disubstituted 1,4-phenylenediamines **23** and alkyl radicals (**140**, **165–168**) and species such as **211–213** rank among various rubber-bound moieties, considered as "nonextractable" nitrogen arising in amine-doped rubbers.

E. Photoantioxidants

Protection of polymers against photochemically induced oxidation in outdoor applications is of key importance. Because of the similarity of intermediates and products in the principal steps of thermal oxidation and photooxidation, formally analogous protection pathways against both oxidations can be exploited. However, photosensitivity of most conventional chain-breaking phenolic stabilizers and of some thioethers limit their performance when applied as single stabilizers. Photostable additives deactivating alkylhydroperoxides and alkylperoxy radicals in photooxidizing polymers are categorized as photoantioxidants and are outlined in this section.

1. Photostable Phenols

If used without photostabilizers or HAS, conventional phenolic antioxidants are rather poor stabilizers in photooxidation of polyolefins or styrenics (48,49). This is due to a rather easy formation of phenoxyls by photolysis or photooxidation, thereby reducing their scavenging capacity in systems with a high production of alkylperoxyls. Substitution of position 4 of hindered phenols by an electron-withdrawing group, such as in **21** ($R = C_{16}H_{32}$, 2,4-di-*tert*-butylphenyl) increases the resistance of the phenolic hydroxy group against photolysis (82). Besides scavenging alkylperoxyls by the general reaction **22a**, phenol **21**, having the aromatic substituent R, is considered to contribute to photostability of polyolefins by a photo-Fries rearrangement, yielding in situ a benzophenone derivative **214**, an effective UV absorber (37) [Eq. (42)].

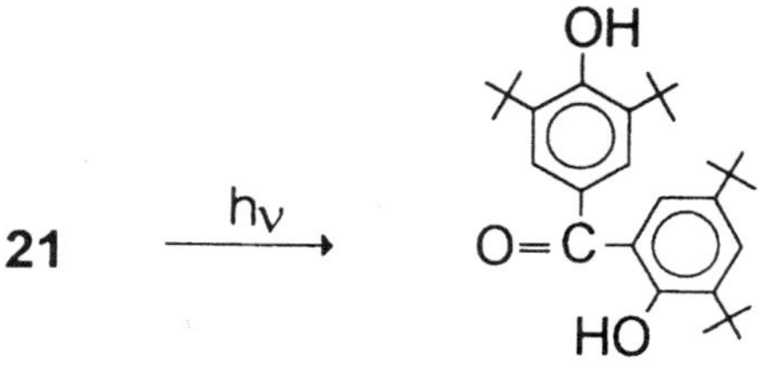

21 $\xrightarrow{h\nu}$ (42)

214

However, aliphatic ester **21** ($R = C_{16}H_{33}$) is a photoantioxidant, comparable with the aromatic analogue. Therefore, the photo-Fries rearrangement is not a necessary condition of the antioxidant activity.

The photoantioxidant effect was also enhanced by a proper substitution of the phenolic part in **215** (n = 1) (83). In spite of the fact that the imidazolidinone moiety has the structure of an HAS, the favorable antioxidant effect was lost in **215** (n = 0) (i.e., in the absence of the electron-withdrawing group).

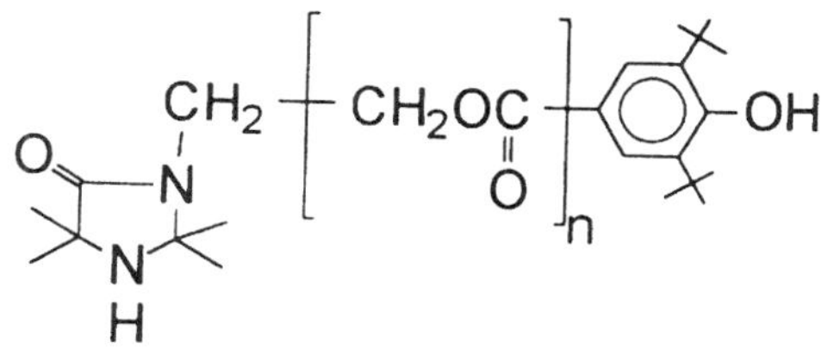

$$\left[CH_2 \overbrace{} CH_2OC \overbrace{} \bigcirc -OH \right]_n$$

215

The phenolic quencher of excited carbonyls **19** (M = Ni) has a pronounced chain-breaking effect and may also be listed among relatively photostable phenols (10); however, no phenolic structure can be fully resistant to a long-term photochemical depletion.

2. Organic Thio Compounds

Complexes of transition metals **36–38** (M = Ni) are photostable hydroperoxide decomposers (see Sect.IV.B.1) and are considered to be effective photoantioxidants (10,38). Transformation products such as **188–190** and their analogues formed from dithiocarbamate **36** or xanthate **38** have been envisaged to arise during photo-processes and impart the peroxidolytic activity. Practical application of **36–38** is limited by polymer discoloration and a lower stability in the polymer matrix.

3. Hindered Amine Stabilizers

Introduction of secondary > NH and tertiary $> NCH_3$ hindered amine stabilizers (HAS) eliminated many problems arising from insufficient photoantioxidant protection of polyolefins. Various derivatives of 2,2,6,6-tetramethylpiperidine (**55, 56, 58–63**) and piperazinone (**57**) have been commercialized. HAS outperform other classes of stabilizers by the complex activity in processes induced by solar and high-energy radiations and thermal oxidation at temperatures not exceeding 120°C (15). Because of their commercial importance, individual pathways of the mechanism have been continuously reevaluated.

Generally accepted pathways of the mechanism are based on a complex of chemical transformations (3,15,56). The primary step includes formation of a nitroxide > NO$^\bullet$ after oxidation of HAS with alkylhydroperoxides or peroxy acids. Complexing of HAS with alkylhydroperoxides [> NH...HOOR] effectively increases HAS concentration in the vicinity of the principal oxidation species responsible for oxidation initiation. Oxidation of secondary amine > NH with alkylhydroperoxides accounts for hydroxylamine > NOH and O-alkylhydroxylamine > NOR [Eqs. (43a,b)], species reacting with alkylperoxyls and forming nitroxide > NO$^\bullet$.

$$2 > NH + ROOH \longrightarrow \begin{cases} > NOH + ROH & (43a) \\ \\ > NOR + H_2O & (43b) \end{cases}$$

Tertiary HAS, mostly $> NCH_3$, are oxidized into salts (formates) of secondary HAS **216** (83) [Eq. (44)]. Salts of secondary HAS and acetic acid **217** are formed together with O-alkylhydroxylamine by oxidation from N-acylpiperidines [e.g., **55** (84); Eq. (45)]. Nitroxides are formed in the following oxidation step from **216** and **217**.

$$> NCH_3 \xrightarrow{\text{ROOH}} \left[> N^+H_2 \cdots {}^-O\overset{\displaystyle O}{\overset{\displaystyle \|}{C}}H \right] \qquad (44)$$

216

$$> NCOCH_3 \xrightarrow{\text{ROOH}} > NOR \; + \; \left[> N^+H_2 \cdots {}^-O\overset{\displaystyle O}{\overset{\displaystyle \|}{C}}CH_3 \right] \qquad (45)$$

217

Formation of nitroxides from various classes of effective HAS [Eq. (46)] explains their equivalency in commercial stabilization.

$$> NH, \; > NCH_3, \; > NOR, \; > NCOCH_3 \xrightarrow{\text{ROOH}} \longrightarrow \; > NO^{\cdot} \qquad (46)$$

The trapping of alkylperoxyls by hydroxylamines and O-alkylhydroxylamines is a key regeneration pathway for nitroxides (15,56). Mechanism of oxidation of a hydroxylamine is a clear process. The regeneration of nitroxides from O-alkylhydroxylamines, considered as the key pathway in the HAS action mechanism, has been discussed many times (15). A realistic explanation accounting for scavenging of alkylperoxyls or peracyl radicals is based on a rather recent product study (85). An intermediary complex **218** is envisaged in the process, resulting in non-peroxidic products, carbonyl compound **219**, and an alcohol or carboxylic acid (Scheme 14). This mechanism explicitly explains regeneration of the nitroxide by depleting both alkylperoxyl (formed in the "hydroperoxide" pathway) and peracyl radicals (reactants in the "peroxy acid" pathway) (85).

Trapping of alkyl radicals by nitroxides accounting for the formation of O-alkylhydroxylamines [Eq. (47)] is a very effective contribution of HAS to the stabilization mechanism (15) and has to compete with self-termination and oxidation of alkyls (4,46).

$$> NO^{\cdot} + R^{\cdot} \longrightarrow \; > NOR \qquad (47)$$

The formed O-alkylhydroxylamine thermolyzes, in an inert atmosphere, to the respective hydroxylamine (a strong chain-breaking antioxidant) and an olefin [Eq. (48a)] (15). Thermolysis performed in air yields nitroxide [see Eq. (48b)].

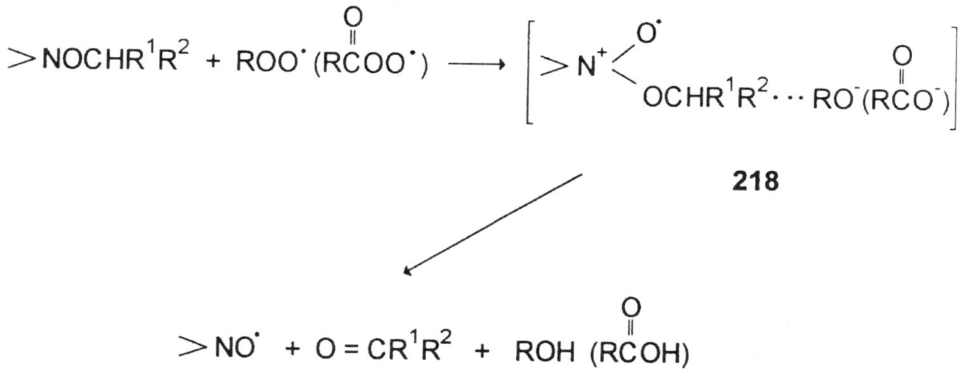

$$>NOCHR^1R^2 + ROO^{\cdot}\,(R\overset{\overset{O}{\|}}{C}OO^{\cdot}) \longrightarrow \left[>N^+ \overset{O^{\cdot}}{\underset{OCHR^1R^2}{<}} \cdots RO^-\,(R\overset{\overset{O}{\|}}{C}O^-) \right]$$

218

$$>NO^{\cdot} + O{=}CR^1R^2 + ROH\,(R\overset{\overset{O}{\|}}{C}OH)$$

219

Scheme 14

O-Alkylhydroxylamines were unambiguously identified in the aged polyolefin matrix doped with HAS and should be considered as the dominant reservoir of nitroxides.

$$>NOR \xrightarrow[N_2]{\Delta} >NOH + \text{olefin} \tag{48a}$$

$$>NOR \xrightarrow[O_2]{\Delta} >NO^{\cdot} + ROO^{\cdot} \tag{48b}$$

Recent analyses of experimental data obtained with HAS-doped polymers revealed the potential importance of charge–transfer complexes [HAS...O$_2$], [Polymer...O$_2$], or [HAS...ROO$^{\cdot}$] in the very early stages of the stabilization mechanism (86).

In spite of the proposed existence of the nitroxide regeneration involving HAS transformation products >NOR and >NOH, the long-term activity of HAS is limited by some depleting processes (15,86). Attempts to improve long-term applications by modification of the known structures has become very attractive today. Because of the relative basicity of secondary and tertiary HAS, difficulties arise in applications in contact with acid species, such as mineral acids arising from dehydrochlorination of PVC, residues of condensation catalysts in coatings, halogenated flame retardants or pesticides, and sulfur-containing hydroperoxide-decomposing antioxidants (15). Mineral acids deactivate HAS by protonation. Salts such as **220** (X = mineral acid residue) are not prone to form nitroxides, the principal intermediate in the HAS mechanism. (Weak carboxylic acids form complexes analogous to **220**; however, they do not deplete HAS.)

Moreover, secondary and tertiary HAS are hardly applicable as stabilizers in polycarbonates (87). Because of their basicity, HAS catalyze hydrolysis of polycarbonates and impair their physical properties. To improve the resistance of HAS against an acid environment, structural modifications of the amino group lowering the basicity were performed. This effort is exemplified by comparison of pK_a values of various HAS (15).

HAS	pK_a	HAS	pK_a
Secondary (> NH)	8.0–9.7	Hydroxylamine (> NOH)	4.3–6.1
Tertiary (> NCH$_3$)	8.5–9.2	O-Alkylhydroxylamine (> NOR)	~4.2
Oligomeric tertiary (> N—)	~6.5	Acylated (> NCOCH$_3$)	2.0

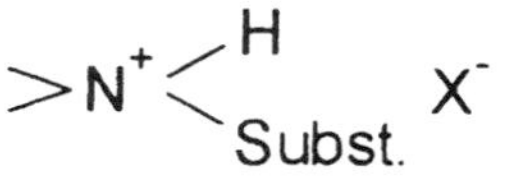

220

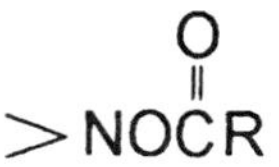

221

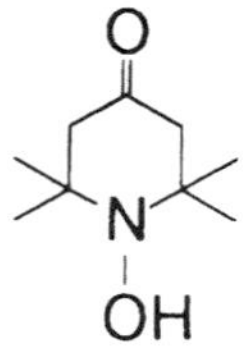

222

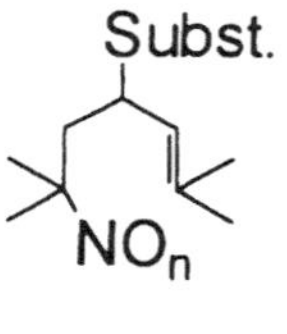

223

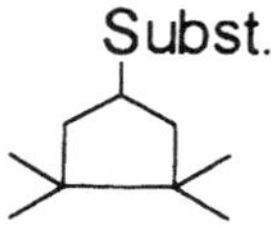

224

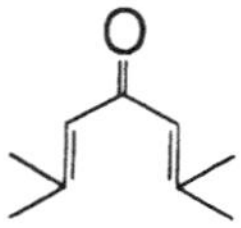

225

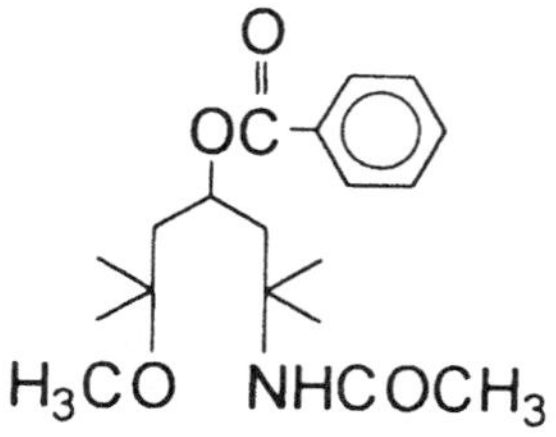

226

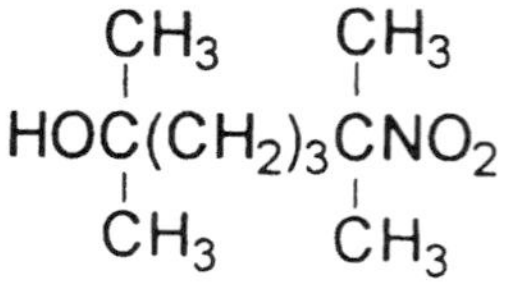

227

Structures **61** and **62** (oligomeric tertiary HAS), *O*-alkylhydroxylamine **56** (R = *iso*–C₈H₁₇), and acylated HAS **55** are examples of commercialized structures applicable in the hostile environment (15,88). The improved resistance of **55** against acid environment is compensated by a reduced performance in weathered polyolefins (88). The essentially nonbasic **56** (R = *iso*–C₈H₁₇) may be considered as a superior stabilizer in polyolefins containing flame retardants, in plasticized PVC, or in acid-catalyzed clearcoats.

Some other irreversible chemical transformations account for the loss of HAS activity (15). This comprises transformation in inactive *N*-acyloxy derivative **221** by recombination of HAS-derived nitroxide and acyl radical generated by photolysis of polymer-bound ketones. Photochemically induced splitting-off of the ester group from position 4 of **56** results in the volatile 4-oxo derivative **222**, which is prone to a subsequent degradation of the piperidine ring. Oligomeric HAS **61** fragments by photolysis at $\lambda > 300$ nm into low molecular weight nitroxides.

Of themselves, HAS do not absorb solar UV radiation. However, the derived nitroxides form excited π–π^* states after absorption of light in the range of 300–320 nm. The piperidine ring fragments, as a consequence of the excitation, and is oxidized. Open-chain nitroso **223** (n = 1) and nitro compounds **223** (n = 2) are formed together with nitrogen-free fragments **224** and **225** (3,15). The nitro compound **223** was also detected in polypropylene following γ-irradiation storage. Besides protonation, hydrogen chloride was reported to induce ring opening of 1-acyl-4-benzoyloxy-2,2,6,6-tetramethylpiperidine, accounting for **226**.

Ozone, the tropospheric oxidizing pollutant, was reported to deplete the tetramethylpiperidine cycle to 2,6-dimethyl-6-nitroheptan-2-ol **227** by an intermediary adduct [$> \mathrm{NH} \ldots \mathrm{O_3}$] (15).

F. Photostabilizers

In 1994, this class for additives accounted for 3% of the total plastics additives market (22). Two principal classes of photostabilizers are used for protection of polymers against harmful effects of the sunlight: absorbers of UV radiation [derivatives of benzotriazole and benzophenone form about 80% of the consumption (23)] and quenchers of excited chromophores (10,39). Because of their commercial importance, photostabilizers attract constant attention. Most commercialized photostabilizers do not act by a single mechanism (40). Elucidation of the chemical behavior of photostabilizers now allows optimization of their structures for particular applications in commodity and engineering plastics and coatings and to explain transformations that deplete the stabilizing effect.

1. UV Absorbers

The UV absorbers compete with the substrate-specific absorption that causes polymer degradation. A short-term photostability may be reached by some fluorescent and photochromic compounds containing stilbene and azobenzene moieties (37). However, the additives suffer from unacceptable color effects and poor inherent photostability.

Effective UV absorbers that have commercial application are colorless compounds characterized by high-absorption coefficients in the spectral range between 300 and 400 nm and by high, inherent photostability (10,50). They absorb sunlight and dissipate, in a harmless way, the energy of absorbed photons within the polymer matrix. A very rapid excited-state intramolecular proton transfer (ESIPT) between parent and cyclohexadienoide structures by means of $O \ldots H \ldots O$ and $O \ldots H \ldots N$ hydrogen tunneling was confirmed (2,10,37,39,41,50,89) for derivatives of 2-hydroxybenzophenone **64** [Eq. (49)], 2-(2-hydroxyphenyl)benzotriazole **65–68** [Eq. (50)], 2-(2-hydroxyphenyl)-1,3,5-triazine **69** [Eq. (51)], 2-(2-hydroxyphenyl) pyrazoles, and 2-(2-hydroxyphenyl)benzoxazoles. ESIPT is responsible for the exceptional inherent photostability of these photostabilizers. Their lifetime is influenced by the microenvironmental structural effects affecting the hydrogen-bonding properties (89). The UV-absorbing capacity remains intact as long as the intramolecular proton transfer is not broken.

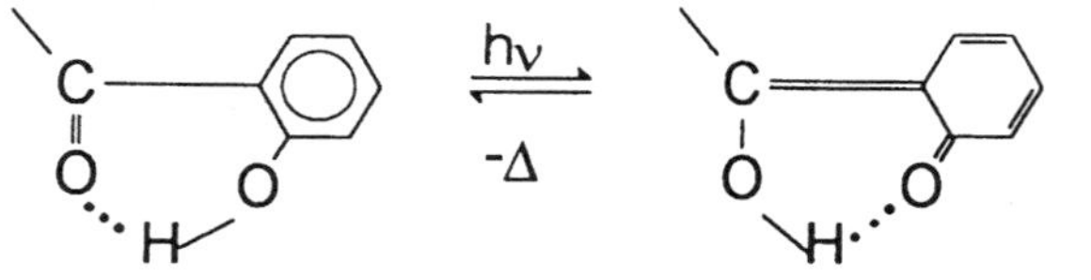

(49)

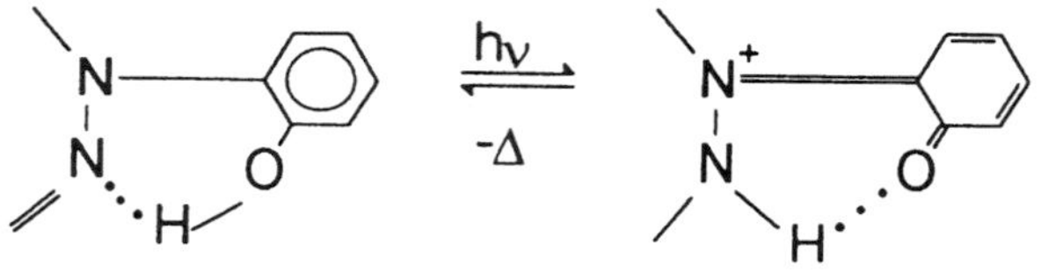

(50)

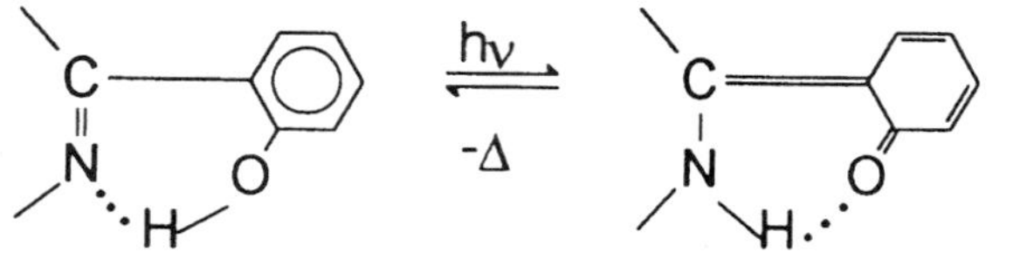

(51)

For oxamide **70**, formation of charge-separated species, Eq. (52), after photoexcitation was postulated as the activity mechanism (2). A similar mechanism was proposed for the 2-cyanoacrylate **71**. Aryl benzoate **72** rearranges under UV radiation into a derivative of 2-hydroxybenzophenone (10,37), acting by the mechanism envisaged for **64** [see Eq. (49)].

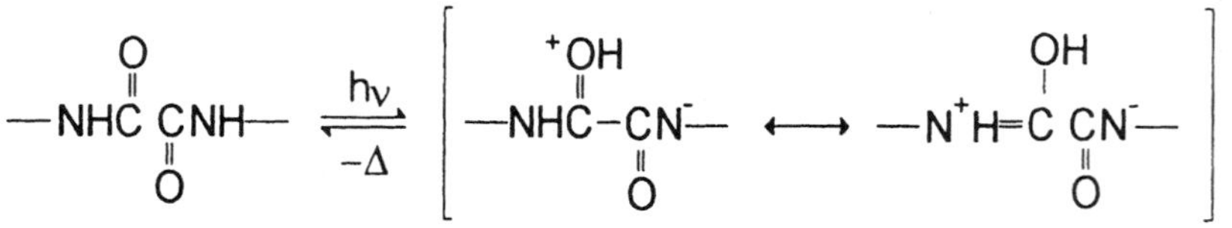

(52)

Ultraviolet absorbers disappear by physical processes (leaching, evaporation) and chemical–photochemical degradation. Physical losses can be minimized by increasing molecular weights of stabilizers (such as in **67, 68**; see also Sec. VII).

During the years of development, the structures of UV absorbers were optimized to reach high absorption close to 300 nm and optimize the inherent physical and chemical stability during long-term protection of polymers. The activity mechanism of UV absorbers has been based on reversible physical processes. They also undergo, however, irreversible chemical and photochemical transformations that cannot be avoided in oxidizing polymer matrices. Photochemical transformations proceed by excited states of UV absorbers (39,41,50,89). The scavenging of alkylperoxyls by photostabilizers that contain phenolic moieties, such as **64–69**, apparently contributes somewhat to the integral performance by the chain-breaking antioxidant effect. This account for oxidative transformations of the phenolic moiety of **64–69**, proceeding by mechanisms characteristic of phenolic antioxidants (see Sec. IV.A.1), resulting in phenoxyls, quinone methides, and substituted cyclohexadienones (42,49) and irreversibly disrupting the H-tunneling mechanisms exemplified in Eqs. (49)–(51). This results in the loss of the photostabilizing power.

Some products of the depleting oxidation of UV absorbers were isolated. Hydroxybenzophenone **64** oxidizes into (alkylperoxy)cyclohexadienone **228**, substituted benzoquinone **229**, and xanthone **230** (30,38). The degradation of **64** is faster in

228   **229**   **230**

easily oxidizing polymers and in polymers that have been subjected to severe processing or that contain a substantial concentration of carbonyl chromophores that arise by degradation (38). Hydroxybenzophenones, therefore, are rather ineffective photostabilizers in high-impact polystyrene or acrylonitrile–butadiene–styrene (ABS) polymers. Formation of intermolecular hydrogen bonds between hydroxybenzophenones and oxo groups of the oxidized polymer matrix **231** increases the lifetime of the excited singlet of the UV absorber (89) and favors degradation of the stabilizer (39). [Alkylperoxy cyclohexadienone **232** was isolated in a model study with benzotriazole **65** (3,30,39).] Other UV absorbers containing phenolic moieties, such as derivatives of benzoxazole, pyrazole, or 1,3,5-triazine, are also prone to depletion by photooxidation, but the products were not identified. It was reported in a more recent study (39,90) that formation of **232** by oxidation of **65** with alkylperoxyls proceeds as a dark reaction. Oxidation of excited **65** results in cleavage of the molecule on the bond *N*–phenyl. Free benzotriazole **233** was isolated as a product. The released phenolic moiety yields nonidentified oxidation products.

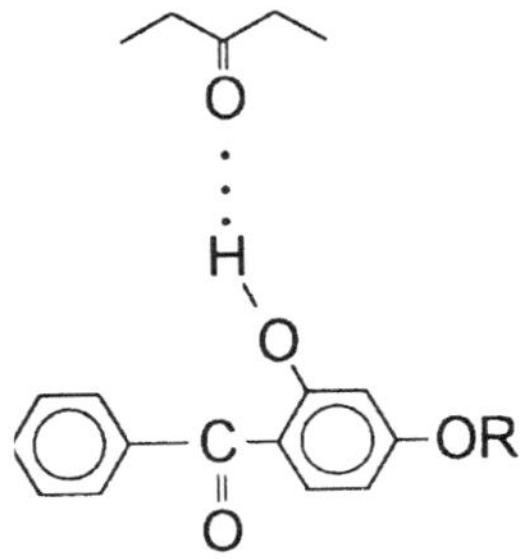

231 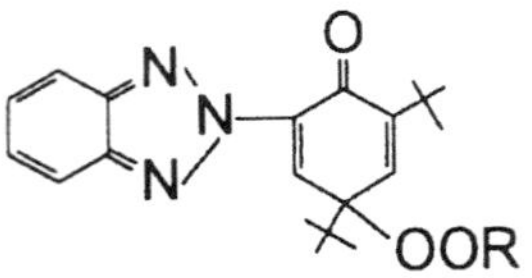   **232** 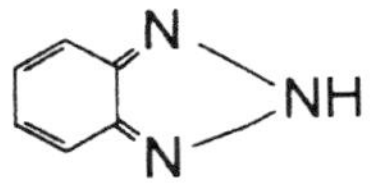   **233**

Because of the reactions with alkylperoxyls, UV absorbers that contain phenolic moieties are deactivated preferentially in rapidly oxidizing polymer matrices (89). The intramolecular H-tunneling mechanism [see Eqs. (49–51)] of UV absorbers is also vulnerable in polar matrices of engineering polymers, which contain functional groups that form strong intramolecular H-bonds with the stabilizer (89), similar to those exemplified in **231** for the system benzotriazole–deeply oxidized polymer. The disruption of the intramolecular hydrogen bond by a polar environment populates the low-lying triplet, resulting in a reactive n,π* excited triplet. This has unfavorable consequences. Excited and intermolecularly bonded UV absorbers are prone to initiate polymer photodegradation (91). A proper molecular architecture may favor the intramolecular H-bonding characteristic of a coplanar structure of the absorber. A bulky substituent (such as a tertiary alkyl or α-methylbenzyl) *ortho* to the phenolic hydroxy group, effectively shields the coplanar intramolecularly fixed form in 2-(2-hydroxyphenyl)benzotriazoles (**65**, **66**, or **68**) (50,89) and prevents formation of a nonplanar and photodegradable intermolecularly bound structure.

It may be extrapolated that the oxidative transformation of the phenolic moiety in **66** and **68** exploits the beneficial chemistry of the propionate-type phenolic antioxidants, including the quinone methide regenerative pathways exemplified by rearrangement of **112** to **113** (see Sec. IV.A.1) and subsequent transformations prolonging the intramolecular H-tunneling, according to Eq. (50), by regeneration of the phenolic moiety.

The mechanisms of oxidative depletion of oxamide **70** and 2-cyanoacrylate **71** are unknown. The interference of free alkylperoxy radicals with the charge-separated species postulated in Eq. (52) was mentioned recently without any mechanistic details (39,92).

Ultraviolet absorbers containing phenolic moieties are subjected to slow radical-induced photoassisted degradation in the course of years of exposure of the stabilizer-doped polymer matrix (39,91,92). The photolysis, accounting for the loss of activity, takes place preferentially in the most radiation-exposed surface layers of weathered polymers. The loss of stabilizers by photochemical processes was reported as a problem specific to glassy polymers and coatings (39). The consequences are observable after a long exposure to sunlight. For example, it was reported that 2-(2-hydroxyphenyl)-1,3,5-triazine (**69**) was inherently more photostable than the physically persistent benzotriazole (**67**) after testing in polycarbonate over long exposure periods (41), although the initial physical persistence and performance of both the stabilizers was comparable. Stabilizers **67** and **69** surpassed 2-hydroxybenzophenone **64** in photoprotection of polycarbonate. The long-term photolysis of **64** yields benzoic acid **234** (39,92).

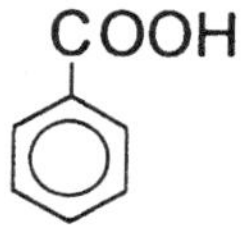

234

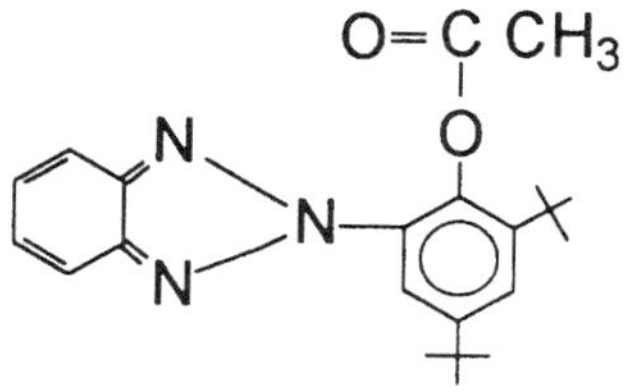

235

The photolytical depletion of UV absorbers has deleterious effects in UV-curable protective surface coatings, which must exhibit a high long-term resistance to sunlight. Added stabilizers interfere with curing and, moreover, lose a part of their stabilizing power after the photocure. The problem may be solved in long-lasting polyurethane–polyacrylate protective coatings by application of a delayed-action UV absorber precursor, an acetylated 2-(2-hydroxyphenyl)benzotriazole **235**. The respective free phenolic derivative, the true strong UV absorber, is generated from **235** during the cure (51). A concentration gradient of the in situ-formed UV absorber from the coating surface to lower layers is formed and provides the strongest protection in the surface layer. The delayed-action additive **235** has a lower detrimental effect on the curing process than conventional UV absorbers.

2. Quenchers of Excited States

Effective quenchers include nickel(II) chelates (10), such as (butylamine)[2,2'-thiobis(4-*tert*-octylphenolato)]nickel(II) **73**, bis(4-decanoyl-3-methyl-1-phenyl-pyrazol-5-ato)nickel(II) **74**, nickel(II) ethyl 3,5-di-*tert*-butyl-4-hydroxybenzyl-phosphonate **19** (M = Ni), or nickel(II) dibutyldithiocarbamate **36** (M = Ni). They deactivate excited carbonyl chromophores Ch* [Eq. (53); Q = quencher] faster than these species can photosensitize degradation of the polymer by energy transfer [see Eq. (9a), Scheme 3].

$$\overset{*}{Ch} + Q \longrightarrow Ch + \overset{*}{Q} \tag{53}$$

Deactivation of Ch* may proceed by a long-range (3- to 10-μm) energy transfer or by a collision transfer to a distance up to 1.5 μm (10). Association of quenchers with excited chromophores enhances the quenching performance.

All nickel chelates **19**, **36**, **73**, and **74** are effective quenchers of excited carbonyls (10,37). Chelates **36** (M = Ni) and **73** also quench singlet oxygen. This effect has been attributed to the presence of the sulfur atom. This structural moiety also imparts a hydroperoxide-deactivating effect (see Sec. IV.B.1) (38).

The quenching of excited states is a physical process. The cooperating scavenging of alkylperoxyls by the phenolic moiety of phosphonate **19** (M = Ni) and hydroperoxide decomposition by **36** (M = Ni) accounts for chemical transformations described in Sections IV.A.1 and IV.B.1, respectively. Quencher **74** was reported to be thermolyzed during polypropylene processing into the corresponding pyrazol-1-yl free radical capable of trapping alkyl radicals (37). A slow photolysis that releases 3,5-di-*tert*-butyl-4-methylphenoxyl (in its mesomeric form **87**, R^1, R^2 = *tert*-butyl; Subst. = H) was reported for phosphonate **19** (M = Ni) (37). The photolyzed fragment **87** dimerizes in ethylenebisphenol **88** (R^1, R^2 = *tert*-butyl; Subst. = H) and oxidizes into discoloring stilbene quinone **110**.

A very high concentration of quenchers (up to 10%) is necessary to reach an effective protection (10). This, together with discoloration problems and environmental objections against nickel, strongly reduces interest in development and commercialization of nickel chelates (22).

G. Heat Stabilizers for PVC

Polyvinyl chloride has a very low thermal stability. Application of heat stabilizers is a prerequisite for commercial processing at elevated temperatures. Heat stabilizers also extend the useful life of PVC products exposed to light.

Heat stabilizers have preventive and curative functions (32). As preventive stabilizers, they should hinder or at least retard PVC dehydrochlorination [see Eq. (2)]. This may be achieved by elimination of the initiation sites in the PVC backbone by substitution or complexation of reactive chlorine atoms or by absorption (chemical binding) of the hydrogen chloride released from the degrading PVC. Elimination of the labile allylic chlorine by substitution with organotin stabilizers **76–78** according to Eq. (54) is of top importance (93,94).

$$\text{(54)}$$

The binding of hydrogen chloride by metal soaps **75** (M = Zn, Ca, Ba) results in metal chlorides (e.g., ZnCl$_2$) and the corresponding free fatty acids (93). Binding by organotin glycolate **78** accounts for **236** and a metal-free fragment **237** (94).

$$(C_8H_{17})_2\, SnCl\, (SCH_2\underset{\underset{O}{\|}}{C}O-(i-C_8H_{17})) \qquad\qquad HSCH_2\underset{\underset{O}{\|}}{C}O-(i-C_8H_{17})$$

236   **237**

Costabilizers such as epoxidized oils (soybean, linseed, sunflower) or esters of oleic acid (e.g., **238**) or synthetic hydrotalcite **239** effectively enhance the binding of the hydrogen chloride. The epoxide **238** is transformed into the respective chlorohydrin **240**.

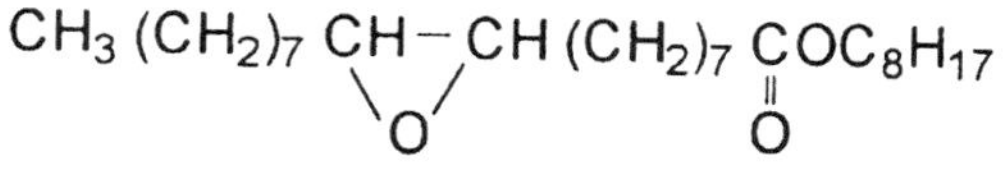

$$CH_3\,(CH_2)_7\,\underset{\displaystyle O}{CH\!-\!CH}\,(CH_2)_7\,\underset{\displaystyle O}{\overset{\displaystyle \|}{C}}OC_8H_{17}$$

$$Mg_{4.5}Al_2(OH)_{13}CO_3 \cdot 3.5H_2O$$

238

239

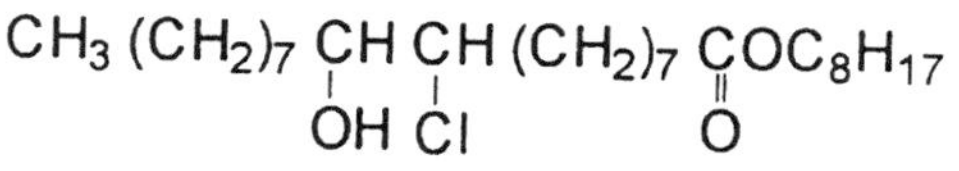

$$CH_3\,(CH_2)_7\,\underset{\displaystyle OH}{CH}\,\underset{\displaystyle Cl}{CH}\,(CH_2)_7\,\underset{\displaystyle O}{\overset{\displaystyle \|}{C}}OC_8H_{17}$$

240

237

$$\underset{\displaystyle \underset{O}{\overset{\|}{SCH_2COR}}}{}$$

76

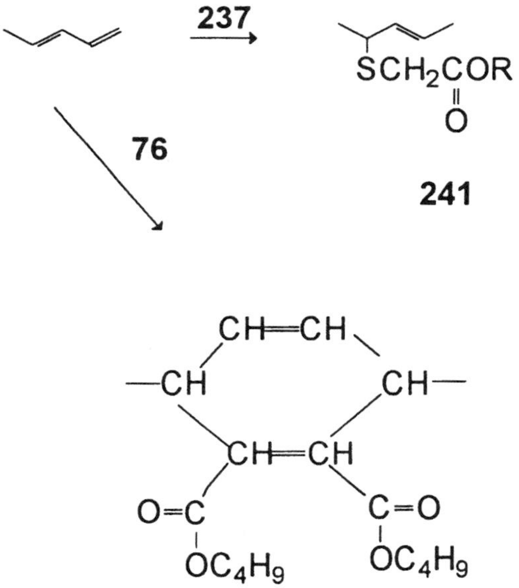

241

242

Scheme 15

In PVC that contains isolated double bonds, the curative mechanism accounts for the reduction in the rate of formation of polyene sequences $-[CH\!=\!CH]_n-$ and deactivation of hydroperoxides (32). Thiol **237**, released from **78**, was reported to stop the growth of polyenes by forming adduct **241** (Scheme 15). Formation of a Diels–Alder adduct **242** with maleate **76** has also been proved (32,94).

Organotin stabilizers that contain sulfur moieties **77**, **78** and their transformation products, such as disulfide **243**, and various S-oxidation products that thermolyze into sulfinic acid **244** having peroxidolytic properties (38), and phosphites, such as alkyl diaryl phosphite **41** or bisphosphite **42**, function in PVC as hydroperoxide-decomposing antioxidants.

The effect of the heat stabilizers **75–78** and common costabilizers **41, 42, 238**, and **239** has been enhanced in PVC stabilization by some polyols, such as pentaerythritol or sorbitol (95), 1,4-dihydropyridine **245** (96) and 1,3-diketones [e.g., benzoyl(stearoyl)methane **246**], and by acid-resistant HAS **55** and **56** ($R = O\text{-}iso\text{-}C_8H_{17}$) (see Sec. V.D).

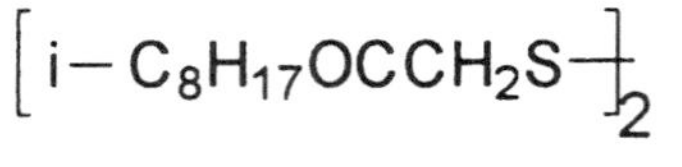

$$\left[i\text{-}C_8H_{17}OCCH_2S \right]_2 \qquad\qquad i\text{-}C_8H_{17}OCS(O)OH$$

243 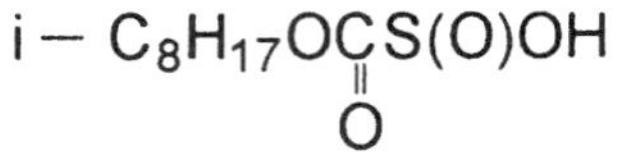   **244**

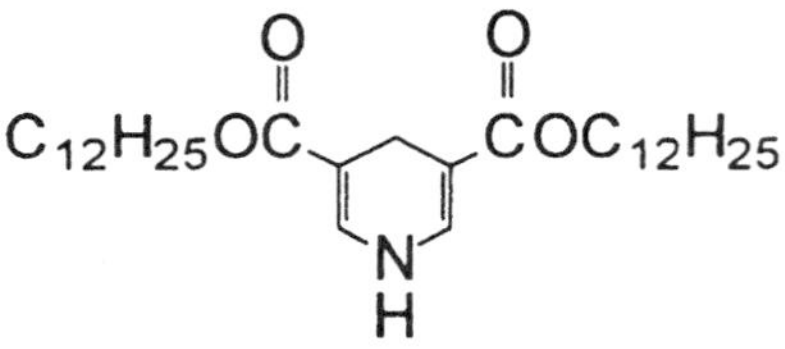

245 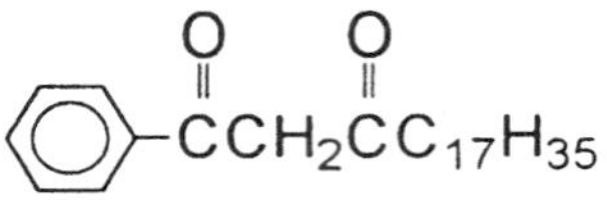   **246**

V. INTERMOLECULAR COOPERATION IN STABILIZER BLENDS

An integral protection against a variable complex of deteriogens that attack polymeric materials during processing and application can be reached by combinations of additives acting by different stabilizing mechanisms (29,77). The chemistry and physics governing the activity of individual stabilizers has been effectively exploited. Mutual protection of individual stabilizers against inactive depletion, and reactions between individual stabilizers and some of their transformation products (77), influence the integral effect. Because of the complementary and supporting effects, additivity or even synergism in the final protection have been reached by successful exploitation of the inherent chemical efficiency of individual stabilizers. Product studies offering a real insight into the cooperative chemistry are rather rare.

A. Combinations of Antioxidants

Synergistic combinations of chain-breaking and hydroperoxide-decomposing antioxidants have been traditionally used in plastics for practical and economical

$$AH + ROO^{\cdot} \longrightarrow A^{\cdot} + ROOH$$

$$\downarrow HD$$

$$HD(O) + ROH$$

Scheme 16

reasons. The complementary activity mechanism is shown in Scheme 16. The hydroperoxide resulting from scavenging of alkylperoxyls is deactivated by a hydroperoxide-decomposing antioxidant (29,77).

Combinations of hindered phenols with aromatic phosphites or phosphonites have been used for effective melt stabilization of polyolefins. It was evidenced by multiple extrusion tests (46) that phosphite **44**, used in excess, reduces consumption of phenol **18** at processing temperatures. Lower amounts of **18** may be added, and more effective long-term stabilization has been reached by the residual **18** during the subsequent polyolefin aging. It was postulated that phosphite **44** protects phenols from destruction by hydroperoxides and converts discoloring quinone methides **86** (arising from phenols) by 1,6-addition in aromatic colorless compounds (e.g., **247**) (29,54). The resistance of polyolefins against discoloration is improved by doping with a phenol–phosphite combination (9).

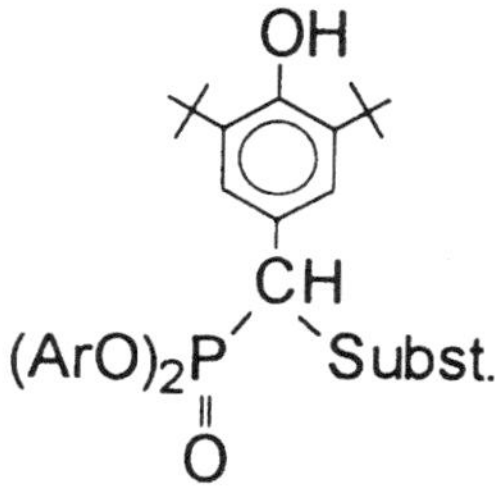

247

A better melt stability of polypropylene was obtained with blends of phenols with a less hydrolytically stable phosphite (79). This indicates that the presence of traces of the hydrogen phosphite **198** may be beneficial and does not break synergism.

The strong synergism of mixtures of phenolics with sulfur-containing stabilizers, **32–34** in particular, has been exploited in the long-term heat stabilization of polyolefins. Thiosynergists are used in approximately threefold concentration excess over phenols (9). Besides the protection of phenolics against depletion by hydroperoxides, the deactivation of (alkylperoxy)cyclohexadienones **91** formed from phenolics at aging temperatures (97) by activated thioethers **34** have been envisaged as a mechanism accounting for the synergism and phenol regeneration (29). A lower sterical hindrance in phenol **12** was reported as beneficial for the synergism with thioethers. In situ deactivation of hydroperoxides resulting from trapping alkylperoxyls in a cage of **248** was proposed to proceed by a hydrogen-associated mechanism (98) [Eq. (55)]. The formed hydroperoxide is reduced in the cage into an alcohol, and the synergist is oxidized in situ in the respective sulfoxide **249**. No free alkylhydroperoxide remains in the matrix.

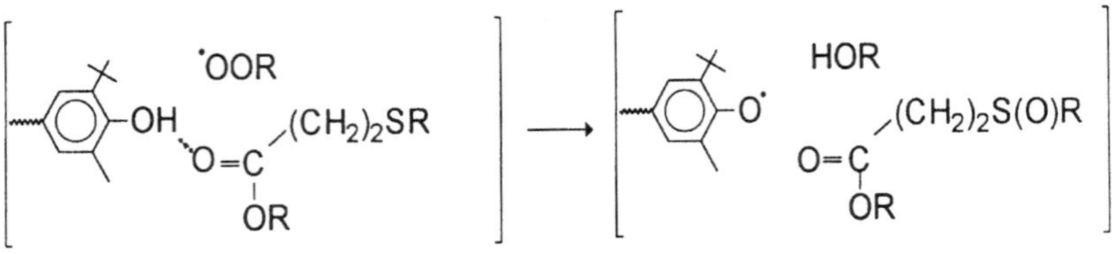

$$\tag{55}$$

248	249

Experiments performed during processing revealed a beneficial effect of the presence of trace amounts (approximately 0.001%) of the thioether **32** ($R = C_{12}H_{25}$) used with 0.05% of the 1 : 1 blend of phenol **18** and phosphite **44** on the melt stability and discoloration resistance of polypropylene (99). The mechanism of the effect was not explained.

A partial regeneration of the stabilizing power of the consumed (transformed) amine component (a low-discoloring substituted diphenylamine **22a**) during the sacrificial process by reaction with phenol **18** was exploited in polypropylene stabilization (58). The aromatic amine $>$NH [Eq. (56)] reacts faster with alkylperoxyls than the phenol. It is, however, regenerated from the formed aminyl $>$N$^{\bullet}$ at the expense of the phenol (AH), serving as a H-donor. Excellent melt and color stability of polypropylene were obtained with a three-component blend of 0.05% amine **22a**, 0.05% phenol **18**, and 0.1% phosphite **44** (100). The regeneration of the amine cannot be completely realized owing to other reactions of the aminyl (oxidation, coupling); see Section IV.A.2.

$$> \text{NH} \xrightarrow{ROO^{\bullet}} > \text{N}^{\bullet} \xrightarrow{\quad AH \quad} > \text{NH} + \text{A}^{\bullet} \tag{56}$$

B. Combinations of Antioxidants with Photostabilizers and Photoantioxidants

Combinations of phenols with UV absorbers and HAS are used for the long-term heat stabilization and photostabilization of polyolefins (29,77). Complementary antioxidant and photostabilizing effects and mutual protection of stabilizers (e.g., UV absorbers protect phenolics against direct photolysis and depletion by excited chromophores; phenols reduce the danger of the chemical breaking of the H-tunneling in UV absorbers) are responsible for excellent results (77). No particular product studies have been published until now.

Chemical evidence of mutual reactivity was revealed in mixtures of phenolic antioxidants with HAS (15,101). HAS-derived nitroxides form a paramagnetic complex with phenols [$>$NO$^{\bullet}$ ...HO–Ar]. Phenols are oxidized in the cage in the next step. Formation of reversible adducts **95** from the cyclohexadienonyl **84b** and HAS derived nitroxyl and of well-defined O-substituted hydroxylamines

96 from the hydroxybenzyl radical **87** and nitroxide (102) was experimentally confirmed. The photoantioxidant effect has been preserved in **96**.

Kinetic measurements in photoxized heptane that was sensitized with a substituted anthraquinone, revealed (77,103) that the (alkylperoxy)cyclohexadienone **91** and the (hydroperoxy)cyclohexadienone **98** do not have a detrimental effect on secondary and tertiary HAS **56** (R = H, CH$_3$). These combinations should be considered in ultimate phases of the locally overoxidized sites in polyolefins. The influence of quinone methides can be extrapolated from the model behavior of stilbenequinone **110**. A combined alkyl-trapping antioxidant and light-screening effects were postulated for **110** (77). Accordingly, compound **110** enhanced the photoantioxidant activity of **56**. It may be stated, from this experimental evidence, that the principal transformation products of phenolics do not have any detrimental effect on HAS.

The effort aiming to minimize polyolefin discoloration caused by stabilizer transformation products concentrates on application of combinations of phosphites (processing stabilizers) with HAS (photoantioxidants and thermal stabilizers at temperatures lower than 120°C), with elimination of phenolic antioxidants (31).

Mixtures of aromatic phosphite **44** with migrating HAS **56** (R = H, CH$_3$) are reported (31,103) as antagonistic in photooxidized polypropylene at low levels of the phosphite. The antagonism was strongly reduced with increasing molar proportions of **44** in the mixture. Aliphatic phosphite **39** exhibited a synergism with the same HAS. The structure of both components of the mixture phosphite–HAS definitely influences the final effect. Combinations, such as phosphite **44** with the high molecular weight HAS **58**, account for an effective long-term heat stability of polypropylene at temperatures not exceeding 130°C and were reported as also desirable for a weathering protection (88). The high integral effect has been explained by a cumulative deactivation of hydroperoxides by both stabilizers and free–radical-scavenging by HAS (77,88). There are no product studies published for the mentioned combinations.

The most recent development in synergistic combinations of stabilizers for polyolefins exploits specific differences in the photoantioxidant and long-term heat-stabilizing effects in mixtures of migrating HAS **55–57** with nonmigrating HAS **58–63** (88,104). For example, the migrating **56** (R = H) is a weaker thermal stabilizer than oligomer **59** (R = –NH-*tert*-C$_8$H$_{17}$) (88). Their mixture has an excellent integral effect. The principle of the complementary effects in combinations of stabilizers differing in molecular weights has been effectively exploited in commercialized blends of HAS **58** and **61**; **59** (R = –NH-*tert*-C$_8$H$_{17}$ and **61**; and **59** (R = –NH-*tert*-C$_8$H$_{17}$) and **56** (22), in application of functionalized terpolymeric acrylate stabilizer **250** that photolyses in migrating HAS-containing fragments having different molecular weights (105); in polymers bearing reactive groups, such as poly(maleic anhydride-*co*-α-methylstyrene); epoxidized polypropylene, polypropylene, ethylene–propylene-diene polymer (EPDM), or poly(styrene–*co*-acrylonitrile) grafted with maleic anhydride; or in polyacrylates with pending epoxy groups functionalized with 4-amino-2,2,6,6-tetramethyl-piperidine **251** (n = 0) or *N*-(2,2,6,6-tetramethyl-4-piperidyl)oxamide **251** (n = 1) and containing residual and migrating **251** (106); or polyolefin photografted with functionalized arylidene malonate **252** containing the migrating unreacted monomer (107).

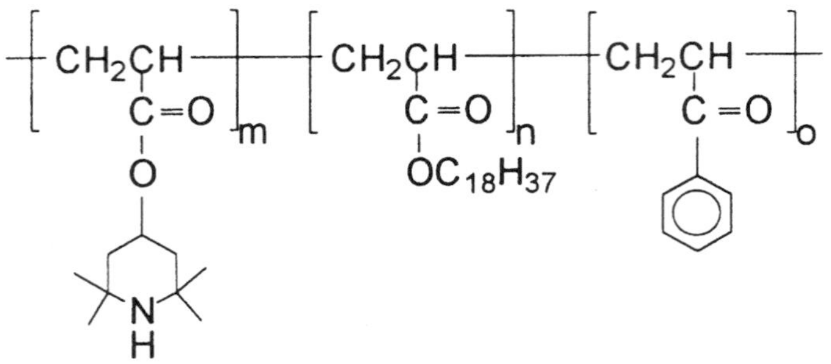

250

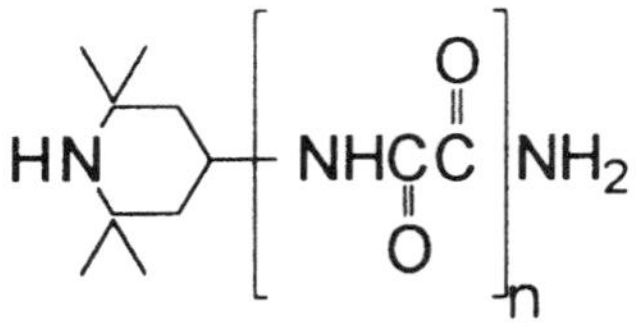

251 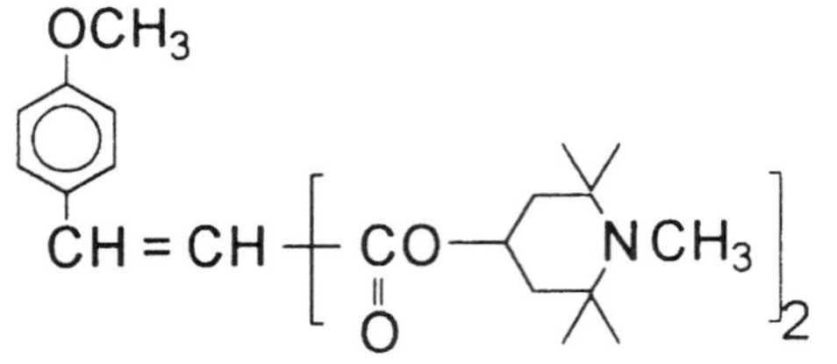   **252**

C. Combinations of UV Absorbers and Photoantioxidants

A very strong stabilizing effect that surpasses the additivity has been reached in plastics and coatings by cooperative combinations of UV absorbers and HAS (29,51). It may be accepted that UV absorbers, the derivatives of benzotriazole **65–68** in particular, prevent the HAS-developed nitroxide from photolysis and the relevant depletion of the piperidine ring (see Sec. IV.E.3), and vice versa, the UV absorbers are protected from breaking of the H-tunneling effect by HAS (39,51,88,91). The favorable integral effect of the HAS–UV absorber combination arose in coatings and in polyolefins containing conventional processing stabilizers (10).

D. Heat Stabilizers and Costabilizers in Poly(vinyl chloride)

Costabilizers effectively enhance the resistance of PVC to thermodegradation when combined with common heat stabilizers; 0.4% of diphenyl decyl phosphite **41** used with 1.2% of the mixed barium–calcium soap **75** (M = Ba and Cd) improves the resistance of PVC against yellowing and retards the development of hydrogen chloride (32). In the stabilizing mechanism, the addition to the conjugated sequence $-[CH=CH]_n-$, binding of HCl and deactivation of PVC-borne hydroperoxides have been envisaged. The stabilizing effect of phosphite **41** increases with its concentration.

Formation of **253** by substitution of the labile chlorine atom on the PVC backbone with epoxide **238**, catalyzed by Cd(II) or Zn(II) ions arising from soap **75** (M = Cd or Zn), and deactivation of $CdCl_2$ or $ZnCl_2$ by **238** were postulated in cooperative blends **238** and **75** (32). Effective protection against yellowing of PVC was attained with a ternary system consisting of 1.2% of Ba/Cd stearate **75** (M = Ca, Cd, R = $C_{17}H_{35}$), 0.4% of phosphite **41**, and 2% of epoxidized soybean oil (32).

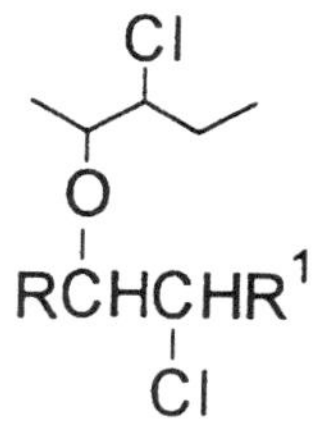

$$
\begin{array}{c}
\overset{\displaystyle Cl}{\underset{\displaystyle |}{|}} \\
\end{array}
$$

Cl
|
O
|
RCHCHR^1
|
Cl

253

Metal chlorides $ZnCl_2$ or $CdCl_2$, formed from metal soaps **75** (M = Cd, Zn) during PVC processing, are complexed with pentaerythritol (32,95). The thermostability of PVC is noticeably improved by this costabilizer or other polyols.

Application of low amounts (about 0.2%) of dihydropyridine **245** or 1,3-diketone **246** accounts for an ecologically improved PVC stabilization. Substitution of cadmium-based soaps **75** (M = Cd) by safer mixed Ba/Zn or Ca/Zn soaps was enabled by using costabilizers **245** and **246**. A good initial color and color retention in thermodegrading PVC was secured (96).

E. Negative Effects in Stabilizer Combinations

Some cooperative effects in stabilizer combinations have negative consequences. Acid transformation products of thiosynergists, such as protonic acids **181** (n = 0–2), **182** (n = 0–2), or **185** or reactive sulfinyl radicals RS˙O have an antagonistic effect on HAS (15,88). Salts or sulfonamides no longer prone to form nitroxides are formed from HAS. The antagonism may be reduced somewhat by calcium stearate and by application of HAS with lower basicity (see Sec. IV.E.3).

The efficiency of secondary HAS can be significantly reduced by adsorption on talc or calcium carbonate used as fillers (88,108). Less basic tertiary HAS **61** was reported as more resistant to the physical deactivation in filled polypropylene.

VI. BIFUNCTIONAL STABILIZERS AND INTRAMOLECULAR COOPERATION

The understanding of principles of the inherent chemical efficiency of monofunctional stabilizers and the necessity of application of their combinations were driving forces in the development of bifunctional stabilizers. An intramolecular autosynergism has been envisaged (2,29). The aim is to synthesize a molecular system with different stabilizing centers having a balanced contribution to the efficiency arising from all individual functional moieties involved. Both active moieties of the bifunctional stabilizer remain in the polymer matrix in the closest proximity and assure autocooperation whenever possible. It was realized that only one of the

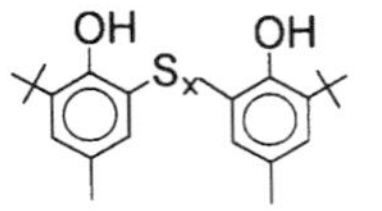

254

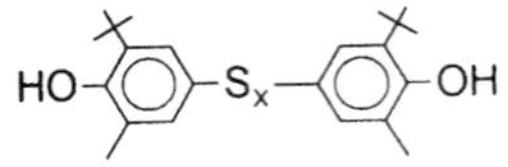

255

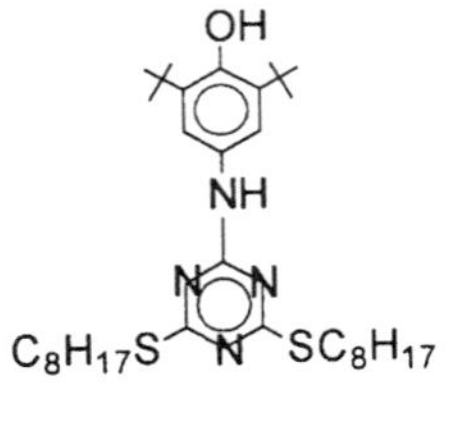

256

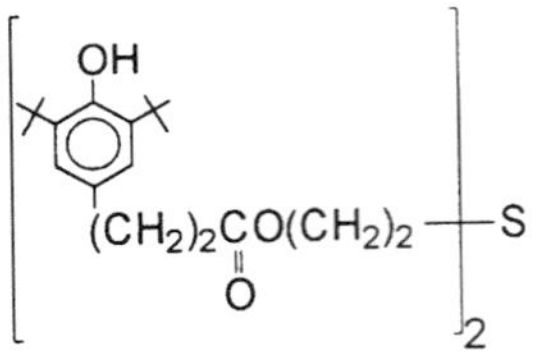

257

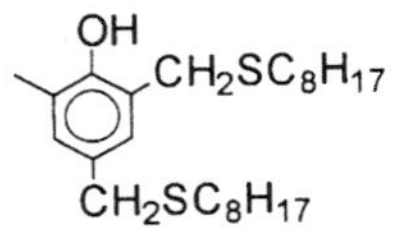

258

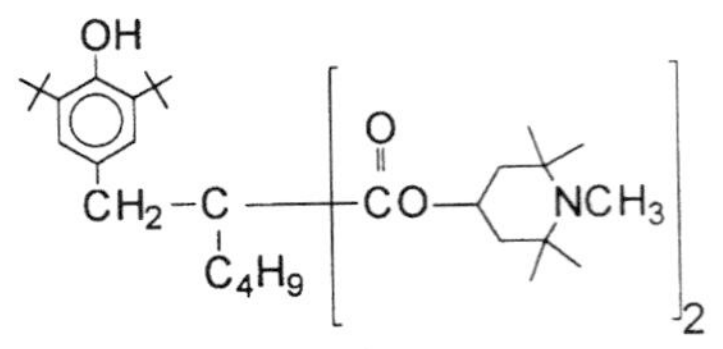

259

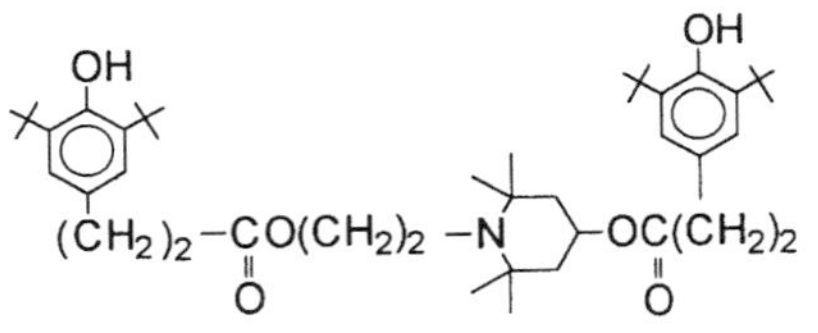

260

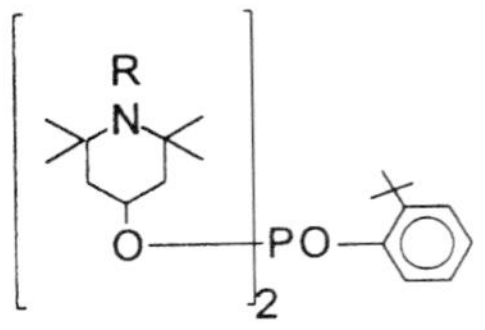

261

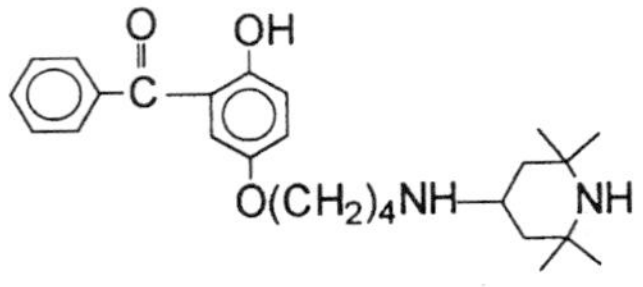

262

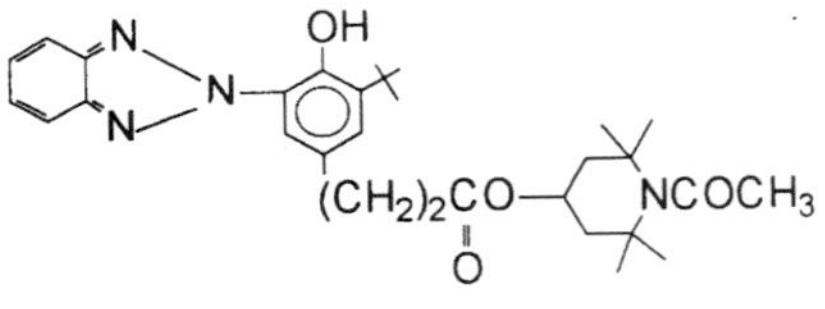

263

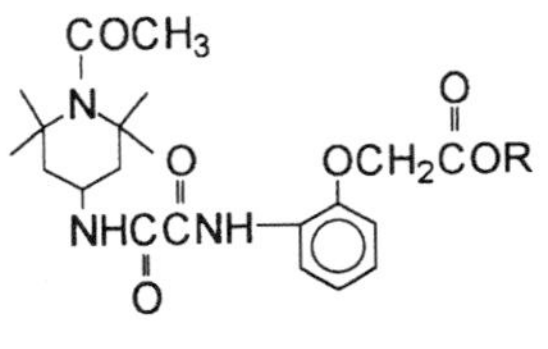

264

functions provides a principal contribution to the final effect, depending on the environmental stress (29,77). The second functional moiety has a supporting or complementary effect. The lack of an equal dual efficiency may also be based on a wrong concentration ratio of the cooperating centers. Examples of different bifunctional stabilizers have been reported in the patent literature. Many compounds were tested, only some of them were selected for commercial development. Compounds **254–266**, as well as metal deactivators **48–50**, **52**, or the nickel(II) salt **19** (M = Ni), are examples of bifunctional stabilizers.

Activity mechanisms have mostly been extrapolated from activities of the relevant monofunctional additives. Product studies explaining the mechanism were performed only with phenolic thioethers **254–258** and **265** and **266** (3,42,44,63). Elucidation of the stabilizing effect of 2- and 4-hydroxybenzyl sulfides **265** and **266** and of different thiobisphenols **254** or **255** (x = 1) and corresponding dithiobisphenols **254** and **255** (x = 2) in model hydrocarbons and polypropylene at temperatures exceeding 100°C revealed an increase in the antioxidant effectivity caused by the presence of the thio moiety, in comparison with analogous phenols containing alkyl groups (e.g., with 2,6-di-*tert*-butyl-4-methylphenol) or alkylidene bridges, such as in 2,2′-methylenebis(4-methyl-6-*tert*-butylphenol). This phenomenon was explained by a concerted cooperation of the phenolic and sulfidic functions. The *para*-substituted additive **266** outperformed at 180°C the respective **ortho**-derivative **265** (44).

Thiobisphenols **254** and **255** are strong antioxidants in polypropylene at 180°C. They produce a pronounced induction period, arising mostly from the

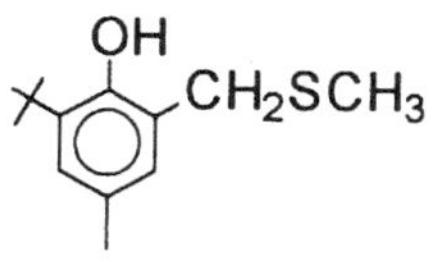

265

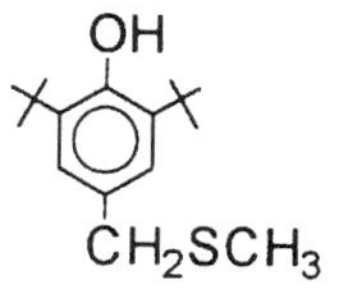

266

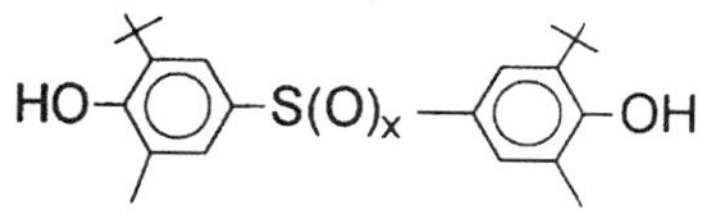

267

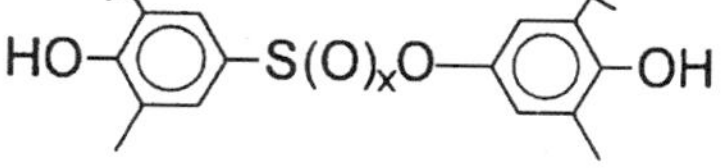

268

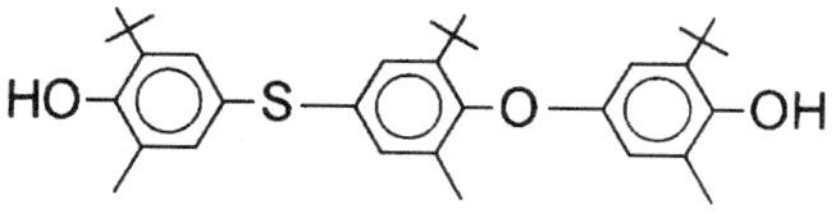

269

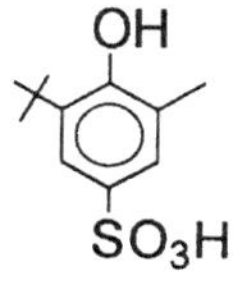

270

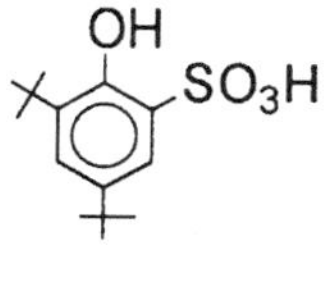

271

chain-breaking effect, and retard the oxidation after the induction period. The retardation effect is due to the thermolysis of primary oxidation products of thiobisphenols that have peroxidolytic properties (109).

Model product studies with thiobisphenols **254** and **255** (x = 1) revealed formation of the related phenoxyls in an alkylperoxyl-rich environment, oxidative transformations of thio bridges, and free radical fragmentation caused by hydroperoxides (42,44). Phenolic sulfoxides and sulfones (e.g., **267**; x = 1 and 2, respectively), are formed, together with the relevant disulfides **254** and **255** (x = 2) and more complicated transformation products, such as 2-*tert*-butyl-4-(3-*tert*-butyl-4-hydroxy-5-methylbenzenesulfonyl)oxy-6-methylphenol **268** (x = 2) or 2-*tert*-butyl-4-[(3-*tert*-butyl-4-hydroxy-5-methylphenoxy)-3-*tert*-butyl-5-methylphenylthiyl]-6-methylphenol **269** (3).

Two distinguished phases are characteristic of hydroperoxide decomposition by **255**: a stoichiometric one (accounting for formation of phenolic sulfoxide or sulfone **267** [x = 1 or 2, respectively] and an overstoichiometric one arising from thermolysis of thermolabile products of the stoichiometric phase, sulfoxide **267** (x = 1; decomposing at 140–180°C) and thiosulfinate **268** (x = 1; unstable at sub-ambient temperatures) (44). The molecule of thiobisphenols **254** and **255** degrades stepwise owing to oxidative and thermolytic reactions. 3-*tert*-Butyl-4-hydroxy-5-methylbenzenesulfonic acid **270** and 3,5-di-*tert*-butyl-2-hydroxybenzenesulfonic acid **271**, sulfur dioxide, phenolic sulfurless fragments (e.g., 3,5,3′,5′-tetra-*tert*-butyl-4,4′-biphenyldiol) and phenolic oxidation products [2,6-di-*tert*-butyl-1,4-benzoquinone **100** and 3,5,3′,5′-tetra-*tert*-butyl-4,4′-diphenoquinone **101** (3,44)] were identified among ultimate products.

The excellent complex performance of phenolic sulfides **256**–**258** is a consequence of the well-balanced intramolecular supporting and cooperative effects. Mechanistic experiments were performed with the partially hindered 2-methyl-4,6-bis[(octylthio)methyl]phenol **258** (63). The (alkylthio)methyl groups in both positions 4 and 6 contribute to its integral stabilizing effect and are a prerequisite of the strong autosynergism observed in the long-term heat aging and dynamic testing of diene-based rubbers. The activity mechanism of **258** accounts for a concerted cage reaction with alkylperoxyls **272** (R = C_8H_{17}), generating a phenoxyl–hydroperoxide complex **273**. An immediate deactivation of the hydroperoxide by the proximity effect of the (alkylthio)methyl group takes place. The latter is oxidized to sulfoxide **274**. The phenoxyl escaping from the cage **273** forms a regioselective coupling product of the type **89** and is oxidized, in the ultimate phase, into a stable quinone methide **275** (63) (Scheme 17).

Phenolic thioethers **256** and **257** are of commercial importance for stabilization of elastomers or elastomer-modified plastics and polyolefins. Some mechanistic information on their chemistry is available. Quinone imine and quinone methide structures, respectively, result from their chain-breaking activity (3,30). Additive **257** combines the favorable mechanistic phenomena of the propionate-type phenolic antioxidants (see Sec. IV.A.1) and of the activated thioethers (see Sec. IV.B.1).

The commercial importance of the high-performance mixtures of phenolic antioxidants with HAS triggered synthesis of various phenolic HAS (15). Compounds **259** and **260** have been commercialized as effective stabilizers against weathering of polyolefins. At a comparable molar basis, the same processing and long-term heat stability of polypropylene was achieved with **260** and **18**.

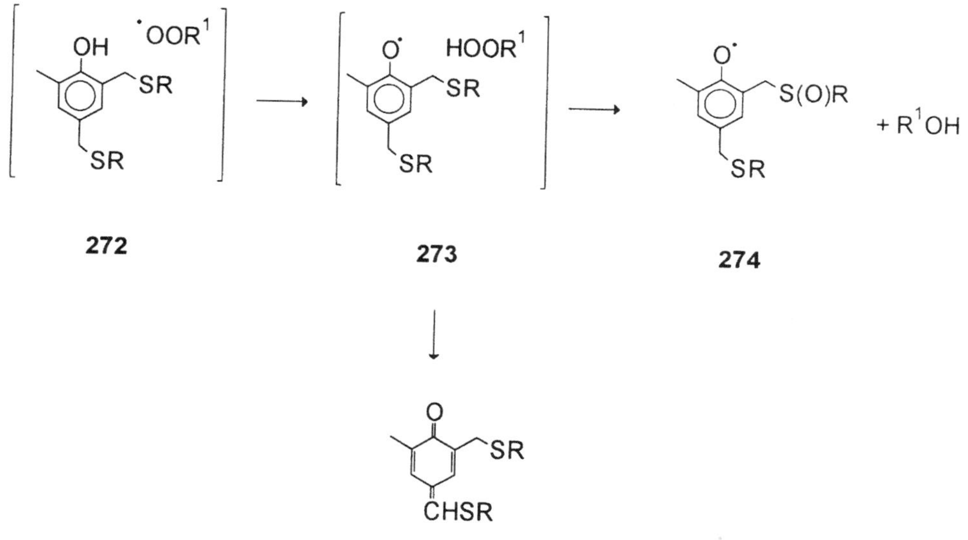

Scheme 17

Stabilizer **260** provided about 70% of the photostability achieved with **56** (R = H) (110).

Data on chemical reactions of phenolic HAS are rare. Additive **259** photolyses and yields the respective phenoxyl. The corresponding aminyl radical was not detected (15). This indicates a preferential reactivity of the phenolic moiety in comparison with the tertiary amine moiety in the alkylperoxyl-rich environment. The regenerative quinone methide→phenol mechanism analogous to the rearrangement of **112**→**113** has been considered for oxidized **260** containing a propionate-type phenolic moiety in its molecule.

Bis(2,2,6,6-tetramethylpiperidin-4-yl) 2,6-di-*tert*-butyl-4-methylphenyl phosphite **261** (R = H), its *N*-methyl derivative **261** (R = CH₃), and various analogous bifunctional HAS–phosphites or phosphonites were tested in polypropylene films as antioxidants at 110°C and photoantioxidants at 30°C (31). The features of autosynergism were reported for **261** when its efficiency was compared with that obtained with a blend consisting of structurally analogous HAS and phosphites. The activity mechanism of **261** includes its oxidation with hydroperoxides, yielding the related phosphate and nitroxide radical. A pronounced hydrolytic resistance was reported for HAS–phosphites (31,103).

HAS do not have any potency to absorb the UV sunlight (15). Much synthetic effort was aimed to modify the HAS structures by introducing effective UV–absorbing moieties. 2-Hydroxy-4-{4-[(2,2,6,6-tetramethylpiperidin-4-yl) amino]butoxy}benzophenone **262** and HAS containing 2-(2-hydroxyphenyl) benzotriazole moiety **263** or oxamide moiety **264** are examples of bifunctional stabilizers designed for polyolefins and coatings. Deactivation of the actinic solar radiation is intramolecularly complemented with the free–radical-scavenging photoantioxidant effect and accounts for mutual intramolecular protection of the moieties and for an integral photostabilization of the polymer. Processing history of polyolefins, enhancing concentration of hydroperoxides or carbonyl groups in the polymer matrix, may reduce the final stabilization effect.

VII. PHYSICAL FACTORS AFFECTING THE INHERENT CHEMICAL ACTIVITY OF STABILIZERS

The low physical persistence limits exploitation of the inherent chemical activity of stabilizers in polymers used under demanding conditions owing to physical losses by volatility and leaching into the environment (18,69). The problem of the physical losses can be solved by application of low-extractable and nonvolatile metal-containing complexes or salts and compounds having higher molecular weights. The approximate ranges of molecular weights of the principal nonpolymeric commercial stabilizers are 200–1200 for phenolic antioxidants, 200–1850 for organic phosphites, 300–450 for aromatic amines, 500–1200 for organic sulfides, 200–900 for UV absorbers, 500–750 for nickel-containing quenchers, 250–2000 for HAS, and 600–750 for PVC heat stabilizers (111). Additives with molecular weights higher than 500 (arbitrarily classified as high molecular weight stabilizers) and oligomers with a proper molecular architecture and molecular weights approximately 3500 (69) generally fulfill requirements for physically persistent high-performing stabilizers.

The optimum molecular weight for HAS is known to change with their application and polymer type (88). HAS that have molecular weights of 500–600 are an optimum option for thick cross-section polyolefins. In thin films or fibers, less volatile HAS with molecular weights of about 2000 are most favorable. Oligomeric HAS, with molecular weights about 3500, have generally more acceptable properties in blending with the host polymer and a broader application spectrum in commodity polymers and coatings than do oligomers with higher molecular weights.

Polymeric stabilizers prepared by polyreactions and reactions on polymers have as yet been exploited only exceptionally (69). They have been considered as substrate-specific and applicable for special purposes. The best example is a terpolymer **276** containing N-(4-anilinophenyl)methacrylamide monomer units and used commercially for semireinforced furnace-black-filled acrylonitrile–butadiene rubber with a low sulfur cure system, and for styrene–butadiene rubbers.

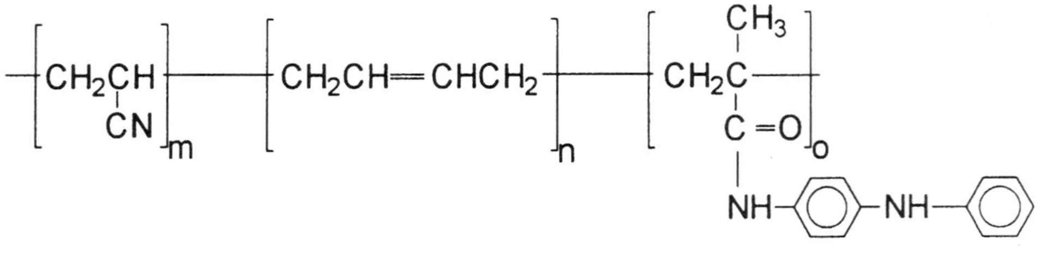

276

In situ formation of polymer-bound stabilizers from reactive functionalized monomers and conventional polymers during processing operations (38,112), copolymerization of functionalized monomers (69), proper functionalization of polymers bearing reactive moieties (69), or photografting of polymers with reactive additives during outdoor application (107) are potential techniques providing combinations of persistent polymeric stabilizers with migrating residual (i.e., unreacted) monomeric stabilizers (see Sec. V.B). The distribution of the bond in stabilizing moieties in functionalized macromolecules is statistic. A minimized migration into the surrounding environment makes polymeric stabilizers safe indirect food additives.

VIII. CONSEQUENCES OF THE STABILIZER CONSUMPTION FOR THE MATERIAL RECYCLING OF PLASTICS

Material recycling of plastics is one of the economically acceptable options for exploitation of industrial scrap and assorted aged postconsumer wastes. The recyclate is characterized by enhanced concentration of photosensitizing polymeric impurities (carbonyl groups, hydroperoxides) and nonpolymeric impurities (metal compounds of various origin, various pigments, printing inks, organic fatty compounds, and transformation products of stabilizers) (1,20). Hence, plastics intended for reuse have an enhanced sensitivity to processing and long-term degradation. Moreover, the residual concentration of processing and long-term stabilizers is mostly lower than the minimal effective level necessary for protection of the degradation-sensitive material. To achieve a durable second-life application of plastics, upgrading by a proper restabilization is necessary (20). Experiments with recyclates revealed the importance of a combination of aromatic phosphite **44** with persistent phenol **18** in reprocessing of high-density polyethylene, polypropylene, and their blends. Phosphite **44** also improves the reprocessing of poly(ethylene terephthalate). Restabilization systems containing phenol **18**, phosphite **44**, and HAS (e.g., **56** [R = H]) are beneficial for recyclates intended for outdoor applications. Pigmented recyclates, such as crates in closed-loop recycling, have a surprisingly high resistance against degradation after restabilization with HAS **56** and UV absorber **65**.

The selection of proper combinations of stabilizers for particular recyclates is a rather delicate problem for smaller-reprocessing units. Prefabricated one-pack master batches with tailor-made combinations of individual additives and impact modifiers are available.

IX. CONCLUSIONS

The plastics, rubbers, and coatings market becomes more competitive. An economical stabilization of polymers without health hazard is the general aim. Trends in stabilization respond to actual problems, such as degradation of polymers arising from more drastic processing conditions or application in aggressive environment, impact of residues of polymerization catalysts of new generations, and degradation of polymer blends and recyclates.

The knowledge of stabilizer chemistry has been exploited profitably to meet increasing demands of the polymer end-use customers and of environmental rules and has contributed to the development of improved stabilizers with an optimized molecular architecture and reduced side effects. A detailed analysis of current toxicological effects arising from properly used polymer stabilizers revealed that the public is protected from intoxication more effectively than from other chemicals in the human environment (113). This trend also remains as the key condition in the future development of polymer stabilizers.

To enhance the acceptability of stabilizers by the public, an effective attempt was made to use synthetic α-tocopherol **277** (synthetic vitamin E, generally recognized as safe for the food contact and as a human medicine) or its analogues for stabilization of plastics in packaging applications (114).

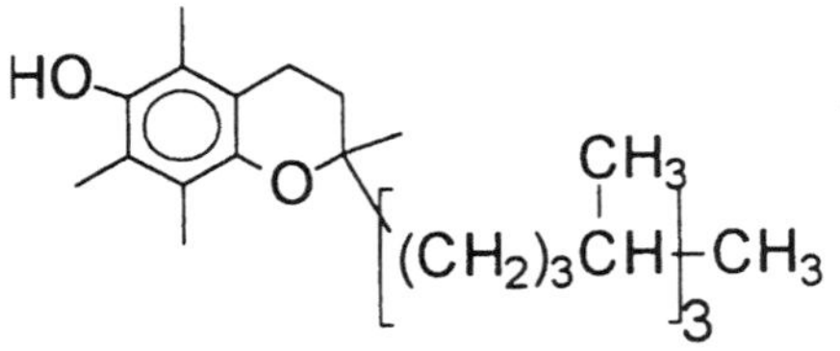

277

ACKNOWLEDGMENTS

The financial support by grants 4050603 from the Grant Agency of the Academy of Sciences of the Czech Republic and KONTAKT-087-1997 from the Ministry of Education, Youth and Sports of the Czech Republic is gratefully appreciated. The authors thank Mrs. D. Dundrová for technical cooperation in the preparation of the manuscript.

REFERENCES

1. J Pospíšil, Z Horák, S Nešpůrek. The origin and role of structural inhomogeneities and impurities in material recycling of plastics. Macromol Symp 135:247–263, 1998.
2. J Pospíšil, S Nešpůrek. Highlights in chemistry and physics of polymer stabilization. Macromol Symp 115:143–163, 1997.
3. J Pospíšil. The key role of antioxidant transformation products in the stabilization mechanisms—a critical analysis. Polym Degrad Stabil 34:85–109, 1991.
4. H Hinsken, S Moss, J-R Pauquet, H Zweifel. Degradation of polyolefins during melt processing. Polym Degrad Stabil 34:279–293, 1991.
5. T Hjertberg, EM Sörvik. Thermal degradation of PVC. In: ED Owen, ed. Degradation and Stabilization of PVC. London: Elsevier Applied Science, 1984, pp. 21–79.
6. M Lazár, J Rychlý. Oxidation of hydrocarbon polymers. Adv Polym Sci 102:189–221, 1992.
7. M Iring, F Tüdös. Thermal oxidation of polyethylene and polypropylene. Effects of chemical structure and reaction conditions on the oxidation process. Prog Polym Sci 15:217–262, 1990.
8. JF Rabek. Polymer Photodegradation. Mechanisms and Experimental Methods. London: Chapman & Hall, 1995, pp. 24–66.

9. F Gugumus. Stabilization of plastics against thermal oxidation. In: J Pospíšil, PP Klemchuk, eds. Oxidation Inhibition in Organic Materials, vol 1. Boca Raton, FL: CRC Press, 1990, pp. 61–172.

10. F Gugumus. Photooxidation of polymers and its inhibition. In: J Pospíšil, PP Klemchuk, eds. Oxidation Inhibition in Organic Materials, vol 2. Boca Raton, FL: CRC Press, 1990, pp. 29–162.

11. NS Allen. Effects of dyes and pigments. In: GC Eastmond, A Ledwith, S Russo, P Sigwalt, eds. Comprehensive Polymer Science, vol 6. Polymer Reactions. Oxford: Pergamon Press, 1989, pp. 529–537.

12. PR Ogilby, M Kristiansen, DO Martire, RD Scurlock, VL Taylor, RL Clough. Formation and removal of singlet ($a^1\Delta g$) oxygen in bulk polymers: events that may influence photodegradation. In: RL Clough, NC Billingham, KT Gillen, eds. Polymer Durability: Degradation, Stabilization and Lifetime Prediction. Washington, DC: American Chemical Society, Adv Chem Ser 249:113–126, 1996.

13. BN Felder. Mechanistic studies of sterically hindered amines in the photooxidation of liquid polypropylene model substances. In: PP Klemchuk, ed. Polymer Stabilization and Degradation. Washington, DC: American Chemical Society, ACS Symp Ser 280:69–86, 1985.

14. RW Layer, RP Lattimer. Protection of rubber against ozone. Rubber Chem Technol 63:426–450, 1990.

15. J Pospíšil. Aromatic and heterocyclic amines in polymer stabilization. Adv Polym Sci 124:87–189, 1995.

16. HHG Jellinek. Reactions of polymers with polluted gases. In: HHG Jellinek, ed. Aspects of Degradation and Stabilization of Polymers. Amsterdam: Elsevier, 1978, pp. 431–500.

17. F Gugumus. Influence of stabilization mechanisms and environment on optimization of UV stability. 3rd International Symposium on Weatherability, Tokyo, 1997.

18. NC Billingham. Location of oxidation in polypropylene. Makromol Chem Macromol Symp 28:145–163, 1989.

19. JR White, NY Rappoport. Stress effects on polymer durability in the oxidative environment. Trends Polym Sci 2:197–202, 1994.

20. J Pospíšil, S Nešpůrek, R Pfaendner, H Zweifel. Material recycling of plastics waste for demanding applications: upgrading by restabilization and compatibilization. Trends Polym Sci 5:294–300, 1997.

21. J-R Pauquet. Antioxidantien. Kunststoffe 86:940–946, 1996.

22. E Kramer. Lichschutzmittel. Kunststoffe 86:948–953, 1996.

23. SJ Ainsworth. Plastics additives. Chem Eng News 70(35):34–55, 1992.

24. BF Greek. Rubber processing chemicals. Chem Eng News 65(20):29–49, 1987.

25. J Pospíšil. Antioxidants and related stabilizers. In: J Pospíšil, PP Klemchuk, eds. Oxidation Inhibition in Organic Materials, vol 1. Boca Raton, FL: CRC Press, 1990, pp. 33–59.

26. R Seltzer, R Ravichandran, RA Patel [Ciba-Geigy AG]. Polyolefin compositions stabilized with long-chain N,N'-dialkylhydroxylamines. Eur Pat Appl EP 323 409, 1989; Chem Abstr 112:21791, 1990.

27. S Yachigo, M Sasaki, Y Takahashi, F Kojima, T Takada, T Okita. Studies on polymer stabilizers: part 1—a novel thermostabilizer for butadiene polymers. Polym Degrad Stabil 22:63–77, 1988.

28. R Pitteloud, P Dubs. Antioxidants for industrial applications. Chimia 48:417–419, 1994.

29. J Pospíšil. Stabilizer mixtures and polyfunctional stabilizers. In: J Pospíšil, PP Klemchuk, eds. Oxidation Inhibition in Organic Materials, vol 1. Boca Raton, FL: CRC Press, 1990, pp. 173–192.

30. J Pospíšil. Chemical and photochemical behaviour of phenolic antioxidants in polymer stabilization: a state of the art report, part I. Polym Degrad Stabil 39:103–115, 1993.

31. I Bauer, WD Habicher, S Korner, S Al-Malaika. Antioxidant interaction between organic phosphites and hindered amine light stabilizers: effects during photooxidation of polypropylene—II. Polym Degrad Stabil 55:217–224, 1997.

32. H Andreas. PVC stabilizers. In: R Gächter, H Müller, eds. Plastics Additives Handbook. 3rd ed. Munich: Hanser Publishers, 1990, pp. 271–325.

33. H Müller. Metal deactivators. In: R Gächter, H Müller, eds. Plastics Additives Handbook. 3rd ed. Munich: Hanser Publishers, 1990, pp. 105–128.

34. MG Chan. Metal deactivators. In: J Pospíšil, PP Klemchuk, eds. Oxidation Inhibition in Organic Materials, vol I. Boca Raton, FL: CRC Press, 1990, pp. 225–246.

35. JF Kuczkowski. The inhibition of oxidative and ozonic processes in elastomers. In: J Pospíšil, PP Klemchuk, eds. Oxidation Inhibition in Organic Materials, vol I. Boca Raton, FL: CRC Press, 1990, pp. 247–290.

36. D Brück. Ozonolyse von ungesättigten Kautschuken und Vulkanisaten—ein Überblick. Kautsch Gummi Kunstst 42:760–770, 1989.

37. JF Rabek. Photostabilization of polymers. London: Elsevier Applied Science, 1990, pp. 42–79, 202–278.

38. S Al-Malaika. Effects of antioxidants and stabilizers. In: GC Eastmond, A Ledwith, S Russo, P Sigwalt, eds. Comprehensive Polymer Science, vol 6. Polymer Reactions. Oxford: Pergamon Press, 1989, pp. 539–578.

39. JE Pickett. Review and kinetic analysis of the photodegradation of UV absorbers. Macromol Symp 115:127–141, 1997.

40. F Gugumus. Re-evaluation of the stabilization mechanisms of various light stabilizer classes. Polym Degrad Stabil 39:117–135, 1993.

41. G Rytz, R Hilfiker, E Schmidt, A Schmitter. Introduction to a new class of high performance light stabilizers and the influence of light stabilizer structure on the polymer life-time. Angew Makromol Chem 247:213–224, 1997.

42. J Pospíšil. Transformations of phenolic antioxidants and the role of their products in the long-term stability of polyolefins. Adv Polym Sci 36:69–133, 1980.

43. K Schwetlick, WD Habicher. Action mechanisms of phosphite and phosphonite stabilizers. In: RL Clough, NC Billingham, KT Gillen, eds. Polymer Durability: Degradation, Stabilization and Lifetime Prediction. Washington, DC: American Chemical Society, Adv Chem Ser 249:349–358, 1996.

44. J Pospíšil. Mechanism of activity and transformations of phenolic sulfides in the stabilization of polyolefins. In: AV Patsis, ed. Advances in the Stabilization and Controlled Degradation of Polymers, vol I. Lancaster: Technomic Publishing, 1989, pp. 55–61.

45. NC Billingham. Physical phenomena in the oxidation and stabilization of polymers. In: J Pospíšil, PP Klemchuk, eds. Oxidation Inhibition in Organic Materials, vol 2. Boca Raton, FL: CRC Press, 1990, pp. 249–297.

46. WO Drake, J-R Pauquet, RV Todesco, H Zweifel. Processing stabilization of polyolefins. Angew Makromol Chem 176/177:215–230, 1990.

47. J Scheirs, J Pospíšil, MJ O'Connor, SW Bigger. Characterization of conversion products formed during degradation of processing antioxidants. In: RL Clough, NC Billingham, KT Gillen, eds. Polymer Durability: Degradation, Stabilization and Lifetime Prediction. Washington, DC: American Chemical Society, Adv Chem Ser 249:359–374, 1996.

48. J Pospíšil. Photooxidation reactions of phenolic antioxidants. In: NS Allen, ed. Developments in Polymer Photochemistry—2. London: Applied Science Publishers, 1981, pp. 53–133.

49. J Pospíšil. Chemical and photochemical behaviour of phenolic antioxidants in polymer stabilization: a state of the art report, part II. Polym Degrad Stabil 40:217–232, 1993.

50. KP Ghiggino. Photostabilization of polymeric materials. J Macromol Sci Pure Appl Chem A33:1541–1553, 1996.

51. C Decker, K Zahouilly, S Biry. In-situ generation of light stabilizers upon photoaging of thermoset- or UV cured-acrylate coatings. Polym Degrad Stabil 47:109–115, 1995.

52. OD Bangee, VH Wilson, GC East, I Holme. Antioxidant induced yellowing of textiles. Polym Degrad Stabil 50:313–317, 1995.

53. J Scheirs, SW Bigger, NC Billingham. Effect of chromium residues on the stability of gas-phase high-density polyethylene produced by supported catalysts. J Polym Sci [A] Polym Chem Ed 30:1873–1889, 1992.

54. J Pospíšil, S Nešpůrek, H Zweifel. The role of quinone methides in thermostabilization of hydrocarbon polymers—II. Properties and activity mechanisms. Polym Degrad Stabil 54:15–21, 1996.

55. F Gugumus. Critical stabilizer concentrations in oxidizing polymers. Polym Degrad Stabil 46:123–140, 1994.

56. J Pospíšil, S Nešpůrek. Chain-breaking stabilizers in polymers: the current status. Polym Degrad Stabil 49:99–110, 1995.

57. J Pospíšil, S Nešpůrek, H Zweifel. The role of quinone methides in thermostabilization of hydrocarbon polymers—I. Formation and reactivity of quinone methides. Polym Degrad Stabil 54:7–14, 1996.

58. J Pospíšil. Chain-breaking antioxidants in polymer stabilization. In: G Scott, ed. Developments in Polymer Stabilization—I. London: Applied Science Publishers, 1979, pp. 1–37.

59. F Tüdös, T Földes-Bereznich. Free-radical polymerization: inhibition and retardation. Prog Polym Sci 14:717–761, 1989.

60. AJ Waring. Cyclohexadienones. In: H Hart, GJ Karabatsos, ed. Advances in Alicyclic Chemistry, vol 1. New York: Academic Press, 1966, pp. 129–256.

61. J Pilař, J Rotschová, J Pospíšil. Antioxidants and stabilizers, 117. Contribution to the transformation mechanisms of 2,6-di-*tert*-butyl-4-methylphenol, a processing stabilizer for polyolefins. Angew Makromol Chem 200:147–161, 1992.

62. EA Altwicker. Chemistry of stable phenoxy radicals. Chem Rev 67:475–531, 1967.

63. H Meier, P Dubs, H Künzi, R Martin, G Knobloch, H Bertermann, B Thuet, A Borer, U Kolczak, G Rist. Some aspects of a new class of sulfur containing phenolic antioxidants. Polym Degrad Stabil 49:1–9, 1995.

64. L Taimr, J Pospíšil. Antioxidants and stabilizers—XXXV. On the character of discolouring compounds formed from 2,2′-methylenebis(4-methyl-6-*tert*-butylphenol) during oxidative degradation of polymers. Angew Makromol Chem 28:13–29, 1993.

65. S Yachigo, F Kojima, M Sasaki, K Ida, S Tanaka, K Inoue. Studies on polymer stabilizers: part IV—prevention of oxidative discoloration. Polym Degrad Stabil 37:107–113, 1992.

66. H Zweifel. Effect of stabilization of polypropylene during processing and its influence on long-term behavior under thermal stress. In: RL Clough, NC Billingham, KT Gillen, eds. Polymer Durability: Degradation, Stabilization and Lifetime Prediction. Washington, DC: American Chemical Society, Adv Chem Ser 249:375–396, 1996.

67. S Yachigo, M Sasaki, T Ishii, S Tanaka. Studies on polymer stabilizers: part III—prevention of NO_x gas discoloration with a new antioxidant. Polym Degrad Stabil 37:99–106, 1992.

68. RE King, C Kuell. Potential impact of catalyst residues on polymer stabilization. Proceedings of the 17th International Conference on Advances in the Stabilization and Degradation of Polymers, Luzern, 1995, pp. 145–165.

69. J Pospíšil. Functionalized oligomers and polymers as stabilizers for conventional polymers. Adv Polym Sci 101:65–167, 1991.

70. J Pospíšil. Aromatic amine antidegradants. In: G Scott, ed Developments in Polymer Stabilization—7. London: Elsevier Applied Science, 1984, pp 1–63.

71. V Ducháček, A Kuta, L Sošková, L Taimr, J Rotschová, J Pospíšil. Antioxidants and stabilizers: part CXII—influence of 3-(*sec*-alkylamino)-6-(phenylimino)-1,4-bis(4-phenylaminophenylimino)benzoquinone on the vulcanization and ageing of natural rubber. Polym Degrad Stabil 29:217–231, 1990.

72. G Knobloch. Influence of processing conditions and additives on the degradation reaction of SIS-block copolymers. Angew Makromol Chem 176/177:333–346, 1990.

73. J Kubo. Radical scavengers from heavy hydrocarbons. Chemtech 10:29–34, 1996.

74. JR Shelton. Organic sulfur compounds as preventive antioxidants. In: G Scott, ed. Developments in Polymer Stabilization—4. London: Applied Science Publishers, 1981, pp. 23–70.

75. DM Kulich, JR Shelton. Organosulfur antioxidants: mechanisms of action. Polym Degrad Stabil 33:397–410, 1991.

76. A Günther, T König, WD Habicher, K Schwetlick. Antioxidant action of organic sulfites—I. Esters of sulphurous acid as secondary antioxidants. Polym Degrad Stabil 55:209–216, 1997.

77. J Pospíšil, S Nešpůrek, H Zweifel, WD Habicher, K Scheim. Cooperative interactions in stabilizer combinations in polymers. Proceedings of the Conference on Long-Term Performance Issues in Polymers: Chemistry and Physics, Boston, 1995, p. 639.

78. C Neri, S Constanzi, RM Riva, R Farris, R Colombo. Mechanism of action of phosphites in polyolefin stabilization. Polym Degrad Stabil 49:65–70, 1995.

79. J Tocháček, J Sedlář. Hydrolysis and stabilization performance of bis(2,4-di-*tert*-butylphenyl)pentaerythrityl diphosphite in polypropylene. Polym Degrad Stabil 50:345–352, 1995.

80. JD Spivack, SD Pastor, A Patel, LP Steihuebel. Bis- and trisphosphites having dioxaphosphepin and dioxaphosphocin rings as polyolefin processing stabilizers. In: PP Klemchuk, ed. Polymer Stabilization and Degradation. Washington, DC: American Chemical Society, ACS Symp Ser 280:247–257, 1985.

81. GJ Klender. Fluorophosphonites as co-stabilizers in stabilization of polyolefins. In: RL Clough, NC Billingham, KT Gillen, eds. Polymer Durability: Degradation, Stabilization and Lifetime Prediction. Washington, DC: American Chemical Society, Adv Chem Ser 249:397–423, 1996.

82. J Pospíšil. Mechanistic action of phenolic antioxidants in polymers—a review. Polym Degrad Stabil 20:181–202, 1988.

83. F Gugumus. Mechanisms and kinetics of photostabilization of polyolefins with *N*-methylated HALS. Polym Degrad Stabil 34:205–241, 1991.

84. T Kurumada, H Ohsawa, T Fujita, T Toda, T Yoshioka. Photostabilizing activity of *N*-acylated hindered amine. J Polym Sci Polym Chem Ed 23:274–2756, 1985.

85. PP Klemchuk, ME Gande, E Cordola. Hindered amine mechanisms: part III—Investigations using isotopic labelling. Polym Degrad Stabil 27:65–74, 1990.

86. F Gugumus. Current trends in mode of action of hindered amine light stabilizers. Polym Degrad Stabil 40:167–215, 1993.

87. GL Gaines. Acceleration of hydrolysis of bisphenol A polycarbonate by hindered amines. Polym Degrad Stabil 27:13–18, 1990.

88. K Kikkawa. New developments in polymer photostabilization. Polym Degrad Stabil 49:135–143, 1995.

89. KP Ghiggino, AD Scully, SW Bigger. Photophysics of hydroxyphenylbenzotriazole polymer photostabilizers. In: E Reichmanis, JH O'Donnell, eds. The Effect of Radiation on High-Technology Polymers. Washington, DC: American Chemical Society, ACS Symp Ser 381:57–79, 1988.

90. MA Dearth, TJ Kornishi. Reaction of benzotriazole ultraviolet light absorber with free radicals. Polym Degrad Stabil 48:121–130, 1995.

91. C Decker. Photostabilization of macromolecular materials by UV-cured protective coatings. In: RL Clough, NC Billingham, KT Gillen, eds. Polymer Durability: Degradation, Stabilization and Lifetime Prediction. Washington, DC: American Chemical Society, Adv Chem Ser 249:319–334, 1996.

92. JE Pickett, JE Moore. Photostability of UV screeners in polymers and coatings. In: RL Clough, NC Billingham, KT Gillen, eds. Polymer Durability: Degradation, Stabilization and Lifetime Prediction. Washington, DC: American Chemical Society, Adv Chem Ser 249:287–301, 1996.

93. A Guyot, A Michel. Stabilization of poly(vinyl chloride) with metal soaps and organic compounds. In: G Scott, ed Developments in Polymer Stabilization—2. London: Elsevier Science Publishers, 1980, pp 89–124.

94. G Ayrey, RC Poller. Organotin stabilizers for poly(vinyl chloride). In: G Scott, ed. Developments in Polymer Stabilization—2. London: Elsevier Science Publishers, 1980, pp 1–52.

95. TV Hoang, A Guyot. The role of polyols as secondary stabilizers for poly(vinyl chloride). Polym Degrad Stabil 12:29–41, 1985.

96. A Michel, TV Hoang, B Perrin, ML Llauro. Synergistic mechanisms of β-diketo derivatives and zinc–calcium soaps in PVC stabilization. Polym Degrad Stabil 3:107–119, 1980/1981.

97. NP Neureiter, DE Brown. Synergism between phenols and sulfides in the stabilization of polyolefins to oxidation. Ind Eng Chem Prod Res Dev 1:236–242, 1962.

98. S Yachigo, M Sasaki, F Kojima. Studies on polymer stabilizers: Part II. A new concept of a synergistic mechanism between phenolic and thiopropionate type antioxidants. Polym Degrad Stabil 35:105–113, 1992.

99. F Gratani, C Neri, GL Laudoni. Influence of small amounts of thioethers on the discoloration of polyolefins during processing. 12th International Conference on Advances in the Stabilization and Controlled Degradation of Polymers [short communication], Luzern, 1990.

100. TM Chucta. An improved stabilizer system for polypropylene. Plast Eng 44(5):73–75, 1988.

101. D Vyprachticky, J Pospíšil, J Sedlář. Possibilities for cooperation in stabilizer systems containing a hindered piperidine and a phenolic antioxidant—a review. Polym Degrad Stabil 27:227–255, 1990.

102. K Murayama. The chemistry of stable N-oxyl radicals. J Synth Org Chem Jpn 29:366–385, 1971.

103. WD Habicher, I Bauer, K Scheim, C Rautenberg, A Lossack, K Yamaguchi. Interactions of stabilizers during oxidation processes. Macromol Symp 115:93–125, 1997.

104. F Gugumus. New trends in the stabilization of polyolefin fibers. Polym Degrad Stabil 44:272–297, 1994.

105. Š Chmela, P Hrdlovič. Reactive oligomeric light stabilizers. In: RL Clough, NC Billingham, KT Gillen, eds. Polymer Durability: Degradation, Stabilization and Lifetime Prediction. Washington, DC: American Chemical Society, Adv Chem Ser 249:473–482, 1996.

106. JL Hahnfeld, DD Devore. Polymer-bound hindered amine light stabilizers for improved weatherability in multi-phase polymer systems. Polym Degrad Stabil 39:241–249, 1993.

107. G Ligner, L Avar. New technologies to tie down UV light stabilizers in polyolefins. Polym Prep 34(2):160–161, 1993.

108. D Vaillant, J Lacoste, J Lemaire. Stabilization of isotactic polypropylene. Problems bond to the interactions of stabilizers with polymers and fillers. J Appl Polym Sci 65:609–615, 1997.

109. I Kúdelka, J Brodilová, M Prusíková, J Pospíšil. Antioxidants and stabilizers: part CVI—mechanism of antioxidant action of dithiobisphenols. Polym Degrad Stabil 17:287–301, 1987.

110. T Toda, T Kurumada. Research and development of hindered amine light stabilizers (HALS). Annu Rep Sankyo Res Lab 35:1–37, 1983.

111. M Minagawa. New developments in polymer stabilization. Polym Degrad Stabil 25:121–141, 1989.

112. S Al-Malaika. Reactive processing and polymer performance. Polym Plast Technol Eng 29:73–86, 1990.

113. J Pospíšil, H-J Weideli. Environmental impacts associated with the application of radical scavenging stabilizers in polymers. Polym Degrad Stabil 52:109–117, 1996.

114. SF Laermer, PF Zambetti. alpha-Tocopherol (vitamin E)—the natural antioxidant for polyolefins. J Plast Film Sheet 8:228–248, 1992.

II
MECHANISMS OF POLYMER DEGRADATION AND STABILIZATION

7

Homogeneous and Heterogeneous Oxidation of Polypropylene

GRAEME A. GEORGE
Queensland University of Technology, Brisbane, Australia

MATHEW CELINA
Sandia National Laboratories, Albuquerque, New Mexico

I. INTRODUCTION

Polypropylene (PP) is today one of the most widely used of all of the commodity polymers. The versatility of this material arises from the cheap petrochemical stocks as raw materials, the sophisticated and efficient catalytic polymerization processes, and the ease of processing of the final polymer as a fabricated article by injection molding or as a textile by spinning (1). This success has occurred in spite of the polymer being one of the most oxidatively unstable of the polyolefins and is a reflection of the extent to which an understanding of the oxidative degradation of the polymer has led to strategies for stabilization (2,3).

Studies of the degradation and stabilization of polypropylene have ranged from macroscopic mechanical property studies of the failure of the polymer article when exposed outdoors, to mechanistic studies of the oxidative pathways in model systems; Fig. 1 summarizes the hierarchy of investigations and methodologies that have been undertaken over the years. The starting point in any study of an actual fabricated polymer is the way the mechanical properties of the material change with time. This may involve simple measurements of tensile strength, elongation at fail and modulus, but in critical applications, it is the change of the fracture toughness, K_{1c} or fracture energy G_{1c} with time of oxidation that is more important (4,5). These latter parameters focus on the significant observation that polypropylene, similar to many polymers, embrittles on environmental aging. The source of this embrittlement may be readily discerned in a scanning electron microscope (SEM) examination of

Observations   **Information Level**

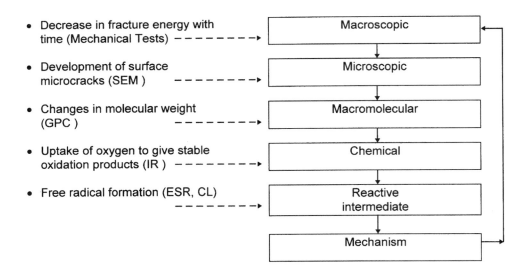

- Decrease in fracture energy with time (Mechanical Tests) – – – – – – – ➤ → Macroscopic
- Development of surface microcracks (SEM) – – – – – – – ➤ → Microscopic
- Changes in molecular weight (GPC) – – – – – – – ➤ → Macromolecular
- Uptake of oxygen to give stable oxidation products (IR) – – – – – – – ➤ → Chemical
- Free radical formation (ESR, CL) – – – – – – – ➤ → Reactive intermediate
→ Mechanism

Figure 1 Hierarchy of investigations of polymer degradation showing the successively higher level of information linking reaction mechanisms to macroscopic properties.

the polymer surface, which reveals a series of microcracks (4,6). The locus of cracking depends on both the internal (fabrication) stresses and applied stresses on the polymer (7), and physical property changes, such as density and crystallinity, indicate that the crack has been the end result of changes to macromolecular properties (8). The most logical macromolecular property to measure is the molecular weight distribution of the polymer, and studies of the changes to number average and weight average molar mass indicate that a chain-scission process has occurred. An extent of scission of one chain scission per molecule may occur within 12 weeks of outdoor exposure of unstabilized polypropylene and produce microcracking and a reduction of fracture toughness of 75% (6,7). The sensitivity of semicrystalline polymers, such as polypropylene, to these changes in polymer molecular weight are a consequence of the degradation being localized in the amorphous region of the polymer. The macromolecules of greatest significance are the tie molecules between the crystalline blocks. Scission of these tie molecules will have a major effect on the ability of the polymer to resist deformation under load. Another consequence of chain scission is that the mobile, lower molecular weight material may recrystallize, because the normal service temperature of polypropylene lies between T_g (0°C) and T_m (165°C). This will produce the changes in density and crystallinity (noted earlier) and so enable microcrack formation (8). Another factor that contributes to microcrack formation is the increase in the polarity of the polymer during environmental degradation. This indicates the importance of the changes to the chemistry of the polymer, as the physical property changes may be considered a consequence of the chemical reactions occurring under the influence of temperature or radiation. Although many polymers, such as the polyamides, are sensitive to moisture, and hydrolytic effects should be considered, with polyolefins it has generally been thought that the chemical changes to the polymer are the result of a free radical oxidative chain reaction (2,9). Evidence for this process has been gained from measurements such as oxygen consumption by the heated polymer (10) and accumulation of hydroperoxide and carbonyl species, as measured by infrared (IR) spectrophotometry (2,9,11). These products are a consequence of the oxidative scission reactions that lower the polymer molecular weight and produce density changes and shrinkage owing to the changes in intermolecular forces. Such has been the importance of understanding these chemical reactions that there have been mechanistic studies using techniques ranging from electron-spin resonance (ESR) spectroscopy (12) to chemiluminescence (13) to understand and control these free radical reactions. The ultimate goals of these studies have been to produce new stabilization strategies for the commercial polypropylene formulations and to be able to predict the safe service lifetime of the polymer (3).

The successful exploitation of polypropylene at high temperatures and in the presence of ultraviolet (UV) radiation has been possible only through the use of a range of stabilization strategies that interrupt the sequence of chemical reactions that oxidize the polymer (3). The prediction of the useful lifetime of a stabilized polymer from short-term accelerated testing has been a technological goal that has, so far, eluded polymer scientists. In principle, from a detailed understanding of the processes shown in Fig. 1, it should be possible to develop a quantitative kinetic model for the oxidation of the polymer. If the causal links from the mechanical properties down to the mechanistic level are understood, then it should be possible to use mechanistic and kinetic information to complete the loop and predict the lifetime.

The tool that has been employed to approach this goal has been chemical kinetics. The theory of free radical chain reaction kinetics in the gas phase or well-mixed liquid state has been long established (14,15). Although molten polymers differ from low molar mass compounds in solution because of persistent entanglements and substantial viscosity effects, the theory for oxidation and stabilization of the polymer melt during processing is consistent with homogeneous chemical kinetics. In this system, the concentration of the reactants may be averaged over the volume to produce a representative value for kinetic modeling. However, as noted before, polypropylene in service is used well below T_m, but above T_g, and the segmental motion of the polymer chains on which the free radicals are formed will be restricted. In addition, the oxidation reactions will be occurring only in those regions in which there is a significant concentration of dissolved oxygen (i.e., the amorphous phase) (2). The kinetic scheme that is appropriate to describe the oxidation of such a system is contentious. At the one extreme it is considered that, because the profiles of oxidation product formation with time show sigmoidal behavior, similar to that for an autoaccelerating chain reaction in solution, then homogeneous chemical kinetics may be employed (16,17). At the other extreme, emphasis is placed on the barriers to the system achieving homogeneity and evidence for highly localized oxidation as a prelude to crack formation in the solid polymer, so that a model for heterogeneous oxidation is considered appropriate (18–20).

In this chapter, it is intended to consider these two approaches to describing the kinetics of oxidation of polypropylene. Particular attention will be paid to the heterogeneous model for oxidation and the extent to which information about degradation mechanisms and stabilization strategies for polypropylene may be deduced from a model in which degradation is considered to initiate in discrete zones within the amorphous region and then spread, similar to an infection, through the polymer (19–21).

II. HOMOGENEOUS OXIDATION OF POLYPROPYLENE

The earliest work related to the established free radical oxidation model was that by Backström on the free radical chain theory for autoxidation (14). This was followed by the detailed studies of Bolland and Gee (15) at the British Rubber Producers Research Association on the oxidation of olefins, which resulted in the current model of radical initiation, propagation, and termination being applied to polymer oxidation (22,23).

A. Free Radical Oxidation Scheme

The radical processes during the thermal or photodegradation of polyolefins are, in principle, identical, with only minor differences owing to variations in the initiation or secondary photochemistry (3,22). Evidence for the reaction scheme and the products that may be formed has been obtained from studies of model compounds as well as analysis of polypropylene at often high extents of oxidation. The determination of specific species (e.g., p-, s-, or t-hydroperoxides) and higher oxidation products, such as peracids) has required specific derivatization reactions coupled to spectroscopic analytical tools, principally infrared spectroscopy (24). As the sensitivity and specificity of these techniques are improved, the possible reaction steps that may occur during the thermal oxidation of unstabilized polypropylene also increase in both number and complexity. In spite of this, the free radical oxidation mechanism has generally been believed to consist of the following steps (22,23).

Initiation

$$\text{Polymer} \quad \rightarrow \quad P^{\bullet} + P^{\bullet} \tag{1}$$

Propagation

$$P^{\bullet} + O_2 \quad \rightarrow \quad POO^{\bullet} \tag{2}$$
$$POO^{\bullet} + PH \quad \rightarrow \quad POOH + P^{\bullet} \tag{3}$$

Chain-branching

$$POOH \quad \rightarrow \quad PO^{\bullet} + {}^{\bullet}OH \tag{4}$$
$$PH + {}^{\bullet}OH \quad \rightarrow \quad P^{\bullet} + H_2O \tag{5}$$
$$PH + PO^{\bullet} \quad \rightarrow \quad P^{\bullet} + POH \tag{6}$$
$$PO^{\bullet} \quad \rightarrow \quad \text{various chain-scission reactions} \tag{7}$$

Termination (leading to nonradical products)

$$POO^{\bullet} + POO^{\bullet} \quad \rightarrow \quad POOOOP (\rightarrow POOP + O_2) \quad (8)$$

$$POO^{\bullet} + P^{\bullet} \quad \rightarrow \quad POOP \quad (9)$$

$$P^{\bullet} + \quad \rightarrow \quad PP \quad (10)$$

In thermal oxidation, initiation [Eq. (1)] results from the thermal dissociation of chemical bonds, whereas in photodegradation, photophysical processes, such as the formation of electronically excited species, energy-transfer processes, or direct photodissociation lead to bond cleavage. The wavelength of maximum sensitivity is 310 nm for PP (3).

The key reaction in the propagation sequences is the reaction of polymer alkyl radicals ($P^{\bullet}$) with oxygen to form polymer peroxy radicals [Eq. (2)]. This reaction is very fast, but can be diffusion-controlled. The next propagation step, [Eq. (3)] is the abstraction of a hydrogen atom by the polymer peroxy radical ($POO^{\bullet}$) to yield a polymer hydroperoxide (POOH) and a new polymer alkyl radical ($P^{\bullet}$). In polypropylene, hydrogen abstraction occurs preferentially from the tertiary carbon atoms as they are the most reactive [Eq. (11)].

$$POO\bullet + CH_2\text{-}\underset{\underset{H}{|}}{\overset{\overset{CH_3}{|}}{C}}\text{-}CH_2\text{-} \longrightarrow POOH + \text{-}CH_2\text{-}\underset{\underset{\bullet}{}}{\overset{\overset{CH_3}{|}}{C}}\text{-}CH_2\text{-} \quad (11)$$

Hydrogen abstraction occurs from secondary carbon atoms in polyethylene [Eq. (12)], and may also occur in polypropylene, but with lower reaction rates.

$$POO^{\bullet} + -CH_2 - CH_2 - \longrightarrow POOH + -CH_2 - \overset{\bullet}{C}H - \quad (12)$$

For polypropylene, intramolecular hydrogen abstraction in a six-ring favorable stereochemical arrangement can preferentially lead to the formation of blocks of hydroperoxides in close proximity (24–27). Infrared studies of polypropylene hydroperoxides showed that more than 90% of these groups were intramolecularly hydrogen-bonded. The results indicated that the peroxide groups were preferentially present in sequences of two or more, which supported intramolecular hydrogen abstraction during the oxidation (25). It was also concluded that the cleavage reaction accompanied the oxidation propagation, rather than the termination reaction. Roginsky (28) also suggested that the reaction of solid polypropylene was dominated by the formation of rather compact blocks of hydroperoxide groups. The relatively rare formation of low molecular weight radicals that can travel over large distances was claimed to result in new blocks of hydroperoxides far from the previous ones.

Chain-branching by thermolysis or photolysis [see Eq. (4)] of polymer hydroperoxides (POOH) results in the formation of very reactive polymer alkoxy radicals ($PO^{\bullet}$) and hydroxyl radicals ($^{\bullet}OH$). The highly mobile hydroxyl in Eq. (5) and polymer alkoxy radical in Eq. (6) can abstract hydrogen atoms from the same or from a nearby polymer chain (3).

Polymer oxy radicals can react further to result in β-scission [see Eq. (13)] the formation of in-chain ketones [see Eq. (14)], or can be involved in termination reactions.

$$
\begin{array}{cc}
\text{CH}_3 & \text{CH}_3 \\
| & | \\
\text{-CH}_2\text{-C-CH}_2\text{-C-} \\
\| \quad | \\
\text{O•} \quad \text{H}
\end{array}
\longrightarrow
\begin{array}{cc}
\text{CH}_3 & \text{CH}_3 \\
| & | \\
\text{-CH}_2\text{-C} & + \quad \text{•CH}_2\text{-C-} \\
\| & | \\
\text{O} & \text{H}
\end{array}
\qquad (13)
$$

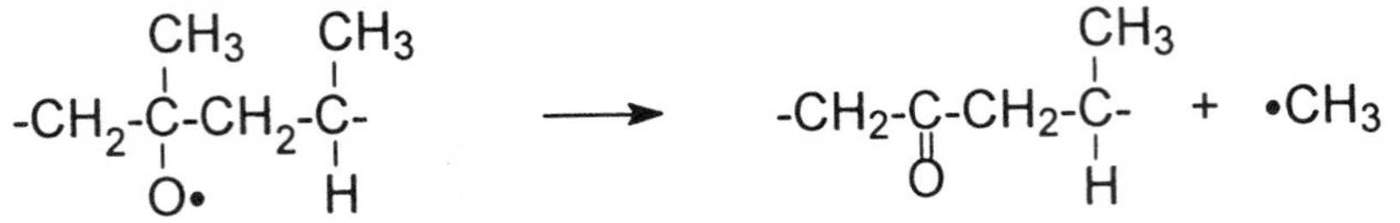

$$
\begin{array}{cc}
\text{CH}_3 & \text{CH}_3 \\
| & | \\
\text{-CH}_2\text{-C-CH}_2\text{-C-} \\
\| \quad | \\
\text{O•} \quad \text{H}
\end{array}
\longrightarrow
\begin{array}{c}
\text{CH}_3 \\
| \\
\text{-CH}_2\text{-C-CH}_2\text{-C-} \quad + \quad \text{•CH}_3 \\
\| \quad\quad | \\
\text{O} \quad\quad \text{H}
\end{array}
\qquad (14)
$$

The termination of polymer radicals occurs by various bimolecular recombinations. When the oxygen supply is sufficient, the termination is almost exclusively via the reaction in Eq. (8). At low oxygen pressure other termination reactions may take place (3). The recombination is influenced by cage effects, steric control, mutual diffusion, and the molecular dynamics of the polymer matrix (3,29,30). In solid polymers, the recombination of polymer peroxy radicals (POO$^{•}$) is subject to the rate of their encounter with each other and is influenced by the intensity of molecular motion (34).

B. Secondary Reactions and Product Formation

1.   Formation of Acids and Peracids

$$
\begin{array}{c}
\text{O} \\
\| \\
\text{-CH}_2\text{-CH}
\end{array}
+ \text{POO•}
\longrightarrow
\begin{array}{c}
\text{O} \\
\| \\
\text{-CH}_2\text{-C•}
\end{array}
+ \text{POOH}
\qquad (15)
$$

Polymer acyl radicals can be produced by hydrogen abstraction from end aldehyde groups, Eq. (15), or by the Norrish type I reaction during photodegradation.
Polymer acyl radicals are easily further oxidized to polymer peracid radicals [Eq. (16)], which can abstract hydrogen to form peracids [Eq. (17)].

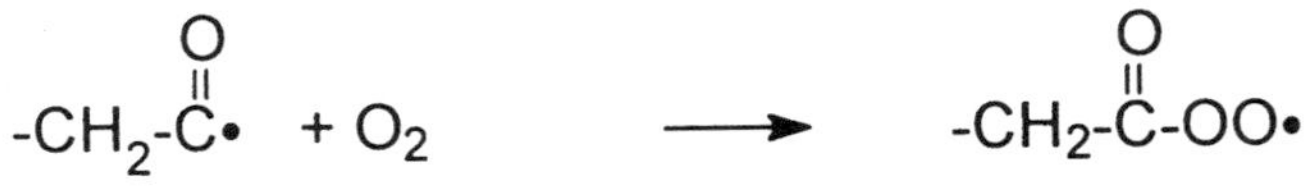

$$
\begin{array}{c}
\text{O} \\
\| \\
\text{-CH}_2\text{-C•}
\end{array}
+ \text{O}_2
\longrightarrow
\begin{array}{c}
\text{O} \\
\| \\
\text{-CH}_2\text{-C-OO•}
\end{array}
\qquad (16)
$$

$$
\begin{array}{c}
\text{O} \\
\| \\
\text{-CH}_2\text{-COO•}
\end{array}
+ \text{PH}
\longrightarrow
\begin{array}{c}
\text{O} \\
\| \\
\text{-CH}_2\text{-C-OOH}
\end{array}
+ \text{P•}
\qquad (17)
$$

The cleavage of peracids can lead to polymer carboxy and hydroxyl radicals [Eq. (18)], with the carboxy radicals again able to abstract hydrogen to form carboxylic groups [Eq. (19)].

$$-CH_2-\overset{\overset{O}{\|}}{C}OOH \quad \longrightarrow \quad -CH_2-\overset{\overset{O}{\|}}{C}-O\bullet \ + \ \bullet OH \qquad (18)$$

$$-CH_2-\overset{\overset{O}{\|}}{C}O\bullet \ + \ PH \quad \longrightarrow \quad -CH_2-\overset{\overset{O}{\|}}{C}-OH \ + \ P\bullet \qquad (19)$$

Alternative sources of acidic species during the oxidation of isotactic polypropylene have been suggested from mass spectrometric analysis of thermal decomposition products from polymer hydroperoxides (31). Acetone, acetic acid, and methanol constitute 70% of the decomposition products, suggesting either a high extent of oxidation involving secondary hydroperoxides, or direct reactions of hydroxyl radicals with ketones (derived through reactions discussed in the next section).

In another related mechanism, the primary alkyl radicals produced by β-scission [see Eq. (13)] may be oxidized further [Eq. (20)] to yield peroxy and carboxylic acids [Eq. (21)] (32).

$$\underset{\underset{H}{|}}{\overset{\overset{CH_3}{|}}{-CH_2-C-CH_2\bullet}} + O_2 + RH \quad \longrightarrow \quad \underset{\underset{H}{|}}{\overset{\overset{CH_3}{|}}{-CH_2-C-CH_2-OOH}} + RH\bullet \qquad (20)$$

$$\underset{\underset{H}{|}}{\overset{\overset{CH_3}{|}}{-CH_2-C-CH_2-OOH}} \longrightarrow \underset{\underset{H}{|}\,\underset{O}{\|}}{\overset{\overset{CH_3}{|}}{-CH_2-C-COOH}} + \underset{\underset{H}{|}\,\underset{O}{\|}}{\overset{\overset{CH_3}{|}}{-CH_2-C-COH}} \qquad (21)$$

Studies of the photoxidation of isotactic PP revealed the formation of peracids in quantities representing 15–20% of the overall peroxide level (33). Solubility measurements indicated that the peracids were of macromolecular nature formed by the oxidation of macroaldehydes, presumably through radical attack on methyl hydrogens, subsequent peroxidation, and disproportionation (33). The analysis of peroxides formed during the oxidation of PP between 50° and 90°C also suggested the presence of different types of peroxides, such as fast-decomposing and slow-decomposing ones. Chemical analysis of the fast-decomposing species showed that these peroxides were peracids (32) formed by the foregoing mechanisms. The difficulty of analysis of these reactive species has meant that unequivocal quantitation of the reactions has not been achieved.

2.   Carbonyl Group Activity

Although the thermal oxidation primarily takes place through the foregoing free radical reaction, some additional reactions can occur in the photodegradation of the polymer. Carbonyl groups, either present in the virgin polymer or formed during the degradation, are sensitive to secondary phototransformations and can be photolysed by the Norrish type I [Eq. (22)] or type II [Eq. (23)] reactions (34).

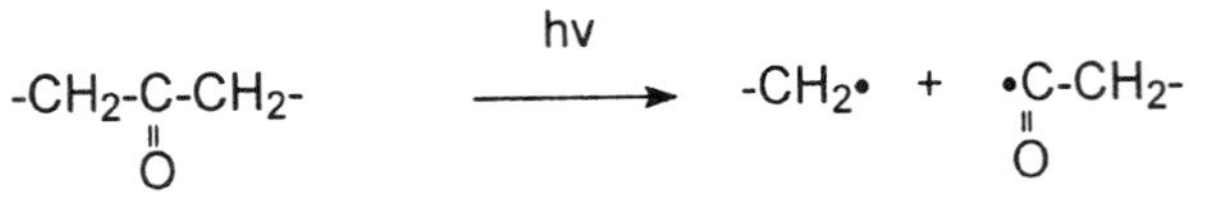

$$-CH_2\text{-}\underset{\underset{O}{\|}}{C}\text{-}CH_2\text{-}\quad\xrightarrow{h\nu}\quad -CH_2\bullet\ +\ \bullet\underset{\underset{O}{\|}}{C}\text{-}CH_2\text{-}\qquad (22)$$

$$\xrightarrow{h\nu}\quad -CH=CH_2 + [CH_2=\overset{OH}{\underset{|}{C}}\text{-}]\ \longrightarrow\ CH_3\text{-}\underset{\underset{O}{\|}}{\overset{O}{\|}}C\text{-}\qquad (23)$$

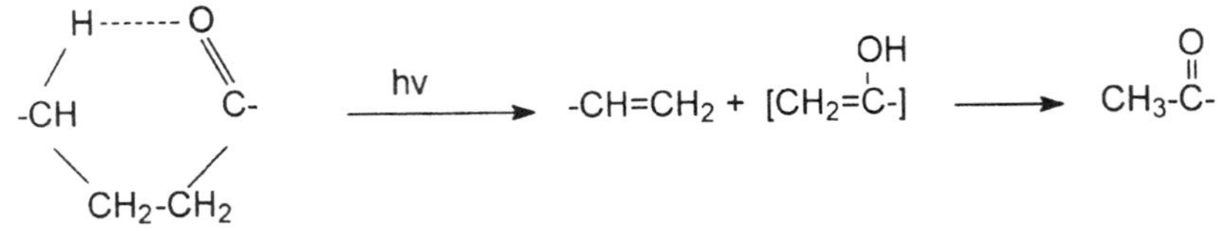

The Norrish type II is essentially a nonradical intramolecular process, dependent on the formation of a six-membered cyclic intermediate, and it requires at least one γ-hydrogen relative to the carbonyl group.

3.   Photodecomposition of Hydroperoxides

The light quanta in solar radiation are energetically sufficient to cleave PO–OH (176 kJ/mol) and P–OOH (293 kJ/mol), but hardly POO–H (377 kJ/mol) bonds. The large difference in the bond dissociation energy between PO–OH and P–OOH means that the formation of PO$^\bullet$ and $^\bullet$OH radicals will be the predominant reaction of photocleavage during irradiation (3). Hydroperoxide groups have a very low molar absorptivity at a wavelength of 340 nm. The O–O bond has no low-lying, stable excited state, and the potential energy surfaces of the first excited state are dissociative. The quantum yield in the near UV is close to 1.0 (3). The photolysis of hydroperoxides under solar irradiation, however, is a slow process. The average lifetime of an hydroperoxide group in a PP film under constant UV irradiation was quoted as approximately 25 h, equivalent to roughly 4–5 days under solar irradiation (35). Photodecomposition of hydroperoxides in polymers may also occur by energy-transfer processes from excited carbonyl or aromatic groups that have a higher molar absorptivity at the wavelengths of terrestrial sunlight (3,34).

C.   Classic Homogeneous Oxidation Kinetics

The starting point in the kinetic analysis of the oxidation of polypropylene is measurement of the extent of oxidation of the polymer as a function of time (Fig. 2). The most common measurement is the uptake of oxygen by the polymer when in a controlled environment (10,36), but it may also involve oxidation product analysis by any of the foregoing techniques.

The fundamental reaction mechanism in Section II.A for the free radical oxidation of hydrocarbons has been used to relate the consumption of oxygen to

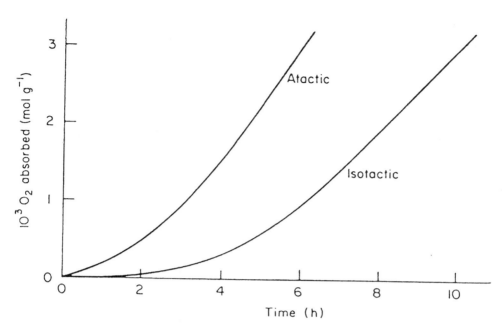

Figure 2 Oxygen uptake curve during the thermal oxidation of polypropylene film at 120°C and the extracted atactic polymer. (From Ref. 10.)

the formation of oxidation products in polypropylene. A kinetic interpretation is based on the steady-state approximation equating the rates of the initiation and termination reactions. With this approach it is possible to derive mathematical equations describing the consumption of oxygen or the formation of specific oxidation products. Simplifying the oxidation of polypropylene to the following reaction sequences [Eq. (1'–6')] the consumption of oxygen may be related to the formation of hydroperoxides at high [Eq. (24)] and at low [Eq. (25)] oxygen pressure (37).

Initiation

$$2\,ROOH \quad \rightarrow \quad RO_2^{\bullet} + R^{\bullet} + H_2O \tag{1'}$$

Propagation

$$R^{\bullet} + O_2 \quad \rightarrow \quad ROO^{\bullet} \tag{2'}$$

$$ROO^{\bullet} + RH \quad \rightarrow \quad ROOH + R^{\bullet} \tag{3'}$$

Termination

$$R^{\bullet} + R^{\bullet} \quad \rightarrow \quad RR \tag{4'}$$

$$R^{\bullet} + ROO^{\bullet} \quad \rightarrow \quad ROOR \tag{5'}$$

$$ROO^{\bullet} + ROO^{\bullet} \quad \rightarrow \quad ROOR + O_2 \tag{6'}$$

Both equations were derived using the steady-state approximation between initiation and termination, with Eqs. (4′) and (5′) being neglected at high oxygen pressure, and Eqs. (5′) and (6′) being redundant at low oxygen pressure.

$$-d[O_2]/dt = k_{3'}(k_{1'}/k_{6'})^{1/2}[ROOH][RH] \tag{24}$$

$$-d[O_2]/dt = k_{2'}(k_{1'}/k_{4'})^{1/2}[ROOH][O_2] \tag{25}$$

A more complex relation, for instance, was presented with the following Eq. (26), obtained with the assumption that the kinetic chain is not too short and that $k_{5'}^2 = k_{4'}k_{6'}$ (r_i = initiation rate) (38).

$$-d[O_2]/dt = \left(k_{2'}, k_{3'}, [PP][O_2]\, r_i^{1/2}\right)/(k_{2'}, k_{6'},^{1/2}[O_2] + k_{3'}, k_{4'},^{1/2}[PP]) \tag{26}$$

One of the obvious features of the oxidation of polypropylene is the formation of hydroperoxides [see Eq. (3′)] as a product. The initiation of the oxidation sequence is usually considered to be thermolysis or photolysis of hydroperoxides formed during synthesis and processing [see Eq. (1)]. The kinetics of oxidation then become those of a branched-chain reaction, as the number of free radicals in the system continually increases with time (i.e., the product of the oxidation is also an initiator). Because of the different stabilities of the hydroperoxides (e.g., p-, s-, t-; isolated or associated) under the conditions of the oxidation, only a fraction of those formed will be measured in any hydroperoxide analysis of the oxidizing polymer. The kinetic character of the oxidation will change from a linear chain reaction, in which the steady-state approximation applies, to a branched-chain reaction in which the approximation may not be valid because the rate of formation of free radicals is not constant and the oxidation demonstrates autoacceleration. Such reactions have long been studied in the high-temperature, gas-phase oxidation of hydrocarbons, and the formation of hydroperoxides as a degenerate branching agent results in the observation of cool flames in which the chain reaction is linear until a critical concentration of hydroperoxide is reached (39). It then becomes branched, an exponentially increasing chain reaction (flame or explosion) ensues that consumes the branching agent, and the system then becomes a linear chain again. This alternating process continues until the fuel is exhausted. By altering the conditions of pressure and temperature, the termination rate may be altered and the system may be held in the explosion (branched) regimen or in the linear regimen in which hydroperoxide accumulates.

Obviously, under the conditions of service, the oxidation of the polymeric hydrocarbon cannot reach the foregoing critical condition, and homogeneous kinetic treatments of polypropylene oxidation involve perturbations of the steady-state approximation. In recent studies, Verdu and collaborators have considered this problem and have developed a "closed-loop" model in which the character of the induction period was explored (16,40). Most recently, the dilemma of using the stationary-state approximation in the solution of an autoaccelerating chain reaction has been addressed (41). Gugumus (17) has also considered the approximations necessary to apply a similar kinetic model to the oxidation of polyethylene. This kinetic approach is considered in detail in Chapter 20.

The application of models involving degenerate branching has been a particular feature of kinetic treatments of polypropylene oxidation by many Russian authors (42). In one such study, it was suggested that under conditions for degenerate branching, the maximum rate of oxygen consumption may be expressed by Eq. (27) (σ = probability of degenerate chain branching), which was derived again by neglecting the reactions in Eqs. (4') and (5') at high oxygen pressure and assuming that the oxidation chain is sufficiently long (43).

$$-\mathrm{d}[O_2]/\mathrm{d}t^{\bullet}_{max} = \delta k'_3[RH]^2/(2k'_6)^{1/2} \qquad (27)$$

In many of these studies, the basic kinetic information is obtained from an oxygen-uptake curve, such as that shown in Fig. 2. If the consumption of oxygen has been measured by a pressure transducer (10,36), then one limitation is the lack of specificity of the measurement. The total pressure in the system will depend on the partial pressure of the products (e.g., carbon dioxide and water) as well as that of oxygen. This may be overcome by undertaking a total analysis of the gases present in the apparatus.

From the slope of the oxygen-uptake curve in the linear region a "steady-state" rate of O_2 consumption has often been used to compare the oxidation of various polymer samples (38,39). Studies have included the variation in the steady-state rate of O_2 consumption with oxygen pressure and film thickness during the photo- or thermal oxidation of polypropylene.

The determination of rate coefficients and kinetic information has been repeatedly obtained from studies of the oxidation of model hydrocarbons in solution at relatively low conversions. Studies of model substances, such as 2,4-dimethylpentane or 2,4,6-trimethylheptane, as well as polypropylene, enabled the evaluation of rate constants for propagation and termination (44). Electron spin resonance (ESR) was applied to evaluate the decay of alkylperoxy radicals in polypropylene, in which a combination of a rapid and slow decay was observed, resulting in information about the termination rates and reaction order (12).

Any attempt to establish precise rate constants for the initiation, propagation, and termination in the autoxidation of PP is complicated, however, because, as pointed out long ago (45), intermediate oxidation products of a saturated hydrocarbon are 10–100 times more susceptible to oxidation than the starting material. Rate constants may apply only to a complex co-oxidation of a highly degraded polymer and its oxidation products (45).

Although the analysis and evaluation of relative oxidation rates and rate coefficients from oxygen uptake and related methods may sometimes be useful, those measurements are fundamentally related to a homogeneous process and, therefore, can yield only average information about the polymer oxidation behavior. The limited mobility of radicals and oxidation products and all related aspects of a nonuniform degradation are not taken into consideration when applying models originally derived to describe the reactions and kinetics occurring in the liquid state or in solution.

III. HETEROGENEOUS OXIDATION OF POLYPROPYLENE

The free radical oxidation scheme of hydrocarbons was primarily aimed at explaining the chemical changes occurring in the material during oxidation. The chemical oxidation in the solid polymer, however, can be complicated by physical features of the material and the oxidation process. The physical ageing of a polymer material is significantly more complex than can be explained by simple chemical reactions. Physical embrittlement can occur despite overall low extents of oxidation. Mechanical properties, such as impact resistance, can be reduced without changes in the visual appearance of the polymer.

A. Microscopic Evidence for Localized Oxidation

The localization of oxidation in polypropylene has been repeatedly investigated using specific-staining techniques in combination with ultraviolet microscopy. Carbonyl groups in oxidized polyolefins were quantitatively reacted with 2,4-dinitrophenylhydrazine (DNPH) (46–48) or 1-dimethylaminonaphthyl-5-sulfonylhydrazine (DNSH) to visualize localized oxidation (10,49,50). Although DNSH, which is fluorescent and so inherently more sensitive, seemed to be very promising, more careful studies revealed a limited solubility or permeability in the polymer so that only the surfaces were stained (50). Sulfur dioxide (SO_2) which reacts quantitatively with hydroperoxides to form alkyl sulfates with a strong absorbance band in the infrared at 1195 cm^{-1} was applied to study the oxidation of polypropylene (51). Localized regions of extensive oxidation appeared as highly discolored areas, particularly around iron impurities. The reaction of gaseous SO_2 with oxidized polyolefins was a convenient method, and limitations in the detection sensitivity were overcome by heating the treated samples to produce conjugated sequences owing to the formation of sulfuric acid and a related charring effect (52). In a completely different approach, oxidized polyolefin samples were immersed in a solution of Sudan III and methylene blue, with the oxidized sites appearing as blue stains and the nonoxidized as pink ones (53). The heterogeneous oxidation of HDPE was also investigated with the DNPH- and SO_2-staining techniques, combined with UV–visible microscopy and spectrophotometry (48).

The UV microphotographs from DNPH-staining experiments on oxidized polypropylene showed the preferential oxidation of spherulitic boundary regions (50). An evaluation of slightly compressed PP films resulted in regions of preferential oxidation, as well as an apparent random distribution of micron-sized oxidation spots, which were attributed to the activity of catalyst residues (50,54).

Various physical aspects of the oxidation process can lead to the established heterogeneous oxidation of solid polymer samples, as discussed in the following.

B. Mobility of Radicals

The limited mobility of radicals in the solid polymer matrix is considered to be one of the main reasons for heterogeneous polymer degradation. Rabek pointed out that a kinetic treatment developed for the liquid phase is not completely valid for solid polymers, for which radical mobility is very restricted (3). Cage recombination of polymer peroxy radicals is considered as the main difference between liquid- and solid-state photooxidation processes (3,26,55). In liquid-state photooxidation, diffusion will quickly randomize radical populations, whereas in the solid state the polymer peroxy radical will separate only by slow segmental diffusion. Buchachenko (56) also emphasized how the kinetics in solid polymers are complicated by molecular motion. All quantitative kinetics of solid polymer reactions are limited by this problem, which makes predictions of polymer lifetimes extremely difficult.

Sufficient evidence has been obtained in the past that the mobility of radicals varies considerably between the crystalline and amorphous phase. Chien and Wang (40) investigated the initiation efficiency of polyolefin oxidation by benzoyl peroxide. They found that initiation efficiencies were as much as ten times lower for semicrystalline polymers than for amorphous ones. The former increased rapidly, approaching the efficiency in the amorphous system when the temperature was raised to the melting point of the polymer. The difference was attributed to the presence of "encumbered" segments and molecules at the phase boundaries, with a higher probability of recombination for the encumbered radicals (40).

Mayo (26), who studied the initiation efficiency in bulk atactic polypropylene and liquid model substances, concluded that a different oxidation behavior was mainly due to changes in the termination constant and the steric effects on inter- and intermolecular propagation. The inhibited oxidation of atactic PP with 2,6-di-*tert*-butyl-*p*-cresol showed that only 16% of the radicals that escaped the cage and were able to propagate chains, could be scavenged. The other 84% reacted with each other before any propagation event occurred, suggesting the vital importance of cage interactions during the oxidation of polypropylene (26).

The photooxidation of PP was discussed by Garton and co-workers, who suggested the importance of secondary cage recombinations (27). The photooxidation of the solid polymer seemed to be greatly affected by the low mobility of radicals. Even when a radical pair escaped the initial polymer cage, the probability of recombination remained high even after several propagation steps. This phenomenon, termed secondary cage recombination, has a pronounced influence on the oxidation kinetics and means that only a few radicals that escape this secondary cage have long enough lifetimes to result in a very high kinetic chain length (27). The yield of radicals from the cage depends on the structure of the polymer and the molecular mobility of the matrix (57).

Similar conclusions were obtained from studies of the decay of peroxy radicals and the thermal decomposition of hydroperoxides in solid polypropylene. The thermal decomposition of PP hydroperoxides consists of two consecutive reactions. An initial faster one shows rates of up to 60 times that of a slower one. The result was claimed to be consistent with an intramolecular radical-induced mechanism for the initial reaction (58). Gijsman et al. also obtained evidence of the oxidation of polypropylene being controlled by different types of hydroperoxides (32). The oxidation rate seemed to be influenced initially by slowly decomposing hydroperoxides followed by the formation of rapidly decomposing hydroperoxides, which indicated the importance of bimolecular decomposition of adjacent hydroperoxides (32). Billingham (54) assumed that the oxidation was localized at the molecular level, leading to the formation of hydroperoxide groups around an initial initiation center that do not diffuse apart, either because of restricted mobility, or because they are on the same chain. Decomposition could then take place in a region of high concentrations, leading to the induced decomposition of adjacent molecules and the destruction of a short length of chain with the evolution of volatiles (54). These volatiles include acetic acid and methanol (31), which was consistent with the high extent of chain scission within highly oxidized domains.

Electron spin resonance (ESR) was applied to multiple studies of the decay of alkyl peroxy radicals in γ-irradiated polypropylene. Roginskii et al. (12) described an initial fast decay, followed by a slower one. Both decays, however, could be described by second-order kinetics. Sohma (58) obtained evidence that those free radicals, which were initially produced uniformly in γ-irradiated PE or PP, decayed preferentially in the amorphous region. This process required a migration of radicals from the original sites into the amorphous material.

Similar studies by Carlsson et al. (59) concluded that the macroperoxy radical reactivity was controlled by mobility factors, and that the radicals may be either highly reactive to propagation or termination, or extremely nonreactive. The proportion of each type of radical appeared to depend on the precise morphology of the sample and other experimental conditions. Both the highly reactive mobile and the less reactive trapped radicals appeared to contribute to the overall degradation (59).

C. Morphology

Polymer morphology plays an integral part in the course of degradation. The oxidation of semicrystalline polymers, such as polyolefins, is generally considered to occur within the amorphous region, which can be treated as a boundary phase of the neighbouring crystalline regions (3). Buchachenko (56) described the material as a microheterogeneous system, with an imperfect submolecular structure in which the crystalline regions alternate with the amorphous ones. This feature leads to a nonhomogeneous distribution of reagents, such as oxygen, oxidation products, and stabilizers, that are concentrated in the amorphous and defective parts of the polymers. These regions also contain the most reactive groups of the macromolecules, such as preoxidized groups and unsaturation. Local concentrations of reagents and, therefore, local rates of chemical reactions, should differ strongly from average ones (56).

Shlyapnikov (60) discussed three levels of irregularities in the solid polymer. Microanisotropy (i.e., differences between directions along and across the polymer chain); topological irregularities (i.e., the existence of end groups and various polymer entanglements); and morphological irregularities (i.e., the existence of relatively large zones of the polymer differing in structure, properties, and chain orientation, such as crystallites and amorphous regions).

Oxygen absorption studies (61) showed that the amount of oxygen reacting in the solid state was apparently inversely proportional to the percentage crystallinity, indicating that the oxidation reaction took place mainly in the amorphous region. The degree of crystallinity and orientation of the sample also has an important influence on the degradation.

Differences in the oxidation kinetics of oriented and isotropic PP were reported by Rapoport et al. (62), who found that orientation of the amorphous phase reduced the yield of hydroperoxides formed. This led to an increase in the induction period and the stability of the material. Similar results were obtained by La Mantia et al. (63), who investigated the reduced photooxidation rates of polypropylene with increasing crystallinity and orientation. Studies of the thermal degradation of oriented polyolefin samples indicated a localization of oxidation in zones of amorphous areas that were isolated from each other by chains in extremely straight conformations. The different oxidation behaviors were attributed to changes in the polymer structure and to reduced oxygen solubility and diffusion coefficients. A proposed model described the drawing of the polymer as leading to an increased isolation of reactive zones in the polymer. This isolation was related to chains in extremely straightened conformation; namely, interfibrillar transition chains, which made the transition free-volume from one microreactor to another one relatively difficult (64).

The thermal oxidation of macro-oriented PP films under load led to a similar conclusion; it was suggested that the differences in the durability of the polymeric specimens were determined by the concentrations of defect zones, rather than by their nature. A proposed model related the stability of the material to the number of centers of oxidative degradation that are similar and cause fracture of the specimen during their growth and linking (65,66).

Studies of the photodegradation of PP samples stabilized with hindered amine light stabilizers (HALS) and carbon black indicated a similar behavior. Whereas severe degradation of the surface layers may quickly lead to surface embrittlement, total failure of the sample also depended on nucleation in the interior of the material. Degradation in the interior occurred at a reduced rate, but was sufficient to alter the failure mode in the HALS-stabilized PP. From the point of fracture properties it seemed beneficial to concentrate the damage into the surface region of the material, rather than allowing it to spread more uniformly through the thickness of the sample (67).

Degradation profiles (i.e., the degree of degradation as a function of depth from the surface) were determined for photodegraded PP and HDPE samples. The shape of the profile was dependent on oxygen diffusion, morphology, and the formation of spontaneous cracks (68). The morphology of HDPE during thermal degradation (69) was analyzed, and it was concluded that the spherulitic structure changed progressively with the degradation. Polarized optical microscopy and small-angle X-ray scattering indicated a decrease in the average lamellar thickness and localized distortions of the spherulites. Solid-state oxidative scission was assumed to occur preferentially at the chin folds where the polymer molecules are strained, resulting in an increase of the crystallinity and a more efficient packing of crystallites formed by shorter, cleaved chains (69).

The analysis of the morphological structure of blends of isotactic and atactic PP showed a considerable proportion of atactic material being present in the spherulite.

It has long been recognized that the atactic material oxidizes more easily than the isotactic fractions (61), and it was also suggested that the spherulite boundaries and interfibrillar regions may be especially vulnerable to oxidation (54). Some studies of the thermal degradation of PP indicated that cracks could pass through the spherulite radii and the actual centers of the spherulites (70), which is consistent with the fact that atactic material is present in the spherulite structures.

D. Solubility and Diffusion of Oxygen

The oxidation of polyolefins below their melting point is a reaction between a gas and a solid and, similar to all such reactions, may be susceptible to rate control by the diffusion of oxygen into the sample, rather than the rate of reaction with the sample.

To oxidize the material, oxygen has to adsorb on the surface and diffuse into the material. In most polyolefins, the amorphous region is more susceptible to oxygen diffusion, and the solubility of oxygen within this region is much higher than in the crystalline fraction.

Studies of the solubility and diffusion of oxygen and other gases in polyethylene were carried out by Michaels and Bixler (71,72). They treated the polymer as a simple two-phase mixture of crystalline and amorphous regions and concluded that the solubility was proportional to the volume fraction of the amorphous material. According to this model even the small helium molecule showed no detectable solubility in the crystallite. It appeared that the basic impenetrable unit in polyethylene was not the spherulite itself, but rather, the lamellar crystallite. Three amorphous regions were distinguished in polyethylene: intralamellar, interlamellar, and interspherulitic (72). The diffusion constants in polyethylene were claimed not only to be dependent on the volume fraction of penetrable amorphous phase, but also on the size, shape, and size distribution of crystallites, as dictated by the prevailing conditions within the polymer during crystal growth. A typical diffusion coefficient of oxygen in polyethylene at 45°C was measured as 12.6×10^7 cm^2/s (72).

Kiryushkin and Shlyapnikov investigated the thermal oxidation of polypropylene in relation to oxygen diffusion (43). It was concluded that a film thickness of less than 40 μm was independent of diffusion, a film thickness of 40–200 μm showed an intermediate, and a sample of more than 200 μm was subject to a diffusion-controlled process. Also, the diffusion influence was temperature-dependent, with a corresponding activation energy of 48 kJ/mol (43). A similar conclusion was presented by Vink (38), who claimed that for films of less than 100 μm, the influence of the diffusion on the photoxoidation process seemed to be negligible.

The importance of a diffusion-limited oxidation during polymer degradation has been repeatedly shown. Diffusion-controlled oxidation occurs when the rate of oxygen consumption within the material is greater than the rate at which oxygen can be replenished by diffusion from the environment (73,74). Gillen et al. carried out detailed studies of the thermal degradation of nitrile rubber specimens ($\sim$2.1-mm thick) and found that the oxidation was diffusion dependent from the earliest ageing times, particularly at high temperatures (125°C).

Similar investigations were carried out by Schoolenberg et al. (68), who studied the UV degradation of 4-mm PP and HDPE samples. The influence of oxygen diffusion was clearly shown by analyzing the depth profile of the degradation. Analysis of microtomed samples by Fourier transform infrared (FTIR) and microfilm tensile testing seemed to be suitable techniques to obtain depth profiles after various oxidation times. It was attempted to relate the oxidation profile to the failure behavior caused by crack formation and propagation using a fracture mechanics approach (4,5).

Studies by Langlois et al. (75,76) of the thermooxidation ageing of cross-linked polyethylene ribbons of 2.2-mm thickness at high temperatures above the melting point also indicated a largely oxygen diffusion-controlled process. Depth profiling was carried out with density and carbonyl index measurements on microtomed polymer sections. Most of the material changes occurred in only a superficial region that could be described by a simple kinetic model of a diffusion-controlled process

(75,76). In general, polypropylene is expected to show diffusion-controlled kinetic behavior similar to that of other polyolefins.

E. Catalyst Residues

Evidence that the thermal oxidation of polypropylene may be initiated by catalyst residues was obtained from degradation studies of slightly compressed PP films using UV microscopy (50). Common PP powder is manufactured by a slurry process that uses second-generation catalyst systems suspended in hydrocarbon solvents (71). The morphological features of the produced PP particles are extremely complex, with different levels of order (78). Highly efficient and improved catalytic systems have resulted in optimized polymerization processes, reduced operating costs, and increased polymer yields per employed catalysts, but have also made catalyst deactivation steps unnecessary. Highly reactive catalyst residues can, therefore, remain in the polymer (79).

The latest "third"-generation catalytic systems have enabled the manufacturing of polymer beads straight from the reactor, rendering compounding and extrusion steps after the polymerization unnecessary, but also leaving active residues within the polymer. Such products have been introduced as "Spheripol" polymers manufactured by a gas-phase polymerization process developed as the Mitsui or Himont technology.

Catalytic efficiencies of 100–1000 kg of polymer per gram of transition metal in second-generation Ziegler–Natta catalytic systems have been reported that leave residual titanium concentrations of 1–20 ppm in the polymer (77). The complex activity of Ti-(IV) compounds, such as n-butyl-o-titanate, n-octadecyl-o-titanate, or titanium tetrastearate in polypropylene was investigated (80). It appeared that these compounds could act as oxidation catalysts at low concentrations and as oxidation inhibitors above critical concentrations, depending on the nature of the ligand (80).

Investigations of the influence of $TiCl_3$-based polymerization catalyst residues on the thermal degradation of polypropylene resulted in curved relations between the induction period of the thermal degradation and the reciprocal temperature, in an Arrhenius-type diagram (81). It appeared that the influence of the catalyst residues depended on the temperature. At temperatures of 130°C almost no differences in the stability of the polymer with different concentrations of titanium were observed. At lower temperatures of 50°C polymers containing 64 or 180 ppm of titanium oxidized with shorter induction periods than polymers with 2 or 8 ppm titanium. As a possible explanation it was assumed that changes in the hydroperoxide decomposition mechanism were responsible. It was suggested that hydroperoxide decomposition occurred primarily thermally at higher temperatures, whereas it was Ti-catalyzed at lower temperatures.

A different study suggested that catalyst residues in polypropylene are difficult to define, for their composition depends not only on the conditions of the polymerization process, but also on subsequent contact with moisture and oxygen (82). Residues of Ziegler–Natta catalysts after the polymerization were described at titanium in its lower valences [Ti(111) with small amounts of Ti(II)] bonded with chlorine and aluminium alkyls. Complex reactions between phenolic antioxidants and titanium compounds were reported, which generally led to the formation of antioxidant titanates by the elimination of hydrogen chloride. The titanates of the antioxidants were seen as being responsible for the discoloration of polypropylene and an overall reduced thermooxidative stability of the material (82).

A similar investigation studying the interaction between metal ions, such as Cr(III) Fe(III), or Cu(II), and various phenolic antioxidants concluded that the metal ions had only minor effects on the unstabilized polymer, but strongly interfered with the antioxidant activity. Beneficial or adverse behaviors, particularly under oven-ageing conditions were reported (83).

The analysis of the thermooxidation of HDPE that contained residual chromium catalyst remnants also revealed a complex interaction between the polymer and catalytic residues (84). Residual chromium seemed to act as a prooxidant in low concentrations (1.6 ppm), depending on the valence state and solubility in the polymer. At higher concentrations (100 ppm) it appeared that oxidative pyrolysis was suppressed, which was related to an ability of the chromium ions to complex macroalkylperoxy radicals, thereby terminating the oxidative chain process and improving the thermal stability of the material (84).

Catalyst residues of currently employed catalytic systems can, therefore, be of wide chemical variety, composition, and activity, and they are able to interfere with the thermooxidative mechanism of the polymer, leading to oxidative initiation centers and reduced stabilities of the polymer (50). In particular, it has been suggested that the residual catalyst particles provide the sites of initiation of the heterogeneous oxidation, with the number of kinetic chains depending on the concentration of catalyst (85).

F. Chemiluminescence

One of the manifestations of the oxidation of polypropylene, as either a liquid or as a solid, is the emission of weak visible light—chemiluminescence—from the onset of oxidation (13). The particular features of this chemiluminescence (CL) have been interpreted as presenting direct evidence for the heterogeneous oxidation of polypropylene (18,19,86), but this interpretation partly depends on the mechanism for production of the emission.

1. Chemiluminescence Mechanisms

Mechanistic studies of CL during the oxidation of hydrocarbons and polymers have been advanced by studies of model compounds, and the spectral analysis of the emission suggests the emitting chromophore may be the triplet state of a carbonyl compound (87,88). If we consider the free radical oxidation sequence of Section II.A and the range of secondary reactions that have been proposed in Section II.B, several possible reactions have been considered to be energetically feasible for the formation of an excited triplet state of carbonyl oxidation product (290–340 kJ/mol).

$$R_2\underset{H}{C}OOH \longrightarrow R_2\underset{H}{C}O\bullet \; + \; \bullet OH \qquad \Delta H = +\,147 \text{ kJ/mol} \qquad (28)$$

$$R_2\underset{H}{C}\text{-}O\bullet \; + \; \bullet OH \longrightarrow R_2C{=}O^* \; + \; H_2O \quad \underline{\Delta H = -471 \text{ kJ/mol}} \qquad (29)$$
$$\Delta H_{total} = -314 \text{ kJ/mol}$$

$$R_2C\text{-}O\bullet + \bullet R \longrightarrow R_2C=O^* + RH \qquad \Delta H = -323 \text{ kJ/mol} \tag{30}$$

(where the first species has an H attached to the carbon)

$$\underset{\begin{smallmatrix}|\\O\\|\\R\end{smallmatrix}}{\overset{R_2\ \ O}{\underset{H}{C}}}\underset{O\bullet}{\overset{O\bullet}{}} \longrightarrow \underset{\begin{smallmatrix}|\\O\\|\\R\end{smallmatrix}}{\overset{R_2\ \ O^*}{\underset{H}{C}}}\overset{O}{O} \longrightarrow R_2C=O^* + O_2 + ROH \tag{31}$$

$$\Delta H = -460 \text{ kJ/mol}$$

Vassil'ev (88) showed that the decomposition of a single hydroperoxide [Eq. (28)] in a cage reaction can yield an excited carbonyl [Eq. (29)], with approximately 314 kJ/mol. Various recombination reactions of alkylperoxy and alkyloxy radicals with alkyl radicals may also lead to excited carbonyls [Eq. (30)]. The exothermic termination reaction of alkylperoxy radicals through a six-membered cyclic intermediate (Russell mechanism) results in the formation of an excited carbonyl, singlet oxygen, and an alcohol [Eq. (31)] (88,89).

The Russell mechanism [see Eq. (31)] requires the participation of at least one primary or secondary alkylperoxy radical. For polypropylene, with a large population of tertiary alkylperoxy radicals as discussed in Section II.A, this should not occur, and it was proposed that β-scission of tertiary alkoxy radicals [Eq. (32)] may lead to the formation of primary alkylperoxy radicals [$R\text{-}CH_2OO\bullet$; Eq. (33)] that may then terminate with a t-peroxy radical (13).

$$\underset{\begin{smallmatrix}|\\O\bullet\end{smallmatrix}}{\overset{CH_3}{\underset{|}{RCH_2\text{-}C\text{-}CH_2R}}} \longrightarrow \overset{CH_3}{\underset{|}{RCH_2\text{-}C}}=O + \bullet CH_2R \tag{32}$$

$$RCH_2\bullet + O_2 \longrightarrow RCH_2\text{-}OO\bullet \tag{33}$$

Thus, CL arises from secondary reactions, rather than from the tertiary alkylperoxy radical reactions of the main oxidation scheme [see reactions in Eqs. (1)–(10)] if the Russell mechanism is applicable to polypropylene. It is this dilemma that has resulted in alternative CL-producing reactions involving t-alkylperoxy radicals to be considered, such as the homolysis of t-hydroperoxides, as shown in Eqs. (28) and (29).

However, in studies in which the concentration of hydroperoxides in polypropylene was analyzed by temperature–ramp chemiluminescence, the signal could be suppressed by a free–radical-scavenging antioxidant (89). Therefore, it appeared unlikely that the emission step involved a cage reaction of the alkyloxy and hydroxyl radical from the hydroperoxide decomposition [see Eq. (29)], but rather, the termination of alkylperoxy radicals [see Eq. (31)].

Other possible CL-producing reactions have been recently reviewed (90), and one of the main problems with a simple Russell mechanism involving s-alkylperoxy radicals was the relatively low quantum yield of CL from polyethylene, compared with polypropylene. The CL-producing reaction in polypropylene may involve a complex sequence of reactions culminating in the termination reaction [see Eq. (31)].

Support for this has been given by recent FTIR emission studies of single particles of polypropylene (91) in which higher oxidation products, such as γ-lactones, were formed at the earliest stages of oxidation. This was consistent with the involvement of primary and secondary alkylperoxy radicals in the CL-producing reaction, and the sequence of reactions leading to CL from polypropylene clearly differ significantly from those in polyethylene. The Russell mechanism also requires that oxygen is produced in an excited singlet state, and there is evidence that the decomposition of polypropylene hydroperoxide produces singlet oxygen (92).

2. Kinetic Applications of Chemiluminescence

Although CL may be observed in both the liquid and solid states, the kinetic methodology generally applied has been one of a homogeneous free radical chain reaction. The intensity of CL, I, is given by the application of the steady-state approximation to the reaction sequence, shown in Eqs. (1')–(6') of Section II.C, so that

$$I = \Phi k_6' [RO_2^{\bullet}]^2 = \Phi r_i \tag{34}$$

where Φ is the quantum yield of CL and r_i is the rate of initiation.

This has been used to model oxidation of polypropylene (90) by using a numerical solution to the reaction sequence and so assess the effect of the magnitude of the rate constants for initiation on the CL–time profile. The CL–time profile has been

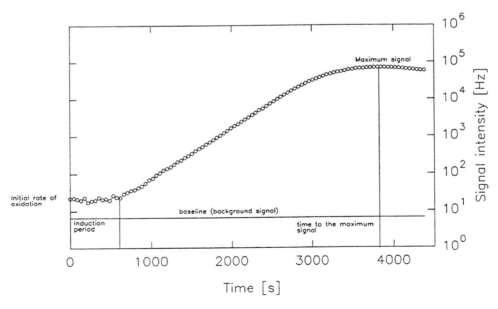

Figure 3 Plot of the intensity of chemiluminescence (CL) from polypropylene powder as a function of the time of oxidation at 150°C. The parameters used in describing the properties of the curve are indicated.

the basic information level from CL experiments and Fig. 3 shows a typical profile from polypropylene powder in oxygen at 150°C. From an examination of these profiles over the temperature range from 90° to 150°C the following features were noted (18,19,86).

1. In unstabilized polypropylene powder there was a measurable steady CL at short times when the overall extent of oxidation was low (the induction period).
2. The "activation energy" for CL intensity in this induction period was identical with that at the maximum after a considerable time of oxidation (112–114 kJ/mol).
3. The activation energy based on induction time was higher (149 kJ/mol), indicating it was independent of the chemical processes leading to CL.
4. By plotting the CL–time curves after the induction period in reduced coordinates up to the maximum [i.e., (I/I_{max}) vs. (t/t_{max})] a single master curve was obtained over the temperature range 90°–150°C.
5. The sigmoidal master curve was interpreted as representing the fraction of the polymer oxidizing at any time, rather than a rate–time curve with kinetic significance.
6. CL imaging (93) and intensity–time profiles of single-reactor particles with controlled separation demonstrated that an oxidizing particle could "infect" an adjacent stable particle, which then commenced emitting CL sooner (86).

These observations could not be reconciled within a homogeneous kinetic scheme and, instead, it was suggested that a heterogeneous model should be applied to the oxidation of polypropylene (18,21). The most significant result in the early studies of CL, performed with very sensitive photon-counting, light-detection equipment, was that the emission of light commenced immediately with the start of oxidation (13,18). This was elaborated on in a later study (94) in which the magnitude of the induction period in any oxidation experiment depended on the sensitivity of the method used to measure it and, therefore, was not a fundamental property of the material. Furthermore, the sensitivity of CL enabled the activation energy to be measured in the induction period (i.e., the samples showed a temperature-dependent emission at zero time, as shown in Fig. 3). That this activation energy was the same as when the polymer was oxidized for a lengthy time period suggested the same chemical processes were occurring throughout the oxidation, and the only difference was that, at short times, the CL intensity was low because only a small fraction of the polymer was oxidizing.

This qualitative heterogeneous oxidation model for CL was able to reconcile many of the apparently anomalous observations on the mechanism for producing CL from polypropylene because, even at short oxidation times, secondary oxidation products would be produced in the small reactive zones. As discussed earlier (see Sect. III.E), these most probably originate around residual catalyst particles (50,85), and then the oxidation may spread similar to an infection through the remainder of the polymer (86). CL-imaging studies of stabilized polypropylene films (95) have shown that the spatial distribution of CL across the surface of a film was nonuniform, with a CL-emission front spreading across the sample. The spatial resolution of CL-imaging systems (e.g., a CCD camera) has not been sufficient to enable direct visualization of the zones of a high reaction rate, and there are fundamental limitations owing to scattering of the emitted light. The average signal from all pixels of the CL images produced an intensity–time profile with similarities to a CL curve of the type shown in Fig. 3, with differences in the intensity beyond the maximum attributable to changes in the spectral distribution of the emission (95). This change in emission intensity profile, depending on the spectral sensitivity of the measuring equipment, reflects a limitation to the use of Eq. (34) to monitor oxidation of the polymer, which is that the quantum yield Φ must be constant throughout the oxidation. Although the foregoing observations (2) and (4) suggest that the mechanism does not change between the induction period and maximum CL, the physical effect of accumulation of colored chromophores that will absorb the emitted radiation may be to produce a CL profile that is thickness-dependent and instrument-specific.

However, the technique has provided strong support for the view (50,54,65,66) that the oxidation of polypropylene is spatially heterogeneous, even within the amorphous region. The method may also be used to provide an understanding of the kinetics of heterogeneous oxidation, as discussed in Section III.H.

G. Scale of Heterogeneity in Polypropylene Oxidation

In the preceding sections there have been successively finer levels of heterogeneity introduced to help explain the experimental observations of the oxidation of solid polypropylene. These may be summarized as follows:

1. Oxidation is localized in the amorphous region of the semicrystalline polymer. When the molten polymer crystallizes, the partitioning of impurities will occur in to the remaining liquid phase which, on solidification, will have a higher specific volume and, consequently, a high solubility of oxygen. Oxidation of the crystalline region, therefore, is considered negligible on the time scale of the degradation of the mechanical properties of the polymer (61).

2. The orientation of polypropylene by drawing further reduces the available sites for oxidation. The oriented chains are equivalent to the crystalline regions of the spherulite in that they have a lower solubility and diffusion coefficient of oxygen. Orientation effectively isolates the amorphous regions from one another, such that oxidation occurs in these smaller volumes of the polymer (65,66).

3. In thick samples, the solubility of oxygen decreases with depth, hence, the reactions become diffusion-controlled in oxygen. This results in a declining extent of oxidation with depth, which may be modeled knowing the diffusion coefficient of oxygen (73,74).

4. Within the amorphous region, oxidation of polypropylene is initiated around catalyst residues and other impurity centers. This results in a high rate of oxidation within discrete zones, whereas other parts of the amorphous region remain unoxidized (10,50,54). The macroscopic oxidation of the polymer corresponds to the spreading of the oxidation front from these zones, resulting in an oxidized region of a size sufficient to nucleate a crack in the polymer (18–20,85,86).

The challenge in the study of the heterogeneous oxidation of polypropylene is to develop a model that encompasses the foregoing observations and allows parameters to be determined that are consistent with the experimental oxidation kinetics. Modeling of a system that incorporates an inert crystalline component and an oxidizable amorphous component has been straightforward, and it is possible to rationalize the oxidation profiles as a function of sample depth using known solubility and diffusion coefficient data for oxygen in polyolefins (73,74). Any model that is developed for the heterogeneous oxidation of the amorphous region must be consistent with the experimental observations made of the oxidation–time profile, as given by oxygen-uptake infrared spectroscopy, or chemiluminescence. This is considered in the following section.

H. Kinetics of Heterogeneous Oxidation of the Amorphous Region

The purpose of a kinetic model is to explain the observations from a wide range of experimental studies of the oxidation of polypropylene as powder, film, or bulk samples and, from this, gain insight into the reaction mechanisms and the way the oxidation may be controlled so that the polymer may be stabilized. The classic homogeneous free radical chain-reaction mechanism has been successful in achieving this (see Sec. II.C) and, although there are conceptual difficulties in understanding how a solid polymer may be described by a physical model developed for liquid hydrocarbons, numerical analysis has allowed construction of computed oxidation–time profiles from which best-fit kinetic parameters may be obtained (16,17,41). It has been argued (96) that the homogeneous kinetic model may be applied to a nonhomogeneous polymer by considering the parameters as average values. However, it has been noted (97), that there is a fundamental problem between microscopic and macroscopic kinetics owing to the measurement of the appropriate concentration terms to appear in the kinetic equations. If oxidation is occurring nonuniformly in the amorphous domains, then these are the kinetic terms that must be evaluated in the rate law using domain concentrations that will be much higher than the mean concentration averaged over the whole polymer. This is not a trivial problem, and it is considered in Section III.I in relation to experimental measurements in heterogeneous systems.

In this section it is intended to summarize those approaches that have been made to a heterogeneous model for the oxidation of polyolefins (19–21,97,98). These have common starting points in which it is considered that oxidation initiates nonuniformly at a few sites in the amorphous region of the polymer. From these sites the oxidation is able to spread (in the absence of a stabilizer, such as a free radical scavenger) so that a progressively larger fraction of the total amorphous region is oxidizing as a function of time. Within any one of these reactive zones, a free radical chain reaction will be taking place at a high local rate, but because there are few of these zones initially, the average rate of oxidation is negligible (i.e., there is an apparent induction period on the macroscopic scale).

1. Monte Carlo Simulation

A highly illustrative approach to heterogeneous oxidation has been provided by using a random-walk model (98) for the spatial propagation of oxidation initiated at catalyst residues. Because diffusion of macroradicals in the solid state is highly restricted, low molecular weight "jogger" radicals are considered the spreading agents. This spreading of the zones by jogger radical microdiffusion occurs at the same time as the extent of oxidation within the original zone also increases. The jogger radical $r^{\bullet}$ is formed from a macroradical with a rate constant k_d and will migrate a distance:

$$R \simeq k_d (D/k_r[\text{PH}])^{1/2} \tag{35}$$

with a characteristic diffusion coefficient D, until it is immobilized by reaction with the polymer [PH] with a rate constant k_r to form a macroradical $P^{\bullet}$ (or $PO_2^{\bullet}$), which again, may then produce oxidation products.

$$r^{\bullet} + \text{PH}(+O_2) \quad \rightarrow \quad rH + P^{\bullet} (PO_2^{\bullet}) \tag{36}$$

The diffusion process has been simulated by a Monte Carlo method, with the following points noted (98):

1. A high initiation rate of 10^{-6} to 10^{-5} mol/kg was used at N initiation sites (ranging from 5 to 100).
2. The diffusion coefficient D was taken as that for gases in polypropylene (i.e., 10^{-5} cm^2/s at 130°C).
3. Within a lifetime t_r of 10^{-2} s, the jogger radical $r^{\bullet}$ moved an average distance of $3\,\mu$m or ten lattice sites before becoming immobilized by the reaction in Eq. (36).
4. At the immobilization site, damage could accumulate and accumulation for a time t_d was considered sufficient to form a microcrack. When the microcrack merged into a percolation cluster, the sample failed by fracture.

This model, while requiring values for several radical lifetimes and related parameters, provided an interesting approach to heterogeneous modeling. It was also possible to examine the role that inhibitors play in such a system. For example, the effectiveness of the inhibitor decreased if there was a substantial concentration of initiating impurities (catalyst residues), and not all of the inhibitor was consumed before the sample failed. Another interesting conclusion was that the durability increased more dramatically if the number of initiation centers were decreased, rather than the rate of radical generation being decreased to the new average value for the same number of initiation centers. This is a consequence of the bimolecular recombination of radicals produced at a higher rate in an initiation zone. These results have been interpreted as indicating that the greatest increase in the durability of polypropylene will be achieved by eliminating the initiation centers by deactivating the catalytic impurities (e.g., by chelation).

2. Epidemic Model

The epidemic model draws on the observation that the oxidation profile from polypropylene (and, indeed, from many other polymers) shows features characteristic of the infectious spreading of a disease through a population. Such a view is supported by microscopic (see Sec. III.A) and chemiluminescence (CL) studies including CL imaging (see Sec. III.F), and it was suggested (97) that the spreading of oxidation could be appropriately modeled by using epidemiological models for the spread of disease (99).

In simple epidemic models, a small number of infected individuals are introduced into a large, fixed population, and the aim is to determine the spread of the infection as a function of time (99). The population, after a given time period, may be divided into three classes: those that have become infected; those that have either recovered from the disease (and are immune) or are dead; and those that are still susceptible to infection.

In applying this epidemic model to the heterogeneous oxidation of a solid polymer the same procedure may be followed. In this case a number of infectious zones are randomly distributed within the polymer. Within each zone are impurities or catalyst residues that result in a chain reaction producing macromolecule scission, formation of volatile oxidation products, and a high local concentration of free radicals. These free radicals are than able to spread from the initial zone by infecting the adjacent polymer. The termination reactions of these radicals will serve to reduce the infectious population, replacing it with dead or oxidized material.

The reactions that are considered to take place are those described earlier in Sections III.A and II.B; that is, the same chemistry that has been invoked in the homogeneous model for polypropylene oxidation, except that within each zone the reaction is occurring at a high rate and produces secondary oxidation products and volatiles.

In the epidemic model three distinct populations in the amorphous region of the polymer may be assigned after a short time of oxidation:

1. The remaining or unoxidized fraction p_r
2. The infectious or oxidizing fraction p_i
3. The dead or oxidized fraction p_d

These populations may be described by a series of coupled differential equations [see Eqs. (37)–(39)] (21), which are developed by noting that oxidation can spread only if an infectious zone has uninfected material available within a contact distance, and this will be proportional to p_r.

$$\begin{aligned} \frac{-dp_r}{dt} &= bp_r p_i \\ &= bp_r(1 - p_r) \end{aligned} \tag{37}$$

where $b(s^{-1})$ is the rate coefficient for spreading and p_d is small compared with p_r at short times of oxidation, such that $p_i = 1 - (p_r + p_d) \simeq 1 - p_r$

$$\frac{dp_d}{dt} = \alpha p_i \tag{38}$$

where α is the rate coefficient for formation of oxidized material from the infectious (free radical) fraction, and this encompasses a range of elementary rate processes for both propagation and termination steps in the oxidation.

$$\frac{dp_i}{dt} = bp_i p_r - \alpha p_i \tag{39}$$

These equations may be solved (21) if the initial infectious fraction p_0 is small, to give the time dependence of the infectious fraction p_i as:

$$p_i = p_r p_0 \exp{(b - \alpha)t} \tag{40}$$

Thus, in this simple model, the spreading of the oxidation through the amorphous region of polypropylene should depend on the initial infectious fraction (i.e., the number of active catalytic initiation centers) and the difference between the spreading rate coefficient b and removal coefficient α.

Although this model may be explored by assuming values for these parameters, they may be explicitly determined from an analysis of the CL profile (21). The only assumption required is that the intensity of CL, I, at any time is given by Eq. (41).

$$I = \phi r p_i \tag{41}$$

where ϕ is the CL quantum efficiency and r is an average rate for termination of peroxy radicals within an infectious zone under the conditions of the oxidation. Equation (41) is identical with Eq. (34), except that it includes the factor p_i, the fraction of the polymer from which CL is being emitted, which is considered to be the infectious fraction.

This relating of CL to the infectious fraction enables a linear function to be obtained from which p_0, α, and b may be calculated from the resulting Eqs. (42) and (43).

$$\ln(I_{max}/I) = \ln(\alpha/b) + (b - \alpha)(t_{max} - t) \tag{42}$$

and

$$\ln[(1 - p_0)/p_0] = b t_{max} + \ln(\alpha/b - \alpha) \tag{43}$$

where I_{max} is the maximum CL intensity after oxidation time t_{max}. To derive Eqs. (42) and (43) it is required that ϕr should not change up to the maximum CL intensity. This is supported by the constancy of activation energies, which is one of the experimental observations noted in Section III.F.1.b. The relations of Eqs. (42) and (43) hold only in the early stages of oxidation (when p_r is very much greater than $p_i + p_d$), but this is sufficient to enable a linear plot to be obtained and values of α, b, and p_0 determined for single particles of polypropylene of different origin (21).

These values, determined in the first 15% of the oxidation are used to simulate the CL curve as representing the change in the infectious fraction over the entire oxidation. Figure 4 shows the comparison of the simulation using Eq. (40) and the parameters given in Table 1 for PPK(1) with the experimental CL curve.

Table 1 also shows the infectious parameters p_0, b, and α for three separate single particles of two different types of polypropylene reactor powders: Hoechst PPK0160 and Polychim A10TB. The following points may be noted:

1. There is wide variation in the value of p_0, the initial infectious fraction, both between and within sample types. This probably reflects the statistical nature of the distribution of residual catalyst and its activity from particle to particle. The values of p_0 indicate that between 0.02 and 2% of the sample is oxidizing at the start of the induction period.

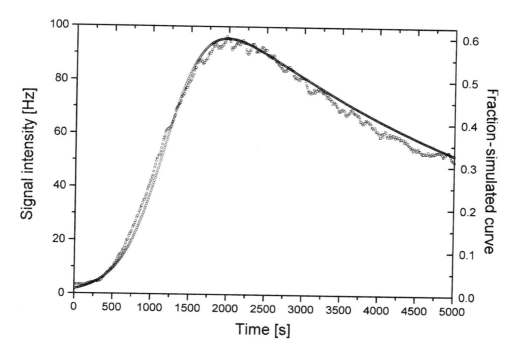

Figure 4 CL from a single particle, PPK(1), of polypropylene compared with the simulated infectious fraction p_i from Eq. (40) using the epidemic model.

Table 1 Infectious Model Parameters for Three Particles of Two Different Types of Polypropylene as Determined by CL Analysis of Single Particles [Eqs. (42) and (43)]

Sample	$p_0 \times 10^3$	$b \times 10^3/s^{-1}$	$\alpha \times 10^4/s^{-1}$
PPK(1)	11.6	3.6	2.2
PPK(2)	23.2	3.3	1.5
PPK(3)	11.3	3.8	3.2
A10TB(1)	0.8	5.2	9.1
A10TB(2)	1.8	4.0	9.6
A10TB(3)	0.2	5.3	12.4

2. The spreading rate coefficient b/s^{-1} is approximately constant within a polypropylene type, and the difference between the values for two different polymer types is sufficiently small to suggest that it is a fundamental property of the polymer.
3. The removal rate coefficient α/s^{-1} is significantly different between the two polymer types and may be more sensitive to the morphology of the particle.

Of some interest in these studies were the simulated profiles for the three separate fractions as a function of oxidation time, as shown in Fig. 5 for the two dif-

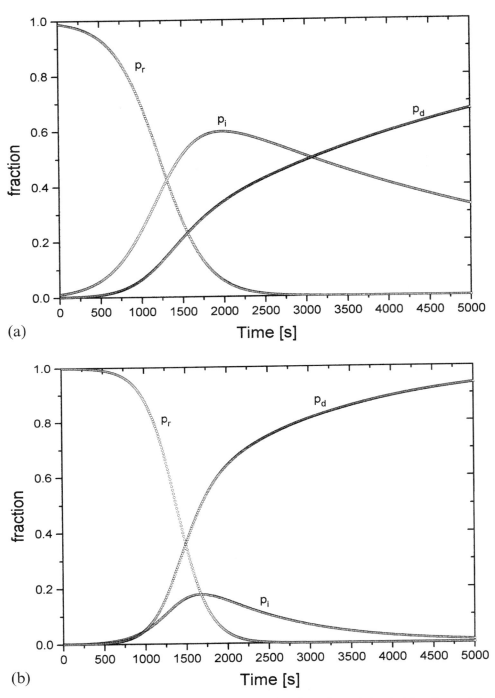

Figure 5   Development of infectious (p_i), oxidized (p_d), and remaining (p_r) fractions of the amorphous region of single particles of (a) PPK(1)(Hoechst) and (b) A10TB(1)(Polychim) polypropylene reactor powder during oxidation at 150°C (in oxygen) based on CL analysis by the infectious spreading model using the parameters from Table 1.

ferent polymer types. The behavior is quite different and reflects the differences in the parameters from Table 1. It should be possible to further test the profiles in Fig. 5 by the measurement of the oxidation product profile (by either infrared absorption spectra, hydroperoxide analysis, or oxygen uptake), but this may be complicated by the known production of volatiles during the oxidation of the polymer (31). However, the following points may be noted from the profiles for the two different polymer types given in Figs. 5a and b.

1. The differences between the two infectious distributions (p_i) is linked to the differences between the initial fraction oxidizing p_0. This is consistent with the importance of the number of active sites, as noted from the different modeling studies of Rapoport (98), discussed in Section III.H.
2. For both polymer types, the peak in p_i occurs when $p_d \sim 0.3$ (i.e., approximately 30% extent of oxidation).
3. The oxidation does not become homogeneous (i.e., $p_r \rightarrow 0$) until well past the maximum in p_i and at least 50% extent of oxidation.
4. The removal parameter α controls the rate of formation of the oxidized fraction p_d compared with the infectious fraction and accounts for the more rapid buildup of p_d in the A10TB samples.

In an extension of the studies of single particles, an investigation was made of the CL from groups of particles because it had earlier been shown that, for polypropylene with a low and unevenly distributed concentration of antioxidant, it was possible for an unstable, oxidizing particle to infect another particle that was located within about 100 μm of it (86,93). When the same analytical method was applied to the group of particles, it was reported (21) that the fit of Eqs. (42) and (43) was much poorer. The model embodied in Eqs. (38)–(40) which yields single spreading and oxidized product formation rate coefficients is too simple to encompass the process of interparticle as well as intraparticle oxidative spreading.

The further development of the epidemiological model requires the consideration of multiple-spreading processes. Of some interest is the effect that an antioxidant may have on the interparticle compared with intraparticle spreading parameters (100). Figure 6 shows CL curves for the oxidation of PPK powder (5 mg) that has been wet-coated with an acetone solution of a hindered phenol antioxidant (Irganox 1010) having the concentration shown (100).

It may be seen from the changed CL profile that the antioxidant behaves as expected in inhibiting the thermal oxidation and producing an increased induction period with increasing concentration. Because of the sensitivity of CL it was possible to determine the infectious parameters by applying Eqs. (42) and (43). These values are given in Table 2, and the following points are noted:

1. There is a steady decrease in b, the spreading parameter with increasing antioxidant (AO) concentration.
2. The removal coefficient α is much reduced, but is independent of AO concentration.
3. The initial infectious fraction p_0 is reduced by a factor of close to 100 by AO.

The value of α is probably controlled by the chemistry of the stabilizer because it represents the rate coefficient for formation of the oxidized material from the

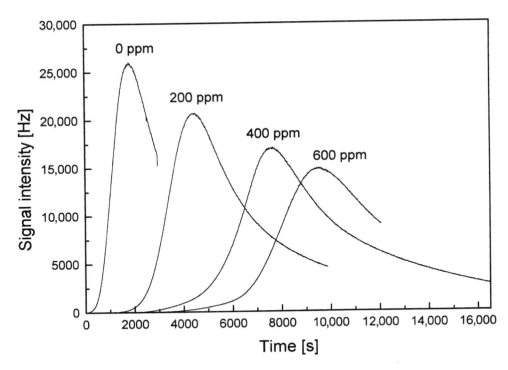

Figure 6 CL oxidation time curves for 5-mg samples of PPK polypropylene reactor powder coated with the indicated concentration of Irganox 1010 hindered phenol antioxidant.

Table 2 Changes in Infectious Spreading Parameters for PPK Powder with Concentration of Added Antioxidant (Irganox 1010)

[Irg 1010]/ppm	$b \times 10^3/s^{-1}$	$\alpha \times 10^4/s^{-1}$	$p_0 \times 10^3$
0	5.60	5.50	2.90
200	3.11	0.77	0.052
400	2.10	0.71	0.029
600	1.39	0.63	0.032

infectious fraction. When the antioxidant is present, then the oxidized fraction p_d remains low.

It might have been expected that the AO would not be able to change a fundamental parameter such as p_0 that is linked to the residual catalyst concentration and other oxidation-initiating centers. However, this may be considered in the light of the mechanisms of hindered phenol antioxidant action in which aggregation of antioxidant at sites containing polar groups may occur. Also, at the reaction temperature, the diffusion coefficient of the AO may be 10^{-7} to 10^{-8} cm^{2}s^{-1} (42) so that migration to active sites may occur. There will, however, be a small number of sites where AO is not present to inhibit effectively, or it may not mechanistically function at that site for chemical reasons. As the AO is consumed at the active site, the oxidation will spread and the role of the AO in the bulk of the resin is to limit that spreading. As shown in Table 2, the spreading coefficient b, continues to decrease with increased concentration of AO.

It is of interest to compare the interpretation of antioxidant behavior with that accepted for a homogeneous free radical oxidation, such as that of a liquid hydrocarbon. The induction period in that case corresponds to a reduction of the kinetic chain length to unity, whereas the antioxidant dominates chain termination and is progressively consumed. The duration of the induction period is controlled by the ratio of the concentration of antioxidant to the rate of initiation. Following consumption of the sacrificial antioxidant, the oxidation proceeds at the same rate as in the absence of stabilizer. Although it has been possible to translate this interpretation to the oxidation of a molten polymer, in solid-state oxidation, the heterogeneity of oxidation requires a different formalism.

The system is no longer well-mixed and the antioxidant is required to both inhibit the widespread onset of oxidation and prevent the spreading of oxidation through a volume of the polymer sufficient to allow the formation of a mechanically weak defect site. In the context of fracture mechanisms, the size of the degradation zone must be kept below the critical crack length, as defined by the critical stress intensity factor K_{1c} of the bulk polymer (4,5). The antioxidant appears to participate, from the results presented here, in both of these ways. By affecting all of the following it may contribute to the environmental durability of the polymer: p_0, the number of potential defect sites; b, the spreading coefficient and thus the rate of growth of defects; and α, the removal coefficient for transforming the infectious fraction of the polymer into oxidized material. It is interesting to compare this with the observations of Livanova (66) in which, in a loaded, oriented polypropylene sample, the AO (present at 0.1 wt% with DLTDP as a synergist) resulted in localization of the oxidation at the initiation sites so that total volatilization of the oxidation products occurred before significant spreading was observed.

3. Simplified Spreading Model

Gugumus (20,97) has discussed the homogeneous and heterogeneous aspects of the oxidation of polyethylene and has emphasized that classic kinetics cannot be applied to oxidation in semicrystalline polymers. As one approach to developing a heterogeneous model (20) he proposed a simplified version of the epidemic model, discussed in the previous section, by considering that the spreading of oxidation followed Eq. (44).

$$\frac{dn}{dt} = rn - sn^2 = rn\left(1 - \frac{n}{N}\right) \tag{44}$$

where n is the number of oxidized domains out of the total number of amorphous domains N in the sample, and r is the rate coefficient for spreading.

This equation was then solved to give an expression Eq. (45) for the time dependence of hydroperoxide concentration, which could be compared with experimental data from infrared spectrophotometry (20,97).

$$n = \frac{N}{1 + \dfrac{N - n_0}{n_0} \exp(-rt)} \tag{45}$$

Where n_0 is the initial number of oxidized domains.

It can be seen that Eq. (44) is identical with Eq. (37), for this simplified model considers the polymer to consist of only two fractions: the oxidized fraction (p_d in the epidemic model) and the unoxidized fraction (p_r in the epidemic model; i.e., $p_r + p_d = 1$).

Thus, in this nomenclature, Eq. (44) becomes:

$$\frac{dn}{dt} = \frac{dp_d}{dt} = rp_d(1 - p_d)$$

$$= bp_r(1 - p_r) = \frac{-dp_r}{dt}, \text{ since } b \equiv r,$$

which is Eq. (37).

This simplified model was able to explain several phenomena in the oxidation of thick samples of low-density polyethylene, including the effect of sample thickness, the role of associated and free hydroperoxides, and the concentration profile of carbonyl groups (20,97). The model has not been applied to the oxidation of polypropylene, and it was noted earlier that the accumulation of oxidation products is mechanistically different for the two polymers (31).

It was also noted by Gugumus that one of the reasons for the success of a homogeneous, classic kinetic model for solid-state polymer oxidation was that the heterogeneous differential Eq. (44) was mathematically identical with the following equation describing the homogeneous rate of formation of hydroperoxides:

$$\frac{d}{dt}[POOH] = a[POOH] - b[POOH]^2 \tag{46}$$

Therefore, it is understandable that both models should give equally good fits to the experimental data; hence, there is no reason (based on data analysis alone) why the heterogeneous model should be preferred. In polypropylene, however, it was noted earlier (see Sec. III.A) that the classic homogenous model could not explain the reason for spatial irregularities in oxidation that ultimately result in crack formation and embrittlement of the polymer. The major difference is that the heterogeneous model "decouples" the experimentally observed oxidation curve (e.g., oxygen uptake) from the chemical kinetic equations for a chain reaction. In the homogeneous model, the oxidation curve is considered to represent the solution of the kinetic equations describing those rate processes (i.e., the macroscopic, measured curve has a direct relation to the chemical kinetics).

In the next section this fundamental problem in the kinetic analysis of macroscopic data for the oxidation of a polymer is considered.

I. The Concentration Problem in Oxidation Kinetics of Solid Polymers

The formal kinetic schemes for oxidative chain reactions (see Sec. II.C) lead to a relation (rate laws) between the rate of oxygen uptake, or rate of oxidation product formation, and concentration of reactants. Again, such rate laws are usually associated with well-mixed systems, such as gases and liquids, for which the concentration profiles may be obtained by a spectrophotometric or other analytical method. When the oxidation kinetics of a heterogeneous solid polymer are measured, it is necessary to consider that a significant fraction of the polymer will not be reacting; hence, the concentration terms that are required for the kinetic equations are those for the amorphous regions that are oxidizing at any point in time. Because of the lack of spatial resolution of most analytical methods; it is not possible to obtain this information; consequently, instead, an average concentration over the sample volume is measured, for example:

1. Infrared transmission spectrophotometry measures absorbance as a function of wavenumber over the area of the beam in the spectral region for oxidation products, and this is converted to concentration by Beer's law or an appropriate calibration curve.
2. Oxygen uptake is averaged over the sample mass and expressed as $mmolO_2/kg$ of polymer.
3. ESR measures the total number of spins per gram of polymer in the cavity to yield average free radical concentrations.

If a polymer is oxidizing heterogeneously with only a fraction of the available number of amorphous domains oxidizing, then to determine the true concentration of reactants and oxidation products it is necessary to probe a sample with a thickness of one amorphous domain using a technique with a spatial resolution of less than one amorphous domain. Because it is considered that the diameters of these domains may be 10–30 nm (18,98), this is then beyond the capabilities of current experimental techniques. In fact, many attempts to demonstrate that heterogeneity in oxidation is not present have used methods with either poor resolution, such as IR spectrophotometry (17), or required high extents of oxidation to detect the changes, such as molecular weight measurements by GPC (101). That both methods failed to detect heterogeneity is, therefore, not surprising.

This fundamental problem in the determination of the formal kinetics for an amorphous domain when one has only access to concentration data for the entire sample volume has been considered by Gugumus (97). He has shown that it was only possible to determine the kinetic law in the amorphous domain from the data for the observation volume if the rate law was first-order. For higher orders, it was necessary to consider further assumptions on the rates of domain oxidation and the reactive volume.

A generalization of the concentration problem in heterogeneous oxidation may be obtained by considering that, in a chemical kinetic treatment of oxidized product formation in a heterogeneous system, it is necessary to obtain the actual local rate of reaction of a microscopic amorphous domain, dC_a/dt, but in practice, it is only possible to obtain a measured rate of a macroscopic volume dC_m/dt. The relation linking these two rates is the volume fraction V_d of the oxidizing polymer in the total volume such that the concentration terms are linked by:

$$C_a = C_m/V_d \qquad (47)$$

that is, because V_d is very much less than 1 especially early in the oxidation, then $C_a \gg C_m$.

The oxidation rate in the amorphous domains is given as the derivative of Eq. (47):

$$\frac{\mathrm{d}\,C_a}{\mathrm{d}t} = \mathrm{d}(C_m/V_d)/\mathrm{d}t \qquad (48)$$
$$= 1/V_d(\mathrm{d}C_m/\mathrm{d}t) - C_m/V_d^2\,(\mathrm{d}V_d/\mathrm{d}t)$$

That is, the true amorphous region oxidation rate will also depend on how V_d changes with time. The principal way that this occurs in the heterogeneous model is by infectious spreading and Fig. 5 shows how the oxidizing fraction changes with time. In the homogeneous model it is assumed that V_d does not change throughout the oxidation so the measured oxidation rate is related directly to the microscopic oxidation rate. In this case:

$$\frac{\mathrm{d}\,C_m}{\mathrm{d}t} = V_d(\mathrm{d}C_a/\mathrm{d}t) \qquad (49)$$

Another extreme example would be if the oxidation rate within an amorphous domain was a constant r such that the change in the measured reaction rate depends only on the rate of spreading of the oxidation through the polymer.

$$\frac{\mathrm{d}\,C_m}{\mathrm{d}t} = (C_m/V_d)\,(\mathrm{d}\,V_d/\mathrm{d}t) + rV_d \qquad (50)$$

It is unlikely that the condition described by Eq. (50) would prevail over the full extent of the oxidation, although it was reported that the infectious spreading (epidemic) model, with a constant rate r of free radical termination in the oxidizing domains, was able to describe the CL profile from oxidizing polypropylene (see Sec. III.H.2). Thus, under these conditions Eq. (50) may be applicable.

Direct measurement of $\mathrm{d}V_d/\mathrm{d}t$ is a challenge to analytical-imaging techniques because the initial amorphous domain dimension may be only 10–20 nm and only a fraction of that may initially be oxidizing. However, it may be possible to image the domains at higher extent of oxidation and, thereby, obtain a direct measure of $(\mathrm{d}V_d/\mathrm{d}t)$. In this way it may be possible to obtain the true concentration of oxidation products in the amorphous domain and thereby determine the kinetic rate laws unambiguously.

To summarize, it is not generally possible to translate data from macroscopic measurements of oxidation product formation (or oxygen uptake) with time of reaction into a kinetic equation that may be used to predict the useful lifetime of the polymer, owing to the heterogeneous oxidative behavior of polypropylene. However, considerable insight may be gained into the processes that ultimately lead to microcrack formation and polymer embrittlement by considering the spatial development of the oxidation that initiates from catalyst particles and related impurity centers.

ACKNOWLEDGMENTS

Work on this chapter commenced while Graeme George was a Visiting Professor at the Ecole Nationale Superieure d'Arts et Metiers (ENSAM), Paris. The author is greatly indebted to Professor J. Verdu and Dr. L. Audouin of ENSAM for their support and valuable discussions of polymer oxidation, and to QUT for the period of study leave.

REFERENCES

1. RB Seymour, T Cheng. History of Polyolefins: The World's Most Widely Used Polymers. Boston: Kluwer Academic, 1986.
2. W Hawkins. Polymer Degradation and Stabilization. Berlin: Springer Verlag, 1984.
3. JF Rabek. Photostabilization of Polymers—Principles and Applications. London: Elsevier Applied Science, 1990.
4. GE Schoolenberg. J Mater Sci 23:1580, 1988.
5. GE Schoolenberg, HDF Meijer. Polymer 32:438, 1991.
6. MM Qayyum, JR White. Polym Degrad Stabil 41:163, 1993.
7. B O'Donnell, JR White. Polym Degrad Stabil 44:211, 1994.
8. JR White, A Turnbull. J Mater Sci 29:584, 1994.
9. M Iring, F Tudos. Prog Polym Sci 15:217, 1990.
10. JB Knight, PD Calvert, NC Billingham. Polymer 26:1713, 1985.
11. DJ Carlsson, DM Wiles. Macromolecules 4:174, 1991.
12. VA Roginskii, Ye L Shania, VB Miller. Polym Sci USSR 20:299, 1978.
13. GA George. Chemiluminescence of polymers at nearly ambient conditions. In: L Zlatkevich, ed. Luminescence Techniques in Solid State Polymer Research. New York: Marcel Dekker, 1989.
14. HL Backström. J Am Chem Soc 51:90, 1929.
15. JL Bolland, G Gee. Trans Faraday Soc 42:236, 1946.
16. L Audouin, V Gueguen, A Tcharkhtchi, J Verdu. J Polym Sci [A] Polym Chem 33:921, 1995.
17. F Gugumus. Polym Degrad Stabil 52:145, 1996.
18. M Celina, GA George. Polym Degrad Stabil 40:323, 1993.
19. M Celina, GA George. Polym Degrad Stabil 50:89, 1995.
20. F Gugumus. Polym Degrad Stabil 52:159, 1996.
21. GA George, M Celina, C Lerf, G Cash, D Weddell. Macromol Symp 115:69, 1997.
22. S Al-Malaika. In: G Allen, J Bevington, eds. Comprehensive Polymer Science. Oxford: Pergamon, 1989. p. 539.
23. N Grassie, G Scott. Polymer Degradation and Stabilization. London: Cambridge University Press, 1985.
24. J Scheirs, DJ Carlsson, SW Bigger. Polym Plast Technol Eng 34:97, 1995.
25. JCW Chien, EJ Vandenberg, H Jabloner. J Polym Sci [A-1] 6:381, 1968.
26. FR Mayo. Macromolecules 11:942, 1978.
27. A Garton, DJ Carlsson, DM Wiles. Makromol Chem 181:1841, 1980.
28. VA Roginski. In: N Grassie, ed. Developments in Polymer Degradation—5. London: Applied Science, 1986, p. 193.
29. A Garton, DJ Carlsson, DM Wiles. J Polym Sci [A-1] 16:33, 1978.
30. A Garton, DJ Carlsson, DM Wiles. Macromolecules 12:1071, 1979.
31. S Commerce, D Vaillant, JL Philippart, J Lacoste, J Lemaire, DJ Carlsson. Polym Degrad Stabil 57:175, 1997.
32. P Gijsman, J Hennekew, J Vincent. Polym Degrad Stabil 42:95, 1993.
33. A Zahradnickova, J Sedlar, D Dastych. Polym Degrad Stabil 32:155, 1991.

34. JF Rabek. Mechanisms of Photophysical Processes and Photochemical Reactions in Polymers. Chichester: Wiley, 1987.
35. DJ Carlsson, A Garton, DM Wiles. In: N Grassie, ed. Developments in Polymer Degradation—1. London: Applied Science, 1979, p. 219.
36. J Scheirs, SW Bigger, NC Billingham. Polym Test 14:211, 1993.
37. L Zlatkevich. J Polym Sci Polym Phys 23:1691, 1985.
38. P Vink. J Appl Polym Sci Appl Polym Symp 35:265, 1979.
39. FR Dainton. Chain Reactions. Methuen, 1956.
40. J Verdu. Macromol Symp 115:165, 1997.
41. J Verdu, S Verdu. Macromolecules 30:2262, 1997.
42. NM Emanuel, AL Buchachenko. Chemical Physics of Polymer Degradation and Stabilization. Utrecht: VNU Science, 1987.
43. SG Kiryushkin, YA Shlyapnikov. Polym Degrad Stabil 23:185, 1989.
44. JCW Chien, DST Wang. Macromolecules 8:920, 1975.
45. FR Mayo. J Polym Sci Polym Lett Ed 10:921, 1972.
46. M Johnson, ME Williams. Eur Polym J 12:843, 1976.
47. DR Burfield, KS Law. Polymer 20:620, 1979.
48. J Scheirs, O Delatycki, SW Bigger, NC Billingham. Polym Int 26:187, 1991.
49. NC Billingham, PD Calvert, JB Knight, G Ryan. Br Polym J 11:155, 1979.
50. NC Billingham, PD Calvert. Pure Appl Chem 57:1727, 1985.
51. P Richters. Macromolecules 3:262, 1970.
52. TJ Henman. In: N Grassie, ed. Developments in Polymer Degradation—6. London: Applied Science, 1986, p. 229.
53. RA da Costa, L Coltro, F Galembeck. Angew Makromol Chem 180:85, 1990.
54. NC Billingham. Makromol Chem Makromol Symp 28:145, 1989.
55. ET Denisov. In: G Scott, ed. Developments in Polymer Stabilization—5. London: Applied Science, 1982, p. 23.
56. AL Buchachenko. J Polym Sci Symp 57:299, 1976.
57. VP Pleshanov, SG Kiryushkin, SM Byerlyant, YA Shlyapnikov. Polym Sci USSR 28:2211, 1978.
58. J Sohma, Pure Appl Chem 55:1595, 1983.
59. DJ Carlsson, CJB Dobbin, DM Wiles. Macromolecules 18:1793, 1985.
60. YA Shlyapnikov. Makromol Chem Makromol Symp 27:121, 1989.
61. WL Hawkins, W Matreyek, FH Winslow. J Polym Sci 41:1, 1959.
62. N Rapoport, AS Goniashvili, MS Akutin, VB Miller. Polym Sci USSR 21:2286, 1980.
63. FP La Mantia, G Spadaro, D Acierno. Polym Photochem 6:425, 1985.
64. LS Shibryaeva, SG Kiryushkin, GE Zaikov. Polym Degrad Stabil 36:17, 1992.
65. NM Livanova, GE Zaikov. Polym Degrad Stabil 36:253, 1992.
66. NM Livanova. Polym Sci Ser A 36:32, 1994.
67. MM Qayyum, JR White. Polym Degrad Stabil 39:199, 1993.
68. GE Schoolenberg, JCM de Bruijn, HDF Meijer. Proceedings 14th International Conference on Advances in Stabilization and Degradation of Polymers, Lucerne, 1992, p. 25.
69. J Scheirs, SW Bigger, O Delatycki. J Polym Sci Polym Phys 29:795, 1991.
70. J Van Schooten. J Appl Polym Sci 4:122, 1960.
71. AS Michaels, HJ Bixler. J Polym Sci 50:393, 1961.
72. AS Michaels, HJ Bixler. J Polym Sci 50:413, 1961.
73. KT Gillen, RL Clough. Polym Degrad Stabil 24:137, 1989.
74. RL Clough, KT Gillen. Polym Degrad Stabil 38:47, 1992.
75. V Langlois, M Meyer, L Audouin, J Verdu. Polym Degrad Stabil 36:207, 1992.
76. V Langlois, L Audouin, J Verdu, P Courtois. Polym Degrad Stabil 40:399, 1993.
77. TE Nowlin. Prog Polym Sci 11:29, 1985.

78. J Wristers. J Polym Sci Polym Phys Ed 11:1601, 1973.
79. KY Choi, WH Ray. J Macromol Sci Rev Macromol Chem Phys C25:57, 1985.
80. O Cicchetti, R De Simone, F Gratani. Eur Polym J 9:1205, 1973.
81. P Gijsman, J Hennekens, J Vincent. Polym Degrad Stabil 39:271, 1993.
82. JE Kresta. Tech Pap Reg Tech Conf SPE:478, 1980.
83. AJ Chirinos-Padron, PH Hernandez, FA Suarez. Polym Degrad Stabil 20:237, 1988.
84. J Scheirs, SW Bigger, NC Billingham. J Polym Sci Polym Chem 30:1873, 1992.
85. LM Livanova, GE Zaikov. Polym Degrad Stabil 57:1, 1997.
86. M Celina, GA George, NC Billingham. Polym Degrad Stabil 42:335, 1993.
87. RE Kellogg. J Am Chem Soc 91:5433, 1969.
88. RF Vassil'ev. Macromol Chem 126:231, 1969.
89. NC Billingham, ETH Then, PJ Gijsman. Polym Degrad Stabil 34:263, 1991.
90. L Matisova-Rychla, J Rychly. Inherent relations of chemiluminescence and thermo-oxidation of polymers. In: RL Clough, NC Billingham, KT Gillen, eds. Polymer Durability. Washington, DC: Am Chem Soc, Adv Chem Ser 249: 175–193, 1996.
91. GA George, M Celina, AM Vassallo, PA Cole-Clarke. Polym Degrad Stabil 48:199, 1995.
92. P Carloni, L Greci, A Marin, G Tosi, A Faucitano. Polym Degrad Stabil 45:415, 1994.
93. M Celina, GA George, DJ Lacey, NC Billingham. Polym Degrad Stabil 47:311, 1995.
94. M Celina, GA George, NC Billingham. Physical spreading and heterogeneity in oxidation of polypropylene. In: RL Clough, NC Billingham, KT Gillen, eds. Polymer Durability. Washington, DC: ACS, Adv Chem Ser 249:149–174, 1996.
95. V Dudler, DJ Lacey, C Kröhnke. Polym Degrad Stabil 51:115, 1996.
96. L Zlatkevich. Polym Degrad Stabil 50:83, 1995.
97. F Gugumus. Polym Degrad Stabil 53:161, 1996.
98. N Rapoport. Proceedings 17th International Conference on Advances in Stability and Degradation of Polymers, Lucerne, 1995, p. 245.
99. JD Murray. Mathematical Biology. Berlin: Springer-Verlag, 1989, p. 610.
100. GA George, M Celina, C Lerf, G Cash, D Weddell. Proceedings 11th IUPAC/FECS International Conference Polymers, Slovakia, July 1996, p. 21.
101. S Girois, L Audouin, J Verdu, P Delprat, G Marot. Polym Degrad Stabil 51:125, 1996.

8

New Mechanisms of Polymer Degradation

**YURII ARSENOVICH MIKHEEV, LUDMILA NIKOLAEVNA GUSEVA and
GUENNADI EFREMOVICH ZAIKOV**

*N.M. Emanuel Institute of Biochemical Physics, Russian Academy of
Sciences, Moscow, Russia*

I. INTRODUCTION

This chapter will analyze the data from the literature on the chain reactions of
polymers from the viewpoint of the theory that has arisen from a knowledge of
the supermolecular structure of noncrystalline polymer domains.

The principle of this organization is outlined in the morphological model used.
By this principle, a noncrystal polymer body represents an aggregate of tightly
packed globules formed by many macromolecules (1–3). Let us remember that electron microscopic investigations of specially prepared glassy polymers usually display
the globules in the 10- to 50-nm–range sizes.

According to the model, polymer chains are packed as a felt (or sponge) in
globules. Thus, they do not reach the extremely dense packing of their units. A definite packing friability of the general chain units in the volume of globules is fixed
because a few other units of the chains participate in the structure formation of
globule coverings, which are composed into a steady continuous spatial skeleton.
The skeletal covering of every globule is composed as a mosaic, consisted of
paracrystalline domains—nanostructures that include up to ten segments of
macromolecules in their approximate parallel packing. Penetrating the polymer
body, this p-skeleton serves as a filigreed reinforcing construction. The fittings play
the simultaneous role of mechanical clips, which do not allow the chains to be packed
into dense static coils in the bulk of globules (i.e., in p-skeleton cells).

The existence of p-domains was discovered for the first time during the analysis
of the wide-angle X-ray scattering on samples of polystyrene, polycarbonate,
polyethylene, polyethylene terephthalate, and natural rubber (caoutchouc) (4).
The p-domains steadily preserve their segmental crystalline regularity and their sizes,

independently of the physical state of the polymers (glass, high elasticity, melting). They were observed visually with the help of electron microscopic methods on thin films of poly(methyl methacrylate) (PMMA) in further experiments, and their sizes were in the ranges of 1.5–3.0 nm (5).

The formation of rigid structural elements, which could be identified as p-domains, was also identified in the investigations of polymer solutions and melts by nuclear magnetic resonance (NMR) and electron spin resonance (ESR) methods. For example, stable rigid aggregates of polymer chains in conditions of high dissolution were observed for solutions of specially labeled PS and PMMA (fragments of inoculated stable nitroxyl radicals served as labels): approximately 75% of dissolvent in PMMA mixture with chloroform and up to 40% of dissolvent in PS mixture with benzene (6). During evaporation of dissolvents at 25°C, rigid elements of the systems present composed a continuous skeleton that fixes up to 20–30% of occluded dissolvent in the gels formed. Thus, each cell of this continuous framework of gel-like polymers seals granules of the sponge, swollen in the solvent.

The formation of swollen chain–sponge granules, encapsulated by p-domains, in the system at the gelatinizing stage ensures that it is often possible to introduce into glassy polymers essential amounts (up to tens of percent) of compounds that cannot dissolve the polymer and that cause no swelling of it (3). Many of substances, such as crystal dibenzoyl peroxide, sterically hindered phenols, and others, are related to this class.

It is important that formed sponge granules of the skeletal body, occluding admixtures, are hermetically isolated from each other by p-domain coverings of globules. This allows one to perform a matrice "conservation" of those substances capable of crystallization or of volatile ones, such as pentane, cyclohexane, heptane, and such, which are applied for obtaining foamy materials from PS and polycarbonate, without separation of essential amounts of them into a separate phase (1–3).

A polymer matrix reinforced by a p-skeleton excludes any possibility of complete collapse of polymer chains to the size of static coils and preserves a significant amount of elastic tension energy. In particular, this is displayed by an abrupt decrease of the internal pressure (and cohesion energy) in a narrow temperature range of polymer transition from the highly elastic state to glass. An abrupt change of the internal pressure, which is experimentally observed at glass transition of PS, PMMA, and polyvinyl acetate, complies with the energy accumulation of up to 40 cal/cm^3 in chains (3,7). Moreover, the method of neutron scattering indicates that the sizes of macromolecular coils in polymer glasses are always 20% higher than those of static coils in Θ-solvents (8). This also indicates that structural porosity of the polymer chains packing is fixed by the p-skeleton.

One can decrease the jump of internal pressure of vitrified samples (7) when treating PS melt by external pressure; thus, it is possible to decrease the size of macromolecular coils, thereby decreasing the empty volume stored in glass. However, the aforementioned changes do not cause an equilibrium state in pressed samples, because their annealing at temperatures even lower than T_g restores the "usual" sizes of macromolecular coils and a balance of elastic forces, which existed initially between the p-skeleton and sponge granules (8).

Concerning the thin structure of granules, which fill the p-skeleton cells, we should mention that the polymer chains, being in a packless state under a measured expansion, form two structural zones differing in the size of micropores (emptiness, reaching 1–2 nm in size) and by the degree of fluctuation dynamics of the chains. Sponge functions, connected with its thin structure, will be considered in detail in the description of the polymer microreactor model. In this section we will simply outline the general idea that structurally caused porosity of chain packing provides the appearance of a colloidal heterogeneity factor in an amorphous polymer. In essence, this factor is a new degree of latitude (9), which primarily differentiates polymer from homogeneous liquid and causes the appearance of new physical and chemical properties. In this sequence, some physical properties, not yet characterized, have been explained (1–3,10) on the basis of properties of a skeleton-stabilized molecular sponge.

Moreover, the important role of skeletal–micellar structure was exhibited in the example of a model reaction of dibenzoyl peroxide (BP) with the polymers, such as relatively rigid-chain polycarbonate and cellulose triacetate (11,12), and flexible chain polypropylene (13), polyethylene (14), polyethylene oxide (15), polyamide PA-548 (16), and polystyrene (17). Importantly, in these investigations, BP possessed the property of an active acceptor of free radicals as well as being an effective initiator of radical transformations. The existence of the aforementioned properties made it possible to apply BP for kinetic sounding of those structural features of the medium that seriously change the kinetic laws and the composition of the reaction products in the polymer matrix, as displayed by investigations (11–17). Processes of all the aforementioned systems were quantitatively described in the ranges of general heterophase mechanisms, which naturally correlates with simple formal kinetics and preserves its force for vitrified, highly elastic and melt states of polymer samples.

Again, many interesting facts were disclosed on the peculiarities of polymer autoxidation when developing our knowledge on the functions of spongy polymer chain micella in chemical processes (18,19). This chapter is designed to demonstrate the validity of new principles in structurekinetic modeling of polymer chain reactions.

II. THE MODEL OF A POLYMER CHAIN–SPONGY MICROREACTOR

In characterizing a chain–spongy micelle as a chemical reactor, the model separates structural zones of two types. For the polymer glass matrix, we take into account that only a small amount of the general polymer chain links is spent for a filigree paracrystal mosaic construction (4). In turn, most of chains fill p-skeleton cells, packed as in sponge micelles (1–3). When combining with sponge granules, polymer chains strive to reach a dense packing, with the highest density of cohesion energy applied as the molecular forces. However, chain units, that participate in paracrystal domains of globule coverings and are composed into a mosaic p-structure, inhibit this process.

It can be supposed that polymer chains, constricted in the middle of a p-skeleton cell, form more or less dense granule–core (v-zones), which split off from p-domain walls of the cell by forming the layer of chain segments that are oriented in the radial direction of the globule–micelle. These border segments play the role of mechanical braces and form a rigid palisade, such that they border a spherical layer of stationary micropores. Micropores themselves must possess the sizes of a thermodynamic segment order, which is in the range of 1.5–3.0 nm for such polymers as PS, PMMA, PVC, PE, and PP (20). They should be related to super-(s)-micropores according to their sizes (21).

In this case, we can say that stationary s-micropores that are formed in a glassy polymer, in turn, form separate s-zones, which are interlayers between paracrystalline coverings of connected globules and intraglobular nucleus granules, filled with small v-micropores.

Glass flexible-chain polystyrene and poly(methyl methacrylate) are related to nonporous sorbents according to their sorption properties (22). We can suppose on this basis that the size of micropores in nucleus granules of sponge micelles must not exceed the thickness of the polymer chains. Actually, direct experiments (23), performed on glassy PMMA, displayed a sharp distribution of micropores by sizes. Most micropores in samples, frozen at 77 K, possessed radii of less than 2.5–3.3 Å, and only 1% of them possessed radii up to 12 Å.

The result obtained (23) complies with the structural model of a sponge micelle, which we roughly presented. Existence of physical surfaces and emptiness, realized as stationary s-micropores and more fluctuationally dynamical v-micropores in micelles, introduces a new degree of latitude into polymers: the factor of colloid dispersity. This factor should be taken into account under consideration of the physicochemical processes connected with functions of spongy micelles. As an example, let us show a logical structure of the approach in which the present model allows us to disclose the physical nature of the phenomenon possessing unexpected and characteristic heat effects of polymer dissolution, particularly in hydrogenated monomers.

It is known that a mixture of low-molecular fractions of polystyrene, poly(methyl methacrylate), or cellulose triacetate with corresponding hydrogenated monomers possesses a zero heat effect, as distinct from glass high molecular polymers, dissolved with heat evolution (24). The investigation stated (25) that the heat amount extracted in a PS mixture with toluene and ethylbenzene decreases rectilinearly with the increase of mixture temperature, from 6 and 7 cal/g (for toluene and ethylbenzene, respectively) at 303 K to ≈ 1 cal/g at 358 K. At higher dissolution temperatures the extracted heat is about 1 cal/g.

An analysis of the known data (25), made it possible to exclude the supposition (24), that the heat release recorded for the foregoing processes is caused by relaxation of packless chains into the dense state, accelerated by the solvent effect. It was also pointed out that the effect is disconnected from the decrease in volume of the mixed components, because the test performed (25) supported the fact, known from the literature, that the amount of released heat was independent of the external pressure under which samples were glassified.

At the same time, the discussion of the experimental material (25) lies in the ranges of a homogeneous model of amorphous polymers. This is probably because this discussion was not finished by construction of an adequate physical picture of the process, whereas a qualitatively obvious description can be made on the basis of the sponge micelle model.

It is evident that sponge dissolution in a hydrogenated monomer causes disappearance of free surfaces of narrow v-micropores as well as supermicropores. In connection with the fact that narrow micropores possess a higher sorption energy, compared with s-micropores (21), a solvent diffused into the polymer should first occupy the system of narrow micropores. In this case, each sponge granule initially transforms to a homogeneous liquid drop of colloidal size, owing to the cohesion equivalence of hydrogenated monomer and monomeric chain units (similar parameters of dissolution). Transformations of tight sponge grains to liquid drops are accompanied by disappearance of free surface v-micropores and a partial decrease of tensile strength of chains in the drops. A corresponding amount of heat Θ_1 is released as a result.

A liquid-viewing drop is formed at the initial stage of micelle saturation by solvent only when surface tension σ provides a spheric form of the grain drop and detaches it from p-domain walls of the rigid skeleton. It is precisely this picture of formation of drops regularly packed in the polymer matrix, which reached the size of tens of nanometers, that was set by small-angle X-ray–scattering method in experiments with PVC containing a large amount of plasticizer (26). However, unlike the plasticizer, which does not dissolve the polymer, the solvent disintegrates the drops and their surfaces, forming a molecular solution. In this case, the potential energy of the total surface S of drops is released as an additional heat $\Theta_2 = \sigma S$.

If we take into account that surface tension of normal liquids decreases linearly with temperature increase (21)

$$\sigma_T = \sigma_0 - \left(\frac{\partial \sigma}{\partial T}\right) \Delta T$$

(here σ_T and σ_0 are surface tension at current temperature and at standard temperature, respectively; $(\partial \sigma / \partial T)$ is the temperature coefficient of surface tension; ΔT is the difference between current and standard temperatures), we can express the total heat amount as follows:

$$\Theta_1 + \Theta_2 = \Theta_1 + S\sigma_T = \Theta_1 + S\left[\sigma_0 - \left(\frac{\partial \sigma}{\partial T}\right)\Delta T\right]$$

It follows from this expression that the total amount of heat, released as a result of initial formation and further disintegration of colloidal drops dispersed in the polymer under the influence of hydrogenated monomer, must decrease by linear law to some value Θ_1 at a mixture temperature increase. Qualitative correlation of this formula with the experiment (25) points out the natural cause of the heat effect, which cannot be explained by the supposition of a homogeneous systems mixture.

If we look at compounds that do not dissolve amorphous polymers (such as crystal dibenzoyl peroxide, sterically hindered phenols, and many hydrocarbons and alcohols) under conditions of direct contact, without participation of general solvent, their molecules are only able to adsorb in the polymer matrix in small amounts, if there is an access to surfaces of structural emptiness. But the same compounds are very often formed with polymer in sufficient amounts (tens of percents) through combined solutions (1–3).

The fact is that during preparation of polymer samples the system reaches the stage of a gel at solvent evaporation. Formed, swollen sponge granules are enveloped in the skeletal structure of the gel, with a large amount of solvent, by thin hermetic coverings from paracrystalline domains (3). The solvent occluded in the gel provides high conformational variability in micelle sponge granules and highly dynamic polymer chains. Consequently, granules can accumulate large amounts of additives that do not dissolve the polymer. They substitute for the solvent in the swollen sponge simultaneously with its evaporation. Finally, they are enveloped in the skeleton cells more or less hermetically, but preserve a degree of molecular dispersion. Thus, the polymer can conserve micellar material with significant degrees of swelling (in accordance with that initially created by the solvent). It also fixes higher dynamic fluctuations of the chain units, if compare with pure glassy polymer.

In occupying the place of a solvent molecule, a molecule of nonsolvent must push apart polymer chains (which would perform more dense packing in absence of an additive) and regulate them in a certain way into a sufficiently large micropore. Formation of such a capsule from polymer chain links is combined with the preservation of a particular excess of the surface energy and a decrease of chain entropy among additive molecules. These phenomena require a compensational increase in entropy of the polymer chains among the capsules formed. Such increase is performed in certain ranges, owing to the decrease of structural tension forces and the increase in the conformational amount of chains in the sponge.

Thus, a nonsolvent molecule introduced into the sponge affects its chains mechanically. In its turn, the sponge affected by the additive performs the opposite effect, producing negative pressure on an additive through capsulating chain links. The end of this force balance, realized with the help of dispersion forces, is that the additive molecule is localized in the volume of the forming micropore and obtains relative freedom for rotation.

To support all this, data from ESR studies confirm the presence of high-frequency rotations of particles of paramagnetic probes in glassy polymers. These data indicate that probe particles are disposed in micropores that provide activation energies and rotation frequencies similar to those in ceolytes with channel diameters of 8–10 Å (27). Also, the data from investigations of plasticized polymers by NMR studies confirm that there is a conformity between the polymer chain structure and that of microporous adsorbents (28). This conformity is not observed for initial polymer glasses.

The study of mobility of additive particles in glassy polystyrene (PS) with the help of the paramagnetic probe tetramethylpyperidinoxyl, allowed determination of the local viscosity of the medium. It changes from 136 to 9 poises in the temperature range of 234–371 K (29). Similar values for the viscosity of liquid glycerin lie in the interval of 266–288 K. These data are remarkable, first of all, by their essential difference from macroviscosity of PS (10^{12} poises at the point $T_g = 373$ K) K) and, second, by their low value for PS instead of glycerin at the low temperature of 266 K (by 2.5 times, approximately). So, narrow v-micropores have to increase their size when they entrap additive particles, providing rotational freedom to additives.

Fluctuation dynamics of the chains in p-skeleton cells also provide a high rate for the transitional migration of particles with low molecular masses. This is undoubtedly confirmed, for example, by experiments on extinguishing the phosphorescence of aromatic admixtures in glassy PMMA by oxygen (30) and small amounts (up to 4%) of methyl methacrylate and methyl isobutyrate, which do not affect the macroscopic viscosity of the polymer (31). The experiments on extinguishing of bisulfate quinine fluorescence by sodium chloride in glassy polyvinyl alcohol (32) are also very effective. The dynamic nature of extinguishment, which even so shortens the short lifetime of singlet-excited paint ($\approx 10^{-8}$ s), testifies to the very high frequency of contacts between additive molecules.

Hence, the translational frequency of additive particle is preserved simultaneously. This is because thermal fluctuational pulses of spongy v-granules suspended on p-skeleton walls stimulate cooperated oscillation of sponge chain units. These cooperated oscillations cause a definite probability that the particles will be translated along the matrix of narrow v-micropores, as the latter also have to increase their size for a short time. Low molecular weight particles thereby obtain suitable freedom for translation in sponge matter and exchange between v- and s-zones.

The functioning of stationary s-micropores and fluctuationally dynamic v-micropores in sponge micelles influence the mechanism of initiated transformations. This is because initiator molecules exist in different steric and energetic states when transferred between s- and v-zones and have to display inequivalent initiating effectiveness in these zones. Actually, the overlapping of force fields from opposite walls is characteristic of narrow v-micropores. In such cases, the initiator is pushed apart into the pore volume and spends general time there, subjected regularly by short-time acts of fluctuationally activated jumps. Although the duration of these acts is significantly short, it does not allow radicals to escape from the initial radical pair, even if it occurs during the occasional decomposition of initiator molecules in the v-zone.

Contrary to narrow v-micropores, initiator molecules obtain significantly more favorable conditions for dissociation in s-micropores. The reason for this is the existence of free space as well as the absence of fields of force overlapping opposite walls. Because of this, the additive particle spends general time adsorbed on the pore surface, not in the volume. Under these conditions radicals of spontaneously formed radical pairs are not subjected to compression, but attack polymer molecules with increased probability under the effect of adsorption forces.

In the reaction of dibenzoyl peroxide (BP) decomposition, which we chose as the model reaction, the decomposition is more rapid in glassy PS than in liquid hydrogenated monomers, despite their difference from PS of extremely low viscosity (17). Here, it is evident that the high mobility of BP molecules in the polymer matrix is provided by the active size fluctuations of v-micropores. It is possible that narrow v-micropores possess some probability for expanding to the sizes of supermicropores. However, such excited pores should differ by a short lifetime in comparison with the characteristic dissociation time of BP molecules.

When moving in micelle pores, BP molecules obtain the ability to dissociate into radicals in the stationary supermicropores, that form border layers between nucleus granules and domain walls of p-skeleton cells. When disposed in s-micropores during this general time, the BP molecule is adsorbed onto the rigidly fixed polymer segments that compose the pore surface. Adsorption forces, acting in such a pore, help the initial radicals attack macromolecules, thereby providing their reorientation and extraction from the initial radical pair. Thus, despite the extremely high macroscopic viscosity of a glassy polymer, the additive initiator acquires the probability of dissociation in the polymer-chain sponge according to a heterogeneous mechanism.

The existence of structural s- and v-zones also affects the polymer radicals, capability to decompose at chain rupture. It is clear that the probability of such an occurrence on the surface of a v-micropore is low, because this requires a simultaneous shift of several chains that fix v-micropore walls under compression from the surface tension of the micelle v-granule. This is possible only at a very high dynamic fluctuation of the chain spongy granules. Considering the data (11–19), such a situation in the v-zone is very rare.

At the same time, the polymer radicals on the s-micropore surfaces develop a definite likelihood of rupturing the chains, which is promoted by the existence of free space and mechanical stretching of s-segments of the polymer chains.

The foregoing features of a heterogeneous chain–sponge micelle, which composes a matrix of noncrystal polymer, make kinetic description of the chain process more complex if compared with a homogeneous reaction. For a polymer, the task of developing an adequate transformation mechanism that demands taking into account the heterogeneity factor, repeats the situation that existed before in investigations of heterogeneous catalysis (33). Now, the current task becomes the development and selection of structural kinetic reaction models.

The stage of corresponding modeling of chain polymer transformations was begun by application of model chain dibenzoyl peroxide reactions with polymers of different structures (such as cellulose triacetate, polycarbonate, polystyrene, polyethylene oxide, or polyamide PA-548; 11–17) and has been continued with reactions of polymer hydroperoxide chain decomposition (34,35).

The current concept of a heterophase reaction is that a complete transformation picture should consist of the following:

1. The scheme of initial reaction chains, spreading in capacious micropores of s-zones in spongy micelles (s-scheme)
2. The scheme of secondary reaction chains in tight v-zone micropores (v-scheme)
3. The scheme of interzone transfer of free radicals (tr-scheme)

If we take these facts into account, we should expect that the features of supermolecular organization of polymer chains are also active in amorphous domains of semicrystalline polyolefins. As a participant in the structure of a solid polyolefin, the spongy micelle is formed and is enveloped by crystallites. These crystallites serve as a rigid skeletal reinforcement, but are capable of restructuring when subjected to the orientational drawing or other factors influencing their structural arrangement. Consequently, all the foregoing items of structure–kinetic modeling must be used in a description of autoxidation processes and transformations of antioxidants applied to polymer material stabilization during processing and exploitation.

A wide application of polymer antioxidants began simultaneously with polymer development; however, development of the scientific ideas on the mechanism of their protective action did not progress from the homogeneous reaction theory until recently (36–38). The study of the model chain reaction of dibenzoyl peroxide with polymers displayed important phenolic antioxidant effects, connected with microheterogeneity of a polymer matrix (11,17,18). A recent review (18) sets forth and analyzes some important, but not yet explained, data from the literature on the mechanism of antioxidant action. The structural kinetic reaction model used for this purpose imparts the physical meaning of an antioxidant functioning in a real heterophase process of polymer aging. It also generally changes its relations to fundamental postulates, which are transferred to polymers from the theory of the homogeneous reactions.

III. THE MECHANISM OF POLYPROPYLENE HYDROPEROXIDE ANAEROBIC DECOMPOSITION

The concept of the heterophase chain reaction is also fruitful for the analysis of the processes related to aging and stabilization of polymers. It allows one to explain the important features of polyolefin autoxidation (19,39) and decomposition of polymer hydroperoxides under conditions of high and low dynamic fluctuation of the polymer-chain sponge (34,35). In this section we consider the mechanism of polypropylene hydroperoxide chain decomposition proceeding in the absence of oxygen, dissolved in the polymer.

Polypropylene (PP) differs from glassy-like polymers by its amorphous phase, which already exists in a high-elastic state at usual temperatures and consists of separate spheres surrounded by crystallites. The crystalline phase usually forms about 60% of mass in isotactic PP ($T_{melt} \approx 438$ K). Its crystallites form a continuous rigid skeleton, imparting the polymer properties of a solid. The oxidation of macromolecules and decomposition of additive reagents proceed in PP samples, that possess such structure, in only highly elastic regions (36,37).

The rigid skeleton of crystallites affects the structure of the noncrystalline PP phase (40); however, characteristic features of chain–micellar sponge composition of amorphous spheres must remain. It is probable that capacious s-micropores in PP will be formed as layers that are also close to the surface of crystallites (40), but the main volume of spongy micelles should be composed of flexible chains as granules possessing more or less narrow v-micropores (Fig. 1). If we suppose that chain–micellar sponge maintains its microheterogeneous structure, even for polyolefins, we should also expect the existence of a specific inequivalence of radical birth rates in s- and v-zones. Polypropylene also possesses s-zones (35,39), in which hydroperoxide (HP in the text, G in formulas) dissociates much faster than in v-zones. The primary reaction chain, initiated in s-zones, induces a chain reaction in v-zones owing to s,v-injection of radicals, performed with low probability. In rigid PP samples, heterophase chemical induction stops at the moment when the HP accumulated in s-zones is completely exhausted.

However, HP decay does not generally stop at this stage because the finished heterogeneous–heterophase stage is changed by a heterogeneous–homophase one, which possesses a lower rate and is connected with v-zones displaying a somewhat higher probability of fluctuational increase of v-micropore size up to s-micropores. This stage determines the frequency of HP dissociation reactions by conditions of polymer sample preparation and residual mechanical tensions acting in the sample.

An adequate s-scheme of HP transformation, formulated on the basis of studies (35,39), is the following:

$$G_s \xrightarrow{k_{0s}} (PO_s^\bullet + HO_s^\bullet) \xrightarrow{PH_s} \delta P_s^\bullet \ ;$$

$$P_s^\bullet \xrightarrow{k_{1s}} R_s^\bullet \quad \text{(polymer chain degradation + products)};$$

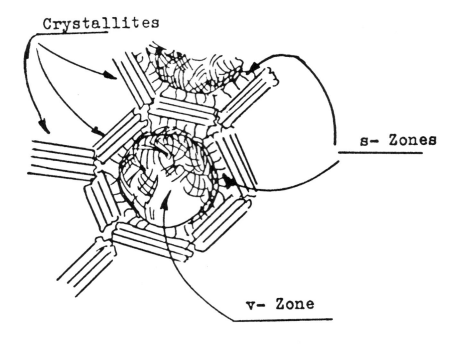

Figure 1   Schematic design of the spongy polymer chain microreactor.

$$R_s^{\bullet} \xrightarrow{k_{2s}} R_{1s}^{\bullet} \quad \text{(end radical isomerization);}$$

$$R_s^{\bullet} + R_{1s}^{\bullet} \xrightarrow{k_{3s}} \text{products.}$$

where PH is the reactional CH-groups of tertiary and secondary carbon atoms in the polymer chains: δ is the coefficient of radical yield on HP dissociation; $P_s^{\bullet}$ is the backbone; and $R_s^{\bullet}$, $R_{1s}^{\bullet}$ are the end macroradicals.

The transfer of free valency from s- to v-zones proceeds with low probability. The corresponding tr-scheme reflects the leading role of the end radicals $R_s^{\bullet}$ in the interzone free-valency relay race:

$$R_s^{\bullet} \xrightarrow{k_m} R_m^{\bullet}; \qquad R_m^{\bullet} \xrightarrow{k_{-m}} R_s^{\bullet}; \qquad R_m^{\bullet} + PH \xrightarrow{k'_{tr}} P^{\bullet}$$

These radicals are capable of entering into the reaction with CH-bonds of v-zones, when exchanging their dislocations in the s-skeleton and in micropore volume ($R_m^{\bullet}$) during their oscillations.

The chain process of HP decomposition, yielding water, proceeds in v-zones owing to this radical transfer. This process is described by the following v-scheme:

$$P^{\bullet} + G \xrightarrow{k_1} PH + POO^{\bullet}; \qquad POO^{\bullet} \rightarrow {}^{\bullet}POOH$$

(β-isomerization of peroxyl to hydroperoxyalkyl);

$$\cdot POOH \rightarrow \varepsilon + HO\cdot$$

(hydroperoxyalkyl decomposition with epoxy compound formation);

$$HO\cdot + PH \rightarrow H_2O + P\cdot; \qquad 2P\cdot \xrightarrow{k_2} \text{ chain termination.}$$

We may obtain an expression for concentration of radicals–translators of valency, applying stationary radical concentration condition to these block schemes:

$$[R_s^\cdot] = \frac{\delta k_{0s} g_s}{2 k_{2s}}$$

(where g_s is the HP concentration in s-zones).

We also obtain the expression for the s,v-transfer rate:

$$V_{tr} = k'_{tr}[PH][R_m^\cdot] = \frac{k_{tr} k_m [R_s^\cdot]}{k_{-m} + k_{tr}} = \frac{\delta k_{tr} k_m k_{0s} g_s}{2 k_{2s}(k_{-m} + k_{tr})}$$

which equals the rate of reactionary v-chain termination $V_{tr} = 2k_2[P\cdot]^2$; the concentration of radicals attacking hydroperoxide:

$$[P\cdot] = \sqrt{\frac{V_{tr}}{2 k_2}}$$

and the rate of chain HP decomposition in v-zones:

$$-\frac{dg_V}{dt} = k_1[P\cdot] g_V = k'_e g_V \sqrt{g_s}$$

By taking into account the current concentration of HP that is dissociating in s-zones, characterized by the exponent

$$g_s = g_{s0} \exp(-k_{0s} t)$$

and consider that the amount of HP in s-zones is low, compared with the amount in v-zones, and that the HP concentration in v-zones is at least equal to that in the amorphous phase ($g_v \approx g$), we obtain a reduced final expression for the HP decomposition rate:

$$-\frac{dg}{dt} = k'_e \sqrt{g_{s0}}\, g \exp(-0.5 k_{0s} t) = k_e g \exp(-0.5 k_{0s} t) \tag{1}$$

Integrating this expression, we obtain Eq. (2), the HP anaerobic decomposition curve:

$$g = g_0 \exp\left\{\left(\frac{k_e}{0.5 k_{0s}}\right)[\exp(-0.5 k_{0s} t) - 1]\right\} \tag{2}$$

Equation (2) is approximated to the first-order equation for significantly small $t \rightarrow 0$:

$$g = g_0 \exp(-k_e t) \tag{3}$$

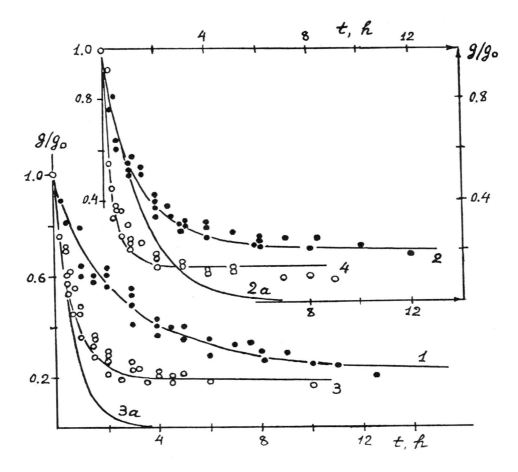

Figure 2 PP-hydroperoxide decay curves for evacuated samples at T values: (1) 383; (2, 2a) 393; (3, 3a) 403; (4) 413 K. The first-order kinetic equation at the rate values of 2.0×10^{-4} and 4.2×10^{-4} s^{-1} were used for the calculation of 2a and 3a curves, respectively.

For large values, $t \rightarrow \infty$, it characterizes a process stop with an unconsumed HP residue:

$$g_{\infty} = g_0 \, \exp\left(\frac{-k_e}{0.5k_{0s}}\right).$$

Theoretical Eq. (2) correlates well with experimental evidence, which is proved by the data in Fig. 2. Curves 1–4 shown in the Fig. 2 were calculated for the values of $T = 383$, 393, 403, and 413 K, respectively. Corresponding values of chain decomposition constants, which equal $k_e \times 10^4 \text{s}^{-1} = 0.865$; 2.00; 4.20; 10.00, respectively, are generalized by Eq. (4):

$$k_e = 3.9 \times 10^9 \exp\left(-\frac{E_e}{RT}\right), \qquad E_e = (100 \pm 5)\text{kJ/mol} \qquad (4)$$

and the determined constants of spontaneous decomposition $k_{0s} \times 10^4 \, \text{s}^{-1} = 1.14$; 2.48; 5.06; 10.00 are generalized as follows:

$$k_{0s} = 6.5 \times 10^8 \exp\left(-\frac{E_{0s}}{RT}\right), \qquad E_{0s} = (93.6 \pm 5) \, \text{kJ/mol} \qquad (5)$$

Curves 2a and 3a, calculated according to Eq. (3), are shown in Fig. 2 for comparison. It is clear that they deviate from the experimental points, whereas theoretical curves 2 and 3 are in complete agreement with the experiment.

A. HP Decomposition Mechanism in the Presence of Dibenzoyl Peroxide

The dependence of the initial rate of HP decomposition (V_0) on dibenzoyl peroxide (BP) concentration was studied (37; p. 104). It displays the existence of a chain component with a quadratic termination of reaction chains:

$$V_0 = k_d g_0 + k_{ind} g_0 \sqrt{k_b c_0} \qquad (6)$$

where g_0 and c_0 are initial concentrations of HP and BP, respectively (c_0 was smaller than 0.05 mol/kg); k_d and k_b are respectively the decomposition rate constants for HP and BP.

The experimentally determined value $k_d = (1.8 \div 1.9) \times 10^{-5} \, \text{s}^{-1}$ (365 K) was the same in the absence of BP and in the extrapolation of the dependence [Eq. (6)] to $c_0 = 0$. It was assumed that the oxidized polymer medium is a homogeneous liquid, and the following description was suggested (37) for the process rate proceeding in the absence of BP:

$$-\frac{dg}{dt} = k_d g + k_e' g^{1.5} \qquad (7)$$

Here, the first term $k_d g$ describes the spontaneous decomposition, and the second the chain decomposition. Meanwhile the chain reaction constant $k_e = 1.9 \times 10^{-5} \, \text{s}^{-1}$, calculated at $T = 365$ K by Eq. (4), equals the k_d constant from Eq. (6). Generally speaking, Eq. (7) is not in accord with the expression in Eq. (2), which neglects the contribution of spontaneous HP decomposition.

Let us consider the situation from the point of view of a heterophase mechanism. Note that BP molecules are rapidly exchanged between zones of chain–sponge micelles:

$$\text{BP}_V \underset{k_{-s}}{\overset{k_s}{\longleftrightarrow}} \text{BP}_s$$

and dissociate in s-zones:

$$\text{BP}_s \overset{k_{bs}}{\longrightarrow} 2r_s^{\bullet}$$

The equilibrium condition of interzone BP exchange is the following:

$$c_S = \frac{k_s}{k_{-s}} c_V = K_e c_V = \frac{K_e}{\beta} c \qquad (8)$$

Here c_s, c_v, and c are the BP concentrations calculated for s- and v-zones and for the amorphous PP phase, respectively; β is the amorphous phase; and K_e is the equilibrium constant.

Because of BP dissociation, the benzoyloxy radicals occur in s-zones of polymer–chain sponge. They carry free valency on oxygen atoms. The reactability of such radicals with the hydrogen of OOH groups is 100 and 1000 times higher than with CH groups (41). This is why the reaction of benzoyloxyl accepting by hydroperoxide should predominate in s-zones at initial stages of the process:

$$r_s^{\bullet} + G_s \rightarrow rH + POO_s^{\bullet}; \qquad POO_s^{\bullet} \rightarrow {}^{\bullet}POOH_s \rightarrow \varepsilon + HO_s^{\bullet};$$
$$HO_s^{\bullet} + PH_s \rightarrow H_2O + P_s^{\bullet}$$

Taking this into account in the s-scheme of HP decomposition considered before, it is easy to obtain an expression for the concentration of radicals that are predominant in the s,v-transfer of valency:

$$[R_s^{\bullet}] = \frac{\delta k_{0s} g_s + 2k_{bs} c_s}{2k_{2s}}$$

and an expression for the s,v-transfer rate:

$$V_{tr} = \frac{k_{tr} k_m [R_s^{\bullet}]}{k_{-m} + k_{tr}} = \frac{k_{tr} k_m (\delta k_{0s} g_s + 2k_{bs} c_s)}{2k_{2s}(k_{-m} + k_{tr})}$$

We now obtain the concentration of radical inductors of HP decomposition based on the v-scheme considered before, and applying the equality condition of the reaction v-chain initiation and termination $V_{tr} = 2k_2[P^{\bullet}]^2$:

$$[P^{\bullet}] = \sqrt{\frac{V_{tr}}{2k_2}}$$

We also obtain the expression for the chain reaction rate:

$$-\frac{dg}{dt} = k_1 g \sqrt{\frac{k_{tr} k_m}{4k_2 k_{2s}(k_{-m} + k_{tr})}} \sqrt{\delta k_{0s} g_s + 2k_{bs} c_s} \qquad (9)$$

Let us mark the number of BP moles in s-zones as $p_s = m_s c_s$, where m_s is the mass of s-zones, for the purpose of transforming the specific BP decomposition rate

$$-\frac{dc_s}{dt} = 2k_{bs} c_s \qquad (10)$$

to the empyric expression (dc/dt). By taking into account that the decrease of BP amount p (in moles) in the sample is equal to the decrease of the mole number in s-zones, we obtain:

$$-\frac{dc}{dt} = -m^{-1}\frac{dp}{dt} = -m^{-1}\frac{dp_s}{dt} = -\left(\frac{m_s}{m}\right)\frac{dc_s}{dt}$$

Substituting the required expressions Eqs. (10) and (8) into this equation, we obtain:

$$-\frac{dc}{dt} = \frac{2k_{bs}m_s K_e c}{\beta^m} = k_b c \qquad (11)$$

The effective rate constant of BP decomposition $k_b = \frac{2k_{bs}m_s K_e}{\beta m}$, which participate in Eq. (11), is $1.2 \times 10^{-4} \text{s}^{-1}$ as it follows from the well-known formula (42)

$$k_b = 8.4 \times 10^{13} \times \exp(-124,000/RT)\text{s}^{-1}$$

This value combines with the fact that parameter $2k_{bs} = k_b \beta m/m_s K_e$ is greater than k_b, because $(m/m_s) \gg 1$ and $K_e < 1$. Consequently, the rate of radical initiation in Eq. (9) by means of BP dissociation is significantly higher than that caused by HP dissociation, for which $k_{0s} = 2.6 \times 10^{-5}\text{s}^{-1}$ at $T = 365$ K, as follows from Eq. (5). In other words, Eq. (9) assumes the following form in a significant range of the initial BP concentrations:

$$-\left(\frac{dg}{dt}\right)_0 = k_{ind} g_0 \sqrt{k_b c_0} \qquad (12)$$

Term (12) is neglected only in the range of very low c_0 values, and Eq. (9) is transformed to the dependence [see Eq. (1)], which characterizes HP chain decomposition with quadratic termination of the reaction v-chains.

The heterophase scheme of the HP reaction stimulated by dibenzoyl peroxide explains the kinetic data of Denisov (37). Moreover, it explains the fact of water formation in the chain process, which was not considered by the homogeneous scheme (37).

IV. THE IONOL EFFECT ON THE PP HYDROPEROXIDE ANAEROBIC DECOMPOSITION

Interesting results were obtained in one of the early studies (43). The features of fast and slow kinetic stages in the general process of HP decomposition (43) are very specific and require special study from positions of heterogeneous mechanisms.

Low molecular syndiotactic PP ($M_n = 13.500$) was applied (43), which was oxidized in a chlorobenzene solution, with an initiator at $T = 373$ K, by bubbling oxygen through the solution to accumulate hydroperoxide.

In carrying out experiments on the decomposition of HP obtained by the aforementioned technique in the absence of oxygen, the authors (43) discovered two kinetic stages of this process: fast and slow. The second one possesses a measurable rate contrary to PP samples oxidized in bulk, which posses the second stage that appears as a kinetic stop (see Fig. 2).

It was shown in experiments (43) that the samples possessed a fast-stage rate that was independent of HP and ionol (4-methyl-2,6-di-*tert*-butylphenol) concentrations. For example, at $T = 408$ K and reagent concentrations of HP/IH (mol/L): 0.0078/0.001; 0.034/0.001; 0.27/0.1; 0.005/0.012, its effective rate constant, $k_{e1} \times 10^4 \text{s}^{-1}$, was 19.8; 15.1; 19.8; 18.2, respectively; and at $T = 393$ K and HP/IH ratios: 0.034/0.000; 0.011/0.001; 0.0078/0.001; 0.27/0.01, $k_{e1} \times 10^4 \text{s}^{-1}$ was obtained as: 5.01; 6.3; 6.41; 6.13, respectively.

At the same time, the formal first-order constant k_{e2}, used for characterizing the second-stage rate, depended on the initial concentration of hydroperoxide g_0, simulating the concentrational dependence of the homogeneous chain reaction. But contrary to this dependence, k_{e2} did not change in the presence of low and moderate ionol concentrations (up to 0.05 mol/L). The second stage lost its chain component only at very high ionol concentration, 0.124 mol/L, and the k_{e2} constant reached its minimum, which was also observed in the absence of ionol at the lowest g_0 concentration.

When discussing the first kinetic stage of HP decomposition the authors (43) suggested a chain mechanism that possesses a specific manner of reaction chain spreading along polymer chains. However, they failed to explain the reason for the decreased amount discovered for rapidly decomposing HP from 0.8÷0.9 to 0.25 under ionol injection up to 0.124 mol/L concentration.

From our point of view, the reaction studied (43) is, indeed, of the chain mechanism. Features of its realization reflected a specific state of the reaction matrix—chain–sponge micelle of oxidized PP–that was more specific. The Local hydroperoxide concentration, g_{s0}, which participates in Eq. (1), is saturated during PP oxidation. In this case, in accordance with other data (35), the constant k_e reaches its maximum (11.0×10^{-4} s^{-1} at $T = 403$ K. Recalculation of this value for 408 and 393 K [taking into account the effective activation energy $E_e = 100$ kJ/mol; Eq. (4)] gives the values of 16.0×10^{-4} s^{-1} and 5.15×10^{-5} s^{-1}, respectively. These values fall within the range obtained for k_{e1} (43). This coincidence suggests that the investigations performed (43) reached the limit of the s-zone content of hydroperoxide at oxidation in solution. Consequently, the rate constant k_e of the initial stage of chain decomposition must not depend on the initial concentration g_0 , calculated for the whole sample [see Eq. (1) and (3)].

In turning to features of the fast stage in presence of ionol, we should remember that ionol significantly decreased the concentrational range in which the fast stage is performed, but it did not change the k_{e1} value. Such a combination may be explained within the framework of the heterophase model by the fact that ionol decreased the amount of transported radicals from s- to v-zones by means of an additional catalytic pathway of decay of HP accumulated in s-zones, but it did not influence the mechanism of radical transfer from s-zones to v-zones.

This situation may be depicted as an s_i-scheme, introducing it to the known s-scheme of the reactions with ionol participation:

$$G_s \xrightarrow{k_{0s}} 0.5\delta(PO_s^{\bullet} + HO_s^{\bullet}); \qquad HO_s^{\bullet} + PH_s \xrightarrow{k_{4s}} P_s^{\bullet};$$

$$PO_s^{\bullet} + IH_s \xrightarrow{k_{5s}} I_s^{\bullet}; \qquad I_s^{\bullet} + G_s \xrightarrow{k_{6s}} POO_s^{\bullet} + IH_s;$$

$$POO_s^{\bullet} + IH_s \xrightarrow{k_{-6s}} I_s^{\bullet} + G_s; \qquad POO_s^{\bullet} + IH_s \xrightarrow{k_{7s}} POOIH_s^{\bullet} \rightarrow POH + IO_s^{\bullet};$$

$$IO_s^{\bullet} + G_s \xrightarrow{k_{8s}} IOH + POO_s^{\bullet}; \qquad POO_s^{\bullet} \xrightarrow{k_{9s}} HO_s^{\bullet} + products,$$

$$P_s^{\bullet} \xrightarrow{k_{1s}} R_s^{\bullet}; \qquad R_s^{\bullet} \xrightarrow{k_{2s}} R_{1s}^{\bullet}; \qquad R_s^{\bullet} + R_{1s}^{\bullet} \xrightarrow{k_{3s}} products$$

Several characteristic features of reagents are displayed in the present scheme. It neglects the reaction of ionol with radicals, $R_s^{\bullet}$ and $R_{1s}^{\bullet}$, because all alkylphenols possess significantly low reactability with polyolefin radicals (36). (Even more active bisphenol 2246 does not influence the concentration of PP radicals, eliminating propylene during PP oxidation induction; 44.)

The scheme considers phenoxyl reaction with HP (37,38) and both directions of POO$^{\bullet}$ radical attack on the phenol molecule, known from the literature: with formation of a phenoxyl and σ-complex. Contrary to liquid solutions, molecular forces provided by micropores make reversible dissociation of σ-complexes difficult in chain–sponge micelle micropores. This gives the opportunity for these complexes to participate in further transformations. In the present case, alcoxyl radical IO$^{\bullet}$ with quinolidic structure occurs from the σ-complex.

It was also taken into account that HO$_s^{\bullet}$ radicals react faster with a polymer because of their high reactivity. Because they are adsorbed as s-micropore walls, they possess less favorable conditions for the reaction with ionol, contrary to PO$_s^{\bullet}$ radicals.

Finally, the reaction of recombination I$_s^{\bullet}$ + POO$_s^{\bullet}$ → POOI is neglected in the s-scheme because of a high local HP concentration in the s-zones in samples (43). The rate of this reaction is supposed to be negligibly low in comparison with the rate of the reaction

$$I_s^{\bullet} + G_s \xrightarrow{k_{6s}} POO_s^{\bullet} + IH_s$$

Applying the condition of steady-state radical concentration to the s$_i$-scheme, it is easy to make sure that it does not change the expression of radical concentration for R$_s^{\bullet}$, that was obtained from the s-scheme of the process without ionol (see Sec. III). The concentration of s-phenoxyls relative to the s$_i$-scheme does not depend on HP$_s$ concentration:

$$[I_s^{\bullet}] = \frac{\delta k_{0s}}{2k_{6s}} + \frac{\delta k_{0s}k_{-6s}[IH_s]}{2k_{6s}k_{9s}} \tag{13}$$

the concentration of s-peroxyls is $[POO_s^{\bullet}] = \delta k_{0s}g_s/2k_{9s}$, and the total rate of s-hydroperoxide decomposition is the sum of several terms:

$$-\frac{dg_s}{dt} = k_{0s}g_s + k_{6s}[I_s^{\bullet}]g_s + k_{8s}[IO_s^{\bullet}]g_s - k_{-6s}[POO_s^{\bullet}][IH_s]$$

$$= k_{0s}g_s + \frac{\delta k_{0s}g_s}{2} + \frac{\delta k_{0s}g_s k_{7s}[IH_s]}{2k_{9s}}$$

We may neglect the second term in this expression comparing it with the first one, which characterizes spontaneous decomposition G$_s$. Integrating this expression and taking into account the interzone equilibrium condition:

$$c_{is} = \frac{K_{ei}}{\beta}c_i$$

(where $c_{is} = [IH_s]$ and $c_i = [IH]$ are the ionol concentrations in the s-zones and noncrystalline PP phase, respectively; K_{ei} is the equilibrium constant; β is the amorphous phase amount), we finally obtain the equation of s-hydroperoxide expenditure in the presence of a chain component, catalyzed by ionol:

$$g_s = g_{s0} \exp\{-(k_{0s} + b_i c_i)t\} \tag{14}$$

Here b_i is a complex constant.

Turning to the mechanism of HP chain decomposition in v-zones, let us remember that, according to the studies of Chien and Jabloner (43), ionol did not influence the k_{e1} value under all studied concentrations c_i, and influenced the k_{e2} value only at significantly high $c_i = 0.124$ mol/kg. Considering this feature, the authors (43) suggest that the reaction chains of HP decomposition proceed by free valency transfer along polymer chains, and any intermolecular acts of P$^\bullet$ radical attack on HP with free peroxyl formation are excluded. The present case displays the following view of the foregoing considered v-scheme:

P$^\bullet$ + G → $^\bullet$POOH → H$_2$O + P$^\bullet$ + products;

2P$^\bullet$ → termination

Here the OH$^\bullet$ radical formed reacts with the neighboring CH group on the same macromolecule. This reflects a significantly porous packing of polymer chains in the structure of the chain–sponge micelles.

Structural porosity (i.e., the increased degree of volume tension of a chain sponge of PP samples oxidized in solution, is stabilized owing to the fixing of stretched chains on the crystallite skeleton. Syndio- and isotactic PP is easily crystallized, and crystallizing polyolefins form no molecular solutions in hydrocarbons at $T < T_{melt}$ (45). But, in any event, they are able to disperse in liquids in certain amounts as colloid solution, although this provides no molecular contacts of the polymer–liquid type. This results as the polymer–polymer type contacts occur in the system, with simultaneous formation of crystal nuclei, or with the paracrystalline domains arranged as a framework (46).

The oxidized PP matrix in samples (43) most likely represented the dispersion of chain–sponge micelles, swollen in chlorobenzene, inside colloid particles.

The stabilization of particular porosity of the chain–sponge micelles in the dried material may result from the skeleton strength of crystallite armor, so the HP decomposition rate was similar for colloidal suspensions in chlorbenzene and in films (43).

The v-scheme of HP decomposition, modified in accordance with the foregoing results (43), does not change the initiation and termination of the reaction v-chains, considered in Section III. In this connection, Eq. (14) may be used in the expression of the chain reaction rate, as shown in Section III:

$$-\frac{dg}{dt} = k_1[P^\bullet]g = k'_e g \sqrt{g_s} = k_e g \exp[-0.5(k_{0s} + b_i c_i)t]. \tag{15}$$

In Eq. (15) the preexponent k_e coefficient possesses the same structure as in Eq. (1), and the exponential multiplicand contains the additional constant $b_i c_i$, connected to the HP chain decomposition inside s-zones.

Integration of Eq. (15) reduces it to the expression of the HP decomposition curve:

$$\ln\left(\frac{g}{g_0}\right) = \frac{2k_e}{k_{0s} + b_i c_i}\left\{\exp[-0.5(k_{0s} + b_i c_i)t] - 1\right\}$$

which transforms to the first-order equation for the initial period of the reaction ($t \to 0$):

$$\ln\left(\frac{g}{g_0}\right) = -k_e t; \qquad g = g_0 \exp(-k_e t)$$

which coincides with Eq. (3) and includes the same constant k_e, independent of the ionol concentration.

The same initial formula transforms to the following expression at significant level of the HP decompodition ($t \to \infty$):

$$\ln\left(\frac{g_\infty}{g_0}\right) = -\frac{2k_e}{k_{0s} + b_i c_i}$$

or

$$g_\infty = g_0 \exp\left[-\frac{2k_e}{k_{0s} + b_i c_i}\right]$$

This characterizes the kinetic stop, which increases with the increasing concentration c_i of ionol in the samples as a result of the catalytic action of ionol on the HP decomposition in s-zones.

A. Features of Homophase HP Chain Decomposition

After the end of the fast stage of the reaction in samples (43), the slow stage began instead of the kinetic stop. The rate of this stage did not change at low and moderate concentrations of ionol (up to 0.05 mol/kg). The existence of relatively low, but reliably measurable, reaction rate and the k_{e2} value for these samples differ from those oxidized in a block, the $g(t)$ curves of which are shown in Fig. 2 and for which the second stage is characterized by the extremely low rate of the process.

Note that the rate of the first stage of samples (43) was also higher (if calculated for equal temperatures) compared with PP, oxidized in bulk. It is evident that the higher chemical activity of suspension-oxidized polymer should be connected with the existence of incomplete packing of polymer chains in the chain–sponge micelle structure.

The porosity of the micellar material, combined with the highly elastic flexibility of polymer chains, increases the level of v-zone dynamics (which accumulate the general amount of hydroperoxide formed during oxidation) for the process of thermofluctuational excitation of v-micropores up to s-micropore size. This causes the increase of excitation frequency and, consequently, the probability of the HP group dissociation inside intensively fluctuating micropores.

If such thermofluctuational provision exists, the second stage of hydroperoxide decomposition occurs in the polymer. This may be characterized as the heterogeneous–homophase stage. Its role becomes predominant after the heterophase s, v-transfer of radicals is over, as a result of HP exhaustion inside the s-zones.

One or another probability of increased fluctuation of v-micropores always exists in the chain–sponge zones, but it is particularly high for a highly elastic or melt polymer state (34,35). At this stage the process of free radical initiation may be presented by the following scheme:

$$G_V \underset{k_{-f}}{\overset{k_f}{\longleftrightarrow}} G_f \overset{k_{0f}}{\longrightarrow} (PO_f^{\bullet} + HO_f^{\bullet}) \rightarrow \delta_f P^{\bullet}$$

in which the heterogeneous acts of excitation, deactivation, and dissociation of HP groups, conjugated with the fluctuations of micropores, are marked by constants k_f, k_{-f}, and k_{0f}; δ_f is the yield of radicals from a fluctuating micropore.

By combining this heterogeneous scheme with the reaction v-scheme, considered earlier in Section III, we obtain the equation for the chain heterogeneous–homophase reaction rate:

$$-\left(\frac{dg}{dt}\right)_c = k_1[P^{\bullet}]g = k_1\sqrt{g^3}\sqrt{\frac{\delta k_{0f}k_f}{2k_2(k_{-f} + k_{0f})}}$$

If we take into account that the occurrence of this reaction is connected with a significantly high contribution of fluctuationally dissociative HP decomposition in v-zones, the total decomposition rate is reduced to

$$-\frac{dg}{dt} = k_d^f g + k_c^f \sqrt{g^3} \tag{16}$$

Here k_d^f and k_c^f are the complex rate constants of nonchain and chain v-reactions, respectively.

The true fulfillment of Eq. (16) was stated (43) by the formal first-order rate constant dependence, typical for this case, on the initial HP concentration:

$$k_{e2} = k_d^f + k_c^f \sqrt{g_0}$$

Let us mention that Eq. (16) is similar in form to Eq. (7) of the homogeneous reaction. However, contrary to the explanation given (37), it characterizes the slow stage of the polymer hydroperoxide decomposition instead of the fast phase.

According to the data (43), the component $k_c^f \sqrt{g_0}$ of the rate is inhibited by a very high ionol concentration only, $c_i = 0.124$ mol/kg. This may be explained by extreme ionol passivity to alkyl polymer radicals $P^{\bullet}$ and to the low probability of its interactions with radicals $PO_f^{\bullet}$ and $HO_f^{\bullet}$, fixed on walls of the thermofluctuationally excited micropore.

V. THE IONOL EFFECT ON THE PP HYDROPEROXIDE DECOMPOSITION UNDER OXYGEN CONDITION

The presence of oxygen dissolved in oxidized PP samples with ionol fundamentally changes the situation that is usual under anaerobic conditions. It puts into first place the heterogeneous–homophase process. This process is limited by a low rate of heterogeneous initiation of radicals in the structural zones of the chain–sponge micelles, where the main amount of PP hydroperoxide, formed during autoxidation, is accumulated. Here, the chain reaction of HP decomposition, initiated in the absence of oxygen, is inhibited by ionol at its very high concentration, because the reactivity of alkyl macroradicals with ionol is negligibly low.

The situation changes in the presence of oxygen. It transforms alkyl radicals to peroxyl radicals and sharply increases the rate of their interactions with ionol and simultaneleously translates the process to the regimen of the catalytic HP decomposition.

An example of such change in the reaction catalytic regimen is from experimental results (47). There the kinetics of HP_1 decomposition in films obtained from melt and then oxidized, and that of HP_2 decomposition in films obtained from colloidal PP oxidized in chlorbenzene, were studied. Ionol injection into the samples induces the occurrence of the oxidation induction period at heating in oxygen ($PO_2 = 1$ atm). The duration of induction periods are significantly different in their dependence on the method of sample preparation and on the hydroperoxide decomposition rates in the induction periods. The highest activity is displayed by films prepared from a polymer that is oxidized in the solution. This should be attributed to increased dynamic fluctuation properties of the chain–sponge micelles in such films (see Sec. IV).

Figure 3 shows the curve (1) of HP_1 expenditure, obtained (47) for the samples possessing low fluctuation dynamics ($T = 403$ K, $g_0 = 0.150$ mol/L, $c_{i0} = 0.013$ mol/L) It is compared with curve 2, displaying phenoxyl expenditure; curves 3 for oxygen absorption; and 4 and 5 for HP and polymer peroxyl accumulation, observed after the end of the induction period.

The amount of oxygen absorbed during the induction period (350 min) does not exceed the sensitivity threshold of an appliance used in the investigation. At this stage the kinetic curve of ionol expenditure (not shown in Fig. 3) and the curve of phenoxyl expenditure are described by the exponential equations:

$$c_i = c_{i0} \exp(-k_i t) \qquad \text{and} \qquad [I^\bullet] = [I^\bullet]_0 \exp(-k_i t)$$

that possess a similar coefficient $k_i = 1 \times 10^{-4} s^{-1}$. Such a combination of kinetic curves means that the condition of the steady-state concentration is set for ionol phenoxyls at the very beginning of the process. It also means that phenoxyls do not participate in bimolecular termination, and do not interact with the oxygen.

The concentration of $POO^\bullet$ radicals is below the ESR spectrometer sensitivity ($\leq 10^{-8}$ mol/L) during the induction period, but it reaches a measurable stationary value $\approx 2 \times 10^{-6}$ mol/L immediately after the end of the induction period.

No empiric function was suggested for the HP_1 expenditure curve (47), and it is still possible to construct a calculated curve (see curve 1 in Fig. 3), taking into account a common 10% error in hydroperoxide concentration determination, which fits the exponential law with the same constant k_i, displayed in the ionol and phenoxyl expenditure:

$$g = g_0 \exp(-k_i t)$$

Here, the initial concentration is set at $g = 0.14$ mol/L. The curve correspondent to this equation, constructed in semilogarithmic coordinates $[\log(g/g_0), t]$, is shown

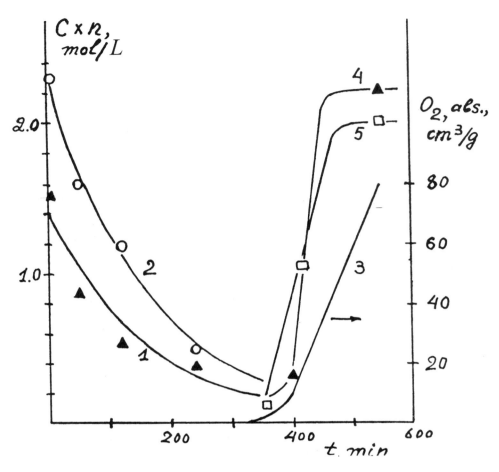

Figure 3 The oxidative kinetics at 403 K and $P_{O_2} = 1$ atm both for the induction period (1, 2) and direct autoxidation of PP (3–5): (1) decay of hydroperoxide; (2) ionol phenoxyl disappearance; (3) oxygen absorption; (4, 5) accumulation of hydroperoxide and free peroxyradicals, respectively. Coefficient n values are: 1, 4–10; 2, 5–1 $\times$ 10^6.

in Fig. 4 (straight-line 1), with the curves of the HP expenditure, obtained in air in the presence of tri-*tert*-butylphenol. Our experiments were performed to clear up the picture of the experimental value scattering caused by their small number presented in the study (47).

Specimens ($\sim$ 1 mg) of fine-grained PP powder for our experiments were oxidized in air at $T = 413$ K until HP accumulation reached concentration $g_0 = 0.14$ mol/kg. Tri-*tert*-butylphenol was injected into the oxidized powder from a chloroform solution, the solvent was evaporated after a 40-min exposure, and its residues were removed by evacuation. HP decomposition was studied by heating the samples in air and applying spectrophotometric modification of the iodometric

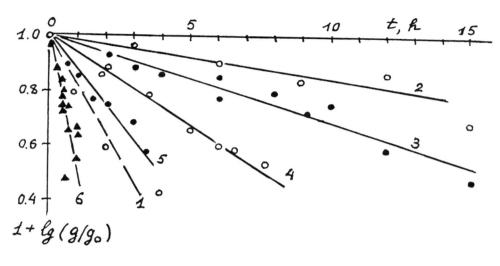

Figure 4 PP hydroperoxide decay plots for PP samples with ionol (1); tri-t-Bu-phenol (2–5) and with no antioxidant (6) in the presence (1–5) and in the absence (6) of O_2 at T values: 2, 372; 3, 383; 4, 393; 1, 5, 6, 403.

analysis of HP (35,39). The experiments displayed overlapping HP expenditure curves at a mass ratio of phenol PP = 1 : 1 and 2 : 1.

The experimental points obtained are shown in Fig. 4 as a series characterized by straight lines (2–5) for $T = 372$ K (2), 383 K (3), 393 K (4), and 403 K (5), respectively. The initial section of the HP decomposition curve at 403 K in the absence of oxygen is also shown for comparison (line 6). From comparison of the dispersion of these points and points, obtained in another study (47), one may conclude that exponential functions could be applied in all cases. Effective rate constants, calculated according to graphs (2–5), equal to $k_i \times 10^5 \mathrm{s}^{-1}$: 0.85, 2.0, 4.26, 8.5; and their generalized expression is $k_i = 6.0 \times 10^7 \exp(-E_i/RT)$, $E_i = 91.5$ kJ/mol.

It is characteristic that at similar to $T = 403$ K HP decomposition with ionol possesses the constant $k_i = 1 \times 10^{-4} \mathrm{s}^{-1}$, insignificantly different from that $(0.85 \times 10^{-4} \mathrm{s}^{-1})$ obtained for tri-*tert*-butylphenol. Both these constants are significantly lower than the constant $(4.2 \times 10^{-4} \mathrm{s}^{-1})$ characterizing HP decomposition in the absence of oxygen and alkylphenol (see Fig. 4, line 6). However, the rate of deceleration mentioned may not be connected to a partial HP reduction owing to chain PP oxidation or interactions of peroxyls with an inhibitor, because the current concentrations: g, c_i, and $[I^\bullet]$, are connected to each other by a proportion form the beginning until the end of the induction period. It is also important that the chain absorption of oxygen is nearly completely suppressed at this stage (see Fig. 3, curve 3). Simultaneous expenditure of HP and ionol, characterized by the same exponential coefficient $\exp(-k_i t)$, occurs at tenfold excess of g_0 over c_{i0}; that is, when the HP formation in the reaction chain termination can by no means compensate the fourfold deceleration of the HP decomposition (see Fig. 4, lines 1 and 6) in the entire measure.

The kinetic law of HP expenditure, applied to the results (47), corresponds to the nonchain homogeneous reaction from a formal point of view. Meanwhile, its effective rate constant occurs in the sequence of chain anaerobic decomposition constants, known for the heterophase stage (35). It is much higher than the rate constant of the nonchain component, observed for heterogeneous–homophase stage. The later equals, at least, $4 \times 10^{-5} \mathrm{s}^{-1}$ at 408 K and $1.5 \times 10^{-5} \mathrm{s}^{-1}$ at 393 K even for fluctuationally active HP_2 samples, whereas HP_1 films possess no measurable value of it at 403 K (see Secs. III and IV).

The important increase of HP decay rate observed in the presence of oxygen, relative to the rate of the second (homophase) stage of the anaerobic reaction, is connected with the alkyl macroradical oxidation to peroxyls. This process does not affect the rate of radical initiation by the mechanism of thermofluctuational excitation of micropores, and it induces sharp acceleration of polymer radical reactions with alkylphenols and leads to an appearance of a new chain reaction of HP decomposition. The introduction of such new reactions was required by the characteristics of the anaerobic HP decomposition in the presence of ionol, and was fixed by the s_i-scheme in Sect. IV. In accordance with the s_i-scheme, the new chain reaction of HP decomposition is induced by alkyloxyls formed from ionol, which possess a six-term carbon cycle structure.

The features of HP decomposition under oxidative conditions, considered in the present section also require the calculation of the new chain reaction of HP decomposition. This conclusion is supported by the independent fact that oxygen injection significantly accelerates the process in HP_2 samples. For example, expenditure of HP_2 and ionol phenoxyls in oxygen at $T = 393$ K ($g_0 = 0.156 \, \mathrm{mol/l}$; $c_{i0} = 0.17$ mol/l) finishes in 20 min, and not in 350 min, as was in the case of HP_1 (47). Despite a lower temperature of HP_2 decomposition, the half period of transformation is 5–10 min instead of 116 min. Under these conditions, the effective rate constant is $23 \times 10^{-4} \, \mathrm{s}^{-1}$ that significantly exceeds the constant of the initial stage of the chain anaerobic reaction for the same samples (see Sec. IV).

Thus, phenomenological action of oxygen changes two macroscopic kinetic stages of the anaerobic process in the chain mechanism by means of significant, but different, acceleration of hydroperoxide decomposition for HP_1 and HP_2, catalyzed by ionol.

The initiation of kinetic v-chains under the oxidative conditions occurs by the homophase pathway, favoured by thermofluctuation excitation of v-micropores, in which hydroperoxide is accumulated. Here, ionol behaves as a good inhibitor of PP autoxidation, intercepting all reaction chains, but it simultaneously participates in the reaction chains of HP decomposition.

Also, the considered regularities of HP decomposition with oxygen and without it (see Sec. IV) exclude any significant alkylphenol molecular reaction with hydroperoxide (and, consequently, the probability of increasing degenerated branchings in the presence of an inhibitor). Moreover, the preservation of ionol catalytic function during the induction period confirms the existence of a definite mechanism of its regeneration.

If we take into account all the aforementioned facts, we may suggest the following v_i-scheme of transformations in v-zones of polypropylene chain–sponge micelles, which start with the thermofluctuation excitation of the v-micropores:

$$G_V \underset{k_{-f}}{\overset{k_f}{\longleftrightarrow}} G_f \overset{k_{0f}}{\longrightarrow} \delta_f P^\bullet ; \qquad P^\bullet + O_2 \rightarrow POO^\bullet ;$$

$$POO^\bullet + IH \overset{k_1}{\longrightarrow} POOIH^\bullet \rightarrow POH + IO^\bullet ;$$

$$IO^\bullet + G \overset{k_2}{\longrightarrow} IOH + POO^\bullet ;$$

$$POO^\bullet + IH \overset{k_3}{\longrightarrow} G + I^\bullet ; \qquad POO^\bullet + I^\bullet \overset{k_4}{\longrightarrow} POOI.$$

Here, propagation of the HP decomposition reaction chains proceeds in the same way as in the previously considered s_i-scheme through the stages of intermediate formation of phenol σ-complex with peroxyl and formation of quinoid alcoxyl (possessing constants k_1 and k_2, respectively). Contrary to the s_i-scheme, the v_i-scheme displays the chain termination performed on ionol. Its rate with the constant k_3 is proportional to the rate of HP decomposition chain propagation, defined by the constant k_1. The existence of this proportion is responsible for ionol providing a catalytic development of the chain HP decomposition, abruptly decreasing peroxyl concentration and reducing the rate of the chain PP autoxidation in the propagating steps to a negligibly low value:

$$POO^\bullet + PH \rightarrow G + P^\bullet ; \qquad P^\bullet + O_2 \rightarrow POO^\bullet$$

Regeneration of the initial ionol (and tri-*tert*-butylphenol) in the present process should be performed from the product of its transformation, quinolic alcohol, in the thermolfluctuation excitation of v-micropores. These reactions proceed as "cryptoradical", or "intracage", ones, resulting in dissociation of quinolic alcohol and further reactions of macromolecule dehydrogenation (18):

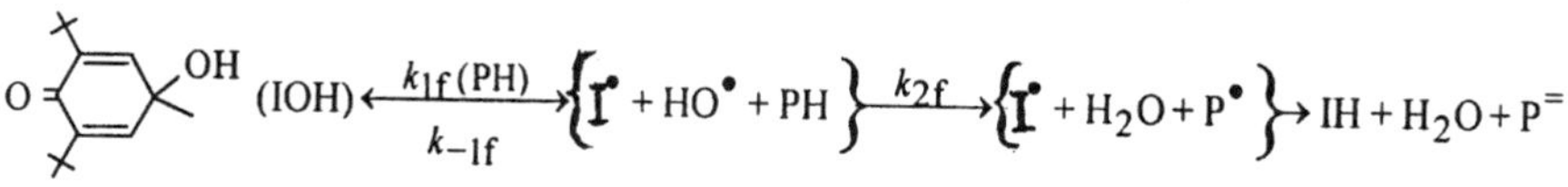

Simultaneous partial loss of phenol should be performed in parallel reactions disproportioning, which proceed in the presence of ionol with formation of methylene quinone and its further transformations.

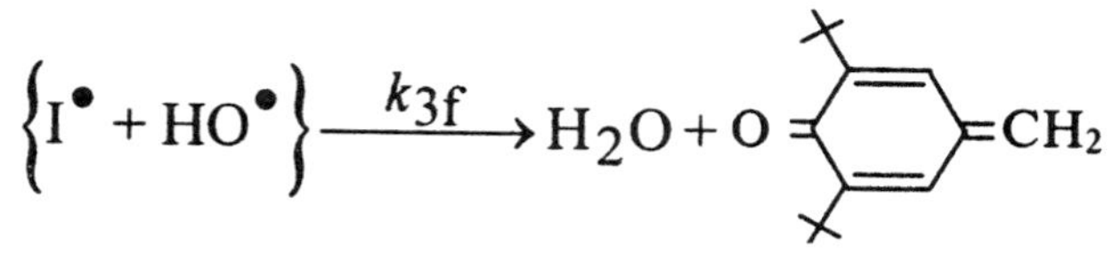

In *tert*-butylphenol, it is performed by detachment of *tert*-butyl radicals with final formation of quinone and other products:

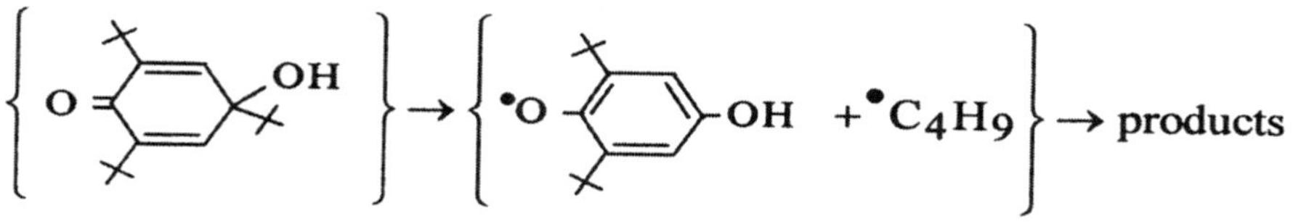

By applying the steady-state concentration condition to active particles, shown in the v_i-scheme (radicals and alcohols IOH), we obtain the rate equation for the reaction chain initiation and termination:

$$\delta_f \alpha_f k_{0f} g = 2k_3 [\text{POO}^\bullet][\text{IH}]$$

Here, δ_f is the radical yield in fluctuational HP decomposition in v-zones;

$$\alpha_f = \frac{k_f}{k_{-f} + k_{0f}}.$$

We also obtain the expression for the HP decomposition rate:

$$-\frac{dg}{dt} = \alpha_f k_{0f} g + k_1 [\text{POO}^\bullet][\text{IH}] - 2k_3 [\text{POO}^\bullet][\text{IH}]$$

$$= \alpha_f k_{0f} g + k_1 \left(\frac{\delta_f \alpha_f k_{0f} g}{2k_3} \right) - \delta_f \alpha_f k_{0f} g = k_i g$$

which characterizes the process as a first-order reaction by HP and a zero-order by alkylphenol:

$$g = g_0 \exp(-k_i t)$$

In accordance with the v_i-scheme, the phenol expenditure rate is

$$-\frac{dc_i}{dt} = k_1 [\text{POO}^\bullet][\text{IH} + k_3 [\text{POO}^\bullet][\text{IH}] - k_2 [\{\text{I}^\bullet + \text{HO}^\bullet + \text{PH}\}]$$

$$= \left[k_3 + \frac{k_1 k_{3f}}{k_{2f} + k_{3f}} \right] [\text{POO}^\bullet][\text{IH}] = \left[\frac{1}{2} + \frac{k_1 k_{3f}}{2k_3 (k_{2f} + k_{3f})} \right] \delta_f \alpha_f k_{0f} g = k_{ii} g$$

Integrating the expression obtained

$$-\frac{dc_i}{dt} = k_{ii} g$$

gives

$$c_i = \left(\frac{k_{ii} g_0}{k_i} \right) \exp(-k_j t) + \text{const}$$

reduced to the following form, which takes into account the initial condition $c_i = c_{i0}$ at $t = 0$

$$c_i = c_{i0} - \left(\frac{k_{ii} g_0}{k_i} \right) [1 - \exp(-k_i t)]$$

In this case, if $c_i = 0$ at $t \to \infty$, we obtain

$$c_{i0} = \frac{k_{ii} g_0}{k_i} \quad \text{and} \quad c_i = c_{i0} \exp(-k_j t)$$

which fits the data (47).

A quasi-equilibrium concentration of phenoxyls in this process is set very fast. Kinetics of their expenditure is described by the theoretical expression

$$[I^{\bullet}] = \frac{k_3}{k_4}c_i = \frac{k_3 c_{i0}}{k_4}\exp(-k_i t)$$

which is fulfilled, since the quasi-equilibrium phenoxyl concentration is set until the end of the induction period of PP oxidation. This theoretical expression, as are the previous ones, adequately displays the experimental results (47). So, we may conclude that the model of the heterogeneous–homophase chain reaction correctly reflects empirical regularities of the oxidative polypropylene HP decomposition in the presence of an antioxidant of the phenolic type.

The mechanism of partial phenol regeneration from the product of its chain degradation, quinolic alcohol, expressed similarly to that of phenol regeneration from product with dibenzoyl peroxide (18), are rather widespread. This is confirmed by an independent experiment with methyl ester of quinolic alcohol (18).

VI. THE MECHANISM OF THE AUTOXIDATION OF POLYOLEFIN NONCRYSTALLINE DOMAINS

A. The Phenomenon

There is a set of attempts in the literature to fit the classic scheme of autoxidation of hydrocarbons onto polyolefins. As a rule, these schemes are based on the idea that a reacting polymer is a homogeneous matter with no singularities, as compared with liquid hydrocarbons. Recently, a new oxidation model, published by Gugumus (48–51), is based on a simple approach that the general process is a superposition of homogeneous and heterogeneous pathways occupying the amorphous domains of polyolefins.

In accord with the model, the oxidized phase is an organization of amorphous domains separated by crystalline interlayers. As in these studies (48–51), the general process, proceeding under the diffusion-unlimited oxygen access, is the result of homogeneous initiation of the reaction chains in the amorphous domains, and the heterogeneous spreading of oxidation from one domain, on reached the steady peroxydation of its matter, to another domain by migration of the molecular weight initiators formed. In addition, the heterogeneous oxidation spreads from primarily highly oxidized sample surfaces into deeper layers, and the rate of oxidation spreading is significantly lower than the rate of developed homogeneous oxidation inside the amorphous domains.

The heterogeneous-spreading model does stress the induction period as the homogeneous reactional stage of oxidation. So, the induction period is the time spent for oxidation of amorphous domains that exist, more or less oxidized, near the surfaces up to the steady peroxidation degree, and then they become the source of low molecular weight peroxides migrants.

The autoaccelerated stage of oxidation is thought to be exponential, because the spreading model is kinetically related to the model of population dynamics and of spreading infectious diseases. This very model requires fulfillment of the exponential law of progression in the initial stage of the processes.

The kinetic equation of OOH group accumulation deduced in the aforementioned studies (48–51) as operative after the induction period is over,

$$\frac{dg}{dt} = a_1 g - a_2 g^2 \tag{17}$$

differ from the common equation, which is used in the literature to characterize the autoxidation reaction of homogeneous hydrocarbons:

$$\frac{dg}{dt} = b_1 g^{0.5} - b_2 g \tag{18}$$

where a_1, a_2, b_1, and b_2 are the constants.

Recently, it was disclosed that these kinetic equations, concerning both the classic reaction model and the spreading model, do not generally fit the experimental data on polymer oxidation. This appears with evidence as one clarification for the existence of a free macroscopic kinetic stage, which is distinguished in its peculiarities (39,52).

For autoaccelerated phase of oxidation, the experimental data from the literature demonstrated an excellent correlation with the parabola equation that rules the process after the induction period is over. The parabolic stage is quite clearly displayed in the general curve (1) of the oxygen absorption exhibited in Fig. 5. This curve was composed (36) for eight carbochain polymers, namely, polyethylene-LD, polypropylene, ethylene copolymer with propylene-13.5%, poly-4-methylpentene-1, polystyrene, synthetic rubber, polyheptene, and polynonene. (Conditions of the experiments for each particular polymer are shown in Ref. 36).

The initial part of the curve 1, recovered in semilogarithmic coordinates (curve 2), is depicted in Fig. 5 for a comparison. It is clearly seen that it includes a linear section characterized by the expression:

$$\ln(N_{O_2}) = \ln(0.009) + 8.0\sqrt{A}(t - t_i) \tag{19}$$

This expression relates to the exponential equation $N_{O_2} = a e^{\alpha x}$, where $a = 0.009 \pm 0.003$ mol/kg is the coefficient obtained by extrapolation to the ordinate axis; $\alpha = 8.0$ kg$^{0.5}$/mol$^{0.5}$; $x = \sqrt{A}(t - t_i)$; t_i is the induction period time.

This equation describes a more narrow passage of curve 1 (0–0.3 mol$^{0.5}$/kg$^{0.5}$) compared with the parabolic equation (0–1.0 mol$^{0.5}$/kg$^{0.5}$). Therefore, it cannot serve as a base for preferable usage of the exponential law. Similarly, any small part of a continuous curve may be approximated by a straight line passage which, however, does not characterize nonlinear sections.

The induction period t_i of polyolefin autoxidation displays a significant scattering of the time values (36,48). In this stage of the process, the polymer samples absorb a definite amount of oxygen. The start of the parabolic autoaccelerated state is induced from the partially oxidized state. The amount of oxygen absorbed during the induction period may be determined by extrapolation of the primary section of generalized curve 1 (see Fig. 5) to the ordinate axis ($t = 0$). However, such extrapolation is quite inaccurate. Therefore, it seems better to use extrapolation of the linear section of curve 2 in Fig. 5, depicted in semilogarithmic coordinates (see Eq. 19). This approach gives the value $a = \Delta[O_2] = 0.009 \pm 0.003$ mol/kg, which adequately relates to all polymers presented in the general curve, as it follows from

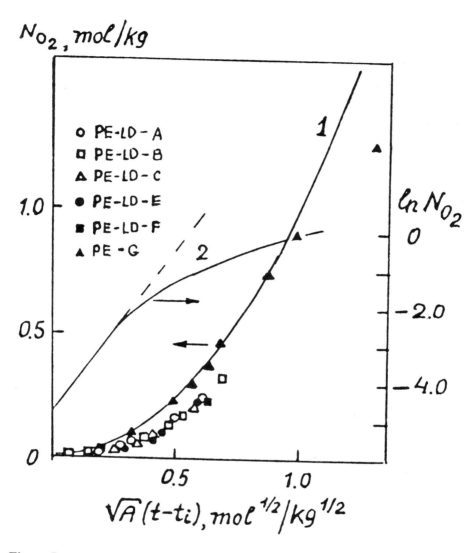

Figure 5 Oxygen absorption at 90°C in thermal oxidation of LDPE as constructed from the data presented by Gugumus: [A–F samples from the reference (48)]; the single master curve (1), presented by Shlyapnikov (36); PE-G points are the same as in the reference (36). Curve (2) is the semilogarithmic plot of curve (1).

Fig. 5. Taking into account this result, the average rate of oxidation (V_i) in the induction period can be characterized by the ratio

$$V_i = \frac{\Delta[O_2]_i}{t_i}$$

The obtained value of $\Delta[O_2]_i$ correlates well with the result of direct determination (obtained in Ref. 54), in which clearly purified PP was oxidized as a finely grained powder under oxygen pressure of $P_{O_2} = 150$ torr (air) and a temperature of 92°C. The manometric instrument used in this study (54), possessed a 1000-time sensitivity over that on which the generalized curve (see Fig. 5) has been obtained.

These experiments displayed the presence of a prolonged primary stage, with oxygen absorption that differed from the aforementioned parabolic curve. This stage assumed $t_i \approx 6000$ s, when 0.008–0.01 mol/kg of oxygen was absorbed. It is characteristic that the oxidation rate observed at this stage was kept constant after a special heating in vacuum at 145°C (during 3 h) and was held to decompose traces of hydroperoxide. Thus, it was shown that oxidation at this stage is not initiated by the traces of OOH groups. When the first stage was finished, the parabolic kinetics of O_2 absorption occurred, strictly fixed in the range of the V_2 rate of $2.0–20.0 \times 10^{-6}$ mol/kg s^{-1} by the presence of directly proportional time-dependent increase of the rate and, consequently, was dependent on the concentration of the hydroperoxide formed.

Thus, in the aforementioned states for polymers presented in Fig. 5 the oxidation in the induction period is initiated by direct oxygen reaction with macromolecules, rather than with hydroperoxide. The conclusion about such a mechanism for the initiation of the process in the primary kinetic stage was confirmed (54).

The oxidation rate

$$V_i = \frac{\Delta[O_2]_i}{t_i} \propto t_1^{-1}$$

changes without saturation in a wider range of oxygen pressures compared with the maximum rate ρ_m. Figure 6 compares curves for the dependence of the rates t_i^{-1} and ρ_m on oxygen pressure, obtained by oxidation of PP films at 90°C (55). It can be seen in Fig. 6 that, after autoxidation, the yield from the induction period to the peroxide initiation function, the rate ρ_m is saturated much faster than in the primary stage in which the initiating function of OOH groups is not yet induced.

A similar picture was observed for initial and maximum rates of oxygen absorption by the oxidized PE melt (56). The results are shown in Fig. 7.

The dependence of t_i on prehistory and the method of polymer sample preparation should also belong to the list of features of the induction period. This very prehistory generally defines scattering of t_i values. Orientation stretching of samples and the existence of structural relaxation in them may make a separate contribution to the phenomenon.

The orientation drawing (its role is discussed in the monograph Ref. 54) increases periods of oxidation induction in PP, HDPE, and poly(methylpentene) films (54,57), but decreases it in LDPE films (57). The higher the degree of drawing of PP films, the lower is the yield of POOH and the higher is the yield of CO groups per 1 g mol of absorbed oxygen (54). The effect of drawing is leveled by annealing of oriented samples and proceeding under the oxidative degradation as a result of structural relaxation.

The effect of structural relaxation on the rate of initial and second-stage oxidation should be taken into account in the kinetic analysis. In the opposite case, an inadequate description of the process is possible. This could be seen in early works (58,59) the results of which were criticized (37). The authors (58,59) have paid no

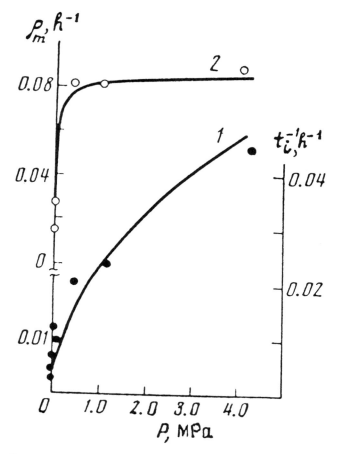

Figure 6 The induction period (1) and maximal rate of carbonyl formation (2) versus oxygen pressure in thermal oxidation of PP films at 90°C. (From Ref. 55.)

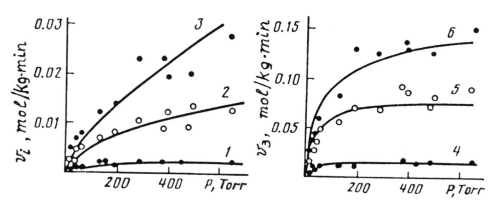

Figure 7 Initial rates (1–3) and maximal rates (4–6) of oxygen absorption in the thermal oxidation of PE melt at 141.7°(1, 4); 157.3°(2, 5) and 165°C (3, 6) versus oxygen pressure. (From Ref. 56.)

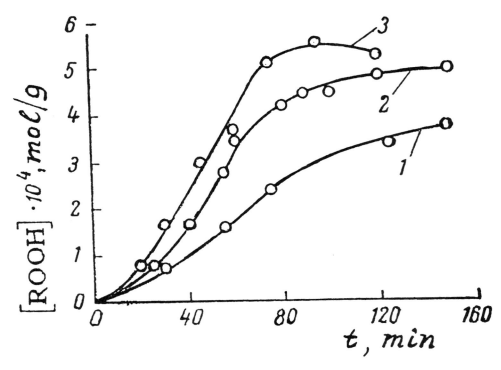

Figure 8 Formation of hydroperoxide in thermal oxidation of PP films at 130°C and $P_2 = 200$ (1); 400 (2); and 600 torr (3). (From Refs. 58, 59.)

attention to the structural features of the samples prepared. The samples were pressed for a short time from isotactic PP powder at 125–130°C (i.e., below the temperature of crystal melting; 165°C). The formation of monolythic material was fixed by a skeleton of nonmelted deformed crystals. The samples were oxidized in the absence of the external pressure and reduced gradually to the primary structure. That process was accompanied by relaxational growth of the molecular–segmental dynamics in the noncrystalline phase. When relating to the mechanical thixotropy (40), this effect resulted the second stage of the autoxidation as significantly prolonged and reached a higher degree of peroxidation (Fig. 8) compared with the samples of isotactic PP, which possessed more stable equilibrium in the supermolecular structure. The samples possessing rather low relaxational dynamics are shown in the reports (60–62).

The results of oxidation of melt-prepared PP films with small spherulites (curves 1 and 3) and with large spherulites (curves 2 and 4) (60) are shown in Fig. 9. The latter (large spherulites) were formed under long heating in vacuum. In this case, the recrystallization did not change the crystallinity. The comparison of the curves in Figs. 8 and 9 (in which concentrations are calculated for the sample mass) clearly shows that the three mentioned methods of film preparation result in significantly different pictures. Contrary to the films with a more dynamic phase (see Fig. 8), more stable films possess a shorter stage of autoacceleration after leaving the induction period (see Fig. 9). These films reach the stage of hydroperoxide decay

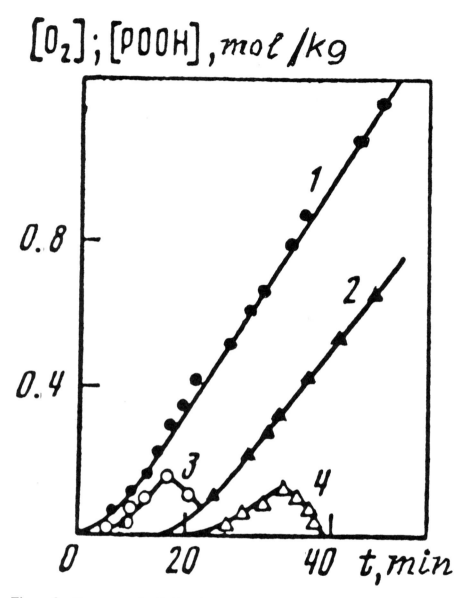

Figure 9 Oxygen uptake (1, 2) and hydroperoxide formation (3, 4) in thermal oxidation of PP films possessing the small-sized (1, 3) and large-sized (2, 4) spherulites; 130°C, $P_{O_2} = 500$ torr. (From Ref. 60.)

much faster, and the low value of the concentration maximum, respectively (see curves 3 and 4). In addition, the recrystallization, induced by the prolonged annealing of the films, results in a significant increase of the induction period (see Fig. 9, curves 2 and 4).

The phenomena mentioned act in the absence of diffusional limitation for oxygen and at the constant part of the amorphous phase in the PP films. They are, in principle, out of the frames of the Gugumus's model and the classic scheme of homo-geneous oxidation of hydrocarbons. The reason for those features should be searched for in the specificities of fine (microheterogeneous) structure of noncrystalline domains formed as the micellar polymer chain–sponge structure. The polymer chain–sponge micelle possesses somewhat intensive fluctuational dynamics to be changed in the structure under the effect of drawing, mechanical thixotropy, and recrystallization. The effect of thermal thixotropy should also be related to that phenomenon (52).

Results (52) indicate the change of primary physical structure of PP by heat treatment which affects the oxidation rate in the induction period with no change in POOH concentration. In addition, the structural change induced makes the poly-mer more sensitive to a structural–relaxational effect under the action of air. The polymer obtains an increased reactivity to oxygen, and PP oxidation is induced at room temperature.

When receiving the maximum rate of oxidation, the process enters the third kinetic stage. One of the important features of the third stage is the zero-order kinetics by hydroperoxide accumulated in a polymeric sample. The reason for the phenomenon was not practically discussed in the literature. Investigators have paid their main attention to the mechanism of the previous autoaccelerated stage, because at this stage polyolefins lose their useful properties. Usually, the problem of the third stage was limited by ranges of the formal approach, and the case was reduced to one for which the rate of oxygen absorption must become a constant at reaching the balance of the rates of chain accumulation and nonchain expenditure of hydroperoxide. This is clearly depicted by equaling zero in Eqs. (17) and (18).

Thus, it was assumed that the rate of oxygen absorption should reach the maxi-mum value only in reaching the maximum of POOH concentration.

However, the situation in the PP samples with stable internal structures indicates the earlier moment of reaching the maximal rate (V_3) of oxygen absorption in relation to the [POOH]$_{max}$ moment. This occasion was commented on (61,62). The pointed kinetic feature (it is also clearly seen in Fig. 9) presents an anomaly from the point of view of schemes that give Eqs. (17) and (18). In addition, it indicates two parallel ways of oxygen uptake in POOH and CO groups. Moreover, the curves in Fig. 9 show one more important feature: the value of V_3 does not significantly change in the steady state when the concentration of hydroperoxide (the agent responsible for branching of the reaction chains) passes its maximum, and then decreases significantly.

The kinetic picture, equal to that presented in Fig. 9, is characteristic not only for PP, but also for oxidizing melts of polyethylene (63), polyethylene oxide (64–66), and polystyrene (67).

The fact that the V_3 value is independent of POOH concentration in the poly-mer is in clear contradiction with the reaction schemes of homogeneous oxidation and the Gugumus model. According to these schemes, upon reaching its limit value, the concentration [POOH] should be practically constant for a long time, because the rate of its decrease is limited by the rate of oxidation of monomeric units. But the latter spend themselves rather thinly in passing through the time period up to when POOH is used up. Therefore, the V_3 value decreases insufficiently. In other words, the decrease of POOH concentration observed in the third stage could be related to neither monomolecular nor to bimolecular reactions by hydroperoxide, as is

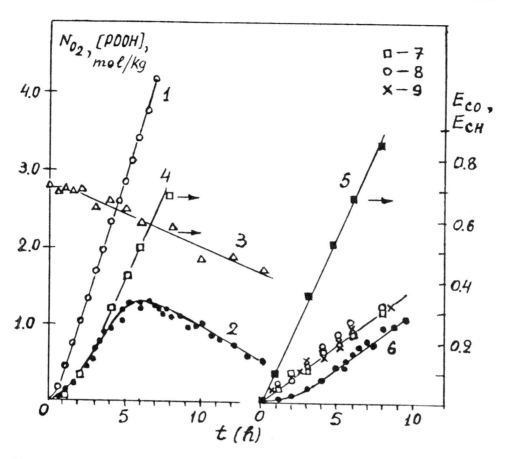

Figure 10   Thermal oxidation of poly(ethylene oxide) melted at 90°C (1–5) and 80°C (6–9) without dibenzoyl peroxide (1–4, 6) and with dibenzoyl peroxide added in concentrations of 0.06 (5); 0.0165 (7); 0.039 (8); and 0.12 mol/kg (9): (1) oxygen uptake; (2, 6–9) hydroperoxide formation; (3) decay of CH-links; (4, 5) carbonyl formation. (From Ref. 19.)

required by Eqs. (17) and (18), but it could be explained by means of the reaction of OOH groups with accumulating oxidation products (64–66).

On the other hand, considering the OOH group concentration decrease after the [POOH]$_{max}$ point (see Fig. 9), it should be taken into account that the afore-mentioned reaction schemes require a simultaneous significant decrease in the rate of oxygen absorption; that is, in contradiction with the data in the Fig. 9.

The foregoing kinetic features are thought to be quite important for kinetic modeling of the process. A direct study of polyethylene oxide (PEO) melt and solid PP oxidation (39,52,66) confirmed the general character of features of the third stage of polymer autoxidation.

Figure 10 shows the curves of oxygen absorption (1), the decrease of optical density of IR absorption of CH bonds (3), the increase of optical density of CO groups (4,5), and the accumulation of hydroperoxide (2,6–9), which characterize oxidation of melted PEO. The curves 5 and 7 through 9 were obtained in the presence of an initiator, dibenzoyl peroxide, added to the PEO melt in concentrations of 0.06, 0.0165, 0.039, and 0.12 mol/kg, respectively.

It follows from the Fig. 10 that the rates of oxygen absorption, expenditure of polymeric CH bonds, and accumulation of CO groups after the end of autoacceleration period do not depend on hydroperoxide concentration. The same range of time displays constant rates of CO and OOH group accumulation in experiments with dibenzoyl peroxide. This indicates that the initiator added loses its effect on the rate of PEO oxidation after the process transfer into the third stage.

The combination of the foregoing kinetic curves was also confirmed in experiments with solid PP and a copolymer of ethylene with propylene (39). As a whole, the third stage of autooxidation may be represented by two consequent phases. The first phase displays a simultaneous increase of OOH and CO group concentrations owing to parallel reactions of oxygen attachment to the polymer. It shows the rates of these processes as independent of the concentration of OOH groups. The second phase of the third stage shows a significant increase of hydroperoxide decomposition, catalyzed by its products. Thus, the concentration of hydroperoxide reaches its limit and then decreases rapidly, but the rate of polymer oxidation remains practically constant.

In addition, the characteristic feature of the third stage is the oxidation remaining the same as the rate obtained under an additive initiator, constant and independent of the concentration of the initiator and current concentration of hydroperoxide. Hence, in the third stage, the autooxidation obtains the kinetic zero order by both hydroperoxide and an additive initiator. Such a combination indicates the situation under the control of hydroperoxide. Hydroperoxide neglects the initiating action of an additive initiator when it reaches a definite concentration, and keeps this function for itself.

If we take into account the aforementioned complex of kinetic features of autooxidation of amorphous polymers, it should be accepted that neither the classic scheme of the liquid-phase oxidation nor the model by Gugumus (48–51), with the attempt to calculate clear heterogeneity of a solid polyolefin, gives any adequate explanation of the features inherent to both solid and melt states.

B. Structure–Kinetic Model of Autooxidation of Polymers

The kinetic modeling of polymer chain reactions requires calculation of fine (microheterogeneous) structural organization of polymeric chains in spongy micelles that fill up amorphous domains in solid polyolefins and polymeric melts. The structure of chain–spongy micelle, shown in Fig. 1, is responsible for creating the s,v-interzone distribution of low molecular additives; namely, initiators, inhibitors, and oxygen. The equilibrium conditions for the interzone exchange is rapidly set and, usually, it does not limit the rate of the chain reactions in the temperature range above room temperature.

Another important result of structural layering of micelle into s- and v-zones is that narrow v-micropores significantly decrease the probability of dissociation of molecules of initiators and OOH groups because of steric and energetic reasons. At the same time, conditions for dissociation and initiation are quite convenient in wide micropores of s-zones. The initiated free radicals transform and decay nearly completely in s-zones. A rather large amount of s-radicals are injected into the matrix of v-zones.

The quite high molecular–segmental dynamics of polymer chain–sponge may significantly increase the frequency of fluctuational formation of supermicropores inside v-zones. The rate of chain initiation in v-zones then grows to excite the chain transformation by the heterogeneous–homophase mechanism. At the same time, this pathway is not generally characteristic for initiation of the process in structurally stable samples. Therefore, the hetegeneous–heterophase model of the process will be discussed in this section (39,52).

The scheme of the primary reaction chain should be written down, at first, as the property of the s-zones of micellar aggregates. Let us take into account that in the absence of hydroperoxide in s-zones the process is initiated by means of direct CH bond interaction with oxygen:

$$PH_s + O_{2,s} \xrightarrow{k_0} \delta_1 P_s^{\bullet}, \qquad P_s^{\bullet} \xrightarrow{k_d} R_s^{\bullet} + R^{=}$$

$$R_s^{\bullet} \xrightarrow{k_{1s}} R_{1,s}^{\bullet}, \qquad R_1^{\bullet} + R_{1,s}^{\bullet} \xrightarrow{k_{2s}} products$$

Here, PH is a chain link; $P_s^{\bullet}$ is the radical carrying free valency on the main polymeric chain; $R_s^{\bullet}$ and $R^{=}$ is a radical and unsaturated group, respectively, formed at ends of broken macromolecules; $R_{1,s}^{\bullet}$ is a product of isomerization of the radical $R_s^{\bullet}$. The particular chemical structure of the mentioned s-radicals of polypropylene were considered in the study (39). The composition of a polyethylene free radical could be easily understood because of the simplicity of its macromolecule.

Let us point out that the present s-scheme fixes only the initial stage of oxidation, when the reaction of macromolecules with oxygen is the source of free radicals.

The process of free valency transfer from the s-zone into the v-zone is quite difficult as a result of low molecular–segmental mobility in the polymer. The process may be carried on only at low frequency and as stimulated by quite movable chain units, the end $R_s^{\bullet}$ radicals, in particular. These radicals should occupy a more or less suitable place in an s-micropore and transform with detachment of a light radical after oxidation:

$$R_s^{\bullet} \underset{k_{-m}}{\overset{k_m}{\longleftrightarrow}} R_m^{\bullet} \qquad R_m^{\bullet} + O_{2,s} \xrightarrow{k_{tr}} ROO_m^{\bullet} (-CH_2OO_m^{\bullet})$$

$$- CH_2OO_m^{\bullet} \rightarrow -\overset{\bullet}{C}HOOH \rightarrow -CHO + HO^{\bullet}$$

By migrating into the v-zone, light $HO^{\bullet}$ radicals initiate the secondary reaction chains with hydroperoxide formation:

$$HO^{\bullet} + PH \rightarrow H_2O + P^{\bullet}, \qquad P^{\bullet} + O_2 \xrightarrow{k_1} POO^{\bullet}$$

$$POO^{\bullet} + PH \xrightarrow{k_2} POOH + P^{\bullet}$$

$$2POO^{\bullet} \xrightarrow{k_3} termination$$

The scheme of the v-reaction should not include dissociation of POOH (the latter is performed only as the hydroperoxide appears in the s-zones). It should be added by parallel chain reaction of formation of CO groups. Additional reactions have been discussed in detail (39). In this review, we point out the formation of an epoxy cycle (E) in polypropylene chains:

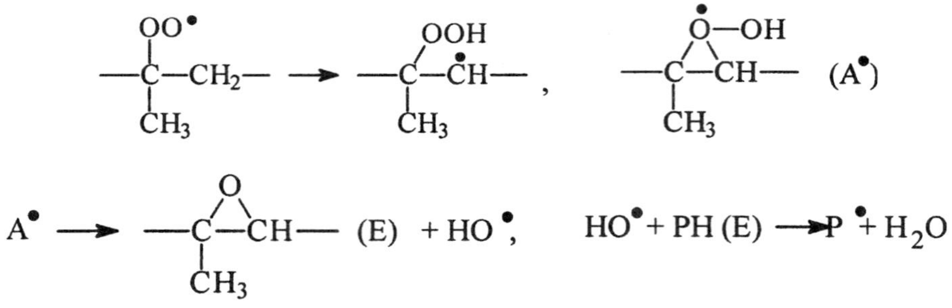

$$
\underset{\underset{CH_3}{|}}{-\overset{OO^\bullet}{\underset{|}{C}}-CH_2-} \;\longrightarrow\; \underset{\underset{CH_3}{|}}{-\overset{OOH}{\underset{|}{C}}-\overset{\bullet}{C}H-} \;,\qquad \underset{\underset{CH_3}{|}}{-\overset{\overset{\bullet}{O}-OH}{\underset{|}{\overset{\diagup\ \diagdown}{C}}\ CH}-}\quad (A^\bullet)
$$

$$
A^\bullet \;\longrightarrow\; \underset{\underset{CH_3}{|}}{-\overset{\overset{O}{\overset{\diagup\ \diagdown}{C}}\ CH}{}-}\ (E)\ +\ HO^\bullet,\qquad HO^\bullet + PH\,(E) \;\longrightarrow\; P^\bullet + H_2O
$$

Simultaneously with POOH, epoxy compound E may pretend to the role of the primary valent-saturated product of oxidation to explain the complementary pathway of the oxygen uptake and chain mechanism of water formation (39). It also explains the noticeable delay in the moment of the CO compound appearance in relation to hydroperoxide. Epoxy compounds isomerize into ketones and aldehydes, and easily interact with free radicals $r^\bullet$, transferring an H atom and attaching $r^\bullet$ (68):

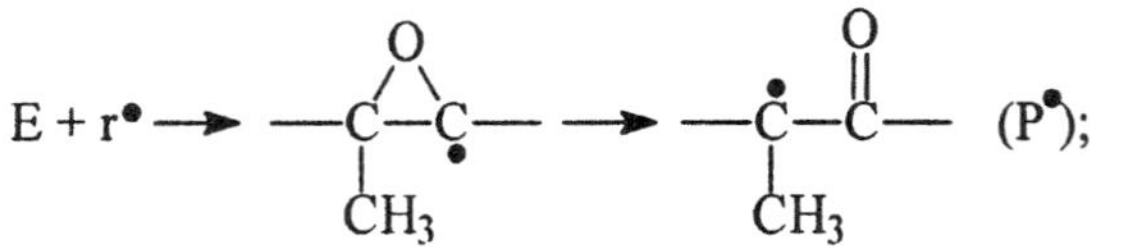

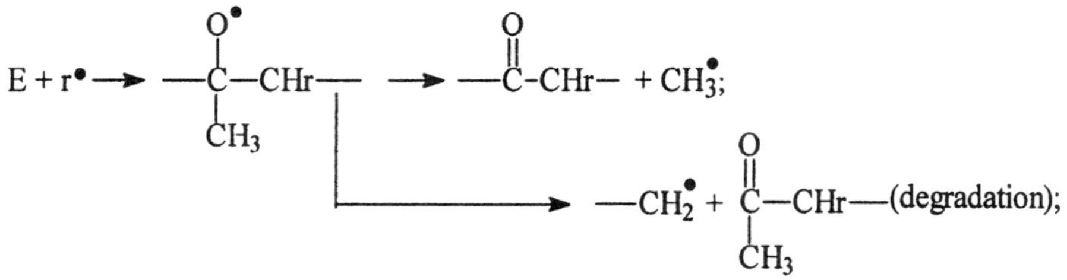

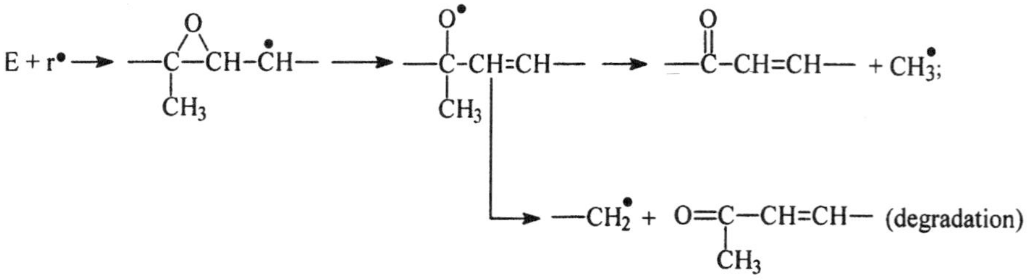

$$
E + r^\bullet \;\longrightarrow\; \underset{\underset{CH_3}{|}}{-\overset{\overset{O}{\overset{\diagup\ \diagdown}{C}}-\overset{\bullet}{C}}{}-} \;\longrightarrow\; \underset{\underset{CH_3}{|}}{-\overset{\bullet}{C}-\overset{\overset{O}{\|}}{C}-}\quad (P^\bullet);
$$

$$
E + r^\bullet \;\longrightarrow\; \underset{\underset{CH_3}{|}}{-\overset{O^\bullet}{\underset{|}{C}}-CHr-} \;\longrightarrow\; -\overset{\overset{O}{\|}}{C}\text{-}CHr- + CH_3^\bullet;
$$

$$
\longrightarrow\; -CH_2^\bullet + \underset{\underset{CH_3}{|}}{\overset{\overset{O}{\|}}{C}-CHr}\text{—(degradation)};
$$

$$
E + r^\bullet \;\longrightarrow\; \underset{\underset{CH_3}{|}}{-\overset{\overset{O}{\overset{\diagup\ \diagdown}{C}}\ CH}{}-\overset{\bullet}{C}H-} \;\longrightarrow\; \underset{\underset{CH_3}{|}}{-\overset{O^\bullet}{\underset{|}{C}}-CH{=}CH-} \;\longrightarrow\; -\overset{\overset{O}{\|}}{C}\text{-}CH{=}CH- + CH_3^\bullet;
$$

$$
\longrightarrow\; -CH_2^\bullet + \underset{\underset{CH_3}{|}}{O{=}C-CH{=}CH-}\ (\text{degradation})
$$

Owing to the reactions mentioned, epoxy compounds in oxidizing polyolefins should rapidly reach a steady concentration, and then the accumulation of CO groups and the oxidative degradation by polymeric alkoxyls proceed in parallel to the processes of POOH formation and oxygen absorption.

Cyclic radical A$^\bullet$, the precursor of E, possesses nearly equal probability of circle disclosure to the structures

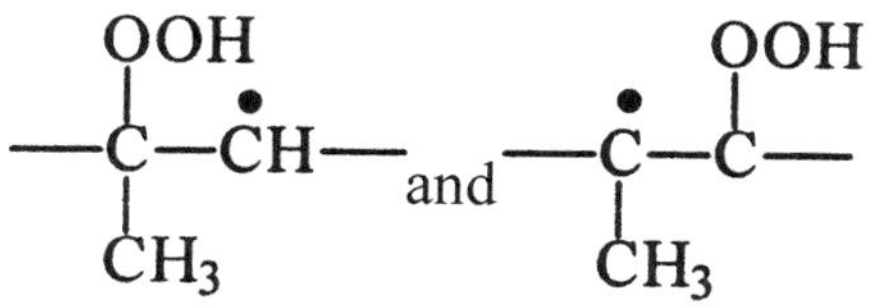

$$-\overset{\displaystyle OOH}{\underset{\displaystyle CH_3}{\overset{|}{\underset{|}{C}}}}-\overset{\displaystyle \bullet}{C}H- \quad \text{and} \quad -\overset{\displaystyle \bullet}{\underset{\displaystyle CH_3}{\overset{}{\underset{|}{C}}}}-\overset{\displaystyle OOH}{\underset{}{\overset{|}{C}}}-$$

In this case, under conditions of a low reaction rate of chain termination and at quite intensive dynamics of sponge chains, a definite rate of chemically induced migration of OOH groups may be provided in the polymer matrix. The totality of the chemical relay race with participation of OOH groups, P$^\bullet$, POO$^\bullet$, and HO$^\bullet$ radicals will provide the appearance of OOH groups in s-zones. Here, a stronger source of initiation will appear in s-zones. However, until that moment the heterophase oxidation is limited by the primary reaction of polymer with oxygen, and the rate of oxygen absorption by the following equation:

$$V_i = k_2[PH][POO^\bullet]_i$$

Here concentration $[POO^\bullet]_i$ appears in the view (52):

$$[POO^\bullet]_i = \left(\frac{\delta_1 k_m k_0 [PH_s]}{2k_3 k_{1s}}\right)^{0.5} \frac{[O_{2,s}]}{\{(k_{-m}/k_{12}) + [O_{2,s}]\}^{0.5}}.$$

By taking into account the expression $[O_{2,s}] = \gamma_s P$ (here γ_s is the constant; P is the oxygen pressure), we obtain the expression for the rate of oxygen absorption:

$$V_i = \frac{a_1 P}{(b + P)^{0.5}} \tag{20}$$

In this case, the induction period shows the following expression:

$$t^{-1} = \frac{V_i}{\Delta[O_2]_i} = \frac{a_2 P}{(b + P)^{0.5}} \tag{21}$$

Parameters a, b, γ_s, and $\Delta[O_2]_i$ are constant in the formulas obtained, if the physical structure of samples does not change significantly during the oxidation. For example, PP films (55) and PE melts (56) reveal the empirical regularities quite well described by the theoretical equation of the heterophase process (52). In particular, the calculated curve 1; in Fig. 6, correlates well with the experimental points (55) (90°C) at used values of coefficients $a_2 = 0.024$ MPa$^{-0.5}$ h^{-1}; $b = 0.1$ MPa [see Eq. (21)]. The efficiency of the theoretical Eq. (20) is displayed in the excellent correlation between calculated curves and the data (56) (see Fig. 7, curves 1–3). In this case, the values of V_i were determined (56) by compositing tangents to the moment $t = 0$ on parabolic sections of curves of oxygen absorption; in other words, when

$V_i = V_{0,2}$. Theoretical curves 1–3 in Fig. 7 were calculated by the formulas:

$$\frac{1 \times 10^{-4}P}{(10 + P)^{0.5}}, \qquad \frac{6.3 \times 10^{-4}P}{(50 + P)^{0.5}}, \qquad \text{and} \qquad \frac{14 \times 10^{-4}P}{(200 + P)^{0.5}}$$

for temperatures 141.7, 157.3, and 165°C, respectively. Here the value of P is given in torr units.

The induction period ends when the samples (possessing stationary micellar structure) absorb a definite amount of oxygen $\Delta[O_2]_i$, and the tranport of OOH group into s-zones begins

$$POOH_v \xrightarrow{k_{vs}} POOH_s$$

under the control of the total penetrability k_{vs}. Hydroperoxide becomes the main source of free radicals resulting the interzone transport:

$$POOH_s \xrightarrow{k_{0s}} \delta P_s^{\bullet}$$

and degenerated heterogeneous–heterophase branching of oxidative chains occurs. Such branching is significantly modified by the occurrence of the heterogeneous reaction of OOH groups, producing no free radicals:

$$POOH_s + POOH_v \xrightarrow{k^{\sharp}} \text{molecular products.}$$

The superposition of the reactions reflect the situation that the s-zone, being the structural border of spongous micelle, stops the spreading of OOH groups in the oxidizing spongy matrix. In this case, probability coefficients k_{vs} and $k^{\sharp}$ of the microheterogeneous matrix possess no sense of elementary rate constants, because they serve in the interzone transition, which probability depends on the structural–dynamic state of the polymer.

The reaction occurring at the meeting of two OOH groups in the s-zones gives no free radicals. This reaction could be presented as follows:

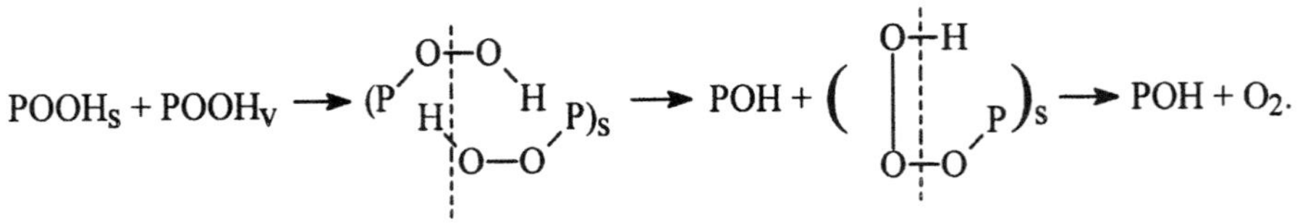

This superposition of heterogeneous actions enables us to write down the following equation for the rate of accumulation of OOH$_s$ groups:

$$\frac{dg_s}{dt} = k_{vs}g_v - k_{0s}g_s - 2k^{\sharp}g_vg_s$$

Here g_s and g_v are concentrations of OOH groups in the s- and v-zones, respectively. This equation easily gives a quasi-stationary concentration of OOH$_s$ groups:

$$g_s = \frac{k_{vs}g_v}{(k_{0s} + 2k^{\sharp}g_v)}$$

and the equation of the radical initiation rate:

$$W_i = \delta\, k_{0s}\, g_s = \frac{\delta\, k_{0s}\, k_{vs}\, g_v}{(k_{0s} + 2k^\sharp\, g_v)} \tag{22}$$

According to Eq. (22), the kinetic step of OOH group inclusion into the free radical initiation is provided by the rate of transfer into s-zones under the initial condition $k_{0s} > 2k^\sharp g_v$, when the equation $W_i = \delta k_{vs} g_v$ is fulfilled. This phase may be significantly prolonged if a micellar structure is rather highly rigid. This may be characteristic for PP samples treated by durable annealing in vacuum to form large spherulites (see Fig. 9). Probably, the reconstruction of crystallites, proceeding under the annealing, promoted polymer chain stretching inside spongy micelles and resulted in obstruction to an OOH group entering the s-zones.

Again, in accordance with the previously considered phenomenology, the drawing of highly crystalline polyolefins significantly increases the induction period. One can say the crystalline reinforcement, formed by drawing, to be a strong mechanical obstacle for the molecular–segmental motion and structural relaxation of fibrillized polymer chains to the initial state of spongy micelles.

Under the "normal" molecular–segmental dynamics of oxidizing samples, the transformation of an OOH group entering s-zones, characterized by Eq. (22), has to begin from the condition

$$W_i = \delta\, k_{0s}\, g_s >> \delta_1 k_0 [PH_s]\, [O_2, s]$$

providing the stage of the parabolic law. Then after, according to Eq. (22) at $k_{0s} < 2k^\sharp g_v$, there appears a new situation; namely, the rate of the radical initiation will reach its kinetically equilibrium value

$$W_i = \frac{\delta\, k_{0s}\, k_{vs}}{2k^\sharp}$$

independently of hydroperoxide concentration. After this moment, the auto-oxidation enters the third stage possessing the rate of oxygen absorption (V_3), dependent on the oxygen pressure as follows (52):

$$V_3 = k_2[PH]\,[PO_2^\bullet]_3 = k_2[PH]\left\{\frac{k_m k_{vs} k_{0s}[O_{2,s}]}{2k_3 k_1 k^\sharp\big([k_{-m}/k_{tr}] + [O_{2,s}]\big)}\right\}^{0.5} = B\left\{\frac{p}{(b+P)}\right\}^{0.5} \tag{23}$$

Here $b = \frac{k_{-m}}{\gamma_s k_{tr}}$, B is the constant.

Equation (23) is algebraically connected to Eqs. (20) and (21). This connection could be easily confirmed by taking the same coefficients b, used in calculations of curves for the graphs in Fig. 6 (curve 1) and Fig. 7 (curves 1–3).

Curve 2 in Fig. 6 was calculated for PP films, oxidizing at 90°C, by the following equation

$$\rho_m = 0.085\left[\frac{P}{b+P}\right]^{0.5}$$

Here $b = 0.1$ MPa; ρ_m is the maximum rate of accumulation of CO groups (55).

Curves 4–6 in Fig. 7 characterize the maximum rate of oxygen absorption by PE melt at 141.7°, 157.3°, and 165°C, and are calculated, respectively, by the following equations

$$V_3 = 1.5 \times 10^{-2} \left[\frac{P}{(19 + P)} \right]^{0.5}; \qquad 8 \times 10^{-2} \left[\frac{P}{(50 + P)} \right]^{0.5};$$

$$16 \times 10^{-2} \left[\frac{P}{(200 + P)} \right]^{0.5}$$

As seen from Figs. 6 and 7, Eq. (23) provides coordination with Eqs. (20) and (21) under the important condition that the dependence of V_3 on P becomes insufficient in the range of oxygen pressure, where t^{-1} value changes significantly.

It should be mentioned once more that coefficients a and b in Eqs. (20), (21), and (23), accepted as constants in the calculations, are not constants if a structural evolution occurs in oxidizing samples. This evolution, even in the absence of a special thermomechanical excitation, is initiated by the oxidative degradation (or cross-linking) of polymer chains and formation of hydrophilic oxygen-containing groups in a hydrophobic matrix. This changes the distribution of elastic forces inside spongy micelles and effects the fluctuational dynamics of chain units.

The inadequacy of the structural–evolutionary processes is the reason for the frequently observed scattering of t_i values and is noncharacteristic for a stable structure. Thus, curves shown in Fig. 8 of OOH group accumulation in the films prepared by pressing of nonmelted powder (58,59), demonstrate the absence of the induction period and much more prolonged phases of hydroperoxide accumulation compared with higher structurally stable samples (see Fig. 9). If we take into account the initial thermomechanical inbalance in films (58,59), it might be suggested that their oxidation was accompanied by structural evolution into an initial nondeformed state, while passing the period of increased fluctuational dynamics. This may be why the frequency of the transferal actions from s-zones to v-zones increases, as well as the fluctuational occurrence of supermicropores in v-zones. Both these processes should increase the rate of hydroperoxide accumulation in v-zones in obscuring the evidence of transition from the first stage of the process into the second and affecting the ratio of POOH and CO group yield. A similar relaxation process is also probable in the thixotropic PP samples excited by an intensive heat pulse (52), and in oriented LDPE films, different from nonoriented ones by the increased oxidation rate (57). Apparently, the low crystallinity of LDPE did not form a resistant crystal reinforcement, contrary to the films from PP, HDPE, and PMP.

In addition, it should be taken into account that the process of heat dissipation when polymer films are cooled, obeys the relaxation mechanism of structural–physical transformations. In this case, at an invariable method of formation of samples, differing only by their thickness, the samples developed more or less layer heterogeneity (69). This heterogeneity affects the microheterogeneous structure of chain spongy–micelles, providing the possibility of one or another influence of the film thickness on the kinetics of brutto process of autooxidation.

Returning to oxidation of samples with normal structure of spongy micelles that do not obey an intensive structural evolution, it should be remembered that the third stage of this chemical process is specified not only by the zero-order kinetics of the hydroperoxide, but also by the zero-order kinetics of the initiator added (39). The mechanism of the phenomenon reflects that the initiator added dissociates in the same s-zones, which are filled by associated OOH_s groups during oxidation. Also, the reactivity of free low molecular weight radicals with hydrogen atoms of a hydroperoxide group is much higher than with CH bonds (41). In this connection, it may be suggested that associated OOH_s groups, treated by the radical-less reaction, serve simultaneously as a specific acceptor of radicals of an additive initiator (39):

$$Y_v \leftrightarrow Y_s; \qquad V_s \rightarrow \{2r_s^{\bullet}\} \quad \text{(radical pair)};$$

$$\{2r_s^{\bullet}\} \rightarrow 2r_s^{\bullet};$$

$$\{2r_s^{\bullet}\} + (POOH \dots HOOP)_s \rightarrow 2rH + POOP + O_2$$

$$\text{or}$$

$$r_s^{\bullet} + (POOH \dots HOOP)_s \rightarrow rH + (POO^{\bullet} \dots HOOP)_s$$

$$r_s^{\bullet} + (POO^{\bullet} \dots HOOP)_s \rightarrow rH + POOP + O_2$$

Polymeric dialkyl peroxide POOP, formed in this manner, is not the initiator (37). That is why, the accumulated hydroperoxide becomes the only initiator of autooxidation in the third stage of the process, which is specified by an "anomalous" zero-order kinetics by OOH groups and additive initiator.

VII. THE REACTIONS OF IONOL DURING THE INDUCTION PERIOD OF INITIATED OXIDATION OF POLYOLEFINS

The material discussed in previous sections allows us to consider one of the fundamental questions of polyolefin stabilization from a new point of view. This question concerns the chemistry of the induction period of inhibited oxidation. This field of science contains much experimental data and theoretical methods, but, as a rule, these methods are in the range of homogeneous reaction models (36–38).

Two main methodological approaches were developed in this field. One of them is based on application of initiators, tested in liquid-phase reactions (37,38). Under conditions of the liquid-phase oxidation, this method allows us to determine the character and features that the influence of the variety of inhibitors have on the rate of oxygen absorption and duration of the induction period, during which inhibitor action is performed and ends.

Another approach studies the process of inhibited polymer oxidation in the absence of a specially introduced additive initiator (36).

The present section of this chapter considers oxidation with an additive initiator, with special attention to regularities of oxygen absorption during the induction period. That oxygen absorption is always observed during the induction period of inhibited polyolefin oxidation has been known for some time, but it has not yet yielded a suitable explanation. Sometimes, it is connected with a low rate of the oxygen absorption in this stage as compared with the stage of direct polymer oxidation (after the end of the induction period). In the task of determining the effectiveness of inhibitors or rate of oxidative chain initiation, the investigators used experimental conditions in which the rate of the inhibitor expenditure did not depend on its concentration. It was supposed that provision of such conditions allows one to neglect the low rate of oxygen absorption during the induction period of the polymer oxidation process (37,38,42). We believe that by using the model of the polymer chain–sponge system, significant improvements in the ideas on the mechanism of oxidative transformations with the participation of inhibitors could be achieved.

In constructing a kinetic model of the oxidation induction period with an initiator, we will use experimental data from the study by Chien and Wang (42), in which polypropylene, poly-(1-butylene), and polymethylpentene were oxidized (oxygen pressure, 1 atm) using dibenzoyl peroxide as the initiator and ionol as the inhibitor. Although the authors (42) have not considered the origin of significant oxygen absorption during the induction period, they have shown the data and detailed characteristics of absorption rates (V) of oxygen accompanied by ionol expenditure rates (V_i) during the induction periods.

The process of oxidation of the partially crystalline polymers mentioned proceeded only in their amorphous regions, and its extent was limited by the initial stage only, which displayed from 0.6 to 2% of BP conversion.

The values $V_i = c_{i0}/t_{ind}$ (here c_{i0} is the initial ionol concentration; t_{ind} is the induction period), measured in a T interval of 344÷378 K, are invariable at the BP concentrations used, as well as values V, which are invariable until abrupt acceleration of the oxidation process, which is connected with the using up of ionol.

The authors clearly proved the absence of the chain BP reaction under oxidative conditions and obtained interesting results on the temperature dependence of the coefficient

$$A = \frac{c_{i0}}{c_0[1 - \exp{(-k_d t_{ind})}]} \qquad (24)$$

where k_d is the rate constant of spontaneous BP degradation. They then used a liquid-phase interpretation of this coefficient and attributed it to the characteristic of the inhibitor effect on the radical yield in a spontaneous BP reaction.

It is easy to explain the coefficient A in another way, comparing ionol and oxygen expenditure rates during the induction period. Results of such comparison are shown in Fig. 11. Here straight lines 1, 3, and 5, constructed in the Arrhenius equation coordinates, characterize the change of oxygen absorption rate in the temperature range studied, and straight lines 2, 4, and 6 characterize that of ionol. The linearity of these plots, which occurred in c_{i0} range of $3.4 \times 10^{-3} \div 72.6 \times 10^{-3}$ mol/L, is the result of independence of these rates on the ionol concentration. The existence of similar slopes of straight-lines for each particular polymer, which characterize oxygen absorption and ionol expenditure, indicates the existence of similar effective activation energies of both macroscopic processes $E_i = E_{O_2}$ (172 kJ/mol for PP, curves 1 and 2, $c_0 = 0.127$ mol/L; 231.0 kJ/mol for poly(1-butylene), curves 3 and 4, $c_0 = 0.376$ mol/L; 231.0 kJ/mol for polymethylpentene, curves 5 and 6, $c_0 = 0.33$ mol/L). When in all temperature ranges studied the number of absorbed oxygen molecules per one expended ionol molecule is constant, it was found that $V/V_i = 3.65$ for PP, 8.0 for poly(1-butylene) and 13.5 for polymethylpentene.

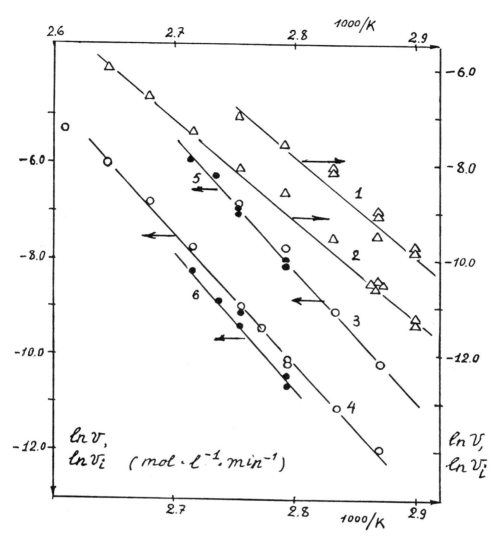

Figure 11 Arrhenius plots for O_2 absorption (1, 3, 5) and ionol disappearance (2, 4, 6) in the induction period of BP-initiated oxidation of polyolefins: (1,2) polypropylene; (3, 4) poly(1-butene); (5, 6) poly(4-methyl-1-pentene).

It is clearly seen that the obtained stoichiometric ratios V/V_i depend on the chemical structure of polymers and reach such a large value that they could not be attributed to oxidation of additive ionol. As a consequence, the process that proceeds during the induction period should be interpreted as initiated chain co-oxidation of alkylphenol and polymer, for which Eq. (24) characterizes the change in the amount of ionol expended in chains of its co-oxidation with the polymer, calculated for a single spontaneously decomposed BP molecule, instead of the influence of ionol and temperature on the radical yield from a "cell" of spontaneously dissociating BP.

In accordance with the data (42), the rate constant of spontaneous BP decomposition is $k_d = 8.4 \times 10^{13} \exp(-124650/RT)$ s^{-1} in PP and poly-(1-butylene), and $7 \times 10^{13} \exp(-123000/RT)$ s^{-1} in polymethylpentene. Such polymers possess activation energy of BP dissociation ($E_d = 123 \div 125$ kJ/mol) significantly lower than that of their co-oxidation with ionol ($E_i = 172 \div 231$ kJ/mol). That is why, the thermal dependence of A (42) is natural for very small depth of BP degradation during time t_{ind} (0.6÷2%):

$$A = \frac{c_{i0}}{c_0[1 - \exp(-k_d t_{ind})]} = \frac{c_{i0}}{c_0 k_d t_{ind}} = \frac{V_i}{V_{d0}},$$

where $V_i = c_{i0}/t_{ind}$ and $V_{d0} = k_d c_0$.

In the structural–kinetic co-oxidation model itself, we take into account the aforementioned material and that there is no hydroperoxide in the systems considered capable of dissociating in v-zones of the chain–sponge matrix according to the mechanism of fluctuational excitation of v-micropores. Also, it should be taken into account that BP molecules dissociate into radicals in s-zones:

$$\text{BP}_v \leftrightarrow \text{BP}_s; \qquad \text{BP}_s \xrightarrow{k_{0s}} 2\delta r_s^{\bullet}; \qquad r_s^{\bullet} \xrightarrow{k_{1s}} P_s^{\bullet} \rightarrow R_s^{\bullet};$$

$$R_s^{\bullet} \xrightarrow{k_{2s}} R_{1s}^{\bullet}; \qquad R_{1s}^{\bullet} + r_s^{\bullet} \rightarrow \text{Rr}$$

The difference between the present s-scheme and that considered previously lies in the existence of s-benzoyloxyl and $R_{1s}^{\bullet}$ macroradical recombination. The corresponding fact was obtained during the study (13) of material balance of phenyl compounds in the BP reaction with PP.

In accordance with the present s-scheme, the steady-state concentration of the radicals, carrying out the interzone s,v-transfer of free valency, is the following:

$$[R_s^{\bullet}] = \frac{\delta k_0 c_s}{k_{25}}$$

and the s,v-transfer reactions themselves should proceed with low frequency, which induces no influence on $[R_s^{\bullet}]$ (this is provided by the chain–sponge skeletal rigidity):

$$R_s^{\bullet} \underset{k_{-m}}{\overset{k_m}{\longleftrightarrow}} R_m^{\bullet}; \qquad R_m^{\bullet} + O_{2s} \xrightarrow{k_{tr}} \text{ROO}_m^{\bullet};$$

$$\text{ROO}_m^{\bullet} + \text{IH} \rightarrow \text{ROOIH}_m^{\bullet} \rightarrow \text{ROH} + \text{IO}^{\bullet}$$

Quinolic alcoxyls, which occur in v-zones according to this pathway, initiate chain oxidation of macromolecules to which ionol is drawn in:

$$\text{IO}^{\bullet} + \text{PH} \xrightarrow{k_5} \text{IOH} + \text{P}^{\bullet}; \qquad \text{P}^{\bullet} + O_2 \rightarrow \text{POO}^{\bullet};$$

$$\text{POO}^{\bullet} + \text{IH} \xrightarrow{k_1} \text{POOIH}^{\bullet} \rightarrow \text{POH} + \text{IO}^{\bullet};$$

$$\text{POO}^{\bullet} + \text{IH} \xrightarrow{k_3} G + I^{\bullet}; \qquad \text{POO}^{\bullet} + I^{\bullet} \xrightarrow{k_4} \text{POOI}.$$

Here, the enumeration of the reaction constants k_1, k_2, k_3, and k_4, assigned to them in the previous section, remains, and preservation of ionol regeneration and transformation reactions is also considered. These reactions are characterized by constants k_{1f}, k_{-1f}, k_{2f}, and k_{3f} of cryptoradical stages of IOH alcohol transformation.

In accordance with the present heterophase scheme, the equation of the reaction v-chain initiation and termination rates is:

$$V_{tr} = \frac{k_{tr}k_m\delta k_{0s}c_s[O_{2s}]}{k_{2s}(k_{-m} + k_{tr}[O_{2s}])} = 2k_3[POO^\bullet][IH] \qquad (25)$$

If we suppose that these chains are significantly long, such that the following approximated equation could be used:

$$k_5[IO^\bullet][PH] \approx k_1[POO^\bullet][IH]$$

we obtain condition of stationary quinolic alcohol concentration:

$$k_1[POO^\bullet][IH] = k_{1f}[IOH][PH]\frac{(k_{2f} + k_{3f})}{(k_{-1f} + k_{2f} + k_{3f})}$$

By expressing [IOH][PH] from this composition and substituting it into the equation of ionol decay rate

$$-\frac{dc_i}{dt} = k_1[POO^\bullet][IH] + k_3[POO^\bullet][IH] - k_{2f}[\{I^\bullet + HO^\bullet + PH\}]$$
$$= (k_1 + k_3)[POO^\bullet][IH] - \frac{k_{2f}k_{1f}[IOH][PH]}{(k_{-1f} + k_{2f} + k_{3f})}$$
$$= (k_1 + k_3)[POO^\bullet][IH] - (1 - \beta)k_1[POO^\bullet][IH]$$

and then neglecting ionol expenditure in the chain termination reaction (with the constant k_3), we obtain [POO$^\bullet$][IH] for the oxygen-saturated system ($k_{-m} \ll k_{tr}[O_{2s}]$) from equation (25). Taking into account this expression, we obtain

$$V_i = -\frac{dc_i}{dt} = \left[\frac{k_{3f}}{k_{2f} + k_{3f}}\right]\left[\frac{k_1k_m\delta k_{0s}c_s}{2k_3k_{2s}}\right]$$

And finally, by taking into account the interzone BP equilibrium [see Eq. (8)], the following reduced expression is obtained:

$$V_i = \beta k_i c$$

where k_i is a complex constant; $\beta = k_{3f}/k_{2f} + k_{3f}$.

The same heterophase scheme under the foregoing conditions displays the following expression of the oxygen absorption rate:

$$V = -\frac{d[O_2]}{dt} = k_5[IO^\bullet][PH] = k_1[POO^\bullet][IH] = \frac{k_1k_m\delta k_{0s}c_s}{2k_3k_{2s}} = k_i c$$

It is clearly seen that the theory of heterophase transformation, in which a polymer obtains significant amounts of oxygen and alkylphenol simultaneously with an initiator, leads to expressions for inhibitor (V_i) and oxygen (V) expenditure rates, independent of their concentrations and proportional to each other:

$$\frac{V}{V_i} = \beta^{-1} = \frac{k_{2f} + k_{3f}}{k_{3f}}$$

The result obtained explains the reason for significant V excess over V_i with the alkylphenol regeneration process. The increase of V/V_i ratio (from 3.65 to 13.5), observed in the sequence PP, poly(1-butylene), polymethylpentene, means growth of the regeneration effectiveness (β^{-1}) and, probably, reflects the increase of the number of reactive hydrogen atoms on surfaces of thermofluctuationally excited v-micropores. The higher the number of hydrogen atoms surrounding $HO^{\bullet}$ and $I^{\bullet}$ radicals, the lower is the probability of mutual disproportioning of the radicals and the higher is the probability of macromolecule dehydrogenation by "cryptoradical" mechanism.

We should especially mention in this section that alkylphenol regeneration must be performed particularly by the "intracage" cryptoradical reaction during the lifetime of the micropore, increased up to the size required. Here, radical escape from such a micropore to the volume of the chain–sponge v-zones may possess a negligibly low rate, which does not affect steady-state v-radical concentrations. Also, the dependence of parameter $A = c_{i0}/c_0[1 - \exp(-k_d t_{ind})]$ on temperature (42), does not relate to the process of diffusional ionol intrusion into a "cage" of spontaneously degradated initiator (which was suggested in Ref. 42). In accordance with ionol and polyolefin chain co-oxidation, this dependence characterizes the stoichiometry of chain co-oxidation. The considered reaction model disclosing the structure of coefficient A is presented by

$$A = \frac{V_i}{V_{d0}} = \frac{\delta \beta k_1 k_m}{2 k_3 k_{2s}}$$

The coefficient β participating in this formula either does not depend, or only weakly so, on temperature. This is indicated by similar temperature dependences of V and V_i for each polymer, studied (42) ($E_{O_2} = E_i$). The main contribution to the temperature dependence of A should be made by the reaction of terminal macroradical $R_s^{\bullet}$ transfer from the skeleton wall structures into the volume of s-micropores (reactions with the constant k_m). The size of the correspondent fragment $R_s^{\bullet} \rightarrow R_m^{\bullet}$ increases in the sequence of PP, poly(1-butylene), polymethylpentene, so it could be expected that the obstacles connected with this motion should become harder. If we take A values (42) and construct the dependence of $\ln A$ on reciprocal temperature (Fig. 12), we will observe the activation energy increase, E_A, from 50 kJ/mol for PP up to 128 kJ/mol for poly(1-butylene), and higher increase for polymethylpentene, as it is in accord with the foregoing discussion.

It is characteristic that the theoretical basis (42) did not allow the authors to construct plots, shown in Fig. 12. In particular, it follows from these plots that the A value becomes 1 at 388 K, and it exceeds 2 at 405 K (see Fig. 12, axis of ordinates: $\ln 2 = 0.693$). It is evident that no more than two free radicals may occur in BP molecule dissociation, so the true realization of the correlation $A > 2$ in the process considered makes senseless the supposition (42) that coefficient A characterizes the effectiveness of the radical yield in oxidative chain initiation.

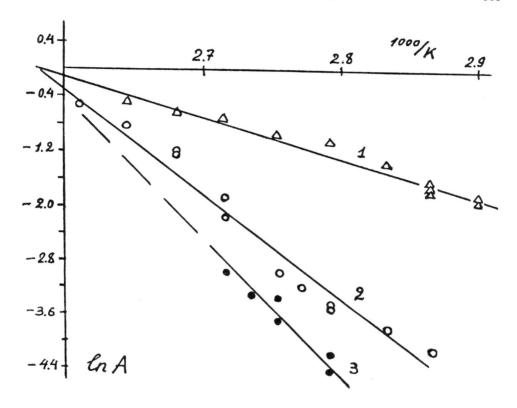

Figure 12 Arrhenius plots for the coefficient A: (1) PP; (2) poly(1-butene); (3) poly(4-methyl-1-pentene).

In the application to polymers, the interpretation of the coefficient f, determined according to methods of homogeneous reactions, should change simultaneously with interpretation of the coefficient A. In the inhibitor method, applied to homogeneous reactions, the coefficient f is calculated by the following formula:

$$f = \frac{W_{i0}}{V_i} = \frac{2\delta V_{d0}}{V_i}$$

Here W_{i0} is the rate of homogeneous reaction of the radical initiation; δ is the coefficient of radical yield in the initiator dissociation.

In this method, the value of f characterizes the number of oxidative chains terminated on a single inhibitor molecule. At the same time, f depends on the reciprocal value of A according to its definition:

$$f \sim \frac{V_{d0}}{V_i} = A^{-1}$$

The moment of inhibitor exhaustion in a polymer, fixed by sharp increase of the oxygen absorption rate, marks the end of co-oxidation period and process transfer to the stage of polymer oxidation, discussed in Section VI.

Application of the homogeneous reaction model to the induction period of polymer oxidation produces specific contradictions. They are, for example, conclusions from the study (70), in which $f = W_{i0}/V_i = 1$ for PP with ionol and dicumylperoxide as initiator ($P_{O_2} = 1$ atm, $T = 388$ K, where W_{i0} is the constant rate of radical initiation, obtained by the inhibitor method; V_i is the ionol expenditure rate). In making sure that, at a constant W_{i0} rate, the inhibitor expenditure rate is also constant and that on this base all chains terminate on the inhibitor, and no side reactions of its expenditure occur, the authors (70) stated that the equation $f = 1$ relates to termination of a single oxidative chain on a single ionol molecule [in accordance with the data (42) for the coefficient $A = 1$ at $T = 388$ K, too].

The authors pointed out (70) that simultaneously there was a discrepancy observed with the liquid-phase value $f = 2$. They also showed the sequence of their reasonings to prove the bimolecular nature of ionol phenoxyl decay for the PP matrix:

$$I^{\bullet} + I^{\bullet} \rightarrow \text{products.}$$

However, according to the material considered in Section V, such a reaction is inconsistent with the true kinetics of phenoxyls.

VIII.   THE CHAIN CO-OXIDATION OF ALKYLPHENOL WITH POLYOLEFIN IN THE ABSENCE OF INITIATOR

A.   The Polymer–Alkylphenol System

In an analysis of the co-oxidation process that proceeds in the absence of initiator, we will use the v-scheme considered in the previous section, taking into account that the rate of radical initiation in v-zones is negligibly small. Under these circumstances s-zones, which are the heterophase sources of radicals, may provide initiation by either direct oxidation of polymer s-chains or oxidation of inhibitor molecules, if the latter possess enough reactivity.

In considering the first stage, we refer to the s-scheme, which allowed us to make a kinetic description of the initial PP autoxidation stage in the absence of initiator and inhibitor (39,52). In accordance with this scheme, the initiation of radicals is performed in supermicropores of chain–sponge s-zones as a result of interactions of polymer chain CH bonds with oxygen adsorbed on s-micropore walls:

$$PH_s + O_{2s} \xrightarrow{k_{0s}^p} \delta P_s^{\bullet}$$

The following transformations of s-macroradicals remain the same as those considered in Sects. II and V:

$$P_s^{\bullet} \xrightarrow{k_{1s}} R_s^{\bullet}; \qquad R_s^{\bullet} \xrightarrow{k_{2s}} R_{1s}^{\bullet}; \qquad R_s^{\bullet} + R_{1s}^{\bullet} \xrightarrow{k_{3s}} \text{products}$$

The expression for radical–mediator concentration, issued from this scheme, is the following:

$$[R_s^{\bullet}] = \frac{\delta k_{0s}^p [PH_s][O_{2s}]}{2k_{2s}},$$

and the process of free valency s,v-translation, in which they participate, becomes:

$$R_s \underset{k_{-m}}{\overset{k_m}{\longleftrightarrow}} R_m^{\bullet}; \qquad R_m^{\bullet} + O_2 \overset{k_{tr}}{\longrightarrow} ROO_m^{\bullet};$$

$$ROO_m^{\bullet} + IH \rightarrow ROOIH^{\bullet} \rightarrow ROH + IO^{\bullet}$$

The scheme of reactional v-chains of alkylphenol and polymer co-oxidation remains with no change, as well as the scheme of cryptoradical reactions of quinolic alcohol IOH transformation, providing partial regeneration of the initial phenol:

$$IO^{\bullet} + PH \overset{k_5}{\longrightarrow} IOH + P^{\bullet}; \qquad P^{\bullet} + O_2 \rightarrow POO^{\bullet};$$

$$POO^{\bullet} + IH \overset{k_1}{\longrightarrow} POOIH^{\bullet} \rightarrow POH + IO^{\bullet};$$

$$POO^{\bullet} + IH \overset{k_3}{\longrightarrow} G + I^{\bullet}; \qquad POO^{\bullet} + I^{\bullet} \overset{k_4}{\longrightarrow} POOI.$$

The present complex of s- and v-schemes fixes nonequivalence of properties of the intrazone macroradicals in relation to oxidation and interaction with inhibitor: a low rate in supermicropores, and a high rate in v-zones. As already mentioned, such selectivity of the reactions is caused by predominant accumulation of low molecular weight compounds (including oxygen and inhibitor) in spongy micelle granules, filled by narrow micropores. An example of such nonequivalence was considered in the description of ionol transformations in s-steps of initiated macromolecule degradation (see Secs. IV and V) and in v-chains of CTA and PC arylation (11,18). We also may note that there are a number of similar data reported in the literature. For example, it is significant that the number of broken polymer chains, observed during 26 h of melt polystyrene oxidation at 473 K ($P = 150$ torr) in the presence of bisphenol 2246 inhibitor and at its high concentration, is practically invariable when the induction period is not yet finished and when oxidation proceeds from the induction period at its low concentrations $c_{ii} \rightarrow 0$ (71). The absence of the inhibitor concentration's influence on polymer degradation rate during the induction period of oxygen absorption also exist in rubber with ionol (72) and in PP with bisphenol 2246: $T = 437$ K, $P_{O_2} = 300$ torr) (44).

In this way the foregoing constructed heterophase scheme of noninitiated co-oxidation of alkylphenol with a polymer fits the general phenomenology of the process during the induction period and allows calculation of the foregoing features of co-oxidation in the heterophase matrix of chain–sponge micelles.

We can easily find the expression of the free valence s,v-transfer rate by applying the condition of the steady-state radical concentration to each block of the total system:

$$V_{tr} = k_{tr}[R_m^{\bullet}][O_{2s}] = \frac{k_{tr}k_m\delta k_{0s}^p[PH_s][O_{2s}]^2}{k_{-m} + k_{tr}[O_{2s}]}$$

which transforms in the oxygen-saturated polymer ($k_{-m} \ll k_{tr}[O_{2s}]$) to

$$V_{tr} = \frac{\delta k_m k_{0s}^p [PH_s][O_{2s}]}{2k_{2s}} \tag{26}$$

This equals the termination rate, which is also the rate of the reaction v-chain initiation

$$V_{tr} = 2k_3[POO^{\bullet}][IH]$$

Under the conditions described, the alkylphenol expenditure rate does not depend on its concentration, but does depend on oxygen pressure:

$$V_i = -\frac{dc_i}{dt} = (k_1 + k_3)[POO^{\bullet}][IH] - (1-\beta)k_1[POO^{\bullet}][IH] \approx \beta k_1[POO^{\bullet}][IH]$$

$$= \frac{\beta k_1 V_{tr}}{2k_3} = \frac{\beta k_1 \delta k_{0s}^p k_m [PH_s][O_{2s}]}{4k_{2s}k_3} = \beta k_i^p P_{O_2} \tag{27}$$

Here $[O_{2s}] = K_{O_2}[O_{2v}] = K_{O_2}P_{O_2}\gamma$; K_{O_2} is the equilibrium constant for oxygen exchange between zones; γ is the Henry constant; $\beta = k_{3f}/k_{2f} + k_{3f}$ is the efficiency of quinolic alcohol expenditure in acts of phenol regeneration (see Secs. V and VII); k_i^p is a complex constant.

In accordance with the scheme, the rate of oxygen absorption

$$V = -\frac{d[O_2]}{dt} \approx k_5[IO^{\bullet}][PH] = k_1[POO^{\bullet}][IH] = k_i^p P_{O_2}$$

exceeds the rate of inhibitor expenditure and does not depend on its concentration c_i.

These particular regularities have been obtained in early stages of investigations of antioxidative stabilization mechanism in experiments with phenols marked as weak inhibitors. Such inhibitors are characterized by more or less strict fulfillment of zero-order kinetic equation by inhibitor, order one by oxygen, and linear dependence of the induction period duration on inhibitor concentration. The list of such inhibitors includes the following substances: 2,6-di(1,1-dimethylhexyl)-4-methylphenol (73,74) (t_{ind} grows linearly from 9.5 min at $c_i = 0$ up to 373 min at $c_i = 0.20$ mol/kg at $T = 473$ K and $P_2 = 300$ torr), 2,4,6-tri-*tert*-butylphenol (75) (the effective activation energy of phenol expenditure $E_i = 36$ kcal/mol in the temperature range from 453 to 473 K). 2,6-di-*tert*-butyl-4-phenylphenol (75) ($E_i = 30$ kcal/mol), bisphenol with spatially separated OH groups—4,4'-methylene-bis(2,6-di-*tert*-butylphenol) (75) (t_{ind} grows linearly and significantly from 9.5 min at $c_i = 0$ up to 250 min at $c_i = 0.025$ mol/kg).

These phenols possess the feature that their expenditure is not accompanied by water formation, which occurrence becomes noticeable at the very end of the induction period. This important fact indicates negligibly low rates of decomposition of peroxyl radicals (with $HO^{\bullet}$ layout) and polymer hydroperoxide. The conclusion about the absence of HP contribution to the process of radical initiation at the stage of polymer–phenol system co-oxidation also follows from the fact that the oxygen and inhibitor expenditure rates are invariable from the beginning until the end of the induction period (76).

In accordance with the foregoing, the reactions of POO• radical and HP group decomposition are absent in the accepted heterophase scheme. That the process is significantly decelerated in the presence of dialkylsulfides may serve as an additional and specific proof of the chain mechanism of alkylphenol co-oxidation with a polyolefin.

B. Co-oxidation Mechanism of the Process Inhibited by Dialkylsulfide

Features of antioxidant transformations in the polymer chain zones of sponge micelles, set forth in the present chapter, make us again turn back to the question about the mechanism of cooperative functioning of phenolic antioxidants and dialkylsulfides. It is common knowledge that the latter possess no significant inhibiting effectiveness, but are capable of a substantial increase in the induction period of polymer co-oxidation with a "weak" phenol, changing the kinetic law of phenol expenditure (75,77).

The ideas about the mechanism of nonadditive combinations of phenol and dialkylsulfide reactions are developed in the literature on the basis of nonradical hydroperoxide decomposition by dialkylsulfide. It is accepted that dialkylsulfide degrades hydroperoxide without a free radial formation and abruptly decreases the rate of degenerated branching of oxidative chains. However, the role of hydroperoxide as branching agent in the co-oxidation stage is not practical because of its low concentration. But even in the samples possessing a high content of pre-liminarily accumulated HP, the branching factor in the presence of alkylphenol is zero (see Secs. IV and V), and in this case, the transformation is reduced to only the oxidative chain HP decomposition.

This means that the mechanism of direct dialkylsulfide and HP interaction, which has not yet been completely described because of its complexity (36), is transformed into an independent problem. In this case, it is rather important that dialkysulfide, which is not the inhibitor of the polymer oxidation itself, inhibits co-oxidation of alkylphenol. Dialkylsulfide extends alkylphenol action by sacrificing itself. This is confirmed, for example, by 2,6-di-(1,1-dimethylhexyl)-4-mehtylphenol and bisphenol 2246 expenditure kinetic curves, obtained at the oxidation temperature $T = 473$ K ($P_{O_2} = 300$ torr) (36,77), shown in Fig. 13. As shown, the expenditure rate of phenols with the mixtures of monophenol/sulfide $=$ 0.16 mol/kg /0.08 mol/kg (curve 2) and bisphenol/sulfide $= 0.08/0.08$ (curve 4) decrease in relation to the rate without sulfide (curve 1 and 3) from the very beginning, when the radical yield in HP dissociation reactions, under the effect of phenol and, also, sulfide, cannot be changed. Such, an influence is absent even at high HP concentration (see Secs. IV and V).

Furthermore, the results from the study (44), according to which the rate of propylene yield in the induction period of the melt PP oxidation ($T = 437$ K, $P_{O_2} = 300$ torr) is constant during 100 min and is similar in the presence of 0.025 and 0.05 mol/kg of bisphenol 2246, and does not change after injection of 0.08 mol/kg dodecylsulfide or dilaurylthiodipropionate into initial mixture. Consequently, the concentration of depolymerizing terminal macroradicals remains constant under these conditions during 100 min, and the process of degenerated branching with HP and phenol may be neglected.

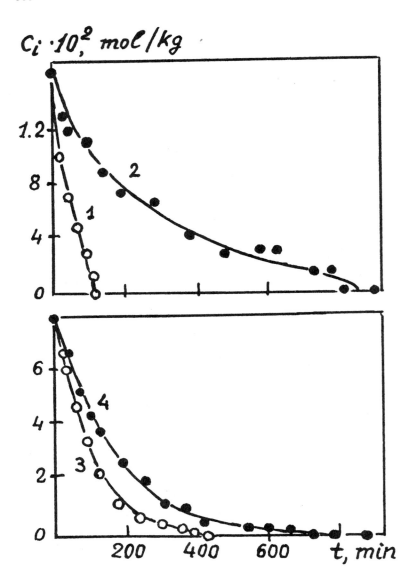

Figure 13 Disappearance of 4-methyl-2,6-di(1,1-di-methylhexyl)phenol (1, 2) and methylene-2,2'-bis(4-methyl-6-t-Bu-phenol) (3, 4) in the induction period of autoxidation ($T = 473$ K; $P_{O_2} = 300$ mmHg) for PP samples, with no (1, 3) and with (2, 4) 0.08 mol/kg didecylsulfide.

At the same time, gas chromatographic analysis (44) showed that dialkylsulfide injection into oxidized PP causes a significant increase of ethylene yield, which indicates a sulfide interaction with polymer peroxyls. Entering the co-oxidation with PP, dialkylsulfides cause no significant increase in the induction period. However, in the presence of a "weak" phenol, they are able to increase the induction period significantly, displaying synergism. The mechanism of this phenomenon may be shown, if we take into account that dialkylsulfides interact with peroxyls (44). Because of this interaction, the decrease of peroxyl concentration occurs, which induces the chain expenditure of alkylphenol.

Dialkysulfide should provide for peroxyl conversion into polymer alcoxyl to carry on such a process. This polymer alcoxyl preferentially reacts with an inhibitor:

$$POO^\bullet + S\overset{R}{\underset{R}{<}} \xrightarrow{k_6} POOS\overset{R}{\underset{R}{<}}\bullet \rightarrow PO^\bullet + O = S\overset{R}{\underset{R}{<}} \ .$$

If we take into account all of the foregoing, the reaction scheme of the polymer v-chains, phenol, and sulfide co-oxidation will transform into the following view:

$$IO^\bullet + PH \xrightarrow{k_5} IOH + P^\bullet ; \qquad P^\bullet + O_2 \rightarrow POO^\bullet ;$$

$$POO^\bullet + IH \xrightarrow{k_1} POOIH^\bullet \rightarrow POH + IO^\bullet ;$$

$$POO^\bullet + IH \xrightarrow{k_3} G + I^\bullet ; \qquad POO^\bullet + I^\bullet \xrightarrow{k_4} POOI$$

$$POO^\bullet + S \xrightarrow{k_6} POOS^\bullet \rightarrow PO^\bullet + O = S$$

$$PO^\bullet + PH \xrightarrow{k_7} POH + P^\bullet ; \qquad PO^\bullet + IH \xrightarrow{k_8} POH + I^\bullet$$

Here S is dialkylsulfide.

The role of dialkylsulfide in the absence of phenol is that it is spent in the reaction with constant k_6, transforming peroxyl into alcoxyl which, in turn, is transformed into oxidizing alkyl $P^\bullet$ in the reaction with constant k_7, finally reducing the $POO^\bullet$ radical. Such sequence of reactions does not change the peroxyl concentration. That is why dialkylsulfide expenditure causes no effect on the polymer oxidation rate. The scheme shows no sulfide transformation with formation of ehtylene, which proceeds either by the displacement mechanism (44),

$$POO^\bullet + R - S - R \rightarrow POO - S - R + R^\bullet ; \qquad R^\bullet \rightarrow C_2H_4 + R_1^\bullet ;$$
$$POOS - R \rightarrow (PO)RSO$$

or as a result of the alcoxyl reaction with the sulfoxide formed:

$$PO^\bullet + R_2S = O \rightarrow PO(R_2)SO^\bullet \rightarrow PO(R)SO + R^\bullet ; \qquad R^\bullet \rightarrow C_2H_4 + R_1^\bullet$$

These processes seem energetically and sterically less favorable if compared with the reaction with the constant k_6 in the scheme. This reaction is followed by the weak peroxide bond break and formation of a strong thio–oxygen bond. On this basis, the process of ethylene formation may be considered as a side effect. Moreover, if that is taken into account, the kinetic image of co-oxidation with phenol will not change as a result of $R_1^\bullet$ radical oxidation to peroxyls which, in turn, participate in further reactions showing no qualitative difference from polymer peroxyls.

In the foregoing co-oxidation v-scheme, phenol interacts with $PO^\bullet$ radicals, thereby terminating oxidative chains with dialkysulfide. This means dialkylsulfide acts as an inhibitor, exhibiting a role it does not primarily possess.

However, at significantly high phenol concentration the oxidative function will prevail owing to the reaction rate increase with the constants k_1 and k_5, making the co-oxidation process independent of dialkylsulfide.

It is easy to obtain the expression for the inhibitor expenditure rate by applying the steady-state radical concentration condition to the v-scheme and the remaining cryptoradical pathway of phenol regeneration:

$$V_i = -\frac{dc_i}{dt} \approx \beta k_1 [POO^\bullet] c_i = \frac{\beta k_1 c_i \, V_{tr}}{2\left(k_3 c_i + \dfrac{k_6 k_8 c_i S}{k_7 [PH] + k_8 c_i}\right)} .$$

This expression is reduced to the same form as the opposite concentrational ratios, $c_i/S \ll 1$ and $c_i/S \gg 1$, showing no dependence on dialkylsulfide concentration S. In the first case, the inequality $k_7[PH] > k_8 c_i$ takes place, indeed, at significantly low concentrations of c_i. It allows one to neglect the second term in the denominator of the main V_i expression fraction. In the second case, $k_7[PH] < k_8 c_i$ at a significantly high concentration c_i, but the inequality $k_3 c_i > k_6 S$ takes place, simultaneously. Both cases display a similar result:

$$V_i = \frac{\beta k_1 V_{tr}}{2 k_3}$$

and the dialkylsulfide effect on the process rate is absent.

At the same time, in the intermediate range of concentrational ratios $c_i/S \approx 1$ when the inequality $k_7[PH] < k_8 c_i$ is combined with that of $k_3 c_i < k_6 S$, the expression of the inhibitor expenditure rate is:

$$V_i = \frac{\beta k_1 V_{tr} c_i}{2 k_6 S}$$

Hence, the kinetics obtains order 1 by phenol at a particular period of time, in accordance with Eq. (26), which discloses the structure of the radical translation rate V_{tr} to v-zones of spongy micelles. Thus, the rate of weak phenol expenditure will decrease because of complete realization of the dialkylsulfide-inhibiting activity effect induced by phenol. Curves 1 and 2, shown in Fig. 13, indicate the true change of the inhibitor expenditure kinetic law, according to theoretical predictions of the heterophase model of the induction period.

C. Features of Dialkylsulfide Co-oxidation with 2,2'-Methylene-bis(4-methyl-6-*tert*-butylphenol)

The effect of dialkylsulfides on the induction period of polyolefin oxidation is lower in the mixtures with bisphenol 2246, which was related to strong inhibitors (36). Deceleration of the bisphenol expenditure, observed in this example, is less effective than with a weak phenol (see Fig. 13, curves 3 and 4). On the one hand, molecules of bisphenol 2246 differ from the foregoing molecules of weak phenols by lower hindrances of OH groups from neighbor alkyl substituents (consequently, by better accessibility for attack of radicals and oxygen molecules). On the other hand, they differ by a specific mutual disposition of phenolic nuclei (18). Owing to these structural features on molecules, the process of direct oxidation with radical formation proceeding in the s-zones of polymer chain sponge, becomes easier for bisphenol:

$$HIIH_s + O_{2s} \xrightarrow{k_{0s}^i} \quad \text{(structure)} \rightarrow \text{(structure)}$$

$$\cdot I\dot{I}HOOH_s \rightarrow HIIOOH \qquad (\cdot I\dot{I}HOOH_s)$$

$$\cdot I\dot{I}HOOH_s \longrightarrow \text{(structure)} \qquad (\cdot IIO_s^{\cdot}) + H_2O$$

$$\cdot IIO_s^{\cdot} + PH_s \rightarrow \cdot IIOH + P_s^{\cdot}; \qquad P_s^{\cdot} \xrightarrow{k_{1s}} R_s^{\cdot};$$

$$R_s^{\cdot} \xrightarrow{k_{2s}} R_{1s}^{\cdot}; \qquad R_s^{\cdot} + R_{1s}^{\cdot} \xrightarrow{k_{3s}} \text{products};$$

$$R_{1s}^{\cdot} + \cdot IIOH_s \rightarrow R^{=} + HIIOH$$

As a result of the self-oxidation bisphenol 2246 becomes the leading oxidation initiator at a significantly high concentration.

Similar conclusions have been reported (44). It was stated that the evolution rate of propylene (PP depolymerization product) is the same at different concentrations of HIIH (0.025 and 0.05 mol/kg), and increases up to its constant value after 100 min, already in the stage of the induction period. The s-scheme displayed explains the constancy of the initial propylene elimination rate by the fact that besides initiating and depolymerizing alkyl $R_{1s}^{\cdot}$ macroradicals, bisphenol leads to their stationary concentration by their combination with bisphenol phenoxyls $\cdot IIOH$, formed in the reactions of radical initiation. In this case, the increase of the propylene-producing rate, up to a constant value during the induction period, may be connected with the dissociation of the peroxide initiator, with the quinolic structure HIIOOH formed from bisphenol, which reaches its stationary concentration.

According to the s-scheme displayed, steady-state concentration of radicals–mediators responsible for the interzone s,v-transfer is the following:

$$[R_s^{\cdot}] = \frac{\gamma c_{ii} P_{O_2}}{2k_{2s}}$$

Here γ is a complex value characterizing the effectiveness of bisphenol action as the initiator.

The free valence transfer from s-zones to v-zones proceeds through the stage of intermediate active quinolic alcoxyl, as in the foregoing phenols:

$$R_s^{\bullet} \underset{k_{-m}}{\overset{k_m}{\longleftrightarrow}} R_m^{\bullet}; \qquad R_m^{\bullet} + O_{2,s} \overset{k_{tr}}{\longrightarrow} ROO_m^{\bullet};$$

$$ROO_m^{\bullet} + HIIH \rightarrow ROO\dot{I}HIH_m \rightarrow ROH + HIIO_m^{\bullet}.$$

On entering the polymer v-zones, this alcoxyl initiates co-oxidation in them.

As known (36,74,78), bisphenol 2246 provides significantly longer induction periods (co-oxidation periods), compared with weak monophenols, and less effective response in the presence of dialkylsulfide. The reason for both of these effects may be connected to a specific bisphenol molecular structure, which causes a high probability of active alcoxyl isomerization to inactive phenoxyl through an intermediate radical of the benzene type:

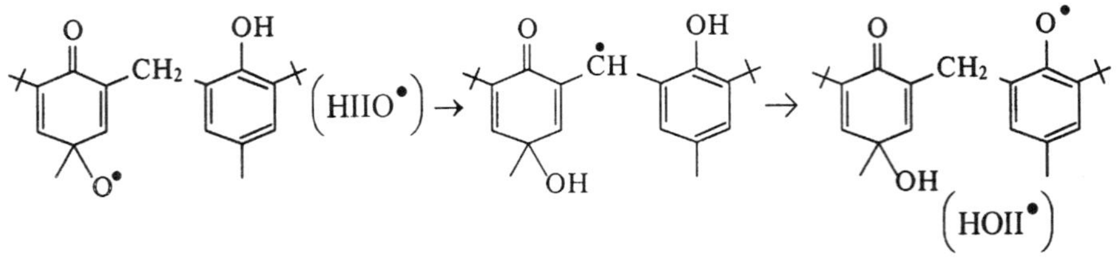

Competition between such isomerization reactions and s,v-translation reactions

$$HIIO_m^{\bullet} \overset{k'_{tr}}{\longrightarrow} HIIO^{\bullet}; \qquad HIIO_m^{\bullet} \overset{k_{is}}{\longrightarrow} HOII_m^{\bullet};$$

$$2HOII_m^{\bullet} \rightarrow products$$

significantly decreases the rate of the reaction v-chain initiation, which becomes the following in the oxygen-saturated system:

$$V_{tr} = k'_{tr}[HIIO_m^{\bullet}] = \frac{\delta_{ii}\gamma c_{ii}P_{O_2}}{2k_{2s}} \tag{28}$$

Here $\delta_{ii} = k_m k'_{tr}/k'_{tr} + k_{is} < k_m$ (contrary to Eq. (26)).

Similar isomerization of phenolic alcoxyls (possessing constant k_9) makes a contribution into the co-oxidation chain termination in polymer v-zones:

$$HIIO^{\bullet} + PH \overset{k_5}{\longrightarrow} HIIOH + P^{\bullet}; \qquad P^{\bullet} + O_2 \rightarrow POO^{\bullet};$$

$$POO^{\bullet} + HIIH \overset{k_1}{\longrightarrow} POO\dot{I}HIH \rightarrow POH + HIIO^{\bullet};$$

$$POO^{\bullet} + HIIH \overset{k_3}{\longrightarrow} G + HII^{\bullet};$$

$$HII^{\bullet} \rightarrow IIH^{\bullet} \text{ [phenoxyl isomerization to } \sigma\text{-complex (18)]}$$

$$POO^{\bullet} + IIH^{\bullet} \rightarrow G + II \text{ (cyclic quinolide)};$$

$$HIIO^{\bullet} \overset{k_9}{\longrightarrow} HOII^{\bullet}; \qquad POO^{\bullet} + HOII^{\bullet} \rightarrow POOIIOH \rightarrow products$$

Here, we suppose that the mechanism of cryptoradical partial regeneration of the initial bisphenol from quinolic alcohol HIIOH remains unchanged, similar to monophenols.

Owing to an additional pathway of the reaction v-chain termination, the process proceeds under the condition of significantly decreased concentrations of peroxyls POO$^\bullet$, determined from conditions of the rate equality of v-chain initiation and termination:

$$V_{tr} = 2k_3[\text{POO}^\bullet]c_{ii} + 2k_9[\text{HIIO}^\bullet] = 2k_3[\text{POO}^\bullet]c_{ii} + \frac{2k_1k_9[\text{POO}^\bullet]c_{ii}}{k_5[\text{PH}]},$$

and as a result of activity loss during the s,v-translation [$\delta_{ii} < k_m$ in Eq. (28)] of the radicals, the bisphenol decay rate decreases significantly compared with the rate of monophenols [see Eq. (27)]:

$$V_{ii} = -\frac{dc_{ii}}{dt} \approx \beta k_1[\text{POO}^\bullet]c_{ii} = \frac{\beta k_1 \delta_{ii} \gamma c_{ii} P_{O_2}}{2k_{2s}\left(2k_3 + \dfrac{2k_1k_9}{k_5[\text{PH}]}\right)}$$

The induction period of the polymer oxidation increases simultaneously, and the potential for realization of synergistic dialkylsulfide cooperation decreases.

Summing up the kinetic picture, the heterophase model of the induction period marks it as a macroscopic stage of chain co-oxidation of an inhibitor and a polymer. Alkylphenol is an active participant of the chain process in this stage, and it is generally expended in the reactions of co-oxidation chain propagation, and not in those of chain termination. This period ends when the inhibitor concentration decreases, and the peroxyl concentration increases so high that the oxygen absorption rate is already determined by the reaction chains of the polymer oxidation (see Sect. VI):

$$\text{POO}^\bullet + \text{PH} \rightarrow \text{POOH} + \text{P}^\bullet; \qquad \text{P}^\bullet + \text{O}_2 \rightarrow \text{POO}^\bullet;$$
$$\text{POO}^\bullet \rightarrow {}^\bullet\text{POOH} \rightarrow \text{HO}^\bullet + \text{products}; \qquad \text{HO}^\bullet + \text{PH} \rightarrow \text{P}^\bullet + \text{H}_2\text{O},$$

and bimolecular acts of the chain termination:

$$\text{POO}^\bullet + \text{POO}^\bullet \rightarrow \text{products}.$$

In this case, the key issue is that the oxidation process with an inhibitor, as well as after its full exhaustion, which does not fall into the regimen with nondegenerated (avalanche-like) branchings of the reaction chains. Transition from the co-oxidation stage to the stage of polymer oxidation proceeds under the condition of the peroxyl steady-state concentration. A similar condition remains at the cooxidation regimen change from zero-order by phenol to an apparent first-order by dialkylsulfide injection into the polymer.

The induction period increase, accompanying the change of the regimen, is often displayed on the $t_{ind}(c_{i0})$ curves with particular sharpness, and phenol concentration at the breakpoint of the curve is interpreted as the critical concentration of the process transition from the nonstationary regimen of branched chain reaction into the stationary oxidation (36). The idea for existence of the stage with nonstationary branchings is stimulated by studies (36) that suggest that the hydroperoxide lifetime is seconds long at relatively high temperatures of approximately 473 K. In this case, the authors pay no attention to the chain heterophase mechanism of the anaerobic HP decomposition.

In the absence of oxygen, the HP decomposition proceeds according to the chain mechanism, and the contribution of spontaneous decomposition into total HP expenditure is negligibly small (see Sect. III). Although at $T \sim 473$ K the HP lifetime in s-micropores is measured in tens of seconds, according to Eq. (5), the dissociative lifetime of the main HP amount localized in narrow micropores of the chain sponge v-zones is substantially higher. Thus, it is significant that a half of polystyrene HP is expended in the absence of oxygen during 12–15 min at $T = 453$ K. In oxidizing polystyrene melt (in the presence of oxygen, the HP decomposition mechanism is generally changed; 39) the hydroperoxide is accumulated together with other oxidation products during 30 and 100 min (67) at $T = 473$ and 453 K, respectively. It is not noticeably decomposed until it enters the reaction with the accumulated products of polymer oxidation. Then, further HP accumulation ends, and its concentration decreases.

The discussions in this chapter show the inadequacy in explaining the "critical" phenol concentration (36). From the point of view of the structural–kinetic reaction model, the breakpoint of the $t_{ind}(c_{i0})$ curve, induced by dialkylsulfide, is stipulated by the phenol expenditure rate decrease during the co-oxidation stage in transition from the zero-order reaction to the first-order one by inhibitor. This occurs at a quasi-stationary peroxyl concentration and in the absence of nonstationary branching of the reaction chains.

The inhibitor expenditure rate is constant in the regimen of the zero-order kinetics:

$$-\frac{dc_i}{dt} = V_i = \frac{c_{i0}}{t_{ind}^i} = \text{const}$$

and the correspondent time of its complete exhaustion (co-oxidation) is

$$t_{ind}^i = \frac{c_{i0}}{V_i}$$

Adding to this value the reaction development time t_0 after inhibitor exhaustion, we obtain dependence of the induction period on the initial phenol concentration:

$$t_{ind} = t_0 + t_{ind}^i = t_0 + (V_i^{-1})c_{i0}$$

The linear dependence $t_{ind}(c_{i0})$ holds for a wide range of monophenol concentrations, studied (73–75). In the presence of dialkylsulfide the $t_{ind}(c_{i0})$ curves display a sharp increase of t_{ind}, starting from some "critical" value $c_{i0,cr}$. A similar situation occurs in bisphenol 2246. The co-oxidation in these systems at $c_{i0} > c_{i0,cr}$ (and $c_{ii0} > c_{ii0,cr}$) conforms to first-order kinetics, characterized by the exponent:

$$c_i = c_{i0} \exp(-k_i t)$$

or

$$\ln\left(\frac{c_{i0}}{c_i}\right) = -k_i t$$

This equation is held until the $c_{i0,cr}$ value is reached (after that, the law of the reaction chain termination changes, and bimolecular termination occurs

$$POO^{\bullet} + POO^{\bullet} \rightarrow products$$

instead of the linear reaction chain termination on an inhibitor molecule).

In this case, the total induction period is presented by the following equation:

$$t_{ind} = t_0 + \left(k_i^{-1}\right) \ln\left(\frac{c_{i0}}{c_{i,cr}}\right)$$

Such dependence is often fulfilled (36).

The present structural–kinetic model of phenol and polymer co-oxidation also allows us to define more exactly the physical nature of monophenol and bisphenol 2246 antagonism at their cooperative injection into polymer (36,73,74). Despite the increase of the total inhibitor content, such systems display the induction period decrease in comparison with its value, characteristic for single bisphenol 2246.

To account for this phenomenon, the authors apply the idea of the specific monophenol and bisphenol effect on the radical yield from a "cage" of spontaneously decomposed hydroperoxide. It is assumed that by trapping one of the radicals from the primary radical pair, monophenol impedes the next radical combination, and the radical yield should consequently increase. At the same time, bisphenol 2246, possessing two functional groups, induces no increase of the radical yield or increases it less than does the monophenol.

Despite the foregoing concept, the material considered in Sections IV and V proves the absence of any significant increase of radical yield from hydroperoxide under the ionol effect.

The reason for this effect is that monophenol increases the initiation rate of the co-oxidation reaction v-chains as a consequence of the increase of free valency injection rate from the s-zones into v-zones of the polymer chain–sponge. By *reaction cage* we mean a general system of supermicropores of structural s-zones, but not the homogeneous environment formed by polymer chain segments.

Interacting with a radical mediator peroxyl, monophenol transforms into a quinolic alcoxyl:

$$ROO_m^{\bullet} + IH \rightarrow ROOIH^{\bullet} \rightarrow ROH + IO^{\bullet}$$

which is incapable of isomerizing into an inactive phenoxyl: it possesses $\delta_i = k_m$, contrary to $\delta_{ii} < k_m$ for bisphenol 2246 [see Eq. (28)].

By competing with bisphenol for the radical s,v-transfer, monophenol increases the rate of the co-oxidation chain initiation in v-zones of the polymer matrix, so it decreases the induction period.

In finishing this section, we should note that sterically hindered monophenols studies displayed no transition to a first-order reaction in the whole range of concentrations studied (75,77). However, it may be that such a transition will become possible for phenols with lower hindrances from alkyl groups and at a significantly high concentration c_i. This will require conditions, favorable for the reversibly dissociating phenol complex with oxygen,

$$IH_s + O_{2s} \leftrightarrow (I^{\bullet} \dots HOO^{\bullet})_s$$

to obtain help from another phenol molecule to escape the free radicals

$$(I^{\bullet} \ldots HOO^{\bullet})_s + IH_s \rightarrow I_s^{\bullet} + HOOIH_s^{\bullet} \, ;$$

$$HOOIH_s^{\bullet} \rightarrow H_2O + IO_s^{\bullet}$$

It should be taken into account that weakening of steric hindrances may induce the occurrence of additional reactions of the co-oxidation chain termination:

$$IO^{\bullet} + IH \rightarrow IOH + I^{\bullet}$$

which may also serve as one reason for synergism in antioxidant mixtures.

IX. CONCLUSION

The principles for application of heterogeneous reaction models to the description of a chain process proceeding in a noncrystalline polymer body matrix were discussed. The basic experimental material required for constructing such models was obtained with the help of the model chain reaction of macromolecule arylation by dibenzoyl peroxide (11–18). Such study of this reaction on a number of polymers, performed in their glassy-like, highly elastic, and melt state, allowed us to clarity its significant kinetic features that differ from the dibenzoyl peroxide reaction with liquid analogues of the polymers studied.

The concepts, worked out with the help of the model reaction, make possible a detailed consideration of the morphological model of supermolecular organization, disclosing the fine structure of polymer globule interior, which is the main structural unit of the noncrystalline polymer spatial skeleton.

The existence of a definite interconnection between microheterogeneous composition of the noncrystalline polymer matrix and the chemistry of polymer transformations was pointed out in the literature with examples of particular processes. However, separate investigations, which displayed only the statements of existence of this interconnection, have not yet been combined into an independent branch of chemical physics of polymers. We may note that the existence of this interconnection was more or less generally manifest in the model of polychromatic (or polychronic; 27) kinetics. However, the methodology for kinetic polychromatism description is reduced to one or another variant of the mathematical formality that applies the most general physical concepts.

Evidently, construction of adequate chemicophysical models of polymer transformations is connected with an improvement of the particular picture of supermolecular organization of the polymer reaction medium. The existence of true microheterogeneous chain–sponge organization of polymer chains significantly increases (if compared with homogeneous liquid) the number of factors capable of affecting a chemical process in any way, and increases the difficulties connected with formulation of the theories. The solution to this problem became possible with formulation and testing of structural–kinetic reaction models (33).

The models of chain processes considered in this chapter account for the calculation of the fine structure of polymer chain–sponge micelles, which possess somewhat dynamic skeletal units. In the present models, the dynamics is connected with the property of micropores that change their size under the influence of thermofluctuations of cooperative units of flexible polymer chains in the structure of a microporous sponge. This is different from the abstract idea of "molecular–segmental mobility."

The structural–kinetic modeling method allowed us to find such important regimes of the chain reactions as heterogeneous–heterophase and heterogeneous–homophase reactions. It also allowed us to improve the concepts of the cryptoradical ('intracage') process, performed inside a micropore, that was thermofluctuationally excited up to a supermicropore size.

At least the construction and analysis of various reaction models make it possible to assume that the general concept of the microheterogeneous polymer chain–sponge chemistry is valid. Broad possibilities remain for further improvement and detailed study of the mechanism of elementary chemical stages that proceeding in a fluctuating sponge body medium.

At the present time we have to refute the idea that the process of polyolefin oxidation, which develops in the autoaccelerated stage, may be interpreted in the frame of the nonstationary, branched chain reaction mechanism.

It was shown that the general amount of phenol is expended in the chain co-oxidation with a polymer in both first-order and zero-order regulation by an inhibitor.

The criterion of the inhibitor method for determination of the oxidative chain initiation rate, which was formulated for homogeneous liquids, also loses its fundamental meaning if applied to polymers.

The factor of the polymer chain–sponge matrix structural morphology raises its own questions connected to determination of the inequivalent rates of the radical initiation in different zones:

1. Quasi-stationary zones consisted of low-dynamic supermicropores.
2. Dynamically more active zones, in which the chain oxidation of the macromolecules is developed.
3. The stage of the interzone radical translation.

The task of development of experimental methods for determination of elementary reaction rate constants, calculated not by homogeneous schemes, and characterization of heterogeneous reactions in the polymer chain–sponge is raised simultaneously.

The present review is a logical continuation of the investigations, performed by the academician N. M. Emanuel and his school, in the branch of science related to the oxidative processes in the condensed phase (27,36,54,79–81).

REFERENCES

1. YA Mikheev, LN Guseva. Some chemical–physical aspects of supermolecular carcass structure of amorphous polymers. Khim Fiz 10:724–733, 1991.
2. YA Mikheev. A new structural model of the amorphous polymer for the description of initiated chain reaction. Int J Polym Mater 16:221–235, 1992.
3. YA Mikheev, LN Guseva, GE Zaikov. On structure effects and entropy factor of compatibility of non-crystalline polymers and liquids. Khim Fiz 15:7–15, 1996.
4. GS Yeh. Current concept of morphology of amorphous polymers. Vysokomol Soedin A21:2433–2446, 1979.
5. VP Lebedev, VD Romanov, DN Bort, SA Arzhakov. Interpretation of electronical–

microscopic pictures of amorphous polymers. Vysokomol Soed A18:85–89, 1976.

6. Z Veksli, WG Miller. The effect of good solvents on molecular motion of nitroxide free radicals in covalently labelled polystyrene and poly(methyl methacrylate). Macromolecules 10:686–692, 1977.

7. EB Bagley, JM Scigliano. The behaviour of the internal pressure of polymers above and below the glass transition temperature. Polym Eng Sci 11:320–325, 1971.

8. VP Privalko. Molecular structure and properties of polymers. Leningrad: Khimia, 1986.

9. DA Fridrikhsberg. Course of Colloid Chemistry. Leningrad: Khimia, 1984.

10. YA Mikheev, LS Pustoshnaya, GE Zaikov. On the mechanism of isothermal evolution of liquid water in a medium consisting of high-elasticity hydrophobic polymers (as exemplified by polymethylmethacrylate). Khim Fiz 15:126–138, 1996.

11. YA Mikheev, LN Guseva, LE Mikheeva, DY Toptygin. The mechanisms of inhibition and catalysis of dibenzoyl peroxide decomposition under the action of ionol in cellulose triacetate. Kinet Katal 28:279–286, 1987.

12. YA Mikheev. Heterophasic mechanism of arylation of cellulose triacetate and polycarbonate by dibenzoyl peroxide. Int J Polym Mater 16:237–259, 1992.

13. YA Mikheev, LN Guseva, DY Toptygin. The catalysis effects in reaction of polypropylene with bibenzoyl peroxide and the structural–kinetic model of polymer. Kinet Katal 28:287–294, 1987.

14. YA Mikheev. Heterophasic mechanism of modification of polyethylene by dibenzoyl peroxide. Khim Fiz 10:715–723, 1991.

15. YA Mikheev, LN Guseva, DY Toptygin. The radical chain mechanism of dibenzoyl peroxide reaction with di-alkyl ethers and polyethylene glycol. The role of the structural microheterogeneity of polymer melt. Khim Fiz 6:251–257, 1987.

16. YA Mikheev. The change of the mechanism of polyamide PA-548 chain reaction with dibenzoyl peroxide near the glass transition temperature. Khim Fiz 8:1110–1117, 1989.

17. YA Mikheev, GE Zaikov. On the functions of molecular sponge in heterophase chain reactions of bulk polymers (on the example of polystyrene arylation by dibenzoyl peroxide). Int J Polym Mater 34:29–57, 1996.

18. YA Mikheev, LN Guseva, GE Zaikov. Heterophase features of chain reactions of macromolecules in non-crystalline polymer matrix with participation of antioxidants. Russ Chem Rev 66:3–33, 1997.

19. YA Mikheev, LN Guseva, GE Zaikov. Analysis of structural–kinetic models of oxidation of polyolefins. Vysokomol Soed B39:1082–1098, 1997.

20. VA Bershtein, VM Egorov. General mechanism of beta-transition in polymers. Vysokomol Soed A27:2440–2453, 1985.

21. YG Frolov. Course of Colloid Chemistry. Surface Phenomena and Dispersed Systems. Moscow: Khimia, 1982.

22. AA Tager, MV Tsilipotkina. The polymer porosity structure and mechanism of sorption. Russ Chem Rev 47:152–175, 1978.

23. GI Lashkov, LS Shatseva, SP Kozel, NS Shelekhov. Reaction of photochemical decay of anthracene dimers in the study of microvoids in solid polymethyl methacrylate. Vysokomol Soed A25:2169–2175, 1983.

24. AA Tager. The effect of the molecular weight of some vitreous polymers on their integral heats of solution. Vysokomol Soed 1:21–28, 1959.

25. FE Filisko, RS Raghava, GSY Yeh. Amorphous structure heat. Temperature dependence of heat of solution for polystyrene in toluene and ethylbenzene. J Macromol Sci Phys B10:371–380, 1974.

26. DM Gezovich, PH Geil. Morphology of the plasticized poly(vinyl chloride). Int J Polym Mater 1:3–8, 1971.

27. NM Emanuel, AL Buchachenko. Chemical Physics of Molecular Degradation and Stabilization of Polymers. Moscow: Nauka, 1988, p. 205.

28. VS Derinovskii, AI Maklakov, MM Sommer, AV Kostochko. The molecular motion in the system of cellulose nitrate—glycerine trinitrate. Vysokomol Soed A16:1306–1311, 1974.

29. VB Stryukov, TV Sosnina, AM Kraitsberg. Investigation of the rotational mobility of iminoxyl radicals (spin probe) in polymers in a wide range of rotation frequencies. Vysokomol Soed A15:1397–1406, 1973.

30. G Shaw. Quenching by oxygen diffusion of phosphorescence emission of aromatic molecules in polymetyl methacrylate. Trans Faraday Soc 63:2181–2189, 1967.

31. RJ Woods, JF Manville. Mobility of phosphorescent solutes in polymethylmethacrylate. Can J Chem 49:515–519, 1971.

32. V Muller, II, Kaletchitz, HGO Bekker, MG Kuzmin, VP Zubov. Kinetics of fluorescence quenching of quinone bisulfate by chlorine ions in polyvinyl alcohol solutions and films. Vysokomol Soed A20:1593–1600, 1978.

33. SL Kiperman. Kinetic models in heterogeneous catalysis. Russ Chem Rev 47:3–38, 1978.

34. YA Mikheev, LN Guseva, LE Mikheeva, SV Kalantsova. The features of the heterophasic decomposition of polymer hydroperoxides related to the high molecular-segmental mobility. Khim Fiz 12:528–537, 1993.

35. YA Mikheev, LN Guseva. The kinetics of anaerobic decomposition of hydroperoxide of solid polypropylene in the heterophase reaction stage. Khim Fiz 15:131–142, 1996.

36. YA Shlyapnikov, SG Kiryushkin, AP Mar'in. Antioxidative Stabilization of polymers. Chichester: Ellis Harwood, 1996.

37. ET Denisov. Oxidation and degradation of the carbochain polymers. Leningrad: Khimia, 1990.

38. VA Roginskii. Phenol Antioxidants: Reactivity and Efficiency. Moscow: Nauka, 1988.

39. YA Mikheev, LN Guseva. The features of the heterogeneous–heterophasic autoxidation of polypropylene. Khim Fiz 12:1081–1096, 1993.

40. GP Andrianova. Physico-Chemistry of Polyolefins. Moscow: Khimia, 1974.

41. VV Lipes. The induced decomposition of alkyl peroxides and hydroperoxides of alkyls and acyls. Kinet Katal 27:1046–1054, 1986.

42. JCW Chien, DST Wang. Autoxidation of polyolefins. Absolute rate constants and effect of morphology. Macromolecules 8:920–928, 1975.

43. JCW Chien, H Jabloner. Polymer reaction. IV. Thermal decomposition of polypropylene hydroperoxides. J Polym Sci A1:393–402, 1968.

44. VS Pudov, BA Gromov, YA Shlyapnikov. The low molecular olefin evolvations in the induction period of the inhibited oxidation of polyolefins. Vysokomol Soed B9:111–113, 1967.

45. DW van Krevelen. Properties of Polymers. Correlations with Chemical Structure. New York: Elsevier, 1972.

46. AA Tager. Physico-Chemistry of Polymers. Moscow: Khimia, 1978.

47. JCW Chien, CR Boss. Polymer reactions. VI. Inhibited autoxidation of polypropylene. J Polym Sci A1:1683–1697, 1967.

48. F Gugumus. Thermooxidative degradation of polyolefins in the solid state. Part 1: Experimental kinetics of functional group formation. Polym Degrad Stabil 52:131–144, 1996.

49. F Gugumus. Thermooxidative degradation of polyolefins in the solid state. Part 2: Homogeneous and heterogeneous aspects of thermal oxidation. Polym Degrad Stabil 52:145–157, 1996.

50. F Gugumus. Thermooxidative degradation of polyolefins in the solid state. Part 3: Heterogeneous oxidation model. Polym Degrad Stabil 52:159–170, 1996.

51. F Gugumus. Thermooxidative degradation of polyolefins in the solid state. Part 4: Heterogeneous oxidation kinetics. Polym Degrad Stabil 53:171–187, 1996.

52. YA Mikheev, LN Guseva. The induction period of autoxidation of polypropylene as the stage of heterophasic process. Khim Fiz 11:964–973, 1992.

53. EL Shanina, VA Roginskii, VB Miller. The kinetics of solid polypropylene autoxidation in the very beginning. Vysokomol Soed B21:892–895, 1979.

54. AA Popov, NY Rapoport, GE Zaikov. Oxidation of Stressed Polymers. Philadelphia: Gordon & Breach, 1991.

55. DE Faulkner. Effects of high oxygen pressure and temperature on the aging of polypropylene. Polym Eng Sci 22:466–471, 1982.

56. T Kelen, M Iring, F Tudos. The kinetics and mechanism of oxidation of polyethylene in melt. Eur Polym J 12:35–39, 1976.

57. LS Shibryaeva, SG Kiryushkin, GE Zaikov. Autoxidation of oriented polyolefins. Int J Polym Mater 16:71–93, 1992.

58. VS Pudov, BA Gromov, EG Sklyarova, MB Neiman. The oxidation of isotactic polypropylene. 1. The kinetics of oxygen uptake. Neftekhimia 3:743–749, 1963.

59. VS Pudov, MB Neiman. The thermo-oxidative degradation of isotactic polypropylene. In: AS Kuzminskii, ed. Ageing and Stabilization of Polymers. Moscow: Khimia, 1966, pp. 5–26.

60. TA Bogaevskaya, BA Gromov, VB Miller, TV Monakhova, YA Shlyapnikov. Effect of supermolecular structure of polypropylene on the oxidation kinetics. Vysokomol Soed A14:1552–1556, 1972.

61. TV Monakhova, TA Bogaevskaya, BA Gromov, YA Shlyapnikov. About some features of autoxidation of solid polypropylene. Vysokomol Soed B16:91–94, 1974.

62. TA Bogaevskaya, TV Monakhova, YA Shlyapnikov. About connection of crystallinity of polypropylene and the oxidation kinetics. Vysokomol Soed B20:465–467, 1978.

63. M Iring, S Laslo-Hedvig, K Barabas, T Kelen, F Tudos. Study of the thermal oxidation of polyolefins—IX. Some differences in the oxidation of polyethylene and polypropylene. Eur Polym J 14:439–441, 1978.

64. YA Mikheev, LN Guseva. About coupled heterophasic reactional pathways in autoxidation of poly(ethylene oxide). Khim Fiz 6:1259–1267, 1987.

65. LN Guseva, YA Mikheev, LE Mikheeva. SV Sukhareva, DY Toptygin. "Solid-phase" features of autoxidation of polyethylene oxide in melt. Vysokomol Soed A30:988–994, 1988.

66. YA Mikheev. Auto inhibition of reactional chain branching and another oxidation features of homologues of glycol. Khim Fiz 8:399–405, 1989.

67. VM Goldberg, VN Esenin, IA Krasotkina. The role of hydroperoxides in thermo-oxidation of polystyrene in melt. Vysokomol Soed A19:1720–1727, 1977.

68. AP Meleshevitch. The mechanism of the free radical reactions of epoxides. Russ Chem Rev 39:444–470, 1970.

69. AM Aryev. The structural thickness effect in the polyethylene film annealing. Plast Massy 2:52–53, 1992.

70. NV Zolotova, ET Denisov. The kinetic characteristics of inhibiting action of phenol in the oxidation of polypropylene. Vysokomol Soed B18:605–608, 1976.

71. VM Goldberg, IA Krasotkina, MM Belitskii, DY Toptygin. A comparison of some methods of estimating the thermo-oxidative stability of polystyrene in the presence and absence of inhibitors. Preprint of International Symposium on Methods of Valuation and Practical Use of Stabilizers and Synergistic Mixtures. Moscow, NIITEKhim, 1973.

72. EM Bevilacqua, ES English. The scission step in Hevea oxidation. J Polym Sci 49:495–505, 1961.

73. YA Shlyapnikov, VB Miller, ES Torsueva. The inhibitor behaviour in the oxidation of polypropylene. Izvest AN SSSR (Ser Khim Nauk):1966–1970, 1961.

74. YA Shlyapnikov, VB Miller, MB Neiman, ES Torsueva. Correlation in the effect of inhibitors in oxidation reactions. I. Alkylphenols. Vysokomol Soed 4:1228–1234, 1962.

75. BA Gromov, VB Miller, MB Neiman, ES Torsueva, YA Shlyapnikov. Mechanism of weak anti-oxidant action in the oxidation of polypropylene. Vysokomol Soed 6:1895–1900, 1964.

76. LN Denisova, ET Denisov. The free radical initiation in reaction $RH + O_2$. VI. Polypropylene, polyethylene. Kinet Katal 17:596–600, 1976.

77. YA Shlyapnikov, VB Miller, MB Neiman, ES Torsueva. Behaviour of inhibitors in oxidation reactions. II. Mixtures of alkylphenols with dodecyl sulfide. Vysokomol Soed 5:1507–1512, 1963.

78. TA Bogaevskaya, NK Tyuleneva, YA Shlyapnikov. Oxidation of polyethylene stabilized by strong phenol anti-oxidant. Vysokomol Soed A23:181–186, 1981.

79. NM Emanuel, GE Zaikov, ZK Maizus. Oxidation of Organic Compounds. Medium Effect in Radical Reaction. Oxford: Pergamon Press, 1984.

80. VA Kritsman, GE Zaikov, NM Emanuel. Chemical Kinetics and Chain Reactions. Historical Aspects. New York: Nova Science, 1995.

81. VL Rubailo, SA Maslov, GE Zaikov. Liquid-Phase Oxidation of Unsaturated Compounds. New York: Nova Science, 1993.

III
CONTROLLED DEGRADATION AND STABILIZATION

9

Polymer Oxidation and Antioxidant Action

EVGUENII T. DENISOV

Institute of Problems of Chemical Physics, Russian Academy of Sciences, Chernogolovka, Russia

LIST OF ABBREVIATIONS

AIBN	Azoisobutyronitrile	NR	Natural resin
APP	Atactic polypropylene	PB	Poly-1-butene
CBA(82/18)	Copolymer butadiene (82%)–acrylonitrile (18%)	PBA	Polybutylacrylate
		PBD	Polybutadiene
CBS(20/80)	Copolymer butadiene (20%)–styrene (80%)	PDMB	Polydimethylbutadiene
		PE	Polyethylene
CEA(40/60)	Copolymer ethylene (40%)–acrylonitrile (60%)	PEA	Polyethylacrylate
		PEMA	Polyethylmethacrylate
CEMA (12/78)	Copolymer ethylene (12%)–methylacrylate (78%)	PFE	Polytetrafluoroethylene
CEP(98/2)	Copolymer ethylene (98%)–propylene (2%)	PIA	Copolymer isoprene–acrylonitrile
		PIB	Polyisobutylene
CEVC	Copolymer ethylene–vinylchloride	PMA	Polymethacrylate
		PMMA	Polymethylmethacrylate
CIBIP	Copolymer isobutylene–isoprene	PMP	Poly-4-methyl-1-pentene
DBP	Dibenzoyl peroxide	PMPD	Polymethylpentadiene
DBPO	Di*tert*-butylperoxalate	PMVE	Polymethylvinyl ether
DCP	Dicumylperoxide	PP	Polypropylene
DIPP	Deuterated isotactic polypropylene	PS	Polystyrene
		PVA	Polyvinylacetate
DLP	Dilaurylperoxide	PVC	Polyvinylchloride
DPE	Deuterated polyethylene	PVCVA	Copolymer vinyl chloride–vinyl acetate
DPP	Dipalmitylperoxide		
HDPE	Polyethylene (high-density)	PVM	Polyvinylmethyl ether
IPP	Isotactic polypropylene	SSR	Stereoregular synthetic resin
LDPE	Polyethylene (low-density)		

I.   CHAIN MECHANISM OF POLYMER OXIDATION

A.   Chain Mechanism of Organic Compounds Oxidation

The low molecular weight analogues of polymers, such as hydrocarbons, alcohols, and esters, are oxidized by dioxygen in the liquid phase through a chain mechanism. The kinetics and mechanism of oxidation of these compounds have been studied in detail (1–15). Free radicals are generated in oxidized RH substances in reactions with dioxygen (see Sec. I.E), as the result of an initiator decomposition (see Sec. I.B), or on decomposition of hydroperoxide ROOH formed by the oxidation of RH (see Sec. I.F), as well as under the action of light. The simplest scheme of chain oxidation of RH includes the following elementary steps (6,10,15).

$$I \text{ (initiator)} \rightarrow 2r^{\bullet} \tag{1}$$

$$r^{\bullet} + RH \rightarrow rH + R^{\bullet} \tag{2}$$

$$R^{\bullet} + O_2 \rightarrow RO_2^{\bullet} \tag{3}$$

$$RO_2^{\bullet} + RH \rightarrow ROOH + R^{\bullet} \tag{4}$$

$$ROOH \rightarrow RO^{\bullet} + HO^{\bullet} \tag{5}$$

$$ROOH + RH \rightarrow RO^{\bullet} + H_2O + R^{\bullet} \tag{6}$$

$$ROOH + CH_2 = CHX \rightarrow RO^{\bullet} + HOCH_2C^{\bullet}HX \tag{7}$$

$$2ROOH \rightarrow RO_2^{\bullet} + H_2O + RO^{\bullet} \tag{8}$$

$$R^{\bullet} + R^{\bullet} \rightarrow RR \text{ (or RH + olefin)} \tag{9}$$

$$R^{\bullet} + RO_2^{\bullet} \rightarrow ROOR \tag{10}$$

$$RO_2^{\bullet} + RO_2^{\bullet} \rightarrow ROH + O_2 + R' = O \text{ or } ROOR + O_2 \tag{11}$$

Principally the same chain mechanism of PH polymer oxidation exists in the liquid and solid phases (16–30). Equations (3) and (4) are chain-propagating and Eqs. (9–11) are chain-terminating steps. Equation (3) is controlled by diffusion, and Eq. (4) is the limiting step of chain propagation in the presence of solved dioxygen at $[O_2] > 10^{-4}$ mol L^{-1} in the liquid. The activation energy of this reaction depends, first, on the enthalpy of the reaction $\Delta H = D(R\text{-}H) - D(ROO\text{-}H)$, where the bond dissociation energy $D(ROO\text{-}H)/kJ$ mol^{-1} is 358.6 and 365.5 for tertiary and secondary hydroperoxides, respectively. The values of the C–H bond dissociation energy, enthalpies, and activation energies of Eq. (4) for several polymer analogues are listed in Table 1. For the rate constants k_p of Eq. (4) for polymers in solution and solid phase see Tables 2 and 3. (For the k_p values in liquid-phase oxidation see Refs. 8,10,11,31–33).

Chain-propagating Eq. (4) proceeds by intermolecular (see foregoing scheme) and intramolecular hydrogen atom abstraction; for example,

$$Me_2C(OO^{\bullet})CH_2CHMe_2 \rightarrow Me_2C(OOH)CH_2C^{\bullet}Me_2 \tag{12}$$

(For the values of rate constants of these reactions see Refs. 10,13.) Oxidation of unsaturated compounds includes the addition of peroxyl radicals to the double bond as a chain propagation.

$$ROO^{\bullet} + CH_2 = CHX \rightarrow ROOCH_2C^{\bullet}HX \tag{13}$$

Table 1 Bond Dissociation Energies of RH (34–36), Enthalpies and Activation Energies of Reaction 4 (37,38) for the Low Molecular Analogues of Polymers

Polymer	RH	D kJ mol^{-1}	ΔH kJ mol^{-1}	E kJ mol^{-1}
PE	MeCH$_2$CH$_2$Me	413	47.5	67.2
PP	Me$_3$CH	400	41.4	63.3
PBD	MeCH=CHCH$_2$Me	345.6	−19.9	48.6
SSR	Me$_2$C=CHCH$_2$Me	337.0	−28.5	45.1
PS	PhCHMe$_2$	357.4	−1.2	43.2
PMA	MeCH$_2$CHMeCOOMe	388.4	29.8	59.2
PMMA	MeCH$_2$CMe$_2$COOMe	404.5	39.0	65.5
PEMA	MeCH$_2$CM$_2$COOCH$_2$Me	398.0	32.5	62.1
PMVE	Me$_2$CHOCHMe$_2$	383.1	24.5	56.4
PVA	Me$_2$CHOC(O)Me	391.5	26.0	58.9

Table 2 Rate Constants of Alkoxyl and Peroxyl Radicals Reactions with C–H Bonds of Polymers and Model Compounds in Liquid Phase

Oxidizing compound	Radical	T (K)	k (L mol^{-1} s^{-1})	Ref.
PE	C$_6$H$_5$(CH$_3$)$_2$CO$_2^{\cdot}$	388	0.12	39
Decane	C$_6$H$_5$(CH$_3$)$_2$CO$_2$	388	$4.8 \times 10^8 \exp(-73/RT)$	9
PP	(CH$_3$)$_3$CO$_2$	363	0.98	40
2,6,10,14-Tetramethylpentadecane	(CH$_3$)$_3$CO$_2$	363	058	40
2-Methylbutane	C$_2$H$_5$(CH$_3$)$_2$CO$_2$	363	$2.0 \times 10^8 \exp(-66/RT)$	41
PP	C$_{10}$H$_{11}$O$_2$	363	3.3	40
2,6,10,14-Tetramethylpentadecane	C$_{10}$H$_{11}$O$_2$	363	2.1	40
PS	C$_6$H$_5$(CH$_3$)$_2$CO$_2$	353	$5.0 \times 10^9 \exp(-72/RT)$	42
Isopropylbenzene	C$_6$H$_5$(CH$_3$)$_2$CO$_2$	353	$5.0 \times 10^6 \exp(-41/RT)$	43
PS	(CH$_3$)$_3$CO$^{\cdot}$	313	26	44
2,4-Diphenylpentane	(CH$_3$)$_3$CO$^{\cdot}$	313	125	44
Isopropylbenzene	(CH$_3$)$_3$CO$^{\cdot}$	313	1020	44
PS	(CH$_3$)$_3$CO$^{\cdot}$	318	26	45
Isopropylbenzene	(CH$_3$)$_3$CO$^{\cdot}$	318	1600	45
PP	(CH$_3$)$_3$CO$^{\cdot}$	318	137	45
2,2,4-Trimethylpentane	(CH$_3$)$_3$CO$^{\cdot}$	318	300	45

(The rate constants of these reactions are given in Refs. 10,14,31–33.)

The motion of a free valence in an oxidizing solid polymer is a complex process. The following three mechanisms are discussed (27,28).

1. The encounter of two free valences is the result of only the diffusion of segments of a macromolecule. Because the radius of segmental diffusion is limited in real time, this mechanism can be effective at a high initiation

Table 3 Rate Constants of Reaction $PO_2^\bullet + PH$ per One C–H Bond in Solid Phase

Polymer	T/K	A (L mol^{-1} s^{-1})	E (kJ mol^{-1})	k_p (300 K) (kg mol^{-1} s^{-1})	Ref.
PE	230–270	2.1×10^2	39	3.4×10^{-5}	46
DPE	270–300	2.4×10^2	43	7.7×10^{-6}	46
IPP	383–413	3.0×10^5	38	2.8×10^{-2}	47
IPP	317–365	9.6×10^{11}	87.8	4.9×10^{-4}	48
IPP	270–310	1.1×10^4	53	6.4×10^{-6}	46
IPP	303	—	—	1.9×10^{-5}	49
DIPP	270–310	1.4×10^4	58	1.9×10^{-6}	46
APP	363–378	1.6×10^7	50.5	2.5×10^{-2}	50
APP	318–336	1.7×10^8	62.8	1.9×10^{-3}	51
PB	363–378	1.1×10^{11}	75	9.5×10^{-3}	50
PS	210–293	3.5×10^8	43	11	52
PS	210–293	1.3×10^{12}	88	6.1×10^{-4}	52
PMMA	250–360	1.0×10^6	50	1.9×10^{-3}	53
PMP	298	—	—	1.8×10^{-3}	54

Table 4 Kinetic Parameters of Free Valence Migration in Polymers: Diffusion Coefficient D and Average Distance r

Polymer, initiation	Macroradical	T (K)	$[P^\bullet]_0 \times 10^6$ or $[PO_2^\bullet]_0 \times 10^6$ (mol L^{-1})	$D \times 10^{18}$ (cm^2 s^{-1})	$r \times 10^{10}$ (m)
PE, γ	$P^\bullet$	363	8.3	2.0	50
PE, γ	$P^\bullet$	343	3.5	3.6	23
PE, γ	$P^\bullet$	343	3.7	2.8	8
PE, γ	$PO_2^\bullet$	363	8.3	10	35
PE, γ	$PO_2^\bullet$	363	8.3	90	15
IPP, mechan.	$PO_2^\bullet$	273	1.7	5.0	75
IPP, mechan.	$PO_2^\bullet$	273	3.3	7.0	65
IPP, mechan.	$PO_2^\bullet$	292	5.0	90	75
IPP, mechan.	$PO_2^\bullet$	292	3.3	80	80
PMMA, γ	$P^\bullet$	301	0.5	0.2	55
PMMA, γ	$P^\bullet$	311	0.5	3.0	55
PMMA, γ	$P^\bullet$	318	0.5	30	20
PMMA, γ	$P^\bullet$	328	0.5	20	15
PMMA, γ	$PO_2^\bullet$	251	5.0	3.0	60
PMMA, γ	$PO_2^\bullet$	273	6.7	50	55

Source: Refs. 55,56.

rate and an intense mobility of polymer segments. Under the conditions of polymer oxidation, this mechanism is possible at the chain length close to unity. Some examples are given in Table 4.

2. The combination of segmental diffusion with the transfer of the free valence to another segment is due to the chemical reaction; for example, $POO^\bullet + PH$ or $POO^\bullet + P'OOH$. The movement of the free valence occurs

Table 5 Rate Constants of Macroperoxyradicals Disproportionation in Solid Polymers

Polymer	T(K)	E (kJ mol^{-1} s^{-1})	log A, A (kg mol^{-1} s^{-1})	$2k_t$ (293 K) (kg mol^{-1} s^{-1})	Ref.
PE	—	77	16	300	63
PE	283–303	90	16.4	3.8	63
PE	364	—	—	1.0×10^{-2}	57
APP	318–336	52	11.0	75	51
APP	363–378	48.5	13.0	3×10^4	50
IPP	273	—	—	3×10^{-3}	64
IPP	292	—	—	5×10^{-2}	64
IPP, [POOH] = 0.1[a]	328–390	119	21.5	4	65
IPP, [POOH] = 0.0025[a]	298	—	—	170	66
IPP, [POOH] = 0.025[a]	298	—	—	12	66
IPP, [POOH] = 0.10[a]	298	—	—	5	66
IPP	299–320	109	20.0	7.52	67
IPP	299–320	110	19.3	1.0	67
IPP	298	—	—	2.2×10^2	68
IPP	298	—	—	2.2×10^3	68
IPP, [POOH] = 0.13[a]	298	—	—	6	68
IPP	333	—	—	0.4	69
IPP	363	—	—	2.3	69
IPP	303	—	—	5.4×10^{-3}	49
IPP	291	—	—	7.2	63
IPP	323–636	103	15.7	4×10^{-3}	70
PB	373	—	—	1.9×10^{-6}	43
PS	265–283	73	14.0	16	71
PS + 5% C$_6$H$_6$	228–283	26	7.5	870	71
PS	248–413	75	14	7	72
PIB	210–245	75	17.3	7×10^3	72
PMMA	278–310	75	12	7×10^{-2}	72
PMMA	273–292	71	12.5	1	72
PMMA	293	—	—	0.36	73
PVA	347–310	75	13	0.7	72
PFE	—	109	11	8×10^{-9}	70
PFE	—	42	5	4×10^{-3}	70
PMP	295–313	92	18.2	107.5	72

[a] $-$[POOH]/mol kg^{-1}.

as the interchange of segmental diffusion with chemical reactions (28). The rate of the free valence transfer depends on the velocity of segmental diffusion and the rate of the chemical reactions. The decay of free radicals in a polymer in a dioxygen atmosphere is accompanied by O$_2$ consumption (4–10 molecules of dioxygen per one radical at room temperature) (57–62). This mechanism seems to be the most probable for the chain oxidation of solid polymers. The values of the rate constants ($2k_t$) of Eq. (11) in solid polymers are collected in Table 5.

3. In the presence of a low molecular weight additive rH that is sufficiently active toward peroxyl radicals, moving the free valence proceeds as the diffusion of free radical r• is formed in the reaction (27,28).

$$POO• + rH \quad \rightarrow \quad POOH + r• \qquad (14)$$

This additive rH should be very active toward peroxyl radicals. It can be hydroperoxide, ROOH, or an antioxidant InH.

B. Initiated Polymer Oxidation

Polymer oxidation is similar to oxidation of low molecular weight analogues in the liquid phase and has several peculiarities caused by the specificity of solid-phase free radical reactions of macromolecules. Several monographs are devoted to this field of chemistry (16–30).

Important characteristics of polymer oxidation were obtained in the study of their initiated oxidation. In the presence of initiator I that generates the chains with the rate $v_i = const.$, the oxidation of polymer PH proceeds with the constant rate v. When the macroradical of the oxidized polymer $P•$ reacts very rapidly with dioxygen (at $[O_2] > 10^{-4}$ mol L^{-1}), the chain termination proceeds mainly by Eq. (11) (see scheme in Sec. I.A), and the oxidation rate is ($k_4 = k_p$ and $2k_t = 2k_{11}$) (28).

$$v = v_i + k_p(2k_t)^{-1/2}[PH]v_i^{1/2} \qquad (15)$$

The oxidation of PH ([PH] is the concentration of monomer fragments, in mol kg^{-1}) proceeds by the chain mechanism when the oxidation rate is not very high (i.e., when $2k_t v_i < k_p^2[PH]^2$). The oxidation rate for very long chains is

$$v = k_p(2k_t)^{-1/2}[PH]v_i^{1/2} = av_i^{1/2} \qquad (16)$$

The values of the a coefficient for different polymers are collected in Table 6. Hydroperoxide is not the only primary product of polymer oxidation: hydroxyl and carbonyl groups are formed simultaneously. The discrepancy between the consumed oxygen $\Delta[O_2]$ and hydroperoxide formed depends on the dioxygen partial pressure P_{O_2} (85).

$$\Delta[O_2]/[POOH] = 1 + BP_{O_2}^{-1} \qquad (17)$$

where $B = k_{3is}/\chi k_3$; χ is Henry's coefficient of dioxygen, k_1 and k_{1is} are defined in Sec. I.A. This dependence is the result of competition between the following reactions:

$$\sim C(OOH)MeCH_2C•Me \sim + O_2 \quad \rightarrow \quad \sim C(OOH)MeCH_2C(OO•)Me \sim \quad (18)$$
$$\sim C(OOH)MeCH_2C•Me \sim \quad \rightarrow \quad \sim C(O•)MeCH_2C(OH)Me \sim \quad (19)$$

Table 6 Values of Kinetic Parameter $a = vv_i^{-1/2} = k_p[PH](2k_t)^{-1/2}$ for Oxidation of Solid Polymers

Polymer	Cristallinity (%)	Initiator	T(K)	a	Ref.
LDPE	65	DCP	389–402	$1.0 \times 10^{10}\exp(-88/RT)$	74
HDPE	75	^{60}Co	318	8.1×10^{-4}	75
HDPE	75	^{60}Co	295	8.5×10^{-4}	76
HDPE	40	DBP	365	5.4×10^{-3}	74
IPP	0	DBP	358–378	$8.5 \times 10^{8}\exp(-71/RT)$	77
IPP	65	DBP, DCP	349–401	$3.0 \times 10^{6}\exp(-57/RT)$	78,74
IPP	60	AIBN	317–365	$4.3 \times 10^{2}\exp(-27/RT)$	79,80
IPP	0	DBP	344–378	$1.1 \times 10^{2}\exp(-26/RT)$	50
IPP	70	^{60}Co	295	9.0×10^{-3}	158
IPP	70	^{60}Co	353	1.9×10^{-2}	67
IPP	70	^{60}Co	343	1.4×10^{-2}	67
IPP	49	DCP	383	6.3×10^{-2}	81
IPP	49	DCP	388	8.0×10^{-2}	81
IPP	50	POOH	383–413	$1.2 \times 10^{4}\exp(-38/RT)$	47
APP	0	DBP	387	0.21	74
APP	0	DBPO	295–318	$2.2 \times 10^{5}\exp(-44/RT)$	51
CEP (98/2)	60	DCP	390	2.2×10^{-2}	78
CEP (96/4)	60	DCP	382–400	$2 \times 10^{9}\exp(-80/RT)$	78
CEP (87/13)	55	DCP	388	3.4×10^{-2}	78
CEP (65/35)	50	DBP	353–373	$3 \times 10^{6}\exp(-58/RT)$	78
CEP (86/14)	14	^{60}Co	318	1.5×10^{-3}	75
CEP (73/27)	5	^{60}Co	318	1.6×10^{-3}	75
CEP (37/63)	0	^{60}Co	318	4.4×10^{-3}	75
PB	—	DBP	349–383	$7.2 \times 10^{9}\exp(-80/RT)$	50
PMP	—	DBP	358	7.8×10^{-3}	50
PMP	—	DBP	368	1.3×10^{-2}	50
PEA	—	DBP	358–378	$7.6 \times 10^{5}\exp(-54/RT)$	77
PBD	—	hv	293	9.9×10^{-2}	82
NR	—	AIBN	353	0.64	83
CBS	—	hv	293	3.2×10^{-2}	82
CBS	—	hv	293	5.8×10^{-2}	82
PVM	—	DBP	348–373	$2.5 \times 10^{8}\exp(-67/RT)$	84

The following values of ratio k_{1is}/k_1 were estimated from data for polymer oxidation.

Polymer	T, K	k_{3is}/k_3	Ref.
PP	365	1.6×10^{-4}	85
PP	298	2.6×10^{-4}	86
PS	463	2.5×10^{-5}	87

The important peculiarity of semicrystalline polymer oxidation is that only the amorphous phase is oxidized (18,25,28).

Table 7   Rate Constants of Initiators Decomposition with Free Radical Generation in Solid Polymers

Polymer	Initiator	T(K)	k_i(s^{-1})	Ref.
PE	AIBN	335–365	$2.0 \times 10^{20} \exp(-176/RT)$	88
IPP	AIBN	345–362	$5.1 \times 10^{17} \exp(-154/RT)$	80
IPP	AIBN	329–364	$3.7 \times 10^{15} \exp(-141/RT)$	91
IPP	AIBN	353–373	$8.5 \times 10^{14} \exp(-134/RT)$	83
PS	AIBN	333–353	$5.5 \times 10^{13} \exp(-128/RT)$	95
PS	AIBN	353	7.8×10^{-5}	89
PE	DBP	365	4.5×10^{-5}	85
IPP	DBP	360	5.9×10^{-5}	92
PE	DCP	384–402	$8.7 \times 10^{15} \exp(-156/RT)$	93
IPP	DCP	378–403	$1.4 \times 10^{14} \exp(-143/RT)$	94
IPP	DCP	388	8.1×10^{-6}	92
IPP	DPP	333–353	$2.1 \times 10^{14} \exp(-127/RT)$	95
PE	DLP	353	4.4×10^{-4}	89
PS	DLP	353	3.5×10^{-4}	89

C.   Decomposition of Initiators and Cage Effect in the Polymer Matrix

A pair of radicals formed from the decomposed initiator molecule collide with each other many times in the condensed phase before their separation, owing to diffusion. This results from the influence of surrounding molecules of a solvent or segments of a macromolecule of polymer that form a cage. Polyolefins are semicrystalline polymers, so the low molecular weight substances penetrate and usually diffuse only in the amorphous phase. Each pair of particles in a polymer is surrounded by the macromolecule segments. The cage formed by polymer segments is more tight than that of the liquid. The probability e for a pair of radicals to go out the cage is considerably less in the polymer than in liquid. The study of the cage effect and molecular mobility in plasticized PP proved the important role not only of the translational but also the rotational diffusion in the fate of a radical pair in the polymer cage (88–90). The initiation rate constant $k_i = 2ek_d$, where k_d are the decomposition rate constants, are given in Table 7. The values of e for a few initiators are given in Table 8. They are sufficiently lower than in the liquid phase (see Ref. 31).

D.   Diffusion of Oxygen and Polymer Oxidation

According to the kinetic scheme of polymer oxidation, the dependence of the oxidation rate v on the dioxygen concentration is nonlinear (see scheme in Sect. I.A). At a high oxygen concentration [O$_2$] when k_3[O$_2$] > k_p[PH], the reaction rate is independent of [O$_2$] (see Eqs. 15 and 17). At the very low [O$_2$], step 3 [see Eq. (15)] limits the chain propagation and the oxidation rate is (6,15)

$$v = k_3[O_2]P = k_3(2k_9)^{-1/2}[O_2]v_i^{1/2} \qquad (20)$$

Table 8 Probability (e) of Free Radical Escape from the Cage into Bulk Volume in Decomposition of Initiators in Polymer Matrix

Polymer	Initiator	T(K)	e	Ref.
PE	AIBN	343	0.006	88
PE	AIBN	353	0.011	88
PE	AIBN	363	0.020	88
PP	AIBN	333	0.016	96
PP	AIBN	353	0.027	96
PP	AIBN	333	0.013	89
PP	AIBN	343	0.017	89
PP	AIBN	353	0.025	89
PS	AIBN	353	0.05	97
PP	AIBN	344	0.14	50
PP	AIBN	378	0.62	50
PBD	AIBN	349	0.018	50
PBD	AIBN	383	0.59	50
PMP	AIBN	358	0.012	50
PMP	AIBN	368	0.05	50
PP	DBP	298	0.014	51
PP	DBP	318	0.061	51
PP	DBP	328	0.15	51
PP	DLP	353	0.05	90

The complete dependence of v on [O_2] at $k_{10} = 2(k_9 k_{11})^{1/2}$ is expressed by Eq. (21) (28)

$$v = a v_i^{1/2} (1 + d[O_2]^{-1})^{-1} \tag{21}$$

where coefficients $a = k_p (2k_t)^{-1/2}[PH]$ and $d = k_3^{-1} k_p [PH](k_9/k_{11})^{1/2}$ (see scheme in Sect. I.A).

In the oxidation of the polymer film, $2l$ thick by dioxygen, the oxidation is accompanied by the O_2 diffusion into this sample. The oxidation rate at the initiation rate v_i depends on the partial pressure P_{O_2}, the diffusion coefficient D, and Henry's coefficient χ for dioxygen and coefficient $a = k_p (2k_t)^{-1/2}$ [PH] according to Eq. (22) (98).

$$P_{O_2} v^{-1} = l^2 (6D\chi)^{-1} + 4P_{O_2}(5a\, v_i^{1/2}) \tag{22}$$

The values of coefficients a are given in Table 6 and the values of D and χ for dioxygen in polymers are shown in Tables 9 and 10. In semicrystalline polymers, dioxygen is dissolved and diffuses in the amorphous phase.

E. Generation of Free Radicals by Dioxygen Reactions

Dioxygen reacts with the C–H bond of any organic compound, with the formation of free radicals (6,9,15).

$$PH + O_2 \quad \rightarrow \quad P^{\bullet} + HO_2^{\bullet} \tag{23}$$

Table 9 Diffusion Coefficients D of Dioxygen in Polymers

Polymer	$T(K)$	$D(cm^2\ s^{-1})$	$D \times 10^7$ (298 K) $(cm^2\ s^{-1})$	Ref.
LDPE	278–328	$0.43\ exp(-35/RT)$	1.7	99
LDPE	—	$0.53\ exp(-37/RT)$	1.7	100
HDEP	278–328	$5.25\ exp(-40.3/RT)$	4.6	99
HDEP	298	4.5×10^{-7}	4.5	101
HDEP	298–340	$2.0\ exp(-38.5/RT)$	5.3	102
PE	278–328	$0.83\ exp(-39/RT)$	12	99
IPP	298–340	$42\ exp(-46.2/RT)$	4.7	102
IPP	298–340	$1.5\ exp(-36.4/RT)$	6.6	102
IPP	403	6×10^{-6}	—	103
IPP	366	7×10^{-7}	—	98
PS	—	$4.8 \times 10^{-4}\ exp(-26/RT)$	0.13	104
PS	—	$0.12\ exp(-34.9/RT)$	1.11	105
NR	298–323	$1.9\ exp(-34.9/RT)$	14	107
NR	298–323	$0.3\ exp(-29.7/RT)$	18	107
SSR	298	17.3	17.3	106
PBD	298–323	$0.095\ exp(-27.2/RT)$	16	106,107
PBD	—	$0.14\ exp(-28.4/RT)$	16	105
PDMB	298–323	$9.2\ exp(-44.3/RT)$	1.5	106,107
PDMB	—	$19\ exp(-46.4/RT)$	1.4	105
CBS	298–323	$0.23\ exp(-29.7/RT)$	14	104
CBA (61/39)	298–323	$42.5\ exp(-48.6/RT)$	1.3	106,106
CBA (61/39)	—	$14\ exp(-45.6/RT)$	1.3	105
CBA (68/32)	298–323	$7.5\ exp(-42.2/RT)$	3.0	106
CBA (68/32)	298–323	$9.8\ exp(-43/RT)$	36	107
CBA (80/20)	298–323	$4.8\ exp(-34/RT)$	52	106,107
CBA (80/20)	298–323	$0.7\ exp(-33.9/RT)$	8.1	105
PEMA	298–358	$0.039\ exp(-31.8/RT)$	1.0	108
PMMA	298	0.03	0.03	109
PVC	—	$41\ exp(-54/RT)$	0.12	110
PMP	298	1.4×10^{-6}	14	112
PMPBD	298–323	$8.5\ exp(-41/RT)$	5.5	106,109
PIA (74/26)	298–323	$70\ exp(-53/RT)$	0.92	106
PVCVA (87/13)	274–361	$1.0\ exp(-44/RT)$	0.13	111
CIBIP (98/2)	298–323	$43\ exp(-49.7/RT)$	0.81	106

In the oxidation of substances with double bonds, dioxygen reacts with these bonds to form peroxyl radicals (14,15).

$$CH_2 = CHX + O_2 \quad \rightarrow \quad C^{\cdot}HXCH_2O_2^{\cdot} \tag{24}$$

Organic compounds with weak C–H bonds ($D_{R-H} < 330\ kJ\ mol^{-1}$) react with dioxygen in trimolecular reactions (6,9,15).

$$RH + O_2 + HR \quad \rightarrow \quad R^{\cdot} + H_2O_2 + R^{\cdot} \tag{25}$$

These reactions are endothermic and proceed slowly. They are important in the

Table 10 Henry's Coefficient of Dioxygen χ (at 298 K) in Polymers

Polymer	Crystallinity	T(K)	$\chi \times 10^8$ (mol kg^{-1} Pa^{-1})	ΔH (kJ mol^{-1})	Ref.
PE	0.73	291–333	1.5	−3.8	113, 114
PE	0.71	290–321	1.4	−2.9	113, 114
PE	0.55	—	1.8	—	113, 114
PE	0.36	291–316	2.4	7.5	113, 114
PE	0.46	291–319	2.2	2.9	113, 114
IPP	0.72	291–323	1.4	—	114
IPP	0.63	291–323	3.4	1.2	114
IPP	0.63	291–323	2.7	1.2	114
APP	0.2	291–323	7.1	10.0	114
PMP	—	—	7.8–15	—	115
PMMA	—	—	2.4	—	116
PEMA	—	308–348	4.1	4.6	108
PVA	—	290–299	2.9	6.7	117
PVC	—	297–363	1.4	1.3	110
PFE	—	303–389	4.6	−7.2	118
PS	—	—	14.7	—	119
NR	—	—	6.8	—	121
CBS(77/23)	—	—	5.7	—	121
CBA(73/27)	—	—	4.1	—	121
PIB	—	—	7.4	—	121

absence of other sources of free radical generation, such as initiators, decay of hydroperoxide groups, and others. Generation of free radicals in polymers is considerably higher than that of liquid analogues of polymers (31) (Table 11).

F. Autooxidation of Polymers and Decomposition of Hydroperoxide Groups

Autooxidation of polymers and hydrocarbons proceeds with acceleration caused by the decay into free radicals to form hydroperoxide groups (18–21,25–28,30). The initial stage of polymer autooxidation develops at long chains according to Eqs. (26) and (27) (28):

$$v = 2\alpha b^2 t \tag{26}$$

$$\Delta[O_2]^{1/2} = \alpha^{1/2} bt \tag{27}$$

where $\alpha = [POOH]/\Delta[O_2]$ at $t = 0$ and $b = ak_i^{1/2}$, k_i is the rate constant of the hydroperoxide decomposition with the generation of free radicals. The values of the b coefficients are outlined in Table 12.

Not all peroxyl radicals are transformed into hydroperoxide groups in polymer oxidation. Macroradicals P$^\bullet$ are transformed in parallel into hydroxyl and carbonyl groups (see Sec. I.B). The values of rate constants of initiation by hydroperoxide

Table 11   Free Radical Generation on Reaction of Polymer with Dioxygen

Polymer	Crystallinity y	% ash	T (K)	$v_{i0} \times 10^7$ (mol kg^{-1} s^{-1})	$k_{i0} \times 10^6$ (L mol^{-1} s-1)	A (L mol^{-1} s-1)	E (kJ mol^{-1})
LDPE	65	0.7	391	0.67	0.70	—	—
LDPE	65	00.4	391	0.20	0.21	—	—
LDPE	65	0.14	391	0.13	0.14	—	—
LDPE	55	0.45	378	0.54	0.57	2.2×10^{10}	117
LDPE	55	0.45	386	2.07	2.18	2.2×10^{10}	117
LDPE	55	0.45	388	3.60	3.80	2.2×10^{10}	117
LDPE	55	0.45	398	8.01	8.43	2.2×10^{10}	117
LDPE	55	0.45	404	12.6	13.3	2.2×10^{10}	117
HDPE	40	—	362	0.51	0.54	6.8×10^{14}	146
HDPE	40	—	367.5	1.19	1.25	6.8×10^{14}	146
HDPE	40	—	372	1.65	1.73	6.8×10^{14}	146
HDPE	40	—	377	3.94	4.14	6.8×10^{14}	146
HDPE	40	—	377	33.45	3.62	6.8×10^{14}	146
IPP	65	0.048	387	0.10	0.12	2.3×10^{7}	92
IPP	65	0.4	380	4.3	5.33	2.3×10^{7}	92
IPP	65	0.4	389	8.2	10.2	2.3×10^{7}	92
IPP	65	0.4	398	11.2	13.9	2.3×10^{7}	92
IPP	65	0.4	405	20.6	25.5	2.3×10^{7}	92
IPP (L. Ph.)	C_6H_5Cl	0.4	385	0.028	0.014	3.4×10^{5}	98
IPP (L. Ph.)	C_6H_5Cl	0.4	389	0.47	0.023	3.4×10^{5}	98
IPP (L. Ph.)	C_6H_5Cl	0.4	394	0.050	0.025	3.4×10^{5}	98
IPP (L. Ph.)	C_6H_5Cl	0.4	399	0.063	0.031	3.4×10^{5}	98
IPP (L. Ph.)	C_6H_5Cl	0.4	403	0.089	0.44	3.4×10^{5}	98

Source: Refs. 92, 122.

Table 12   Kinetic Parameters of Polymer Autoxidation ($2b$) in Trichlorbenzene Solution [Eqs. (26) and (27)]

Polymer	T (K)	$2b$(mol$^{1/2}$ L$^{-1/2}$ s^{-1})	E(kJ mol^{-1})	Ref.
APP	413	8.3×10^{-5}	113	123
APP	408	8.8×10^{-5}	127	124
PP	408	4.4×10^{-5}	113	125
HDPE	439	6.2×10^{-5}	155	123
PBA	533	5.3×10^{-5}	150	123
IPP	403	3.6×10^{-4} s^{-1}	84	123
PIB	344	3.2×10^{-4} s^{-1}	117	123

groups k_i are measured by the kinetics of autoxidation and by the technique of acceptors of free radicals (28) (Table 13).

Free radicals are produced from hydroperoxide groups as the result of different reactions. In addition to the cleavage of hydroperoxides at O–O bonds, the following

Table 13 Rate Constants of Hydroperoxide Group Decomposition into Free Radicals Measured by Kinetics of Autoxidation and by Free Radical Acceptor Method

Polymer	Conditions of oxidation	T (K)	$k_i(s^{-1})$	Ref.
PE	C_6H_5Cl, 383 K	351–403	$1.6 \times 10^{14} \exp(-146/RT)$	126
PE	Solid phase, 363 K	365	5×10^{-6}	85
IPP	$C_6H_5CH(CH_3)_2$, 383 K	392–403	$2.5 \times 10^{12} \exp(-134/RT)$	126
IPP	C_6H_5Cl, 383 K	365–387	$2.4 \times 10^{6} \exp(-79/RT)$	126
IPP	Solid phase, 363 K	365–387	$7 \times 10^{7} \exp(-92/RT)$	85
IPP	Solid phase, 358 K	322–370	$3.1 \times 10^{11} \exp(-119/RT)$	126
IPP	Solid phase	393	4.6×10^{-5}	127
IPP	Solid phase	403	8.1×10^{-5}	127
IPP	Solid phase	413	1.4×10^{-4}	127
IPP	Solid phase	383	1.2×10^{-4}	47
IPP	Solid phase	393	2.5×10^{-4}	47
IPP	Solid phase	403	3.5×10^{-4}	47
IPP	Solid phase	413	6.0×10^{-4}	47
IPP	Solid phase	365	1.4×10^{-5}	128
IPP	Solid phase	383	2.6×10^{-6}	126
IPP	Solid phase	393	8.9×10^{-6}	126
IPP	Solid phase	383	2.0×10^{-5}	126
IPP	Solid phase	393	4.6×10^{-5}	126
IPP	Solid phase	403	6.2×10^{-5}	126
IPP	Solid phase	348	1.3×10^{-6}	85
IPP	Solid phase	365	5.0×10^{-6}	85
IPP	Solid phase	387	2.8×10^{-5}	85
IPP	Solid phase	403	6.0×10^{-5}	129
PMP	Solid phase	430	1.2×10^{-5}	130
HDPE	Melt	430	1.0×10^{-4}	131

reactions proceed in the oxidizing polymer (28):

$$POOH + \sim CH=CH\sim \quad \rightarrow \quad PO^\bullet + \sim CH(OH)C^\bullet H \sim$$
$$\sim CMe(OOH)CH_2C(OH)Me \sim \quad \rightarrow \quad \sim CMe(O^\bullet)CH_2C(O^\bullet)Me \sim +H_2O$$
$$\sim CH(OOH)CH_2C(O) \sim \quad \rightarrow \quad \sim CH(O^\bullet)CH_2C(O^\bullet)(OH) \sim$$
$$\sim CH(OOH)CH_2C(O)OH \quad \rightarrow \quad \sim CMe(O^\bullet)CH_2C(O)O^\bullet + H_2O$$

Two types of hydroperoxide groups are formed by PP oxidation and they differ in thermal stability (28,126,132). Single OOH groups are more stable than adjacent ones. The latter are decomposed by the reaction.

$$\sim CMe(OOH)CH_2C(OOH)Me \sim \quad \rightarrow \quad \sim CMe(OO^\bullet)CH_2C(O^\bullet)Me \sim +H_2O$$

The rate constants of the decay (k_d) and efficiency e of free radicals initiation ($k_i = 2ek_d$) for hydroperoxide groups of PE and PP are given in Table 14. The order way for decomposition of hydroperoxide groups is their reaction with alkyl macroradicals (induced decomposition) (85).

$$P^\bullet + POOH \quad \rightarrow \quad POH + PO^\bullet$$

Table 14 Rate Constants of Polymer Hydroperoxide Groups Decay

Polymer	T (K)	Medium	k_d (s^{-1})	k_i/k_d	Ref.
HDPE	373–408	C_6H_5Cl	$1.1 \times 10^8 \exp(-104/RT)$	0.2	133
HDPE	393–408	C_6H_5Cl	$1.2 \times 10^8 \exp(-113/RT)$	—	133
LDPE	383–403	C_6H_5Cl	$8.5 \times 10^{15} \exp(-146/RT)$	0.02	126
LDPE	373–393	$C_6H_5Cl + InH$	$1.9 \times 10^8 \exp(-84/RT)$	0.15	126
HDPE	333–373	Solid phase	$2.2 \times 10^8 \exp(-92/RT)$	—	134
HDPE	373–393	Solid phase	$6.2 \times 10^{17} \exp(-162.5/RT)$	—	135
HDPE	413–443	Solid phase	$2.5 \times 10^{14} \exp(-146/RT)$	—	136
HDPE	430	Solid phase, O_2	2.6×10^{-3}	—	131
PE	403	Solid phase	4.7×10^{-4}	—	137
PE	403	Solid phase, O_2	5.3×10^{-3}	—	137
APP	385–395	C_6H_5Cl	$2.5 \times 10^{15} \exp(-113/RT)$	—	138
APP	363–393	Solid phase	$9.5 \times 10^{11} \exp(-113/RT)$	—	139
APP	363–393	Solid phase	$2.3 \times 10^{12} \exp(-115/RT)$	—	140
IPP	398	C_6H_5Cl	6×10^{-4}	0.07	126
IPP	392–407	C_6H_6	$3.8 \times 10^{10} \exp(-109/RT)$	0.03	126
IPP	393–413	Solid phase	$7.5 \times 10^{10} \exp(-109/RT)$	—	141
IPP	393	Solid phase	$(2.6 \div 3.1) \times 10^{-4}$	—	142
IPP	403	Solid phase	$(6.7 \div 8.7) \times 10^{-4}$	—	143
IPP	403	Solid phase	11.5×10^{-4}	—	144
IPP	—	Solid phase	$1.3 \times 10^{10} \exp(-100/RT)$	—	145
IPP	—	Solid phase	$1.7 \times 10^{12} \exp(-129/RT)$	—	145
IPP	—	Solid phase	$2.2 \times 10^{10} \exp(-104/RT)$	0.14	146
IPP	403	Solid phase, O_2	8.7×10^{-4}	—	127
IPP	398	Solid phase	1.8×10^{-4}	0.03	126
PMP	—	Solid phase	$2 \times 10^7 \exp(-84/RT)$	0.04	147
PMP	—	Solid phase	$2.4 \times 10^8 \exp(-96/RT)$	—	147
PMP	403	Solid phase	7.0×10^{-4}	—	130
CEP (95/5)	403	Solid phase	3.7×10^{-4}	—	137
CEP (76/24)	403	Solid phase	1.2×10^{-4}	—	137
PS	353–413	C_6H_5Cl	$1.2 \times 10^{21} \exp(-82.5/RT)$	—	148
PS	463	Solid phase	6.4×10^{-4}	—	87
PS	463	Solid phase	6.8×10^{-3}	—	87
PS	453–473	Solid phase	$8.1 \times 10^{13} \exp(-150.7/RT)$	—	136
PDB	353–403	Solid phase	$1.1 \times 10^3 \exp(-46.5/RT)$	—	149
SR	353–403	Solid phase	$2.2 \times 10^7 \exp(-77/RT)$	—	149
CEVC (5/95)	403	Solid phase	3.1×10^{-4}	—	137
CEVC (89/11)	403	Solid phase	3.5×10^{-4}	—	137

The rate of induced decomposition is

$$v_{ind} = k_i^{1/2} k_{ind}[POOH]^{3/2} \tag{28}$$

(The values of $k_{ind} = k(P^\bullet + POOH)(2k_9)^{-1/2}$ are given in Ref. 28. For the values of k_d see Table 15.)

Table 15 Kinetic Parameters of Induced Decomposition of Hydroperoxide Groups in Solid Phase

Atmosphere	Polymer	T (K)	$k_d \times 10^4$ (s^{-1})	$k_i \times 10^5$ (s^{-1})	e	$k_{ind} \times 10^2$ (kg$^{1/2}$ mol$^{-1/2}$ s$^{-1/2}$)
N$_2$	IPP, OOH group (adjacent)	387	1.4	2.8	0.10	13.0
N$_2$	IPP, OOH group (adjacent)	365	0.21	0.5	0.12	4.5
O$_2$, 10^5 Pa	IPP, OOH group (adjacent)	387	1.4	2.8	0.10	13.6
O$_2$, 2.6 × 10^4 Pa	IPP, OOH group (adjacent)	365	0.19	0.5	0.13	12.0
O$_2$, 10^5 Pa	IPP, OOH group (adjacent)	365	0.21	0.5	0.12	2.9
O$_2$, 10^5 Pa	IPP, OOH group (adjacent)	365	—	0.5	—	3.0
N$_2$	IPP, OOH group (single)	387	0.10	0.21	0.11	9.6
N$_2$	IPP, OOH group (single)	365	0.039	0.017	0.02	3.0
N$_2$	PE	365	0.05	0.5	0.5	0.36
O$_2$, 10^5 Pa	PE	365	0.05	0.5	0.5	0.81
N$_2$	PE	393	4.6	33	0.37	11.0

Source: Ref. 85.

Table 16 Kinetic Parameters of Macromolecular Degradation in Oxidized Polymers

Polymer	T (K)	Conditions of oxidation	$(v_s/v) \times 100$	$k_s \times 10^3$ (s^{-1})	Ref.
LDPE	378	0.36 mol L^{-1} in C$_6$H$_5$Cl	2.0	3.2	163
LDPE	388	0.36 mol L^{-1} in C$_6$H$_5$Cl	1.7	5.0	163
LDPE	398	0.36 mol L^{-1} in C$_6$H$_5$Cl	1.3	8.1	163
LDPE	388	0.36 mol L^{-1} in C$_6$H$_5$Cl	1.6	4.8	166
LDPE	388	Solid phase	1.1	—	164
IPP	393	0.12 mol L^{-1} in C$_6$H$_5$Cl	1.6	3.8	165
IPP	393	0.12 mol L^{-1} in C$_6$H$_5$Cl	1.3	3.1	166

The autoxidation of solid polymer films and particles is local. The oxidation begins in some microvolumes and then develops in the surrounding volume. This has been proved by the chemiluminecence technique (150–156).

As the result of polymer oxidation, the following oxygen-containing groups are formed along with polymer chains: hydroperoxide (18–28), peroxide (75,85,157), hydroxyl (18,28,158), and carbonyl (16–22,76,159–161). The following groups are formed as the products of hydroperoxide group decomposition: hydroxyl, carbonyl, ester, carboxyl, and lactones (162,163).

G. Polymer Degradation in Oxidation

Polymer oxidation is accompanied by degradation that is the result of the decomposition of macroradicals with the C–C bond scission. The following three types of macroradicals are formed in oxidizing polymers: peroxyl, alkoxyl, and alkyl. Each of them is decomposed.

1. Peroxyl radicals of PE and PP are decomposed with the C–C bond scission and double C–C bond production (164–167). The rate of this degradation of polymer v_s is proportional to the rate of oxidation v. The following mechanism was proposed (167).

$$\sim CMe(OO^{\bullet})CH_2CHMe \sim \quad \rightarrow \quad \sim C(O)Me + CH_2 = CMe \sim + HO^{\bullet}$$

The kinetic parameters of such decomposition are seen in Table 16.

2. Alkoxyl radicals are formed during the polymer oxidation in Eq. (11) (see Sec. I.A) and as the result of thermal and induced decomposition of hydroperoxide groups (see Sec. I.F). When Eq. (11) is the main one in the alkoxyl radical production, the stationary concentration is $[PO^{\bullet}] \sim v_i$. As the result, the rate of the polymer degradation is $v_S \sim v_i$. The examples of this degradation are shown in Table 17. (For the absolute rate constants of low molecular weight $RO^{\bullet}$ radical decay see Refs. (10,32,33). Alkoxyl radicals are very active in reactions of hydrogen atom abstraction, and the degradation rate depends on the competition between these reactions (the rate constants of the abstraction reaction; 31–33).

Table 17 Kinetic Parameters (Ratio v_s/v_i) of C–C Bond Scission of Macroperoxyradicals at Initiated Oxidation of Polymers in Chlorobenzene Solutions

Polymer, copolymer	T (K)	[Polymer] (mol L^{-1})	$(v_s/v_i) \times 10^2$	Ref.
PS	348	0.48	7.2	168
PS	393	0.70	7.5	169
Isobutylene-styrene	393	0.91	7.5	169
Isobutylene-p-chlorostyrene	393	0.56	6.2	169
Isobutylene-o-chlorostyrene	393	0.56	5.6	169
Isobutylene-p-methylstyrene	393	0.69	7.1	169
Isobutylene-o-methylstyrene	393	0.69	5.7	169

3. Alkyl macroradicals react with dioxygen very rapidly, and the reaction is limited by dioxygen diffusion. At the slow diffusion of O_2 into the bulk of polymer this reaction proceeds slowly, and radicals P$^\bullet$ recombine. This process of cross-linking increases the molecular weight of the polymer. However, the decay of alkyl macroradicals predominate with an increase in temperature. The decomposition of the end macroradicals produces the monomer (16). The activation energy of depolymerization is close to its enthalpy.

II. PECULIARITIES OF BIMOLECULAR REACTIONS IN THE POLYMER MATRIX

Interaction of two particles in the liquid phase proceeds in a cage formed by small molecules, so that all common orientation of the pair of particles are energetically equivalent owing to the high flexibility of surrounding molecules. This cage may be regarded as "soft". The formation of the transition state of the bimolecular reaction in a soft cage of nonpolar liquid does not require additional energy for reorganization of surrounding molecules.

   In the polymer matrix, each particle, or a pair of particles, is surrounded by polymer segments. They are connected with C–C bonds and, hence, form a "rigid" cage. The different common orientations of the pair of particles are energetically unequal. Therefore, a pair of reagents in a polymer matrix requires additional energy to take the orientation necessary to react in a rigid cage. This peculiarity of the bimolecular reaction in the polymer matrix describes the rigid cage model (170–172). According to this model, a preexponential factor of the bimolecular reaction in the polymer matrix includes the factor $\exp(-U_{or}/RT)$, where U_{or} is the additional energy for the proper orientation of reacting particles. The energy of orientation U_{or} correlates with the energetic barrier of rotation U_r of one of particles ($U_{or} = mU_r$, $m < 1$). The rate constant of a bimolecular reaction in the polymer matrix k_s is expressed through the liquid-phase rate constant k_l and the ratio of frequency of rotation of one of the reagents in the liquid (v_l) and solid (v_s) phases.

$$k_s = k_1(v_i/v_s)^{m-0.5} I_0[0.5 \ln(v_1/v_s)] \tag{29}$$

where $I_0(x)$ is the Bessel function of imaginary argument. The frequencies of

Table 18 Physical Parameters of the Polymer Matrix's Influence on the Bimolecular Reaction of 4-Benzoyloxy-2,2,6,6-tetramethylpiperidinoxyl with 2,6-Di-*tert*-butylphenol

Polymer	T (K)	$k_1 \times 10^3$ (L mol^{-1} s^{-1})	v_l/v_s	U_0 (kJ mol^{-1})	m	U_{or} (kJ mol^{-1})
IPP	313	1.27	235	14.1	0.73	10.0
PE	323	2.10	112	12.6	0.55	6.9
IPP	323	2.10	232	14.6	0.63	9.2
PE	333	3.36	92	12.5	0.52	6.5
IPP	333	3.36	160	14.0	0.50	7.0
PS	333	3.36	480	17.0	0.56	9.5

Source: Ref. 172.

rotations of the nitroxyl radical 4-benzoyloxy-2,2,6,6-tetramethylpiperidineoxyl and values of rate constants in its reaction with 2,6-di-*tert*-butylphenol are shown in Table 18.

The potential barrier U_{or} depends on the common volumes of two reagents V. The dependence has the linear form (at $V^{\ddagger} \geq V^*$) (177):

$$RT \ln(k_1/k_s) = g(V^{\ddagger} - V^*) \tag{30}$$

Coefficients g and V^* were the following for a line of reactions of substituted piperidinoxyls with a substituted phenol.

T/K	303	333	362
$g \times 10^2/kJ$ cm^{-3}	2.50	2.23	1.98
V^*/cm^3 mol^{-1}	127	110	117

The frequency of nitroxyl radical (2,2,6,6-tetramethyl-4-benzoyloxypiperidinoxyl) rotation v_s in IPP was equal to $\log v_s/s^{-1} = 14.9 - 2090/T$.

III. PHENOLIC ANTIOXIDANTS

A. Mechanism of Action of Antioxidants Reacting with Peroxyl Radicals

The compounds that inhibit oxidation of hydrocarbons in the liquid phase may be classed into four groups based on the mechanism of this inhibition: 1) inhibitors that terminate chains through reactions with peroxyl radicals (phenols, aromatic amines, thiophenols; see Secs. III.A and III.B); 2) inhibitors that terminate chains through reactions with alkyl radicals (stable nitroxyl radicals, quinones, quinoneimines; see Sec. III.C); 3) agents that decompose peroxides without generating free radicals (sulfides, phosphites, metal thiophosphates, and carbamates) (178–183); and 4) agents that consume dioxygen very rapidly and so prevent the polymer oxidation (184). The following reactions take place in the system at induction of phenol ArOH, which reacts with the peroxyl radical (185–190).

Table 19 Formulas for the Rate v and Kinetics of Oxygen Consumption $\Delta[O_2]$ (t) of Inhibited Oxidation of Organic Compounds

Limiting steps (Eq.)	v (mol L^{-1} s^{-1})	$\Delta[O_2](t)$ (mol L^{-1})
(4)(31)	$v_i(1 + k_4[RH]/fk_{31}[InH])$	$v_i t - k_4 k_{31}^{-1}[RH]\ln(1 - t/\tau)$
(4)(31)(33)(35)	$2k_4 [RH]^{3/2} (k_{35}v_i/fk_{31}k_{33} [InH])^{1/2}$	$2v_0\tau\, F(t)$
(4)(31)(32)(33)	$k_4 [RH] (k_{32}v_i [ROOH]/fk_{31}k_{33} [InH])^{1/2}$	$v_0^2\, \tau^2 [ROOH]_0^{-1}\, F(t) + 2v_0\, \tau F(t)$
(4)(31)(36)	$k_4 [RH](v_i + 2k_{36} [ROOH][InH])/fk_{31} [InH]$	$2v_0\tau F(t)$
(4)(31)(37)	$k_4 [RH](v_i + 2k_{37} [O_2][InH])/fk_{31} [InH]$	$2v_0\tau\, F(t)$
(4)(31)(38)	$3k_4\, v_i\, [RH]/fk_{31} [InH]$	$-k_4 k_{31}^{-1} [RH]\ln(1 - t/\tau)$
(4)(31)(39)	$k_4 [RH] (k_{39}v_i/fk_{31}k_{33} [InH])^{1/2}$	$2v_0\tau\, F(t)$

$\tau = f[InH]_0/v_i;$ $\quad F(t) = 1 - (1 - t/\tau)^{1/2};$ $\quad v_0 = v_i k_4[RH]/fk_{31}[InH]_0.$
Source: Ref. 28.

$$RO_2^{\bullet} + ArOH \rightarrow ROOH + ArO^{\bullet} \tag{31}$$
$$ArO^{\bullet} + ROOH \rightarrow ArOH + RO_2^{\bullet} \tag{32}$$
$$RO_2^{\bullet} + ArO^{\bullet} \rightarrow ArOOR\ or\ ROH + Q \tag{33}$$
$$ArO^{\bullet} + ArO^{\bullet} \rightarrow products \tag{34}$$
$$ArO^{\bullet} + RH \rightarrow ArOH + R^{\bullet} \tag{35}$$
$$ArOH + ROOH \rightarrow products \tag{36}$$
$$ArOH + O_2 \rightarrow ArO^{\bullet} + HO_2^{\bullet} \tag{37}$$
$$OArOOR \rightarrow OArO^{\bullet} + RO^{\bullet} \tag{38}$$
$$ArO^{\bullet} \rightarrow Q + r^{\bullet} \tag{39}$$
$$ArO^{\bullet} + O_2 \rightarrow Q + HO_2^{\bullet} \tag{40}$$

The kinetics of oxidation by antioxidant depends on some of these reactions, which can be named as key reactions. The number of key reactions depends on the structure of the inhibitor and oxidation conditions. For example, quinolide peroxides are formed in the reaction of 2,4,6-trialkylphenoxyl with RO_2 see Eq. (33) (188) and quinone Q is produced on reactions of peroxyl radicals with phenols having free *ortho*- or *para*-positions. Equation (39) proceeds only with *par-a*-alkoxy-substituted phenoxyls (188). Only semiquinone radicals take part in Eq. (40). The formulas for the oxidation rate and kinetics of dioxygen consumption in the different mechanisms of the antioxidant action at the constant initiation rate are collected in Table 19.

An important distinction of autoxidation of hydrocarbon or polymer is that its product (hydroperoxide) is the initiator, which causes a progressively increasing initiation rate in the course of the reaction. The rate of acceleration, in turn, depends on the rate of chain oxidation (i.e., there is a type of a positive-feedback between autoinitition and autoxidation reactions). First, the more effective is the chain termination by the inhibitor, the slower is its rate of consumption and the longer is the effective inhibition period τ, whereas in an initiated chain reaction with $v_i = const$, it is independent of the InH effectiveness. Second, autoxidation of RH can be inhibited not only by compounds that terminate chains, but also by com-

Table 20 Formulas for $[ROOH] = F([InH])$ at Inhibited Hydrocarbon Oxidation in Nonstationary Regimen; $x = [InH]/[InH]_0$, $[InH] = [InH]_0$ at $t = 0$

Key steps (Eq)	[ROOH]	a
(4)(31)	$[InH]_0 (1 - x - a \ln x)$	$k_4 [RH]/k_{31} [InH]_0$
(4)(32)(33)	$a([InH]_0^{1/2} - [InH]^{1/2}$	$2k_4 [RH] (fk_{32}/k_5k_{31}k_{33})^{1/2}$
(4)(31)(33)(35)	$a([InH]_0^{1/2} - [InH]^{1/2})^{2/3}$	$(9fk_4^2k_{35}/k_5k_{31}k_{33})^{1/3} [RH]$
(4)(31)(36)	$[InH]_0 (b(1 - x) - a \ln x)$	$k_4 [RH]/k_{31} [InH]_0$, $b = 2 k_4k_{36} [RH]/fk_5k_{31}$
(4)(31)(32)	$a [InH]_0(- b \ln x)$	$2 k_4 [RH]/k_{31} [InH]_0$
		$b^{-1} = 1 + f/4(1 + k_5 (1 + a)/k_{37} [O_2])$
(4)(31)(38)	$a [InH]_0(- \ln x)$	$2 k_4 [RH]/k_{31} [InH]_0$
(4)(33)(39)	$a\{([InH]_0^{1/2} - [InH]^{1/2}\}^{2/3}$	$(9fk_4^2k_{39}[RH]^2/k_5k_{31}k_{33})^{1/3}$
(4)(34)(40)	$[InH]_0\{b(1 - x) - a \ln x\}$	$k_4[RH]/k_{31}[InH]_0$
		$b^{-1} = 1 + [k_5k_9(1 + a)[InH]_0]^{1/2}/(k_{15}[O_2])$

Source: Ref. 28.

pounds that decompose ROOH (178–183). Third, critical phenomena are often observed in inhibited autoxidation experiments, which must be attributed to the aforementioned feedback effect (28).

Because ROOH is decomposed during autoxidation, the oxidation may follow either of two regimens, a non–steady-state or quasi–steady-state relative to ROOH (15,25,28). In the non–steady-state process, the ROOH is stable and almost no decomposition is observed during the induction period, i.e., its decomposition rate constant $k_3 < \tau^{-1}$. Obviously, this regimen appears owing to specific conditions of inhibited oxidation, which depend on the structure and reactivity of RH, ROOH, and InH. As oxidation of RH and consumption of InH are interrelated processes, the O_2 absorption rate may be quantitatively expressed in terms of the consumption rate of InH in the system. For each possible mechanism of inhibited oxidation, $[RO_2^\bullet]$ can correlate with [InH] and [ROOH]. For a particular mechanism, the correlation has its specific form. Table 20 contains formulas relating the degree of oxidation with consumed InH. The calculation was based on the following assumed conditions (1) $\Delta[O_2] = [ROOH]$ and (2) the rate of initial chain generation $v_{i0} \ll k_3[ROOH]$.

At a sufficiently high temperature, or in the presence of an ROOH decomposer, ROOH rapidly dissociates and, therefore, the oxidation regimen quickly becomes quasi-steady for the ROOH concentration, with the decomposition rate equal to the rate of its formation. However, the ROOH concentration will tend to increase, because as InH is consumed, the inhibition effect declines and the ROOH formation rate increases. A necessary condition for the quasi–steady-state process is the inequality $k_d \tau \gg 1$, where k_d is the overall rate constant of the ROOH consumption by all possible routes, including dissociation to radicals, decomposition to molecular products, and decomposition under the attack of free radicals. The change from the non–steady-state to the quasi–steady-state condition is related to the induction period τ, which depends on the InH type and concentration. The transition from one inhibitor to another often manifests itself in transitions from one type of autoxidation process to another and to various other critical phenomena (28). What we mean by the critical effects in inhibited autoxidation of RH is that under a certain critical InH concentration $[InH]_{cr}$, a sharp change in the τ vs. [InH] relation takes

Table 21 Formulas for Kinetic Parameters of Hydrocarbon Autoxidation as Chain Reaction in Quasi-stationary Regimen[a]

Key steps (Eq.)	v	$[InH]_{cr}$	$[ROOH]_s$
(4)(31)	$k_4[RH]/fk_{31}[InH]$	$\beta k_4[RH]/fk_{31}$	$\beta v v_{i0}/k_5(1-\beta v)$
(4)(33)(35)	β^{-1}		$\beta^2 k_4 k_{35}[RH]^3/fk_5 k_{31} k_{33}[InH]$
(4)(32)(33)	$k_4 k_{32}^{1/2}[RH](fk_5 k_{31} k_{33}[InH])^{-1/2}$	$\beta^2 k_4^2 k_{32}[RH]^2/fk_5 k_{31} k_{33}$	$v_{i0}/fk_5\{([InH]/[InH]_{cr})-1\}$
(4)(31)(36)	$k_4[RH]/fk_{31}[InH]$	$\beta k_4[RH]/fk_{31}$	$\beta v v_{i0}/(1-\beta v)(k_5+k_{36}[InH])$
(4)(31)(37)	$k_4[RH]/fk_{31}[InH]$	$\beta k_4[RH]/fk_{31}$	$\beta v(v_{i0}+k_{37}[O_2][InH])/$ $/k_5(1-\beta v)$
(4)(31)(38)	$k_4[RH]/fk_{31}[InH]$	$\beta k_4[RH]/fk_{31}$	$2\beta v v_{i0}/k_5(1-2\beta v)$
(4)(33)(39)	β^{-1}		$\beta^2 k_4^2 k_{39}[RH]^2/fk_5 k_{31} k_{33}[InH]$
(4)(34)(40)	$k_{15}[O_2](2k_9 v_{i0})^{-1/2}$ $\times (2+k_4 k_{31}^{-1}[RH][InH]^{-1})$	$k_{31}(2k_9 v_{i0})^{1/2}(\beta k_4 k_{15}[RH])^{-1}$ $-2k_4^{-1}k_{31}[O_2][RH]^{-1}$	$\beta v v_{i0}/k_5(1-\beta v)$

[a] Chain length v; critical concentration of inhibitor $[InH]_{cr}$; and quasi-stationary concentration of hydroperoxide $[ROOH]_s$. The following symbols are used: $\beta = k_5/k_d$ and v_{i0} is rate of free radical generation on reaction of RH with oxygen.
Source: Ref. 28.

Table 22 Formulas for Induction Period τ of Inhibited Oxidation of Hydrocarbons in Quasi-stationary Regimen[a]

Key steps (Eq.)	τ/τ_0	x
(4)(31)(33)	$1-x^{-1}(1+\ln x)$	$fk_{31}[InH]_0/\beta k_4[RH]$
(4)(33)(35)	$1-x^{-1}\ln(1+x)$	$fk_{31}k_{33}[InH]_0 v_{i0}/\beta^2 k_4^2 k_{35}[RH]^3$
(4)(31)(36)	$[InH]_0/[InH]_{cr} - a \ln x$	$(k_{31}k_{36}[InH]_{cr}+b)[InH]_0/(k_{31}k_{36}[InH]_0+b)[InH]_{cr}$ $a=\beta k_5[RH][(f[InH]_0)(k_4^{-1}k_5 k_{31}+\beta k_{36}[RH]);$ $b=k_5 k_{31}+\beta k_4 k_{36}[RH]$
(4)(31)(37)	$cx^{-1}(2-b/a)\{$arc tan $[c(2ax+b)]-$ arc tan $[c(2a+b)]\}$	$fk_{31}[InH]_0/\beta k_4[RH],$ $a=fk_{31}k_{37}[O_2][InH]_{cr}^2,\ b=v_{i0}k_{31}[InH]_{cr},$ $c^{-1}=b[4(f-1)k_{37}[O_2][InH]_{cr}v_{i0}^{-1}-1]^{1/2}$
(4)(32)(33)	$1-x^{-1}\{1+\ln(2-2x^{1/2}+x)+$ arc tan$(x^{1/2}-1)$	$fk_5 k_{31}k_{33}[InH]_0/\beta^2 k_4^2 k_{32}[RH]^2$
(4)(33)(39)	$1-x^{-1}\ln(1+x)$	$fk_{31}k_{33}[InH]_0 v_{i0}/\beta^2 k_4^2 k_{39}[RH]^2$

[a] Symbols are the following: $\tau_0 = f[InH]_0 v_{i0}^{-1}$; $\beta = k_5/k_d$; v_{i0} is the rate of free radical generation on reaction of RH with oxygen.
Source: Ref. 28.

place (i.e., $d\tau/d[InH]$ for $[InH] > [InH]_{cr}$ is much greater than $d\tau/d[InH]$ for $[InH] < [InH]_{cr}$). The decay of hydroperoxide groups is sufficiently rapid at elevated temperatures that the condition $k_3 \gg \tau^{-1}$ for $[InH] > [InH]_{cr}$ is satisfied. Table 21 contains formulas for chain lengths v, $[InH]_{cr}$, and the quasi–steady-state hydroperoxide concentration $[ROOH]_s$ for different inhibited oxidation mechanisms. Table 22 also contains formulas for the induction periods of inhibited autoxidation. The induction period was calculated as the time of InH decrease from $[InH]_0$ to $[InH]_{cr}$ or to zero.

B. The Reactivity of Phenols and Phenoxyls

Peroxyl radicals react rapidly with O–H bonds of phenols [see Eq. (31)] with low activation energy (185). This high reactivity of phenols toward $RO_2^{\bullet}$ is the result of a very low triplet repulsion in the transition state $ROO...H...OAr$. The reactivity of an individual phenol depends on its O–H bond dissociation energy (see Ref. 185). The solvent forms a hydrogen bond with phenol that lowers its reactivity (9,11,12). The reaction of the peroxyl radical with phenol proceeds slower in the polymer matrix than in the liquid phase (Table 23). This results from the influence of the polymeric rigid cage on the bimolecular reaction. The effectiveness of the inhibiting action of phenols depends not only on their reactivity, but also on the reactivity of phenoxyls formed (see scheme in Sect. III.A). Equation (33) leads to chain termination and proceeds rapidly in hydrocarbons (see Ref. 185) as well as in polymers (Table 24). Quinolide peroxides produced in this reaction by sterically hindered phenoxyls are unstable at elevated temperatures (188). (For the rate constants of their decay see Ref. 185.) The reaction of sterically hindered phenoxyls with hydroperoxide groups proceeds slowly in solution (see Ref. 185) and in the polymer matrix (Table 25).

C. Effectiveness of Phenols as Inhibitors of Polymer Oxidation

The effectiveness of the antioxidant action is characterized by the induction period τ at fixed oxidation conditions. Another parameter is the critical concentration of antioxidant $[InH]_{cr}$ (see Tables 21 and 22). This value is found experimentally as the point on the $\tau - [InH]_{cr}$ curve, so that τ slightly depends on $[InH]$ at $[InH] < [InH]_{cr}$, and dependence becomes strong at $[InH] > [InH]_{cr}$. Some values of $[InH]_{cr}$ and τ are shown in Table 26.

The effectiveness of the antioxidant depends not only on its reactivity, but also on its molecular weight, which affects the rate of the antioxidant loss by evaporation. The following example illustrates this dependence. Antioxidants 2,6-di-*tert*-butylphenols, with *para*-substituents of the general structure $ROCOCH_2CH_2$, were introduced into Decalin and polypropylene films that were oxidized by dioxygen at 403 K (200). The experimentally measured values were the following: the induction period τ_1 of Decalin oxidation, induction period τ_2 of PP film oxidation in an atmosphere of dioxygen, induction period τ_3 of PP film oxidation in the flow of dioxygen, and time of evaporation of half of phenol $t_{1/2}$ in a nitrogen atmosphere at $T = 403$ K. The results follow:

R	CH_3	C_6H_{13}	$C_{12}H_{25}$	$C_{18}H_{37}$
$MW/g\ mol^{-1}$	292	362	446	530
$t_{1/2}/s$	1,010	13,000	229,000	2,376,000
$\tau_1 \times 10^{-3}/s$	90	83	72	72
$\tau_2 \times 10^{-3}/s$	342	1,123	1,512	720
$\tau_3 \times 10^{-3}/s$	7.2	7.2	7.2	594

The following correlation was established for IPP oxidation at $T = 473$ K inhibited by phenols between the induction period τ and the rate constant of the reaction of the same phenols with cumylperoxyl radicals k_7 in cumene at $T = 333$ K

Table 23 Rate Constants of Reaction of $PO_2^{\bullet}$ with Phenols in Solid Polymer and $RO_2^{\bullet}$ in Liquid Phase

Phenol	PH	T (K)	$k_7 \times 10^{-3}$ (L mol^{-1} s^{-1}) (polymer)	$k_7 \times 10^{-4}$ (L mol^{-1} s^{-1}) (cumene)	Ref.
2,6-Di-*tert*-butyl-4-methylphenol	IPP	388	5.6	2.5	191
2-6-Di-*tert*-butyl-4-methylphenol	IPP	353	3.5	3.5	79
2,6-Di-*tert*-butyl-4-methylphenol	SSR	353	13.5	3.5	192
2,6-Di-*tert*-butyl-4-methoxyphenol	IPP	353	8.1	20	79
2,6-Di-*tert*-butyl-4-*tert*-butoxyphenol	IPP	353	3.9	8.3	79
2,4,6-Tri-*tert*-butylphenol	IPP	353	3.2	2.5	79
2,6-Di-*tert*-butyl-4-phenylphenol	IPP	353	1.8	3.3	79
2,6-Di-*tert*-butyl-4-chlorophenol	IPP	353	2.7	1.4	79
2,6-Di-*tert*-butyl-4-acetylphenol	IPP	353	0.55	0.42	79
2,6-Di-*tert*-butyl-4-benzoylphenol	IPP	353	0.95	0.30	79
2,6-Di-*tert*-butyl-4-cyanophenol	IPP	353	0.61	0.28	79
2,6-Di-*tert*-butyl-4-diphenylmethylphenol	IPP	353	1.28	—	79
2,6-Di-*tert*-butyl-4-(2'-methoxyoxoethyl)phenol	IPP	353	1.02	—	79
2,6-Di-*tert*-butyl-4-octadecyloxyoxoacetyloxyphenol	IPP	353	1.03	—	79
α-Naphthol	IPP	388	4.2	40	191
2,2'-Methylenebis(4-methyl-6-*tert*-butyl)phenol	IPP	388	11.2	65	191
2,2'-Methylenebis(4-methyl-6-*tert*-butyl)phenol	IPP	353	13.3	65	192
4,4'-Methylenebis(2,6-di-*tert*-butyl)phenol	SSR	353	12.0	4.4	192
2,4,6-Tri-(3',5'-di-*tert*-butyl-4-hydroxyphenyl)phenol	SSR	353	9.1	—	192
2,6-Di-*tert*-butyl-4-(1'-isobutyl)ethylylphenol	IPP	388	22	6.3	191
Tetra(methyl-3,5-di-*tert*-butyl-4-hydroxyphenyl)methane	SSR	353	12.2	8.4	192
2,6-Dimercapto-4-*tert*-butylphenol	IPP	388	1.2	0.94	193

Table 24 Rate Constants of Cross-Recombination of 1-Cyano-1-methylethylperoxyl Radical with *p*-Substituted 2,6-Di-*tert*-butylphenoxyl Radical in Solid IPP (k_s) and Liquid (Benzene) (k_1) Phase

p-Substituent	$k_s \times 10^{-6}$ (L mol^{-1} s^{-1})	$k_1 \times 10^{-8}$ (L mol^{-1} s^{-1})	k_1/k_s
OCH$_3$	1.2	7.5	625
OC(CH$_3$)$_3$	5.9	3.8	64.4
C(CH$_3$)$_3$	2.9	8.5	293
C$_6$H$_5$	1.6	7.3	456
Cl	6.7	1.0	15
COC$_6$H$_5$	4.1	0.96	23.4
CN	24	2.4	10
CH$_2$CH$_2$COOCH$_3$	6.0	8.0	133
CH$_2$CH$_2$COOC$_{18}$H$_{37}$	6.5	—	—
CH(C$_6$H$_5$)$_2$	21	—	—

Source: Ref. 79.

Table 25 Rate Constants for Reaction of *para*-Substituted 2,6-Di-*tert*-butylphenoxyl Reaction with Hydroperoxide Groups of IPP

Phenoxyl	k_{-7} (353 K) (L mol^{-1} s^{-1})	log A, A (L mol^{-1} s^{-1})	E (kJ mol^{-1})	k_s/k_1 (353 K)
2,6-[(CH$_3$)$_3$C]$_2$-C$_6$H$_3$O$^\bullet$	0.29	9.4	67	0.13
2,6-[(CH$_3$)$_3$C]$_2$-C$_6$H$_3$O$^\bullet$	1.49	10.4	67	0.67
2,6-[(CH$_3$)$_3$C]$_2$-4-(CH$_3$)COC$_6$H$_3$O$^\bullet$	8.9×10^{-3}	15.0	115	1.5×10^{-2}
2,6-[(CH$_3$)$_3$C]$_2$-4-(CH$_3$)$_3$CC$_6$H$_3$O$^\bullet$	1.7×10^{-2}	12.8	98	4.3×10^{-3}
2,6-[(CH$_3$)$_3$C]$_2$-4-C$_6$H$_5$COC$_6$H$_3$O$^\bullet$	4.4×10^{-2}	12.5	94	2.2×10^{-3}
2,6-[(CH$_3$)$_3$C]$_2$-4-CNC$_6$H$_3$O$^\bullet$	0.14	16.0	115	4.0×10^{-4}

Source: Refs. 174 and 196.

Table 26 Critical Concentrations of Phenols and Induction Periods of IPP Oxidation at 473 K and a Dioxygen Partial Pressure of 10^4 Pa

Inhibitor	[InH]$cr \times 10^3$ (mol kg^{-1})	$\tau \times 10^{-3}$ (s)
2,2'-Methylenebis(4-methyl-6-*tert*-butyl)phenol	1.2	14
Di(3,5-di-*tert*-butyl-4-hydroxybenzyl)sulfide	2.0	13
1,2-Thiobis(4-methyl-6-*tert*-butyl)phenol	3.0	15
2,2'-Methylenebis(4-chloro-6-*tert*-butyl)phenol	4.0	2.7
2,2-Thiobis(4-chloro-6-*tert*-butyl)phenol	4.0	6.0
4,4'-Methylenebis(2,6-di-*tert*-butyl)phenol	4.0	7.5

Source: Refs. 197–199.

(201):

$$\tau/s = 1.44 \times 10^4 - 2.88 \times 10^4 \, k_7^{-1} \tag{41}$$

The diffusion coefficients and solubility of phenols in polymers are given in Tables 27 and 28.

Table 27 Diffusion Coefficients of Inhibitors D in Polymers

Inhibitor	Polymer	T (K)	D (cm^2 s^{-1})	Ref.
2,6-Di-*tert*-butyl-4-methylphenol	HDPE	296–347	$6.92 \times 10^3 \exp(-73.4/RT)$	202
2,6-Di-*tert*-butyl-4-methylphenol	HDPE	347–363	$1.82 \times 10^9 \exp(-109.9/RT)$	202
2,6-Di-*tert*-butyl-4-methylphenol	IPP	333–383	$5.02 \times 10^3 \exp(-96.4/RT)$	203
2,2′-Methylenebis(4-ethyl-6-*tert*-butyl)phenol	HDPE	296–363	$1.74 \times 10^8 \exp(-102.8/RT)$	202
2,2′-Methylenebis(4-ethyl-6-*tert*-butyl)phenol	APP	363–473	$2.88 \times 10^8 \exp(-114.8/RT)$	204
2,2′-Methylenebis(4-methyl-6-α-methylcyclohexyl)phenol	HDPE	296–363	$2.63 \times 10^7 \exp(-100.0/RT)$	202
Octadecyl-3-(3′,5′-di-*tert*-butyl-4′-hydroxyphenyl)propionate	HDPE	325–363	$1.41 \times 10^9 \exp(-108.3/RT)$	202
1,1,3-Tris(5′-*tert*-butyl-4′-hydroxy-2′-methylphenyl)butane	HDPE	325–363	$2.88 \times 10^3 \exp(-71.3/RT)$	202
1,1,3-Tris(5′-*tert*-butyl-4′-hydroxy-2′-methylphenyl)butane	HDPE	296–363	$1.70 \times 10^6 \exp(-92.7/RT)$	202
1,1,3-Tris(5′-*tert*-butyl-4′-hydroxy-2′-methylphenyl)butane	IPP	373–423	$1.58 \times 10^4 \exp(-93.3/RT)$	205
Bis(3,5-di-*tert*-butyl-4-hydroxyphenyl)ethoxycarbonylethylsulfide	PMP	413–473	$0.08 \exp(-59.4/RT)$	205
Bis(3,5-di-*tert*-butyl-4-hydroxyphenyl)ethoxycarbonylethylsulfide	HDPE	296–347	$1.02 \times 10^{14} \exp(-143.2/RT)$	202
Pentaerythrityl tetrakis[3-(3′,5′-di-*tert*-butyl-4′-hydroxyphenyl)propionate]	HDPE	347–363	$1.38 \times 10^4 \exp(-77.8/RT)$	202
Pentaerythrityl tetrakis[3-(3′,5′-di-*tert*-butyl-4′-hydroxyphenyl)propionate]	HDPE	296–363	$9.12 \times 10^8 \exp(-115.7/RT)$	202
Pentaerythrityl tetrakis[3-(3′,5′-di-*tert*-butyl-4′-hydroxyphenyl)propionate]	IPP	298	8.0×10^{-13}	206
Pentaerythrityl tetrakis[3-(3′,5′-di-*tert*-butyl-4′-hydroxyphenyl)propionate]	PBD	343–373	$0.52 \exp(-41.1/RT)$	207
Pentaerythrityl tetrakis[3-(3′,5′-di-*tert*-butyl-4′-hydroxyphenyl)propionate]	CBA(82/18)	343–373	$4.57 \exp(-52.4/RT)$	207
Pentaerythrityl tetrakis[3-(3′,5′-di-*tert*-butyl-4′-hydroxyphenyl)propionate]	CBA(74/26)	343–373	$0.44 \exp(-50.3/RT)$	207
Pentaerythrityl tetrakis[3-(3′,5′-di-*tert*-butyl-4′-hydroxyphenyl)propionate]	CBA(60/40)	343–373	$11.5 \exp(-54.5/RT)$	207
2-Hydroxy-4-methoxybenzophenone	HDPE	343–363	$1.9 \exp(-50.2/RT)$	208,209
2-Hydroxy-4-methoxybenzophenone	LDPE	343–363	$3.02 \times 10^2 \exp(-72.8/RT)$	208,209
2-Hydroxy-4-methoxybenzophenone	IPP	353–383	$6.76 \times 10^2 \exp(-76.1/RT)$	208,209
2-Hydroxy-4-octyloxybenzophenone	HDPE	313–343	$3.3 \times 10^2 \exp(-65/RT)$	210,211
2-Hydroxy-4-octyloxybenzophenone	IPP	323–383	$1.00 \times 10^4 \exp(-87.0/RT)$	208,209

Table 28 Solubility of Inhibitors (mol% kg^{-1}) in Polymers

Inhibitor	Polymer	T (K)	Solubility (mol% kg^{-1})	Ref.
2,6-Di-*tert*-butyl-4-methylphenol	HDPE	300–342	$6.76 \times 10^7 \exp(-43/RT)$	25
2,6-Di-*tert*-butyl-4-methylphenol	IPP	303–344	$1.29 \times 10^6 \exp(-33/RT)$	212
2,6-Di-*tert*-butyl-4-methylphenol	PBD	295	18.0	213
2,6-Di-*tert*-butyl-4-methylphenol	SSR	295	16.0	213
2,6-Di-*tert*-butyl-4-methylphenol	CBA(82/18)	295	24.8	213
2,6-Di-*tert*-butyl-4-methylphenol	CBA(74/26)	295	20.1	213
2,6-Di-*tert*-butyl-4-methylphenol	CBA(60/40)	295	13.1	213
2,4,6-Tri-*tert*-butylphenol	HDPE	303	$6.03 \times 10^5 \exp(-90.6/RT)$	25
2,4,6-Tri-*tert*-butylphenol	IPP	303	$1.90 \times 10^7 \exp(-43.2/RT)$	212
2,6-Di-*tert*-butyl-4-phenylphenol	IPP	313–373	$1.29 \times 10^6 \exp(-36.9/RT)$	212
2,6-Di-*tert*-butyl-4-phenylphenol	PMP	303–323	$6.61 \times 10^{13} \exp(-81.3/RT)$	25
2,6-Di-*tert*-butyl-4-(methoxyoxo-2'-ethyl)phenol	HDPE	303–363	$2.40 \times 10^8 \exp(-50.3/RT)$	25
2,6-Di-*tert*-butyl-4-(methoxyoxo-2'-ethyl)phenol	IPP	303–333	$2.19 \times 10^7 \exp(-43.1/RT)$	212
2,6-Di-*tert*-butyl-4-(methoxyoxo-2'-ethyl)phenol	PMP	303–333	$6.03 \times 10^{15} \exp(-97.6/RT)$	25
2,2'-Methylenebis(4-methyl-6-*tert*-butyl)phenol	HDPE	303–353	$5.50 \times 10^3 \exp(-27.6/RT)$	25
2,2'-Methylenebis(4-methyl-6-*tert*-butyl)phenol	HDPE	296–263	$2.19 \times 10^3 \exp(-15.9/RT)$	202
2,2'-Methylenebis(4-methyl-6-*tert*-butyl)phenol	IPP	303–353	$1.48 \times 10^7 \exp(-47.8/RT)$	212
2,2'-Methylenebis(4-methyl-6-*tert*-butyl)phenol	IPP	323–373	$5.13 \times 10^4 \exp(-26.8/RT)$	214
2,2'-Methylenebis(4-methyl-6-*tert*-butyl)phenol	PMP	303–363	$5.62 \times 10^7 \exp(-55.3/RT)$	25
2,2'-Methylenebis(4-methyl-6-*tert*-butyl)phenol	PBD	295–253	$2.40 \times 10^5 \exp(-29.0/RT)$	213
2,2'-Methylenebis(4-methyl-6-*tert*-butyl)phenol	SSR	295–253	$4.68 \times 10^4 \exp(-26.0/RT)$	213
2,2'-Methylenebis(4-methyl-6-*tert*-butyl)phenol	CBA(82/18)	—	12.7	213
2,2'-Methylenebis(4-methyl-6-*tert*-butyl)phenol	CBA(74/26)	—	14.4	213
2,2'-Methylenebis(4-methyl-6-*tert*-butyl)phenol	CBA(60/40)	—	10.9	213

Compound	Polymer	Temperature (K)	Rate expression	Ref.
4,4'-Thiobis(6-*tert*-butyl-3-methyl)phenol	HDPE	295–263	$1.29 \times 10^{11} \exp(-80.9/RT)$	202
2,2'-Methylenebis(4-chloro-6-*tert*-butyl)phenol	IPP	313–373	$3.47 \times 10^{4} \exp(-28.5/RT)$	212
2,2'-Methylenebis(4-ethyl-6-*tert*-butyl)phenol	HDPE	296–263	$8.91 \times 10^{3} \exp(-26.8/RT)$	202
2,2'-Methylenebis(4-methyl-6-α-methylcyclohexyl)phenol	HDPE	296–263	$8.13 \times 10^{4} \exp(-32.8/RT)$	202
2,2'-Methylenebis(4-methyl-6-α-methylcyclohexyl)phenol	IPP	313–363	$1.15 \times 10^{4} \exp(-27.6/RT)$	212
2,2'-Methylenebis(4-methyl-6-α-methylcyclohexyl)phenol	PBP	343–373	$2.19 \times 10^{4} \exp(-23.0/RT)$	207
2,2'-Methylenebis(4-methyl-6-α-methylcyclohexyl)phenol	CBA(82/18)	343–373	$1.29 \times 10^{5} \exp(-27.2/RT)$	207
2,2'-Methylenebis(4-methyl-6-α-methylcyclohexyl)phenol	CBA(74/26)	343–373	$1.55 \times 10^{5} \exp(-27.2/RT)$	207
2,2'-Methylenebis(4-methyl-6-α-methylcyclohexyl)phenol	CBA(60/40)	343–373	$4.90 \times 10^{4} \exp(-27.2/RT)$	207
2,2'-Methylenebis(4-methyl-6-*tert*-butyl)phenol	HDPE	296–263	$1.58 \times 10^{5} \exp(-38.5/RT)$	202
2,6-Di-*tert*-butyl-4-(2'-carboxyethyl)phenol	HDPE	296–322	$3.98 \times 10^{7} \exp(-27.2/RT)$	202
2,6-Di-*tert*-butyl-4-(2'-carboxyethyl)phenol	HDPE	325–363	$9.33 \times 10^{6} \exp(-72.8/RT)$	202
1,1,3-Tris(5'-*tert*-butyl-4'-hydroxy-2'-methylphenyl)butane	HDPE	295–363	$2.00 \times 10^{6} \exp(-47.3/RT)$	202
Bis(3,5-di-*tert*-butyl-4-hydroxyphenyl)-etoxycarbonylethylsulfide	HDPE	323–373	$4.57 \times 10^{5} \exp(-49.7/RT)$	214
2,4,6-Tris(3',5'-di-*tert*-butyl-4'-hydroxybenzyl)mesitylene	HDPE	296–347	$1.00 \times 10^{9} \exp(-61.6/RT)$	202
2,4,6-Tris(3',5'-di-*tert*-butyl-4'-hydroxybenzyl)mesitylene	HDPE	347–363	$6.03 \times 10^{6} \exp(-46.9/RT)$	202
2,4,6-Tris(3',5'-di-*tert*-butyl-4'-hydroxybenzyl)mesitylene	HDPE	296–363	$1.10 \times 10^{3} \exp(-28.1/RT)$	202
2,4,6-Tris(3',5'-di-*tert*-butyl-4'-hydroxybenzyl)mesitylene	IPP	323–373	$3.89 \times 10^{8} \exp(-62.9/RT)$	214
Pentaerythrityl tetrakis[3-(3',5'-di-*tert*-butyl-4'-hydroxyphenyl)propionate]	HDPE	296–362	$3.98 \times 10^{8} \exp(-62.4/RT)$	202
Pentaerythrityl tetrakis[3-(3',5'-di-*tert*-butyl-4'-hydroxyphenyl)propionate]	IPP	343–413	$3.02 \times 10^{3} \exp(-23.9/RT)$	25
Octadecyl-3,5-bis-*tert*-butyl-4-hydroxybenzenepropionate	PP	323	0.10	215
Octadecyl-3,5-bis-*tert*-butyl-4-hydroxybenzenepropionate	PP	353	0.17	215
Octadecyl-3,5-bis-*tert*-butyl-4-hydroxybenzenepropionate	CEP	323	0.21	215
Octadecyl-3,5-bis-*tert*-butyl-4-hydroxybenzenepropionate	CEP	353	0.51	215

IV. ACCEPTORS OF ALKYL RADICALS

A. Polymer Oxidation Inhibited by Nitroxyl Radicals and Quinones

Stable nitroxyl radicals are formed as active intermediates from sterically hindered piperidines (hindered amine light stabilizers; HALS) (216–223). Nitroxyl radicals $AmO^{\bullet}$ react very rapidly with alkyl radicals (Table 29).

$$AmO^{\bullet} + R^{\bullet} \quad \rightarrow \quad AmOR \tag{42}$$

This reaction is the chain termination step in polymer oxidation. The rate of polymer oxidation v in the presence of $AmO^{\bullet}$ and $k_7'[P^{\bullet}][AmO^{\bullet}] > 2k_t[POO^{\bullet}]$ obeys Eq. (43) (228).

$$v = k_1[O_2]v_i/k'_7[AmO^{\bullet}] \tag{43}$$

The same kinetics is observed for polymer oxidation retarded by benzoquinone (228). The values of ratio k_1/k_7' are shown in Table 30.

B. Cyclic Chain Termination by Nitroxyl Radicals

The HALS appeared to be effective light stabilizers owing to mechanisms of cyclic chain termination when one radical or molecule of inhibitor terminates a few oxidative chains. The following five mechanisms of cyclic chain termination were described.

1. Nitroxyl radical $AmO^{\bullet}$ produces hydroxamic ether AmOP in the reaction with the alkyl macroradical $P^{\bullet}$. The hydroxamic ether formed has weak C–H bonds that are attacked by peroxyl macroradicals with regeneration of the initial nitroxyl radical (228–235).

$$POO^{\bullet} + AmOCH_2CHME\sim \quad \rightarrow \quad POOH + CH_2 = CMe \sim +AmO^{\bullet}$$

2. Hydroxamic ether has a relatively weak C–O bond, which is cleaved at elevated temperature to produce the pair of radicals $AmO^{\bullet}$ and $P^{\bullet}$ in one cage. Because of the disproportionation, hydroxylamine is formed, which is easily attacked by the peroxyl radical to produce the nitroxyl radical (236–240).

$$AmOP \rightarrow [AmO^{\bullet} + P^{\bullet}] \quad \rightarrow \quad AmOH + CH_2 = CMe\sim$$
$$POO^{\bullet} + HOAm \quad \rightarrow \quad AmO^{\bullet} + POOH$$

3. The N–O bond of hydroxamic ether is also thermally unstable. The homolysis of this bond initiates the following cascade of reactions (241, 242).

$$AmOP \quad \rightarrow \quad Am^{\bullet} + PO^{\bullet}$$
$$Am^{\bullet} + POO^{\bullet} \quad \rightarrow \quad AmO^{\bullet} + PO^{\bullet}$$
$$AmO^{\bullet} + P^{\bullet} \quad \rightarrow \quad AmOP$$

Table 29 Rate Constants of Alkyl Radicals Reactions with Nitroxyl Radicals

Alkyl radical	Nitroxyl radical	T (K)	k (1 mol^{-1} s^{-1}) or log $k = A - E/\theta$	Ref.
$C^{\bullet}H_3$	$cyclo\text{-}[N(O^{\bullet})C(CH_3)_2CH=C(CONH_2)C(CH_3)_2]$	298	7.8×10^8	224
$CH_3(CH_2)_7C^{\bullet}H_2$	$cyclo\text{-}[N(O^{\bullet})C(CH_3)_2(CH_3)_3C(CH_3)_2]$	270–317	$10.4 - 7.5/\theta$	224
$(CH_3)_3C^{\bullet}$	$cyclo\text{-}[N(O^{\bullet})C(CH_3)_2(CH_2)_3]$	293	7.6×10^8	225
$(CH_3)_3C^{\bullet}$	$1,2\text{-}cyclo\text{-}[C(CH_3)_2N(O^{\bullet})C(CH_3)_2] - C_6H_4$	293	8.8×10^8	225
$(CH_3)_3CC^{\bullet}H_2$	$cyclo\text{-}[N(O^{\bullet})C(CH_3)_2(CH_2)_3C(CH_3)_2]$	293	9.6×10^8	225
$cyclo\text{-}[CH(C^{\bullet}H_2)C(CH_3)_2(CH_2)_2]$	$1,2\text{-}cyclo\text{-}[C(CH_3)_2N(O^{\bullet})C(CH_3)_2] - C_6H_4$	333–398	$10.0 - 2.5/\theta$	226
$cyclo\text{-}[CH(C^{\bullet}H_2)(CH_2)_2]$	$1,2\text{-}cyclo\text{-}[C(CH_3)_2N(O^{\bullet})C(CH_3)_2] - C_6H_4$	333–398	$7.4 - 4.6/\theta$	226
$CH_2=CH(CH_2)_3C^{\bullet}H_2$	$1,2\text{-}cyclo\text{-}[C(CH_3)_2N(O^{\bullet})C(CH_3)_2] - C_6H_4$	333–398	$9.7 - 3.8/\theta$	226
$CH_2=CHC(CH_3)_2C^{\bullet}H_2$	$1,2\text{-}cyclo\text{-}[C(CH_3)_2N(O^{\bullet})C(CH_3)_2] - C_6H_4$	353	$8.9 - 0.4/\theta$	226
$CH_2=CHCH_2OCH_2C^{\bullet}(CH_3)_2$	$1,2\text{-}cyclo\text{-}[C(CH_3)_2N(O^{\bullet})C(CH_3)_2] - C_6H_4$	353	$9.2 - 2.9/\theta$	226
$CH_2=CHCH_2OC^{\bullet}H_2$	$1,2\text{-}cyclo\text{-}[C(CH_3)_2N(O^{\bullet})C(CH_3)_2] - C_6H_4$	353	$9.5 - 1.3/\theta$	226
$CH_2=CHCH_2OCH_2C^{\bullet}H_2$	$1,2\text{-}cyclo\text{-}[C(CH_3)_2N(O^{\bullet})C(CH_3)_2] - C_6H_4$	333–398	$9.6 - 2.1/\theta$	226
$CH_2=CHCH_2OCH_2C^{\bullet}HCH_3$	$1,2\text{-}cyclo\text{-}[C(CH_3)_2N(O^{\bullet})C(CH_3)_2] - C_6H_4$	353	$8.5 - 2.1/\theta$	226
$C_6H_5C^{\bullet}H_2$	$cyclo\text{-}[N(O^{\bullet})C(CH_3)_2(CH_2)_3C(CH_3)_2]$	233–306	$9.3 - 3.7/\theta$	225
$C_6H_5C^{\bullet}H_2$	$1,2\text{-}cyclo\text{-}[C(CH_3)_2N(O^{\bullet})C(CH_3)_2] - C_6H_4$	293	5.5×10^8	225
$C_6H_5C^{\bullet}HCH_3$	$cyclo\text{-}[N(O^{\bullet})C(CH_3)_2(CH_2)_3C(CH_3)_2]$	293	1.6×10^8	225
$C_6H_5C^{\bullet}(CH_3)_2$	$cyclo\text{-}[N(O^{\bullet})C(CH_3)_2(CH_2)_3C(CH_3)_2]$	293	1.2×10^9	225
$(C_6H_5)_2C^{\bullet}CH$	$cyclo\text{-}[N(O^{\bullet})C(CH_3)_2(CH_2)_3C(CH_3)_2]$	293	4.6×10^7	225
$(C_6H_5)_2C^{\bullet}CH_3$	$cyclo\text{-}[N(O^{\bullet})C(CH_3)_2(CH_2)_3C(CH_3)_2]$	293	4.6×10^7	225
$1\text{-}C^{\bullet}H_2 - C_{10}H_7$	$cyclo\text{-}[N(O^{\bullet})C(CH_3)_2(CH_2)_3C(CH_3)_2]$	293	8.2×10^7	225
$2\text{-}C^{\bullet}H_2 - C_{10}H_7$	$cyclo\text{-}[N(O^{\bullet})C(CH_3)_2(CH_2)_3C(CH_3)_2]$	293	5.7×10^7	225
$C^{\bullet}H_2OH$	$cyclo\text{-}[N(O^{\bullet})C(CH_3)_2CH_2CH(C(O)NH_2)C(CH_3)_2]$	298	4.6×10^8	224
$C^{\bullet}H_2OH$	$cyclo\text{-}[N(O^{\bullet})C(CH_3)_2CH=CH(C(O)NH_2)C(CH_3)_2]$	298	4.6×10^8	224
$(C_6H_5)_2C^{\bullet}OH$	$cyclo\text{-}[N(O^{\bullet})C(CH_3)_2CH=C(C(O)NH_2)C(CH_3)_2]$	295	5.0×10^7	227
$C^{\bullet}H_2CH_2OH$	$cyclo\text{-}[N(O^{\bullet})C(CH_3)_2CH=CH(C(O)NH_2)C(CH_3)_2]$	298	4.7×10^8	224

Solvent: H_2O (224); isooctane (225); $cyclo\text{-}C_6H_{12}$ (226); 1-propanol (227); $\theta = 2.3RT$.

Table 30 Relative Rate Constants of Inhibitors Terminating the Chains on Reactions with Alkyl and Peroxyl Radicals (see Sec. III.A)

Inhibitor	T/K	$2k_7/k_p$	$2k_7'/k_1$
4-Benzoyloxy-2,2,6,6-tetramethylpiperidinoxyl	387	0	9.5×10^{-2}
p-Benzoquinone	387	0	7.6×10^{-2}
Anthracene	387	1.3×10^{-1}	3.6×10^{-3}
2,6-Dinitrophenol	387	9.0×10^{-2}	4.8×10^{-3}
1,1'-Bis(3,5-di-*tert*-butyl-4-on-2,5-cyclodienylydene)	366	8.4×10^{-1}	9.3×10^{-2}
1,2-Bis(3',5'-di-*tert*-butyl-4-on-2',5'-cyclohexadiene) ethanediylydene	366	6.3×10^{-1}	1.8×10^{-2}

Source: Refs. 224–226.

4. Hydrogen peroxide is formed as the product of decay of adjacent hydroperoxide groups of oxidized PP (see Sec. I.F). Radicals $HO_2^{\bullet}$ are produced in an exchange reaction of $POO^{\bullet}$ with hydrogen peroxide. They have the reductive activity and reduce the nitroxyl radical into hydroxylamine. The latter react readily with peroxyl radicals forming the nitroxyl radical (243–245).

$$POO^{\bullet} + H_2O_2 \rightarrow POOH + HOO^{\bullet}$$
$$POO^{\bullet} + HO_2^{\bullet} \rightarrow POOH + O_2$$
$$POO^{\bullet} + AmOH \rightarrow POOH + AmO^{\bullet}$$

This mechanism of cyclic chain termination was also observed in the oxidation of PP retarded by quinones (246).

5. The nitroxyl radical effectively terminates the chains in hydrocarbon (ethylbenzene) oxidation in the presence of hydrogen peroxide and organic acid HA, according to the following cyclic mechanism (247,248).

$$AmO^{\bullet} + HA \Leftrightarrow AmOH^{+}, A^{-}$$
$$ROO^{\bullet} + AmOH^{+}, A^{-} \rightarrow ROOH + AmO^{+}, A^{-}$$
$$AmO^{+}, A^{-} + H_2O_2 \rightarrow AmOH + HA + O_2$$
$$ROO^{\bullet} + AmOH \rightarrow AmO^{\bullet} + ROOH$$

The same phenomenon was observed in the same system with quinone instead of nitroxyl (249).

The values of inhibition coefficient f in polymer oxidation in the presence of piperidinoxyls and quinones were the following (231,232,243).

Inhibitor	Polymer	T/K	f
2,2,6,6-Tetramethyl-4-benzoyloxypiperidineoxyl	PP	387	14
2,2,6,6-Tetramethyl-4-stearyloxypiperidineoxyl	PP with POOH	388	40
Stilbenquinone	PP with POOH	365	20

REFERENCES

1. JL Bolland. Q Rev 3:1, 1949.
2. L Bateman. Q Rev 8:147, 1954.
3. KU Ingold. Chem Rev 61:563, 1961.
4. WO Lundberg, ed. Autooxidation and Antioxidants, vols 1, 2. New York: Interscience, 1962.
5. IV Berezin, ET Denisov, NM Emanuel. The Oxidation of Cyclohexane. Oxford: Pergamon Press, 1966.
6. NM Emanuel, ET Denisov, ZK Maizus. Liquid Phase Oxidation of Hydrocarbons. New York: Plenum, 1967, p. 71.
7. JQ Betts. Rev Chem Soc 26:265, 1971.
8. JA Howard. Adv Free Radical Chem 4:49, 1972.
9. ET Denisov, NI Mitskevich, VE Agabekov. Liquid–Phase Oxidation of Oxygen-Containing Compounds. New York: Consultants Bureau, 1977, p. 7.
10. T Mill, DG Hendry. In: Comprehensive Chemical Kinetics, vol 16. Amsterdam: Elsevier, 1980, p. 13.
11. ET Denisov. In: Comprehensive Chemical Kinetics, vol 16. Amsterdam: Elsevier, 1980, p. 125.
12. NM Emanuel, GE Zaikov, ZK Maizus. Oxidation of Organic Compounds. Effect of Medium. Oxford: Pergamon, 1984, p. 85.
13. RV Kucher, IA Opeida. Cooxidation of Organic Compounds in Liquid-Phase. Kiev: Naukova Dumka, 1989, p. 43 [in Russian].
14. MM Mogilevich, EM Pliss. Oxidation and Oxidative Polymerization of Unsaturated Compounds. Moscow: Khimiya, 1990, p. 53 [in Russian].
15. ET Denisov, GI Kovalev. Oxidation and Stabilization of Jet Fuels. Moscow: Khimiya, 1990, p. 25 [in Russian].
16. N Grassie. Chemistry of High Polymer Degradation Processes. London: Butterworths, 1956.
17. CH Bamford, CFH Tipper, eds. Comprehensive Chemical Kinetics, vol. 14. Amsterdam: Elsevier, 1980, p. 425.
18. G Scott. Atmospheric Oxidation and Antioxidants. Amsterdam: Elsevier, 1965.
19. MB Neiman. Agening and Stabilization of Polymer. New York: Consultants Bureau, 1965.
20. L Reich, SS Stivala. Autoxidation of Hydrocarbons and Polyolefines. New York: Marcel Dekker, 1969, p. 188.
21. G Geiskens, ed. Degradation and Stabilisation of Polymers. London: Applied Science Publishers, 1975.
22. B Ranby, JF Rabek. Photodegradation, Photooxidation and Photostabilization of Polymers. London: John Wiley & Sons, 1975.
23. HHG Jellinec, ed. Aspects of Degradation and Stabilization of Polymers, Amsterdam: Elsevier, 1978.
24. JF McKellar, NS Allen. Photochemistry of Man-made Polymers. London: Applied Science Publishers, 1979.
25. YA Shlyapnikov, SG Kiryushkin, AP Mar'in. In: Antioxidative Stabilization of Polymers. Moscow: Khimiya, 1986, p. 50 [in Russian].
26. AA Popov, NY Rapoport, GE Zaikov. Oxidation of Oriented and Strained Polymers. Moscow: Khimiya, 1987, p. 25 [in Russian].
27. NM Emanuel, AL Buchachenko. Chemical Physics of Molecular Degradation and Stabilization of Polymers. Moscow: Khimiya, 1988, p. 154 [in Russian].
28. ET Denisov. Oxidation and Degradation of Carbonchain Polymers, Leningrad: Khimiya, 1990, p. 1–287 [in Russian].

29. ET Denisov, In: G Scott, ed. In Developments in Polymer Stabilisaton—5. London: Applied Science Publishers, 1982, p. 23.
30. J Verdu. Macromol Symp 115:165, 1997.
31. ET Denisov. Liquid-Phase Reaction Rate Constants. New York: Plenum, 1974.
32. Landolt-Bornstein. In: H Fischer, ed. Numerical Data and Functional Relationships in Science and Technology, New Series, Group II: Atomic and Molecular Physics, vol. 13. Radical Reaction Rates in Liquids, New York: Springer-Verlag, 1984.
33. Landolt-Bornstein. In: H. Fischer, ed. Numerical Data and Functional Relationships in Science and Technology, New Series, Group II: Atomic and Molecular Physics, vol. 18, Radical Reaction Rates in Liquids. New York: Springer-Verlag, 1997.
34. W Tsang. In: A Greenberg, J Liebman, eds. Energetics of Free Radicals. New York: Blackie Academic & Professional, 1996.
35. ET Denisov. Zh Fiz Khim 67:2416, 1993.
36. ET Denisov. Zh Fiz Khim 68:29, 1994.
37. ET Denisov, TG Denisova. Kinet Katal 34:199, 1993.
38. ET Denisov, TG Denisova. Kinet Katal 35:338, 1994.
39. PA Ivanchenko, ET Denisov, VV Kharitonov. Kinet Katal 12:492, 1971.
40. E Niki, Y Kamiya. Bull Chem Soc Jpn 48:3226, 1975.
41. TG Degtyareva. Mechanism of oxidation of 2-methylbutane. Dissertation, Institute of Chemical Physics, Chernogolovka, 1972.
42. AF Guk, SP Erminov, VF Tsepalov. Kinet Katal 13:86, 1972.
43. IS Gaponova. Kinet Katal 12:1137, 1971.
44. L Dulog, KH David. Makromol Chem B177:1717, 1976.
45. E Niki, Y Kamiya. J Org Chem 38:1403, 1973.
46. ER Klinshpont, BK Milinchuk. Vysokmol Soed B17:358, 1975.
47. JCW Chien, CR Boss. J Polym Sci A-1 12:3091, 1967.
48. VA Roginskii, EL Shanina, VB Miller. Vysokmol Soed A20:265, 1978.
49. LM Postnikov, G Geskens. Vysokmol Soed B28:89, 1986.
50. JCW Chien, DSF Wang. Macromolecules 8:920, 1975.
51. E Niki, C Decker, F Mayo. J Polym Sci Polym Chem Ed 11:2813, 1973.
52. VA Radzig, MM Rainov. Vysokmol Soed A18:2022, 1976.
53. PY Butyagin, IV Kolbanov, AM Dubinskaya, MU Kislyuk. Vysokmol Soed A26:2265, 1968.
54. LS Shibryaeva, NY Rapoport, GE Zaikov. Vysokmol Soed A28:1230, 1986.
55. WY Wen, DR Johnson, M Dole. J Phys Chem 78:1798, 1974.
56. PY Butyagin. Vysokmol Soed A16:63, 1974.
57. E Lawton, R Powell, J Balwitt. J Polym Sci 32:257, 1958.
58. P Neudorf. Kolloid Z 224:132, 1968.
59. VA Radtsig. Vysokmol Soed A28:777, 1986.
60. B Loy. J Phys Chem 65:58, 1961.
61. H Fischer, KH Hellwege, P Neudorf. J Polym Sci A1:2109, 1963.
62. P Neudorf. Kolloid Z 224:25, 1968.
63. L Davis, C Pampilo, T Chiang. J Polym Sci Polym Phys Ed 11:841, 1973.
64. ER Klinshpont, VK Milinchuk. Khim Vysok Energii 3:74, 1969.
65. VA Roginskii, VB Miller. Dokl Akad Nauk SSSR 215:1164, 1974.
66. VA Roginskii, EL Shanina, VB Miller. Vysokmol Soed A24:189, 1982.
67. NY Rapoport, AS Goniashvili, MS Akutin, VB Miller. Vysokmol Soed A20:1432, 1978.
68. AL Margolin, AE Kordonskii, YV Makedonov, VY Shlyapintokh. Vysokmol Soed A29:1067, 1987.
69. F Czocs. J Appl Polym Sci 27:1865, 1982.
70. YS Lebedev, YD Tsvetkov, VV Voevodskii. Kinet Katal 1:496, 1960.

71. GP Vskretchyan, YS Lebedev. Izv Akad Nauk SSSR Ser Khim 6:1378, 1976.
72. PY Butyagin, AM Dubinskaya, VA Radtsig. Uspekhi Khim 38:593, 1969.
73. SI Oniski, I Nitta. J Polym Sci 38:441, 1959.
74. YB Shilov, ET Denisov. Vysokmol Soed A11:1812, 1969.
75. C Decker, F Mayo, H Richardson. J Polym Sci 11:2875, 1973.
76. NY Rapoport, AS Goniashvili, MS Akutin. Vysokmol Soed A23:393, 1981.
77. AV Tobolsky, PM Norling, NH Frick, M Yu. J Am Chem Soc 86:3925, 1964.
78. YB Shilov, ET Denisov. Vysokmol Soed A14:2385, 1972.
79. VA Roginskii. Oxidation of polyolefines inhibited by sterically hindered phenols. Dissertation, Institute of Chemical Physics, Chernogolovka, 1983.
80. EL Shanina, VA Roginskii, VB Miller. Vysokmol Soed A18:1160, 1976.
81. AV Kirgin, YB Shilov, ET Denisov, AA Efimov. Vysokmol Soed A28:2236, 1986.
82. VB Ivanov, SG Burkova, YL Morozov, VY Shlyapintokh. Vysokmol Soed B20:852, 1978.
83. VV Pchelintsev, ET Denisov. Vysokmol Soed A25:781, 1983.
84. PM Norling, AV Tobolsky. Rubber Chem Technol 39:278, 1966.
85. YB Shilov, ET Denisov. Vysokmol Soed A19:1244, 1977.
86. NY Rapoport, AS Goniashvili, MS Akutin, VB Miller. Vysokmol Soed A20:1652, 1978.
87. VM Goldberg, VN Esenin, GE Zaikov. Vysokmol Soed A28:1634, 1986.
88. AP Griva, LN Denisova, ET Denisov. Vysokmol Soed A18:219, 1977.
89. AP Griva, LN Denisova, ET Denisov. Zh Fiz Khim 59:2944 1985.
90. AP Griva, LN Denisova, ET Denisov Zh Fiz Khim 58:557, 1984.
91. BE Krisyuk, EN Ushakov, AP Griva, ET Denisov. Dokl Akad Nauk SSSR 227:630, 1984.
92. AP Griva, ET Denisov. Kinet Katal 17:1465, 1976.
93. YB Shilov. Kinetics and mechanism of initiated oxidation of polypropylene, Dissertation, Institute Chemical Physics, Chernogolovka, 1979.
94. YB Shilov, ET Denisov. Vysokmol Soed A20:1849, 1978.
95. ON Karpukhin, TV Pokholok, VY Shlyapintokh. Vysokmol Soed A13:22, 1971.
96. VA Roginskii, VA Shanina, VB Miller. Dokl Akad Nauk SSSR 227:1167, 1976.
97. NV Zolotova, ET Denisov. Vysokmol Soed A11:946, 1969.
98. ET Denisov, YB Shilov. Vysokmol Soed A15:1196, 1983.
99. AS Michaels, HJ Bixler. J Polym Sci 50:393, 1961.
100. Polymer Handbook, vol 6. New York: Wiley, 1979, p. 43.
101. RS Giberson. J Polym Sci A2:4965, 1964.
102. SG Kiryushkin, BA Gromov. Vysokmol Soed A14:1715, 1972.
103. SG Kiryushkin, VP Filipenko, VT Gontkovskaya. Karbonchain Polymers. Moscow: Nauka, 1971, p. 182 [in Russian].
104. Y Watanade, T Shirota. Bull Agency Ind Technol 2:149, 1981.
105. J Crank, Diffusion in Polymers. New York: Academic Press, 1968.
106. G Amerongen. J Polym Sci 5:307, 1950.
107. R Kosiyanov, R Gregor. J Appl Polym Sci 26:629, 1981.
108. V Stannet, JL Williams. J Polym Sci C 10:45, 1965.
109. G Shaw. Trans Faraday Soc 63:2181, 1967.
110. BP Tikhomirov, HB Hopfenberg, V Stannets, JL Williams. Makromol Chem 118:177, 1968.
111. CA Kumins, J Roteman. J Polym Sci 55:683, 1961.
112. H Yasuda, K Rosengren. J Appl Polym Sci 14:2834, 1970.
113. TV Monakhova, SG Kiryushkin, TP Belov. Vysokmolek Soed B18:90, 1972.
114. SG Kiryushkin. Dissertation, Institute Chemical Physics, Moscow, 1975.
115. H Yasuda, KJ Rasengren. J Appl Polym Sci 14:2839, 1970.

116. OM Yariev, MA Askarov, AI Dzhalilov, YA Shlyapnikov. Uzb Khim Zh 1:42, 1978.
117. P Meares. J Am Chem Soc 76:3415, 1954.
118. RA Pasternak, MV Christensen, J Heller. Macromolecules 3:366, 1970.
119. OE Yakimchenko, IS Gaponova, VM Goldenberg. Izv. AN SSSR Ser Khim 2:354, 1974.
120. AS Michaels, WR Vieth, JA Barrie. J Appl Phys 34:1, 1963.
121. GJ Van Amerongen. Rubber Chem Technol 37:1065, 1964.
122. LN Denisova, ET Denisov. Kinet Katal 17:596, 1976.
123. L Dulog, E Radlmann, W Kern. Makromol Chem 60:1, 1963.
124. CEH Bawn, SA Shaudri. Polymer 9:123, 1968.
125. L Dulog, E Radlmann, W Kern. Makromol Chem 80:67, 1964.
126. NV Zolotova, ET Denisov. J Polym Sci 9:3311, 1971.
127. VS Pudov, BA Gromov, ET Sklyarova, MB Neiman. Neftekhimiya 3:743, 1963.
128. EL Shanina, VA Roginskii. Vysokmolek Soed B20:145, 1978.
129. NY Rapoport, VB Miller. Vysokmolek Soed A18:2343, 1976.
130. NY Rapoport, LS Shibryaeva, VB Miller. Vysokmolek Soed A25:831, 1983.
131. M Iring, T Kelen, F Tudos. Makromol Chem 175:467, 1974.
132. JCW Chien, EJ Vandenberg, H Jabloner. J Polym Sci Part A-1 6:381, 1968.
133. JCW Chien. J Polym Sci Part A-1 6:375, 1968.
134. NM Emanuel, SG Kiryushkin, AP Mar'in. Dokl Akad Nauk SSSR 275:408, 1984.
135. AM Tolks. Kinetics and mechanism of hydroperoxide groups decay in polymers. Dissertation, Polytechnic Institute, Riga, 1973.
136. VM Goldberg, VI Esenin, IA Krasochkina. Vysokmolek Soed A19:1720, 1977.
137. IG Latyshkaeva, GN Belov, TA Bogaevskaya, YA Shlyapnikov. Vysokmolek Soed B24:70, 1982.
138. SS Stivala, V Chavla, L Reich. J Appl Polym Sci 17:2739, 1973.
139. Z Manyasek, D Beren, M Michko. Vysokmolek Soed 3:1104, 1961.
140. VV Dudorov, AL Samvelyan, AF Lukovnikov. Izv AN Arm SSR Ser Khim 15:1104, 1962.
141. AS Kuzminskii, ed. Ageing and Stabilization of Polymers. Moscow: Khimiya, 1966 [in Russian].
142. TV Monakhova, TA Bogaevskaya, BA Gromov, YA Shlyapnikov. Vysokmol Soed B16:91, 1974.
143. SG Kiryushkin, YA Shlyapnikov. Dokl Akad Nauk SSSR 220:1364, 1975.
144. SG Kiryushkin, YA Shlyapnikov. Vysokmol Soed B16:702, 1974.
145. JCW Chien, H Jabloner. J Polym Sci Part A-1 6:393, 1968.
146. VS Pudov, MB Neiman. Neftekhimiya 2:918, 1962.
147. VS Pudov. Kinetics and mechanism of solid polymer degradation and stabilization, Dissertation, Institute Chemical Physics, Moscow, 1980.
148. L Dulog, KH David. Makromol Chem 53B:72, 1962.
149. EA Il'ina, SM Kavun, ZI Tarasova. Vysokmol Soed B17:388, 1975.
150. NC Billingham. Makromol Chem Makromol Symp 28:145, 1989.
151. P Richters. Makromolecules 3:262, 1970.
152. NC Billingham, PD Calvert. Pure Appl Chem 57:1727, 1985.
153. M Celina, GA George. Polym Degrad Stabil 40:323, 1993.
154. M Celina, GA George, NC Billingham. Polym Degrad Stabil 42:335, 1993.
155. M Celina, GA George, NC Billingham. Adv Chem Ser 249:159:1996.
156. GA George, M Celina, M Ghaemy. Abstracts, International Conference on Advances in the Stabilization and Degradation of Polymers, Luzern, Switzerland, 1992, p. 75.
157. C Decker, F Mayo. J Polym Sci Polym Chem Ed 11:2847, 1973.

158. NY Rapoport, AS Goniashvili, MS Akutin, VB Miller. Vysokmol Soed A19:2211, 1977.
159. JP Luondo. J Polym Sci 42:139, 1960.
160. M Iring, T Kelen, F Tudos, ZS Laslo-Hedvig. J Polym Sci Polym Symp 57:89, 1976.
161. M Iring, ZS Laslo-Hedvig, K Bavabas. Eur Polym J 14:439, 1978.
162. J Lacoste, Y Deslandes, P Black, DJ Carlsson. Polym Degrad Stabil 49:21, 1995.
163. J Lacoste, DJ Carlsson, S Falicki, DM Wiles. Polym Degrad Stabil 34:309, 1991.
164. PA Ivanchenko, VV Kharitonov, ET Denisov. Vysokmol Soed A11:1622, 1969.
165. PA Ivanchenko, VV Kharitonov, ET Denisov. Kinet Katal 13:218, 1972.
166. TG Degtyareva, NF Trofimova, VV Kharitonov. Vysokmol Soed A20:1873, 1978.
167. NF Trofimova, VV Kharitonov, ET Denisov. Dok Akad Nauk SSSR 253:651, 1980.
168. NM Zalevskaya, IA Opeyda, RV Kucher. Vysokmol Soed B20:453, 1978.
169. ZA Sadykov, SR Kusheva. Vysokmol Soed B22:403, 1980.
170. ET Denisov. Zh Fiz Khim 49:2473, 1975.
171. ET Denisov. Izv AN SSSR Ser Khim 1:51, 1976.
172. ET Denisov, AP Griva. Zh Fiz Khim 53:2417, 1979.
173. AP Griva, ET Denisov. Dokl Akad Nauk SSSR 219:640, 1974.
174. AP Griva, ET Denisov. J Polym Sci Polym Chem Ed. 14:1051, 1976.
175. AP Griva, LN Denisova, ET Denisov. Dokl Akad Nauk SSSR 232:1343, 1977.
176. AP Griva, LN Denisova, ET Denisov. Kinet Katal 19:309, 1978.
177. SD Grinkina, AP Griva, ET Denisov, VD Sen', IK Yakuschenko. Kinet Katal 34:245, 1993.
178. VJ Voigt. Die Stabilisierung der Kunststoffe gegen Licht und Warme. Berlin: Springer-Verlag, 1966.
179. SK Ivanov. In: G Scott, ed. Developments in Polymer Stabilisation—3. London: Applied Science Publishers, 1980, p. 55.
180. LR Shelton. In: G Scott, ed. Developments in Polymer Stabilisation—4. London: Applied Science Publishers, 1981, p. 55.
181. K Schwetlic. In: G Scott, ed. Mechanism of Polymer Degradation and Stabilisation. London: Elsevier Applied Science, 1990, p. 23.
182. DG Pobedimskii, NA Mukmeneva, PA Kirpichnikov. In: G Scott, ed. Developments in Polymer Stabilisation—2. London: Applied Science Publishers, 1981, Chap. 4.
183. DM Shopov, SK Ivanov. Reaction Mechanism of Inhibitors Peroxide Decomposers. Sofia Publ. House BAS, 1988.
184. GP Gladyshev, YA Ershov, OA Shustova. Stabilization of Thermostable Polymers. Moscow: Khimia, 1979, p. 1–271 [in Russian].
185. ET Denisov. Handbook of Antioxidants. Boca Raton, FL: CRC Press, 1995, p. 1–174.
186. ET Denisov, IV Khudyakov. Chem Rev 87:1313, 1987.
187. WL Hawkins. Polymer Stabilisation. New York: Wiley, 1972.
188. J Pospisil. In: G Scott, ed. Developments in Polymer Stabilisation—1. London: Applied Science Publishers, 1978, p. 1.
189. G Scott. In: G Scott, ed. Developments in Polymer Stabilisation—4. London: Applied Science Publishers, 1981, p. 1.
190. P Klemchuk, J Pospisil. Oxidation Inhibition of Organic Materials vols 1 and 2. Boca Raton, FL: CRC Press, 1990.
191. NV Zolotova, ET Denisov. Vysokmol Soed B18:605, 1976.
192. VV Pchelintsev, ET Denisov. Vysokmol Soed A25:1035, 1983.
193. NV Zolotova, ET Denisov. Vysokmol Soed A24:420, 1982.
194. DG Pobedimskii, VA Kurbatov, PA Kirpichnikov. Vysokmol Soed A18:2650, 1976.
195. LL Gervits, NV Zolotova, ET Denisov. Vysokmol Soed A17:2112, 1975.
196. VA Roginskii, VB Miller. Dokl Akad Nauk SSSR 213:642, 1973.

197. NM Livanova, NC Vasileiskaya, DV Muslin. Izv Akad Nauk SSSR Ser Khim 5:1074, 1972.
198. NC Vasileiskaya, NM Livanova, VB Miller. Izv Akad Nauk SSSR Ser Khim 11:2614, 1972.
199. NM Livanova, NC Vasileiskaya, VV Samarina. Izv Akad Nauk SSSR Ser Khim 5:1074, 1972; 8:1672, 1973.
200. N Grassie, G Scott. Polymer Degradation and Stabilisation. Cambridge: University Press, 1985.
201. AA Kharitonova, GP Gladyshev, VF Tsepalov. Kinet Katal 20:245, 1979; 23:1020, 1982.
202. JY Moisan. Eur Polym J 16:979, 1980.
203. MB Neuman. In Progress in Polymer Chemistry. Moscow: Nauka, 1969, p. 396.
204. IA Tutorskii, IS Yurovskaya, BS Grishin. Vysokmol Soed B21:25, 1979.
205. RA Jackson, SRD Oldland, A Pajazkowski. J Appl Polym Sci 12:1297, 1968.
206. NC Billingham, PD Calvert. In: G Scott, ed. Developments in Polymer Stabilisation—3. Applied Science Publishers, 1980, p. 165.
207. BS Grishin, IA Tutorskii, IS Yurovskaya. Vysokmol Soed A20:1967, 1978.
208. M Dubini, O Cicchetti, GP Vicario, E Bua. Eur Polym J 3:473, 1967.
209. M Dubini, P Parrini, GP Vicario, E Bua. Eur Polym J 4:419, 1968.
210. JF Westlake, M Johnson. J Appl Polym Sci 19:319, 1975.
211. JF Westlake, M Johnson. J Appl Polym Sci 19:1745, 1975.
212. IA Shlyapnikova, AP Mar'in, GE Zaikov, YA Shlyapnikov. Vysokmol Soed A27:1736, 1985.
213. LS Feldshtein, AS Kuzminskii. Kauchuk I Rez 10:16, 1970.
214. NC Billingham, PD Calvert, AS Manke. J Appl Polym Sci 26:3543, 1981.
215. NC Billingham, OJ Hoad, F Chenard, DJ Whiteman, Macromol Symp 115:203, 1997.
216. VY Shlyapintokh, VB Ivanov. In: G Scott, ed. Developments in Polymer Stabilisation—3. London: Applied Science Publishers 1982, p. 41.
217. PP Klemchuk. In: PP Klemchuk, ed. ACS Symposium Series 280 Polymer Stabilization and Degradation. Washington, DC: Am Chem Soc, 1985, p. 1.
218. EG Rozantsev, ES Kagan, VD Sholle, VB Ivanov, VA Smirnov. In: PP Klemchuk, ed. ACS Symposium Series 280 Polymer Stabilization and Degradation. Washington, DC: Am Chem Soc, 1985, p. 11.
219. T Toda, T Kurumada, K Murayama. In: PP Klemchuk, ed. ACS Symposium Series 280, Polymer Stabilization and Degradation. Washington, DC: Am Chem Soc, 1985, p. 37.
220. HK Muller. In PP Klemchuk, ed. ACS Symposium Series 280, Polymer Stabilization and Degradation. Washington, DC: Am Chem Soc, 1985, p. 55.
221. T Kelen, F Tudos, G Balint, A Rockenbauer. In: PP Klemchuk, ed. ACS Symposium Series 280 Polymer Stabilization and Degradation. Washington, DC: Am Chem Soc, 1985, p. 109.
222. F Gugumus. Polym Degrad Stabil 34:205, 1991.
223. F Gugumus. In: G Scott ed. Mechanism of Polymer Degradation and Stabilisation. London: Elsevier Applied Science 1990, p. 169.
224. S Nigam, K-D Asmus, RL Willson. J Chem Soc Faraday Trans 1:2314, 1976.
225. J Chateauneuf, J Lusztyk, KU Ingold. J Org Chem 53:1629, 1988.
226. ALJ Beckwith, VW Bowry. J Org Chem 53:1632, 1988.
227. LL Koroli, VA Kuzmin, IV Khudyakov. Int J Chem Kinet 16:379, 1984.
228. YB Shilov, ET Denisov. Vysokmol Soed A16:1736, 1974.
229. YB Shilov, ET Denisov. Vysokmol Soed A16:2313, 1974.
230. YB Shilov, ET Denisov. Vysokmol Soed A26:1753, 1984.
231. YB Shilov, RT Battalova, ET Denisov. Dokl Akad Nauk SSSR 207:388, 1972.

232. ET Denisov. Russ Chem Rev 65:505, 1996.
233. ET Denisov. In: PP Klemchuk, ed. ACS Symposium Series 280 Polymer Stabilization and Degradation. Washington, DC: Am Chem Soc, 1985, p. 87.
234. ET Denisov. Polym Degrad Stabil 25:209, 1989.
235. ET Denisov. Polym Degrad Stabil 34:325, 1991.
236. DJ Carlsson, A Garton, DM Wiles. In: G Scott, ed. Developments in Polymer Stabilization-1. London: Applied Science Publishers, 1979, p. 219.
237. KB Chakraborty, G Scott. Polymer 21:252, 1980.
238. H Berger, TA Bolsman, DM Brower. In: G Scott, ed. Developments in Polymer Stabilization-6. London: Applied Science Publishers, 1983, p. 219.
239. G Scott. J Polym Sci Polym Lett Ed 22:553, 1984.
240. DJ Carlsson. Pure Appl Chem 55:1651, 1983.
241. S Korcek, RK Jensen, M Zinbo, JL Gerlock. In: H Fischer, ed. Regeneration of Amine in the Catalytic Inhibition of Oxidation in Organic Free Radicals. Berlin: Springer, 1988, p. 95.
242. RK Jensen, S Korcek, M Zinbo, JL Gerlock. J Org Chem 60:5396, 1995.
243. AV Kirgin, YB Shilov, ET Denisov, AA Efimov, VV Pavlikov. Kinet Katal 31:58, 1990.
244. ET Denisov. Russ Chem Bull 45:1870, 1996.
245. ET Denisov. Kinet Katal 38:259, 1997.
246. YB Shilov, ET Denisov. Vysokmol Soed A29:1359, 1987.
247. VI Goldenberg, ET Denisov, NA Ermakova. Izv Akad Nauk SSSR Ser Khim 4:738, 1990.
248. ET Denisov. In: 12th International Conference on Advances in the Stabilization and Controlled Degradation of Polymers, Luzern, Switzerland, 1990. p. 77.
249. VI Goldenberg, NA Ermakova, ET Denisov. Izv Akad Nauk SSSR Ser Khim 1:79, 1995.

10

Biodegradation of Polymers in Historical Perspective Versus Modern Polymer Chemistry

ANN-CHRISTINE ALBERTSSON
Royal Institute of Technology, Stockholm, Sweden

I. BIODEGRADATION OF POLYMERS IN A HISTORICAL PERSPECTIVE

In striving for a sustainable society, polymers with well-defined lifetimes are required. The focus of the 1930s on materials that were so much better than natural materials in the sense of their lifetime, shifted in the early 1980s to polymers that were biodegradable. The first environmental approach was taken when the oil crisis of the 1970s initiated development of materials filled with what is now called renewable polymers (e.g., starch). In some respect we are now closing the loop. This chapter will give an exposé of biodegradation of polymers in a historical perspective. It also summarizes recent biodegradable polymers relative to chemical, physical, and degradation behavior.

The role of microorganisms in the spoilage of solid materials and the development of protective chemical biocides initiated studies acquiring the title of "biodegradation." Their status was recognized by the adoption of standard specifications [e.g., American Society for Testing and Materials (ASTM) D 1924 (1963)], fungal overgrowth specification. The term *biodegradation*, previously adopted in connection with disposal of water-soluble materials, such as surfactants, was used and the search for biodegradable polymers began.

Already in 1946 Brown (1) claimed that products such as poly(vinyl acetate), melamine formaldehyde polymers, and other synthetic resins were resistant to fungal growth. Hueck (2) and Wesel (3) divided plastics into two groups: one with good and one with poor resistance to microorganisms.

421

More than 300 types of polymers and elastomers were tested from the microbial point of view (4). Growth in tests on media containing paraffins were compared with those grown on media lacking an easy carbon source, and polyethylenes, poly(vinyl chlorides), and polystyrenes were found to be rather resistant (4). Two hundred fourteen various materials widely used in electronics and electrotechniques were tested to determine their resistance to superficial growth of fungi (5). A review of the resistance of plastics to microorganisms evaluated in different countries by different methods was reported by Dolezel (6).

Different materials underwent soil burial tests, and it was found that the copolymerization of acrylic acids, acrylamide, or vinyl acetate with ethylene and propylene increased the biodegradability of these polyolefin polymers, whereas unmodified polyolefins were almost completely resistant (7).

Connolly (8,9) summarizes some of the trends in the behavior of plastic materials after 8 years of soil exposure. Soil burial is a very realistic test, but it is rather complex and difficult to control.

Products formulated with many different kinds of additives, and the individual constituents, were tested for their ability to support growth of bacteria, yeasts, actinomycetes, and fungi. Simple polymeric materials were inert, but some more complex products, such as plasticized poly(vinyl chloride) (PVC), supported the growth (10).

The biodegradability of packaging plastics was analyzed for the U.S. Environmental Protection Agency (EPA) (11). With the ASTM method D-1924-63, 31 samples of commercial plastics were compared. Almost all of these widely used commercial polymeric materials were resistant to degradative attack by the specific strains of test fungi recommended in this method, which, in fact, were cellulose-degrading fungi.

Detailed review articles, such as those of Kaplan (12) and Küster (13), threw more light on the complex problem of biodegradation. Küster was certainly more optimistic about the possibility of biodegradation in many polymers. When looking for support for this idea in his earlier work, he concluded that even the basically inert polymers were attacked by microorganisms, probably by means of enzymatic activity (14,15).

Lazar (16), in Romania, approached the problem in a new way. For a great number of fungi he tested their enzymatic activity individually for each species. He also noted the age of the culture, the media used in the tests, and the presence of possible microbiocides in the plastics.

II. BIODEGRADATION OF HYDROLYZABLE POLYMERS

Polyurethanes based on polyether or polyester diols have been tested by several workers, and the general conclusion is that the polyester polyurethanes are considerably more susceptible to hydrolytic attack than polyether polyurethanes (17–24). Darby and Kaplan (20) tested 100 polyurethanes and reported that the microbial susceptibility of the polyethers was related to the number of adjacent methylene groups in the polymer chain. At least two such groups were required for a significant attack to occur, whereas the presence of side chains on the diol moiety of the polyurethane reduced susceptibility (25,26).

The fungal resistance of polyamides has also been widely discussed. Bomar (27) investigated the surface growth on polyamides, and Dayal et al. (28) isolated *Penicillium janthinellum*, which was capable of degrading synthetic fabrics, such as nylon, Terylene, and Orlon. Pavlov and Akopdzhanyan (29) interpreted the deterioration of polyamides as an abiotic oxidation or hydrolysis, rather than as a microbial action.

Ennis and Kramer (30) isolated special bacteria capable of degrading polymeric and low molecular weight amides. The following year they reported (31) a special technique for testing the biodegradability of nylons and related polyamides.

Esters of most kinds are generally rather easily accessible to chemical or enzymatic hydrolysis and it was indeed found, according to expectations, that the polyester-based polyurethanes were susceptible to microorganisms (11,17,20, 32–36). The polyesters were reexamined in the hope of a better understanding of specific structural configurations in a polymer that favors biodegradability. The only types of synthetic high polymer found to be enzymatically degradable were those having aliphatic ester linkages in the main chain. Fields et al. continued the work with polycaprolactones (37–40). Polyester structure, molecular weight distribution, and crystal morphology were examined for their possible relation to susceptibility by biodegradation. When polyester samples with increasing molecular weight were exposed to the attack of *Aureobasidium pullulans*, the test material lost weight, but in an inverse relation to molecular weight. Also, when polymers with a broad molecular weight fraction were degraded by this "black yeast," the low molecular weight fraction was preferentially utilized. Diamond et al. (41) suggested that the relative degradability of polyester films is related to the carbon chain spacing between ester groups. Suzuki et al. (42) also investigated the degradation of polycaprolactone and other polyesters, applying a *Pencillium* fungal species, but not in soil (34) nor with *Aureobasidium pullulans* (38). This mold almost completely degraded even a high molecular weight (MW) polycaprolactone (MW 25,000).

III. BIODEGRADATION OF INERT POLYETHYLENE

In several surveys it is claimed that polyethylene is an inert polymer with good resistance to microorganisms (43–46). A fungal growth may occur on the surface of polyethylene (47–50), and Connolly (8,9) found that some insects attacked low-density polyethylene (LDPE). A change in tensile strength for LDPE after biological exposure was shown by Kestelman et al. (51,52).

Investigations in natural surroundings or in corresponding model experiments were often based on lower fungi—more explicitly the molds—and performed in accordance with the ASTM recommendations, thus relying on observation and estimation of the extent of visible spreading of mycelial growth. On the other hand, degradation experiments based predominantly on bacteria, simulated submerged conditions found in marine, estuarian, or sewage environments.

Potts and co-workers (33,34) investigated the relation between the molecular weight of different alkanes as well as polyethylenes and the observable corresponding fungal growth. Haines and Alexander (53) showed that normal alkanes containing up to 44 carbon atoms were metabolized by microorganisms. In addition, they also concluded that failure of earlier investigators to demonstrate microbial decomposition of high molecular weight polyethylene glycol (PEG) may have resulted from the absence of an appropriate microorganisms or conditions in the biological oxygen demand (BOD) test system not conclusive to the particular activity (54). Tsuchi et al. (55) have also examined the microbial degradation of polyethylene oligomers, especially the effect of dispersion on the microbial degradation of waxes. Mixed with softener (squalene) and ultrasonically treated at 95°C, n-tetratetracontane was finely dispersed into the culture medium and could then be metabilized by *Acinetobacter*.

The generally observed behavior of polyethylenes, as well as the growing interest in finding biodegradable plastic materials, led to a demand for a "biodegradable polyethylene," and many different products of this type appeared during the beginning of the 1970s (56–59). Despite the older tests on biodegradation of polyethylene and other synthetic polymers, there are still many unanswered questions about biodegradation. The degradation changes in polyethylene (PE) take place very slowly, and the detection and monitoring of the events are difficult. However, by using a ^{14}C-labeling technique we have been able to study microbial and oxidative effects in the degradation of PE (60). A limited microbial conversion of ^{14}C in PE to $^{14}CO_2$ is measured when the substrate is incubated with some soil fungi. Further studies have shown that the biodegradation of PE is affected by exposure to ultraviolet (UV) irradiation, by morphology and surface area, antioxidants, and various oxidation promotors and enhancers, as well as by the polymeric structure and molecular weight (61–63). We have mixed 10% $C_{32}H_{66}$ with ^{14}C-labeled high-density PE (HDPE) which gave a film containing an additive known to be degraded by fungi (64). Over a period of 2 years the degradation of this film was followed and compared with pure PE film containing ^{14}C using a $^{14}CO_2$ scintillation method. The liberation of $^{14}CO_2$ from the HDPE film was slightly enhanced by the additive in the first year, but somewhat retarded thereafter. Thus, the biodegradation of PE was not stimulated by the addition of the compatible biodegradable additive. This was probably because of the short period during which the samples were studied. We have suggested that PE biodegradation in soil consists of three distinct phases. In the first phase, there is a slow and constant degradation rate. The second phase consists of decline in degradation rate, as determined by mineralization studies with labeled material. The third and final phase is characterized by a return to a slow and constant degradation rate that could lead to the final destruction and mineralization of the material (65). During synthesis and degradation of polyesters we found similar degradation phases (66,67).

IV. MEANS OF INCREASING BIODEGRADABILITY

Degradable polymers are materials with inherent sensitivity to different degradation factors. A degradation is performed by physical, chemical, mechanical, or biological agents. To answer the need of ultimate disposability or recyclability of plastic materials, great efforts have been made to develop degradable polymers, thus prohibiting the irreversible buildup of these materials in the environment.

Several means are used to achieve degradability. The repetitive chain of the polymer is either hydrolyzable, oxidizable, or thermolabile. Some polymers are sensitive to biodegradation by enzymes or microorganisms or both, whereas others in later stages of deterioration are biodegraded. Addition of biopolymers to synthetic polymers renders the materials more susceptible to autoxidation owing to the porous matrix left after degradation of the biopolymeric additive.

The methods that have emerged for degradable polymers are the following:

1. Photooxidizable degradable polymers
2. Directly biodegradable polymers
3. Autoxidizable polymers that then biodegrade
4. Water-soluble polymers
5. New forms of natural polymers

The combining of several modes of degradation in a material is a successful way to render a plastic degradable. Package materials sensitive to photooxidation and biodegradation will be easily assimilated and constitute part of other organic waste, later becoming soil humic materials (compost).

Several of the degradable materials are produced by renewable resources, in contrast with the oil-based production of the bulk polymers polyethylene, polypropylene, and other such. In principle it is also possible to produce polyethylene from renewable resources (e.g., from ethanol obtained from agricultural products) instead of using oil.

Photooxidative materials have a larger market, especially in agricultural applications (mulch films), whereas truly biodegradable polymers in general have so far had a limited economic value, at least for packages. A higher price is acceptable for packages used for biomedical products than for foodstuffs.

A. Photodegradable Bulk Polymers

The major approaches to preparing photodegradable polymers are by synthesis of sensitized copolymers or use of additives to photosensitize the polymers.

Guillet has formulated a copolymer of vinyl ketones and styrene and ethylene in which the ketone groups are coordinated with the main polymer chain (68). The obtained materials—trademarked as Ecolyte—were shown to biodegrade, and this degradation was followed using respirometry (69–71). Adding carbonyl compound groups to inert bulk polymers should be a successful way to obtain degradable polymers, as the carbonyl readily absorbs UV light.

It is customary to use light to initiate free radical chain reactions in polymerization and oxidation processes. Scott discusses two examples of this type: molecules in which excited states behave as radical-initiating species (carbonyl compounds), and molecules that dissociate to free radicals (peroxides and transition metal salts) (72). By using these processes, it is possible to control the oxidative breakdown of polymers.

An additive that acts as photoinitiator for the oxidation of the polymer substrate is an interesting solution for creating degradable polymers (73–76). The nontoxic organosoluble acetyl acetonates of these transition metals are photo-prooxidants, but the transition metal compounds interfere during polymer processing by accelerating the rate of hydroperoxide thermolysis (76). The material has found wide use in applications for different agricultural products, such as mulch films and packages (77). The degradation mechanism is a photooxidation, leaving a hydrophilic surface that subsequently biodegrades in combination with further oxidative processes.

Dithiocarbamates are effective processing and heat stabilizers for polyolefins. Their photoantioxidant activity depends on the metal ion, for which iron and manganese complexes are the least stable. These metal complexes show a well-characterized induction period that increases with their concentration in the polymer (78). The sulfur ligand–iron complexes exhibit a high level of antioxidant activity during processing and storage and in the early stages of exposure to light; after this initiation a very fast photooxidation occurs. Contrary to the iron and manganese complexes, the nickel complexes have a stabilizing effect, and combining the two types of complexes gives the possibility of creating materials with a predetermined degradation time.

An oxidized polymer is always more brittle and more hydrophilic than a nonoxidized one and is thus more prone to biodegrade. The hydrophobic surface of most synthetic polymers is often the major obstacle to biodegradation. On the other hand, most enzymes adhere very closely to polymeric surfaces. Low-density polyethylene (LDPE) that was degraded in a solution containing surfactant showed an increased degradation rate compared with samples without surfactants (79). A concentration of 0.5% nontoxic surfactant polyoxyethylene (20)-sorbitan monooleate (Tween 80) in LDPE increased the degradation of the LDPE in a solution of microorganisms (80).

The effect of UV irradiation on a polymer can be modified by using protective or sensitizing additives. The increased degradation of a polyethylene film containing a UV-sensitizer was reported (65,81). The degradation curves are characterized by a straight-line progression in the first 100 days of observation before declining. The trend was occasionally reversed to a progressive increase in $^{14}CO_2$ liberation in some of the long-term experiments and suddenly increased the degradation rates. This degradation proceeds in three distinct steps: a first step, with a constant increasing degradation rate; a second step; with a lower degradation rate; and a third step, during which the degradation rate increases again (65). The same three steps can be observed for several degrading polymers besides PE.

A closely related method of imparting photodegradability is by copolymerizing ethylene with carbon monoxide (82,83). This is a commercial process used by Dow Chemical, DuPont, and Union Carbide, primarily for "six-pack" connectors for beverage containers. With current catalytic systems, the copolymer can be made only with LDPE (84); carbon monoxide is copolymerized at levels less than 2% with ethylene in a specially fitted high-pressure reactor to give E/CO. Because the degradation depends on chain scission at the site of each carbonyl, blends of a concentrated copolymer with homopolymer to the same total CO concentration do not degrade as effectively and are not generally used (83).

B. Starch–Plastic Composites

Biopolymers, such as cellulose, starch, and lignin, constitute the major part of living matter. Starch is the principal food reserve polysaccharide in the plant kingdom and is found in numerous sources, such as corn, potato, rice, wheat, barley, oat, and tapioca. The two main components of starch are amylose and amylopectin, which consist of glucose units linked by glycosidic bonds (85). Amylose is principally linear, whereas amylopectin contains branches. Animals, plants, and microorganisms are able to utilize starch as a source of energy. Microorganisms produce various starch-hydrolyzing enzymes, such as amylase, and their wide distribution assures the biodegradation of starch in nature.

The extensive occurrence of starch, together with its biodegradability and low price, has led to the exploitation of starch as a means of enhancing the biodegradability of otherwise biologically inert polymers. Research on starch–plastic composites began in the 1970s, and several different technologies are currently being studied. The starch can be used in its natural granular form as a biodegradable filler, or in the gelatinized or destructurized form in starch-based compositions. Other possibilities are starch graft copolymers having thermoplastic branches (e.g., starch-g-poly(methyl acrylate) (PMA) or starch-g-polystyrene (86). There are also products that are completely starch-based.

1. Starch as a Biodegradable Filler

The main problem associated with the use of starch as a filler is its hydrophilic nature and the generally hydrophobic nature of the polymer matrix. Starch particles from different plants have varied sizes and shapes, making it possible to choose the best-suited type of starch for each use. The starch particle size limits the amount to be incorporated if the composition is to be blown to a thin film, but this is not such a limitation for sheets or injection-molded articles. For very thin films, small–particle-sized starches, such as rice, may be needed, whereas wheat and cornstarch give good results with most blown films.

Griffin has invented formulations based on granular starch typically containing 6–12% starch by weight. In his early patents (87,88) the starch was surface-treated to improve the compatibility between the starch and the synthetic polymer matrix. The first product on the market was shopping bags made of polyethylene with approximately 7% starch incorporated in the form of a master batch denoted Ecostar. In 1988 Griffin (89) patented an improved process in which starch is used together with a thermoplastic elastomer that functions both as a compatibilizer and as a prooxidant. It is licensed to the American company ADM (Archer Daniels Midland Co.; world license), and is sold under the name Polyclean. Another patent in this area is a recent one by Willett (90) on a composition that includes granular starch and a copolymer carrier resin.

Biodegradation of the starch filler hollows out the polymer matrix and is associated with loss of mechanical properties, increased permeability, and a greater surface/volume ratio. This facilitates further abiotic degradation processes. If the degradation takes place in aqueous environments, the starch particles can swell and cause disruption of the polymer surface. Furthermore, enzymatic breakdown of the starch particles results in the polymer surface being covered with a film of starch degradation products, thus providing an intermediate carbon source for the invading microorganisms, as well as overcoming the hydrophobic nature of the surface. If the starch level exceeds 41% by weight, continuous biodegradation can take place, leaving behind a disintegrated polymer matrix. Wool and co-workers (91,92) have applied this so-called percolation theory in degradation studies to the connectivity of the starch particles in the polymer matrix at different starch concentrations. Griffin and Nathan (93) found that enzyme diffusion can take place through very thin films of polyethylene, suggesting that a somewhat larger portion of the starch than that calculated from the percolation theory can be accessible to continuous enzymolysis.

Specially treated starch, has other than in polyethylene, has been tested as an additive in polypropylene, polystyrene, polyurethane, ethylene vinyl acetate copolymer, poly(vinyl chloride) and ABS (94). Starch has also been incorporated into polymers which themselves are inherently biodegradable. These studies include the work done by Tanna et al. (95) and by Holland et al. (96,97) showing the greater degradation of matrices of hydroxybutyrate–hydroxyvalerate copolymers when polysaccharides are used as fillers. Tanna et al. observed that the addition of 5–10% starch accelerated the degradation of the matrix by almost 15% under aerobic composting conditions.

Utilization of starch as a sole additive is inferior to compositions containing additional additives (e.g., prooxidants). The incorporation of additional pro-degradant additives provides the potential for degradation by abiotic mechanisms, such as photo- and thermooxidation. Extensive degradation studies on starch-filled polyethylene, with and without prooxidants, have been performed by Albertsson et al. (98–100). Examples of other publications in this area are those of Ianotti et al. (101), Pometto et al. (102–104), Maddever and Chapman (105), and Sung and Nikolov (106).

Polyethylene–starch blends are also susceptible to macrobiodeterioration (i.e., degradation caused by organisms larger than bacteria or fungi, such as insects). Woodlice are able to consume starch-modified polyethylene even when offered alternative food (107). When offered both normal and starch-modified polyethylene they clearly preferred the modified material.

2. Starch-Based Materials

Starch intended for starch-based compositions is most often used in its gelatinized or destructurized state. These two terms refer to basically the same condition of starch and differ merely in the production process. The result of gelatinization or destructurization is disruption of the granular starch structure accompanied by liberation of amylose and amylopectin. Early efforts were mostly directed to starch–poly(vinyl alcohol) (PVA) compositions that could be cast from aqueous suspensions and dried to flexible films (108). These films had a high water sensitivity (e.g., too high for agricultural mulch applications) that led to evaluation of other compositions to enhance the properties. Films laminated with poly(vinyl chloride) (PVC) were developed. Several techniques of producing starch-based materials have been patented since the 1970s. Starch in its gelatinized form is used in formulations developed by Otey and Westhoff (109). Together with the complexing agent ethylene acrylic acid copolymer (EAA) and ammonia, it was possible to incorporate up to about 60% gelatinized starch in milled or blown films. The films exhibited good water resistance and were flexible without added plasticizers. The work was expanded to include polyethylene (PE) in the formulation (110). In starch–EAA blends, the starch forms an inclusion complex with the EAA, which explains the partial miscibility between the two polymers. The complex is insoluble. Usually, an aqueous solution of urea and ammonium hydroxide is added, and a typical formulation contains 40% starch, 25% EAA, 10% PE, and 10% urea as the solid components. Ammonium hydroxide is added as a plasticizer because starch films are brittle owing to the relatively high T_g and lack of sub-T_g main-chain relaxation. The material can be blown to film or injection-molded.

Bastioli et al. (111–115) have patented numerous technologies based on destructurized starch, typically incorporating more than 60% starch by weight together with synthetic copolymers (e.g., those of ethylene–acrylic acid). Urea and ammonia can also be included in these formulations. They are marketed under the Mater-Bi trademark and produced by the Italian company Novamont (part of the Montedison Co). Mater-Bi is claimed to have both low-modulus applications, such as films in diapers, bags for compost, and holders for six-can beverages, as well as high-modulus applications (e.g., disposable containers). Other formulations are based on 60–98% starch and often contain some modified polysaccharide or polysaccharide derivative (116–122). Other ways of obtaining high starch levels have been reported by Suominen et al. (123–125), who in several patents have claimed to have overcome the constraints on starch loadings caused by particle size, thereby allowing 10–60% of the biopolymer to be incorporated. In addition, microbes, in the form of spores, are incorporated. Vanderbilt and Neeley (126) have invented blends comprising ethylene–carbon monoxide copolymers, polysaccharide (preferentially cornstarch), and optionally unsaturated oil. The blends are claimed to be both photo- and biodegradable. Toms (127) has patented a composition applicable to backsheets in diapers, sanitary napkins, and such. Films made of destructurized starch, EAA, or ethylene vinyl alcohol copolymers and an aliphatic polyester such as polycaprolactone are stated to be liquid impervious.

The foregoing inventions serve as examples of the most abundant approaches to utilizing starch in the design of materials possessing enhanced degradability. The actual time required for biodegradation of any starch-containing plastic in the environment depends on several highly variable environmental factors and will be difficult to predict accurately unless some defined standard degradation procedure is employed.

C. Microbial Polyesters

Polyhydroxybutyrate is one of the naturally occurring polyesters that are produced by microorganism. Microorganisms use renewable sources, such as glucose, to produce the polymer. Prokaryotic organisms, for example, bacteria and cyanobacteria, accumulate poly(3-hydroxybutyrate) [P(3HB)] as inclusions in the cytoplasmic fluid, amounting to 30–80% of the cellular dry weight when their growth is limited by the depletion of an essential nutrient such as nitrogen, oxygen, phosphorus, sulfur, or magnesium. The polyester functions either as a carbon or energy reserve, or as a sink for excess reducing equivalents in the microorganism. Zeneca Inc. (now part of Monsanto) developed a controlled-fermentation process in which polyhydroxyalkanoate (PHA) copolymers are produced by feeding bacterial monocultures with a variety of carbon substrates. With this process a copolymer of 3-hydroxybutyrate and 3-hydroxyvalerate [P(3HB-co-3HV)], is produced commercially by *Alcaligenes eutrophus* from propionic acid and glucose (128).

This copolymer, which contains 0–20% hydroxyvaleric acid, is known under the tradename Biopol, sold by Zenca Ltd (but now owned by Monsanto) may be under a new name. The homopolymer P(3HB) is relatively stiff and brittle, but becomes more flexible and tougher as the HV content increases. The microbial polyesters are—unlike other biological polymer, such as proteins and polysaccharides—thermoplastics, with a melting temperature close to 180°C (128).

The microbial polyesters have attracted industrial attention as environmentally, degradable thermoplastics for a wide range of agricultural, marine, and medical applications. The products of microbial polyesters, such as films and fibers, are degraded in soil, sludge, or seawater. Under optimum conditions, the degradation rate is extremely fast. Some microorganisms, such as bacteria and fungi, excrete extracellular P(3HB) depolymerases that hydrolyze P(3HB) and its copolymers into the dimers or monomers in the vicinity of the cells, and the resulting products are absorbed and utilized as nutrients. The extracellular P(3HB) depolymerase is able to hydrolyze a highly crystalline product of pure P(3HB); this is distinct from the properties of intracellular P(3HB) depolymerase, which can hydrolyze only native P(3HB) granules of an amorphous elastomer. Some microorganisms, such as fungi, an incapable of accumulating P(3HB), but they do excrete the extracellular depolymerases (128).

The polymer chains of microbial polyesters are hydrolyzed in water without enzymes at a very slow rate. An involvement of simple hydrolytic chain scission in the degradation processes of microbial polyesters may be important in certain biological environments (in vivo) as implant materials (128). The hydrolysis rate is enhanced by increasing the temperature or by making the hydrolysis medium more alkaline. The pH effect has to be determined if the material is to be used inside the body, where large pH changes occur. The enhancement of degradation rate as the hydrolytic medium becomes more alkaline is possible partly because of the increasing solubility of the degradation products at higher pH (129). Mainly, 3-hydroxybutyrate, 3-hydroxyvalerate, and crotonic acid are formed as degradation products during alkali hydrolysis of P(3HB) and P(3HB)-co-P(3HV).

Examples of medical applications for the microbial polyesters include controlled drug release, surgical sutures, and bone plates. All of these potential uses depend on the biocompatibility and slow resorption of microbial polyesters in biological environments (in vivo). The ultimate biodegradation product, (−)-3-hydroxybutyric acid, is a normal metabolite in human blood, and P(3HB) shows negligible oral toxicity. Because human tissue does not contain P(3HB)-utilizing bacteria and fungi, simple hydrolytic degradation is an important process for the biodegradation of P(3HB) (130).

Other applications for biodegradable polyesters include agriculture mulching films, fishing nets, packaging films, bottles, and containers (128). The ready sorption of the PHA latex by fibrous constructs, such as paper or nonwoven goods, also suggests applications such as binder, coating material, or barrier. These applications in fibrous constructs become particularly attractive when the natural biodegradability is considered (131).

Although there is strong evidence that physiological degradation within the body is substantially dependent on a conventional hydrolytic process, the degradation behavior of fabricated devices cannot be reasonably predicted simply on the basis of the properties of the unprocessed starting material. This is because processing conditions used in polymer fabrication have a marked potential effect on such properties as molecular weight and crystallinity of the material (132). PHAs are melt-unstable, which may cause problems during processing. The main decomposition of bacterial PHB and its copolymers occurs at temperatures of about 250°C and results in the formation of thermal decomposition products of the HB and HV repeating units (133). The molecular weight and molecular weight distribution decreased with increasing processing temperature and decreasing extrusion rate (134,135). In addition, changes in the morphology and atomic positioning of the oxygen atoms (polyester functional group) occurred during processing (134).

The synthetic polyesters are less susceptible to enzymatic degradation than the bacterially produced polyesters. In contrast to the isotactic bacterial copolyesters that have a random stereosequence, the synthetic polyesters are blocky and only partially stereoregular (136–138). The [S]-stereoblocks halt the enzymatic degradation and thus make it difficult for the depolymerase to penetrate into the surface and access the available [R]-stereoblocks (139).

The physical properties and biodegradability of microbial polyesters may be regulated by blending with synthetic or natural polymers. P(3HB) is miscible, to some degree, with a number of functional polymers, such as nylon, polyethyleneoxide, polyvinylacetate, polyvinylchloride and polysaccharides, because of the hydrogen bonding. The rate of hydrolytic degradation of the P(3HB)–polysaccharide blends is dramatically affected by the presence of polysaccharides. The 3-hydroxyalkanoate comonomers are internal plasticizers for P(3HB) and improve its mechanical properties. Its external plasticizers may have a dramatic effect on the physical properties of P(3HB) (128).

V. DEFINITION OF BIODEGRADATION

The term *biodegradation* has been used to encompass events taking place both in the natural environment and in the living human body. The environmental degradation by microorganisms, such as bacteria or fungi, is probably not comparable with the degradation under physiological conditions imposed by living bodies. In the field of sutures, bone reconstruction, and drug delivery, the term biodegradation may simply imply hydrolysis. On the other hand, for environmentally degradable plastics, the term biodegradation may mean fragmentation, loss of mechanical properties, or sometimes degradation through the action of living organisms.

Interpretations and discussions in the field of biodegradable polymers are, and will continue to be, polarized for a while for at least four reasons:

1. Definitions and nomenclature in the field are sometimes still vague and not generally accepted.
2. The field covers rather wide interdisciplinary aspects, ranging from mechanics and physical chemistry, to enzymology and microbial taxonomy, which will seldom be covered by one researcher or even by groups of workers.
3. All experimentation is difficult and, consequently, often imperfect because such work must be related to rather complex and extremely long-term phenomena in a wide range of natural environments that are intrinsically variable.
4. Immediate acceptance of interpretations of apparently opposite character will be more difficult if it interferes with patents production, or marketing interests.

In natural environments there are many different degradation modes that synergistically combine to degrade polymers. This deterioration and degradation of polymeric structures in nature, which are the basic mechanisms ensuring recycling of the elements of the biosphere, may be called environmental degradation. Biodegradation might be better used as a term only when it is necessary and to distinguish clearly between the action of living organisms and other degradation modes (e.g., photolysis). A general definition that excludes degradation modes other than biodegradation may be "transformation and deterioration of polymers solely by living organisms (including the microorganisms and/or enzymes excreted by them)" (140).

The accessibility of a polymer to degradative attack by living organisms has no direct relation to its origin, but merely to its molecular composition and architecture. Complex macromolecules, such as lignin and asphalt, show great inertness, despite being biopolymers or natural polymers, respectively. On the other hand, synthetic polymers with intermittent ester linkages (e.g., polyester polyurethanes) are readily accessible to the biodegradative action of esterases, despite their usual enzymatic specificity. In addition, solitary examples of an extremely uniform chain, characteristic of a synthetic polymer molecule (for instance, polyethylene with 100–1000 or more carbon atoms), certainly occur sporadically in nature as an artifact, intermixed throughout with countless types of other aliphatic and aromatic macromolecules of higher petroleum derivates. How then should biodegradation of polymers or any origin be defined? Preferentially in three ways, all feasible theoretically, but unlikely to occur alone in nature as a single phenomenon (141,142):

1.   A biophysical effect, such as mechanical damage of a material by the swelling and bursting effect of growing cells
2.   A secondary biochemical effect, resulting from the excretion of substances other than enzymes from cells, which might act directly on the polymer or by changing the pH or redox conditions of the surroundings
3.   Direct enzymatic actions, leading to splitting or oxidative breakdown of the material.

The macroscopically or mechanically observable consequences of these three effects might be expressed in many ways, as corrosion, abrasion, deterioration, cracking, decrease in tensile strength and so on.

Fungi and bacteria can also use plasticizers and fillers as a source of nutrient, and this might accelerate the aging of the plastics (96). Growth of a pure culture of one defined microbial strain on a specified accessory substance in a plastic material is unlikely to induce the ribosomal production of an entirely different enzyme in the same strain, directed toward the main molecular species in this particular plastic product. Such a mechanism is not in agreement with existing knowledge on the mechanism of enzyme induction, which is just a triggering of an existing gene to produce a number of molecules of the corresponding enzyme.

It is still more futile to hope that the induction of biosynthesis of yet unknown enzymes might eventually occur in the case of polyethylenes. There is instead a synergism between biodegradation and environmental degradation. The mechanism for the biodegradation of polyethylene is comparable with that of paraffin. We have shown that in biodegradation of polyethylene, abiotic steps also contribute to the total degradation (143). There is always a synergistic effect between photooxidative degradation and biodegradation (81,141,143,144). In environmental degradation of polyethylene. UV light or oxidizing agents are so much faster than the biodegradation. The final end products of a total mineralization are carbon dioxide and water (81,143), but before that ever happens many low molecular weight degradation products are evolved and may take part in the catabolisms of nature (145).

VI. PRESENT STANDARDS FOR TESTING DEGRADABLE POLYMERS

As we have summarized before, there are a few traditional test methods for the testing of biodegradability (e.g., 140):

1. Visual inspection of mycelium growth on the polymer surface
2. Quantitative estimation of microbial growth
3. Quantitative estimation of the weight loss of the polymer
4. Measurement of changes in polymer properties, such as molecular weight changes, changes in functional groups, crystallinity, tensile strength, or a combination thereof.

The currently available biodegradation testing methods for polymers are really only screening methods in laboratory-simulated environments. The best that can be achieved is an indication of the potential of the material to degrade. In addition, the environmental concentration of residual fragments must be measured (i.e., environmental interaction), which is a key factor in an environmental safety assessment (ESA). An ESA is a systematic way to evaluate the ecological hazards of biodegradation products, to measure their concentrations, and to ensure that the concentrations are below the level at which they are toxic to plant and animal life (no observable effect concentration; NOEC) (146). The condition for acceptability in the environment may be expressed as environmental concentration (EC) <NOEC where EC is a function of the rate and extent of biodegradability (BD):

$$EC = fn(BD)$$

Thus, determining the rate and extent of biodegradation of a polymer and identifying the mechanism of biodegradation and the metabolites produced enables ESA prediction.

The ASTM, CEN, and ISO standards and definitions on degradable polymers are available today. They give guidelines on the subject of biodegradability and compostability. The means of addressing handling of degradable polymers are soil burial disposal, incineration, recycling energy and materials recovery, composting, and biodegradation in aerobic and anaerobic atmospheres.

The ASTM subcommittee D-20.96 (on environmentally degradable plastics) is constantly working on the development of standards for degradable plastics. The sections under D-20.96 are:

Biodegradable (D-20.96.01)
Photodegradable (D-20.96.02)
Chemically degradable (D-20.96.03) (hydrolytic and oxidative)
Environmental fate (D-20.96.04)
Terminology (D-20.96.05)
Classification and marketing (D-20.96.06)

Details about ASTM and its activities can be obtained from ASTM headquarters, 1916 Race Street, Philadelphia, PA 19103-1187, USA.

VII. ENVIRONMENTAL INTERACTION OF POLYMERS

There are several different types of low molecular weight compounds that can be present in polymers. It is possible to detect traces of monomers, solvents and initiators, different additives, and degradation products. The low molecular weight compounds interact with the environment and may create problems and change the properties of the materials. The long-term properties are affected (e.g., leakages of additives will shorten the lifetime of the plastics).

Degradable polymers evolve low molecular weight compounds during degradation. Some are present in the materials from the beginning and can migrate and diffuse to various extent in the polymers during use, and they are eventually released from the polymeric surface to the environment. Some of the additives, which may in part be inorganic compounds, evolve only after offset of the degradation and add to the total evolution of low molecular weight products.

The degradation products of polymers range in number from one or two (e.g., in PLA or PGA) to several hundreds (e.g., in PE and PP). The combination of synthetic polymers and biopolymers added to the number of degradation products further complicates the degradation product pattern.

Gas chromatography (GC) and liquid chromatography (LC), together with mass spectrometry (MS) are the methods most suitable for detecting and identifying low molecular weight compounds in polymers. It is often necessary to develop special extraction procedures when analyzing polymers for the presence of smaller molecules. We have studied abiotic degradation of LDPE and the breakdown products by GC. We have earlier shown that it was possible to monitor the biodegradation of casein used as an additive in some current products (147,148).

The degradation products of accelerated-aged LDPE, containing iron dimethyl dithiocarbamate and carbon black used as degradable mulch films in agricultural materials, are acetaldehyde, methanol, acetone, and butanol, among others. Hydrocarbons, ketones, carboxylic acids, and dicarboxylic acids are formed during early stages of photooxidation, and during prolonged photooxidation, additional oxidation of ketones and monocarboxylic acids occurs (149). Dicarboxylic acids and ketoacids are also identified as secondary oxidation products and a zip depolymerization mechanism by back-biting through a cyclic transition state has been proposed (149).

The degradation products of biodegraded starch-filled LDPE differ from those identified in the corresponding abiotically degraded ones. Short-chain carboxylic acids are readily found in abiotically aged samples, whereas in the starch-filled LDPE that is degraded with microorganisms, these acids are missing. This we explained by the mechanism for the biodegradation of PE presented in 1987, which proposed bacterial assimilation of short-chain carboxylic acids (143). For PE, initial photooxidation before biodegradation, is necessary to create carbonyl compounds regularly along the polymer backbone. We recently demonstrated differences in the degree of degradation and products formed for LDPE with starch and prooxidant aged in either air or water (150). Mono- and dicarboxylic acids were formed in both environments, whereas ketones were identified only in air. Photo- and thermooxidation of LDPE with either starch or prooxidants or photosensitizers identified both oxygen-containing compounds and unoxidized hydrocarbons

(151,152). An efficient method to trap very volatile photooxidation products using closed vials was developed in that context (151).

The screening of low molecular weight compounds present and formed in the materials during long-term use and plastic waste handling is equally important for degradable polymers and for inert ones. To evaluate useability and waste treatment, the interaction with the environment of polymer materials must be known, and this is best achieved through careful analysis of the degradation products (e.g., Refs. 98,99,153,154). Three different observations are part of the monitoring of low molecular weight compounds in polymers, predicting the interaction of polymers with the environment. These are 1) the product pattern (i.e., the type of products identified and the pattern of these products); 2) the amount of products (nano- to milligrams) formed; and 3) the possible harmful effects of the compounds (e.g., nontoxic or natural and safe metabolite). In addition the cumulative effects must be analyzed. It is quite obvious that both natural and synthetic polymers may produce harmful degradation products. Therefore, identification and quantification of the products formed in every separate case are necessary. This gives us the basis to evaluate the interaction of the degradation products with the environment (156,157).

VIII. CONCLUDING REMARKS

Today, great efforts are being made to make existing nondegradable polymers biodegradable, photodegradable, or hydrolyzable by chemical modification or by the inclusion of additives (e.g., sensitizers and biopolymers) and by developing methods making it possible to use natural biodegradable polymers [e.g., poly(β-hydroxybutyrate)]. Renewed interest in several hydrolyzable polymers (e.g., polyesters, polyanhydrides, and polycarbonates) has resulted in new materials suitable for medical devices. In the future, new degradable polymers should be able to participate in the metabolism of nature (i.e., degraded by bioconversion) (157). Interest will be focused on the directly biodegradable materials (PLA, PCL, PCA, and others), on modified biopolymers (cellulose, starch, and soya protein derivatives), and on biodegradable blends (e.g., starch and polyolefins) (155). Many of the degradable polymers are still very expensive solutions for packaging, diapers, disposable tableware, and mulch films, but in biomedical connections the cost is of minor importance. In the primary step, a new material can be expensive, but in large-scale production, even these complex (bio)degradable polymers are usable for applications such as packagings. The use of degradable polymers is dependent on the knowledge of the interaction of these materials with the environment. This interaction is governed by the type and amount of small molecules evolving from the materials by diffusion and migration (156,157).

REFERENCES

1. A Brown. Mod Plast 23:189, 1946.
2. HJ Hueck. Plastics Oct:419, 1960.
3. CJ Wesel. SPE Trans. 4:93, 1964.
4. A Schwartz. Dissertation, Deutsche Acadmie Wissenschaften Berlin, K1 Chem Geol Biol 5:1, 1963.

5. R Wasserbauer. Technol Digest 9:3, 1967.
6. B Dolezel. Br Plast 40:105, 1967.
7. HE Worne. Plast Technol July:23, 1971.
8. RA Connolly. Bell Syst Technol J 51:1, 1972.
9. RA Connolly. In: H Walters, EH Heuck-van der Plas, eds. Biodeterioration of Materials. London: Applied Science Publishers, 1972, p. 168.
10. ES Pankhurst, MJ Davies, HM Blake. In: H Walters, EH Heuck-van der Plas, eds. Biodeterioration of Materials. London: Applied Science Publishers, 1972, p. 76.
11. J Potts, RA Clendinning, WB Ackart. An investigation of the Biodegradability of Packaging Plastics, US Environmental Protection Agency (EPA) publication EPA-R2-72-046, 1972.
12. AM Kaplan. First Intersect. Congr Int. Assoc. Microbiol. Soc, Preprints, Tokyo, Sept 1974.
13. E Küster. IUPAC Symp Long-Term Properties of Polymer Materials, Stockholm, Aug 1997.
14. A Azadi-Bakshs. Dissertation, Justus Liebig Universität, Giessen, 1972.
15. E Küster, A Azadi-Bakshs. Degradability of Polymers and Plastics, Preprints. London: Plastic Institute, Nov 1973, p. 16/1.
16. V Lazar. Int Biodeterior Bull 11:16, 1975.
17. DM Evans, I Levisoh. PATRA (UK) Printing Lab Rep 71, 1966.
18. GA Kanavel, PA Koons, RE Lauer. Rubber World 154(5):80, 1966.
19. ZT Ossefort, FB Testroet. Rubber Chem Technol 39:1308, 1966.
20. RT Darby, AM Kaplan. Appl Microbiol 16:900, 1968.
21. AM Kaplan, R Darby, M Greenberger, MR Rogers. Dev Ind Microbiol 9:201, 1968.
22. P Edmonds, JJ Cooney. Appl Microbiol 16:426, 1968.
23. DM Evans, I Livisohn. Int Biodeterior Bull 4:89, 1968.
24. HG Hedrick, MG Crom. Appl Microbiol 16:1826, 1968.
25. FBG Jones, T Le Champion Alsumard. Int Biodeterior Bull 6:119, 1970.
26. MP Rogers, AM Kaplan. Int Biodeterior Bull 7:15, 1971.
27. M Bomar. Plast Kaut 4:287, 1957.
28. HM Dayal, KL Maheswari, PN Agarwal, SS Nigam. J Sci Ind Res 21:356, 1962.
29. NN Pavlov, ZA Akopdzhanyan. Plast Massy 5:72, 1968.
30. DM Ennis, A Kramer. Lebensm Wiss Technol 7:214, 1974.
31. DM Ennis, A Kramer. J Food Sci 40:181, 1975.
32. JE Potts, RA Clendinning, WB Ackart, WD Niegisch. Tech Pap Reg Technol Conf Soc Plast Eng, Oct:63, 1972.
33. JE Potts, RA Clendinning, WB Ackart, WD Niegisch. Polym Preprints 13:629, 1972.
34. JE Potts, RA Clendinning, WB Ackart, WD Niegisch. In: J Guillet, ed. Polymers and Ecological Problems. New York: Plenum Press, 1973, p. 61.
35. JE Potts, RA Clendinning, WB Ackart. Conference on Degradability of Polymers and Plastics. London: Institute of Electrical Engineering, 1973, p. 12/1.
36. JE Potts, RA Clendinning, S Choen. Soc Plast Eng Tech Pap 21:567, 1975.
37. RD Fields. Dissertation Cornell University Microfilms, 74-10.873, 1973.
38. RD Fields, F Rodriguez, RK Finn. Polym Preprints 14:1244, 1973.
39. RD Fields, F Rodriguez, RK Finn. J Appl Polym Sci 18:3571, 1974.
40. RD Fields, R Rodriguez. In: JM Sharpley, AM Kaplan, eds. Proceedings of the 3rd International Biodegradation Symposium. London: Applied Science Publishers, 1976, p. 753.
41. MJ Diamond, B Freedman, JA Garibaldi. Int Biodeterior Bull 11:127, 1975.
42. Y Tokiwa, A Tadano, T Suzuki. J Ferment Technol 54:603, 1976.
43. PL Steinberg. Bell Syst Technol J 40:1369, 1961.
44. W Summer. Corrosion Technol 11:19, 1964.

45. C Wessel. SPE Trans 4:193, 1964.
46. HE Worne. Plast Technol July:23, 1971.
47. B Dolezel. Br Plast 40:105, 1967.
48. F Demmer. Mater Org (Berlin) 3:19, 1968.
49. WH Heap, SH Morrell. J App Chem 16:1826, 1968.
50. JD Campos de Alaniz, AA Solari. Bioquim Clin 7:171, 1973.
51. WN Kestelman, VL Jarovenko, EI Melnikova. In: AH Walters, EH Hueck-van der Plas, eds. Biodeterioration of Materials, vol. 2. London: Applied Science Publishers, 1972, p. 61.
52. VN Kestelman, VL Yarovenko, EI Melnikova. Int Biodeterior Bull
53. JR Haines, M Alexander. Appl Microbiol 28:1084, 1974.
54. JR Haines, M Alexander. Appl Microbiol 29:621, 1975.
55. A Tsuchi, T Suzuki, F Fukuoka. Rep Ferment Res Inst 55:10, 1980.
56. B Baum, RD Deanin. Polym Plast Technol Eng 2(1), 1973.
57. L Binder, S Sultan. Report to the Swedish Board for Technical Development (STU), 71-290/U, 1973.
58. RG Caldwell. Tappi 58:64, 1975.
59. JD Cooney, DM Wiles. Soc Plast Eng Tech Pap 20:420, 1975.
60. A-C Albertsson. J Appl Polym Sci 22:3419, 1978.
61. A-C Albertsson, ZG Banhidi, LL Beyer-Ericsson. J Appl Polym Sci 22:3434, 1978.
62. A-C Albertsson, B Ranby. Appl Polym Symp 34:423, 1979.
63. A-C Albertsson, ZG Banhidi. J Appl Polym Sci 25:1655, 1980.
64. A-C Albertsson, B Ranby. In: JM Sharpley, AM Kaplan, eds. Proceedings of the 3rd International Biodegradation Symposium. London: Applied Science Publishers, 1976, p. 743.
65. A-C Albertsson, S Karlsson. J Appl Polym Sci 35:1289, 1988.
66. A-C Albertsson, O Ljungquist. Acta Polym 39:95, 1988.
67. T Mathisen, A-C Albertsson. J Appl Polym Sci 38:591, 1989.
68. PH Jones, D Prasad, M Heskins, MH Morgan, JE Guillet. Environ Sci Technol 8:919, 1974.
69. JE Guillet, TW Regulski, TB McAnenay. Environ Sci Technol 8:923, 1974.
70. LR Spencer, M Heskins, JE Guillet. In: JM Sharpley, AM Kaplan, eds. Proceeding of the 3rd International Biodegradation Symposium, London: Applied Science, 1976, p. 753.
71. JE Guillet. Polym Mater Sci Eng 63:946, 1990.
72. G Scott. Arabian J Sci Eng 13:605, 1988.
73. G Scott. In: J Guillet, ed. Polymers and Ecological Problems, New York: Plenum Press, 1973, p. 27.
74. G Scott. J Oil Colour Chem Assoc 56:521, 1973.
75. MU Amin, G Scott. Eur Polym J 10:1919, 1974.
76. G Scott. Polym Age 6:54, 1975.
77. D Gilead, G Scott. Br. Patent 1 586 344, 1978.
78. D Gilead, G Scott. In: G Scott, ed. Developments in Polymer Stabilization, vol 5. London: Applied Science, 1982. p. 71.
79. S Karlsson, O Ljungquist, A-C Albertsson. Polym Degrad Stabil 21:237, 1988.
80. A-C Albertsson, C Sares, S Karlsson. Acta Polym 44:243, 1993.
81. A-C Albertsson, S Karlsson. Prog Polym Sci 15:177, 1990.
82. R Johnson. In: Proceedings of Symposium on Degradable Plastics. Society of the Plastics Industry, Washington, DC, June 10, 1987, p. 6.
83. GM Harlan. In Proceeding of Symposium on Degradable Plastics. Society of the Plastics Industry, Washington DC, June 10, 1987, p. 14.
84. MH Levy. Convert Packag p. 136, 1987.

85. RL Whistler, EF Paschall, eds. Starch: Chemistry and Technology, vol 1.
86. EB Bagley, GF Fanta, RC Burr, WM Doane, CR Russell. Polym Eng Sci 17:311, 1977.
87. GJL Griffin. US Pat 4 016 117, 1977.
88. GJL Griffin. US Pat 4 021 388, 1977.
89. GJL Griffin. Int Pat WO 88/9354, 1988.
90. JL Willett. US Pat 5 087 650, 1992.
91. JD Peanasky, JM Long, RP Wool, J Polym Sci Polym Phys Ed 29:256, 1991.
92. SM Goheen, RP Wool. J Appl Polym Sci 42:2691, 1991.
93. GJL Griffin, PS Nathan. J Appl Polym Sci Appl Polym Symp 35:475, 1979.
94. GJL Griffin, SA Hashemi. Iran J Polym Sci Technol 1:45, 1992.
95. ST Tanna, R Gross, SP McCarthy. Polym Mater Sci Eng 67:294, 1992.
96. M Yasin, SJ Holland, AM Jolly, BJ Tighe. Biomaterials 10:400, 1989.
97. SJ Holland, M Yasin, BJ Tighe. Biomaterials 11:206, 1990.
98. A-C Albertsson, C Barenstedt, S Karlsson. Polym Degrad Stabil 37:163, 1992.
99. A-C Albertsson, C Barenstedt, S Karlsson. J Appl Polym Sci 51:1097, 1994.
100. A-C Albertsson, S Karlsson. Polym Mater Sci Eng 67:296, 1990.
101. G Ianotti, N Fair, M Tempesta, H Neibling, F-H Hsieh, R Mueller. In: SA Barenberg, JL Brash, R Narayan, AE Redpath, eds. Degradable Materials. Boca Raton, FL: CRC Press, 1990, p. 425.
102. AL Pometto III, B Lee, KE Johnsson. Appl Environ Microbiol 58:731, 1992.
103. KE Johnsson, AL Pommetto III, ZL Nikolov. Appl Environ Microbiol 59:1155, 1993.
104. KE Johnsson, AL Pometto III, L Somasundaram, J Coats. J Environ Polym Degrad 1:111, 1993.
105. WJ Maddever, GM Chapman. Plast Eng 45:31, 1989.
106. W Sung, ZL Nikolov. Ind Eng Chem Res 31:2332, 1992.
107. GJL Griffin, K Tarverdi. In: TA Oxley, S Barry, eds. Macrobiodeterioration of Thermo-plastics: Starch Filler as an Attractant for Woodlice in Biodeterioration 5. New York: Wiley, 1983.
108. FH Otey, AM Mark, CL Mehltretter, CR Russell. Ind Eng Chem Prod Res Dev 13:90, 1974.
109. FH Otey, RP Westhoff. US Pat 4 133 784, 1979.
110. FH Otey, RP Westhoff. US Pat 4, 337181, 1982.
111. C Bastioli, R Lombi, G Del Tredici, I Guanella. Eur Pat Appl 0 400 531, 1990.
112. C Bastioli, V Belloti, L Del Giudice, G Del Tredici, R Lombi, A Rallis. Int Pat WO 90/10671, 1990.
113. C Bastioli, L Del. Giudice, R Lombi. Int Pat WO 91/02024, 1991.
114. C Bastioli, G Del Tredici. Int Pat WO 91/02025, 1991.
115. C Bastioli, V Bellotti, G Del Tredici, R Lombi, A Montino, R Ponti. Int Pat WO 92/19680, 1992.
116. J Silbiger, J-P Sachetto, DJ Lentz. Eur Pat Appl 0 408 501 A2, 1991.
117. J-P Sachetto, J Silbiger, DJ Lentz. Eur Pat Appl 0 408 502 A2, 1991.
118. J Silbiger, DJ Lentz, J-P Sachetto. Eur Pat Appl 0 408 503 A2, 1991.
119. J-P Sachetto, J Rehm. Eur Pat Appl 0 407 350 A2, 1991.
120. J-P Sachetto, J Silbiger, DJ Lentz. Eur Pat Appl 0 409 781 A2, 1991.
121. DJ Lentz, J-R Sachetto, J Silbiger. Eur Pat Appl 0 409 782 A2, 1991.
122. J-P Sachetto, J Silbiger, DJ Lentz. Eur Pat Appl 0 409 783 A2, 1991.
123. HL Suominen, J Melartin, K Karimo. Int Pat WO 89/10391, 1989.
124. HL Suominen. US Pat 5118725, 1991.
125. HL Suominen. US Pat 5133 909, 1991.
126. JJ Vanderbilt, CM Neeley. Int Pat WO 90/15843, 1990.
127. D Toms. Int Pat WO 93/00399, 1993.
128. Y Doi. Microbial Polyesters. New York: VCH, 1990.

129. SJ Holland, AM Jolly, M Yasin, BJ Tighe. Biomaterials 8:289, 1987.
130. T Saito, K Tomita, K Juni, K Ooba. Biomaterials 12:309, 1991.
131. RH Marchessault, CJ Monasterios, P Lepoutre. In: EA Dawes, ed. Novel Biodegradable Microbial Polymers. Dordrecht: Kluwer Academic, 1990, p. 97.
132. M Yasin, SJ Holland, BJ Tighe. Biomaterials 11:451, 1990.
133. H Mimoto, E Ota. Sen'i Gakkaishi 47:89, 1991.
134. R Renstad, S Karlsson, A-C Albertsson, PE Werner, M Westdahl. Polym Int 43:201, 1997.
135. R Renstad, S Karlsson, A-C Albertsson. Polym Degrad Stabil 57:331, 1997.
136. S Bloembergen, DA Holden, TL Bluhm, GK Hamer, RH Marchessault. Macromolecules 22:1663, 1989.
137. S Bloembergen, DA Holden, RH Marchessault. Polym Preprints 29:594, 1988.
138. N Yoshie, M Sakurai, Y Inoue, R Chujo. Macromolecules 25:2046, 1992.
139. JJ Jesudason, RH Marchessault, T Saito. J Environ Polym Degrad 1:89, 1993.
140. A-C Albertsson, S Karlsson. In: SA Barenberg, et al, eds. Degradable Materials: Perspectives, Issues and Opportunities. Boca Raton, FL: CRC Press, 1990, p. 263.
141. A-C Albertsson. In: A Patsis, ed. Advances in Stabilization and Degradation of Polymers, vol. I. Lancaster, PA: Technomic Publishing, 1989, p. 115.
142. A-C Albertsson. Dissertation. The Royal Institute of Technology, Stockholm, 1977.
143. A-C Albertsson, S-O Andersson, S Karlsson. Polym Degrad Stabil 18:73, 1987.
144. A-C Albertsson. Eur Polym J 16:623, 1980.
145. A-C Albertsson, S Karlsson. ACS Symp Ser 433:60, 1990.
146. G Swift. In: International Science Workshop on Biodegradable Plastics and Polymers, Osaka, Japan. 1993, p. 41.
147. S Karlsson, ZG Banhidi, A-C Albertsson. J Chromatogr 442:267, 1988.
148. S Karlsson, ZG Banhidi, A-C Albertsson. Mater Struct RILEM 22:163, 1989.
149. S Karlsson, M Hakkarainen, A-C Albertsson. Macromolecules 30:7721, 1997.
150. M Hakkarainen, A-C Albertsson, S Karlsson. J Appl Polym Sci 66:959, 1997.
151. F Khabbaz, A-C Albertsson, S Karlsson. Polym Degrad Stabil 61:329 (1998).
152. F Khabbaz, S Karlsson. (manuscript).
153. A-C Albertsson, C Barenstedt, S Karlsson. Acta Polym 45:97, 1994.
154. S Karlsson, C Sares, R Renstad, A-C Albertsson. J Chromatogr A 669:97, 1994.
155. A-C Albertsson, S Huang. Degradable Polymers, Recycling and Plastic Waste Management. New York: Marcel Dekker, 1995.
156. S Karlsson, A-C Albertsson. JMS Pure Appl Chem A32:599, 1995.
157. S Karlsson, A-C Albertsson. Macromol Symp 118:733, 1997.

11

Assessment of Biodegradability in Organic Polymers

ANTHONY L. ANDRADY

Research Triangle Institute, Research Triangle Park, North Carolina

I. INTRODUCTION

The global demand for plastics has consistently increased in the recent years and presently stands at about 135 million tons annually. It was back in mid-1980s that the world production of plastics exceeded that of steel. For a material with a short history of about half a century of commercial use, this is a major achievement that illustrates the versatility and the cost-effectiveness of plastics in a myriad of applications. Interestingly, the world demand for plastics has increased linearly with the population, allowing the estimation of the resin production in year 2010, when the global population is expected to level off, to be about 165 million metric tons per year. The consumption of plastics in various applications is not uniform in all parts of the world. There is a wide disparity in the per capita consumption of plastics in affluent countries, such as Japan or the United States, and developing countries, such as India or China. The differences reflect mainly those in the lifestyles in the respective countries. The more affluent nations rely on disposable single-use plastic packaging more extensively, and use more plastic building materials in construction. The exceptional mechanical properties and the weatherability of adequately stabilized formulations and the low cost of plastics, compared with competing materials, have resulted in rapid growth of plastics' use even in the developing world. Even a small change in per capita consumption in a populous country, such as China, can amount to very large changes in the absolute volume of resin demand annually.

Most plastic products have a limited lifetime and invariably end up as municipal solid waste. With those used for packaging applications, such as Styrofoam cups or polyethylene straws, the useful lifetime of only a few minutes is very much shorter than the available lifetime based on the ability of the material to continue to perform the intended task. This leads to more packaging plastics in the solid waste stream as well as in urban litter. As litter in particular is a problem directly related to population density, the present trend of migration of population to urban centers and the popularity of megacities will contribute to this problem in the future. Given the consumption patterns, one might expect plastic wastes to be a particular problem in two settings: urban centers in affluent countries and those in highly populated developing countries. Signs are already evident that the disposal of plastic wastes and litter in these types of communities is rapidly becoming a concern.

Why do plastics, which constitute only about 7% or less by weight of the municipal solid waste stream, cause such disproportionate concern? Over 50% of the solid waste stream is paper-based and the same concerns are rarely expressed about paper. The main reason has to do with the biological inertness of plastics; in general, synthetic polymers do not undergo microbial-mediated breakdown in the environment. Therefore, except for the small fraction of plastics that is incinerated, all the plastic resin that was ever produced still remains intact somewhere, most probably in landfills. Although durability is certainly an excellent attribute of materials during their useful lifetime, the same is not true from a waste-management perspective. From an environmental standpoint, this amounts to slowly converting a fraction of the limited crude oil reserves on earth into xenobiotic material that is, for all practical purposes, excluded from the carbon cycle in the short term. With wood, paper, leather, and cotton textiles the discarded waste material (if allowed enough moisture and a microbial environment) breaks down into small organic molecules and finally into carbon dioxide and water. These products can be used by other members of the ecosystem, for instance, to create biomass; hence, they are effectively recycled. Possibly owing to the very short history of plastics in the biosphere, microbial species with physiological apparatus needed to carry out such a digestion have not yet evolved. It is this single feature of biodegradation that distinguishes plastics from most (glass, for instance, also falls into this category) other components in the municipal solid waste or litter stream.

This chapter will examine the biodegradability of synthetic organic polymers, with emphasis on the approaches to assessment and quantification of biodegradability.

II. DEFINITIONS

Several attempts have been made to define the terms *degradation* and *biodegradation* within the context of environmental considerations. Biodegradation might be conveniently defined as "a chemical change in polymer facilitated by living organisms, usually microorganisms" (1). This definition is somewhat restricted, however, in that it excludes chemical breakdown processes on polymer substrates mediated by either isolated natural enzymes or synthetic man-made enzymes. However, this fits in well with the classification of degradation reactions shown in Figure 1, and has been adopted by recent reviewers (2). The scheme defines degradation as a chemical change that is brought about by one or more of several agents acting on the polymer. Depending on the predominant factor, the degradation can be classified into various subcategories. Biodegradation can often involve several steps, and the Japanese Biodegradable Plastics Society definitions recognizes this by requiring

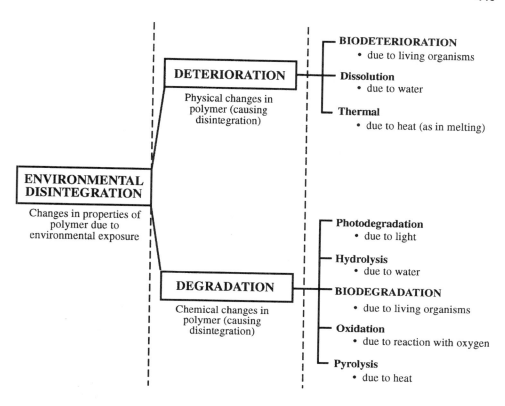

Figure 1 Definitions of degradability.

"at least one step" in the degradation process to be mediated by naturally occurring microorganisms. A clear distinction, however, needs to be made between true biodegradation and biologically mediated disintegration or volume reduction of polymers, which does not amount to degradation (see Fig. 1) and is referred to as *biodeterioration*. The attack of polyethylene by insects (3,4), for instance, belongs to this latter category. In spite of the "damage" suffered by the polymer, the predominant change is physical, and the indigestible polymer is merely reduced in particle size at the end of process. Several such "biodeteriorable" polymer compositions that were physical blends of polyethylene with starch have been commercially introduced. On environmental exposure the starch fraction is expected to biodegrade rapidly, weakening the polyethylene matrix to such an extent that it fragments. Although technologies that could blend anywhere from about 6–60% of starch into a extrusion-blown thin film were developed, their biodeteriorability characteristics fell short of expectations. In low-starch formulations, limited accessibility of the starch granules to microbial flora, restricted facile biodegradation, limiting it to only about 10% of the available starch (4–6). In high-starch formulations, the starch domains were sufficiently interconnected to allow a more complete degradation of the starch, but such films often exhibited poor mechanical properties. Adequate connectivity between starch domains to allow biodegradation is expected only at high levels of the additive, around 30 wt% of starch in the blend (7).

A practical definition must reach beyond a clarification of the chemical process and hopefully assist a user in distinguishing an environmentally acceptable biodegradable polymer from a regular plastic material. Such definitions require a further elaboration of the simple definition to include at least two additional considerations: a) A benchmark rate of biodegradation that is deemed to be environmentally acceptable. The biodegradation rates of various polymers might then be compared with the benchmark rate in an effort to quantify their biodegradability; and b) some means of addressing the toxicity of by-products and residues produced by the degrading polymers. Adopting a definition for practical purposes, particularly to delimit those polymers that are "environmentally biodegradable" is more complicated and several attempts have been made in this direction (8).

As all polymers are invariably biodegradable (the term "biodegradable polymer" is a misnomer—all organic polymers do invariably breakdown, although at an extremely slow rate), some guidelines are needed to recognize biodegradation rates fast enough to obtain some environmental benefit. From a solid-waste management point of view, a slow biodegradation that reduces the volume of solid plastics by a fraction of a percentage each year is of little value. Adoption of a suitable benchmark, however, has not been easy. With no agreed benchmark of a "reasonable" rate of biodegradation as yet available, the use of natural biopolymers as standard biodegradable materials (9) has been suggested. Because of their rapid rates of breakdown, regenerated cellulose or filter paper (10), wood pulp (11), and even whole leaves might be used as reference materials in biodegradation studies. The environmental justification here is that these natural biopolymers, such as lignocellulosics, have been a part of the ecosystem for a very long duration, and a biodegradation rate approaching that of biopolymers, therefore is, likely to be environmentally acceptable. The requirement that the biodegradation process does not lead to the production of toxic by-products is also important. Xenobiotic materials are known to yield toxic products on biodegradation (12,13). For instance, the notion of an environmentally acceptable biodegradable polymer has been discussed (14). This definition requires the polymer to be either completely mineralized or partially mineralized, but not producing any environmentally harmful residue. The DIN 103.2-1993, the German Working Group definition of biodegradable polymers, also includes the requirement of complete biodegradation.

Interestingly, none of the reported definitions of environmental biodegradability refer to the microbial mechanism of the breakdown. At least, in theory, the possible fate of a synthetic polymer in a biotic environment could include a) its complete or partial biodegradation by microorganisms; b) cometabolism of the substrate; or c) the incorporation of the breakdown products of the substrate in natural humic polymers. In cometabolism a polymer substrate is biodegraded only in the presence of a second substrate, but the organisms do not derive any energy from the polymer. Cometabolism of organic compounds is common and is a significant breakdown mechanism in soil environments. The requirement in some definitions of biodegradation that the microorganisms utilize the substrate (for energy and carbon) preclude cometabolic breakdown from the definition.

A distinction also needs to be made between aerobic an anaerobic biodegradation. In general the aerobic processes predominate in the presence of air. The process of complete biodegradation of a polymer or any organic material can be represented by the following simple chemical process.

$$C_xH_yO_z + (2x - z + y/2)\,O \longrightarrow x\,CO_2 + y/2\,H_2O \qquad (1)$$
$$C_xH_yO_zN_p + (2x - z + (3p - y)/2)\,O \longrightarrow x\,CO_2 + (3p - y)/2\,H_2O + pNH_3 \quad (2)$$
$$\text{for } 3p > y$$
$$C_xH_yO_zN_p + (2x - z + (y - 3p)/2)\,O \longrightarrow x\,CO_2 + (y - 3p)/2\,H_2O + pNH_3 \quad (3)$$
$$\text{for } 3p < y$$

In environments, such as in deep-sea sediments, anaerobic digesters, marshy soil depleted of oxygen, or even modern landfills, however, alternative chemical routes that do not rely on oxidation are available for biodegradation. Such anaerobic biodegradation processes may yield a different mix of products that often include methane. But the original material, the substrate, nevertheless, is broken down into simple, mostly gaseous, products. With polymers such as polyethylenes that biodegrade extremely slowly, copolymerization or other means of introducing polar biodegradable functionalities into the polymer chain can induce biodegradation in the polymer.

III. FACTORS AFFECTING BIODEGRADABILITY

Biodegradability of a polymer is essentially determined by its physical (in the case of biodegradation as a solid substrate) and chemical characteristics. The availability of functional moieties that can participate in enzyme-mediated degradation reactions on the polymer is the crucial requirement. For instance, 2 methylene 1,3-dioxepane can be used as a comonomer with styrene, or a ketene acetal might be used with ethylene comonomer. This approach increases the biodegradability of the co-polymer (15). The lower molecular weight polyethylenes generated during biodegradation are more likely to undergo faster breakdown than the unaltered polyethylene. Alternatively, block copolymers of a synthetic (and essentially recalcitrant) polymer with a biopolymer, such as a polysaccharide, might be used to achieve the same end. A segmented block copolymer can be made by reacting low molecular weight amylose or cellulose blocks that have terminal hydroxy groups with a synthetic prepolymer (such as a polyether) with reactive end groups using an appropriate diisocyanate. Gilbert et al. used a multistep reaction sequence to produce several such biodegradable block copolymers (16–18).

A. Long Chain-like Molecular Geometry

Long chain-like molecular geometry (and the consequent high molecular weights) of polymers need not necessarily preclude biodegradation. With those enzymatic systems that act on chain ends in exo-type reactions (19), the higher molecular weights do retard biodegradation, as the bulk concentration of the reactive terminal groups decreases with increasing chain length biodegradation (20–22). Lower average molecular weights also discourage the formation of crystalline domains that are generally difficult to biodegrade within the bulk polymer. Amorphous regions are preferentially biodegraded in synthetic polymers (23–26) as well as in biopolymer lignocellulose (27). However, biopolymers (such as cellulose or chitin), as well as some synthetic polymers (e.g. polycaprolactone), are readily biodegradable despite the long-chain geometry.

B. Structural Complexity

The biodegradability of a polymer generally implies the existence of a set of micro-organisms able to utilize the polymer substrate as a carbon and energy source. Because this needs to be accomplished with minimum expenditure of energy by these organisms, the more complex polymers that would require numerous enzyme-mediated steps for their breakdown is not an energy-effective choice (28). Several species of microorganisms acting in concert to produce the required ensemble of enzymes may not be available in the microenvironment of interest. Higher levels of structural complexity of a polymer, therefore, leads to recalcitrance in the environment. Persistence of soil humic acids, naturally occurring biopolymers in soil, is attributed to their structural complexity (29).

C. Hydrophilicity

Synthetic polymers that are soluble in water, such as poly(vinyl alcohol) (30) poly(acrylic acid) (31), and polyethers (32) tend to be more biodegradable than water-insoluble polymers of comparable molecular weight and structural complexity. Also, increasing the hydrophilicity of a polymer by chemical modifi-cation promotes its biodegradability (33). The functional groups that impart water-solubility also apparently contribute to ready biodegradability of these polymeric systems. This is not surprising, as the presence of a dissolved substrate in reasonable concentrations would induce the production of necessary enzymes within the microorganisms relatively more easily than would a solid insoluble substrate.

IV. APPROACHES TO ASSESSMENT OF BIODEGRADABILITY

In aerobic biodegradation of polymers a set of enzymes secreted by microorganisms facilitate extracellular degradation processes that progressively reduce the molecular weight and the structural identity of macromolecules. Generally, the high polymer is first reduced to a "low polymer" by a combination of chain scission and removal of repeat units from chain ends of the polymer. This leads to fragmentation or dissol-ution of the polymer. Further enzymatic action yields oligomeric fragments and simple organic compounds that are intermediates of the biodegradation process. If biodegradation is allowed to continue, complete mineralization would ensue, with the transformation of the organic polymer into carbon dioxide, water, and perhaps other products, such as methane. The microbial community that facilitates the pro-cess derive energy and simple organic compounds for anabolic processes from the reaction sequence to foster their own growth and reproduction. A typical example of this sequence of events is found in the biodegradation of cellulose, the most abundant biopolymer. Several enzymes act synergistically in the break-down of cellulose in a series of hydrolysis reactions (34,35). The reactions occur mainly in the amorphous regions of the biopolymer matrix. *endo*-Cellulases cause

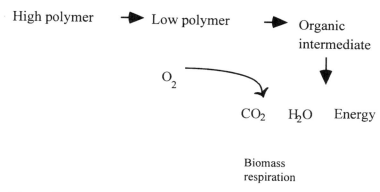

Figure 2 Hypothetical stages in the biodegradation of a high polymer.

random chain scission, whereas *exo*-cellulases act at terminals of cellulose chains splitting off cellobiose units that, in turn, are hydrolyzed by β-glucosidase. Lignin, often found associated with cellulose in substrates such as wood or unbleached paper, can biodegrade through specific oxidative pathways. Relevant enzymes (such as lignases, laccase, and alcohol oxidase) have been reported (36,37). The role of white-rot fungi in causing environmental biodegradation of lignin is well known (38).

The sequence of changes that generally accompany environmental biodegradation of polymers is given in Figure 2. The high polymer initially breaks down into a low-polymer or oligomers, some of which might be water-soluble. Further breakdown of these yield organic intermediates that are further reduced to inorganic products. The pathway suggest several possible approaches available for detection and measurement of biodegradation in polymers.

Although each individual approach is valid, they measure different aspects of the biodegradation process, and the one selected for use needs to be based on the definition of biodegradation adopted by the experimenter as well as the end-use of the measurement. In general, the standard tests used to assess biodegradation attempt not only to qualitatively ascertain that the material biodegrades, but also to quantify the rate at which the process occurs under the conditions used in the test. Because of the numerous factors that can influence the process, it is usual to express the latter as a relative measurement. Therefore, experiments generally include a reference material (preferably a polymer such as cellulose powder or cellophane) as well as a control environment with no sample or reference material.

Biodegradability of a given polymer can be discussed only in relation to a) the criterion (or definition) adopted in its assessment, and b) the nature of microbial environment used for the purpose. This second issue is crucial and will be discussed in detail later in this chapter. It is obvious that the rate of biodegradation of a readily biodegradable polymer obtained in a given environment is determined by the density and biodiversity of the microorganisms in that environment. Generalization of experimental data (particularly the description of a polymer as being "biodegradable" without qualification) can be misleading and contribute to confusion in the literature.

A. Monitoring the Accumulation of Biomass

An indirect, but acceptable indication of biodegradation of a substrate is the increase in biomass in its immediate environment. The microorganisms essentially use the polymer as a source of carbon as well as energy, undergoing both somatic growth and reproduction at the expense of the substrate. Some of the early tests on biodegradation of polymers employing a single species of microorganism relied on this particular approach. A sample of the polymer is incubated in a pure culture of the test microorganism for a time period, and the surface coverage by the biomass achieved is rated on a scale of 0–4 in American Society for Testing and Materials (ASTM) G21, G29, and ASTM D1924 tests.

An improvement over visual assessment of the growth is by simple measurement of the biomass using turbidimetry (39). Alternatively, either direct microscopy (40) or standard plate-count techniques might be used to quantify the increase in biomass. The disadvantage of the latter technique is that it selectively detects some fraction of the microflora that can thrive on the particular culture medium employed. Biomass might also be quantified by subsequent mineralization (and measurement of the CO_2 evolved) (41) or by sophisticated chemical methods, such as the analysis of culture media for adenosine triphosphate (ATP) (42) that is indicative of live biomass.

The main difficulty with this general approach is the variability of the conversion factor (called the yield) of substrate carbon into biomass carbon. If the yield is accurately known, the extent of substrate conversion can be simply calculated from the increase in biomass yield. However, the yield of biomass is both substrate-specific and species-specific (43) and, for instance, can vary from values as low as 10% for lignins to 60% for saccharides or amino acids, for consortia of soil microbes (44). Yield factors for even single species acting on polymer substrates are not known. With microbial consortia, the situation is even more complicated, with species competing for the substrate, and the yield itself possibly changing with the duration of incubation. A practical difficulty is encountered in studying the biodegradation of multicomponent systems such as plasticized polymers. The low molecular weight plasticizer may often undergo biodegradation preferentially, and the increase in biomass can be mistaken as evidence of the polymer matrix (as opposed to the plasticizer alone) undergoing biodegradation.

B. Monitoring the Depletion of Polymer or the Demand for Oxygen

Biodegradation must invariably result in the loss of substrate from the system, but loss of polymer from the environment is never a good indication of complete biodegradation (or mineralization) of the material. Biodegradation might reduce the polymer molecular weight to an extent to make it merely water-soluble. The resulting loss of substrate is due to only the initial stages of the biodegradation process. However, when the interest is on this phase of the process (the removal of high polymer substrate from the system), monitoring depletion of polymer can be an effective approach.

Depletion of polymeric substrate is measured using a simple technique such as monitoring the weight loss during biodegradation (10,45). Experimental difficulties, such as the loss of sections of the polymer substrate by fragmentation and the possible close association of the biomass with the polymer (leading to overestimate of weight), make this an approximate technique. These result in poor correlation of data on weight loss and those from other measures of biodegradation (46,47), particularly with weak substrate such as paper or foamed polymers. Where finely divided polymer powder is used the "clear zone" technique (48,49) on agar plates might be used for preliminary assays of biodegradability. The technique is similar to that use in screening antibiotic activity of chemicals using mixed cultures of microbes. Alternatively, an analytical technique, such as spectroscopy (7), and even thermogravimetry (50), might be used to study loss of substrate during biodegradation. The success of using any analytical technique depends both on the specificity of the method for the particular substrate in question and the detection limits obtained under the experimental conditions.

A second important reactant in the aerobic biodegradation processes is oxygen. Not only are the biodegradation pathways often oxidative, but also, oxygen is needed by the biomass for growth and reproduction. Oxygen demand by a biodegrading polymer, therefore, is used to quantify the process (51,52). Theoretical oxygen demand (TOD) for a pure substrate can be obtained by considering the mineralization chemistry [see Eqs. (1)–(3) given earlier]. The chemical oxygen demand (COD) experimentally observed (perhaps in a complete breakdown of the substrate by a strong oxidizing agent) is in the range of 90–100% of the TOD, depending on the substrate. The biochemical oxygen demand (BOD) is the oxygen used up by the biotic system in the process of biodegradation of the substrate. Standard tests measure short-duration BOD during a 5- or 7-day period of activity. Although this correlates well with the extent of substrate breakdown, the newly formed biomass and synthetic compounds in the cells, survive in the medium for a much longer time.

C. Monitoring the Products of Reaction

Whereas initial stages of biodegradation of a polymer may yield simple organic intermediates, complete biodegradation involves further biooxidation of the product into carbon dioxide and water. For instance in the biodegradation of cellulose some cellobiose and glucose is observed in the reaction mixture during early reaction, but carbon dioxide is the main product of aerobic breakdown. Although, in principle, it is possible to analytically monitor the formation of the intermediate, it is the mineralization (conversion of the organic carbon into CO_2) that is often of interest. Quantifying the water of reaction is beset with experimental difficulties, and it is of little interest. Several techniques available for quantifying biodegradation of polymers rely on monitoring the evolution of CO_2 during the process. Some of the popular standard test protocols for establishing aerobic biodegradability of polymers, published by the ASTM (Committee D20.96), MITI Japan, and OECD (Test 304-A) are based on this approach.

The test is much easier and more meaningful when applied to those polymers that can biodegraded rapidly (in a matter of weeks, as opposed to in several years or decades). With recalcitrant polymers, such as polyethylenes, radiotracer techniques are used to quantify the minute amounts of CO_2 formed over very long time periods (53,54). Only about 0.1%/year of the carbon that makes up the polyethylene is transformed into CO_2 by biodegradation, even under the best laboratory conditions. With biopolymers, such as cellulose or chitosan, and some synthetic poly-

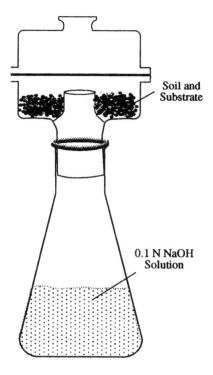

Soil and
Substrate

0.1 N NaOH
Solution

Figure 3   A novel design for a biometer flask, as described by Andrady. (From Ref. 1.)

mers, such as the aliphatic polyesters, the biodegradation rates are rapid and
titrimetric estimation of CO_2 is possible (10,55).

The test is typically carried out in a respirometer flask (also called a biometer
flask) that carries a known mass of the finely divided polymer substrate, a biotic
medium (usually soil inoculated with microbial cultures), and an aliquot of a dilute
alkali to absorb the gas produced during the process. This alkali solution is removed
periodically and titrated against a standard acid to estimate the amount of CO_2
absorbed by it. Care must be taken to ensure that there is adequate oxygen in
the flask at all times to support the biotic environment and that the titrations
are carried out before the sorbant alkali is totally neutralized (56). With most designs
of respirometer flasks, daily titration is recommended with biopolymer substrates.
Figure 3 shows a recent design of such a biometer flask suitable for the purpose
reported by Andrady (1). This design is convenient, as the top (desiccator-shaped)
part can be removed and the alkaline sorbant titrated in the flask itself. With
the other older design the alkali is removed from the flask through a syringe needle
for titration in a separate titration flask. During the removal of the liquid, fresh
air is allowed to enter the flask by opening the stopcock and the cover of the
drying tube attached to the flask. With the Andrady flask (1) aeration is automatic
as the lower part of the apparatus is removed for titration and a fresh flask with
a measured amount of alkali is substituted for it. For rigorous work involving
slow-degrading polymers this switching of flasks might be carried out in a glove
box of CO_2-free air. Even with moderately degradable polymers, however, changing

the flasks daily in ambient air does not affect the reproducibility of the results. An alternative design uses a continuous stream of air to remove the CO_2 formed from the biodegradation chamber and bubble into one or more sorption bottles containing alkali for later titrimetric analysis (57). The exit airstream can also be analyzed using a flow-through cell and an infrared detector for CO_2.

The test might be carried out in water or soil medium. In either case a suitable volume of activated sludge, to obtain complete mineralization of the sample in about a month, is used as an inoculum. About 10 mL/L is recommended for aqueous systems and about 5 mL/50 g for use with soil. The inoculum must be used the same day it is collected and kept aerated until use. Also added to the medium is sufficient urea and potassium hydrogen phosphate (0.1 and 0.05% of weight of polymer substrate to be added) to fortify it and to promote faster microbial growth. The soil water content is adjusted to 60% of its maximum water-holding capacity, and the system is incubated at 32°C for a day. About 50 g of the soil is transferred into the upper chamber of the biodegradation flask assembly (see Fig. 3). A small quantity (usually 0.1 g/50 g soil) of the plastic, preferably in a finely divided form, is mixed into the soil in test flasks. A blank flask with no polymer mixed with the soil, and a flask with a reference material such as cellulose (cellophane) is also generally used.

A known volume of 0.1 N KOH is placed in the bottom of the flask, and it is closed firmly with the upper part, using a very thin film of high-vacuum grease to ensure air-tightness. Under incubation, biodegradation proceeds yielding CO_2 that flows down from the upper chamber into the bottom of the flask to dissolve in the alkali. The amount of CO_2 reacted with the alkali can be determined by titration against a standard solution of hydrochloric acid. Although the precision in titrimetry is quite good, it is still desirable to carry out the test in triplicate (making a total of nine flasks for a single substrate test). The amount of CO_2 liberated in the test flask is corrected for background levels given by the blank determination. The test is continued until 3–4 days of near zero net differences in CO_2 is obtained between the blank and the test flasks. Typical data for a sample of bleached paperboard is shown in Fig. 4.

The cumulative CO_2 evolution curve is convex over the horizontal time axis and reaches a plateau. Sometimes a small lag-time of c (days) is observed before any gas evolution is detected. Thereafter, gas evolution proceeds rapidly over a period of several months until the cumulative curve reaches the asymptote value of a. The shape of the curve suggests that it might be quantified using the following empirical equation.

$$y = a\{1 - \exp(-k(t - c)\}$$
$$= 0 \text{ for } t < c \tag{4}$$

where y is the percentage mineralization of the substrate carbon, and k is a rate coefficient. To carry out the curve-fitting, the carbon content of the substrate needs to be determined for the substrate. Fitting the experimental data with this empirical curve yields a value for k (days^{-1}), and the plateau of the curve gives y_{max}, the maximum percentage conversion of the substrate carbon achieved during the experiment. Most natural polymers and readily biodegradable compounds reach a value of $y_{max} \geq 60\%$ in about 28–30 days. The value of k (days^{-1}) and y_{max} might be used

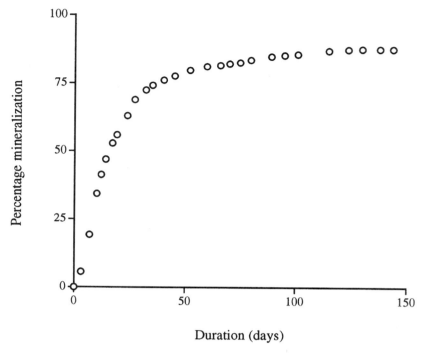

Figure 4 Carbon dioxide evolution curve for a sample of shredded bleached paperboard substrate: soil medium, at 25°C; inoculum: sewage sludge.

to characterize the biodegradability of the polymer substrate. These numbers might be compared with those obtained for the reference material to obtain a relative appreciation of the polymer biodegradability.

In interpreting the data, several possible sources of inaccuracy need to be taken into consideration. Substrates containing S and N in their chemical structure will produce ammonia and sulfur oxides, in addition to CO_2, during biodegradation. These will interfere with the acid–base titration procedure employed to estimate the CO_2 evolved. There is also the chance that some fraction of the CO_2 remains sorbed in the soil itself and is not available for reaction with the alkali in the biometer flask. At the end of the experiment some acid is added into the soil to liberate any CO_2 bound in the form of inorganic carbonates in the soil water, to take this error into account. It is important to appreciate that the rate of biodegradation obtained, and in complex substrates even the $y_{max} = a$ value, can depend on the concentration and the nature of microbial inoculum used in the experiment. The values of k therefore, are not absolute values and are meaningful only as relative measures in comparison with those of reference materials, such as cellulose powder or cellophane.

D. Monitoring Changes in the Properties of the Polymer

This approach is similar to monitoring the depletion of polymer reactant, except that it focuses on the initial stages of biodegradation, as opposed to the final mineralization stage. During early stages of breakdown of a polymer there is little noticeable volume reduction. However, the degradative reactions result in changes in certain physical and chemical characteristics of the polymer. These changes are reflected in the properties of the material, such as modulus, strength, solubility, or molecular weight. It is precisely these changes in properties that are critical from an applications standpoint. During the service life of a polymer, it is desirable to minimize such changes. A wide choice of properties that change on bio-degradation are available, but one that is relevant to the end use of the particular polymer material being studied is often adopted.

A commonly used valuable measure of degradation is the change in the average molecular weight of the polymer (26,57,58). A lowering of the average value and a general broadening of the distribution of molecular weight of the polymer are indicative of degradation. These changes might be detected using gel permeation chromatography or viscometry. Also widely used to monitor the progress of biodegradation are changes in mechanical properties of polymer films (59). Even low levels of degradation, particularly degradative changes resulting from the scission of polymer chains, cause very significant changes in mechanical properties, such as the tensile properties (60–63) or the dynamic mechanical properties (64).

Commonly used standard test methods for quantifying biodegradability of polymers (and organic compounds in general) are mainly based on measurements of oxygen consumed, carbon dioxide evolved, or the depletion of dissolved organic content in the system. As the results from a biodegradability test depends very much on the particular approach adopted, it is important to include the details of the approach used in measurement when reporting such data. Ideally, the biodegradation data should include the molecular weight of the polymer, the particular test used in establishing biodegradability, the physical form of polymer used (fiber, powder, film, or beads), the reference materials employed, as well as origin, nature, and biomass characteristics of the inoculum used in the experiment. Table 1 summarizes the key parameters employed in the standard tests used to quantify biodegradability.

V. CRITICAL CONSIDERATIONS IN THE ASSESSMENT OF BIODEGRADABILITY

A crucial component of any experiment designed to assess the biodegradability of a material is the biotic component. The simplest experiments use an enzyme in place of biomass; therefore, they are somewhat easier to control. With most studies, however, live biomass, either cultures of a single species or a sample of a microbial consortium isolated from nature, is used to inoculate the system. This is particularly true of studies aimed at establishing environmentally biodegradability of polymers. The particular inoculum used depends on the objectives of the experiment and must relate to the specific environment of interest. In general, several different environments are of practical interest. The biotic composition of these environments can be very different and, in most instances, the breakdown of polymers obtained in these environments is the result of a concerted effect of several different degradative processes. Table 2 summarizes the more important of these environments.

Table 1 Summary Details for Six Biodegradability Tests Based on OECD Guidelines

Test method	[S]	Units[a]	[I]	Units	Cells/L	Analysis	pH/T°C
Die-away (301 A)	10–40	mg DOC/L	<100	mL/L	$10^7 - 10^8$	Dissolved organic carbon	7.4/22
CO$_2$ evolution (301 B)	10–20	mg/L	<100	mL/L	$10^7 - 10^8$	Carbon dioxide evolution	7.4/22
Respirometry (301 F)	100	mg/L	<100	mL/L	$10^7 - 10^8$	Oxygen consumption	7.4/22
Modified screening test (301 E)	10–40	mg DOC/L	0.5	mL/L	10^5	Dissolved organic carbon	7.4/22
Closed bottle (301 D)	2–10	mg/L	<5	mL/L	$10^4 - 10^6$	Dissolved oxygen	7.4/22
MITI (1) (301 C)	100	mg/L	30	mg/L	$10^7 - 10^8$	Oxygen consumption	7.0/25

[a] Milligrams of suspended solids per liter.
DOC, dissolved organic carbon; [S] substrate concentration; [I] inoculum concentration.
Source: Ref. 69.

Table 2 Biotic Environments Relevent to Biodegradation Tests

Environment	Nature of biota	Mechanisms[a]	ASTM test methods
1. Natural			
Garden soil	Aerobic	B,P,O,H	
Subsurface soil	Aerobic/anaerobic	B,O,H	
Marine floating	Aerobic	B,P,H,O	D 5437-93
Coastal sediment	Aerobic	B,O,H	
Deep-sea	Anaerobic, barophilic	B,H	
2. Man-made			
Landfills	Anaerobic	B,H	
Compost systems	Aerobic, thermophilic	B,O,H	D 5338-93
Wastewater system	Aerobic	B,O,H	D 5271-92
Anaerbic digesters	Anaerobic	B,H	D 5209-91, D 5210-92

[a] B, biodegradation; P, photodegradation; H, hydrolysis; O, thermooxidative degradation.

The importance of adequate characterization and control of the biotic elements in biodegradation testing cannot be overstated. Early tests to qualitatively assess biodegradation relied on pure cultures of microorganisms. Although the biotic components in these tests were well characterized, the findings were rarely applicable to any real environments where stable consortia of microorganisms participate in the biodegradation process. To be able relate the test results to the biotic processes taking place in real environments, it is crucial to duplicate, to the extent possible, the microbial makeup those environments in the laboratory. The recent tests for biodegradability attempt to address this issue by employing mixed microbial inocula (such as garden soil or activated sludge), yet fall short in their attempts at characterization of the inoculum.

Several key problems are associated with sampling and maintaining a microbial consortium in a laboratory environment for the purpose of carrying out biodegradation tests.

1. Owing to the heterogeneity of the microbial populations and their uneven distribution (especially in soil environments), representative sampling is often very difficult to achieve. Particularly, with small samples of soil, it is unlikely that the inoculum sampled from the soil is typical of the environment. In addition to variations in the horizontal plane (soil surface), there is also a vertical distribution of species in soil and water environments. Particularly, in seawater or freshwater environments the attenuation of light in the surface layers results in a marked change in microbial composition with depth. In collecting surface samples of soil or water, the role of the biota in lower layers of the ecosystem is often excluded from the experiments. Furthermore, there can be cyclic fluctuations of the population composition at a given environment, even within a single day. Sampling soil or sediment at different times of the day, therefore, may result in significantly different consortia of organisms.

2. The mere act of separating a soil or sewage sample from its outdoor microenvironment and maintaining it in a laboratory culture container, even with adequate aeration and control, can easily lead to changes in the biocomposition

of the sample. Containment of a sample of marine and estuarine water, for instance, results in the drastic changes in the specific heterotrophic activity of the population and a selective increase in the fraction of bacteria showing an active electron transport system (65,66). Apparently, select members of the consortium better adapted to life under containment thrived under the conditions, changing the biotic makeup as well as the hetertrophic potential of the sample quite rapidly.

3. In some biodegradation tests, nutrients, such as inorganic N and P compounds, are added to the soil in an effort to promote a rich growth of microorganisms. This is often necessary to ensure a high enough concentration the organisms and, consequently, biodegradation at a rapid enough rate for each measurement. The practice, however, must alter the equilibrium composition of the consortium, allowing certain species, which are better adapted to utilizing the nutrients under these conditions, to thrive at the expense of others.

Unfortunately, the same also applies to the substrate, especially where the substrate is water-soluble. The presence of a biodegradable substrate must favor the growth and reproduction of those organisms that are capable of enzymatically breaking down the substrate, leading to an altered population. The opposite effect is when a substrate (particularly at a high concentration) poisons or retards the growth of microorganisms. Changes in osmotic pressure, owing to soluble intermediates of biodegradation or toxicity of these same compounds, easily alter the composition of microbial consortia. Although these changes do occur in the real environments as well, their effect on the microbial population is unlikely to be as pronounced as in laboratory experiments, because of the facility for free exchange of microbiota (as well as the organic intermediates or products) with the surrounding soil or water media.

4. Preadaptation of the microbial consortium to specific types of polymer substrates is also a concern (67). *Adaptation* is defined as the alteration of a microbial community that enables it to easily or rapidly biodegrade specific test compounds. For instance, activated sludge could have microbes especially adapted to degradation of paper or cellulose. This might be brought about by various mechanisms such as enzyme induction in cells, gene-transfer mutation in the organisms, or population-level compositional changes. In biodegradation of xenobiotic substrates, the microbial populations preadapted to those substrates showed rates of breakdown that were 1000-fold higher than those for unadapted populations (68). Enrichment culture of microorganisms would probably result in some extent of preadaptation, and the use of cultured consortia in biodegradation studies must be undertaken with caution. The more serious concern is unintentional use of preadapted population during sampling from natural environments. Either discharge of materials or other pollution may increase the local concentration of a particular substrate in soil or sediment environments for a duration long enough to obtain such preadaptation. At the time of collection it is difficult, if not impossible, to establish that the population is preadapted.

5. All changes that take place when a polymer is exposed to a biotic environment are not necessarily biologically mediated. Hydrolytic and thermooxidative processes can initiate degradation processes that are then continued by biological processes. This is particularly true of polymers that are able to photo- or thermooxidatively change into more polar hydrophilic materials that, in general, are more amenable to enzymatic attack. This synergistic interplay of the biotic

and abiotic factors that are important in outdoor environments does not play the same role in standard biodegradation tests carried out in the dark at constant temperature.

VI. CONCLUSIONS

Biodegradation is an important contributing process in the environmental breakdown of polymeric materials. Mineralization of the polymer substrate into simple compounds represents natural the endpoint of the biodegradation process. However, depending on the intent of the investigator, other stages of the process, such as removal of high polymer from the system, or the change in mechanical properties typical of polymers, might be more pertinent endpoints. This leads to several approaches for measurement of biodegradability. Of the various approaches available, the monitoring of oxygen use, or of carbon dioxide production, is popularly used to quantify aerobic biodegradation. The use of a biometer to titrimetrically quantify the gas evolution provides a particularly convenient means of establishing the rate and extent of carbon conversion during the breakdown of readily biodegradable polymers. An empirical curve-fitting procedure allows the calculation of an empirical rate constant for the carbon conversion process, but it is only a relative measure. It is useful to the extent that it allows the rate of carbon conversion during biodegradation of a polymer compared with that of a reference material or compound. Although the inclusion of a reference material in biodegradation experiments is crucial, it does not always remove the uncertainties caused by the use of poorly characterized atypical consortia of microorganisms.

The assessment of biodegradation of polymers is presently a difficult experimental undertaking. There are many unanswered issues, particularly those relating to the control of microbial composition used in measurements. The availability of different criteria for establishing biodegradation and the reported data that do not always clearly indicate the criterion adopted and its justification, add to the difficulty. However, with ongoing research in several laboratories around the world, the understanding of polymer biodegradation processes is rapidly improving and may yield more robust assessment methods fairly soon.

REFERENCES

1. AL Andrady, J Macromol Sci Rev Macromol Chem Phys C34: 25–76, 1994.
2. A Calamon-Decriaud, V Bellon-Maurel, F Sylvestre. Adv Polym Sci 135: 208, 1998.
3. RA Connolly. In: H Walters, EH Huech-van der Plas, eds. Biodeterioration of Materials. London: Applied Science Publishers, 1972, p. 168.
4. RP Wool, JS Peanasky, JM Long, SM Goheen. In: SA Barenberg, JL Brash, R Narayan, AE Redpath, eds. Degradable Materials. Perspectives, Issues, and Opportunities. Boca Raton, FL: CRC Press, 1990, p. 515.
5. RG Austin. In: SA Barenberg, JL Brash, R Narayan, AE Redpath, eds. Degradable Materials: Perspectives, Issues and Opportunities. Boca Raton, FL: CRC Press, 1990, p. 237.
6. G Iannotti, N Fair, M Tempesta, H Neibling, FH Hsieh, R Mueller. In: SA Barenberg, JL Brash, R Narayan, AE Redpath, eds. Degradable Materials; Perspective, Issues and Opportunities. Boca Raton, FL: CRC Press, 1990, p. 425.
7. SM Goheen, RP Wool. J Appl Polym Sci 42:2691, 1991.

8. RM Ottenbrite, AC Albertsson. In: M Vert, J Feijen, A Albertsson, G Scott, E Chellini, eds. Biodegradable Polymers and Plastics. Cambridge, UK: Royal Society of Chemistry, 1992, p. 73.

9. AL Andrady. ASTM Standardization News Oct:46, 1988.

10. V Coma, Y Couturier, B Pascat, G Bureau, S Guilbert, JL Cuq. In: M Vert, J Feijen, A Albertsson, G Scott, E Chellini, eds. Biogradable Polymers and Plastics. Cambridge UK: Royal Society of Chemistry, 1992, p. 242.

11. AL Andrady, VR Parthasarathy, Ye Song. Tappi 75(4), 1992.

12. JM Wood. Science 183:1049, 1974.

13. R Bartha, HAB Linke, D Pramer. Science 161:582, 1968.

14. G Swift. In: Y Doi, K Fukuda, eds. Biodegradable Plastics and Polymers. New York: Elsevier, 1994, pp. 228–236.

15. WJ Bailey, V Kuruganti, JS Angle. In: JE Glass, G Swift, eds. Agricultural and Synthetic Polymers. Biodegradability and Utilization, vol 433. Washington, DC: ACS Symposium Series, 1990, p. 149.

16. K-S Lee, RD Gilbert. Carbohydr Res 88:162, 1981.

17. MM Lynn, VT Stannett, RD Gilbert. J Polym Sci Polym Chem Ed 18:1967, 1980.

18. SL Kim, VT Stannett, RD Gilbert. J Macromol Sci 7:101, 1979.

19. E Kuhlwein, F Demmer. Kunstoffe 57(3):183, 1967.

20. Y Toikawa, T Suzuki. Agric Biol Chem. 42:1071, 1978.

21. RD Fields, F Rodriguez, RK Finn. J Appl Polym Sci 18:3571, 1974.

22. RD Fields, F Rodriguez. In: JM Sharpley, AM Kaplan, eds. Proceedings of the Third International Biodegradation Symposium. Barking, UK: Applied Science, 1976, p. 775.

23. AL Andrady, Y Tropsha. Unpublished data, 1993.

24. WJ Cook, JA Cameron, JT Bell, SJ Huang. J Polym Sci Polym Lett Ed. 19:159, 1981.

25. Y Doi, Y Kumagai, N Tanahashi, K Mukai. In: M Vert, J Feijen, A Albertsson, G Scott, E Chellini, eds. Biodegradable Polymers and Plastics. Cambridge, UK: Royal Society of Chemistry, 1992, p. 193.

26. SJ Huang, C Marci, M Roby, C Benedict, JA Cameron. ACS Symp Ser 172:471, 1981.

27. KKY Wong, KF Deverell, KL Mackie, TA Clark, LA Donaldson. Biotechnol Bioeng 24:447, 1988.

28. RL Tate. In: Soil Organic Matter—Biological and Ecological Effects. New York: John Wiley & Sons, 1987, p. 158.

29. RJ Swaby, JN Ladd. In: RA Silow, ed. The Use of Isotopes in Soil Organic Matter Studies. Oxford: Pergamon Press, 1966, p. 153–157.

30. M Shimao, N Kato. In: International Symposium on Biodegradable Polymers Abstracts. Tokyo: Biodegradable Plastics Society, 1990, p. 80.

31. RW Lenz. Adv Polym Sci 107:28, 1993.

32. F Kawai. In: JE Glass, G Swift, eds. Agricultural and Synthetic Polymers. Biodegradability and Utilization, ACS Symp Ser 433:110, 1989.

33. KW March, CR Widevuur, WL Sederel, A Bantjes, J Feijen. Biomed Mater Res 11:405, 1977.

34. G Halliwell, M Griffin. Biochem J 128:1183, 1973.

35. LER Berghem, LG Petterson. Eur J Biochem 37:21, 1973.

36. M Tien, TK Kirk. Science 221:661, 1983.

37. M Shimada, T Higuchi. In: DN-S Hon, N Shiraishi, eds. Wood and Cellulosic Chemistry. New York: Marcel Dekker, 1992, pp. 557; 625.

38. TK Kirk, E Schultz, WJ Connors, LF Lorenz, JG Zeikus. Arch Microbiol 117:277, 1978.

39. D Gilmore, RC Fuller, R Lenz. In: SA Barenberg, JL Brash, R Narayan, and AE Redpath, eds. Degradable Materials: Perspectives, Issues and Opportunities. Boca Raton, FL: CRC Press, 1990, p. 481.

40. H Seki. Appl Microbiol 26:318, 1973.

41. DS Jenkinson, DS Powlson. Soil Biol Biochem 8:209, 1976.
42. BS Asumus. Biol Sol 14:8, 1971.
43. WJ Payne. Annu Rev Microbiol 24:17, 1984.
44. JA Van Veen, JN Ladd, MJ Frissel. Plant Soil 76:257, 1984.
45. JE Potts. In: Encyclopedia of Chemical Technology. 2nd ed. New York: Wiley Interscience, 1984, p. 626.
46. AL Andrady, VR Parthasarathy, Ye Song. Tappi 74(7):185, 1991.
47. K Hardie. Int Biodeterior Bull 17:1, 1963.
48. FP Delafield, M Doudoroff, NJ Palleroni, CJ Lusty, R Contopoulos. J Bacteriol 90:1455, 1965.
49. CJ Lusty, M Doudoroff. Proc Natl Acad Sci USA 56:960, 1966.
50. AL Andrady, S Nakatsuka. J Appl Polym Sci 45:1881, 1992.
51. Y Yakabe, M Kitano. In: Y Doi, K Fukuda, eds. Biodegradable Plastics and Polymers. New York: Elsevier Science, 1994, pp. 331.
52. S Matsumara, T Tanaka. J Environ Polym Degred 2:89, 1994.
53. AC Albertsson, C Barnstedt, S Karlsson. J Environ Polym Degrad 1:241, 1993.
54. JE Guillet. Adv Chem Ser 169:1, 1978.
55. J Gu, D Eberiel, SP McCarthy, RA Gross. J Environ Polym Degrad 1:281, 1993.
56. R Bartha, D Pramer. Soil Sci 100:68, 1965.
57. H Eya, N Ewaki, Y Otsuji. In: Y Doi, K Fukuda, eds. Biodegradable Plastics and Polymers. New York: Elsevier, 1994, pp. 337–344.
58. J Mergaert, A Wouters, J Swings, K Kersters. In: M Vert, J Feijen, A Albertsson, G Scott, and E Chellini, eds. Biodegradable Polymers and Plastics. Cambridge, UK: Royal Society of Chemistry, 1992, pp. 267.
59. A Nakayama, N Kawasaki, I Arvanitoyannis, N Yamamoto. In: Y Doi, K Fukuda, eds. Biodegradable Plastics and Polymers. New York: Elsevier, 1994, p. 557.
60. L Tilstra, D Johnsonbaugh. J Environ Polym Degrad 1:257, 1993.
61. DF Gilmore, S Antoun, RW Lenz, RC Fuller. J Environ Polym Degrad 1:269, 1993.
62. DF Gilmore, RC Fuller, B Schneider, RW Lenz, N Lotti, M Scandola. J Environ Polym Degrad 2:49, 1994.
     REFERENCES MISSING
65. RT Wright, AW Bourquin, PH Pritchards, eds. Microbial Degradation of Pollutants in the Marine Environment (USEPA, Gulf Breeze, FL, 1979), Vol. 600/9-79-012, 1979, p. 119.
66. LH Stevenson. Microbiol Ecol 4:127, 1978.
67. JC Spain, PA Van Veld. Appl Environ Microbiol 45:428, 1983.
68. JC Spain, PH Pritchard, AW Bourquin. Appl Environ Microbiol 40:726, 1983.
69. AL Andrady. In: JE Mark, ed. Physical Properties of Polymers Handbook. American Institute of Physics, 1996, p. 625–637.

12

Environmental Degradation of Polymers

PETER P. KLEMCHUK

Institute of Materials Science, University of Connecticut, Storrs, Connecticut

I. INTRODUCTION

Increasing amounts of municipal solid waste (MSW), decreasing landfill capacity for disposal, slow degradation rates of MSW in sanitary landfills, slow degradation of plastic litter in the environment, generated intense interest in degradable plastics in the last quarter century. Injuries and deaths of animals, birds, fish, and other sea creatures from encounters with plastic litter received considerable publicity, as did the findings that even readily biodegradable materials, such as food and paper, degraded hardly at all in sanitary landfills in 20–30 years. What could be the expectations for plastics to degrade in such landfills?

The amounts of plastics in MSW were relatively small in the 1960s and early 1970s. Presently, plastics are 9–10% of MSW by weight, considerably more by volume, and growing. The increasing plastic contents of MSW and their slow rates of degradation in the environment have caused examination of means to decrease the problem. Of more interest are recycling, chemical recovery by pyrolysis and hydrolysis, incineration with energy recovery, source reduction, and degradable technologies. Of these, degradable plastics were hoped to be a quick "fix" for complex solid waste problems. That expectation still has not been realized (1). However, photodegradable plastic technologies are being used successfully for agricultural mulches and six-pack retaining rings.

At the outset, in the 1970s and 1980s, attention was concentrated on photodegradable plastics and mixtures of biodegradable starch with polyethylene. The former found practical use as agricultural mulches and in six-pack retaining rings. The latter were advanced as biodegradable compositions for use in packaging,

461

including grocery bags. In time, recognition grew that photodegradable and bio-degradable were not synonymous. Photodegradable plastics filled important needs, mainly by reducing large articles to small, less visible pieces, through the action of sunlight and air. The resultant smaller pieces were not degradable further in the environment within reasonable time periods through the action of micro-organisms—bacteria, fungi, yeasts, and algae—and their enzymes (2–5). Moreover, mixtures of starch and polyethylene came to be recognized as only partially biodegradable—starch degraded readily, the resultant polyethylene "lace" hardly at all; new technologies were needed for truly biodegradable plastics and completely environmentally degradable plastics.

As industrial organizations committed resources to the development of degradable plastics, the American Society for Testing and Materials (ASTM) played a key role by developing standardized test methods for evaluating degradable plastics, by helping remove confusion about the photo/bio dichotomy and, in doing so, provided the means for a better understanding of degradable plastics.

Degradable plastic technologies have evolved from early concentration on photodegradable materials to the present interest in all degradation technologies, photo-, bio-, oxidative, or hydrolytic, all together constituting environmental degradation. Photodegradable plastics of the 1970s and 1980s proved to be highly effective in breaking down into small particles under the action of sunlight. Realization soon developed that particles from photodegradation of photo-degradable plastics were degraded further by biological processes only at very low rates, because they did not meet essential compositional requirements for biodegradation. That realization generated concerns about releasing large amounts of photodegraded plastic particles into the environment and led to searches for new biodegradable plastics that could be made available at acceptable cost. The evolution of degradable plastic technology has reached the point that several companies are making progress in developing new biodegradable polymers that degrade in relatively short, acceptable times. These developing compositions still cost more than commercial packaging plastics, such as polyethylene (PE), polypropylene (PP), polystyrene (PS) or poly(vinyl chloride) (PVC), but some are finding niche markets. In time, some of the newly developed materials are expected to be more cost competitive. In some instances, increased stability during high-temperature processing needs to be provided for candidate biodegradable polymers.

The environmental safety of degradation products of degradable plastics is viewed to be important. Several toxicological test methods have been adapted from tests used to assess the safety of new chemical substances, to assess the environmental safety of degradation products of plastics. This chapter reviews the environmental degradation technology of polymers and the means for safety assessment of products from degraded plastics.

A. Mechanisms for the Degradation of Polymers in the Environment

An ASTM committee has defined a *degradable plastic* as "a plastic designed to undergo a significant change in its chemical structure under specific environmental conditions, resulting in a loss of some properties that may vary as measured by standard test methods appropriate to the plastic and the application in a period of time that determines its classification" (6). The definition can be applied to all means of polymer degradation including photodegradation, thermooxidation, hydrolysis, and biodegradation.

Autooxidation is responsible for thermooxidative degradation of organic materials in the absence of light. Thermoplastic articles usually contain sufficient hydroperoxides to initiate autooxidation. Hydroperoxides are formed in polymers during melt processing at high temperatures, during manufacturing, and during fabrication. Trace amounts of oxygen in processing equipment are sufficient to generate small amounts of oxidation-initiating hydroperoxides. Autooxidation and photooxidation have many common features: free radical reactions; initiation from homolysis of hydroperoxides, thermally in the case of autooxidation, photolytically in the case of photooxidation; oxidation propagation by peroxy radicals; termination by radical combination, notably by combination of peroxy radicals (Schemes 1 and 2). Autooxidation can be initiated also by reactions of hydroperoxides with catalytic amounts of transition metal cations, notably iron, cobalt, and copper (Scheme 3).

Natural photodegradation is the process by which sunlight, through photooxidation and direct bond cleavage, reduces the molecular weight of polymers so that plastic articles become brittle and disintegrate. *Biodegradation* is the process whereby bacteria, fungi, yeasts, algae, and their enzymes consume a substance as a food source so that its original form disappears. Care must be taken in discussing degradable polymers to define which process is involved for a particular degradable polymer. All commercial-packaging polymers are not biodegradable owing to high

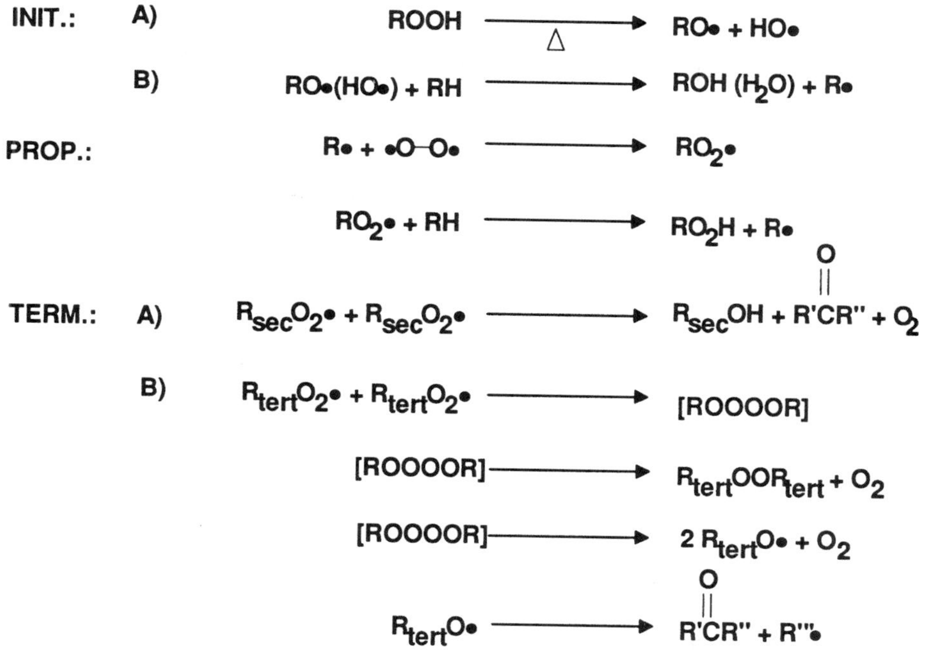

INIT.:   A) $ROOH \xrightarrow{\Delta} RO\bullet + HO\bullet$

          B) $RO\bullet(HO\bullet) + RH \longrightarrow ROH\,(H_2O) + R\bullet$

PROP.: $R\bullet + \bullet O\text{--}O\bullet \longrightarrow RO_2\bullet$

                      $RO_2\bullet + RH \longrightarrow RO_2H + R\bullet$

TERM.:   A) $R_{sec}O_2\bullet + R_{sec}O_2\bullet \longrightarrow R_{sec}OH + R'\overset{\overset{\displaystyle O}{\|}}{C}R'' + O_2$

          B) $R_{tert}O_2\bullet + R_{tert}O_2\bullet \longrightarrow [ROOOOR]$

                      $[ROOOOR] \longrightarrow R_{tert}OOR_{tert} + O_2$

                      $[ROOOOR] \longrightarrow 2\,R_{tert}O\bullet + O_2$

                      $R_{tert}O\bullet \longrightarrow R'\overset{\overset{\displaystyle O}{\|}}{C}R'' + R'''\bullet$

Scheme 1   Thermooxidative degradation of polymers.

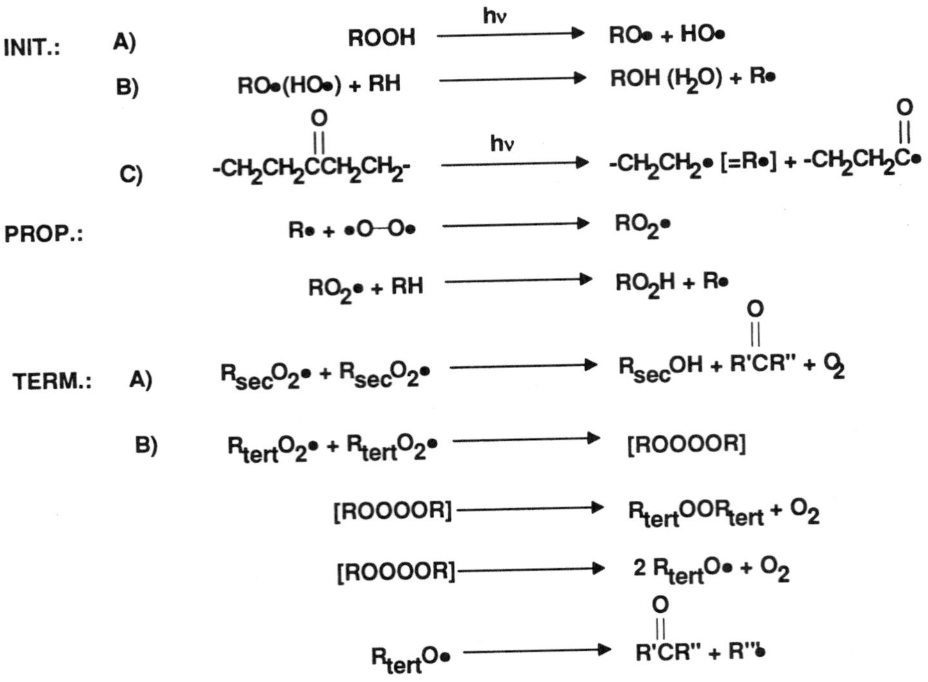

INIT.: A)   ROOH $\xrightarrow{h\nu}$ RO• + HO•

        B)   RO•(HO•) + RH $\longrightarrow$ ROH (H$_2$O) + R•

                     $\overset{\displaystyle O}{\overset{\displaystyle \|}{}}$ $\overset{\displaystyle O}{\overset{\displaystyle \|}{}}$
        C) -CH$_2$CH$_2$CCH$_2$CH$_2$- $\xrightarrow{h\nu}$ -CH$_2$CH$_2$• [=R•] + -CH$_2$CH$_2$C•

PROP.:   R• + •O–O• $\longrightarrow$ RO$_2$•

                 RO$_2$• + RH $\longrightarrow$ RO$_2$H + R•

                                                                        $\overset{\displaystyle O}{\overset{\displaystyle \|}{}}$
TERM.: A)   R$_{sec}$O$_2$• + R$_{sec}$O$_2$• $\longrightarrow$ R$_{sec}$OH + R'CR'' + O$_2$

        B)   R$_{tert}$O$_2$• + R$_{tert}$O$_2$• $\longrightarrow$ [ROOOOR]

             [ROOOOR] $\longrightarrow$ R$_{tert}$OOR$_{tert}$ + O$_2$

             [ROOOOR] $\longrightarrow$ 2 R$_{tert}$O• + O$_2$

                                                 $\overset{\displaystyle O}{\overset{\displaystyle \|}{}}$
             R$_{tert}$O• $\longrightarrow$ R'CR'' + R''•

Scheme 2 Photooxidative degradation of polymers.

ROOH + Me$^{+(+)}$ $\longrightarrow$ RO• + Me$^{++(+)}$ + HO$^-$

ROOH + Me$^{++(+)}$ $\longrightarrow$ ROO• + Me$^{+(+)}$ + H$^+$

_____

OVERALL:

   2ROOH $\longrightarrow$ RO• + ROO• + H$_2$O

Me = Fe, Cu, Co

Scheme 3 Catalysis of autooxidation initiation by transition metals.

molecular weights (2–4), to substituents that prevent biodegradation by the enzymatic fatty acid oxidation mechanism (2–5), or to rigid structures (7). Although there has been a tendency to presume polymers in the environment degrade virtually completely by biodegradation, in most instances, photodegradation is the primary degradative process. This chapter will cover mainly photodegradable and biodegradable technologies for polymers.

II. PHOTODEGRADABLE TECHNOLOGY

A. Introduction

Organic materials degrade through photodegradation (including photooxidation and weathering, the combined action of sunlight, air, heat, rain, and pollutants), biodegradation, autoxidation, and hydrolysis. However, because most presently available, commercial polymers are not responsive to biodegradation by bacteria, fungi, yeasts, or enzymes, photodegradation, photooxidation, and autoxidation are the main pathways for their degradation.

Weathering is particularly severe for organic materials because it combines the photophysical and photochemical effects of ultraviolet radiation (UV) photons with the oxidative effects of atmospheric oxygen and the hydrolytic effects of water. UV photons of terrestrial sunlight contribute importantly to the photodegradation of organic materials because they are sufficiently energetic to break some bonds directly and to initiate the breaking of others through photosensitized processes and photooxidation. All commercial polymers with carbon backbones are prone to degradation by weathering to varying degrees. The large molecules (high molecular weight components) among the distribution of molecules in polymers are the important contributors to the strength of the polymers.

Low degrees of chain scissioning of the large molecules, which have the highest probability for chain scissioning, are sufficient to cause severe losses of physical properties.

As the polymer industry evolved, considerable effort was expended to understand, at a molecular level, the degradation of polymers during weathering, and means were found to retard it. Ultraviolet-absorbing additives and trappers of oxidation-propagating free radicals were developed that permit the use of commercial polymers for articles that must last many years in direct weathering. The knowledge gained from efforts to stabilize polymers during weathering has been put to use in developing photodegradable polymer technology. The main approaches to make photodegradable polymers have been to insert photosensitive ketonic groups on polymer backbones and in pendant groups and to use photosensitizing additives, which increase the rate of photooxidation.

This section provides an assessment of photodegradable polymer technology and attempts to put into perspective the various aspects of this complex subject. Photodegradable polymers are viewed as meaningful for some applications, such as agricultural mulches and to reduce unsightly and wildlife-harmful litter. They are not solutions to municipal solid waste problems but offer attractive properties in their own right.

B. Polymers and Photodegradation

Photooxidation, one of the two major mechanisms involved in the photodegradation of polymers has as its main features the following: oxidation initiation by photolysis of hydroperoxides, oxidation propagation by chain reactions, and termination through bimolecular reactions of alkyl peroxy radicals, as shown in Scheme 2. Nearly all commercial thermoplastic polymers are sensitive to photooxidation, and when exposed in the environment they photooxidize and eventually lose physical properties. A second major contributor to the photodegradation of polymers is

ketone photolysis, which proceeds through two major reactions called Norrish I and
Norrish II, as shown in Scheme 4. Ketones are introduced onto the backbones of
polymers by photooxidation. On exposure to light, these ketone functionalities
absorb photons of appropriate energy, break carbon–carbon bonds, and scission
the polymer backbone.

In Norrish I reactions, a carbon–carbon bond is broken to form two
carbon-centered free radical species: an alkyl radical and a carbonyl radical. Very
often carbon monoxide is lost from the carbonyl radical and two terminal
carbon-centered alkyl free radicals are generated. These react readily with oxygen
to yield alkylperoxy radicals and propagate the photooxidation. In Norrish II
reactions, carbon–carbon bonds on polymer backbones, which are β to ketones,
are broken. The scission of backbones by both the Norrish I and Norrish II reactions
is significant because polymers depend on relatively large molecules for their
strength. When the large molecules are broken down, as by ketone photolysis processes, the resultant smaller molecules do not contribute as effectively to the strength
of specimens, which eventually undergo significant loss of physical properties and
become brittle. The ASTM methods in Table 1 can be used to investigate the photo-
and oxidative degradability of polymers.

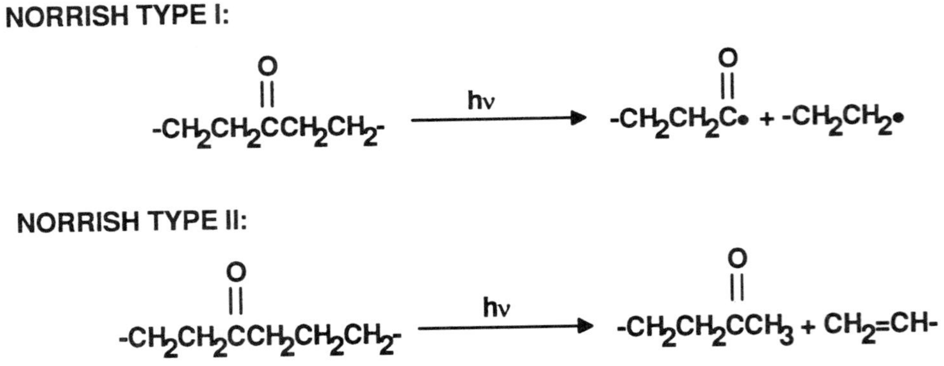

NORRISH TYPE I:

$$-CH_2CH_2CCH_2CH_2- \xrightarrow{h\nu} -CH_2CH_2C\bullet + \bullet CH_2CH_2\bullet$$

NORRISH TYPE II:

$$-CH_2CH_2CCH_2CH_2CH_2- \xrightarrow{h\nu} -CH_2CH_2CCH_3 + CH_2=CH-$$

Scheme 4 Ketone photolysis.

Table 1 ASTM Standard Test Methods for Photo- and Oxidative Degradation Testing of
Polymers

ASTM test number	Test method
D3826	Degradation End Points Using a Tensile Test
D5071	Operation of a Xenon Arc-Type Exposure Device
D5208	Operation of a Fluorescent Ultraviolet and Condensation Device
D5272	Outdoor Exposure Testing of Photodegradable Plastics
D5437	Weathering of Plastics Under Marine Floating Exposure
D5510	Heat-Aging of Oxidatively Degradable Plastics
G26	Operating Light-Exposure Apparatus (Xenon-Arc Type)
G53	Operating Light- and Water-Exposure Apparatus (Fluorescent UV/Condensation Type)

Table 2 contains data for the thermal oxidation of high-density polyethylene film samples investigated by oxygen uptake (8). Oxygen consumption, elongation, molecular weight, and carbonyl absorbance were measured periodically for the samples. The degree of chain scissioning was calculated from the number-average molecular weight. As can be seen in Table 2, an average of one scission per molecule was sufficient to cause these samples to lose all elongation and become brittle. That occurred because, although the chain scissions averaged only one, the large molecules were scissioned several times, smaller ones not at all, and as a result the large molecules were degraded to small molecules of much reduced strength. Because physical integrity and strength are contributed by the large molecules of polymers, as found here, relatively few scissions of large polymer molecules can cause marked reduction in strength.

C. Making Polymers Photodegradable

Two main approaches have been taken to make polymers photodegradable. One has been to introduce ketonic functionalities on backbones and on the α carbons of pendant groups. This approach takes advantage of the sensitivity of such ketones to undergo the Norrish I and Norrish II reactions on absorption of UV photons—chain scissioning is a usual consequence (see Scheme 4). The other approach is to use one of several photosensitizing additives to initiate photooxidation of the polymer, as a result of which chain scissioning also takes place. Both approaches have been used commercially for photodegradable polymers. Table 3 is a summary of the main patented approaches to photodegradable plastics.

1. Copolymer Approaches to Photodegradable Polymers

Ketones are introduced onto backbones and in pendant groups of polymers to enhance photodegradability. That is done by the copolymerization of ethylene and carbon monoxide at high pressures and by the copolymerization of ethylene or styrene with vinyl ketones, notably methyl vinyl ketone. Both types of copolymers photodegrade by the Norrish I and Norrish II mechanisms (Scheme 5).

Ethylene–carbon monoxide (ECO) copolymers have been known for over 40 years. They are prepared by using carbon monoxide as a comonomer in the high-pressure polymerization of ethylene, the process used to prepare low-density polyethylene (LDPE). Carbon monoxide is an ideal comonomer for ethylene, for the two add at about the same rate to growing copolymers (R. J. Statz, personal

Table 2 Correlation of HDPE Properties and Oxygen Uptake; 5-Mil Unstabilized HDPE Films; Oxygen; 100°C

Elongation % retention	$\overline{Mn}$	$\overline{Mw}$	Avg. scission per molecule	CO ABS	ROOH ABS	Oxygen uptake mmol/mmol HDPE
100	8,270	151,000	0	0	0	0
93	6,460	137,000	0.28	0.01	0.0001	0.04
50	4,900	23,000	0.63	0.24	0.0049	1.0
0	4,100	14,000	1.1	0.48	0.0100	2.1

HDPE, high-density polyethylene.

Table 3 Photodegradable Polymer Technology: Patents

Technology	Patentee	Date	Patent no.
Et.–CO copolymers	Bayer	8/27/41	Ger 863,711
Et.–CO copolymers	DuPont	1/24/50	U.S. 2,495,286
Vinyl ketone copolymers	J Guillet	8/21/73	U.S. 3,753,952
Vinyl ketone copolymer ctgs.	J Guillet	5/21/74	U.S. 3,811,931
Vinyl ketone copolymers	J Guillet	12/10/74	U.S. 3,853,814
Photodegradable master batches	J Guillet	1/14/75	U.S. 3,860,538
Aromatic ketone additives	Bio-Degradable Plastics, Inc.	6/10/75	U.S. 3,888,804
Metal complex sensitizers (Fe)	G Scott	10/17/78	U.S. 4,121,025
Ti/Zr complexes + subs. Benzophenones	Princeton Polymer Laboratories, Inc.	1/22/85	U.S. 4,495,311
Metal complex stab.–sensit. (Ni–Fe)	Scott–Gilead	7/24/84	U.S. 4,461,863
		5/28/85	U.S. 4,519,161

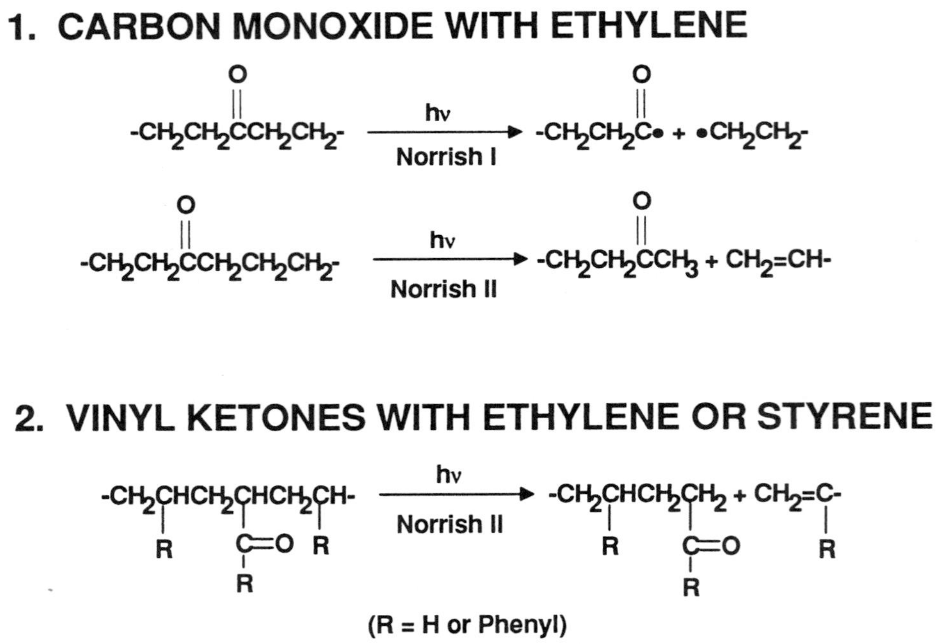

1. CARBON MONOXIDE WITH ETHYLENE

2. VINYL KETONES WITH ETHYLENE OR STYRENE

(R = H or Phenyl)

Scheme 5 Copolymer approaches to increasing photodegradation of packaging plastics.

communication, 1988). The copolymer composition is determined by the relative partial pressures of the two monomers in the polymerization reactor. The copolymers have physical and thermal stability properties that are comparable with those of normal low-density polyethylene (9). Copolymers can be made with equal molar ratios of ethylene and carbon monoxide. However, films with lower carbon monoxide contents are normally used for packaging applications. Dow, Dupont, and Union Carbide, all have manufactured ethylene–carbon monoxide copolymers with about 1% carbon monoxide for photodegradable applications. (Dow is likely the only current source of the copolymer in the United States.)

The rate of photodegradation of ECO copolymers is a function of the carbon monoxide content—the higher the carbon monoxide content, the faster the photodegradation. Films, 2.0 mil thick, with 13% carbon monoxide lost their physical properties in about a half day of exposure outdoors in Orange, Texas; with 1% carbon monoxide, the properties were lost in about 2 days. Normal low-density polyethylene required more than 9 months exposure to achieve the same degree of degradation (9).

The other copolymer approach to the manufacture of photodegradable polymers is the copolymerization of vinyl ketones with various monomers: ethylene, styrene, methyl methacrylate, vinyl chloride, among others (10,11). These copolymers also have physical properties that are comparable with the homopolymers and rates of photodegradation that are a function of the carbonyl content of the copolymers. With copolymers of vinyl ketones and ethylene, the Norrish I mechanism does not lead to direct chain scission, but generates radicals that initiate photooxidation of the polymer (Scheme 6). The Norrish II mechanism leads to direct chain scission, and it is that mechanism that is responsible for the direct photodegradation of copolymers. However, because in current practice, ketone-containing copolymers are added as master batches to homopolymers to sensitize photodegradation, the Norrish I mechanism is responsible for the bulk of photodegradive activity (12).

The use of commercially available copolymer master batches is favored over in-house copolymerization to provide photodegradable polymers based on Guillet's technology. The effectiveness of the master batches has contributed to that being a preferred route. One may view these master batches as additives for the photodegradation of polymers. In a study with Ecolyte polystyrene 210 (a styrene–vinyl ketone copolymer), added at several concentrations to homo-

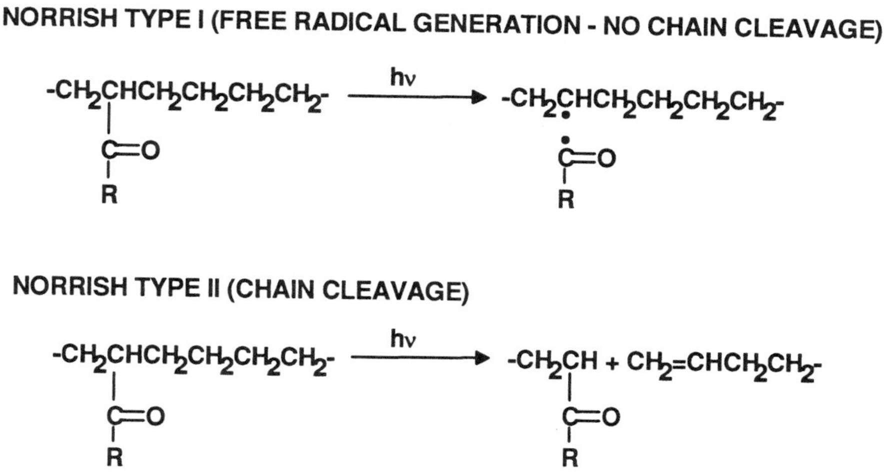

NORRISH TYPE I (FREE RADICAL GENERATION - NO CHAIN CLEAVAGE)

$$-CH_2CHCH_2CH_2CH_2CH_2- \xrightarrow{h\nu} -CH_2\overset{\bullet}{C}HCH_2CH_2CH_2CH_2-$$

$$\underset{R}{\overset{|}{C}}=O \qquad\qquad \underset{R}{\overset{|}{C}}=O$$

NORRISH TYPE II (CHAIN CLEAVAGE)

$$-CH_2CHCH_2CH_2CH_2CH_2- \xrightarrow{h\nu} -CH_2CH + CH_2=CHCH_2CH_2-$$

$$\underset{R}{\overset{|}{C}}=O \qquad\qquad \underset{R}{\overset{|}{C}}=O$$

Scheme 6 Photolysis of ethylene-vinyl ketone copolymers.

polystyrene, and the compositions exposed outdoors in southwestern Ontario, the rate of degradation was dependent on the amount of master batch in the composition. Exposures were started in October and continued for 20 weeks. The 100% copolymer lost half its initial tensile strength in less than 2 weeks. The 20% blend took 6 weeks to the same result. Blends with 10, 5, and 0% of the master batch suffered losses of 35, 12, and 2%, respectively, of initial tensile strength in 20 weeks. The same work showed seasonal variation in the rate of degradation of the copolymer. However, rates of degradation at different seasons were comparable when evaluated on the basis of hours of bright sunshine—temperature had little effect on the rate of photodegradation (12).

Some idea of the susceptibility of various ketonic photodegradable polymers can be gained from the relative quantum yields for the loss of carbonyl groups in ethylene–vinyl ketone copolymers: quantum yields were 1 for copolymer with 1% carbon monoxide, 4.1 for copolymer with 2% methyl vinyl ketone and 8.3 for copolymer with 2% methyl isopropentyl ketone (13). Significant molecular weight reductions took place on exposure of the copolymers of methyl vinyl ketone to UV light (14).

Absorption of UV light by aliphatic ketones, low molecular weight compounds and polymers, is characterized by small molecular extinction coefficients, usually less than 150, with absorption maxima in the vicinity of 280 nm (14,15). Thus, absorption is relatively slight at the tail end of terrestrial sunlight, and yet the number of UV photons that are absorbed by polymers with ketonic groups is sufficient to cause severe reduction in physical strength and embrittlement in only a few days of exposure to sunlight.

2. Additive Approaches to Photodegradable Polymers

Aromatic ketones are effective photosensitizers for the photodegradation of polymers. Benzophenone, which in the photoexcited state exists as a diradical, is capable of extracting hydrogen from polymeric substrates and initiating photooxidation of the polymer (Scheme 7). Patents covering a variety of other aromatic ketones have been issued. The low molecular weight of benzophenone makes it quite volatile and not suitable for most polymer applications. Another approach to impart photodegradability to polymers is the use of titanium or zirconium complexes with substituted benzophenones as additives. The combination of the metal complex and the aromatic ketone significantly increased the rate of photodegradation of polymers (16).

Several metal complexes are effective sensitizers of polymer photodegradation. Ferric dialkyldithiocarbamates, which were covered in a U.S. patent, are among them (17). This initial approach with a ferric compound as a photosensitizer has evolved to the use of a combination of nickel and ferric dialkyldithiocarbamates

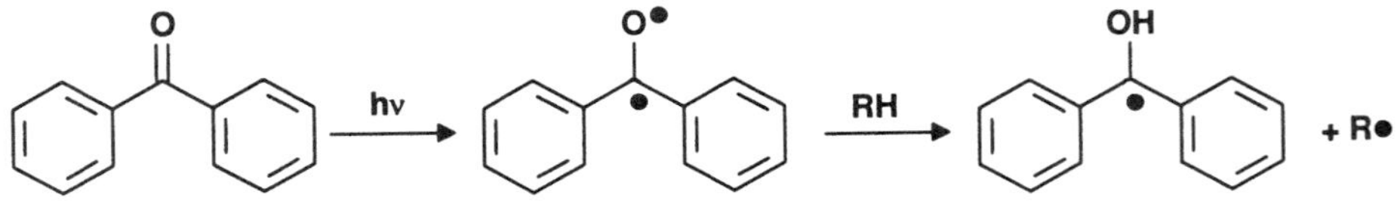

Scheme 7   Initiation of photooxidation by benzophenone.

(18). Although other compounds retard the photoactivating effect of ferric dimethyldithiocarbamate, only nickel diethyldithiocarbamate provided the ideal behavior required for time-controlled photodegradable agricultural film. In addition to control of photodegradation, the combination of ferric and nickel dimethyldithiocarbamates provided effective melt stabilization during processing (19). By varying the ratios of the nickel and iron compounds, the stabilities of polyethylene films can be adjusted over a considerable range of exposure times to meet the needs of various field crops.

D. Uses of Photodegradable Polymers

1. Packaging and Disposables

Much of the pressure for degradable polymers has come from desires to overcome unsightly collections of packaging trash that have been discarded by uncaring persons in the environment. When discarded on land, the plastic trash in mainly unsightly, but when discarded in the oceans it also causes injuries and deaths to animals, birds, fish, and other sea creatures that ingest or otherwise become involved with the litter. Photographs of animal and bird encounters with plastic articles and photographs of mounds of plastic bottles and other plastic trash washing up on beaches have made us very much aware of the problems associated with marine dumping of solid waste. In the past, much dumping at sea was done by ships, but agreements among the major seafaring nations have controlled dumping of solid waste at sea.

The high interest in degradable polymers for packaging and disposables to minimize litter has resulted in the availability of several commercial products. Dow, Dupont, and Union Carbide have offered ECO copolymers with usually about 1% carbon monoxide. These copolymers are now being used to make six-pack retaining rings for beverage containers, which undergo photodegradation when placed in sunlight for just a few days. Many states and Congress have legislated that the six-pack retaining rings should be made of photodegradable plastic. ECO copolymers also have the potential for other applications for which rapid photodegradation is desired, such as in degradable polyethylene bags. They are regulated by the FDA for food contact uses as a result of a petition from Dow (20).

Photodegradable copolymers of ethylene or styrene with methyl vinyl ketone or methyl isopropenyl ketone, based on Guillet technology have been offered commercially under the Ecolyte trade name. Ecolyte Atlantic, Inc. of Baltimore, Maryland, presently has supplied photodegradable Ecolyte copolymer master batches for use in sensitizing the photodegradation of plastics. These master batches are added at concentrations of 2.5–5% to polyolefin and polystyrene packaging plastics and disposables.

In addition to the copolymer approaches, photodegradable polymers based on photosensitizing additives have been offered by several companies, including Ampacet, Ideamasters, Princeton Polymer Laboratory, and Rhône-Poulenc. These include the use of ferric, cerric, and other metal compounds, benzophenones, and mixtures of substituted benzophenones with titanium or zirconium chelates to sensitize photodegradation. Interest in photo- and biodegradable shopping bags is increasing, and new offerings have become available using these additive technologies.

Recycling of packaging plastics, also in conjunction with deposit laws, so that plastic containers are returned and reused for useful articles is growing in interest as an alternative to degradable polymers. Effective, large-scale recycling of plastics promises to become a meaningful way to reduce litter and to reduce the amounts of plastics in municipal solid waste that are disposed by incineration and landfilling. However, some questions have arisen as to whether photodegradable polymers are compatible with recycling because significant amounts of photodegradable polymers in postconsumer plastics could reduce lifetimes of articles made from recovered materials.

2. Agricultural Mulches

Photodegradable polymers have found a niche as agricultural mulches. These plastic film mulches help plants grow, and when no longer needed, they photodegrade in the fields, thereby avoiding the cost of removal from fields and disposal. The plastic films are desirable because they conserve moisture, reduce weeds, and increase soil temperatures, thereby boosting the rate of plant growth.

Polymers used for this purpose usually contain light-sensitizing additives that cause the polymers to undergo photodegradation. A particularly unique system consists of a mixture of ferric and nickel dialkyldithiocarbamates, the ratio of which is adjusted to provide protection for specific growing periods. Thus, the lifetime is timed so that once a crop is grown, the polyethylene mulch will begin to photodegrade. This system, based on Scott–Gilead (18) technology, was first used in Israel and presently is being used in several countries, including the United States (Plastigone from Ideamasters), for agricultural mulches. It is also being advanced for other degradable polymer applications. A combination of substituted benzophenones and titanium or zirconium chelates is another additive system that have been offered for this application (21).

These photosensitizing additive systems function mainly to photodegrade low-density and linear low-density polyethylene (LLDPE) films and result in the formation of small particles of polymer film. The resultant particles are sufficiently small so that once photodegradation has taken place they are no longer visible in the soil. Some marketers of these degradable polymers have made claims for the biodegradation of the fragments in the soil. However, although some studies (22,23) have shown biodegradation of ^{14}C-labeled polyethylene and polystyrene can take place at measurable rates, the results are not encouraging that complete biodegradation would take place in less than 50 years.

The environmental safety of the degraded polymer films and ecological considerations of the accumulation of polymer particles in soil over many years have been investigated in a short-term study. The main focus in the study was the effect of nickel, which is phytotoxic, on the growth of lettuce and green peppers in soil containing photodegraded mulch film fragments. Greenhouse- and field-grown vegetables were grown on soil that contained amounts of photodegraded low-density polyethylene mulch films equivalent to 30 years of repetitive use. Analyses for dithiocarbamate residues in the vegetables showed none to be present and analyses for nickel showed no consistent differences between treated and control samples of crops and soils (24). The nickel content of the photodegradable mulch used in the study was 0.46 ppm, which is much lower than nickel levels in mineral soils, 10–140 ppm. Similar results had been reported from an earlier study, and it was concluded that either the nickel had been released in unavailable form to the plants, or that the plastic had not degraded sufficiently to release the nickel (25). The results of these studies suggest the potential phytotoxicity and environmental effect of the photodegradable film related to release of nickel and dithiocarbamate are likely very low.

Ennis, in reviewing the environmental toxicology of photodegradable plastics, concluded the evaluation of the toxic potential of degradable plastics was an evolving discipline and, as degradation products are identified, the toxic potential of their interactions with the environment will become better understood (26). He predicted there will not be any increased toxicity of the breakdown products of degradable plastics in comparison with regular plastic products.

III. BIODEGRADABLE TECHNOLOGY

A. Introduction

In biodegradation, microorganisms, bacteria, fungi, yeasts, algae, and their enzymes, consume a substance as a food source so that its original form disappears. Under appropriate conditions of moisture, temperature, and oxygen availability, biodegradation of truly biodegradable substances can be a relatively rapid process—food, leaves, organic garden refuse, and such can be composted in a few months. Biodegradation in 1–3 years, at most, is a reasonable target for complete assimilation and disappearance of a truly biodegradable plastic article.

A landmark study on the biodegradability of packaging plastics was carried out by Potts and associates of the Union Carbide Corporation for the Environmental Protection Agency (EPA) (2–4). That study was carried out according to the provisions of ASTM method G21-70, *Determining Resistance of Synthetic Polymeric Materials to Fungi*. In the test, specimens being tested for biodegradability were placed on agar gel containing all nutrients for supporting the growth of organisms, with the exception of carbon, and a suspension of fungi was sprayed onto the surface. The fungi used in the test were *Aspergillus niger, Penicillium funiculosum, Chaetomium globosum, Gliocaldium vireus*, and *Aureobasidium pullulans*. The inoculated test samples were incubated at 82–86°F and at greater than 85% relative humidity for a minimum of 21 days. The growth of organisms was recorded each week. The degree of fungal growth received a numerical rating according to the following observations:

Observed growth	Rating
None	0
Traces of growth (10%)	1
Light growth (10–30%)	2
Medium growth (30–60%)	3
Heavy growth (60% to complete coverage)	4

The study revealed that all commercial thermoplastics, with few exceptions, are quite immune to biodegradation. Exceptions are aliphatic polyesters and aliphatic polyurethanes derived from aliphatic esterdiols, which were readily biodegraded under the test conditions.

B. Biodegradability of Hydrocarbons

Of great significance to the biodegradability of plastics is the finding from the Union Carbide study that straight-chain hydrocarbons were biodegradable only up to a molecular weight of about 500. Beyond 500, straight-chain hydrocarbons did not support the growth of fungi. Also significant is the finding that hydrocarbon branching prevented the organisms, from utilizing them as food. The effect of molecular weight on biodegradability was emphasized in studies with pyrolyzed high-density polyethylene and pyrolyzed low-density polyethylene. Only when the molecular weights of the pyrolyzed (535°C) polymers were lowered from about 60,000–120,000 to less than 3,000 were polymer samples able to support growth of the fungi. Because polymers are not molecules of single molecular weights, but a distribution of molecular weights, it is not surprising that at an average molecular weight of 2,000 or 3,000 some growth of fungi took place. It is likely that some polyethylene fragments of molecular weight 500 or less were included in the distribution of polymer molecular weights.

C. Biodegradability of Polymers

Copolymers of ethylene with the following monomers were not assimilated by fungi according to the ASTM test procedure: vinyl acetate, vinyl alcohol, acrylic acid, ammonium acrylate, ethyl acrylate, carbon monoxide, aconitic acid, itaconic acid, and lauryl acrylate. Moreover, copolymers of ethylene with several unsaturated vegetable oils, including castor, linseed, safflower, soybean, neats foot, peanut, rapeseed, olive, corn, and oleic acid, gave negative results in the ASTM test. Early efforts to make polyethylene biodegradable, by starch–polyethylene compositions, are summarized in Table 4. Polyethylene in the starch compositions was not significantly more biodegradable than the neat polymer.

Synthesized polystyrene was not biodegradable under these test conditions, even at molecular weights as low as 600. The pyrolysis of high molecular weight polystyrene at high temperatures did not produce fractions that were capable of supporting growth of the organisms. Efforts to make polystyrene biodegradable

Table 4 Starch-Based Biodegradable Plastic Technology

Technology	Patentee	Date	Patent no.
Polyethylene, starch, unsat. fatty acid or ester	Coloroll Ltd.	4/5/77	4,016,117
Polyethylene, etherified/esterifed starch	Coloroll Ltd.	5/3/77	4,021,388
Ethylene acrylic acid copolymer, starch	U.S. Dept. Agric.	1/9/79	4,133,784
Ethylene acrylic acid copolymer, gelatinized starch	U.S. Dept. Agric.	6/29/82	4,337,181

by attaching biologically active end groups and by copolymerizing with other monomers were unsuccessful.

Aliphatic polyesters, but not aromatic ones, were highly biodegradable in the
Union Carbide study (Table 5). Polycaprolactone was highly biodegradable in that
study, but polyethylene terephthalate was not. The biodegradability of
polycaprolactone was used to advantage in the development of biodegradable planting containers for tree seedlings. These containers permitted machine planting of
tree seedlings at a rapid rate, with the polyester container biodegrading soon after,
permitting the tree roots to penetrate the surrounding soil. The weight loss of these
containers, 95% in 12 months of burial (3), can serve as a useful benchmark to assess
the biodegradability of other polymers.

In another investigation, the biodegradability of photodegraded polymers,
polypropylene, low-density polyethylene, and Ecolyte PS, a copolymer of polystyrene and a vinyl ketone, was studied with sewage sludge and soils (27). The
biodegradation studies were carried out in a closed system, and the rate of oxygen
consumption was used as the measure of biodegradation. The data from this study
(Table 6) show the polypropylene and low-density polyethylene samples to be
slightly biodegradable, with the polyethylene somewhat more so. The Ecolyte PS

Table 5 Biodegradability of Polyesters

Polyester	Reduced viscosity	Growth rating
Caprolactone polyester	0.7	4
Pivalolactone polyester	0.1	0
Polyethylene succinate	0.24	4
Polytetramethylene succinate	0.59	1
Polytetramethylene succinate	0.08	4
Polyhexamethylene succinate	0.91	4
Polyhexamethylene fumarate	0.25	2
Polyhexamethylene fumarate	0.78	2
Polyethylene adipate	0.13	4
Polyethylene terephthalate	High	0
Polycyclohexanedimethanol terephthalate	High	0
Polybisphenol A carbonate	High	0

Table 6 Biodegradation of Photodegraded Polymer Samples in Soil and Sewage Sludge

Polymer	Degradation method	MW	Biodegradation[a]	
			Sewage sludge (150 h)	Garden soil (70 days)
PP	UV	2,200	1%	4.2%
LDPE	Thermoxid	2,300	2%	9%
Ecolyte PS	UV	15,000	None	None

[a] Based on percentage oxygen consumed.

was virtually nonbiodegradable. Another study with prephotooxidized ^{14}C-labeled Ecolyte PS (with 5% vinyl ketone) showed less than 1% of the radiocarbon was liberated as [14]CO$_2$ in about 2 months contact with soil or sewage sludge (28). The Ecolyte PS used in that study had been photodegraded to molecular weights of 14,000–15,000.

Guillet and co-workers reported that several species of plants were able to utilize the carbon in photodegraded ^{14}C-labeled Ecolyte polystyrene that was contained in a soil mixture on which the plants were grown (29). The plants grown in a closed terrarium were reported to be able to utilize about 5% of the carbon in the polystyrene in 6 months.

Albertsson and Karlsson examined the biodegradation of ^{14}C-labeled low-density polyethylene (LDPE) in soil (22). Samples that were prephotodegraded for 42 days evolved significantly more [14]CO$_2$ over a period of 10 years than samples that had not been previously photodegraded (Table 7). Samples of LDPE with a UV sensitizer were also somewhat more biodegradable than the samples that did not contain a UV sensitizer. At best, a little more than 8% [14]CO$_2$ was evolved in 10 years.

These studies with soil and sewage sludge disclosed definite, but low rates of biodegradation. However, the degree of biodegradation was nowhere near that achieved with the polycaprolactone reported by Potts and co-workers (2,3). Unless the polymers are first photodegraded to low molecular weights, virtually no biodegradation can take place. Even so, the degree of biodegradation after photodegradation to relatively low molecular weights is not encouraging for complete biodegradation of the polymer samples in 1–2 years. It is not possible to estimate how long it would take for the samples of Guillet, Albertsson, and others to completely biodegrade, but the reported results suggest that more than 50 years would probably be required. More recent work has shown that polyethylene residues from photodegradation biodegrade more readily with select organisms when examined in a carbon-starved medium (30). However, it remains to be seen whether biodegradation will be fast enough (1–3 years under carbon-starved conditions) to qualify as true biodegradation.

D. Biodegradation Mechanisms

In reviewing the biochemistry of fatty acid and polyester biodegradation it is not surprising that the commercial packaging polymers show limited biodegradability. In fatty acid biodegradation, enzymes oxidize unsubstituted β-carbon atoms to a carbonyl group, and a two-carbon segment is removed from the fatty acid as acetyl

Table 7 Biodegradation of ^{14}C-Labelled LDPE in Soil

Sample	No. days photodegradation	% [14]CO$_2$ evolved in 10 yr
LDPE	0	0.2
LDPE + UV sensitizer	0	1.0
LDPE	42	5.7
LDPE + UV sensitizer	42	8.4

coenzyme A (Scheme 8). Biodegradation of the resulting fatty acids shortened by two carbon atoms proceeds in a similar manner, with two-carbon segments being removed at a time until the fatty acid is completely degraded. The acetyl coenzyme A, which is removed from the fatty acid, enters into the citric acid cycle and becomes a source of energy for the organism. Substituents other than hydrogen on the β-carbon prevent a molecule from entering into this biodegradation scheme. Thus polypropylene, with a methyl group; polystyrene, with a phenyl group; and PVC, with a chlorine atom, all would not be expected to undergo this biodegradation path. Aliphatic polyesters are susceptible to enzymatic hydrolysis and easy assimilation of the acid and alcohol fragments from hydrolysis. In addition to the problem with substituents on β-carbons, the rigidity of large polymer molecules can prevent access by active sites of enzymes and frustrate interaction (7).

E. Prospects for Biodegradable Plastics

The work of Potts, Guillet, Albertsson, and others show that biodegradability of the major commercial-packaging plastics is quite limited and takes place only to slight degrees. Polypropylene, polystyrene, and polyvinyl chloride are not degradable in the fatty acid biooxidation process, as previously mentioned, because the enzymes required for the biooxidation are not able to assimilate molecules with substituents other than hydrogen. Polyethylene is virtually the only viable candidate among

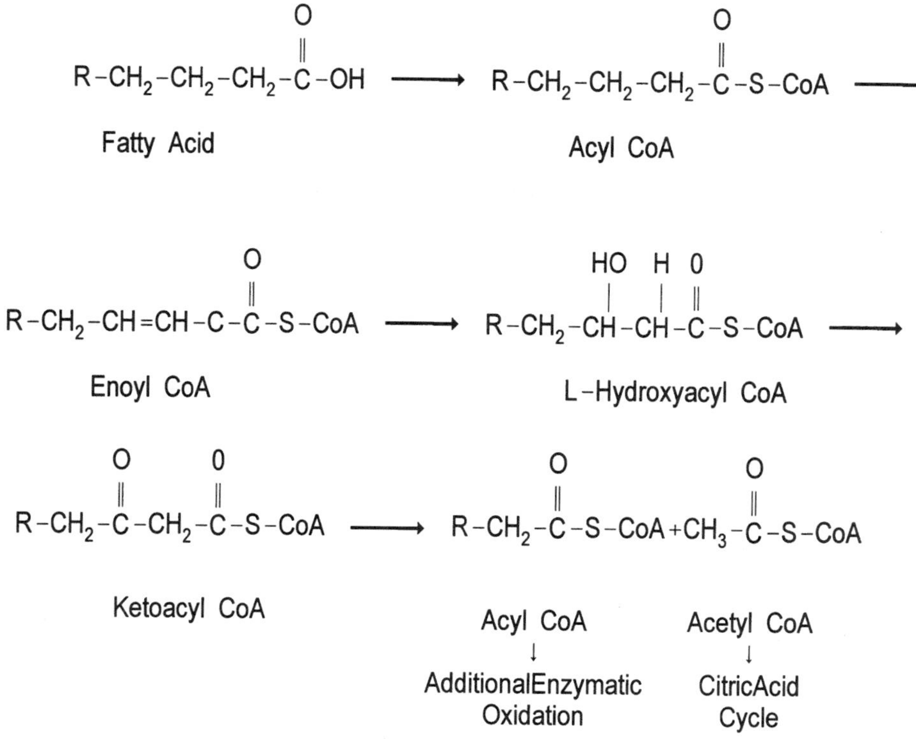

Scheme 8   Biodegradation via enzymatic fatty acid oxidation.

the commercial-packaging plastics for biodegradation, but even with polyethylene, as with all hydrocarbons, there is a molecular weight limitation, and the molecular weights of the polymer samples have to be reduced significantly before any reasonable degree of biodegradation can take place. Other polymers, aliphatic polyesters in particular, are susceptible to reasonable rates of biodegradation, but they are not, at least at present, large-volume–packaging polymers for plastics. Copolymers of 3-hydroxybutyrate and 3-hydroxyvalerate are biodegradable, but because of cost, are not viewed yet as large-volume–packaging polymers. They are expected to achieve certain market niches where cost is not a stringent limitation.

Starch-filled polyethylene was an early approach to biodegradable plastics technology (see Table 4). The starch biodegrades relatively rapidly when in contact with suitable organisms, but that leaves biodisintegrated polyethylene of still significantly high molecular weight that biodegrades very little. Some approaches to starch-filled systems include the incorporation of autooxidizable substances, such as unsaturated oils and sensitizers (see Table 4) and prooxidant iron compounds (31). However, it has not been shown that these lead to significantly increased rates of polyethylene biodegradation after the starch has biodegraded.

Biodegradation was hoped to further degrade the polymer fragments formed from photodegradation of photodegradable polymers. Virtually all of the technologies referred to under the photodegradable approaches would depend on biodegradation to consume the polymer fragments resulting from photodegradation. Because the biodegradation of polymers with substituents other than hydrogen on the main chain is exceedingly slow with existing enzymes, only photodegraded polyethylene would be susceptible to reasonable rates of biodegradation. However, data such as in Table 7, that show less than 10% biodegradation of UV-exposed polyethylene took place in 10 years is not encouraging for complete biodegradation of photodegraded fragments in reasonable time.

Because biodegradation of photodegraded plastic particles of commercial-packaging plastics takes place slowly, a build up of fragments in soil and in the environment might be expected. Although fragments that consist of linear, unsubstituted hydrocarbons could be expected to biodegrade if their molecular weights were low enough, the fate of larger molecular weight fragments of those and other packaging polymers is unknown. Thus, it is conceivable that harmful products might build up in soils over an extended time. This matter requires further exploration before massive amounts of photodegraded polymers are contributed to the environment.

Several polymer producers, recognizing the need for truly biodegradable packaging plastics, have undertaken research to invent and develop such products. Table 8 lists some of the most active polymer systems being investigated. These efforts should provide some polymer offerings that can be expected to be truly biodegradable. However, commercial acceptance of these products will require improved stability during high-temperature processing for some and acceptable pricing for all to realize large-volume applications. The pricing to date of poly(hydroxybutrate) (PHB) and poly(hydroxyvalerate) (PHV) has prevented this truly biodegradable polymer composition from achieving large-volume uses.

The ASTM has developed test methods to assess biodegradability of plastic formulations under a variety of exposure conditions. Table 9 lists methods for aerobic, anaerobic, marine, and composting conditions. Biodegradation under con-

Table 8 Biodegradable Polymer Technologies Under Active Investigation

Polymer	Company
Polylactic acid	Cargill, Ecochem, Biopak, Mitsui Toatsu
Poly(caprolactone)	Union Carbide, Solvay
Poly(aspartic acid)	Rohm & Haas
Poly(hydroxybutyrate and hydroxyvalerate) (PHB, PHV)	Monsanto
Cellophane	Flexel
Cellulosics	Rhône-Poulenc, Eastman
Aliphatic polyesters	Showa High Polymer
Poly(vinyl alcohol)	Rhône-Poulenc, Air Products, Kuraray, Hoechst
Poly(ethylene glycol)	Union Carbide, Dow
Poly(ethylene oxide)	Planet Technologies
Starch-based polymers	Novamont
Starch foam	National Starch

Table 9 ASTM Biodegradation Test Methods

ASTM test number	Test environment	Test measurement
D5209	Aerobic sewage sludge	Volume CO_2 evolved
D5210	Anaerobic sewage sludge	Volume CO_2/CH_4 evolved
D5247	Aerobic with specific organisms	Polymer molecular weight
D5271	Aerobic activated sewage sludge	Volume O_2/CO_2 evolved
D5338	Aerobic controlled composting	Volume CO_2 evolved
D5437	Marine floating conditions	Physical properties
D5509	Simulated compost	Physical properties
D5511	Anaerobic biodegradation	Volume CO_2/CH_4 evolved
D5512	Simulated compost	Physical properties
D5525	Simulated landfill	Physical properties
D5526	Accelerated landfill	Volume CO_2/CH_4 evolved

ditions with adequate moisture and oxygen produces carbon dioxide, the determination of which measures the rate of degradation. When moisture is adequate, but oxygen is excluded, methane is the major product of biodegradation. Its rate of formation also is a measure of the rate of biodegradation. ASTM D5526, an accelerated landfill test, is a useful screening test to determine biodegradability characteristics of polymers. In that test, cumulative biogas production over 300 days for control compositions containing cellulose, a positive control, is compared with the result for the test material.

F. Degradation by Composting

Composting has come under consideration as the pressure for disposal of solid waste becomes greater, because it is a natural way to dispose of biodegradable wastes. Thus, garbage, garden residues, and grass would be candidates for composting in a comprehensive solid waste disposal program that separates materials for disposal by different methods. Composting of biodegradable plastics would be an interesting way to dispose of such materials. Unfortunately, most commercial packaging polymers are still not suitable for composting because they do not biodegrade at sufficiently high rates. Although attractive, this method is not a near-term solution for the existing problems with disposal of plastics. Composting of biodegradable plastics is a disposal method of the future when new biodegradable packaging plastics become available. In the meantime, progress is being made in identifying candidate polymers and methods for biodegradation. Cellulose acetate films with a degree of substitution less than 2.5 and blends of cellulose acetate with aliphatic polyesters biodegraded acceptably rapidly by composting (32,33). Cellulose acetate with a degree of substitution less than 2.2 compared favorably, relative to rate of biodegradation, with poly(hydroxybutyrate-*co*-valerate) and polycaprolactone (33). Surprisingly, in that same composting study, various polylactic acids biodegraded considerably slower than the three rapidly biodegrading polymers.

G. Applications of Biodegradable Plastics

Biodegradable plastics can satisfy a number of different uses. They have been used in sutures for surgical procedures for many years. That serves to emphasize that a high-priced market niche can be met by biodegradable plastics. Biodegradable plastics are of interest for packaging, agricultural mulches, and agricultural planting containers. Ultimate biodegradability, as in composting, is of interest because it would permit biodegradable plastics to be combined with other biodegradable materials and converted into useful soil-improving materials. The prospect of having truly biodegradable plastics available for packaging is appealing, for that would open up possibilities for disposal of those materials by composting.

Considerable interest exists for plastic articles to be biodegradable in landfills. However, this is not a realistic expectation because once a landfill is covered with soil, limited oxygen and water penetration reduces degradation of even usually rapidly biodegrading materials to exceedingly slow rates. Thus, even a highly biodegradable plastic article would not be expected to degrade in a reasonable time in such a landfill. In short, the biodegradation of biodegradable plastic materials in municipal solid waste landfills is not a realistic expectation. On the other hand, that plastics articles survive in landfills for long periods provides a degree of security in the sense that products from the degradation of the plastics are not leaching from the landfill into the environment.

IV. ENVIRONMENTAL SAFETY ASSESSMENT

Biodegradability testing is only one component of environmental safety assessment (ESA), not an end in itself. If the polymer is truly biodegradable, polymer fragments from these processes will undergo mineralization, complete biodegradation. The toxicological properties of biodegrading media, volatiles, residues, or fragments need to be determined before the polymer involved can be deemed to be biodegradable without any environmental safety concerns. Assessment of environmental safety includes estimating environmental concentrations and the no-effect-concentrations of the test materials. Optimally, the no-effect concentration should be larger than the environmental concentration. Rapid and complete biodegradation predict unlikely environmental disturbance (1). Slow or partial biodegradation indicates questionable safety and the need for an ESA.

Solid and liquid samples from the ASTM tests can be examined for ecotoxicity by several methods. The ASTM recommends conducting ASTM Test Method D 5526, *Standard Test Method for Determining Anaerobic Biodegradation of Plastic Materials Under Accelerated Landfill Conditions.* The reactor atmosphere is analyzed periodically for methane and carbon dioxide from which the percentage of biodegradation relative to the positive control, cellulose, is calculated. The residual material in the anaerobic digester is subjected to ecotoxicity testing.

The European Economic Community (EEC) has developed a base set of ecotoxicity tests. Included are a Test for Ready Biodegradability based on CO_2 evolution, an Acute Toxicity Test with *Daphnia magna* (water fleas), an Acute Toxicity Test with *Brachydanio Rerio* (zebra-fish), a Green Algae Inhibition Test, a Wastewater Bacteria Inhibition Test, and an Abiotic Degradation Test, in which hydrolysis is measured as a function of pH. Many of these test methods have been adapted from test methods that are used to assess the toxicological properties of new chemicals.

The Organization for Economic Cooperation and Development (OECD) provides test methods for assessing environmental toxicity. OECD Test Method 301B measures ready biodegradability by means of CO_2 evolution. OECD 301B is a modified Sturm test, the objective of which is to determine if a test substance is readily degradable in an aerobic, aqueous medium by measuring the quantity of CO_2 evolved in the dark at 22°C. The inoculum is activated domestic sewage sludge. Reference substances, glucose, aniline, and sodium acetate must yield more than 60% CO_2 in 28 days. Seven test vessels are used for each test compound: Samples 1 and 2 contain inoculum and the test material at 20 mg C/L; samples 3 and 4 contain inoculum only; sample 5 contains inoculum and a reference substance at 20 mg C/L; sample 6 contains a sterilizing agent and the test material at 20 mg C/L; sample 7 contains inoculum, and test material and reference substance each at 10 mg C/L. The CO_2 evolved in the dark at 22°C is trapped and measured. Reported results are the percentage of theoretical CO_2 evolution.

OECD 202 determines the acute toxicity of materials to the waterflea (*Daphnia magna*) under static conditions. The objective is to determine the 24- and 48-h median effective concentration (EC_{50}) values for *Daphnia* that are less than 24-h old. A range-finding toxicity test is carried out by varying concentrations to produce nontoxic to acute toxic results. The range-finding test information is used to set up the definitive test with a geometric series of at least five concentrations of the test material. Four replicates of each of the concentrations with five water fleas per container are used. The dead or immobilized water fleas are used to calculate EC_{50} for the test substance.

A test method to rapidly screen test substances for possible effects on aerobic microbiological treatment plants has been adapted from a test method to evaluate the toxicity of chemical substances in chemical plant effluents. The inoculum is activated domestic sewage sludge. In the test procedure, two control samples and five concentrations of the test substance are examined. The respiration rate at 20°C is monitored over 10 min with a polarographic oxygen electrode connected to a potentiometric recorder. The measurement provides the EC_{50} for the test substance.

These are some of the most useful test methods for assessing environmental safety. Others are available to meet special requirements. Still others might need to be developed to satisfy special circumstances. The existing test methods that have been reviewed here might be modified to meet the special requirements. A method to assess low molecular weight degradation products from degradable polymers by (head-space)-gas chromatography–mass spectrometry (GC–MS) has been described (34).

V. CONCLUSIONS

Environmental degradation of polymers has evolved over many years to mean the involvement of all degradation mechanisms—photo-, photooxidative, thermo-oxidative, biological, and hydrolytic—to achieve mineralization, complete biodegradation, in less than 2 years. Perceived disadvantages of photodegradation and photooxidation for polymer waste disposal have resulted in increased emphasis on biodegradation. The combination of photodegradation to reduce plastic items to small pieces and biodegradation for mineralization is attractive for polymers in the environment.

Several effective technologies have been developed for photodegradable polymers. They have proved to be a meaningful approach for beverage can loops, which will rapidly photodegrade and reduce prospects for injuries and deaths to creatures; in agricultural mulches as a practical means of improving the growing of various plants; and in plastic bags. None is being used yet commercially for bottles or food containers. It is questionable how important photodegradable polymers are for applications other than the current ones. The major objective appears to be to relatively quickly remove plastic waste from the environment. However, that objective needs to be balanced with the disadvantages posed by photodegradable polymers in terms of soil accumulation of polymer fragments, potential inadvertent contamination of nonphotodegradable polymers, and contamination of recycled plastics (31), an important means for reducing amounts of plastics in municipal solid waste. In these instances, photodegradable polymers can have adverse effects on photostability.

Photodegradable polymers are meaningful for some applications, but they are not a solution to the municipal solid waste problem. Recycling, incineration with energy recovery, and composting are more meaningful approaches to help solve the solid waste problem. Modern incineration technology permits the safe disposal of plastic materials that cannot be recycled. The considerable heat content of plastics makes them ideal for energy recovery during incineration as well as for assisting in the incineration of substances with low heat energy (garbage, for example). Composting can be a useful means for the disposal of biodegradable polymers.

Although considerable progress has been made in identifying and developing biodegradable polymers, as yet, none is cost competitive with commercial-packaging polymers. The shift in emphasis from photodegradable to biodegradable polymers is, in itself, evidence of progress. Natural polymers are among those being investigated for this use. We hope truly biodegradable packaging polymers will become available in the near future and continue to grow in importance. Cost–performance considerations of biodegradable polymers will dominate their growth for some time. In addition, assurance of environmental safety will continue to be a factor for new degradable polymer introductions.

Test methods of ASTM, EEC, and OECD, among others, have been developed to assess the environmental safety of degradable polymers and to ensure that degradation products from candidate degradable materials will not have a negative effect on the environment.

REFERENCES

1. G Swift. Opportunities for environmentally degradable polymers. J Macromol Sci Pure Appl Chem A32:641–651, 1995.
2. JE Potts, RA Clendinning, WB Ackart. Environmental Protection Technology Series, Report EPA-R2-72-046, Office of Research and Monitoring, U.S. Environmental Protection Agency, Washington, DC, 1972.
3. JE Potts, RA Clendinning, WB Ackart, WD Niegisch. In: JE Guillet, ed. Polymers in Ecological Problems. New York: Plenum Press, 1973. p 61.
4. JE Potts. Environmentally degradable plastics. In: Encyclopedia of Chemical Technology, Supplemental Volume. 3rd ed. New York: John Wiley & Sons, 1984, pp 628–668.
5. PP Klemchuk, Degradable plastics: a critical review. Polymer Degrad Stabil 27:183–202, 1990.
6. ASTM Standard Method D883, Standard Terminology Relating to Plastics.
7. SJ Huang, M Bitritto, KW Leong, J Paolisko, M Roby, JR Knox. The effects of some structural variations on the biodegradability of step-growth polymers. Adv Chem Ser 169:205–214, 1978.
8. PP Klemchuk, P-L Horng. Perspectives on the stabilization of hydrocarbon polymers against thermo-oxidative degradation. Polym Degrad Stabil 7:131–151, 1984.
9. RJ Statz, NC Dorris. Photodegradable polyethylene. Proceedings of Symposium on Degradable Plastics, The Society of the Plastics Industry, Washington, DC, 1987, pp 51–55.
10. JE Guillet. U.S. Patent 3, 753, 952, 1973.
11. JE Guillet. U.S. Patent 3, 853, 814, 1974.
12. SA May, EC Fuentes, N Sato. The weathering characteristics of Ecolyte polystyrene: Part 1—A photodegradable copolymer masterbatch and blends with polystyrene. Polym Degrad Stabil 32:357–368, 1991.
13. SKL Li, JE Guillet. Photochemistry of ketone polymers XIV. Studies of ethylene copolymers. J Polym Sci Polym Chem Ed 18:2221–2238, 1980.
14. JE Guillet. Polymers with controlled lifetimes. In: J.E. Guillet, ed. Polymers and Ecological Problems. New York: Plenum Press, 1973, pp 1–25.
15. GH Hartley, JE Guillet. Photochemistry of ketone polymers. I. Studies of ethylene–carbon monoxide copolymers. Macromolecules 1:165–170, 1968.
16. DE Hudgin, T Zawadski. Novel additive system for photodegrading polymers. In: SS Labana, ed. Ultraviolet Light Induced Reactions in Polymers. Washington, DC: ACS Symposium Series 25, 1976, pp 290–306.
17. G Scott. Polymer compositions. U.S. Patent 4,121,025, 1978.
18. G Scott, D Gilead. Controllably degradable polymer compositions. U.S. Patent 4,461,863, 1984. Agricultural processes using controllably degradable polymer composition film U.S. Patent 4,519,161, 1985.

19. S Al-Malaika, AM Marogi, G Scott. Mechanisms of antioxidant action: time-controlled photoantioxidants for polyethylene based on soluble iron compounds. J Appl Polym Sci 31:685–698, 1986.

20. FDA. 21 Code of Federal Regulations 177.1312. Food and Drug Administration Federal Register 57:32421–32423, 1992.

21. DE Hudgin, T Zawadski. Novel additive system for photodegrading polymers. U.S. Patent 4,495,311, 1985.

22. A-C Albertsson, SJ Karlsson. The three stages in degradation of polymers—polyethylene as a model substance. Appl Polym Sci 35:1289–1302, 1988.

23. JE Guillet. Environmental studies on photodegradable plastic packaging. Polym Mater Sci Eng 63:946–949, 1990.

24. DW Wolfe, CA Bache, DJ Lisk. Analysis of dithiocarbamate and nickel residues in lettuce and peppers grown in soil containing photodegradable plastic mulch. J Food Safety 10:281–286, 1990.

25. HG Taber, RS Ennis. Plant uptake of heavy metals from decomposition of Plastigone photodegradable mulch. Proceedings Twenty-first National Agricultural Conference, Orlando, 1989, pp 47–52.

26. RS Ennis. Controlled degradation of plastics: environmental toxicology. In: SH Hamid, MB Amin, AG Maadhah, eds. Handbook of Polymer Degradation. New York: Marcel Dekker, 1992, pp 383–408.

27. PH Jones, D Prasad, N Heskins, NH Morgan, JE Guillet. Environ Sci Technol. 8:919, 1974.

28. JE Guillet, TW Regulski, TV McAneney. Environ Sci Technol 8:923, 1974.

29. JE Guillet, N Heskins, LR Spencer. Division of Polymeric Materials Science and Engineering. ACS, Preprints, June 1988, p 80.

30. J Lemaire, R Arnaud, P Dabin, G Scott, S Al-Malaika, S Chohan, A Fauve, A Maaroufi. Can polyethylene be a photo(bio)degradable synthetic polymer? J Macromol Sci Pure Appl Chem A32:731–741, 1995.

31. S. Al-Malaika, S Chohan, M Coker, G Scott, R Arnaud, P Dabin, A Fauve, J Lemaire. A comparative study of the degradability and recyclability of different classes of degradable polyethylene. J Macromol Sci Pure Appl Chem A32:709–730, 1995.

32. RA Gross, J-D Gu, D Eberiel, SP McCarthy. Laboratory-scale composting test methods to determine polymer degradability: model studies on cellulose acetate. J Macromol Sci Pure Appl Chem A32:613–628, 1995.

33. CM Buchanan, DD Dorschel, RM Gardner, RJ Komarek, AW White. Biodegradation of cellulose esters: composting of cellulose ester–diluent mixtures. J Macromol Sci Pure Appl Chem A32:683–697, 1995.

34. S Karlsson, A-C Albertsson. Degradation products in degradable polymers. J Macromol Sci Pure Appl Chem A32:599–605, 1995.

13

Medical Polymers and Diagnostic Reagents

NAIM AKMAL

Union Carbide Technical Center, South Charleston, West Virginia

ARTHUR M. USMANI

ALTEC USA, Indianapolis, Indiana

I. INTRODUCTION

Without polymers, our lives would be very difficult, if not impossible. Polymers are used in homes, offices, schools, space, underground, underwater, and in fact, everywhere. Polymers occupy a very important place in our daily lives.

Advances in materials science and technology, notably polymers, have resulted in many successful implants, therapeutic devices, and diagnostics assays in the medical field. The selected construction materials must be nontoxic and stable under biological conditions. In addition to polymers (fibers, films, rubber products, molded plastics, coatings, polymeric carbon, and composites), other biomaterials include metals, ceramics, and natural products.

Polymers are by far the most important and dominant biomaterials because they can be manufactured in a wide range of compositions. Their properties can be regulated by suitable composition and process modifications. Furthermore, they can be easily molded into simple as well as complex shapes. Examples of polymeric biomaterial applications include arterial grafts, cardiac pacemakers, heart valves, heart-assist devices, contraceptive devices, implants, catheters, sensors, medical adhesives, blood bags, contact lenses, artificial skin, artificial hearts, heart–lung blood oxygenators, artificial ears, artificial blood, dental and orthopedic cements, sutures, vascular grafts, intrauterine devices, bone and hip joints, dialysis membranes, drug-delivery devices, immunodiagnostics, and dry reagents diagnostics devices.

More than 22.7 billion kg (50 billion lb) of polymer plastics were consumed in the United States in 1998, of which about 907 million kg (2 billion lb) were used in medical applications.

485

Currently, drug and medical packaging account for about 37% of all medical plastics used, followed by general medical and surgical supplies (25%), fluid administration (21%), and medical parts (16%). Low-cost commodity polymers, such as polyethylene, poly(vinyl chloride) (PVC), polystyrene, and polypropylene, dominate the resins used for medical applications. Higher-cost polymers, such as acrylonitrile–butadiene–styrene (ABS), styrene–acrylonitrile (SAN) acrylics, nylons, and polycarbonates are low-volume materials. Polyurethanes, polyesters, and silicones are other types of resins that find medical applications.

Today's medicine is placing demands on technology for new materials and new application techniques. Polymers are thus finding ever-increasing usage in biomedical and life-saving devices. Furthermore, the aging American population would also require an increase in health care products during the next decade. Therefore, use of plastics in medical applications is expected to increase at an annual rate of about 5%.

The application of polymeric materials in medicine is a specialty, with a range of specific applications and requirements. Although the volume of polymers used in medicine is a small fraction of total polymer production, the dollar amount spent annually on prosthetic and biomedical devices exceeds 18 billion dollars in the United States. These applications include more than a million dentures, about 500 million dental fillings, about 8 million contact lenses, over a million hip, knee, or finger replacement joints, about a half million plastic surgery procedures (breast, reconstruction), about 30,000 heart valves, some 65,000 pacemakers, and millions of feet of sutures. Moreover, about 45,000 people use an artificial kidney (hermodialysis) regularly and about 100,000 bypass operations, often using polymers, are done each year.

Polymeric materials used in medicine are somewhat limited to silicones, polyethylenes, plasticized PVCs, acrylics, polyurethanes, polyethylene terephthalate, and Teflon. However, a large number of polymers are still under investigation. In the past, commercial-grade polymers were tried in medical applications. This is not a good approach, because commercial commodity polymers are formulated with additives, such as plasticizers and stabilizers. Currently, biomedical polymers are manufactured specifically for this purpose. This trend will continue, and new biomedical polymers, based on known polymers, will appear in the marketplace.

Most biomolecules present in the human body consist of proteins and nucleic acid macromolecules. Therefore, it makes sense to partially or completely replace tissues or organs with synthetic polymers when and if required. Biomedical applications of polymers are described in some recent books (1–7). Currently, conventional forming techniques, such as injection molding, extrusion, blow molding, and vacuum forming, are most widely employed techniques. However, new-processing techniques should be researched and adapted as routes to improved performance in medical applications.

In general, degradation considerations in medical polymers are similar to polymers used in other applications. Medical reagents that utilize enzymes (also polymeric in nature) need special consideration for deactivation and degradation. However, there are some unique environmental elements, such as body fluids, that must be considered in medical polymer applications. Disposable medical products made from polypropylene are readily degraded by high-energy irradiation encountered during radiation sterilization. In this chapter, we will describe sources of degradation, medical polymers, and their applications.

II. DEGRADATION

Polymers, whether man-made or natural, degrade during use, with progressive loss of properties such as appearance, mechanical strength, and overall integrity. Degradation, caused by the service environment, produces chemical reactions or physical changes, eventually leading to product failure. Some polymers degrade more rapidly than others. An understanding of the cause and source of these degradations will help find stabilization methods on a scientific basis, rather than by trial and error.

In general, degradation involves chemical alteration of the polymer by the service environment. Frequently, but not always, alterations caused by the environment reduces performance of the polymeric materials.

A. Mechanisms

Extensive work on degradation of many polymers has been done and key degradative reactions have been identified (8–15). From these sets of reactions, it is possible to accurately anticipate the major causes of degradation of any new or modified polymer system.

The origin of all degradative processes is the initial bond-breaking reaction. The primary bond-scission reactions may end the extent of deterioration, but may also give rise to secondary chemical reactions, leading to more bond scission, recombination, disproportionation, or substitution reactions. Examples of bond-cleavage processes important in medical polymers are shown in Table 1.

Bond cleavage to produce a free radical pair can be induced by radiation energy (ultraviolet, gamma, x-ray, electron beam), mechanical action, or heat. Many polymers containing heterogeneous structures in the backbone undergo ionic cleavage, either base- or acid-catalyzed hydrolysis. The third form of the primary scission reaction is the bond rearrangement to give molecular products.

Radicals from the primary processes may take part in secondary chemical reactions, such as self-reactions (cross-linking, branching, recombination, disproportionation), addition reactions, depolymerization, and chain reactions involving oxygen.

All polymers undergoing stretching, grinding, milling, sheeting, or any type of shearing process will produce free radicals owing to the fracture of the main chain. In melt blending, it may be difficult to separate the degradative effects of heat and mechanical degradation.

Polymer degradation caused by ionization radiation (gamma sources) is common in medical polymers, as radiation sterilization of medical equipment and nonwoven materials is routinely practiced in hospitals. The high-energy radiation interacts with polymers, producing a cascade of fast physical and chemical processes, leading to the generation of free radicals and other products. The sensitivity of the polymer to radiation depends on the polymer structure (Table 2) (16–18). Polymers with aromatic groups are relatively resistant to gamma radiation, whereas

Table 1   Bond Scission Reactions

Degradative reactions[a]	Source	Polymers
	UV, γ, x-ray, eB, short-wave	All polymers
	Heat, mechanical irradiation	
	UV	Amides, urethanes, ester, ketones
	Base-catalyzed hydrolysis	Amides, urethanes
	Acid-catalyzed hydrolysis	Amides, urethanes, esters, siloxanes
	Thermal	Polyolefins
$RO\cdot \ + \ -\overset{|}{\underset{|}{C}}-H \ \longrightarrow \ ROH \ + \ -\overset{|}{\underset{|}{C}}\cdot$	Thermal	All oxidizing polymers
$ROO\cdot \ + \ -\overset{|}{\underset{|}{C}}- \ \longrightarrow \ ROOH \ + \ -\overset{|}{\underset{|}{C}}\cdot$		
$-Cl\cdot \ + \ -\overset{|}{\underset{|}{C}}-H \ \longrightarrow \ HCl \ + \ -\overset{|}{\underset{|}{C}}\cdot$	Thermal	PVC

[a] R, alkyl or H; X and Y, hetero- or C atom.

Table 2   Polymer Degradation by Gamma Radiation in Air

Polymer	Oxygen uptake[a]	Average kinetic chain length	Dose rate (Mrad/h)	Residual elongation at break	
				%	Dose (Mrad)
PS	1	5	0.015	50	53
			0.36	86	2.5
	11	1.9	0.015	50	2.5
HDPE			0.015	50	1.5
	45	9	0.36	77	2.5
PP			0.015	50	1

[a] Oxygen molecules/100 eV of radiation absorbed at 0.06 Mrad/h.

aliphatic-based ones deteriorate substantially, mainly owing to postirradiation thermal oxidation.

The radiation degradation of polyolefins in both air and in a vacuum has been investigated in detail (19–21). There are two phases in this degradation. The first is the rapid chemical modification of the polymer during the actual radiation exposure. The second phase is a slow postirradiation, occurring later during storage or use in air. Discoloration from degradation is obvious, along with progressive brittleness. Postirradiation deterioration may be due to the presence of longlived macroradicals resulting from gamma irradiation. It has been proposed that these radicals are trapped in the crystalline domain of the polyolefins. The macroradicals move very slowly into the amorphous region where they react with oxygen and begin an oxidative chain. Alternatively, slow decomposition of hydroperoxides and some other peroxidic products lead to postirradiation deterioration of polyolefins, specifically polypropylene.

In the United States, the sterilization dose of gamma irradiation is about 2.5 Mrad. Therefore, polypropylene and other polyolefins used in medical applications must be stabilized before sterilization. Degradation both during gamma radiation and during storage can be prevented by phenolic and piperidyl stabilizers. The sterilization of nylon 6 sutures in the air reduces their mechanical properties only marginally (22). Aromatic polymers are not prone to much degradation by gamma radiation, whereas polyethylene oxide is highly sensitive to oxidation (16).

B. Sources of Degradation

Scission of bonds is most important in polymer degradation. Input of many forms of energy can cause this scission. Bond scission is the main reason for degradation, and the combined effects of other elements are depicted in Fig. 1. The partially degraded polymer most often degrades faster owing to its sensitivity toward the elements. Degradation by internal human body elements is discussed later.

C. Biodegradation

Conventional applications of polymers are based on their relative inertness to biodegradation. In the past, this has led to biodegradation studies for preventing attacks on polymers by bacteria and fungi. Most polymers are bioresistant, and advance research is currently being undertaken to study biodegradibility of polymers for their easier disposal. Furthermore, biodegradable polymers find applications in sutures, surgical implants, and controlled release of drugs.

Testing for biodegradation involves traditional soil burial, as well as microbial degradation with cultured microorganisms and purified enzymes. In vitro testing of polymers for biomedical applications is generally performed in media such as blood serum and isolated issues; for example, chick embryo liver tissue (23) and the yolk sac (24). Generally, in vitro degradation studies are done before in vivo experiments to reduce time as well as cost. Both kinds of testing give a combined result of enzymatic and nonenzymatic degradation. For in vivo testing, the polymer device is implanted for the test time; it is then explanted and evaluated for property loss, such as tensile or flexural strength.

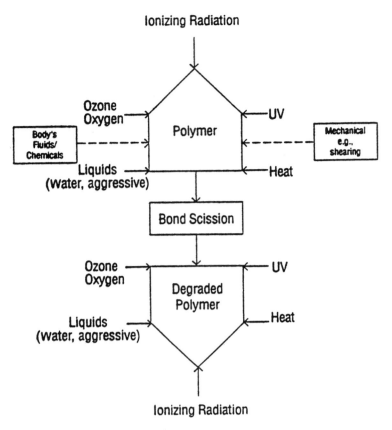

Figure 1 Sources of polymer degradation.

D. Polymer Biocompatibility

In biomedical polymers, biocompatibility—a term not easy to define—is extremely important. Biocompatibility comprises the interactions of living body tissues, body chemicals, and body fluids, including blood, with an implanted or contacting material, such as a polymer. Ultimately, biocompatibility refers to the human body interactions with a biomedical device. In some cases, interactions are desirable, whereas in others inertness is required. The interactions could be either the actions of the body materials on the polymer (e.g., degradation, impairment of function), or the action of the alien polymeric material on the body (25,26). Needless to say, each polymer–body material interaction must be evaluated individually. However, theses interactions can be treated under three broad classes. These are polymer stability, tissue–fluid interactions, and blood compatibility. They are discussed in the following.

1. Polymer Stability

The polymer should be substantially stable during the lifetime of the prosthetic or medical device. However, erodible controlled-release drug systems, sutures, and resorbable bone plates are a few exceptions. Thus, very few polymers can be used safely as implanted biomaterials or devices. Polymers that rely on low molecular weight additives (e.g., plasticizers) for achieving their properties will not be suitable for internal applications. They may find applications in external prosthetic devices, such as the ear and the finger. In the external application, the polymer stability encountered will be similar to the other polymer applications. Aesthetic considerations, specifically discoloration, is more important in external prosthetic devices. In such applications, stabilizers, plasticizers, and other additives that do not produce allergy or toxic reactions are used.

The human body contains a host of enzymes and chemicals that can produce polymer degradation. Polymers with ester or amide linkages will hydrolyze slightly. Implanted nylons lose about half of their tensile strength after 17–24 months, whereas polyesters are more stable (27,28). Polyethers are hydrolytically more stable than polyesters; therefore, polyurethanes based on polyethers are more stable in the internal environment of the human body. Some polyether polyurethane ureas show very little degradation. Silk and cotton degrade very rapidly (28,29). In addition to the hydrolytic degradation promoted by physiological factors, degradation resulting from bond-scission pathways, discussed earlier, can take place. Polyethylene can lose up to 30% of its original tensile strength after 17 months of implantation (30). Polymers such as polyethylene and Teflon can become brittle owing to cross-linking after implantation.

Adsorption as well as absorption of various body chemicals (e.g., triglycerides and steroids) by the polymer can alter polymer flexural properties. In heart valves, but not in nonvascular implants, silicone rubbers can degrade by absorption (25). Furthermore, polymer implants are also subject to abrasion and stress under normal use. Examples are dental devices and prosthetic joints. In the physiological environment, all polymeric materials will show chemical and mechanical degradation. However, polymeric materials will survive better than metal and ceramic materials.

2. Tissue–Fluid Interactions

The body's exact response to an implant depends not only on the polymer structure, but also on the form (foam, fiber, film), shape, implant movements, and the location within the human body. Body response can vary from a benign acceptance of the implant, to a rejection. In outright rejection, the body may try to expel the implant by chemical reactions through phagocytic or enzymatic activity. With all implants, polymeric or nonpolymeric, the body forms a fibrous membrane capsule around the implant (31).

Nonpolar polymers (e.g., polytheylene and Teflon) show minimal tissue in comparison with polar polymers (e.g., nylons and polyacrylonitrile). This may be due to the historical interaction with hydrolyzed products from polymers containing amide or ester linkages, as well as arising from immunological interactions. With increasing tissue response, inflammation increases, producing numerous macrophages around the implant. Polar polymers also produce undesirable adhesion to tissues.

A smooth, rounded shape will give less interactions than a rough-edged shape for a given type of material. In general, surface smoothness reduces undesirable tissue adhesion. Within a class, a powdered polymer will give very high tissue interactions owing to its large surface area. Implant movement will also increase the interaction.

Low molecular weight polymers produce significant tissue interactions. Low weight additives, because of their mobility and solubility in body chemicals, produce adverse tissue reactions. Exclusion of additives, removal of residual monomers, and even polymerization catalysts are, therefore, advantageous for implant applications.

The determination of biocompatibility of polymers is a difficult task. Animal testing is being widely used and is both expensive and not fully reliable. Some promising preanimal testing schemes have been developed and are being used extensively.

3. Blood Compatibility

In many extracorporeal devices, blood contact invariably occurs, thus requiring the polymer to be biocompatible with blood. Blood clotting can be prevented by anti-coagulants (e.g., heparin). In other cases, the blood-containing material must be blood compatible. The mechanism of blood clotting caused by polymeric materials is not yet well understood. Very few polymeric materials have good blood compatibility. These include some hydrogels, some polyether urethane ureas, and some materials made by affixing biologically inactivated natural tissue to the polymer surface.

III. MEDICAL POLYMERS AND THEIR APPLICATIONS

Application of synthetic polymers as biomaterials in medicine, has been extensively pursued in the last several decades. A *biomaterial* is a "nonviable material used in medical devices intended to interact with the biological system." Other biomaterials, such as ceramics, metals, glasses, modified biopolymers, and composites, have also been applied in medicine. Because of the close contact of biomaterials with biological systems, interaction and biocompatibility of synthetic polymers are of prime concern for selection.

In general, the following minimum standards must be met when choosing a biomaterial:

1. The biomaterial must be biocompatible (tissue, cell, and blood compatible). It should not cause immunological response in the host, and should be nontoxic and noncarcinogenic.
2. The biomaterial must not release any components into the biological system unless intended.
3. The mechanical properties of the biomaterial must satisfy the intended usage for the expected life of the device.
4. The biomaterial must be nondegradable in the biological system unless the degradation is intended and desirable.

Four major classes of polymers used as biomaterials and their applications will be briefly described.

A. Polyurethanes

Polyurethane polymers with biomedical applications are mostly polyether urethanes and are used in a variety of products. These polymers show high tensile strength, excellent elastomeric properties, and are resistant to biodegradation.

Polyurethane elastomers have been used in heart-assist devices, blood bags, hemodialysis tubing, surgical prostheses, and catheters. Polyurethane foams, with high flexibility and open cell structures, are used in medical bandages. These foams when impregnated with silicone are used in blood oxygenation devices. Rigid polyurethane foams have been used in artificial legs and arms.

Other applications of polyurethane polymers include artificial hearts, ligament replacements, heart valve prostheses, vascular graft protheses, and coating for blood compatibility.

B. Acrylic Polymers

Acrylic and methacrylic polymers, such as poly(hydroxyethyl methacrylate) and poly(acrylic acid) have been used in contact lenses, bone cements, and restorative and filling materials in dental applications. Recent improvements in oxygen permeability of acrylic–silicone lenses and comfort of hydrogel lenses have reduced the use of metacrylic polymers in contact lenses. Bone cements are used to anchor artificial joints in the modullary cavity. They consist mainly of methyl methacrylate monomer and the initiator, which when mixed together form a putty. The putty is used in knee and hip joints and other orthopedic devices, and on polymerization it hardens to form a cement-type connection.

C. Polymeric Hydrogels

Polymeric hydrogels are water-insoluble, highly hydrophilic, cross-linked materials that swell extensively in water at physiological conditions. Hydrogel polymers include mostly poly(hydroxyethyl methacrylate), poly(vinyl alcohol), and poly-acrylamide. Polymeric hydrogels are used in coatings on catheters, sutures, intrauterine devices (IUDs), and blood-compatible materials. These polymers are also used in a variety of other biomedical applications, such as hemodialysis membranes, breast tissue substitutes, burn dressings, particulate antibody carriers, dental fillings, controlled drug-release systems, and contact lenses. Various polymers other than hydrogels are used in controlled-release systems include polyesters, polyacids, polyaminothiazole, and cellulose.

D. Silicone Polymers

The silicane group of biomedical polymers comes from polydimethylsiloxane, also known as silicone rubber, which can be cross-linked to modify its physical and mech-anical properties. These polymers can be synthesized in very pure form for artificial skin, joint replacement, vitreous replacement, artificial heart and assist materials, plastic surgery implants, contact lenses, corneal bandages, and different types of catheters.

A large number of other synthetic polymers with specific applications in diag-nostic devices, disposable medical materials, packaging, and so on, are also exten-sively used.

Applications of biomedical polymers are summarized in Table 3. The list is by no means comprehensive.

Table 3   Medical Applications of Polymers

Application	Comments and polymers
Orthopedic	
Cast, braces, and bone repair	Replace conventional heavy plaster with molded polyethylene, polypropylene lined with flexible polyester or polyurethane foam. Molding matches contours of patients.
	Degradable poly(lactic acid) as resorbable bone plate.
Joint replacement	For finger joint, silicone rubber. Excellent durability does not promote bone resorption or damage. Also used is poly(1,4-hexadiene).
Hip joint	Metal–metal joints unsatisfactory owing to corrosion and friction. All ceramic fractures. All plastic is not strong. HDPE used for socket and metal balls.
Knee joint	HDPE–metal.
Knee cap	Poly(dimethylsiloxane).
Joint cement	Acrylic copolymer/MMA/redox initiator.
Artificial limb	No internal body contact. Central metal shaft with lightweight plastic foam.
Cardiovascular	
Pacemaker	Wires/electrodes are plastic coated and embedded in plastic for protection from body fluids.
Heart valve	Natural polymeric materials; also silicone rubbers, Teflon, polyacetals. These polymers are not truly biocompatible; therefore, long-term anticoagulant treatment is required.
Blood vessel	Flexible plastic tubes in coronary bypass. Woven Dacron or foamed Telfon for vascular prostheses; neither is blood-compatible. For blood vessel less than 4 mm, no satisfactory replacement. Polyurethane urea, polyester and hydrogel are promising.
Heart-assist devices	Mechanical device to relieve damaged heart. Intra-aortic balloon pump—polyether urethane urea (PEUU).
	Ventricular assist—PEUU, silicone rubbers, PVC, Dacron.
Artificial heart	PEUU, biolized poly(1,4-hexadiene). Silicones, plasticized PVC, and natural rubber are thrombogenic and weak.
Dental	
Filling and denture	Acrylic, polyurethane, vinyl acetate–vinyl chloride–acrylic, vulcanized rubbers, epoxies, filled thermoset acrylics.
Crown bridge	Silicones, polysulfides, polyethers, and alginates.
Artificial organs	
Kidney, hemodialyzer	Silicone, PVC tubing, and cellulosic or polyacrylonitrile membrane.
Lung	No implantable artificial lung: extracorporeal oxygenator—polydimethylisiloxane.

Table 3 Continued

Application	Comments and polymers
Artificial pancreas	Infusion pump outside body—silicone rubbers.
	Nylon wall encapsulated beta cells to produce insulin on demand.
Eyes	
Contact lens, hard	Acrylic copolymers.
Contact lens, soft	Hydroxyethyl methacrylate or N-vinyl pyrrolidone, cross-linked.
Intraocular lens implant	Acrylics, hydrogels.
Craniofacial	
Cartilaginous, nose, ear	Silicone rubbers, polyethylene, Telfon.
Alveolar ridge	Reconstructed by polymethyl methacrylate, Teflon/carbon composite.
Cranial bone	Reconstruction by polyethyl methacrylate, nylon.
Soft tissues	
Breast prosthesis	Polydimethylsiloxane, polyurethanes.
Facial plastic surgery	Polydimethylsiloxane, polyurethanes.
Hernia repair	Polyethylene.
Breast replacement	Silicone rubber gel in silicone bag.
Nonphysiological testicle replacement	Silicone rubber gel in silicone bag.
Drug-release systems	
Implantable osmotic pump	Ethylene vinyl acetate membrane.
Biodegradable releasing polymer	Maleic anhydride copolymers, hydrogels with degradable cross-links.
Encapsulated drugs	Silicones, polyesters, nylons, hydrogels, polyurethanes.
Others	
Skin replacement	Composite of collagen–glycosaminoglycan layer (1.5 mm) covered with thin silicone rubber.
Polymeric drugs	Divinyl ether–maleic anhydride cyclic alternating co-polymers, caboxylated acrylics, vinyl analogues of nucleic acid.
Dry diagnostic chemistry	Vinyl, acrylic emulsions.
Imunodiagnostics	Monodisperse polystyrene latex.

IV. MEDICAL DIAGNOSTIC REAGENTS

Purified enzymes are widely used in medical diagnostic reagents and in the measurement of analytes in urine, plasma, serum, or whole blood. There has been a steady growth of dry chemistry during the past three decades. It has surpassed wet clinical analysis in the number of tests performed in hospitals, laboratories, and homes because of its ease, reliability, and accuracy.

Enzymes are specific catalysts that can be derived from plant and animal tissues; however, fermentation continues to be the most popular method. Enzymes are extensively used in diagnostics, immunodiagnostics, and biosensors. They measure or amplify signals of many specific metabolites. Purified enzymes are expensive, and their use for a large number of analytes can be costly. This is the main reason for the increasing use of reusable immobilized enzymes in clinical analyses.

Wet chemistry methods for analysis of body analytes (e.g., blood glucose or cholesterol) require equipment and trained analysts. Millions of persons with diabetes check their blood glucose levels and are able to obtain results in a matter of a few minutes. However, science has not yet invented an insulin delivery system that can respond to the body's senses. Injected insulin does not automatically adjust; therefore, the dose required to mimic the body's response must be adjusted daily or even hourly, depending on the diet and physical activity. Self-monitoring of blood glucose levels is essential for diabetics. This has become possible over the last 30 years owing to the advent of dry chemistry (32–39). Accurate monitoring of blood glucose levels by an expectant woman will enable her to have a normal pregnancy and give birth to a healthy child. Athletes with diabetes can self-test their blood glucose levels to avoid significant health problems. Dry chemistries are useful not only for diabetes, but also for patients with other medical problems. They are also used in animal diagnosis, food evaluation, fermentation, agriculture, and environmental and industrial monitoring (32).

We will now describe the principles and biochemical reactions involved in diagnostic reagents. Dry chemistry construction and recent advances will also be addressed. For completeness we shall describe biosensors.

A. Biochemical Reactions for Assaying Cholesterol and Glucose

1.   Cholesterol

$$\text{Cholesterol esters} + H_2O \xrightarrow{\text{cholesterol.esterase}} \text{cholesterol} + \text{fatty acids}$$

$$\text{Cholesterol} + O_2 \xrightarrow{\text{cholesterol.oxidase}} \text{cholest-4-en-3-one} + H_2O_2$$

$$H_2O_2 + \text{chromogen} \xrightarrow{\text{peroxidase}} \text{dye} + H_2O$$

Notes

1. Endpoint followed by dye formation.
2. Amount of oxygen consumed can be measured amperometrically by an oxygen-sensing electrode.
3. The H_2O_2 produced by cholesterol oxidase requires phenol to produce dye. A popular alternative step is to substitute *p*-hydroxybenzenesulfonate for phenol in the reaction with pyridine nucleotide. The subsequent reaction is as follows:

$$H_2O_2 + \text{ethanol} \xrightarrow{\text{catalase}} \text{acetaldehyde} + 2H_2O$$

$$\text{Acetaldehyde} + \text{NAD(P)}^+ \xrightarrow{\text{aldehyde.dehydrogenase}} \text{acetate} + H^+ + \text{NAD(P)H}$$

4. Free cholesterol is determined, if cholesterol esterase is omitted.

2. Glucose

1. $\text{Glucose} + O_2 + H_2O \xrightarrow{\text{glucose .oxidase}} \text{gluconic acid} + H_2O_2$

    $H_2O_2 + \text{chromogen} \xrightarrow{\text{peroxidase}} \text{dye} + H_2O_2$

    Note: Formation of dye.

2. $\text{Glucose} + O_2 + H_2O \xrightarrow{\text{glucose .oxidase}} \text{gluconic acid} + H_2O_2$

    Note: The rate of oxygen depletion is measured with an oxygen electrode. Two additional steps prevent the formation of oxygen from H_2O_2 and are as follows:

    $H_2O_2 + \text{ethanol} \xrightarrow{\text{catalase}} \text{acetaldehyde} + 2H_2O$

    $H_2O_2 + 2H^+ + 2I^- \xrightarrow{\text{molybydate}} I_2 + 2H_2O$

B. Dry Chemistry

1. History

Dry- or solid-phase chemistry can trace its origins back to the ancient Greeks. Some 2000 years ago, copper sulfate was an important ingredient in tanning and preservation of leather. At that time, dishonest traders were adulterating valuable copper sulfate with iron salts. Pliny has described a method for soaking reeds of papyrus in a plant gall infusion or a solution of gallic acid that would turn black in the presence of iron. This is the first recorded use of the dry chemistry system. In 1830, filter paper, impregnated with silver carbonate, was used to detect uric acid qualitatively.

Self-testing (e.g., measuring ones own weight and body temperature) is likewise an old concept. Elliot Joslin advocated, some 80 years ago, that diabetics should frequently monitor their glucose level by testing their urine with Benedict's qualitative test (40). When insulin became available in the early 1920s, self-testing of urine became necessary. In the mid-1940s, Compton and Treneer compounded a dry tablet, consisting of sodium hydroxide, citric acid, sodium carbonate, and cupric sulfate (41). This tablet, when added to a small urine sample and boiled, resulted in the reduction of blue cupric sulfate to a yellow or orange color, if glucose was present. Glucose urine strips, impregnated and based on enzymatic reactions using glucose oxidase, peroxidase, and an indicator were introduced about 1956 (42). Dry chemistry blood glucose test strips coated with a semipermeable membrane to which whole blood could be applied and wiped off were introduced in 1964. Currently, low-cost, lightweight, plastic-housed reflectance meters with advanced data management capabilities are available and are being extensively used.

Impregnated dry reagents have a coarse texture, with high porosity and an uneven large pore size, resulting in a nonuniform color development in the reacted strip. In the early 1970s a coating film-type dry reagent was developed by applying an enzymatic coating onto a plastic support. This gave a smooth, fine texture and, therefore, uniform color development (43). Nonwipe dry chemistries, wherein the devices handle the excess blood by absorption or capillary action, have appeared in the marketplace during the last 5 years.

Discrete multilayered coatings, developed by the photographic industry were adapted in the late 1970s to coat dry reagent chemistry formats for clinical testing (44,45). Each zone of the multilayered coating provides a unique environment for sequential chemical and physical reactions. These devices consist of a spreading layer, separation membrane, reagent, and reflective zone, coated onto the base support. The spreading layer wicks the sample and applies it uniformly to the next layer. The separation layer can hold back certain components (e.g., red blood cells) allowing only the described metabolites to pass through to the reagent zone, which contains all the necessary reagent components. Other layers may be incorporated for filtering, or eliminating the interfering substrates.

2. Background

Enzymes are essential in dry chemistries. In a typical glucose-measuring dry reagent, glucose oxidase (GOD) and peroxidase (POD) enzymes, along with a suitable indicator, (e.g., 3,3',5,5'-tetramethylbenzidine; TMB), are dissolved or dispersed in a latex or water-soluble polymer. This coating is then applied to a lightly pigmented plastic film and dried to obtain thin dry film. The coated plastic can be cut to a suitable size, 0.5 by 0.5 cm and mounted on a plastic handle. The user applies a drop of blood on the dry reagent pad and allows it to react for about 60 s or less. The blood can be wiped off manually by a swab or by the device itself. The developed color is then read by a meter or visually compared with the predesignated printed color blocks to determine the precise glucose level in the blood. Thus, dry chemistries are highly user-friendly.

Dry chemistry test kits are available in thin strips that are usually disposable. They may be either film-coated or impregnated. The most basic diagnostic strip consists of a paper or plastic base, polymeric binder, and reacting chemistry components consisting of enzymes, surfactants, buffers, and indicators. The diagnostic coatings or impregnation must incorporate all reagents necessary for the reaction. The coating can either be single or multilayer. A list of analytes, enzymes, drugs, and electrolytes assayed by dry chemistry diagnostic test kits is given in Table 4 (46,47).

Dry chemistry systems are being widely used in physician's offices, hospital laboratories, and by millions of patients worldwide. They are used for routine urine analysis, blood chemistry determinations, and immunological and microbiological testing. The main advantage of dry chemistry technology is that it eliminates the need for reagent preparation and many other manual steps common to liquid reagent systems, resulting in greater consistency and reliability of test results. Furthermore, dry chemistry has a longer shelf stability; therefore, its use helps reduce wastage of reagents. Each test unit contains all the reagents and reactants necessary to perform the assays.

Dry chemistry tests are used for the assay of metabolites by concentration or by activity in a biological matrix. In general, reactive components are present in excess, except for the analyte being determined. This is done to make sure that the reactions will be completed quickly. Other enzymes or reagents are used to drive the reactions in the desired direction (48). Glucose and cholesterol are the most common measured

Table 4 Dry Chemistry Diagnostic Test Kits

Substrate	Enzymes	Drugs	Electrolytes
Glucose		Phenobarbitone	
Urea		Phenytoin; theophyline	Sodium ion
Urate		Carbamazepine	Chloride
Cholesterol (total)	Alkaline phosphate (ALP)		CO_2
Triglyceride	Lactate dehyrogenase (LDH)		
Amylase (total)	Creatine kinase MB isoenzyme (CK-MB)		
Bilirubin (total)	Lipase		
Ammonium ion			
Creatinine			
Calcium			
Hemoglobin			
HDL cholesterol			
Magnesium (II)			
Phosphate (inorganic)			
Albumin			
Protein in cerebrospinal fluid			

analytes. The biochemical reactions in dry chemistries for many analytes, including glucose and cholesterol, have been described earlier.

3. Components of the Dry Chemistry System

The basic components of typical dry reagents that utilize reflectance measurement are a base support material, a reflective layer, and a reagent layer that can either be single- or multilayered. We will now describe the functions of the various building blocks. The base layer serves as a building base for the dry reagent and is usually a thin, rigid thermoplastic film. The reflective layer, usually white pigment-filled plastic film, coating, foam, or paper, reflects those components not absorbed by the chemistry to the detector. The reagent layer contains the integrated reagents for a specific chemistry. Typical materials are paper matrix, fiber matrix, coating film, as well as combinations thereof.

An example of a single-layer coating reagent that effectively excludes red blood cells (RBC) is shown in Fig. 2. Here, an emulsion-based coating containing all the reagents for specific chemistry, is coated onto a lightly filled thermoplastic film and dried. For glucose measurement, the coating should contain GOD, POD, and TMB. It may also contain buffer for pH adjustment, minor amounts of ether–alcohol-type organic coalescing agent, and traces of a hindered phenol-type antioxidant to serve as a color–signal-ranging compound. Whole blood is applied and allowed to react for 60 s or less; excess blood is wiped off; and the developed color is read visually or by a meter. Paper-impregnated–type dry chemistry is shown in Fig. 3.

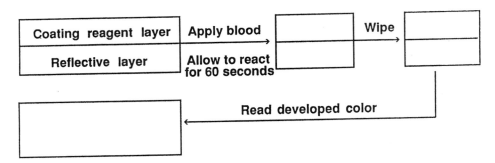

Figure 2   Coating film-type dry chemistry.

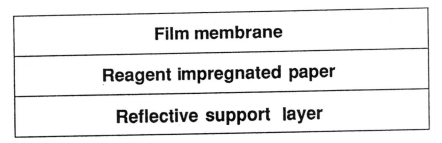

Figure 3   Paper impregnated and overcoated-type dry chemistry.

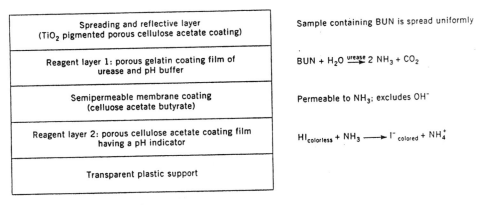

Figure 4   Example of multilayered coated dry chemistry for blood urea nitrogen.

The schematic of a typical multilayered coating dry reagent, for example, blood urea nitrogen (BUN), is shown in Fig. 4. It consists of a spreading and reflective layer. A sample containing BUN is spread uniformly on this layer. The first reagent layer is a porous-coating film containing the enzyme urease and a buffer (pH 8.0). Urease reduces BUN to NH_3. A semipermeable membrane coating allows NH_3 to permeate while excluding OH^- from the second reagent layer. The second reagent layer is composed of a porous-coating film containing a pH indicator, wherein the indicator color develops when NH_3 reaches the semipermeable coating film. Typically, such dry reagents are slides (2.8×2.4 cm) with an application area of 0.8 cm², and the spreading layer is about 100-mm thin.

4. Application of Diagnostic Technology in Monitoring Diabetes

Frequent measurement of blood glucose to manage diabetes is one of the most important applications of diagnostic reagents. Researchers are making new inroads in understanding diabetes, whereas medicinal chemists are exploring new drugs to treat the disease and its complications. At the same time, biotechnologists are discovering better ways to manage the disease.

It has been estimated that there are about 15 million diabetics in the United States, but in only half of them has the disease been diagnosed. More than 1.5 million diabetics are treated with injected insulin and the rest with a weight-loss program, diet, and oral antidiabetic drugs (e.g., sulfonylureas; tolbutamide, tolazamide, chloropropamide, glipizide, glyburide).

The current U.S. market for drugs to control blood glucose is about 1 billion dollars, equally divided between insulin and antidiabetic drugs (49). Insulin will grow by about 10% annually, whereas, the antidiabetic drug market will shrink by about 3%. The blood–glucose-monitoring market is about 750 million dollar, in the United States, and is expected to grow at a rate of 10% annually.

5. Diagnostic Polymers and Coatings

In most dry chemistries, polymers account for more than 95% of the strips. Polymers, therefore, are important and must be selected carefully. Usmani and co-workers have linked polymer chemistry to biochemistry, and that has led to a better understanding of and improvement in dry chemistries. General considerations of polymers in dry chemistries are described in the following paragraphs.

The polymer binder is necessary to incorporate the system's chemical components in the form of either a coating or impregnation. The reagent matrix must be carefully selected to mitigate or eliminate nonuniformity in reagents' concentrations owing to improper mixing, settling, or nonuniform-coating thickness. Therefore, aqueous-based emulsion polymers and water-soluble polymers are being extensively used. Table 5 provides a list of commonly used matrix binders (50,51).

Polymers must be carefully screened and selected to avoid interference with the chemical reactions. The properties of polymers (e.g., composition, solubility,

Table 5 Common Matrix Binders

Emulsion polymers	Water-soluble polymers
Acrylic	Polyvinyl alcohol
Polyvinyl acetate: homo- and copolymers	Polyvinyl pyrrolidone; highly hydroxylated acrylic
Styrene acrylics	Polyvinylethylene glycol acrylate
Polyvinyl propionate: homo- and copolymers	Polyacrylamide; hydroxyethyl cellulose
Ethylene vinyl acetate	Other hydrophilic cellulosics
Lightly cross-linkable acrylics	Various copolymers
Polyurethanes	

viscosity, solid content, surfactants, residual initiators, film-forming temperature, and particle size) should be carefully considered (52). In general, the polymer should be a good film-former with good adhesion to the support substrate. Furthermore, it should have no or only minimal, tack for handling purposes during manufacturing of the strip. The coated matrix or impregnation must have the desired pore size and porosity to allow penetration of the analyte being measured, as well as have the desired gloss, swelling characteristics, and surface energetics. Swelling of the polymer binder by absorption of the liquid sample may or may not be advantageous, depending on the system. Emulsion polymers have a distinct advantage over soluble polymers, owing to their high molecular weight, superior mechanical properties, and potential for adsorbing enzymes.

Polymeric binders used in multilayered coatings include various emulsion polymers, gelatin, polyacrylamide, polysaccharides (e.g., agarose), water-soluble polymers (e.g., polyvinyl pyrrolidone, polyvinyl alcohol, copolymers of vinyl pyrrolidone and acrylamide), and hydrophilic cellulose derivatives (e.g., hydroxyethyl cellulose and methyl cellulose).

6. Waterborne Coatings

The waterborne type is by far the most important coating film in use today because enzymes are water soluble. During the late 1980s we researched a tough coating film in which whole blood was allowed to flow over the coating film. This is shown in Fig. 5. The RBC must roll over and not stick to the coating film. Almost all types of emulsions and water-soluble polymers were investigated: a styrene–acrylate emulsion was the most suitable. A coating containing 51 g styrene–acrylate emulsion (50% solids), 10 g linear alkylbenzene sulfonate (15 wt% in water), 0.1 g GOD, 0.23 g POD, 0.74 g TMB, 10.0 g hexanol, and 13.2 g 1-methoxy-2-propanol gave a good dose–response, as seen in Fig. 6. This coating film gave a linear dose–response, in transmission mode, up to 1500 mg/dL of glucose. The coating was slightly tacky and was highly residence- and time-dependent.

To mitigate or eliminate tack, ultrafine mica was effective. This made possible the rolling and unrolling of coated plastic for automated strip construction. In devices wherein blood residence time is regulated, the residence–time-dependent coating will be of no problem. Coating film dependency creates a serious problem where blood residence time cannot be regulated by the strip. To make the coating film residence–time-independent, continuous leaching of color from films during exposure time is essential. Enzymes and the indicator can leach rapidly near the blood–coating film interface and develop color that can be drained off. The color formed on the coating film is then independent of the exposure time. This gave a coating film with minimum time dependency. A low molecular weight polyvinyl pyrrolidone (e.g., PVP K-15 from GAF or PVP K-12 from BASF) gave coatings with no residence time dependency. Such a coating consists of 204 g of styrene–acrylate emulsion (50% solids), 40 g linear alkybenzene sulfonate (15% solution in water), 0.41 g GOD (193 U/mg), 1.02 g POD (162 U/mg), 2.96 g TMB, 7.4 g PVP K-12, 32 g micromica C-4000, 20 g hexanol and 6 g Igepal CO-530. The nonionic surfactant serves as a surface modifier to eliminate RBC retention.

Excellent correlation was found when results at 660 nm and 749 nm were correlated with a reference hexokinase glucose method. The dose–response was excellent up to 300 mg/dL glucose. In general, waterborne coatings do not lend themselves to ranging by antioxidants.

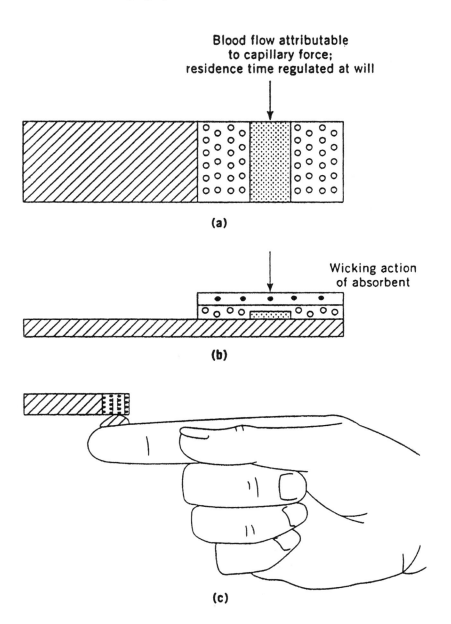

Figure 5   Touch and drain dry chemistry construction: (a) dry-coated surface, (b) cross-section of dry-coated surface, adhesive, and cover piece, (c) contact with blood drop results in blood filling the cavity. After desired reaction time, blood is drained off by touching end of the cavity with absorbent material.

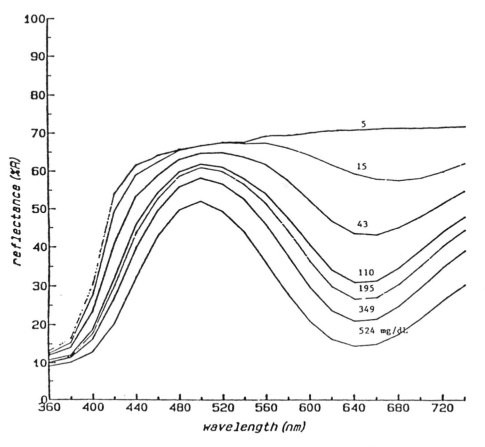

Figure 6   Dose–response of waterborne-coating film for 120-s residence time.

7. Nonaqueous Coatings

For the past 25 years, dry reagent coatings have exclusively been waterborne owing to the belief that enzymes function only in a water medium. Recently, nonaqueous enzymatic coatings for dry chemistries have been researched, developed, and refined by Usmani et al. (47,53). The RBCs will not adhere to such chemical constructions, and they gave a quick endpoint. These coatings also gave superior thermostability. Furthermore, these coatings could be easily ranged by antioxidants, whereas waterborne coatings are very difficult to range. These researchers synthesized nonaqueous hydroxylated acrylic diagnostic polymers, with good hydrophilicity and hydrogel character. The enzymes, (e.g., GOD and POD) are insoluble in organic solvents, but become extremely rigid and can be dispersed with ease. Dispersions of less than 1 mm were made using an Attritor mill or a ball mill. To prepare nonaqueous coatings, polymer solution, TMB, mica, surface modifiers, and solvents were added to the enzyme dispersion and mixed slightly on a ball mill. The ranging compound can be added later. The weight percentage composition of a typical nonaqueous coating useful for a low-range blood glucose measurement is 33.29

hydroxylethyl methacrylate–butyl methacrylate–dimethylaminoethyl methacrylate (65/33/2) polymer (40% solid), 2.38 TMB, 1.17 GOD, 2.68 POD, 3.28 sodium dodecyl benzene sulfonate, 26.53 xylene, 26.53 1-methoxypropanol, and 1.96 cosmetic grade C-4000 ultrafine mica. Many surfactants and surface modifiers were investigated that eliminated RBC retention. Antioxidants that function as ranging compounds in these nonaqueous systems include 3-amino-9-(aminopropyl)-carbazole dihydrochloride, butylated hydroxy toluene (BHT), and a combination of BHT–propyl gallate. These ranging compounds are effective in ranging a compound to the TMB indicator molar ratio of 1 : 2.5 to 1 : 20.

The long-term stability of the nonaqueous coatings films under elevated temperature and moderate humidity has been reported to be better than aqueous coatings. Furthermore, color resolution and sensitivity of reacted nonaqueous-coating films were excellent, as demonstrated by curve fitting.

8. Molded Dry Chemistry

In general, most enzymes are very fragile and sensitive to pH, solvent, and elevated temperatures. Typical properties of select diagnostic enzymes are given in Table 6. The catalytic activity of most enzymes is reduced dramatically as the temperature is increased. Common enzymes used in diagnostics (e.g., GOD and POD) are almost completely deactivated at close to 65°C in solid form or as a water solution. Despite wide and continued use of such enzymes in diagnostics for more than 40 years, limited or no thermal analysis work on these biopolymers have been reported until recently (54,55). Recent differential scanning calorimetric (DSC) analysis results, indicating glass transition temperature (T_g), melting temperature (T_m), and decomposition temperatures (T_d), are shown in Table 7. Below T_g, the enzymes are in a glassy state and should be thermally stable. Around T_g, onset of the rubbery state begins, and the enzyme becomes prone to thermal instability. When the

Table 6 Typical Properties of Select Diagnostic Enzymes

	Cholesterol oxidase (CO)	Cholesterol esterase (CE)	Glucose oxidase (GOD)	Peroxidase (POD)
Source	*Streptomyces*	*Pseudomonas*	*Aspergillus*	Horseradish
EC no.	1.1.3.6	1.1.13	1.1.3.4	1.11.1.7
Molecular weight	34,000	300,000	153,000	40,000
Isoelectric point	5.1 ± 0.1 and 5.4 ± 0.1	5.95 ± 0.05	4.2 ± 0.1	—
Michaelis constant	4.3×10^{-5} M	2.3×10^{-5} M	3.3×10^{-2} M	—
Inhibitor	Hg^{2+}, Ag^+	Hg^{2+}, Ag^+	Hg^+, Ag^+, Cu^{2+}	CN^-, S^{2-}
Optimum pH	6.5–7.0	7.0–9.0	5.0	6.0–6.5
Optimum temperature	45–50°C	40°C	30–40°C	45°C
pH stability	5.0–10.0 (25°C, 20 h)	5.0–9.0 (25°C, 20 h)	4.0–6.0 (40°C, 1 h)	5.0–10.0 (25°C, 20 h)

Table 7   DSC Analysis of Select Diagnostic Enzymes

Enzyme	Source	$T_g(°C)$	$T_m(°C)$	$T_d(°C)$
Cholesterol oxidase	*Nocardia*	50	98	210
Cholesterol oxidase	*Streptomyces*	51	102	250
Cholesterol esterase	*Pseudomonas*	43	88	162
Glucose oxidase	*Aspergillus*	50	105	220
Peroxidase	Horseradish	50	100	225

enzymes melt at about T_m, all the tertiary structures are destroyed, thus making the enzyme completely inactive. The presence of chemicals can considerably influence enzyme stability.

The redox centers ($FAD/FADH_2$) in the GOD enzyme can conduct electrons and are catalytically relevant. To keep or sustain the enzyme's activity, the redox centers must remain intact. The bulk of the enzyme, which is polymeric in composition, is an insulator; altering it, will not reduce the enzyme's catalytic activity. Recently, an enzymatic compound containing GOD, POD, TMB, a linear alkylbenzene sulfonate, and polyhydroxy ethyl methacrylate was molded between 105 and 150°C. This gave a good response to glucose (58). Molding at 200°C resulted in enzyme deactivation. A mechanism has been proposed whereby the enzymes are protected by the tight PHEMA coils. These researchers further suggest that molding of strips using RIM may lead to useful chemistries, including biosensors, in the future.

B.   Biosensors

Biosensors and DNA probes are a few new trends in diagnostic reagents. Work is currently underway on the near-infrared method (NIR), which will be a reagentless system as well as noninvasive. Prospects for the reagentless NIR method for glucose and other analytes is uncertain at this time.

In the early 1960s, Clark investigated a promising approach to glucose monitoring in the form of an enzyme electrode that used oxidation of glucose by the enzyme GOD (56). This approach has been incorporated and used in a few clinical analyzers for blood glucose determination. There are about three detection approaches in the glucose enzyme electrode (Fig. 7). Oxygen consumption is measured in the first method. A reference, nonenzymatic electrode, however, is required to provide an amperometric signal. The second approach detects H_2O_2 , but requires an applied potential of about 650 mV and an inside permselective membrane. In the third-generation biosensor, advantage is taken of the fact that the enzymatic reaction happen in two steps. The GOD enzyme is reduced by glucose, and then the reduced enzyme is oxidized by an electron acceptor (i.e., a mediator), most specifically, a redox polymer. Direct electron transfer between GOD and

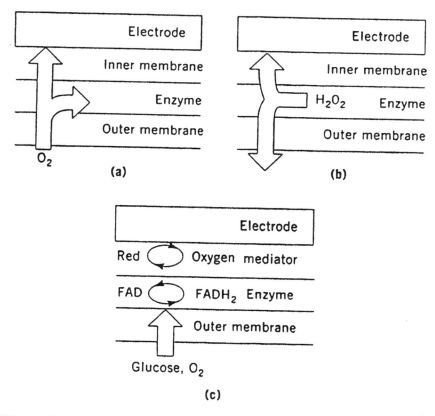

Figure 7 Detection methods for glucose enzyme electrode based on (a) oxygen, (b) peroxide, and (c) a mediator.

the electrode occurs extremely slowly; therefore, an electron acceptor mediator is required to make the process rapid and effective (57).

During the past several years, much work has been done on exploration and development of redox polymers that can rapidly and efficiently shuttle electrons. Several research groups have "wired" the enzyme to the electrode with a long-chain polymer having a dense array of electron relays. The polymer penetrates and binds the enzyme, and is also bound to the electrode. Heller and his group have done extensive work on the Os-containing polymers. They have made a large number of such polymers and evaluated their electrochemical characteristics (58). Their most stable and reproducible redox polymer is a poly(4-vinyl pyridine) to which Os(bpy)$_2$Cl$_2$ has been attached to one-sixth of the pendant pyridine groups. The resultant redox polymer is water-insoluble. To make it water-soluble and biologically compatible, Heller et al. have partially quaternized the remaining pyridine pendants with 2-bromoethylamine. This polymer is now water-soluble, and the newly introduced amine groups can react with a water-soluble epoxy (e.g., polyethylene glycol diglycidyl ether) and GOD to produce a cross-linked biosensor-coating film. Such coating films produce high-current densities and a linear response to glucose up to 600 mg/dL. The synthesis and application of osmium polymers have been refined by Usmani and co-workers (59). Researchers have also successfully made osmium monomers that should also shuttle electrons, much like the polymer version (63,64). A schematic depiction of osmium polymer and polymer–GOD hydrogel films useful in biosensors is shown in Fig. 8 (60).

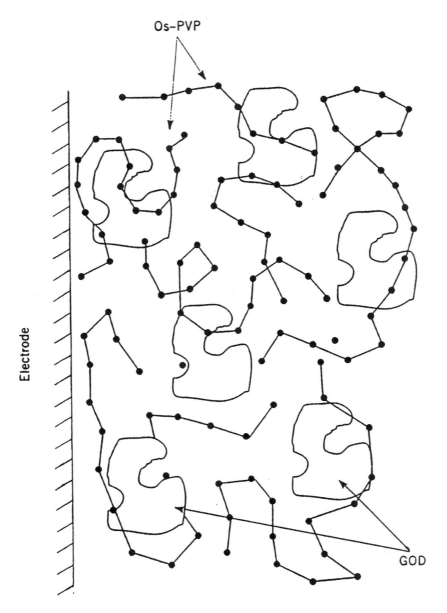

Figure 8   Structure of Os-conducting redox polymer and depiction of polymer–GOD hydrogel film on a biosensor.

Okamoto and his group used a flexible polymer chain, in contrast, to secure relays (61,62). Their polymers provide communication between GOD's redox centers and the electrode. No mediation occurred when ferrocene was attached to a nonsilicone backbone. Their ferrocene-modified siloxane polymers are stable and nondiffusing. Therefore, biosensors based on these redox polymers gave a good response and are superior in stability. Recently, commercial electrochemical microbiosensors (e.g., Exactech [Medisense] and silicone-based 6^+ system [i-Stat]) have appeared in the marketplace. These new technologies will certainly affect rapid blood chemistry determinations by the turn of this century. A typically important example is that of blood glucose determination on very small blood volumes (<5 μL) obtained by a fingerstick. This is possible because detection instruments can be designed more compactly than opto-electronic systems. Certainly, there is a market for small, disposable electrochemical tests in the emergency room, surgical and critical care units, as well as homes.

In a biosensor we seek to compete effectively with GOD's cofactor— oxygen—for the transfer of redox equivalents from the active site to the electrode surface. A good choice for the artificial electron relay depends on a molecule's ability to reach the reduced $FADH_2$ active site, undergo fast electron transfer, and then transport this to the electrodes as rapidly as possible. Surridge and co-workers have studied electron-transport rates in an enzyme electrode for glucose, using interdigitated array electrodes (63). In addition, Surridge has proposed the following mechanism in osmium polymer–GOD biosensor films.

$$GOD(FAD) + G \xrightarrow{k_1} GOD(FAD).G \xrightarrow{k_2} GOD(FADH_2) + GL$$

$$GOD(FADH_2) + 2Os\,(III) \xrightarrow{k_3} GOD(FAD) + 2Os\,(II) + 2H^+$$

$$Os(II)_1 + Os(III)_2 \xrightarrow{k_e} Os(III)_1 + Os(II)_2$$

$$Os(II)_2 \xrightarrow{fast} Os(III)_2 + e^-$$

where:

Os(II)	= reduced form
OS(III)	= oxidized form
G	= glucose
GL	= glucolactone

Schuhmann has recently suggested that the next generation of amperometric enzyme electrodes will be based on immobilization techniques that are compatible with microelectronic mass-production processes and will be easy to be miniaturized (64). Integration of enzymes and mediators simultaneously should improve the electron-transfer pathway from the active site of the enzyme to the electrode. In this work, Schuhmann deposited functionalized conducting monomers on electrode surfaces aiming for covalent attachment or entrapment of sensor components. Conductive polymers (e.g., polypyrrole, polyanaline, and polythiophene) are formed at the anode by electrochemical polymerization. For integration of bioselective compounds or redox polymers into conductive polymers, functionalization of conductive polymer films is essential. In Fig. 9, schematic representation of an amperometric biosensor with the enzyme covalently bound to a functionalized conductive

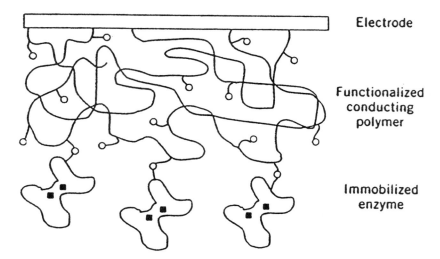

Figure 9 Schematic representation of enzyme covalently bound to a functionalized conductive polymer.

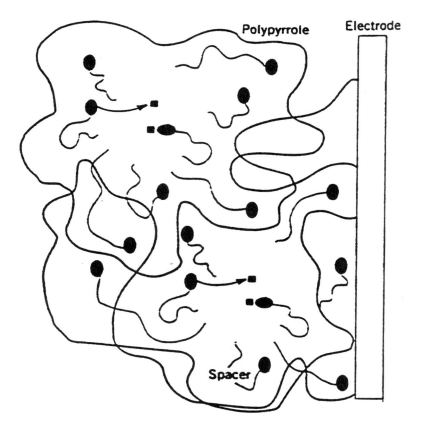

Figure 10 Entrapment of mediator-modified enzymes with a conductive polymer film.

polymer (e.g., β-amino(polypyrrole), poly[N-(4-aminophenyl)-2,2'-dithienyl]pyrrole is shown. Entrapment of ferrocene-modified GOD within the polypyrrole is shown in Fig. 10. Finally, schematic representation of an amperometric biosensor, with a pyrrole-modified enzyme copolymerized with pyrrole is presented in Fig. 11.

There is a pressing need for an implantable glucose sensor for optimal control of blood glucose concentration in diabetics. A biosensor providing continuous readings of blood glucose will be most useful at the onset of hyper- or hypoglycemia, enabling a patient to take corrective measures. Furthermore, incorporating such a biosensor into a closed-loop system with a microprocessor and an insulin infusion pump could provide automatic regulation of the patient's blood glucose. Morff et al. used two novel technologies in the fabrication of a miniature electroenzyme glucose sensor for implantation in the subcutaneous tissues of humans with diabetes (65). They developed an electrodeposition technique to electrically attract GOD and albumin onto the surface of the working electrode. The resultant enzyme–albumin layer was cross-linked by butraldehyde. They also developed a biocompatible-polyethylene glycol–polyurethane copolymer to serve as the outer membrane of the sensor to provide differential permeability of oxygen relative to glucose and thus avoid the oxygen deficit encountered in physiological tissues.

V. CONCLUDING REMARKS

Polymers have found wide acceptance in medicine, and this trend will continue to grow. Biocompatibility and polymer stability in the body's environment are not yet fully understood. A fusion technology and medical science could provide an understanding that may lead to development of new biomedical polymers.

A new emerging area is the synthesis of protein using gene-splicing technology. Such polymers may find application in artificial skin and wound dressing by

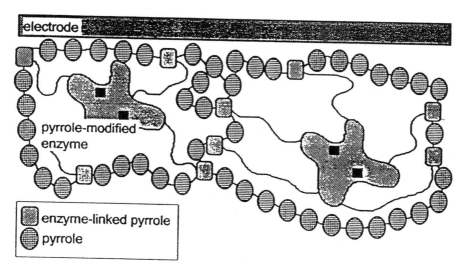

Figure 11 Schematic representation of an amperometric biosensor with a pyrrole-modified enzyme copolymerized with pyrrole.

attracting cells that promote the healing process. Thus, engineered protein polymers in extremely pure form could become important. New biomedical polymers made by chemical methods will also emerge into the marketplace. Needless to say, polymers made in biological systems are much more expensive than those made by chemical methods. Dry reagents and biosensors applications will take a quantum jump during the next few years.

REFERENCES

1. CG Gebelein. Prosthetic and biomedical devices. In: Kirk-Othmer Encyclopedia of Chemical Technology Vol 19, 3rd ed. New York: Wiley, 1982.
2. SL Cooper, NA Pappas. eds. Biomaterials: Interfacial Phenomena and Applications, ACS Symp. 199. Washington DC: American Chemical Society, 1982.
3. CG Gebelein, FF Kobiltz, eds. Biomedical and Dental Applications of Polymers. New York: Plenum Press, 1981.
4. EP Goldberg, A Nakajima eds. Biomedical Polymers: Polymeric Materials and Pharmaceuticals for Biomedical Use. New York: Academic Press, 1980.
5. J Andrade, ed. Hydrogels for Medical and Related Applications. ACS Symp. 31. Washington, DC: American Chemical Society, 1976.
6. HP George, ed. Biomedical Applications of Polymers. New York: Plenum Press, 1975.
7. RL Kronenthal, Z. Oser, E Martin. Polymers in Medicine and Surgery. New York: Plenum Press, 1975.
8. WL Hawkins, ed. Polymer Stabilization. New York: Wiley, 1972.
9. B Randy, JF Rabek. Photodegradation, Photooxidation and Photostabilization of Polymers: Principles and Applications. New York: Wiley, 1975.
10. Ultraviolet Light Induced Reactions in Polymers, ACS Symp. 57. Washington, DC: American Chemical Society, 1976.
11. N Grassie, ed. Developments in Polymer Degradation, Vol. 1. London: Applied Science Publishers, 1977; Vol 2, 1979; Vol. 3, 1981; Vol. 4, 1982; Vol. 5, 1984.
12. DL Allara, WL Hawkins, eds. Stabilization and Degradation of Polymers, Advanced Chemistry Series 169. Washington DC: American Chemical Society, 1978.
13. Durability of Macromolecular Materials, ACS Symp. 95. Washington, DC: American Chemical Society, 1979.
14. JF McKellar, NS Allen. Photochemistry of Man-Made Polymers. London: Applied Science Publishers, 1979.
15. HHH Jellinek, ed. Degradation and Stabilization of Polymers. Amsterdam: Elsevier, 1983.
16. C Decker. J Appl Polym Sci 20:3321, 1976.
17. H Wilkis, G Dich, E Gaubo HJ Leugering, R Rosinger. Colloid Polym Sci 259:818, 1981.
18. SV Nablo, JE Hipple. Soc Plast Eng Brookfield Center, CT, 1972.
19. M Dole, ed. The Radiation Chemistry of Macromolecules. New York: Academic Press, 1973.
20. C Decker, FR Mayo. J Polym Sci Polym Ed 11:2847, 1973.
21. TS Dunn, JL Williams. J Ind Irrad Technol 1:133, 1983.
22. W Szymanski, G Lerke, B Rymian. Agnew Makromol Chem 99:75, 1981.
23. KE Williams, EM Kidston, F Beck, JB Lloyd. J Cell Biol 64:113, 1975.
24. AF Hegyeli. J Biomed Mater Res 7:205, 1973.
25. LL Hench, EC Ethridge. Biomaterials, An Interfacial Approach. New York: Academic Press, 1982.
26. J Black. Biological Performance of Materials: Fundamentals and Biocompatibility. New York: Marcel Dekker, 1981.

27. RI Leininger. Plastics in Surgical Implants. Philadelphia: ASTM, 1965.
28. RW Postlethwait. Ann Surg 171:892, 1970.
29. BD Halpern. Ann NY Acad Sci 146:193, 1968.
30. RI Leininger, V Mirkovitch, RE Beck, PG Andrus, WJ Kolff. Trans Am Soc Artif Intern Organs 10:237, 1964.
31. MB Habel. Biomater Med Devices Artif Organs 7:229, 1979.
32. B Walter, Anal Chem 55:449A, 1983.
33. AH Free, HM Free. Lab Med 15:1595, 1984.
34. TK Mayer, NK Kubasik. Lab Manage 43, 1986.
35. JA Jackson, ME Conard. Am Clin Prod Rev 6:10, 1987.
36. AF Azhar, AD Burke, JE DuBois, AM Usmani. Polym Mater Sci Eng 59:1539, 1989.
37. E Diebold, M Rapkin, AM Usmani. Chem Technol 21:462, 1991.
38. MT Skarstedt, AM Usmani. Polym News 14:38, 1989.
39. RS Campbell, CP Price. J Intern Fed Clin Chem 3:204, 1991.
40. EP Joslin, HP Root, P White, A Marble. The Treatment of Diabetes, 7th ed. Philadelphia, 1940.
41. WA Compton, JM Treneer. U.S. Patent 2,387,244, 1945.
42. AH Free, CE Adams, ML Kercher. Clin Chem 3:163, 1957.
43. HG Rey, P Rieckmann, H Wielinger, W Rittersdorf. U.S. Patent 3,630,975, 1971, assigned to Boehringer Mannheim.
44. TL Shirey. Clin Biochem 16:147, 1983.
45. EP Przybylowicz, AG Millikan. U.S. Patent 3,992,157, 1976, assigned to Eastman Kodak.
46. HE Spiegel. Kirk-Othmer Encyclopedia of Chemical Technology, 3rd ed., New York: Wiley, 1985.
47. AM Usmani. Diagnostic Biosensor Polymers. ACS Symp Ser 5562, 1994.
48. NW Tietz. Fundamental of Clinical Chemistry. Philadelphia: WB Saunders, 1976.
49. SC Stinson. C&EN, Sept 30:635 1991.
50. AM Usmani. Biotechnology Symposium, 1992.
51. BJ Bruschi. U.S. Patent 4,006,403, 1978, assigned to Eastman Kodak.
52. W Scheler. Makromol Chem Symp 12:1, 1987.
53. AF Azhar, AM Usmani, AD Burke. J DuBois-Bousamra, ER Diebold, MC Rapkin, MT Skarstedt. U.S. Patent 5,260,195, 1993, assigned to Boehringer Mannheim.
54. JE Kennamer, AM Usmani. J Appl Polym Sci 42:3073, 1991.
55. JE Kennamer, AD Burke, AM Usmani. Biotechnology and Bioactive Polymers. New York: Plenum, 1992.
56. LC Clark, C Lyons. Ann NY Acad Sci 102:29, 1962.
57. G Reach, GS Wilson. Anal Chem 64:381A, 1992.
58. BA Gregg, A Heller. J Phys Chem 95:5970, 1991.
59. AM Usmani, D Deng. Unpublished work, Boehringer Mannheim, 1992.
60. AM Usmani, NA Surridge, ER Diebold. Unpublished work, Boehringer Mannheim, 1992.
61. L Boguslavsky, P Hale, T Skotheim, HI Karan, H Lee, Y Okamoto. Polym Mater Sci Eng 64:322, 1991.
62. HI Karan, HL Lan, Y Okamoto. Diagnostic Biosensor Polymers. ACS Symp Ser 556:169, 1994.
63. NA Surridge, ER Diebold, J Chang, GW Neudeck. Diagnostic Biosensor Polymers. ACS Symp Ser 556:47, 1994.
64. W Schuhmann. Diagnostic Biosensor Polymers. ACS Symp Ser 556:110, 1994.
65. KW Johnson, DJ Allen, JJ Mastrototaro, RJ Morff, RS Nevin. Diagnostic Biosensor Polymers. ACS Symp Ser 556: 84, 1994.

14

Drug Release from Polymers: Quantitative Approach

ALEXANDER YAKOVLEVICH POLISHCHUK and GUENNADI EFREMOVICH ZAIKOV

N.M. Emanuel Institute of Biochemical Physics, Russian Academy of Sciences, Moscow, Russia

I. INTRODUCTION

The quantitative approach to regulation of multicomponent transport in polymers and polymer blends is developed, with the scope of application to polymer systems of controlled-release drugs. The approach is based on the theory of multicomponent transport in polymers of different hydrophilic character and integrates the modeling of chemical, physical, morphological, and transport properties of polymeric materials. To govern the kinetics of the release of low molecular weight species from polymers, the methods of controlled modification of matrix are proposed, which comprise the following:

- Alteration of hydrophilic–hydrophobic balance in a polymer by controlled polymer degradation
- Controlled load of fillers and additives
- Application of multilayer systems, and others

The fundamental aspects of multicomponent transport in polymers are discussed with reference to eventual application of the theory into practice.

The understanding of general regularities of multicomponent transport in polymers gives an opportunity to alter its kinetics and the rate. This chapter deals with the fundamental approaches to such an alteration in the systems of controlled drug delivery. In recent years new original methods have been developed using the following approaches:

1. The synthesis of novel polymeric materials with regulated hydrophilic–hydrophobic balance
2. The alteration of this balance by the controlled modification of polymers or by use of fillers and additives
3. The use of multilayer systems

The methods used to regulate the behavior of the controlled-delivery devices are not limited by the foregoing approaches. There are numerous other methods serving many specific purposes. We should mention some of them such as magnetism-activated drug delivery, ultrasound-activated drug delivery, pH-activated drug delivery, and ion-activated drug delivery, which are not considered in this chapter. The reader can find the details of their principles in the brilliant monograph edited by Lee and Good (1) and other monographs published during last decade (2–7). Also we do not discuss here the technological aspect of the design of devices, although this problem is very important for engineers.

Our aim is not to review all possible methods to regulate the controlled release, but to show how the theory of multicomponent transport can be applied to the development of these methods. In the meantime, we could not pass up the problem of biocompatibility of devices implanted in the body. In the final paragraph of the chapter we show how this very important problem can also be considered in terms of transport phenomena in polymers.

II. MODELS OF SIMULTANEOUS TRANSPORT OF SOLVENT AND SOLUTES IN POLYMERS

A. Hydrophilic Polymers

The general principles of simultaneous transport of water and low molecular weight solutes in hydrophilic polymers include consideration of the processes of water diffusion in and solute diffusion from the polymer matrix, the increase of volume of the test piece (swelling), the changes in molecular mobility (plasticization and relaxation), and the mechanical properties of the polymer.

The simplest mathematical description of these processes was proposed by Peppas and coauthors (8). In assumption of constant diffusion coefficients of both components (D_w, D_s) they wrote

$$\frac{\delta C_w}{\delta t} = \frac{D_w \delta^2 C_w}{\delta x^2} \qquad 0 < x < 1 \tag{1}$$

$$\frac{\delta C_s}{\delta t} = \frac{D_s \delta^2 C_s}{\delta x^2} \qquad 0 < x < 1 \tag{2}$$

where C_w, C_s are concentrations of water and solute in polymer, respectively.

During the time of swelling t the thickness of the sample increases from the initial value l_0 to the current value 1, and the front of the solvent penetrates to the depth $X(t)$ (Fig. 1). This defines specific initial and boundary conditions,

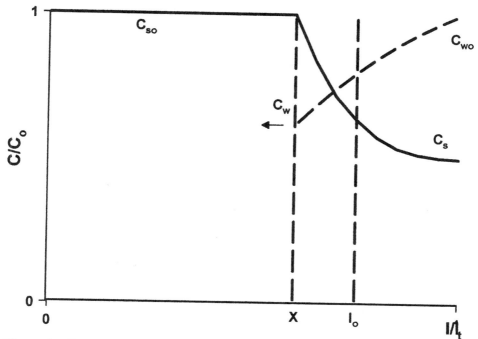

Figure 1   Concentration profile in a polymeric system for controlled release.

which are

$$C_s(X, t) = C_{s1} \qquad C_s(l, t) = 0; \qquad C_w(l, t) = C_{wo}$$

$$\frac{\delta C_s(0, t)}{\delta x} = 0; \qquad \frac{\delta C_w(X, t)}{\delta x = 0}$$

$$C_s(x_20) = C_{so}; \qquad C_w(x, 0) = 0$$

where C_{wo} is the water concentration in polymer in equilibrium and C_{so} is the initial concentration of solute in polymer. The physical meaning of C_{s1} is less clear, although its relation to the solubility of solute in polymer and in water is obvious.

The foregoing model can reproduce some of the known kinetic curves of the solute desorption. One such case occurs when the kinetics of release follows water uptake $(D_w \ll D_s)$ (6).

The so-called phenomenological approaches to describing transport phenomena in polymers are based on the search of the best mathematical fitting of experimental kinetic curves, rather than on real physicochemical processes occurred in the material. These approaches give an opportunity to obtain analytical solutions of diffusion equations (6,9) and to calculate parameters, thereby allowing one to compare the kinetics and the rate of the transport of species in the polymer. However, the aforesaid parameters generally have uncertain physical meaning. This is the main disadvantage of phenomenological approaches, as a whole. For example, we applied one of the phenomenological approaches in an attempt to describe the simultaneous transport of water and dioxidine in copolymers of vinylpyrrolidone with butyl methacrylate (VP/BM) and methyl methacrylate (VP/MM) (10). Therein we assumed that water sorption follows the equation

$$\frac{\delta C_w}{\delta t} = \left(\frac{\delta}{\delta x}\right)\left(\frac{D_w \delta C_w}{\delta x}\right) - \frac{v \delta C_w}{\delta x}$$

where v is the velocity of local volume transfer, controlled by swelling stresses in the matrix.

Although we obtained reasonable agreement between theoretical and experimental results, and could make qualitative conclusions about the regularities of the behavior of the aforesaid systems, the uncertain meaning of parameters did not permit us to extend this model quantitatively to other systems. Therefore, the analysis of real processes in a swelling matrix is required for an explanation of the behaviors of a system: hydrophilic polymer–water–solute.

The Petropoulos model of "differential swelling stress" (11,12) is one of those that generalize the description of the main processes occurring in swelling polymers. This model has been extended (13,14) for a case of simultaneous transport of solvent and solutes. The model describes the transport of solvent by the diffusion equation of a standard form.

$$\frac{\delta C_w}{\delta t} = \left(\frac{\delta}{\delta x}\right)\left(\frac{D_w \delta C_w}{\delta x}\right) \qquad 0 < x < 2l \tag{3}$$

where

$$D_w = D_w \exp(k_{w1} C_w + k_{w2} f) \tag{4}$$

coupled with an equation describing the buildup and relaxation of the local differential swelling stresses f along the plane of the film

$$\frac{\delta f}{\delta t} = \frac{(G_o - G_{oo})\delta s}{\delta t} + \frac{\delta(sG_{oo})}{\delta t} + \frac{(G_o - G_{oo})^{-1}\delta(G_o - G_{oo})}{\delta t} - \beta(f - sG_{oo}) \tag{5}$$

The change in the film area is given by

$$A(C_w = A(0)(1 + k_s C_w) \tag{6}$$

Because the actual area of a thin film is constrained to a uniform value $\mathbf{A}(t)$ the creation for local strains is described by the expression

$$s = \frac{\mathbf{A}(t)}{A(x, t)} - l \tag{7}$$

and corresponding local stresses f add up to zero net overall stress.

The plasticizing action of solvent and the opposite action of the solute could be described by exponential dependence of the instantaneous and long-term elastic modules on solvent and solute concentrations

$$\begin{aligned}
G_o &= G_{o0} \exp(-k_{g1} C_w + k_{g2} C_s) \\
G_{oo} &= G_{oo0} \exp(-k_{g1} C_w + k_{g2} C_s)
\end{aligned} \tag{8}$$

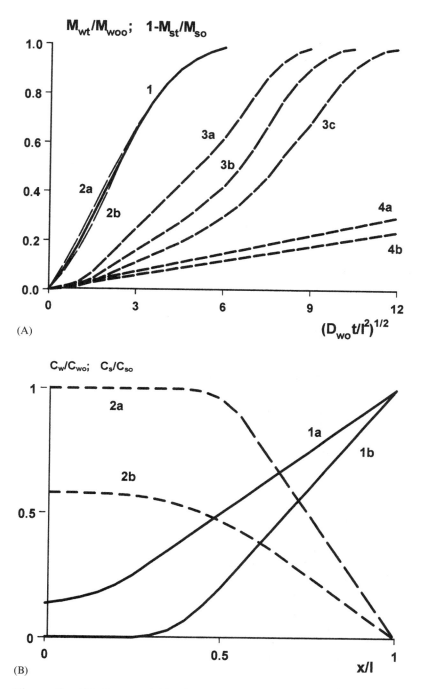

Figure 2 (A) Computed kinetic curves of water uptake curve (1) and solute desorption curves (2–4): (1) water uptake; (2) $D_s(C_{wo})/D_{wo} = 10$; $D_{so}/D_{wo} = 1000$ (a), 100 (b); (3) $D_s(C_{wo})/D_{wo} = 1$; $D_{so}/D_{wo} = 1000$ (a), 100 (b), 10 (c); (4) $D_s(C_{wo})/D_{wo} = 0.1$; $D_{so}/D_{wo} = 100$ (a), 10 (b). (B) Distribution of solvent (1) and solute (2) in polymer at (a) $(D_{wo}t/l^2)^{1/2} = 8$ and (b) 12 in Fig. 2A.

kinetics (curves 3a–c), the rate of which increases with the increase of k_{w3}. The kinetic regularities remain unchanged. Desorption of low molecular weight solute with small values of $D_s(C_{wo})/D_{wo}$ (of an order of 0.1) becomes Fickian because the water uptake is completed at an early stage (curves 4a–b). The influence of k_{w3} on the rate of desorption is minor for this example. The profiles of C_w (curves 1a and b) are diffuse and seem quite linear, whereas profiles of C_s (curves 2a and b) are quite steep (see Fig. 2B).

For high values of $D_s(C_{wo})/D_{wo}$ (on order of 10) the solubility of solute affects neither the kinetics nor the rate of its desorption. Such an effect becomes visible for moderate values of $D_s(C_{wo})/D_{wo}$ (on order of 1), although it relates only to the rate of desorption (Fig. 3). which is higher for more soluble compounds (curves ac). The kinetic regularities remain unchanged. Such a regularity of solute desorption in swelling polymers is in contrast with the solute desorption from a nonswelling matrix, which is considered later. The profiles of C_w are not affected by solubility and remain linear, whereas the steep character of C_s profiles develops as solubility of solute reduces (see Fig. 3B).

The second group of variations involved parameters concerned with the dependence of diffusion coefficient of water (Fig. 4). The acceleration of water sorption, coupled with minor changes in the shape of the kinetic curve is the main result of the strong dependence of D_w on C_w (curve 2 in comparison with curve 1). An increase of k_{w2} decelerates water uptake owing to the strong dependence of D_w on f (curve 3 against curve 1), and the kinetic curve becomes markedly more S-shaped.

Because C_w and f affect D_w in opposite directions their effects tend to cancel at low M_{wt} (overall water uptake) values (curves 1 and 4). However, there is substantial difference between the weak (curve 1) and strong (curve 4) dependencies of D_w on C_w and f at high M_{wt} values. Acceleration of water sorption is the consequence of the increase of β_0 (curve 5) and the shape of the curve becomes Fickian (at the same time the profile of water in the film remains sharp because of strong dependence of D_w on C_w). Both, the kinetics and the rate of solute release are governed by the corresponding features of water uptake (whatever the underlying causes may be) at high $D_s(C_{wo})/D_{wo}$ values (curves labeled a). In contrast to this, for moderate $D_s(C_{wo})/D_{wo}$ values (curves labeled b), the rate of the desorption of solute is affected by changes in the rate of water uptake, but the kinetics of release remains the same. Variations of mechanical properties have been considered in the form of a strong dependence of elastic modules on the concentrations of solvent and solute (Fig. 5). A five-orders decrease of elastic modules, which was used to represent strong dependence on concentration of water, causes almost case II diffusion of water until $M_{wt}/M_{woo} = 0.6$, whereas nonlinear uptake was observed at higher values of M_{wt}/M_{woo} (curve 1). The strong dependence of elastic modules on solute concentration is characterized by Fickian water uptake (curve 2). The solubility of solutes affects only the rate of water sorption whatever its kinetics may be (curves a–c). For obvious reasons this influence is much more considerable for a strong dependence of elastic modules on solute concentration (curves 2a–c).

The kinetics of water sorption that correspond to a moderate value of β_0 and a weak dependence of β on concentrations of water and solute look Fickian (curve 1 of Fig. 6). Because of solute desorption, deceleration of water uptake was observed for the strong dependence of stress relaxation on concentration of solute, in comparison

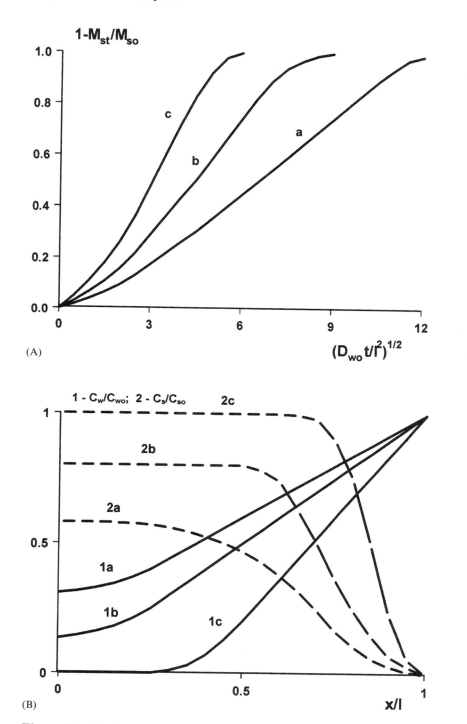

Figure 3 (A) Computed kinetic curves of the release of solutes of different solubility: $D_s(C_{wo})/D_{wo} = 1$; $K_s = 0.1$ (a), 0.3 (b), 1 (c). (B) Distribution of solvent (1) and solute (2) in polymer at $(D_{wo}t/l^2)^{1/2} = 8$ in Fig. 3A. The labels are the same.

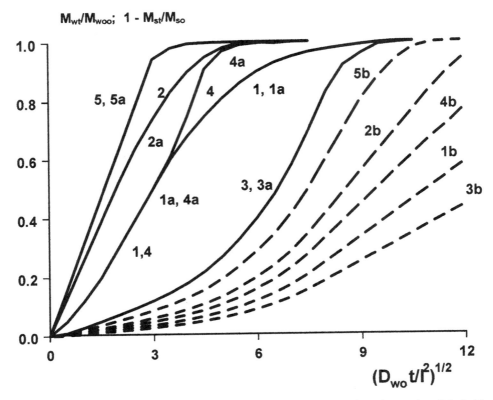

Figure 4 Computed kinetic curves of water uptake (1–4) and solute desorption (labeled by letters): $D_s(C_{wo})/D_{wo} = 10$ (a), 1 (b);
(1) $D_w(C_{wo}, 0)/D_{wo} = e$; $D_w(0, f_{max})/D_{wo} = \exp(f_{max})$; $\beta_o = 1$;
(2) $D_w(C_{wo}, 0)/D_{wo} = 20{,}000$; $D_w(0, f_{max})/D_{wo} = \exp(f_{max})$; $\beta_o = 1$;
(3) $D_w(C_{wo}, 0)/D_{wo} = 20{,}000$; $D_w(0, f_{max})/D_{wo} = \exp(10f_{max})$; $\beta_o = 1$;
(4) $D_w(C_{wo}, 0)/D_{wo} = e$; $D_w(0, f_{max})/D_{wo} = \exp(10f_{max})$; $\beta_o = 1$;
(5) $D_w(C_{wo}, 0)/D_{wo} = e$; $D_w(0, f_{max})/D_{wo} = \exp(10f_{max})$; $\beta_o = 100$.

with the same initial value of β corresponding to unfilled dry polymer (curves 4 and 2). Strong dependence of β on concentration of solvent causes further deceleration of water uptake at lower values of M_{wt} (curve 3). At higher values of M_{wt}, deviation of the kinetic curve of water sorption above the Fickian one was observed, and this leads to higher velocity of water uptake in comparison with when there is strong dependence on the content of solute.

With a strong dependence of relaxation frequency on the concentration of solute, the effect of solubility on the kinetics of water sorption is most essential. Figure 7 illustrates how solubility of the second compound alters either the kinetics or the rate of water sorption. The calculations have been made for values K_s of 0.1 (a), 0.3 (b) and 1 (c).

The similar approach, which takes proper account of the simultaneous uptake of a liquid leachant by a polymer matrix and the consequent release of a bioactive or other solute incorporated therein, was applied by Petropoulos and associates

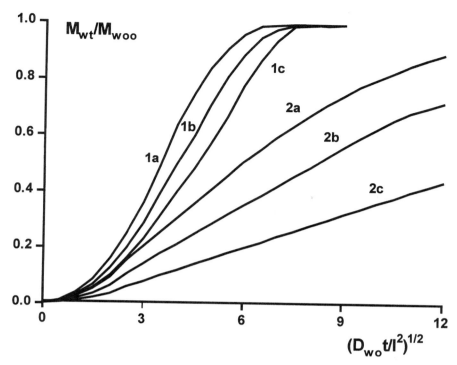

Figure 5 Computed kinetic curves of water uptake (1–2): (1) $G_i(0, C_{so})/G(C_{wo}, 0) = 100,000$; (2) $G_i(0, C_{so})/G(C_{wo}, 0) = 20$; $K_s = 0.1$ (a); 0.3 (b); and 1 (c).

(18,19) to predict the kinetics of release of several leachants under various conditions. They showed that the foregoing model can be easily modified or extended, in accordance with the information available, about any particular system. It is, therefore, expected that the generalized model of differential swelling stresses is proved useful as a basis for the design of monolithic controlled-release devices of this type or for the evaluation of the leachability of low-level and medium-level wastes "immobilized" in polymeric matrices.

B. Moderately Hydrophilic and Hydrophobic Polymers

Even for hydrophobic polymers, the diffusion of solutes is, in general, affected by thermodynamics and kinetics of water uptake.

Guo (21) described the effects of water on the properties of ethylcellulose films. The change of density, morphology, porosity, and drug transport properties of ethylcellulose films with the water content in the polymer solutions were investigated. Because of the strong hydrogen-bonding property of water, the process of dissolving and mixing of ethylcellulose in a solvent (ethanol) was dramatically changed when a nonsolvent (water) was added in this system. As the solubility parameter difference between water and the rest of the components and the evaporating rate difference between ethanol and water could cause the phase separation of polymer solution during the film-forming process, the porosity of ethylcellulose films

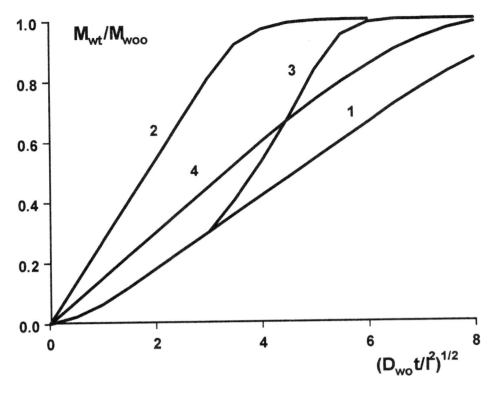

Figure 6   Computed kinetic curves of water uptake (1–4): $D_s(C_{wo})/D_{wo}=1$;
(1) $\beta_0=1$; $\beta(0,\ 0)/\beta(C_{wo},\ 0)=\exp(-1)$; $\beta(0,\ C_{so})/\beta(C_{wo},\ 0)=e$;
(2) $\beta_o=100$; $\beta(0,\ 0)/\beta(C_{wo},\ 0)=\exp(-1)$; $\beta(0,\ C_{so})/\beta(C_{wo},\ 0)=e$;
(3) $\beta_o=1$; $\beta(0,\ 0)/\beta(C_{wo},\ 0)=0.01$; $\beta(0,\ C_{so})/\beta(C_{wo},\ 0)=e$;
(4) $\beta_o=1$; $\beta(0,\ 0)/\beta(C_{wo},\ 0)=\exp(-1)$; $\beta(0,\ C_{so})/\beta(C_{wo},\ 0)=100$.

increased with water content in the polymer solutions. The film density and the drug
permeability of ethylcellulose films decreased and increased with water content in the
polymer solutions, respectively; and these results were very consistent with the
porosity increase of the ethylcellulose films (21).

One of the main aims of practical application of a theory of transport in
polymers is to alter the hydrophilic–hydrophobic balance in the polymer.
Accordingly, the principle difference between polymers should not be searched
for in the equilibrium concentration of water, but in the effect of the limited water
sorption on the kinetics and rate of release. This effect is visible, to a certain extent,
for all polymers that sorb less than 10% of water in equilibrium, and even for a
dozen of highly swelling copolymers and polymer blends with hydrophobic
fragments, domains, and such. The load of hydrophilic fillers and additives also does
not cancel the influence of the polymer's nature. For the aforesaid and many other
reasons we combine the foregoing polymers in one group to consider in this
paragraph. Our consideration is based on the comprehensive analysis of these systems made by Petropoulos (18,19).

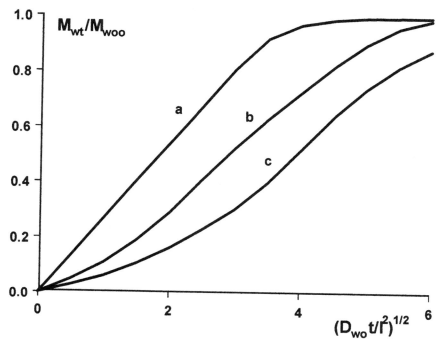

Figure 7 Computed kinetic curves of water uptake in presence of solute of different solubility: $D_s(C_{wo})/D_{wo} = 1$; $\beta_o = 1$; $\beta(0, 0)/\beta(C_{wo}, 0) = \exp(-1)$; $\beta(0, C_{so})/\beta(C_{wo}, 0) = 100$; $K_s = 0.1$ (a); 0.3 (b); 1 (c).

The simplified theoretical model for elution of a solute from an initially dry hydrophilic polymer matrix with simultaneous inhibition of water [see Eqs. (3) and (4)] may be applied for less hydrophilic polymers (22,23). However, the applicability of the model of Korsmeyer, Peppas, and their colleagues (22,23) is severely limited by certain simplification.

First, the equilibrium water uptake of the matrix tends to increase as the amount or water solubility of the embedded solute increases (24). Second, the total concentration of solute (C_s) at any point in the matrix is, in general, composed of a mobile or "dissolved" part (concentration C_{ss}), and an immobile or "dispersed" part (concentration $C_s - C_{ss}$). C_{ss} is the increasing function of the water content of the matrix. Third, and largely as a consequences of the foregoing points, the driving force of diffusion cannot be properly formulated in terms of concentration gradients (25). Accordingly, Petropoulos proceeded to formulate a more general and rigorous treatment, in which proper account is taken of all the foregoing points. He formulated the dimensionless diffusion equation in terms of chemical potential gradient ($\delta\mu/\delta x$) for both of the diffusion species

$$\frac{\delta C_w}{\delta t} = \left(\frac{\delta}{\delta x}\right)\left(\frac{m_{Tw} C_w \delta\mu_w}{\delta x}\right)$$

$$\frac{\delta C_s}{\delta t} = \left(\frac{\delta}{\delta x}\right)\left(\frac{m_s C_s \delta\mu_s}{\delta x}\right) \tag{15}$$

where m_T is the "thermodynamic" mobility coefficient. Because $\mu = RT \ln a$, Eq. (15) becomes

$$\frac{\delta C_w}{\delta t} = \left(\frac{\delta}{\delta x}\right)\left(\frac{D_{wT}S_w \, \delta a_w}{\delta x}\right)$$
$$\frac{\delta C_s}{\delta t} = \left(\frac{\delta}{\delta x}\right)\left(\frac{D_s S_s \, \delta a_s}{\delta x}\right) \tag{16}$$

where $D_s = m_T RT$ and $S_i = C_i / a_i$ are thermodynamic diffusion and sorption coefficients. The thermodynamic diffusion coefficient is not identical with the "Fick diffusion coefficient," unless $S = $ constant.

Therefore Eqs. (3) and (12) should be replaced by Eq. (16) with boundary conditions

$$a_w(0, t) = 1; \qquad \delta a_w(l, t)/\delta x = 0; \qquad a_w(x, 0) = 0$$
$$\frac{\delta a_s(l, t)}{\delta x} = 0; \qquad a_s(0, t) = 0; \qquad a_s(x, 0) = a_{s0} \tag{17}$$

In Eq. (17) a_w is given by the relative water vapor pressure. The experimental water sorption isotherms in hydrophobic and moderately hydrophilic polymer matrices indicate that S_w usually tends to increase with a_w and also with the salt load C_s (18,19,26). Petropoulos et al. suggested representing this dependence semi-quantitatively by the relation [Eq. (18)].

$$S_w = \frac{C_w}{a_w} = (K_{w1} + K_{w2}a_w)(1 + K_{w3}C_s) \tag{18}$$

In the meantime, we found that water solubility in poly(hydroxybutyrate) containing potassium or sodium chloride is better described by exponential dependence

$$S_w = K_{w1} \exp(K_{w2}a_w + K_{w3}C_s) \tag{19}$$

Obviously both Eqs. (18) and (19) may be further modified to represent any other particular system more accurately. Hence, the local a_w value may be found at any point of the matrix, during the diffusion process from the local C_w following Eq. (18) or (19).

For the solute we obtain $S_s = C_{ss}/a_s$. The sorption isotherm relating C_{ss} to the concentration of solute in aqueous solution may be written in form of Eq. (13). For low hydrophilic polymers, one should expect that k_s is progressively less than 1 as C_w is reduced. If any experimental information on k_s behavior is available Eq. (13) allows evaluation of C_s for that particular case. Thus we can obtain

$$S_s = \frac{C_s}{a_s} = k_{ss}C_w V_w c_{ss}^o = K_s C_w \tag{20}$$

Since Eq. (20) copies Eq. (14) we can use Eq. (20) to represent the saturation value of C_{ss} at any point within the matrix.

Both diffusion coefficients D_w and D_s also must be formulated as functions of C_w in general forms [see Eqs. (4) and (11), respectively]. In comparison with hydrophilic polymers, here we may expect the weak dependence of the diffusion coefficient of water on swelling stress ($k_{w2} \sim 0$). Although the general expression for dependence of D_w on C_w [see Eq. (4)] remains the same; for hydrophobic and moderately hydrophilic polymers, the value of k_{w1} may be positive, negative, or close to zero. The concentration dependence of D_s on C_w is always assumed to follow the "free volume theory" of Yasuda et al. (16). Therefore, the effect of solute on water sorption in nonswelling polymers is defined, mostly, by the dependence of the solubility of water on the concentration of solute, which in addition, presents in these polymers either in dissolved or dispersed forms. This general difference between swelling and nonswelling polymers defines further differences in regularities of multicomponent transport.

Following the foregoing theoretical consideration, the major factors that affect the kinetics and rate of water sorption and solute release must be

1. Solubility of components in polymer
2. Presence of third components, such as fillers, additives, stabilizers, impurities, which can change (sometimes, dramatically) the hydrophobic–hydrophilic balance in the matrix.

All these factors are considered in the following, with references to corresponding prediction of the foregoing model.

1. Effect of Solubility

We were modeling the kinetics of water sorption in polymers of structural heterogeneity using Eqs. (15)–(17) and (19) with constants K_{w1}, K_{w2}, and K_{w3} corresponding to the system of poly(hydroxybutyrate) (PHOxB)–water–KCl (NaCl). The computed isotherms of water sorption are shown in Fig. 8, and the kinetics of water uptake and drug release is illustrated in Fig. 9. If we increase the initial load of salt, both processes accelerate, and starting from the certain value of C_s the kinetic curve of water sorption passes maximum ($w1=w5$) and anomalous character of release develops ($s2s5$). The equilibrium water concentration was equal to

$$C_w = a_w K_{w1} \exp(K_{w2}a_w + K_{w3}C_{sim})$$

where C_{sim} is the concentration of the immobile part of the solute. This is also in agreement with experimental results that we obtained for the system of VP–MM copolymer–water–dioxidine (15).

Petropoulos did his computations using Eq. (18) instead of (19), and obtained the same qualitative results. He mentioned the intensification of the tendency toward an S-shaped salt elution curve with the increase of its load. He refers the results of computations to the results reported by Lee in (27).

The simple equation for a description of the release of solute of limited solubility was suggested by Higuchi (28). This equation is well known as Higuchi kinetics. Using the nomenclature of the current report we can write the Higuchi dependence of the concentration of the released compound on time as

$$\frac{C_s}{C_{so}} = 2\left(\frac{D_{sF}t}{\pi l^2}\right)^{1/2}$$

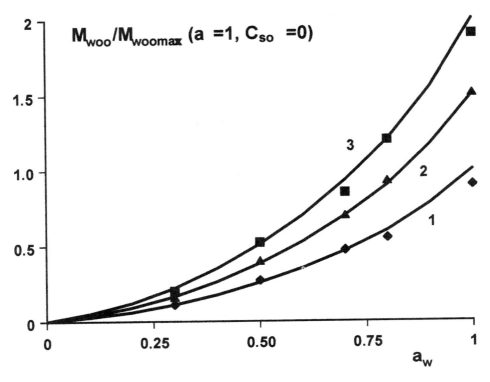

Figure 8 Computed and experimental (points) isotherms of water sorption in PHOxB containing KCl: $C_s = 0$ (1); 0.5 (2); 0.8 (3) mol/dm³.

with an apparent diffusion coefficient

$$D_{sF} = \pi D_s (1 - K_s) K_s / 2$$

[the meaning of K_s is the same as in Eq. (14)].

Therefore, one of the main problems occurs with whether or not the imbibed water (when $C_w = C_{wo}$) can dissolve the whole of the initial salt load of the polymer film. Petropoulos (18) did corresponding computations for $k_s = \exp(-k_{ss}/C_w)$. He assumed that k_s is progressively lower as C_w is reduced. For computations more specifically representative of the cellulose acetate–NaCl–water system he put $k_s = 0.2$ and varied K_s within the realistic frame of 0.13–0.013. The results illustrated in Fig. 10 show that S-shaped curve is predicted for higher K_s value (curve a, which corresponds to higher solubility or lowest solute load), but this S-shaped curve tends to disappear as K_s is lowered (curve b), and finally the desorption curve obeys the Higuchi kinetics (curve c).

This result was obtained for moderate values of $D_s(C_{wo})/D_{wo}$ (of order of 1) and is in contrast with the regularities of solute desorption from a swelling matrix, for which, at equivalent values of $D_s(C_{wo})$ and D_{wo}, the solubility affects only solute desorption. Desorption of low molecular solute with small values of $D_s(C_{wo})/D_{wo}$ (of

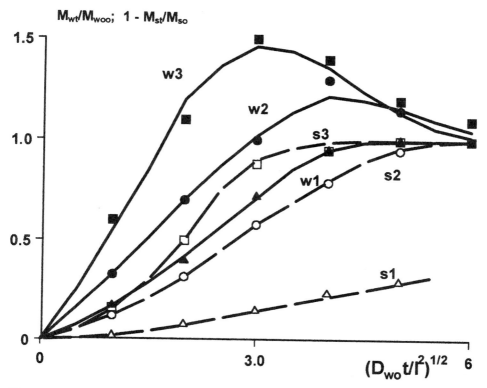

Figure 9 Computed and experimental (points) kinetic curves of water sorption and release of potassium chloride from PHOxB films: $C_s = 0$ (1); 0.5 (2); 0.8 (3) mol/dm^3.

order of 0.1) becomes Fickian either for swelling or nonswelling polymers because water uptake is completed at an early stage.

2. Effect of Fillers and Additives

The effect of the third component depends on its own solubility in water. Obviously, the corresponding equation should be added in the system [see Eqs. (3)–(17) or (3)–12)]

$$\frac{\delta C_f}{\delta t} = (\delta/\delta x)\left(\frac{D_f S_f \, \delta a_s}{\delta x}\right) \tag{21}$$

and the dependencies of solubility of water and, in general, elastic modules and frequency of relaxation should contain the third contributing factor.

The simplest cases are when the third component just makes the aforesaid dependencies stronger or weaker. Then the main regularities described in the foregoing remain unchanged. There is, however, a remarkable possibility to alter, in principle, the hydrophilic character of the polymer.

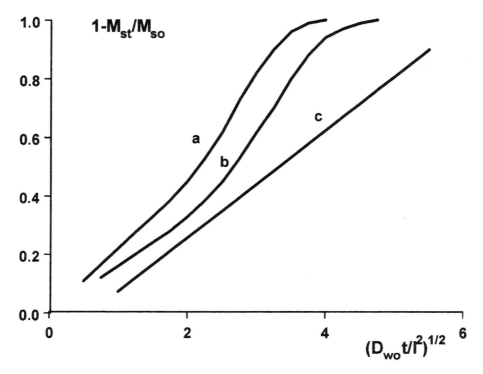

Figure 10   Computed salt elution curves: $K_s = 0.13$ (a), 0.066 (b), and 0.013 (c).

The effect of plasticizer was modeled by the addition to system Eq. (21) and the replacement of Eq. (19) by

$$S_w = \frac{C_w}{a_w} = K_{w1} \exp(K_{w2}a_w + K_{w3}C_s + K_{w4}C_f)$$

and the effect of concentration of this filler has been investigated. The results are shown in Figs. 11 and 12. Up to certain value the increase of C_f accelerates water sorption (see curves w1, and w2 in Fig. 12) and the release of solute (see curves s1 and s2 in Fig. 12), and their kinetics and shape of the water sorption isotherm (see curves 1 and 2 in Fig. 11) remain unchanged. Then, however, remarkable changes come in (curves w3w6, s3s6 in Fig. 12). This effect may be explained by the increase of the solubility of solute. Accordingly, the contribution of solute in the solubility of water becomes negligible, and this changes the shape of water isotherm (see curves 3–6 in Fig. 11). Further acceleration of solute release also occurs with the change in its kinetics.

III.   APPLICATIONS

A.   Regulation of Polymer Structure

The alteration of hydrophilic–hydrophobic balance in copolymers by variation of the ratio between monomers or substitution of monomers is the obvious way to

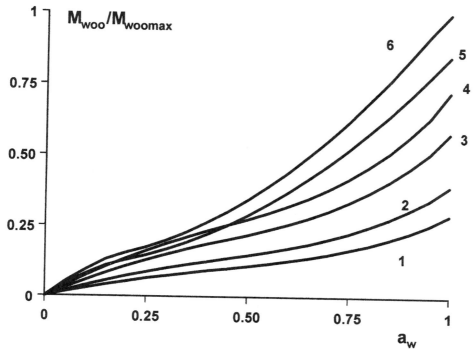

Figure 11 Computed water isotherms in polymer containing highly and lower-soluble solutes: $C_F/C_s = 0.1$ (1), 0.2 (2), 0.3 (3), 0.5 (4), 0.8 (5), and 1 (6).

regulate the kinetics of water sorption and further release of the solute. Although the obvious does not mean the simplest, the possibility of such a regulation is the reason for the wide application of copolymers, polymer composites, and blends in the devices of controlled release.

We applied the theoretical approach that has been described in previous sections for the explanation of the behavior of several copolymers. For example, films of vinylpyrrolidone–butyl methacrylate (VP–BM) and vinylpyrrolidone–methyl methacrylate (VP–MM) copolymers were investigated (15). Table 1 gives a list of their characteristics. The copolymers contained hydrophilic (N-vinylpyrrolidone) and hydrophobic groups that enable the thermodynamics and the kinetics of water sorption to be varied.

Table 1 and Fig. 13 show that equilibrium concentration of water and the rate of water sorption are the functions of the content of hydrophilic groups. The existence of the linear part of the kinetics curves of water sorption may be explained in terms of the "generalized model of differential swelling stresses," which has been reported in Section II [see Eqs. (3)–(11)]. The expected kinetics of the release of antiseptic drug, dioxidine, has been obtained for the films of all aforesaid N–VP copolymers. As pointed out in Section II for the high relative value of the diffusion coefficient of solute $[D_s(C_{wo})/(D_{wo}) \geq 10]$, a drug release should follow water uptake. This feature is observed for VP/BM copolymer and, according to the kinetics of water sorption, the kinetic curve of dioxidine release out of this co-

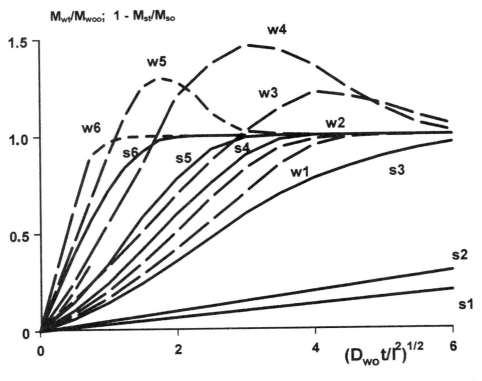

Figure 12 Computed kinetics of water sorption in and solute release from polymer filled with the third component: $C_F/C_s = 0.1$ (1), 0.2 (2), 0.3 (3), 0.5 (4), 0.8 (5), and 1 (6).

Table 1 Characteristics of Copolymers of N-Vinylpyrrolidone

	Copolymers		
Characteristics	VP–BM	VB–MM-1	VB–MM-2
Nitrogen content (%)	5.2–6.3	2.8–3.6	4.4
Relative viscosity	1.15–1.29	1.18–1.29	1.18–1.24
Relative strain/(%)	250–300	80–100	4–6
Content of N–VP groups (g/g)	0.7	0.4	0.2
$C_{wo}/(g/g)$	0.46	0.25	0.07
$D_{wo}/(10^{-8} \text{ cm}^2/\text{s})$	7.5	4.2	2.0
$D_s(C_{wo})/(10^{-8} \text{ cm}^2/\text{s})$	50	6.2	0.4

polymer has a steady-state part (Fig. 14). In agreement with the model, the rate of release decreases as the foregoing ratio reduces, showing non-Fickian diffusion for moderate $D_s(C_{wo})/D_{wo}$ values and the Fickian one for lower $D_s(C_{wo})/D_{wo}$ values, which were obtained for VP–MM-1 and VP–MM-2 copolymers, respectively (see Table 1).

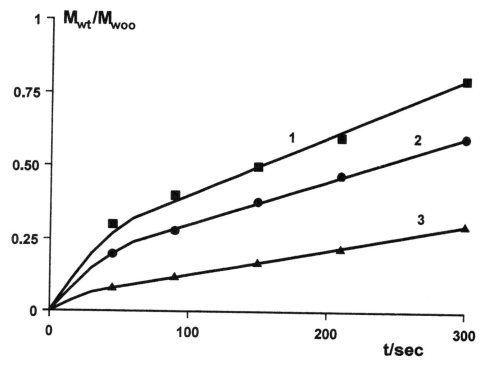

Figure 13   Kinetics of water sorption by copolymers of N-vinylpyrrolidone–VP–BM (1), VP–MM-1 (2), and VP–MM-2 (3). $1 = 100\ \mu m$.

Then, the model and experimental findings were applied for the preparation of fibers for surgery with programmed release of dioxidine. The good agreement between calculated kinetics of release from these fibers and those obtained in the clinical experiments is illustrated in Fig. 15.

The similar approach may be applied for a regulation of drug release from reservoir devices, in which the steady-state release cannot be obtained for the negative exponential dependence of the diffusion coefficient of water on its concentration in the coating (29). We obtained this result experimentally (Fig. 16, curve 1) when studying the release of albuterol (salbutamol) salt of benzoic acid from the reservoir coated by cellulose diacetate (CDA) (29). The concentration dependence of the diffusion coefficient of water in CDA is quantitatively described by the equation

$$D_{w1} = 3.5 \times 10^{-8} \exp(-27 C_w)\ [\mathrm{cm}^2/\mathrm{s}] \tag{22}$$

where C_w is the concentration of water in the polymer of the dimension of [g/g]. Accordingly, the rate of albuterol release passes through the sharp maximum that is usually undesirable. Thus, we considered other cellulose derivatives to be more suitable for this device. Cellulose acetyl fluorate (CAF) and its blend with cellulose triacetate (CTA) are characterized by two opposite processes: the swelling of polymer matrix, causing the acceleration of transport of water; and formation of clusters, slowing it down. Isotherms of water sorption (Fig. 17) indicated slight plasticization

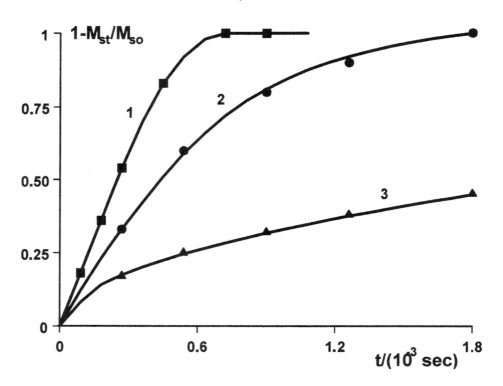

Figure 14 Kinetics of drug release from films of VP–BM (1), VP–MM-1 (2), and VP–MM-2 (3) copolymers.

of these polymers. The analysis of these isotherms showed that the bimodal distribution of water in the matrix should be expected in CAF and CAF + CTA with the number of water molecules in a cluster equal to 10 and 15, respectively. The main consequence of such a behavior of these polymer–water systems is the moderate increase of the diffusion coefficient of water, which may be described by linear equations.

$$D_{w2} = 8 \times 10^{-10} + 1.1 \times 10^{-7}\, C_w \; [cm^{-2}/s] \qquad (23)$$

$$D_{w3} = 5.6 \times 10^{-9} + 3.3 \times 10^{-7}\, C_w \; [cm^2/s] \qquad (24)$$

for CAF and CAF + CTA, respectively.

Experimental results confirmed the model prediction (which is described in Ref. 29). The steady-state flux of albuterol solution was reached for both coatings, and the time interval of this steady-state part was defined by constants involved in Eqs. (23) and (24). Particularly, the CAF + CTA coating provided steady-state release for a longer time (see Fig. 16, curve 3). The rate of release was higher for CAF (see Fig. 16, curve 2) owing to the higher value of water solubility in this polymer.

Polyurethane (PU) microspheres were prepared by a novel and simple method using the condensation polymerization technique (30). Microspheres of different

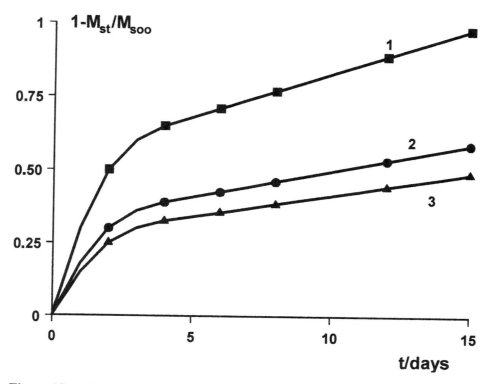

Figure 15 Kinetics of drug release from coatings of nylon fibers: VP–BM (1), VP–MM-1 (2), and VP–MM-2 (3) fibers.

morphological characteristics were prepared using tolylene 2,4-diisocyanate (TDI) and methylene diphenyl diisocyanate (MDI). These microspheres were fully characterized by thermal analysis, scanning electron microscopy, and spectral techniques. MDI-containing microspheres were more porous than TDI-containing microspheres. The infrared spectra indicated the complete utilization of the isocyanate groups in the synthesis of microspheres. Bromothymol blue (BTB) was used as a model drug for the in vitro release studies from the spheres. The results indicated that BTB released much faster using the MDI spheres than from the TDI spheres.

Nuclear magnetic resonance (NMR) microscopy was used to monitor the formation of the gel layer in hydrating hydrophilic polymer tablets. Such tablets were used in the controlled delivery of drugs, for which the rate and extent of the swelling of the outer gel layer critically influenced the kinetics of drug release. Tablets were hydrated in distilled water at 37°C and then imaged at discrete time intervals using a 500 MHz microscope. The growth of the gel layer was clearly observed in time sequences of radial and axial sections. Axial images showed some interesting dimensional changes, with the gel at the flat surface of the tablet becoming concave. This was probably a reflection of the occurrence of uniaxial stress relaxation as hydration proceeds. Diffusion- and T_2-weighted images provided evidence that the water in the gel layer was more strongly bound close to the dry core of the tablet than at the more fully hydrated outer surface. In images of tablets containing

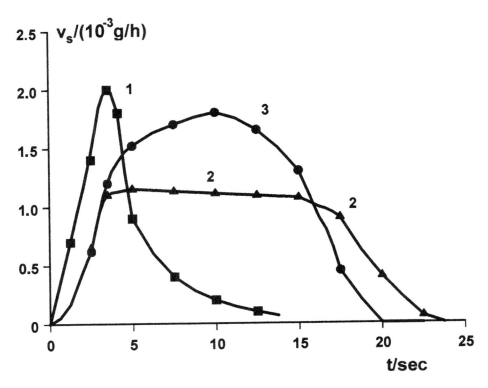

Figure 16 Kinetics of the release of albuterol (salbutamol) benzoate from the reservoir coated by CDA (1), CAF (2), and CAF + CTA (3).

diclofenac, disruption of the gel layer occurred primarily from the flat surfaces of the tablet, whereas the distribution of particles could be seen in tablets doped with insoluble calcium phosphate (31).

Injectable collagen is a concentrated dispersion of phase-separated collagen fibers in aqueous solution. The structure and properties of collagen fibers are defined by the magnitudes of electrostatic and hydrophobic attractive forces between neighboring collagen molecules within collagen fibers. The structure and mechanical properties of collagen fiber dispersions were studied by dynamic rheological measurements and by polarized microscopy (32). Rheological measurements were performed therein over pHs ranging from 6 to 9 and over temperatures ranging from 283 to 298 K. At higher pHs, the fiber dispersions possessed more rigid fibers and stronger interfiber attractive forces. This response is argued to result from changes in the ionization of amino acid side chains, which result in larger net electrostatic attractive forces. Raising the temperature caused fibers to become rigid through enhanced hydrophobic attractive forces. Gels formed by lower-pH–higher-temperature fiber dispersions possess different properties than gels formed at higher pHs and lower temperatures.

Sah and Chien prepared a biodegradable microreservoir-type microcapsule for the controlled release of a model protein bovine serum albumin (BSA). This was incorporated into the microcapsules with a high efficiency of 96.1%. The

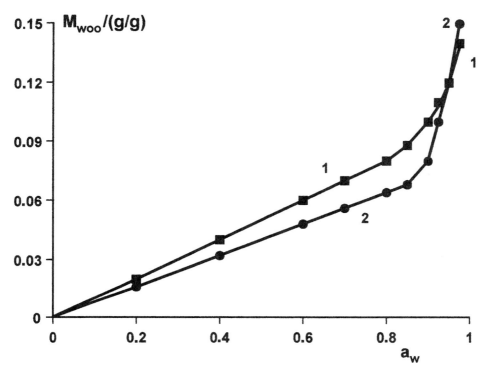

Figure 17 Water sorption isotherms for CAF (1) and CAF + CTA (2) films.

encapsulation did not cause any changes in the molecular weight or conformation of BSA, which was proved by biochemical analyses, such as gel electrophoresis, circular dichroism, and high-performance liquid chromatography (HPLC). The compositions and the fabrication technique of microcapsules were closely related to the release of BSA from the microcapsules and their degradation. Depending on microcapsule formulations, the in vitro release profile of BSA was either monophasic or biphasic. The microcapsules provided various release rates. It was also possible to control the delay before the initiation of BSA release and the total duration of its delivery (33).

Jou and Huang investigated the behavior of blends of rod-like pyromellitic dianhydride–benzidine (PMDA-B) and semiflexible 6F-dianhydride-phenyl-endiamine polyimides (6FDA-PDA) with several different compositions. According to the results from x-ray diffractometry and Fourier transform infrared (FTIR) spectrometry, these two polyimides were incompatible when mixed at room temperature for 20 min. When mixed at 50°C for 40 h, the polymers, thermodynamically incompatible, became compatible owing to exchange reactions. Bending beam diffusion experiments showed that the diffusion of moisture in these films belongs to case I. The diffusion coefficient D_w was 10^{-10} cm^2/s in PMDA-B and 1.65×10^{-9} cm^2/s in 6FDA-PDA. In the blends, D_w increased with increasing content of 6FDA-PDA. The slow diffusion in PMDA-B was attributed to its highly crystalline structure and relatively small interchain spacing. The diffusion of moisture was faster in the compatible films. Regardless of compatibility, diffusion in all the blends was much slower than in pure 6FDA-PDA. This was attributed to the comparatively small average interchain spacing of the blends (34).

The mass transport of two different compounds through polydimethylsiloxane (PDMS)–silica films was investigated by Dahl and Sue (35) for qualitative demonstration of how this coating system can alter the release of various compounds. Various ratios of PDMS elastomer and silica were used to coat monodisperse particle-sized pellets layered with an ionizable compound (tartrazine) and a nonionized compound (acetaminophen). The 2 : 1 PDMS–silica composition, containing the polyethylene glycol (PEG) 8000 pore former, allowed mainly pore transport through void spaces in the PDMS films. Both compounds rapidly diffused through the film as a result of the solubilization and subsequent removal of the PEG 8000 from the film matrix. As the PDMS–silica ratios in the films changed from a 1 : 1, to a 2 : 1, to a 4 : 1 (all without PEG 8000) coating formulation, the differences in release rate between acetaminophen and tartrazine changed. The lower ratio of PDMS–silica allowed much faster tartrazine diffusion compared with acetaminophen. As the ratio increased from 1 : 1 to 2 : 1 the two compounds were released at similar rates. When the ratio reached 4 : 1, acetaminophen was released significantly faster than tartrazine. This study demonstrated that utilization of this polymer system offers a useful tool to modify release rates of ionic and nonionic drug substances.

Crystalline poly-(−)-3-hydroxybutyrate (PHB) and its copolymers with poly-3-hydroxyvalerate [P(HB–HV)] are biodegradable polyhydroxyalkanoates with potential application in controlled drug-delivery systems. Matrices of PHB and P(HB–HV) copolymers containing a model drug, methyl red, were prepared by solvent casting and melt processing. Drug release from P(HB–HV) copolymer matrices produced by any given fabrication technique was dependent on copolymer composition. Progressively faster rates of drug release were obtained on increasing HV content. This could be explained by the different crystallization behavior of P(HB-HV) copolymers. Evidence from polarized light microscopy of copolymer spherulites suggested that copolymer films with increasing HV content exhibited morphologies with decreasing abilities to entrap the model drug (36).

The aggregation between hydrophobic–hydrophilic block copolymer unimers in water was analyzed by Petrak (37). Based on the molecular weight and composition of the block polymer, the number of unimers in each particulate aggregate, the size of the particle, and the size of the particle core, the average distance between the solvated chains attached to the micelle core was calculated. The distance between the terminally attached chains was compared with the Flory radius $R(f)$ which was related to the extent that the chains were forced to stay in the "brush" conformation. By selecting the proper structure of block copolymer, micelles composed of water-soluble polymer chains were prepared with a "near ideal" saturated surface.

Various theoretical frameworks of analysis of drug transport in block and graft copolymers were examined by Harland and Peppas (38), and the importance of partially or totally impermeable domains was underlined therein. The drug transport in and its release from block or graft copolymers containing hydrophilic and hydrophobic domains were governed by the size and shape of the domain and thermodynamic interactions between hydrophilic and hydrophobic components. This heterogeneous structure can be used to the advantage for the delivery of hydrophilic or hydrophobic drugs at desirable rates. On the basis of their theoretical approach Harland and Peppas synthesized a series of hydrophilic–hydrophobic copolymers and characterized them for possible applications in drug delivery. Such systems were prepared by random, graft, or block copolymerization of various monomers, followed by selective hydrolysis. The copolymer composition was investigated using NMR spectroscopy, whereas their molecular weight distribution was determined by gel permeation chromatography. Thermodynamic and physical properties, such as transition temperatures and degrees of swelling were measured to elucidate the effect of phase separation on the formed network structure. Microdomain formations, balance of chemical properties, and network characteristics affected the partitioning and transport of theophylline and myoglobin in these swollen networks. Finally, the nonlinear network swelling, selective drug partitioning, and control of drug permeation through and release from copolymer membranes were correlated with the presence of hydrophobic and hydrophilic microdomains in the copolymer networks for a selected number of copolymers studied (39).

Copolymers of 2-hydroxyethyl methacrylate with several multiethylene glycol dimethacrylates were prepared by bulk copolymerization in the presence of AIBN as an initiator. The content of the dimethacrylate varied between 30 and 50 mol%. Additional copolymers were prepared by solution polymerization in the presence of water or ethanol. All samples were swollen in water up to thermodynamic equilibrium, and their dynamic swelling behavior was studied as a function of time. It was concluded that the mechanism of water transport in these moderately and highly cross-linked polymers was a coupled relaxation and diffusion and that the relaxational contribution to the overall transport depended on the chain length of the multiethylene glycol dimethacrylate (40).

The permeability of solutes may be regulated by the formation of liquid crystal zones in a polymer membrane (6). As reported (41), polysiloxanes with substituted side groups are able to form structures with a phase transition temperature in the interval of 45–65°C. Figure 18 shows the effect of a mezogene phase of polysiloxanes on the rate of diffusion of salicyclic acid in a wide temperature interval, including that of phase transition. The increase of the content of a liquid crystal phase induces the reduction of the diffusion coefficient of solute, thereby providing the possibility to govern the rate of its ejection.

B. Controlled Alteration of Hydrophobic–Hydrophilic Balance in Polymers

1. Controlled Polymer Heating

Although a synthesis of a polymeric material of optimal composition is the most spread method for the regulation of multicomponent transport in the controlled-delivery devices, there are more accurate ways of controlled modification of polymer structure. One of them is thermal modification. Such an approach has been developed to govern the kinetics and rate of drug release from a polymer matrix (42). The water-soluble, alternating copolymer of maleic anhydride and vinyl acetate (MAn–VAc) has been isothermally degraded at 250°C to obtain stable residues, for which their insolubility in water develops as the time of degradation increases. The multicomponent system of partly degraded polymer, water, and drug

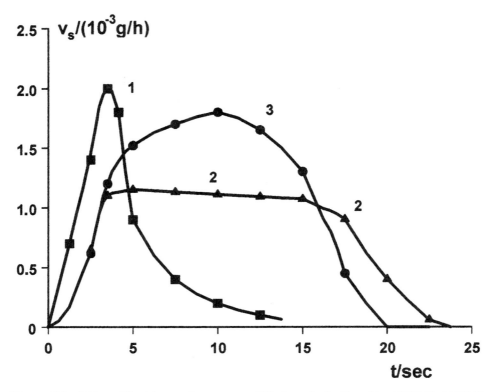

Figure 18   Effect of liquid crystal content on diffusion of salycilic acid at 80°C (1), 70°C (2), 60°C (3), 50°C (4), 40°C (5), 30°C (6), and 20°C (7).

Table 2   Conditions for Copolymerization of Maleic Anhydride (MAn) with Vinyl Acetate (VAc) at 60°C

Total concentration of monomers	2.5 M
Total volume	85 mL
Initial mole fraction of MAn	0.5
Initiator (AIBN) concentration	10^{-2} M
Conversion	20%

has been studied theoretically and experimentally. The details of polymerization of copolymer of MAn–VAc copolymer are presented in Table 2.

The thermal degradation of this copolymer has been studied by thermal volatilization analysis (TVA) (43). Formation of conjugated double bonds and hydroxyl groups, as well as the development of insolubility in water owing to intermolecular dehydration are the most essential rearrangements occurring under isothermal investigation at 205°C. Because it is related to the time of degradation the data obtained show that the first stage of degradation basically consists of acetic acid loss (Scheme 1) followed by subsequent elimination of carbon dioxide from the MAn (Scheme 2) units remaining in the chain.

The CO_2 loss leads to the development of conjugation and, finally, to the total breakdown of the anhydride ring (Scheme 3).

The gradual insolubility develops partly because of the reduction in flexibility of the backbone, but also as a result of cross-linking by intermolecular dehydration involving OH groups produced in some of the structures (Scheme 4).

The concentration dependence of the diffusion coefficient of water in undergraded and partly degraded MAn–VAc copolymer (Table 3) confirms the different hydrophilic characters of these materials. These dependencies put in the model of Eqs. (3)–(11) were used for the quantitative prediction of the regularities of controlled release of several drugs.

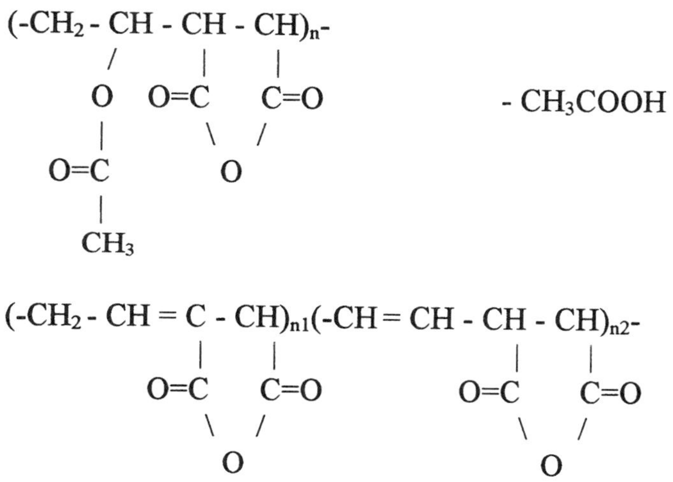

```
(-CH₂ - CH - CH - CH)ₙ-
       /     |    |
      O   O=C   C=O           - CH₃COOH
      |     \   /
    O=C      O
      |
     CH₃
```

```
(-CH₂ - CH = C - CH)ₙ₁(-CH = CH - CH - CH)ₙ₂-
              |   |              |    |
            O=C   C=O          O=C    C=O
              \   /              \    /
               O                  O
```

Scheme 1

```
(-CH₂ - CH = C - CH)ₙ₁(-CH = CH - CH - CH)ₙ₂-
              |   |              |    \
            O=C   C=O          O=C    C=O         - CO₂
              \   /              \    /
               O                  O
```

```
(-CH₂-C=CH
     |  |
     C - CH)ₙ₃(-CH=CH-CH-CH)ₙ₂(-CH=C-CH₂
     ||              |   |         |  |
     O             O=C   C=O      C=C)ₙ₅(-CH=C-CH
                     \   /         |        |   ||
                      O           H-O      CH-C)ₙ₆-
                                                |
                                               H-O
```

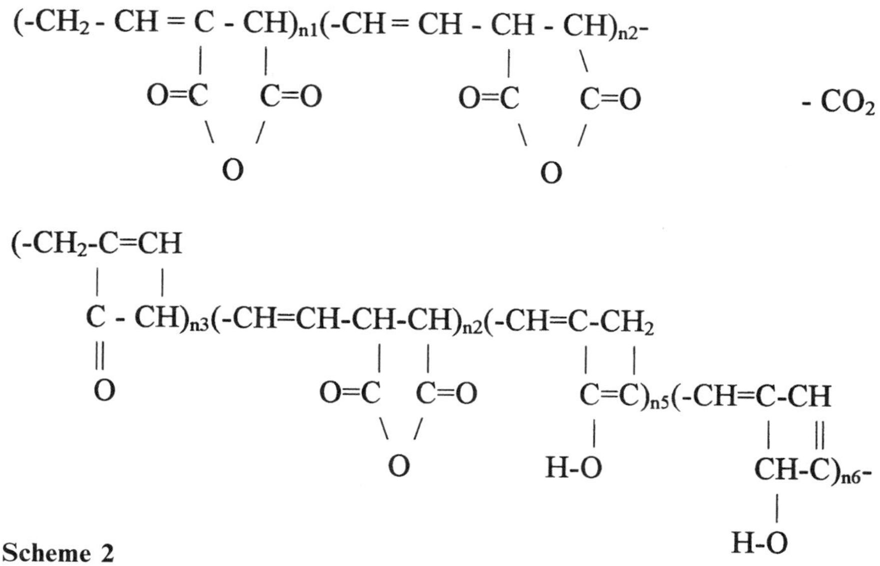

Scheme 2

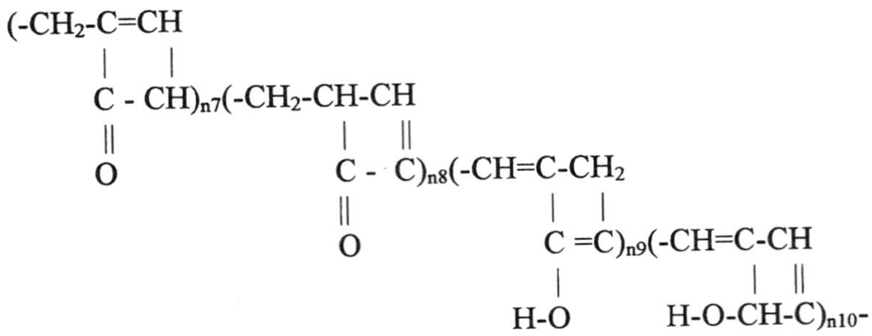

```
(-CH₂-C=CH
     |    |
     C - CH)n7(-CH₂-CH-CH
     ||              |    ||
     O               C - ·C)n8(-CH=C-CH₂
                     ||             |  |
                     O              C =C)n9(-CH=C-CH
                                    |        | ||
                                    H-O    H-O-CH-C)n10-
```

Scheme 3

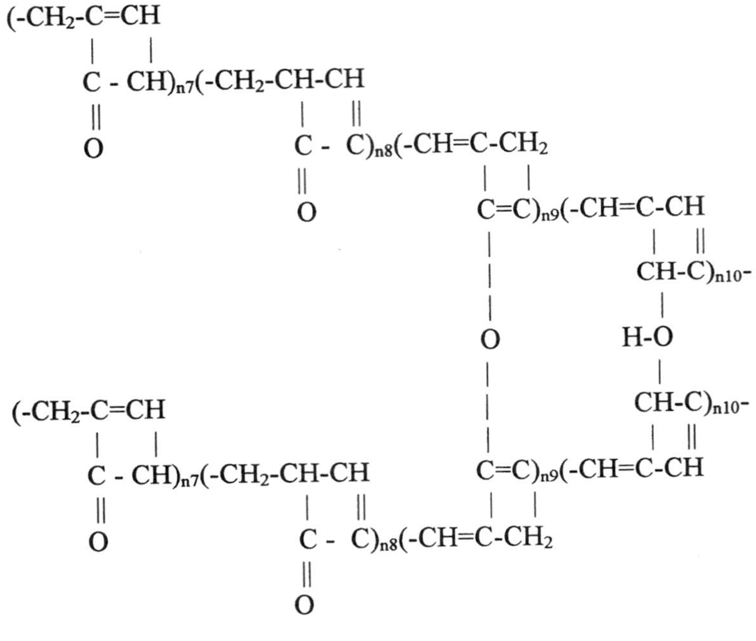

```
(-CH₂-C=CH
     |    |
     C - CH)n7(-CH₂-CH-CH
     ||              |    ||
     O               C - C)n8(-CH=C-CH₂
                     ||            |  |
                     O             C=C)n9(-CH=C-CH
                                   |         | ||
                                   |        CH-C)n10-
                                   |         |
                                   O         H-O
                                   |         |
(-CH₂-C=CH                         |        CH-C)n10-
     |    |                        |         | ||
     C - CH)n7(-CH₂-CH-CH          C=C)n9(-CH=C-CH
     ||              |    ||       |  |
     O               C - C)n8(-CH=C-CH₂
                     ||
                     O
```

Scheme 4

Table 3 Parameters Concerned with the Concentration Dependence of the Diffusion Coefficient of Water in Initial and Partly Degraded MAn–VAc Copolymer

$$D_w = D_{wo}(f)\exp(k_{w1}C_w)$$

Parameter	Degradation time (min)						
	0	5	10	15	20	30	40
$D_{wo}/(10^{-8}\text{ cm}^2/\text{s})$	5.5	5.9	5.9	7.5	9.0	2.9	1.0
k_{w1}	5.3	1.8	0.3	0.12	0.02	−1.4	−2.1

Among the parameters responsible for drug release from the system under study, the diffusion coefficient and solubility in water are the most important. The permeability of the partially degraded copolymer of MAn–VAc to the different drugs has been studied (44). Table 4 gives a list of these drugs and their solubility in water, and Table 5 illustrates the dependencies of the diffusion coefficients of these drugs on the time of degradation. The diffusion coefficient in the initial copolymer corresponds to the diffusion coefficient of the drug in pure water, and the dependence, as a whole, is the dependence of the diffusion coefficient of the drug on the concentration of water in the polymer, which should be put in the model [see Eq. (11)].

The effect of deceleration of drug release by controlled degradation of MAn–VAc copolymer has been predicted through calculations and proved by experimental data (Fig. 19).

The highly hydrophilic residues of MAn–VAc copolymer are characterized by high values of $D_s(C_{wo})/D_{wo}$. For these coatings the expected result for drug release to follow water uptake was obtained. Neither solubility nor mobility of drug affect

Table 4 Solubility of Furan-Containing Drugs in Water (Concentration of the Saturated Solution)

Drug	Chemical formula	$c_s^o/(\text{mg/g})$
Furacilin, dissociated	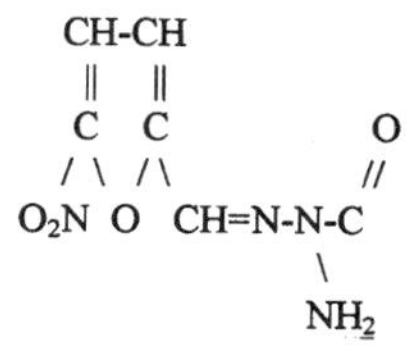 CH-CH ‖ ‖ C C O / \ /\ // O_2N O CH=N-N-C \ NH_2	5
Furagin, potassium salt	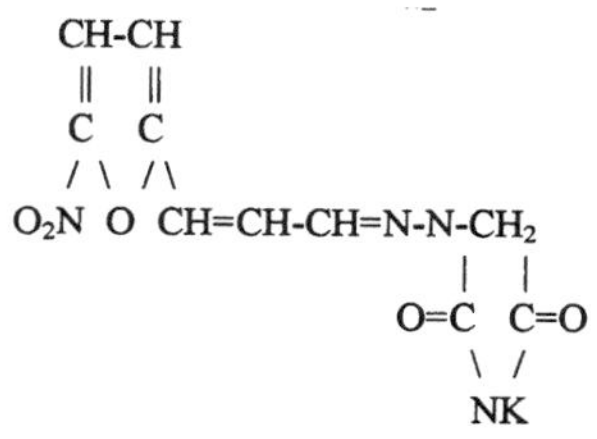 CH-CH ‖ ‖ C C / \ /\ O_2N O CH=CH-CH=N-N-CH_2 ∣ ∣ O=C C=O \ / NK	2
Furacilin, associated	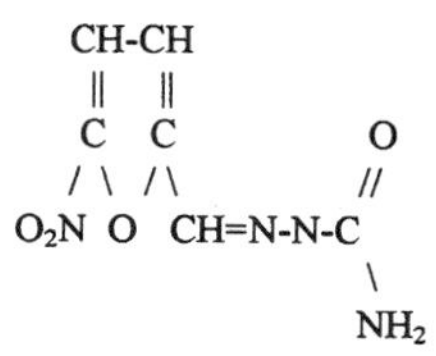 CH-CH ‖ ‖ C C O / \ /\ // O_2N O CH=N-N-C \ NH_2	0.25
Furadonin	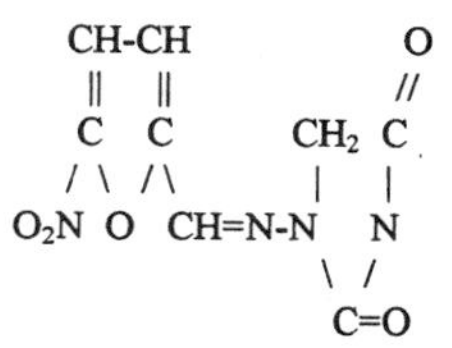 CH-CH O ‖ ‖ // C C CH_2 C / \ /\ ∣ ∣ O_2N O CH=N-N N \ / C=O	0.125

Table 5 Dependence of the Diffusion Coefficient of Drugs on Time of Thermal Degradation and the Corresponding Value of Overall Water Uptake

Degradation time/min		0	5	10	15	20	30	40
$C_{wo}/(g/g)$		oo	5.7	2.4	1	0.5	0.12	0.09
No.	$c_s^0/(mg/g)$	$D_s(C_{wo})/(10^{-7} \text{ cm}^2/s)$						
1	5	10	8.6	6.2	4.3	1.2	0.25	0.11
2	2	8.0	6.8	5.8	3.8	1.2	0.30	0.14
3	0.25	8.0	6.7	5.4	3.2	0.8	0.12	0.06
4	0.12	7.5	6.5	5.7	4.0	1.5	0.37	0.19

the kinetics of its release, except minor variations near equilibrium. This is most probably due to the release of the rest of drug, which is distributed in the bandage itself.

The contribution of the solubility becomes visible for coatings that contain from 20 to 50% water in equilibrium, and the most remarkable for moderately swelling coatings that contain no more than 10%. For the first group of the coatings (20–50%) the solubility of drug influences mostly the rate of release, whereas its kinetics remains essentially unchanged and includes the linear (case II) section in the kinetic curve. We should emphasize, however, that this is not an effect of solubility only, but also of the mobility (relative diffusion coefficient) of the drug.

For low-swelling coatings (<10%) the solubility affects both the rate and the kinetics of the release, and the latter is the effect of solubility only. The decrease of solubility finally leads to Higuchi kinetics as the theory predicts.

Therefore, we obtain evidence that the simultaneous control of chemical structure of a polymer and of its transport properties gives a good opportunity to develop polymer systems for which required kinetics and rate of drug release can be obtained.

Thermoresponsive hydrogels of poly(*N*-isopropylacrylamide-*co*-butyl methacrylate) [poly(IPAAm–*co*-BMA)] are capable of swelling–deswelling changes in response to external temperature. As poly(IPAAm–*co*-BMA) gels swell larger at a lower temperature, the degree and rate of the swelling could be controlled by temperature without altering the chemical structure. Therefore, drug release profiles were remarkably changed by alternation of temperature. The release profiles of indomethacin from poly(IPAAm–*co*-BMA) were observed to be zero-order at 20°C. Okuyama and coauthors explained this release (45) in terms of a case II diffusion mechanism that indicates relaxation of polymer chains with swelling was rate-determining. At 10°C, release demonstrated a sigmodal profile. The acceleration of drug release was due to a rapid increase in swelling, with disappearance of the glassy core, which had constrained swelling. The regulation of the water-uptake process by changing external temperature remarkably affected drug release and resulted in several different release profiles.

Thermoresponsive poly(*N*-isopropylacrylamide-*co*-alkyl methacrylate) gels are capable of "on–off" regulation of drug release in response to external temperature changes because a gel surface skin, formed with increasing temperature, stops

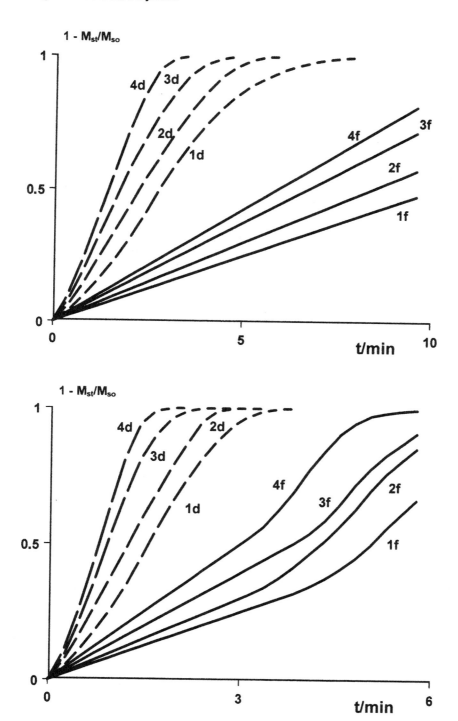

Figure 19 Kinetics of release of dioxidine (d) and furacilin (f) from polyacrylic emulsion at different initial ratio between drugs loaded: $C_{s1o}/C_{s2o} = 10 : 2$ (1); $5 : 5$ (2); $2 : 4$ (3); $1 : 5$ (4); $C_{s3o} = 0$ (top), 0.04 (bottom) g/g.

drug release from the gel interior. In this gel-shrinking process, observation of bubble formation on the surface indicates that pressure is induced within the gel. This pressure may result in an outward convection of water. Therefore, the drug must be released not only by diffusion, but also by convective transport. Yoshida et al. (46) created a drug-release model for this shrinking process using a tortuous pore model and simulated four decreasing patterns of release rate for different induction patterns of pressure. Experiments using indomethacin could match simulated release patterns by changing the chemical structure of polymer and thermal gradient. These changes induce different pressure fluctuations within gels and affect the release pattern from the gel "on" state to the "off" state.

The mechanism of thermal regulation of release in hydrogels with a phase diagram having low critical temperature of mixing (LCTM) was reported (6). When the thermally reversible gel passes LCTM under heating or cooling, the polymer matrix is contracted or expanded. This results to the desorption of the previously sorbed compounds (47,48) or their extra sorption. This mechanism works when toxins are being ejected from physiological medium (48) or enzymes are being immobilized in a gel (49). Figure 20 illustrates this effect for release of the model compound (methylene blue) from hydrogels of 1 : 1 poly(N-isopropylacrylamide-co-methacrylic acid) with LCTM near 40°C (48).

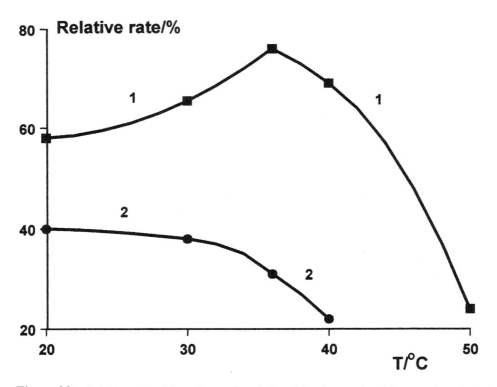

Figure 20   Relative rate of the release of methylene blue from poly-N-isopropylacrylamide gel in citric phosphate buffer (pH = 7.4) (1) or in deionized water (2).

Aging can modify polymer structure at the molecular, macromolecular, or the morphological level and, thus, induce changes in the mechanical properties. Stiffness is generally not modified for nonrubbery materials, except for mass transfer (solvent plasticization or plasticizer loss) in amorphous polymers or phase transfer (crystallization or crystal destruction) in semicrystalline polymers. The most significant modulus changes occur in the radiochemical aging of semicrystalline polymers the amorphous phase of which is the rubbery state. Yield properties generally vary in the same way as stiffness. Physical aging at $T < T_g$ can lead to a significant increase in the yield stress. Very general features can be observed for rupture properties, for instance 1) only ultimate elongation E is a pertinent variable in kinetic studies of aging involving tensile testing and related methods; 2) the amplitude of E variation for a given degradation conversion is considerably higher for initially ductile materials than for brittle ones; and 3) the rupture envelope $\sigma = f(\varepsilon)$ (i.e., the ultimate stress) is often very close to the initial tensile curve, except for rubbery materials undergoing predominant cross-linking. The mechanisms of ultimate property changes have been reviewed (50). A kinetic approach was proposed for the very important case of heterogeneous, diffusion-controlled aging.

Michelle and Vittoria (51) measured a sorption and diffusion of dichloromethane vapor in poly(aryl ether ether ketone) films after different aging times. In the fresh samples, the diffusional behavior was characterized by two stages: in the first stage the diffusion coefficient D was weakly dependent on concentration, and transport occurred mainly through a frozen system; whereas in the second stage, a steep dependence of D on concentration was found. The aged samples also showed an intermediate stage, characterized by an anomalous or non-Fickian sorption behavior. Furthermore, the diffusion coefficients of the aged samples were lower than those of the fresh samples, in agreement with a reduced fractional free volume. Aging at a temperature slightly lower than the glass transition temperature decreased the sorption at low activity, leading to the suggestion that ordered domains, impermeable to the vapor at low activity, were formed at this temperature. The process of solvent-induced crystallization on the amorphous samples was investigated, and it was found that crystallization is induced after an activity 0.7. Both the fresh and the aged samples showed the same behavior, although the level of crystallinity attained was higher for the latter samples (51).

The hydraulic permeation of water through heat-treated polyvinyl alcohol membranes was investigated by Yang and Chu (52). The transport of water in a heat-treated polyvinyl alcohol membrane follows the mechanism of the solution diffusion equation. The activation energy of water diffusivity increased, and water diffusivity decreased with increase in the period of heat treatment. The compressibility of polyvinyl alcohol membranes decreased with increase in the period of heat treatment.

2. Controlled Load of Solute, Fillers, and Additives

Although there is a principal possibility of governing the water sorption in the solute release from the polymer matrix by adjusting the load of fillers and additives, and some theoretical and experimental studies have been done in this area (see Sec. III.B.3), the application of this method has so far received little discussion. This is, most probably, due to the toxic character of many chemicals usually used as additives. For this reason their possible release together with the main components is not desirable and, mainly, water-soluble ingredients are used to regulate the kinetics and the rate of release.

The effect of additives on the release of drugs from polyacrylic emulsion have been studied for its possible application as skin bandages. We studied the simultaneous release of drugs of different solubility and the effect of a water-soluble ingredient, sodium alginate, on the drug transport. For this particular system the effect of additives (including drugs) on the equilibrium concentration of water in the polymer is described by the equation

$$C_{w0} = C_{wo}^0 \exp(K_{w1} C_{s1} + K_{w2} C_{s2} + K_{w3} C_{ss})$$

where C_{wo}^0 is the equilibrium concentration of water in unfilled polymer ($C_{wo}^0 = 0.5$ g/g), C_{s1}, C_{s2}, C_{s3}, are the concentration of dioxidine, furacilin, and sodium alginate in the polymer, respectively, K_{w1}, K_{w2}, K_{w3} are the thermodynamic constants equal to 1, 3.6, 12 L/mol, respectively. In the meantime, the kinetics of water sorption and its rate remains unchanged ($D_w = 2.3 \times 10^{-7}$ cm^2/s) for variations of the drug load in a wide range, and minor changes are observed if sodium alginate is incorporated in the emulsion.

Because the kinetics and rate of solute release is affected by water concentration in the polymer, the programmed regulation of release is possible as it is shown in Fig. 19a. The main regularities observed are the following. In the absence of sodium alginate the diffusion of dioxidine may be described in terms of Fickian diffusion

$$\frac{\delta C_{s1}}{\delta t} = \left(\frac{\delta}{\delta x}\right)\left(\frac{D_{s1} \delta C_{s1}}{\delta x}\right)$$

where

$$D_{s1} = 9.2 \times 10^{-6} \exp(-2.4/C_{wi}) \, [\text{cm}^2/\text{s}]; \qquad D_{s1}(C_{wo}) = 6.3 \times 10^{-8} \, [\text{cm}^2/\text{s}]$$

The release of the drug of limited solubility (furacilin) follows Higuchi kinetics with an effective diffusion coefficient

$$D_{s1} = 5 \times 10^{-11} \exp\left(-\frac{2.4}{C_{wi}}\right) \, [\text{cm}^2/\text{s}]$$

The load of the water-soluble filler (sodium alginate) accelerates the release of both drugs and alters the kinetics of the desorption of furacilin. The dependence of the diffusion coefficient of dioxidine changes in presence of sodium alginate to

$$D_{s1} = 9.2 \times 10^{-6} \exp\left(-\frac{2.4}{C_{wi}}\right) \, [\text{cm}^2/\text{s}]; \qquad D_{s1}(C_{wo}) = 5 \times 10^{-7} \, [\text{cm}^2/\text{s}]$$

showing that the load of water-soluble ingredient is one of the approaches to the regulation of the release kinetics of incorporated solutes.

Schroen et al. (53) studied problems encountered with an emulsion–membrane bioreactor. In this reactor, enzyme (lipase)-catalyzed hydrolysis in an emulsion was combined with two in-line separation steps. One was carried out with a hydrophilic membrane, to separate the water phase, the other with a hydrophobic membrane, to separate the oil phase. In the absence of enzyme, sunflower oil–water emulsions with an oil fraction between 0.3 and 0.7 could be separated with both membranes operating simultaneously. However, two problems arose with emulsions containing lipase. First, the flux through both the hydrophilic and the hydrophobic membranes decreased with exposure to the enzyme. Second, the hydrophobic membrane showed a loss of selectivity demonstrated by permeation of both the oil phase and the water phase through the hydrophobic membrane at low transmembrane pressure. These phenomena were explained by protein (i.e., lipase) adsorption to the polymer surface within the pores of the membrane. It was proved that lipase was present at the hydrophilic membrane and that this, in part explains the flux decrease of the hydrophilic membrane. To prevent the observed loss of selectivity with exposure to protein, the hydrophobic polypropylene membrane (ENKA) was modified with block copolymers of propylene oxide (PO) and ethylene oxide (EO). These block copolymers act as a steric hinderance for proteins that come near the surface. The modification was successful: After 10 days of continuous operation, the minimum transmembrane pressure at which water could permeate through an F 108-modified membrane was 0.5 bar, the same value as that observed in the beginning of the experiment. This indicated that less of selectivity owing to protein adsorption may be prevented by modification of the membrane (53).

Solid dispersions of carbamazepine with polyvinylpyrrolidone–vinylacetate copolymer at different loading ratios were prepared to study the influence of this copolymer on the solubility and dissolution rate of the drug. The greatest increase in the dissolution rate of carbamazepine was obtained from solid dispersion at the 1 : 4 (w/w) drug/polymer ratio. Furthermore, the influence of pH on dissolution was studied. The data indicate that the release profile of the drug was not modified by a change in pH. Several kinetic models were applied in an attempt to describe the mechanism of drug dissolution. Physical characterization of the prepared systems was carried out by differential scanning calorimetry (DSC), x-ray diffractometry, and wettability studies (54).

The mechanical properties of ethylcellulose films were determined to evaluate effects of polymer, plasticizer, and dispersed solid. The concentration of propylene glycol exhibited no significant effect on the tensile strength, whereas that of diethyl phthalate did. Both plasticizers also had significant effects on the film elongation. Theoretical calculations revealed that the release of diltiazem hydrochloride from ethylcellulose film-coated pellets could be described by the combination of constant and nonconstant activity source diffusion-controlled model. A dramatic modification in drug release characteristics was observed after the film-coated pellets had been compressed into tablets. However, the increase in compression force within the working range (181.8, 272.7, and 363.6 kg) distinctly slightly affected the release from pellets being compressed. These behaviors may not be caused by flaws or failures within the film, but instead, by the appreciable alterations in some physical properties of the film itself under pressure. The electron photomicrographs confirmed such a hypothesis (55).

Adhesive dispersion-type transdermal drug delivery (a-TDD) systems, consisting of a monolayer of drug-loaded adhesive matrix, were developed from three types of silicone-based pressure-sensitive adhesives. The adhesive polymers were tailored such that two of them were lipophilic (Bio-PSA(R) X7-2920 and Dow

Corning(R)-355 Medical Adhesive) and one was relatively hydrophilic (E8086(R) adhesive). Three steroids—progesterone, testosterone, and hydrocortisone—were used as model penetrants, and their release from the a-TDD systems and permeation through skin were investigated. The adhesive properties of these systems were studied, and the partial and total solubility parameters of these adhesive polymers were determined. The release of steroid molecules was observed to be a complex function of the physicochemical properties of the drug and polymer. The adhesiveness, as determined from a standard peel test; indicated that incorporation of the drug in higher drug-loading doses results in a loss of adhesiveness. The results suggest that the chemical nature of the polymer is an important consideration when studying such adhesives for transdermal drug delivery (56).

3. Multilayer Systems

Many materials that are widely used in medicine, agriculture, and industry are not suitable enough when their application for controlled delivery of chemicals (mainly drugs) is required. On the other hand, there are dozens of polymers with appropriate transport characteristics which, however, cannot be applied themselves as fibers, bandages, or such, because of their lack of mechanical and other properties. In these cases, the development of multilayer systems solves this problem. In the meantime, the application of multilayer systems is not only a technological problem. It is one of the ways to regulate the kinetics and the rate of release.

We particularly studied the multilayer system that included nylon coated by a blend of MAn–VAc copolymer with drug (42). The hydrophilic–hydrophobic balance in the second (copolymer) layer was varied by controlled isothermal heating, as described in Section III.B.1. Redistribution of drug between two polymers defines the multistage process of its further release, which includes desorption from moderately hydrophilic polyamide being in contact with solvent, following by the transport in swelling or dissolving MAn–VAc copolymer.

By taking into account these stages, we used the following equations to describe multicomponent transport in each layer

$$\frac{\delta C_{wi}}{\delta t} = \left(\frac{\delta}{\delta x}\right)\left(\frac{D_{wi}\delta C_{wi}}{\delta x}\right) \qquad 0 < x < 1 \tag{25}$$

with

$$D_{wi} = D_{woi}\exp(k_{w1i}C_w + k_{w2i}f) \tag{26}$$

and

$$\frac{\delta C_{si}}{\delta t} = \left(\frac{\delta}{\delta x}\right)\left(\frac{D_{si}S_{si}\delta C_{si}}{\delta x}\right) \qquad 0 < x < 1 \tag{27}$$

with

$$D_{si} = D_{soi}\exp(-k_{w3i}/(C_{wi} + k_{w4i}) \tag{28}$$

$$S_{si} = \frac{C_{si}}{a_{si}} = k_{ssi}C_{wi}V_{wi}c_s^o = K_{si}C_{wi} \tag{29}$$

where the main definitions are the same as in Section II in which the generalized swelling stress model has been reported and index i corresponds to the number of layers (1 or 2).

The boundary and initial conditions for this multilayer system were selected as follows ($x = 0$ corresponds to the nylon–water interface; $x = 1$ corresponds to the nylon–MAn–VAc interface, and $x = L$ corresponds to MAn–VAc–water interface).

$$D_{w1}(\delta C_{w1}(l, t)/\delta x) = D_{w2}(\delta C_{w2}(l, t)/\delta x),\ C_{w1}(l, t) = \alpha_w C_{w2}(l, t) \tag{30}$$

$$D_{s1}(\delta C_{s1}(l, t)/\delta x) = D_{s2}(\delta C_{s2}(l, t)/\delta x),\ C_{s1}(l, t) = \alpha_s C_{s2}(l, t) \tag{31}$$

$$C_{w1}(0, t) = C_{w1o}; \qquad C_{w2}(L, t) = C_{w2o} \tag{32}$$

$$C_w(x, 0) = 0; \qquad 0 < x < L \tag{33}$$

$$C_s(x, 0) = 0; \qquad 0 < x < 1 \tag{34}$$

$$C_s(x, 0) = C_{s0}; \qquad 1 < x < L \tag{35}$$

where α_w and α_s are the distribution coefficients between two layers for water and solute, respectively.

The effect of the first layer on the kinetics of release of a model drug is visible near equilibrium, mainly, for highly swelling coating, as illustrated in Fig. 21 in comparison with Fig. 22. These variations become minor and disappear as the

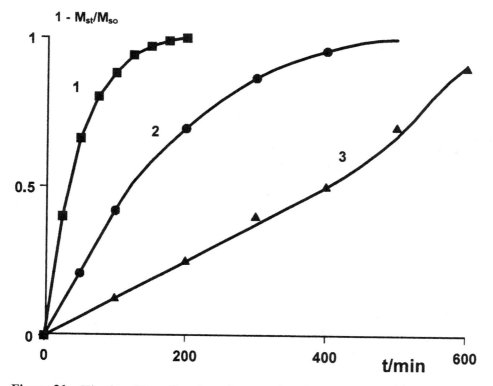

Figure 21 Kinetics of furacilin release from a polyamide textile bandage through films of partly degraded MAn–VAc copolymer after heating for 5 (1); 15 (2); and 30 (3) min.

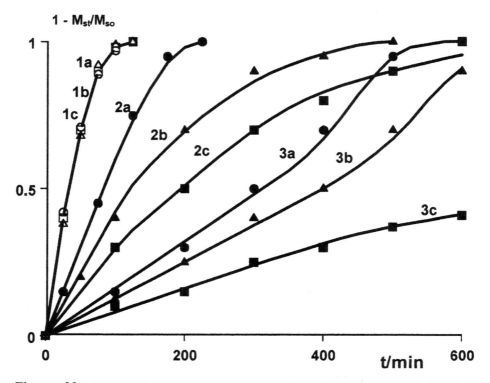

Figure 22 Computed and experimental kinetics of the release of furagin (a), furacilin-associated (b), and furdonin (c), from partly degraded MAn–VAc copolymer after heating to 5 (1); 15 (2); and 30 (3) min.

hydrophobicity of the MAn–VAc coating (owing to the programmed degradation) develops.

The aforesaid example relates to the systems when the second layer initially contains the solute. These systems are applied mostly in transdermal devices or other kinds of skin bandages (44,57).

If the solute is initially loaded in the first layer we must substitute the initial conditions [see Eqs. (33)–(35)] by

$$C_w(x, 0) = 0; \qquad 0 < x < L \tag{36}$$

$$C_s(x, 0) = C_{s0}; \qquad 0 < x < 1 \tag{37}$$

$$C_s(x, 0) = 0; \qquad 1 < x < L \tag{38}$$

The composition and thickness of the second layer were varied to alter the kinetics of the release of dioxidine from the aforementioned VP–MM films. The agreement between theoretical and experimental data (Fig. 23) showed that the transport of water and solute is controlled by the diffusion in the layers and is not substantially affected by the interlayer processes. The fundamental results were then applied for the development of VP–MM coatings for fibers with the programmed release of dioxidine.

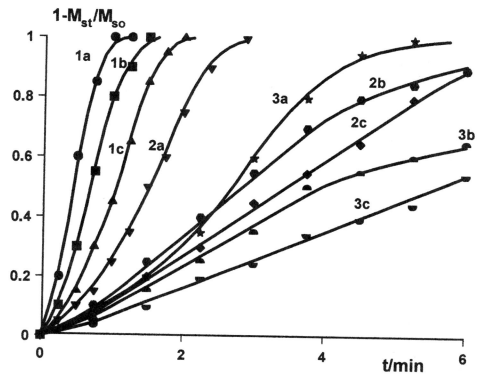

Figure 23 Kinetics of dioxidine release from the films of VP–MM-1 films coated by the second layer of VP–BM (1); VP–MM-2 (2); and polyethoxycyanoacrylate (3). Thickness of the second layer = 50 (a), 80 (b), 120 (c) μm.

A new method of preparing composite poly(vinyl alcohol) (PVA) beads with a double-layer structure was developed (58) and involved a stepwise saponification of suspension polymerized poly(vinyl acetate) (PVAc) beads and subsequent stepwise cross-linking of the PVA core and shell with glutaraldehyde. This process results in PVA beads with thin, highly cross-linked outer shells and lightly cross-linked inner cores of different degrees of cross-linking. In addition to structural characterization of the polymer based on equilibrium swelling measurements, the kinetics of water swelling and drug release from these beads were studied at 37°C using acetaminophen and proxyphylline as model drugs. The results showed that the outer shell functions as a rate-controlling membrane on increasing its cross-linking ratio X above 0.47. This aspect is reflected in the observed diffusional time lags and constant-rate regions during swelling and drug release. From the observed time lags, the diffusion coefficient of water through the outer PVA shell with a high cross-linking ratio of $X = 0.5$ was estimated to be at least six times higher than that of acetaminophen and proxyphylline. In addition, drug diffusion coefficients in the lightly cross-linked PVA core appeared to be at least ten times larger than that in the highly cross-linked outer shell. At lower shell cross-linking ratios ($X < 0.4$), the diffusional time lags appeared to be absent and the diffusion profiles are apparently first-order (Fickian) in nature (58).

Fu and colleagues reported (59) that solvent transport in multilayer thin film structures can induce damaging stresses. As a model for such systems, they focused on the connection between solvent transport in a thin, supported film and the induced bending curvature of the film–substrate combination. They developed a simple mechanical model to calculate the bending curvature based on the transport-induced stresses and used a phenomenological moving boundary description of the non-Fickian solvent transport, which is often found in glassy polymers. As an application of the model, experimental data for a polyimide (PI)–quartz–N-methyl-2-pyrrolidinone (NMP) system involving significant swelling (15–20%) of the PI film were analyzed. The analysis showed that the measured bending during the transport of NMP in the PI film compared well with that predicted based on an "intermediate," non-Fickian diffusion mechanism of NMP, consistent with the finding obtained from a laser interferometric study.

A new multilayer tablet device was proposed (60) for the extended release of drugs at a constant rate. It consisted of the application, during tableting, of compressed barrier layers on one or both surfaces of a hydrophilic matrix containing the active ingredient. The barriers can limit the core hydration process by maintaining the planar surfaces of the tablet covered during dissolution. Moreover, the coatings reduce the extent of the core surface exposed to the interaction with the dissolution medium and to drug release. As a result, the release rate was reduced and the kinetics shifted toward constant drug release. The effects of the coating application on the release patterns of three different matrix cores containing trapidil or sodium diclofenac as model drugs were tested. The matrix cores were designed to provide different release mechanisms and extents.

Latex film matrix systems, with a nonuniform drug distribution, were prepared (61) for a coating process. In this process a drug concentration gradient in the coating dispersion was generated by the programmable pumping of the latex into a drug reservoir which contained the drug and latex. The film matrix formed by the dispersion would have a built-in drug concentration gradient as the coating process proceeded. A mathematical model was developed to describe the concentration change of the active ingredient in the coating dispersion, as a function of the spraying rate of this dispersion, the pumping rate of the latex into the drug reservoir, and the initial drug concentration in the dispersion. The applicability of this process was demonstrated by the controlled in vitro dissolution of acetaminophen from ethylcellulose latex film matrices formed on glass beads. The release profile of the active ingredient from the systems changed as the drug concentration gradient profile in the matrix was altered, and a higher drug concentration gradient in the matrix resulted in a slower release rate and a more linear release profile (61).

Wan and Lai developed (62) a new model system for controlled-release using multilayer-coated granules prepared by consecutively spraying aqueous solutions of diphenhydramine and HCl, dissolved in methylcellulose (MC) of various viscosity grades, onto lactose granules in a fluidized bed. The in vitro drug release profiles of the coated product were a function of the polymer viscosity grade and the sequential order of application of the polymeric component layers on the lactose granules. The permutation and the mass fraction of the component layers on the multilayer-coated granules were altered to provide different drug release rates. Although the equilibrium hydration and swelling of each component layer influenced drug diffusion to some extent, the overall release was not entirely governed by it. The drug-release profiles were also modulated by the dissolution characteristics, particularly the particle size of the recrystallized drug that is embedded in the polymeric coat. This approach on controlled-release drugs provides an outline for further work in this field (62).

Physical mixtures of an acrylate–methacrylate copolymer (in fine powder) and salicyclic acid (drug model) in a ratio of 1 : 4 were compressed directly (i.e., without further granulation) to matrix (nondisintegrating) tablets to serve as core models. A special technique was used to form an aqueous-based coating system consisting of a water-insoluble copolymer and sucrose in varying proportions. Film coatings were applied on the matrix cores to a thickness of 75 μm (SEM). Drug-release rates increased as a linear function of the initial sucrose content in the film coatings. The amount of sucrose leached from the film coatings into the leaching fluid also increased linearly and in parallel with the increase in release rates. The release profiles obtained with cores with coating thickness of 75 μm were generally without a time lag. Doubling the coating thickness introduced a time lag (preceding the onset of release) of about 2.5, 1.3, or 0.5 h for polymeric coatings without sucrose, or with sucrose 12.5% w/w, or 25% w/w, respectively. Coatings with a higher sucrose content did not display the release time lag. Overall release rates were hardly affected by the increase in coating thickness (63).

A good example was recently given (64) of a combined application of many approaches described in the foregoing. These investigators developed a multiparticulate, sustained-release formulation of theophylline and evaluated it in vitro. The formulation comprised spherical pellets of high drug loading, coated with a rate-controlling membrane. The pellets were prepared using an extrusion spheronization method, whereas coating was performed with an aqueous dispersion of ethylcellulose using a fluidized bed coating technique. When ethylcellulose was used alone as the coating polymer, the drug-release profile was unsatisfactory, but could be improved by incorporating a coating additive. Several additives were evaluated, and methylcellulose of high viscosity grade was found most satisfactory. The in vitro theophylline release was relatively linear and pH independent, and could be varied in a predictable manner by manipulating the coat thickness. In addition, when the coated pellets were subjected to additional thermal treatment, the drug release was stable after storage for one year (64).

C. Regulation of Biocompatibility of Polymeric Materials

The application of polymer devices in medicine, especially for their implantation in the body, requires the analysis of biocompatibility of polymeric materials. In connection with this problem the study of the adsorption of proteins on a nonphysiological surface is attracting the attention of many researchers.

The report of Griesser (65) described a work using low-pressure gas plasma techniques for the fabrication of surfaces designed for compatibility with biological tissue and fluids. The smoothness of the surfaces was investigated using scanning, tunneling microscopy (STM), and the wettability was assessed by measurement of air–water contact angles. Plasma surface treatment rendered fluorocarbon polymers wettable by polar liquids, but the wettability decreased with time during storage of the samples in air. Plasma polymerization, on the other hand, allowed the deposition of thin coatings that were more stable with time. STM of plasma polymers, which were coated with a sputtered film of platinum or gold, showed that these coatings were smooth and continuous. The attachment and growth of human umbilical artery endothelial cells was used to assess the biocompatibility of several plasma polymer surfaces. There was some correlation between the ability of a surface to grow cells and the wettability of that surface with polar liquids, but significant differences in cell growth on surfaces with very similar hydrophilicity indicated that other factors are at least as important (65).

Analysis of products formed in the end-capping step during the synthesis of two block copolyurethanes being studied as biomedical materials showed the presence of dimeric soft segments and free diisocyanate. In this standard two-step synthesis, the presence of these compounds leads to block copolyether–urethane–ureas containing sizeable amounts of dimeric hard and soft segments. These standard copolymers were compared in terms of IR spectra, stress–strain properties, and dynamic mechanical properties with their pure analogues that contain no dimeric segments (66).

Studies on protein adsorption and platelet adhesion indicate that polymer surfaces can selectively adsorb proteins from whole blood in vivo, and the composition of this adsorbed protein layer influences the adhesion of platelets, thereby determining the thrombogenicity of the polymer. The microphase structure of the block copolyurethanes often seems to confer a selectivity in albumin adsorption, which appears to make these materials relatively nonthrombogenic. Analyses of the synthetic procedures for preparing these block copolyurethanes indicated that the standard two-step, synthetic procedure introduces sizeable amounts of dimeric soft and hard segments into the copolymers. FTIR and ESCA were used to determine the bulk and surface morphology of these standard block copolyurethanes as well as their pure $(AB)_n$ analogues, to understand protein binding to their surfaces. A study of these materials as vascular grafts indicated that the bulk properties (i.e., the compliance of the graft) can also affect graft patency under in vivo conditions (67).

Surface design aimed at reduced adhesion and preserved functions of platelets is of great importance for extracorporeal devices. Matsuda and Ito (68) explored a coating technique using hydrophilic–hydrophobic block copolymers on a hydrophobic poly(acrylonitrile) (PAN) hemodialyzer. The hydrophilic block of copolymers was composed of either poly(methoxy polyethylene glycol methacrylate) or poly(dimethyl acrylamide), and the hydrophobic block was poly(methyl methacrylate). The copolymers were coated on the dialyzer membrane by means of a solution-coating method. On coating, the hydrophobic block of the copolymers was anchored on a PAN membrane, and the hydrophilic block oriented toward the blood–material interface. This was deduced from water wettability measurements. Significantly reduced transmembrane stimulation of platelets was observed, which was evaluated by determining the intracellular calcium ion concentration of platelets eluted through treated hollow fibers. This suppression was enhanced as the relative fraction of the hydrophilic block of the copolymers increased. Furthermore, the number of platelets adhering to the copolymer-coated PAN membrane was drastically reduced. Thus, coating of the hydrophilic–hydrophilic block copolymers provided better biocompatibility on a hydrophobic PAN dialyzer (68).

Materials with unique complex of the properties of elasticity and tensile strength processed from the natural rubber latex (NRL) were modified to improve their blood compatibility. ESCA, ATR-IR, UV-spectroscopy, and other methods were used for the analysis of modification, structure, and properties of material. Thromboresistant properties were studied by in vivo and ex vivo tests. The blood compatibility was raised after purification of the material from nonrubber components by means of two-stage extraction, changing the physicochemical parameters of the surface.

Two different principles were used in the process of modification: (1) coating by thin polyurethane (PU) coverings with improved thromboresistant properties—thus, the problem of providing high adhesion interaction of covering with the latex base was solved; (2) heparin surface immobilization—higher efficiency of modification of latex material in gel form without preliminary protein adsorption—was shown. Modification allowed an increase in blood compatibility of the latex materials while, at the same time, preserving their elasticity and tensile strength (69).

The adsorption of plasma proteins is the primary act of thrombus formation on the polymer surface in contact with blood (70). The most convincing version appears to be that the activation of platelets on the proteinated surface is stimulated by the partial rearrangement in structure of the protein that adsorbed on the surface (71,72). Earlier we have shown (72–75) that the mechanism of adsorption constitutes a complex multistage process, including diffusion of the protein globules to the polymer surface and the subsequent stages of formation of structurally rearranged and native protein layer. The correlation between the kinetic parameters of adsorption and the biocompatibility of polymers have been established either for materials giving a good (76) or an unsuitable (77) account of themselves in medical practice. Then, the approach was extended to test and alter the biocompatibility of materials widely used in medicine, particularly for design of drug-delivery systems. These are segmented polyetherurethanes and siloxanecarbonates (78). To provide this purpose the diffusion–kinetic model was modified to be relevant to the process of adsorption onto the surface of block copolymers, and FTIR-spectroscopy with accessories of attenuated total reflection (FRIR-ATR) was applied to express evaluation of the amount of adsorbed protein.

Earlier we have shown (72–75) that the equilibrium protein layer on the surface of hydrophobic and moderately hydrophilic polymers incorporates two adsorption layers of different structure, architecture, and intramolecular bond energy. The first (termed irreversible, or firmly adsorbed) consists of molecules subjected to structural rearrangement and is monomolecular; this layer interacts directly with active centers of the surface. The structure of the second (reversible) layer is close to the structure of native protein. Macromolecules contained on this layer are adsorbed by active centers of molecules that have undergone structural rearrangement. Interaction of native globules are neglected because of the absence of such interaction under normal conditions in solution. The formation of an irreversible layer onto the surface of block copolymer may be presented in the form of second- and first-order parallel and step-by-step reactions

$$A_v + \Theta_{11} \underset{k_{121}}{\overset{k_{111}}{\rightleftharpoons}} A_{s11} \overset{k_{11}}{\longrightarrow} A_{s11}{}^*$$

$$A_v + \Theta_{12} \underset{k_{122}}{\overset{k_{112}}{\rightleftharpoons}} A_{s12} \overset{k_{12}}{\longrightarrow} A_{s12}{}^* \tag{39}$$

where A_v is the molecule in solution coming into contact with the surface; Θ_{1i} is the active center of the surface corresponding to the i-block; A_{sli} is the native molecule which either undergoes structural rearrangement, or desorbs into the solution, A_{sli}^* is the molecule that sustained structural rearrangement.

Similarly, a second-order equation may describe the adsorption of reversible layer:

$$
\begin{aligned}
A_v + \Theta_{21} &\xrightleftharpoons[k_{221}]{k_{211}} A_{s21} \\[2mm]
A_v + \Theta_{22} &\xrightleftharpoons[k_{222}]{k_{212}} A_{s22}
\end{aligned}
\tag{40}
$$

where Θ_{2i} is the active center on the surface of the corresponding layer, A_{s2i} is the molecule incorporate in the reversible layer.

The system of kinetic equations corresponding to the scheme [see Eqs. (39) and (40)] takes the form

$$
\frac{dC_{s1}}{dt} = \sum_{i=1}^{2} k_{11i} C_v(0, t)(C_{sli}^{oo} - C_{sli} - C_{sli}^*) - (k_{12i} + k_{1i}^*)C_{sli}
\tag{41}
$$

$$
C_{s1} = \sum_{i-1}^{2} C_{sli} + C_{sli}^*
\tag{42}
$$

$$
\frac{dC_{sli}^*}{dt} = k_{1i}^* C_{sli}
\tag{43}
$$

$$
\frac{dC_{s2}}{dt} = \sum_{i=1}^{2} k_{21i} C_v(0, t)(N_i C_{sli}^* - C_{s2li}) - k_{22i} C_{s2i}
\tag{44}
$$

$$
C_{s2} = \sum_{i=1}^{2} C_{s2i}
\tag{45}
$$

Here C_{s1} is the overall concentration directly on the surface, C_{sli}, C_{sli}^*, and $C_v(0, t)$ are the concentrations of molecules A_{sli}, A_{sli}^*, and A_v, respectively (surface coordinate $x = 0$); C_{sli}^{oo} is the maximum concentration of irreversible layer corresponding to protein adsorption by all active centers of i-sort of the surface, C_{s2i} is the concentration of molecules A_{s2i}, N is the average number of active centers of Θ_{21}, k_{ijk} $(i, j, k = 1, 2)$ are the rate constants.

The rate of adsorption of proteins depends not only on polymer–globular interactions, but also on hydrodynamic conditions of the experiment (73,74,76). This is explained by the fact that concentration $C_v(0, t)$ is determined by the transfer in the diffusion boundary layer near the surface. According to the diffusion–kinetic model (74), protein concentration in a diffusion layer of a thickness l conforms to the equation

$$
\frac{\delta C_v}{\delta t} = \frac{D\delta^2 C_v}{\delta x^2}
\tag{46}
$$

with boundary conditions

$$
-\frac{\bar{D}\delta C_v}{\delta x} = \frac{R\mathrm{d}C_v}{\mathrm{d}t} \qquad x = 1
$$
$$
\frac{D\delta C_v}{\delta x} = \frac{\mathrm{d}C_{sp}}{\mathrm{d}t} \qquad x = 0 \qquad\qquad (47)
$$
$$
C_{sp} = \sum_{i=1}^{2} C_{s1i} + C_{s1i}^* + C_{s2i}
$$

where $\bar{D}$ is the diffusion coefficient of protein in solution; C_v is the protein concentration in the zone of mixing; $C_{sp} = C_{s1} + C_{s2}$, R is the typical size of an experimental cell.

Constants in the system of Eqs. (41)–(47) may be divided into two groups. One group (rate constants, N, C_{s1}^{oo}) directly characterizes the polymer–protein system, on which the properties depend, the second (l, C_v, R) is determined by experimental conditions and is given by a researcher. Coefficient D is a characteristic of a given protein and may be found from independent experiments. Controlling adsorption by changing flow rate (and, therefore, l) and the initial concentration of protein in solution, the necessary data may be obtained for calculating rate constants, N and C_{s1}^{oo}, and the prediction of properties of the polymer–protein system as a whole.

The general consequences from the model, which are essential for further analysis of particular polymers, are that

1. All kinetic curves of irreversible adsorption tend to the limit (C_{s1}^{oo}), which is altered by the ratio of monomers (m) (Fig. 24).
2. The total equilibrium concentration C_s^{oo} also depends on the ratio of monomers in copolymer through a corresponding dependence of N (Fig. 25).

Two kinds of block copolymers were tested with application of the aforesaid approach. The first, segmented polyetherurethane (PEU) (trade mark VITUR-0134), was synthesized on the basis of copolymer of polytetra-methylene-glycol, methane diisocyanate, and ethylene diamine. The ratio of monomers also varied in the second copolymer of polysiloxane carbonate (PSC). Human serum albumin (Sigma) contained 99% conversion to dry substance. The concentration of protein solution in Tirade-3 buffer (pH = 7.4) varied within the range of 0.01–5 g/L. Other essential parameters of proteins should be indicated as 69,000 molecular weight and 6×10^{-7} cm^2/s diffusion coefficient in buffer solution. Special experimental order is required for accurate measurements of parameters involved in the model [see Eqs. (41)–(47)].

Parameters of reversible adsorption must be measured first from the desorption experiments. The kinetics of desorption of a reversible layer from the surface of polymers were examined at various rates of solvent flow Q, initial bulk concentrations of protein, and different monomer ratio in block copolymers (as shown in Fig. 26, for example, of polysiloxane carbonate). The total time of desorption related to a certain Q value as it follows the diffusion–kinetic model (77). All kinetic curves tend to one limit, which corresponds to the concentration of irreversible layer under a given ratio of monomers. In the meantime, the variation of this ratio alters

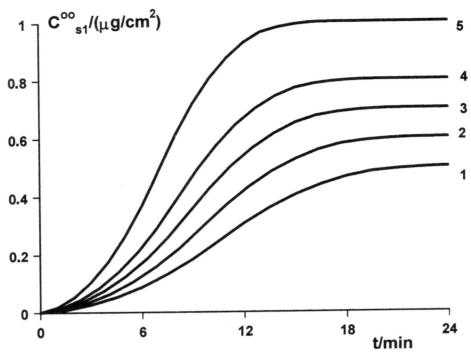

Figure 24 Computed kinetics of irreversible adsorption of proteins on the surface of block copolymers: $m = 0$ (1); $1 : 3$ (2); $1 : 1$ (3); $3 : 1$ (4); $1 : 0$ (5). $C_{s11}^{oo} = 0.5$, $C_{s12}^{oo} = 1$.

such a limit. The examination of kinetic curves at different protein concentrations in the initial solution showed that each curve is characterized by its own equilibrium value of C_{sp}, which also depends on the ratio of monomers in the block copolymer. The rate constants of the reversible protein layer were calculated on the basis of the desorption kinetics, and results are presented in Tables 6 and 7. Other parameters that are also collected in these tables have been calculated from the kinetic curves of irreversible adsorption (Fig. 27 illustrates the example of polyetherurethane). Calculations have been made on the basis of comparison of theoretical curves (lines) and experimental data (points).

   The tables also contain the effective parameters that could be calculated using a previous unmodified model (72). The validity of the modified approach was then checked by the comparison of theoretical and experimental results of overall albumin adsorption on the surface of polyetherurethanes with different ratios of hard and soft segments (see Table 6; Fig. 28). Because of the good coincidence of theoretical and experimental data, we can draw the following general conclusions. The increase of the soft segment in the chemical structure of polyurethanes causes the decrease of the total amount of irreversibly adsorbed protein accompanied with the shift of the reaction for formation of the reversible layer to adsorption. Although the number of reversible binding centers depends slightly on the kind of segment, the contribution of this number itself in the thermodynamics and kinetics of adsorption is negligible in comparison with the chemical structure of these centers, followed

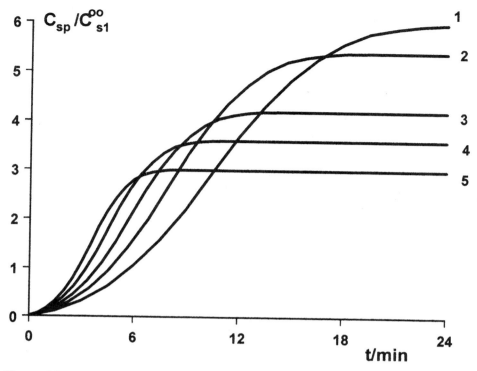

Figure 25 Computed kinetics of reversible adsorption of proteins on the surface of block copolymers: $m = 0$ (1); $1 : 3$ (2); $1 : 1$ (3); $3 : 1$ (4); $1 : 0$ (5). $N_1 = 5$, $N_2 = 2$.

by structural rearrangement of macromolecules of the corresponding irreversible layer. In other words, the effect of the kind of segment on the values of rate constants is much more essential. The deactivation of the centers of the reversible adsorption before the moment of contact between platelets and proteinated surface improves the biocompatibility of the polymer. The worsening of the biocompatibility of the polymer coincides with a shift of equilibrium in the reaction of formation of the reversible layer toward desorption.

The alteration of the polymer's biocompatibility by the controlled variation of the ratio between monomers was also shown for an albumin–polysiloxane carbonate system (see Table 7; Fig. 29). The results illustrate the expected effect of the siloxane fraction of the copolymer on the quantity of irreversibly adsorbed protein. The most considerable effect is that for which the ratio of copolymer components has on the reversible adsorption kinetics. The larger extent of structural rearrangement in proteins stimulates protein–protein interactions. This intensifies the reversible adsorption and increases the equilibrium constant of the reversible layer. Once this opposite effect of structural rearrangement on kinetics of adsorption exists, an optimal ratio of monomers should be expected from the point of polymer biocompatibility. For in vitro conditions, this optimal ratio should be expected at values of 50 : 50 for polysiloxane carbonate and 60 : 40 for polyetherurethane.

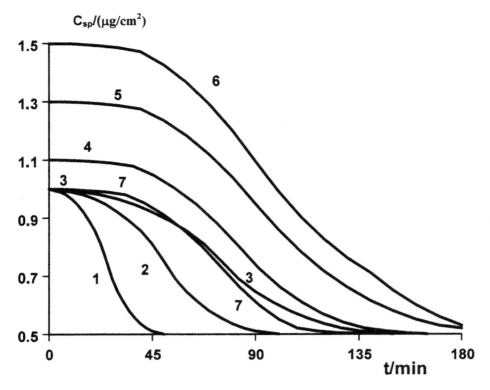

Figure 26   Kinetics of desorption of albumin from the PSC surface: $Q = 10$ (1, 4–7); 30 (2); 60 (3) mL/min. $C_v^0 = 1$ (1); 2 (4); 5 (5) g/L. $m = 3 : 1$ (1); 1 : 1 (6); 1 : 3 (7).

Table 6   Physicochemical Parameters of Albumin Adsorption on the Surface of Segmented Polyetherurethane

Parameter	Hard segment	Soft segment	Effective value at soft/hard segment ratio of		
			72 : 28	60 : 40	56 : 44
$k_{11}/(\text{mL}/\text{g*s})$	50	30	35	40	45
$k_{12}/(\text{L}/\text{s})$	0.05	0.01	0.02	0.25	0.03
$k_1^*/(\text{L}/\text{s})$	0.03	0.03	0.03	0.03	0.03
$k_{21}/(\text{mL}/\text{g*s})$	30	40	40	40	35
$k_{22}/(\text{L}/\text{s})$	0.4	0.1	0.1	0.3	0.35
N	2	4	3.8	3	2.6
$C_{sl}^{oo}/(\text{Mkg}/\text{cm})$	0.8	0.6	0.6	0.65	0.7

Therefore, the foregoing diffusion–kinetic model allows the reversibility factor to be controlled with time. This opens a way to improving biocompatibility of copolymers used in medicine.

Table 7 Physicochemical Parameters of Albumin Adsorption on Siloxane-Containing Polymers

Parameter	Siloxane fragment	Carbonate fragment	Effective value at siloxane fraction of		
			0.65	0.50	0.25
$k_{11}/(\text{mL}/\text{g*s})$	25	30	25	28	30
$k_{12}/(\text{L}/\text{s})$	0.001	0.05	0.02	0.025	0.003
$k_1^*/(\text{L}/\text{s})$	0.03	0.03	0.03	0.03	0.03
$k_{21}/(\text{mL}/\text{g*s})$	10	40	22	25	32
$k_{22}/(\text{L}/\text{s})$	0.014	0.1	0.04	0.06	0.08
N	4	2	3.5	3	2.8
$C_{s1}^{oo}/(\text{Mkg}/\text{cm})$	0.5	0.6	0.52	0.55	0.6

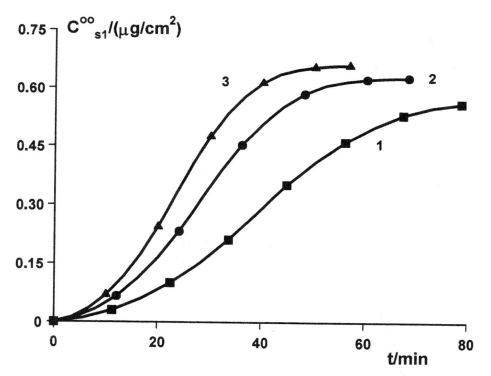

Figure 27 Kinetics of irreversible adsorption of albumin on the PEU surface: $Q = 10$ mL/min. $C_v^0 = 1$ g/L; $m = 3 : 1$ (1); $1.5 : 1$ (2); $1.25 : 1$ (3).

IV. CONCLUDING REMARKS

Since the 1970s polymeric-controlled delivery of drugs and of other low molecular weight compounds has become an important area of research, development, and production. In this short time, several devices with programmed kinetics and rate of release have progressed from the university or institution laboratory to the

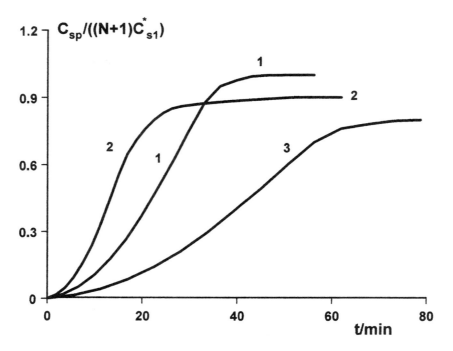

Figure 28 Kinetics of overall albumin adsorption on the PEU surface: $l = 100\ \mu m$; $C_v^0 = 45$ g/L; $m = 72 : 28$ (1); $60 : 40$ (2); $56 : 44$ (3).

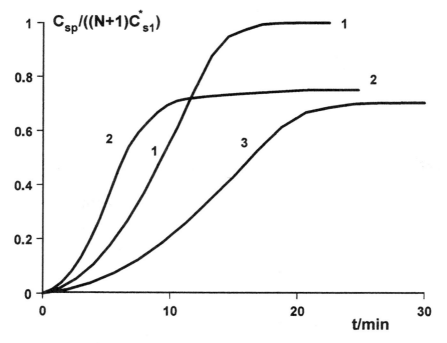

Figure 29 Kinetics of overall albumin adsorption on the PSC surface: $l = 100\ \mu m$; $C_v^0 = 45$ g/L; $m = 75 : 25$ (1); $50 : 50$ (2); $25 : 75$ (3).

hospitals and agricultural fields, and the shops. Although newer and more specific polymeric controlled-delivery systems continue to be developed, the understanding of the mechanism of multicomponent transport becomes more and more important for the design of particular devices. The theories discussed in this chapter represent fundamental approaches that were developed to describe the regularities of multicomponent transport in polymers. As related to the systems for controlled release, these regularities may be used to design devices more properly corresponding to the practical requirements. The successful application of fundamental research will be a significant challenge that may provide new acceleration of the investigations in this important and fruitful area of our life.

ACKNOWLEDGMENTS

The authors thank their colleagues from the states of ex-USSR who provided them with materials for experimental investigations with particulars to Dr. T. Khanlarov (Azerbaijan State University, Baku), Professor G. E. Krichevskii (Textile Progress Ltd., Moscow, Russia), Dr. N. D. Oltarzhevskaya (Institute of Textile Industry, Moscow, Russia), and Dr. E. D. Zagreba (Institute of Microbiology, Riga, Latvia). The authors address their special thanks to Dr. Ian C. McNeill (the University of Glasgow, UK), Dr. J. H. Petropoulos ("Demokritos" National Research Center, Athens, Greece), and Professor G. Camino (the University of Turin, Italy) for the excellent opportunity to work in their laboratories and for their very useful comments.

MAIN DEFINITIONS*

Independent Variables

$$t = \text{time}$$
$$x = \text{direction of diffusion}$$

Dependent Variables

$$a_\text{w} = \text{activity of water in material}$$
$$a_\text{s} = \text{activity of solute in material}$$
$$C_\text{w} = \text{concentration of water in the material}$$
$$M_\text{wt} = \int_0^l C_\text{w}(x, t)\text{d}x = \text{overall water uptake}; \quad M_\text{woo} = \int_0^l C_\text{w}(x, \infty)\text{d}x$$
$$C_\text{s} = \text{concentration of solute in the material}$$
$$M_\text{st} = \int_0^l C_\text{s}(x, t)\text{d}x = \text{overall amount of solutes}; \quad M_\text{so} = \int_0^l C_\text{s}(x, \infty)\text{d}x$$
$$C_\text{ss} = \text{concentration of the mobile (dissolved in water) part of solute, referred}$$
$$\text{to the whole material}$$
$$c_\text{ss} = \text{concentration of solute in water}$$
$$C_\text{f} = \text{concentration of fillers in the material}$$
$$C_\text{v} = \text{bulk concentration of proteins}$$

---

* Other definitions are given in the text.

C_{s1} = surface concentration of proteins at the first layer
C_{s1}^* = concentration of proteins undergone conformational changes at the first layer
C_{s2} = surface concentration of proteins at the second layer
C_{sp} = overall surface concentration of proteins
f = mechanical stress
s = swelling strain

Variable Parameters

β = frequency of stress relaxation
D_w = diffusion coefficient of water in the material
D_s = diffusion coefficient of solute in the material
E_o, E_{oo} = initial and final elastic modules
S_w = solubility coefficient of water in the material
S_s = solubility coefficient of solute referred to the whole material

Constants

β_o = frequency of stress relaxation of the dry material
C_{wo} = equilibrium concentration of water in the material
C_{so} = initial concentration of solute in the material
C_{f0} = initial concentration of fillers in the material
c_{ss}^o = concentration of saturated solution of solute in water (solute solubility)
C_{s1}^{oo} = maximum concentration of irreversible layer of proteins
D_{wo} = diffusion coefficient of water in dry and unstressed material
D_{so} = diffusion coefficient of solute in pure water
E_{o0}, E_{oo0} = initial and final elastic modules of dry material
K_s = apparent solubility constant
l, L = thickness of polymer slab, film, or membrane
S_{wo} = solubility coefficient of water in dry and unfilled matrix
V_w = molar volume of water

REFERENCES

1. PI Lee, WR Good, eds. Controlled-Release Technology: Pharmaceutical Applications, ACS Symp Ser 348, 1987.
2. MA El-Nokaly, DM Piatt, BA Charpentier, eds. Polymeric Delivery Systems: Properties and Applications. ACS Symp Ser 520, 1993.
3. RL Dunn, RM Ottenbrite, eds. Polymeric Drugs and Drug Delivery Systems. ACS Symp Ser 469, 1991.
4. SW Shalaby, CL McCormick, GB Butler, eds. Water-Soluble Polymers: Synthesis, Solution Properties, and Applications. ACS Symp Ser 467, 1991.
5. NA Plate, LI Valuev. [Polymer Contact with the Living Body]. Issue 8. Moscow: Znanie, 1987 [in Russian].
6. AL Iordanskii, TE Rudakova, GE Zaikov. Interaction of Polymers with Bioactive and Corrosive Media. Utrecht: VSP, 1994.
7. M Gillespie. 3i Technology Forecast. Issue Number 2. Drug delivery systems into 1990s. 3i Group Plc., 1990.

8. NA Peppas, R Gurny, E Doelker, P Buri. Modelling of drug diffusion through swellable polymeric systems. J Membr Sci 7:241–253, 1980.

9. D Hershey. Transport Analysis. New York: Plenum Press, 1973.

10. AY Polishchuk, LA Zimina, RY Kosenko, AL Iordanskii, GE Zaikov. Diffusion-activation laws for drug release from polymer matrices. Polym Degrad Stabil 31:247–254, 1991.

11. JH Petropoulos, PR Roussis. Influence of transverse differential swelling stresses on the kinetics of sorption of penetrants by polymer membrance. J Membr Sci 3:343–356, 1978.

12. JH Petropoulos. Application of the transverse differential swelling stress model to interpretation of case II diffusion kinetics. J Polym Sci Polym Phys Ed 22:183–189, 1984.

13. AY Polishchuk, GE Zaikov, JH Petropoulos. The modelling of simultaneous transport of solvent and low-molecular compounds in swelling polymers. Int J Polym Mater 19:1–14, 1993.

14. AY Polishchuk, GE Zaikov, JH Petropoulos. General model of transport of water and low-molecular solute in swelling polymers. Int J Polym Mater 25:1–12, 1994.

15. AY Polishchuk, LA Zimina, AL Iordansky, GE Zaikov, RY Kosenko, SI Belykh. Sorption and kinetic laws of the drug release from polymer materials. Polym Sci USSR 32:2203–2209, 1990.

16. H Yasuda, CE Lamaze, A Peterlin. Diffusive and hydraulic permeabilities of water in water-swollen polymer membranes. J Polym Sci A2 9:1117–1131, 1971.

17. AY Polishchuk, GE Zaikov, JH Petropoulos. General aspects of multicomponent transport in swelling polymers. Proceedings of 4th International Conference on Polymeric Materials, "Polymat' 94," Imperial College, London, 1994.

18. SG Amarantos, KG Papadokostaki, JH Petropoulos. Characterization of Radioactive Waste Forms. Final Report for the Commission of the European Communities, Luxembourg, 1991.

19. JH Petropoulos, KG Papadokostaki, SG Amarantos. General model for the release of active agents incorporated in swellable polymeric matrices. J Polym Sci Polym Phys 30:717–725, 1992.

20. AL Iordansky, AY Polishchuk, RY Kosenko, NN Madyuskin, OV Shatalova, LL Razumova, GE Zaikov. Interrelation between structural–morphological and diffusion–kinetic processes in hydrophilic and moderately hydrophilic polymers. Int J Polym Mater 16:195–212, 1992.

21. JH Guo. Investigating the effect of water on the porosity of polymer film for controlled drug delivery. Drug Dev Ind Pharm 20:2467–2477, 1994.

22. RW Korsmeyer, NA Peppas. Solute and penetrant diffusion in swellable polymers. Proceedings of the International Symposium on Controlled Release of Bioactive Materials, New York, 10:141–143, 1983.

23. RW Korsmeyer, SR Lustig, NA Peppas. Solute and penetrant diffusion in swellable polymers. I. Mathematical modelling. J Polym Sci Polym Phys 24:395–408, 1986.

24. A Apicella, HB Hopfenberg. Water swelling behavior of an ethylene–vinyl alcohol copolymer in presence of sorbed sodium chloride. J Appl Polym Sci 27:1139–1148, 1982.

25. JH Petropoulos, PP Roussis. Non-Fickian diffusion anomalies through time lags. Some time-dependent anomalies. J Chem Phys 47:1491–1496, 1967.

26. AY Polishchuk, LA Zimina, GE Zaikov, JH Petropoulos, VMM Lobo. Simultaneous transport of water and low-molecular weight electrolytes in moderately hydrophobic polymers. Khim Fizi [Chem Phys] 16:110–117, 1997.

27. PI Lee. Dimensional change during drug release from a glassy hydrogel matrix. Polym Commun 24:45–47, 1983.

28. T Higuchi. Mechanism of sustained-action mediation. Theoretical analysis of rate of release of solid drugs dispersed in solid matrices. J Pharm Sci 52:1145–1149, 1963.

29. Al Iordansky, AY Polishchuk, LP Razumovsky. Transport of water through cellulose derivatives used as the cover of osmotic systems for drug release. Indian J Chem 31A:366–368, 1992.

30. KL Shantha, KP Rao. Drug release behavior of polyurethane microspheres. J Appl Polym Sci 50:1863–1870, 1993.

31. R Bowtell, JC Sharp, A Peters. NMR microscopy of hydrating hydrophilic matrix pharmaceutical tablets. Magn Reson Imaging 12:361–365, 1994.

32. J Rosenblatt, B Devereux, DG Wallace. Injectable collagen as a pH-sensitive hydrogel. Biomaterials 15:985–995, 1994.

33. Sah, YW Chien. Evaluation of a microreservoir-type biodegradable microcapsule for controlled release proteins. Drug Devel Ind Pharm 19:1243–1263, 1993.

34. JH Jou, PT Huang. Compatibility effect on moisture diffusion in polyimide blends. Polymer 33:1218–1222, 1992.

35. TC Dahl, IIT Sue. Mechanism to control drug release from pellets coated with a silicon elastomer aqueous dispersion. Pharm Res 9:398–405, 1992.

36. S Akhtar, CW Pouton, LJ Notarianni. The influence of crystalline morphology and copolymer composition on drug release from solution constant melt-processed P (HB-HV) copolymer matrices. J Controlled Release 17:225–234, 1991.

37. K Petrak. Biocompatible particles based on block-copolymer aggregates for intravascular administration. J Bioactive Compat Polym 8:178–187, 1993.

38. RS Harland, NA Peppas. Drug transport in and release from controlled delivery devices of hydrophilic/hydrophobic block and graft copolymers. Eur J Pharm Biopharm 39:229–233, 1993.

39. RS Harland, NA Peppas. Hydrophilic/hydrophobic block and graft copolymeric hydrogels: synthesis. J Controlled Release 26:157–174, 1993.

40. LY Shieh, NA Peppas. Solute and penetrant diffusion in swellable polymers. XI. The dynamic swelling behaviour of hydrophilic copolymers containing multiethylene glycol dimethacrylates. J Appl Polym Sci 42:1579–1587, 1991.

41. H Loth, A Euschen. Diffusion flux control by liquid–crystalline side-chain polysiloxane elastomer foils. Macromol Chem Rapid Commun 9:35–38, 1988.

42. AY Polishchuk. Models of simultaneous transport of water and drugs in hydrophilic polymer systems. Int J Polym Mater 25:37–47, 1994.

43. IC McNeill, AY Polishchuk, GE Zaikov. Thermal degradation studies of alternating copolymers: I—maleic anhydride–and vinyl acetate. Polym Degrad Stabil 37:223–332, 1992.

44. GE Krichevsky, AY Polishchuk, LB Savilova, ND Oltarzhevskaya, AL Iordansky. New technology to obtaining textile materials for prolonged action. Tekstil'naya Khim [Textile Chemistry] 1:91–100, 1992.

45. Y Okuyama, R Yoshida, K Sakai, T Okano, Y Sakurai. Swelling controlled zero order and sigmoidal drug release from thermoresponsive poly(N-isopropylacrylamide-co-butylmethacrylate) hydrogel. J Biomater Sci Polym Ed 4:545–556, 1993.

46. R Yoshida, K Sakai, T Okano, Y Sakurai. Drug release profiles in the shrinking process of thermoresponsive poly(N-isopropylacrylamide-co-alkylmethacrylate) gels. Ind Eng Chem Res 31:2339–2345, 1992.

47. AS Hoffman, A Afrassiabi, LC Dong. Thermally reversible hydrogels. In: IUPAC Macromolecules' 86 Meeting. Oxford: Butterworth-Heinemann, 1986, pp 65–70.

48. AS Hoffman, A Afrassiabi, LC Dong. Thermally reversible hydrogels: II. Delivery and selective removal of substances from aqueous solutions. J Controlled Release 4:213–222, 1986.

49. LC Dong, AS Hoffman. Thermally reversible hydrogels: III. Immobilization of enzymes for feedback reaction control. J Controlled Release 4:223–227, 1986.

50. J Verdu. Effect of ageing on the mechanical properties of polymeric materials. J Macromol Sci Pure Appl Chem A31:1383–1398, 1994.

51. A Michele, V Vittoria. Transport properties of dichloromethane in glassy polymers. Polymer 34:1898–1903, 1993.

52. MH Yang, TJ Chu. Hydraulic penetration of water through water-swollen polymer membrane: application of heat treated poly(vinylalcohol) membrane. Polym Test 12:97–105, 1993.

53. CGPH Schroen, MC Wijers, MA Choen-Stuart, A van der Padt, K van't Riet. Membrane modification to avoid wettability changes due to protein adsorption in an emulsion/membrane bioreactor. J Membr Sci 80:265–274, 1993.

54. G Zingone, F Rubessa. Release of carbamazepine from solid dispersions with polyvinylpyrrolidone/vinylacetate copolymer (PVP/VA). STP Pharm Sci 4:122–127 (1994).

55. MP Patel, M Braden. Heterocyclic methacrylates for chemical application. Biomaterials 12:645–648, 1991.

56. RD Toddywala. Effect of physicochemical properties of adhesive on the release, skin permeation and adhesiveness of adhesive-type transdermal drug delivery system (a-TDD) containing silicon-based pressure sensitive adhesives. Int J Pharm 76:77–89, 1991.

57. LB Savilova, ND Oltarzhevskaya, AY Polishchuk, AL Iordansky, GE Krichevsky. [Diffusion of drugs from bandages obtained by sealing, into the environment] (Russian) In: Development of Novel Textile Medical Products. Moscow: Pre-print of the Research Institute of Textile Materials, 1992, pp 42–47.

58. CJ Kim, PI Lee. Composite poly(vinylalcohol) beads for controlled drug delivery. Pharm Res 9:10–16, 1992.

59. TZ Fu, CJ Durning, HM Tong. Simple model for swelling-induced stresses in a supported polymer thin film. J Appl Polym Sci 43:709–721, 1994.

60. U Conte, L Maggi, A Lamanna. Compressed barrier layers for constant drug release from swellable matrix tablets. STP Pharm Sci 4:107–113, 1994.

61. LC Li, YH Tu. Latex film matrix system with a concentration gradient for controlled drug delivery. Drug Dev Ind Pharm 17:2041–2054, 1991.

62. LSC Wan, WF Lai. Multilayer drug-coated cores: a system for controlling drug release. Int J Pharm 81:75–88, 1992.

63. RS Okor, S Otimenyin, I Ijen. Coating of certain matrix cores with aqueous-based system of acrylate, a water-insoluble copolymer and drug release profiles. J Controlled Release 16:349–354, 1991.

64. KH Yuen, AA Deshmukh, JM Newton. Development and in-vitro evolution of a multiparticulate substained-release theophylline formulation. Drug Dev Ind Pharm 19:855–874, 1993.

65. HJ Griesser, RC Chatelier, TR Gengenbach, ZR Vasic, G Johnson, JG Steele. Plasma surface modification for improved biocompatibility of commercial polymers. Polym Int 27:109–117, 1992.

66. TLD Wang, DJ Lyman. Morphology of block copolyurethanes. IV. Effects of synthetic procedure on chain structure. J Polym Sci Polym Chem 31:1983–1995, 1993.

67. DJ Lyman. Bulk and surface effects on blood compatibility. J Bioact Compat Polym 6:283–295, 1991.

68. T Matsuda, S Ito. Surface coating of hydrophilic–hydrophobic block copolymers on a poly(acrylonitrile) haemodialyser reduces platelet adhesion and its transmembrane simulation. Biomaterials 15:417–422, 1994.

69. N Kislinovskaya, ID Khodzhaeva, SP Novikova, NB Dobrova. High elastic blood compatible material. Intern J Polymeric Mater 17:131–141, 1992.

70. GE Zaikov, AL Iordansky, VS Markin. Diffusion of Electrolytes in Polymers. Utrecht: VSP, 1988.

71. SD Bruck. Problems and artefacts in the evaluation of polymeric materials for medical uses. Biomaterials 1:103–107, 1980.

72. AL Iordansky, AY Polishchuk, GE Zaikov. Structural and kinetic aspects of blood plasma proteins adsorption on the surface of hydrophobic polymers. JMS Rev Macromol Chem Phys 23C:33–59, 1983.

73. AY Polishchuk, AL Iordansky, GE Zaikov. Diffusion–kinetic model of plasma proteins adsorption onto hydrophobic polymer surface. Dokl USSR Acad Sci 264:1431–1435, 1982.

74. AY Polishchuk, AL Iordansky, GE Zaikov. Kinetic aspect of the blood plasma proteins adsorption onto polymer surface. Khim Fiz [Chem Phys] 1:1268–1278, 1982.

75. AY Polishchuk, AL Iordansky, GE Zaikov. Kinetics of adsorption of proteins on the surface of hydrophobic polymer. Polym Sci USSR 26:1068–1076, 1984.

76. AY Polishchuk, VL Vladimirov, AL Iordansky, GE Zaikov, OG Fortunatov, AV Trezvova, EV Il'yakov. Ratio of reversible and irreversible protein adsorption layers on the surface of segmented polyester urethane. Polym Sci USSR 27:1327–1331, 1985.

77. AY Polishchuk, AL Iordansky, GE Zaikov. [Kinetic analysis of globular proteins adsorption onto surface of hydrophobic medical polymers] [in Russian]. Kompositsionnye Mater [Composites] 28:77–83, 1986.

78. AY Polishchuk, GE Zaikov. Testing of the thrombus-resistant properties of elastomers. Proceedings of the International Conference "Rubber'94," Moscow, 1994.

IV
WAVELENGTH SENSITIVITY
OF POLYMERS

15

Wavelength Sensitivity of Photodegradation of Polymers

AYAKO TORIKAI

Nagoya University, Nagoya, Japan

I. INTRODUCTION

Many kinds of polymeric materials have been developed and used in various industrial fields, and these amounts are increasing. It is said that the amounts of polymeric materials will be over 50% of total materials (including metals and ceramics) in the 21st century. These materials are used under the influence of terrestrial sunlight or other artificial light sources. Most polymeric materials have chromophores in the wavelength region of terrestrial sunlight. As a result, the exposure of these materials to sunlight or other light sources causes chemical changes in them, and these microscopic changes induce the deterioration of polymeric materials. Fundamental studies on photodegradation of polymers are required to develop photostable polymers and to estimate the desirable lifetime of the materials. More directly, it is important to develop photodegradable polymers and to solve the polymer litter problem.

In 1974, it was reported that partial depletion of the stratospheric ozone layer was caused by chlorofluorocarbons (1). The depletion of stratospheric ozone was actually observed in the ozone layer over Antarctica in the 1980s (2). The partial ozone depletion may lead to the enhancement of shorter-wavelength solar radiation reaching the earth. This may lead to acceleration of photodegradation of polymeric materials, resulting in a decrease of useful lifetimes in the outdoors. Because degradation of polymers is dependent on the irradiation wavelength, the changes in the wavelength and the amount of solar radiation by partial ozone depletion may affect the service life of polymeric materials.

The action spectrum, which represents the efficiency of light-induced damage of a polymer per one-incident photon, expressed as a function of irradiation wavelengths, gives crucial information on spectral sensitivity of the material. From analysis of the action spectrum of a specified polymer, it is possible to clarify the spectral region that will damage the polymer. An action spectrum is useful to know the wavelength sensitivity of a material and then, for example, to select suitable photostabilizers. The initial step of photodegradation of polymers is the photon absorption at a specified wavelength, depending on the polymer structure, impurities, or abnormal bonds. Consequently, the action spectrum offers important information with which to clarify the degradation mechanism of polymers.

Many studies have been reported on the photodegradation of polymers (3). These studies were carried out using a radiation source, such as 254 nm from a low-pressure mercury lamp, polychromatic radiation from various kinds of radiation sources, or filtered radiation through cut-on filters. Monochromatic radiation with high intensity had not yet been obtained with a conventional irradiation technique.

In 1980, a large spectrograph was built at the National Institute for Basic Biology (NIBB) in Okazaki, Japan and named as Okazaki Large Spectrograph (OLS). By using the OLS it becomes possible to study the wavelength sensitivity of polymers, as the OLS is the most efficient and the largest apparatus in the world for monochromatic radiation.

In this chapter, I will introduce the outline of OLS and the studies on the wavelength sensitivity of polymeric photodegradation by irradiation with monochromatic radiation using this spectrograph.

II. MONOCHROMATIC RADIATION SOURCES

Usually, studies on photodegradation of polymers have been carried out by using polychromatic radiation or radiation with a specific wavelength width obtained applying cut-on filters. As mentioned in the previous section, the action spectrum offers a lot of information on the photodegradation of polymers. A large-scale, powerful spectrograph that enables simultaneous monochromatic irradiations on many samples at different wavelengths is useful to carry out the irradiations efficiently and to obtain action spectra of polymers.

A. Large Spectrograph

The first large spectrograph was built in the mid-1940s in the United States. After this, several large spectrographs were constructed in the United States (1956, 1970), France (1964), and Japan (1971) (4). These spectrographs have been used to determine action spectra for vaious photobiological phenomena. With these large spectrographs, little has been known about the action spectra of whole plants because the fluence rate in the ultraviolet (UV), near-ultraviolet, and infrared (IR) regions were too low to obtain a reliable action spectra, and to make detailed studies of photoreactions requiring a high fluence rate. A more powerful spectrograph is expected to overcome these problems.

B. The Okazaki Large Spectrograph

A large-scale, powerful, and computer-controlled spectrograph was built at the National Institute for Basic Biology, Okazaki Japan in 1980 and named Okazaki Large Spectrograph (OLS). The OLS has been developed and improved for photobiological experiments and used by many researchers not only in Japan, but also in other countries. The spatial arrangement of the OLS is shown in Fig. 1

The spectrograph has two xenon short-arc lamps (30 kW and 6 kW) as its radiation sources. One of these sources is used according to the fluence rate (photon energy). Radiation from the source was dispersed into a spectrum by using a double-blazed plane grating with 1200 lines/mm. A spectrum covering UV, visible, and near-infrared region (250–1000 nm) is projected onto a 10-m–long focal curve.

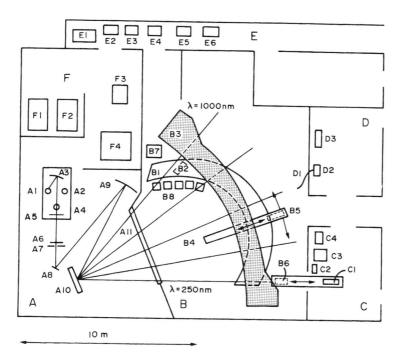

Figure 1 Spatial arrangement of the Okazaki Large Spectrograph: A, monochromator room; B, optical fiber room; C, sample box preparation room; E, microcomputer room; F, power supply room. A1, 300-kW Xe short-arc lamp; A2, 6-kW Xe short-arc lamp; A3, rotatable condensing mirror; A4, medium-pressure mercury lamp for wavelength calibration; A5, shutter; A6, heat-absorbing filter; A7, entrance slit; A8, plane mirror; A9, condensing mirror; A10, double-blazed plane grating; A11, window; B1, focal curve stage; B2, sample box; B3, x-axis frame; B4, y-axis frame; B5, arm; B6, origin; B7, interface for entrance slit control; B8, interface for control panel; C1, trolley; C2, control panel; C3, CRT (cathode ray tube) terminal; C4, printer; D1, optical fiber bundle (11 m long); D2, optical fiber outlet unit; D3, panel for monitoring; E1, 16-bit host microcomputer; E2, data typewriter; E3, CRT terminal; E4, printer; E5, NC (numerical control device) for the automatic carrier system; E6, NC interface; F1, air-cooling unit for lamps; F2, power supply for lamps; F3, control panel for lamps; F4, water-cooling unit for lamps. (From Ref. 5.)

A detailed description of this spectrograph is reported by Watanabe et al. (5). The wavelength dispersion is about 0.8 nm/cm, and the half-width at specific wavelength is about 6 nm (incident slit width, 50 nm) and 1 nm (slit width, 5 nm). The beam is focused at each wavelength by using a surface mirror (20 × 10 cm). The stability of the source is continuously monitored during irradiations at two preselected wavelengths. A long-term variation of about 2% in the photon intensity over a 10-h period was observed in this irradiation system. The photon intensity at each sample position is measured by using a specially made photon density meter for this spectrograph. The photon fluence rate expressed by photon number at an interval of 10 nm in the wavelength is shown in Fig. 2.

Irradiations with monochromatic radiation of any desired wavelength between 250 and 1000 nm can be made by placing the samples at appropriate positions on a 10-m–long focal curve. The schematic representation of the irradiation procedure is shown in Fig. 3.

The first work applying the OLS on polymer studies was performed by our group in 1989 (6). Recently (in this last 2 or 3 years), three other groups in Japan have begun exposure experiments on polymers using the OLS.

C. The Future Plan for the Okazaki Large Spectrograph

The OLS is the largest monochromatic radiation irradiator in the world and has been operating since the early 1980s at the National Institute for Basic Biology (NIBB) in

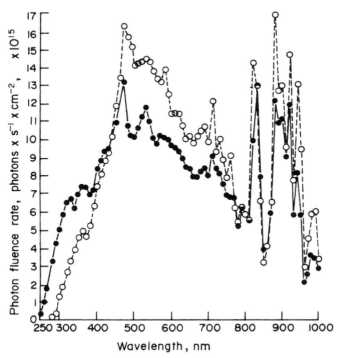

Figure 2 Photon fluence rate on the focal curve (●——●) and at the outlet of the optical fiber bundle (○----○). (From Ref. 5.)

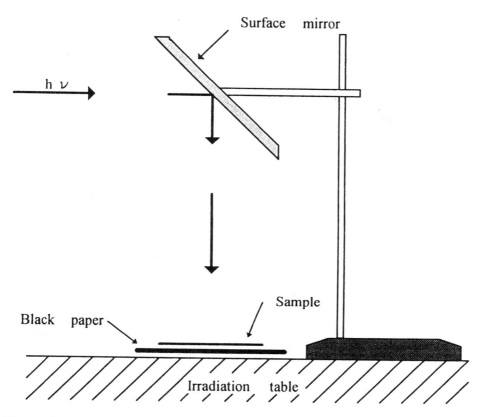

Figure 3 Schematic representation of sample irradiations.

Japan. The spectrograph has been used actively for interuniversity—in Japan and partially international—cooperative programs. For more advanced studies, a more powerful and effective large spectrograph is needed. A symposium on "New Prospects of Photobiology and the Future Plan of the Okazaki Large Spectrograph" was held in November 1996 at NIBB. In this symposium, the future plan of the large spectrograph, which is called as the super spectrograph, has been proposed. The spectrograph will be designed to obtain high-intensity, monochromatic radiation. A paper presented at this symposium, indicated that the major structure of the super spectrograph will be a collection and simultaneous use of nearly 60 newly developed high-power tunable lasers that cover 200–1000 nm and provide at least one or two orders of magnitude higher intensities (after beam expansion to a 5-cm diameter) than the OLS. If the super spectrograph is constructed in the near future, the exposure experiments with monochromatic radiation will be conducted more actively and efficiently.

I have introduced monochromatic radiation from the OLS in this section and will discuss the studies obtained by using the OLS. The studies on wavelength sensitivity by using radiations having wavelength width are given in Chapter 16 of this volume.

III. WAVELENGTH SENSITIVITY OF POLYMERS

Generally the photodegradation processes of polymers are very complex because various types of impurities, additives, or abnormal bonds in polymers absorb UV radiation. Moreover various factors affect the photodegradation processes of polymers: 1) the molecular weight and its distribution; 2) the processing technique; 3) the temperature maintained during processing; 4) the mechanical tension during the preparation of the film; 5) the density; 6) the extent and distribution of crystallinity; 7) the structure on the surface of the film; 8) the size of a crystallite; 9) the boundary region between crystalline lamellae (tie molecule); 10) any defects within the crystal cores; and 11) the orientation or mobility of the chains. Numerous problems are involved in investigating the exact mechanism for polymer degradation. Because the first step of photodegradation of polymers is the photon absorption by chromophores existing in the polymer, it is important to clarify the wavelength sensitivity of the polymer. This means to clarify what reactions occur and by which wavelength radiation. If we can specify the wavelength that induces the damage on polymers, the photodegradation mechanism of the polymer can be considered. Under these considerations, the author and collaborators have been studying the wavelength sensitivity of photodegradation of polymers. It is important to confirm the specific wavelength that causes the degradation in polymers and to clarify the reaction mechanisms involved in such photodegradation.

Recent progress of wavelength sensitivity of photoinduced reaction of individual polymers is reviewed in the following section. Attention is especially focused on the studies carried out using the OLS.

A. Synthetic Polymers

The wavelength sensitivities of photodegradation of synthetic polymers that are widely used in various industrial fields have been studied. Poly(vinyl chloride), polymethyl methacrylate, polycarbonate, polystyrene, polyolefins, and others are included in this section.

1. Polyvinyl Chloride

a. *Action Spectrum of Rigid PVC Formulations*

Poly(vinyl chloride) (PVC) is used mainly as siding or door and window frames in the building industry. This material undergoes photoyellowing on exposure to sunlight. Photodegradation of PVC results in the formation of conjugated double bonds and carbonyl groups.

Andrady et al. used the yellowness index (YI) as a measure to monitor photodegradation of PVC (6,7). Rigid PVC formulations containing 0, 2.5, 5 parts per hundred (phr) of titanium oxide showed wavelength-dependent yellowing or photobleaching in the range of irradiation wavelengths between 280 and 500 nm. Yellowing was observed in irradiation at 280, 300, 320, and 340 nm, whereas photobleaching occurred with the irradiation of wavelength at 400 and 500 nm. The light-induced yellowing can be expressed quantitatively as linear plot of $\ln(\Delta \text{YI}/\text{photon})$ versus the irradiation wavelength having the following intercept a and gradient b, where r is the correlation coefficient for the data.

$$\ln(\Delta \text{YI}/\text{photon}) = a + b\,\lambda\,(\text{nm})$$

0 phr of TiO_2 $a = -31.1$ $b = -0.048$ $r = 0.99$
2.5 phr of TiO_2 $a = -30.6$ $b = -0.058$ $r = 0.98$
5 phr of TiO_2 $a = -27.9$ $b = -0.073$ $r = 0.99$

The numerical value of the gradient b (nm^{-1}) quantifies the dependence of yellowing on the wavelength of irradiation. The plots describing the action spectra for yellowing of PVC formulations are shown in Fig. 4.

Irradiations at 400 and 500 nm caused photobleaching rather than photoyellowing of PVC formulations. The yellowing of PVC under the terrestrial sunlight is the net result of competing yellowing and photobleaching reactions. The mechanism of photobleaching is not fully understood. It has been reported by several authors that quenching of polyenyl radicals by oxygen and possible reaction of polyenes with hydrogen chloride products formed during photoinduced dehydrochlorination occurs.

b.   Accelerated Degradation of PVC

Pure PVC has no chromophores in the wavelength region of terrestrial sunlight. The photodegradation may start from the photon absorption by chromophores, such as impurities or abnormal bonds. We have investigated accelerated degradation of PVC by introducing chromophores having the absorption in UVB region into PVC (8). This study is useful to find a method to degrade polymer waste materials that are accumulating in the environment.

PVC that was preirradiated with radiations of shorter wavelengths was irradiated with wavelength radiation longer than 290 nm. If the accelerated degradation occurs, this method will become an excellent tool for the polymer waste prob-

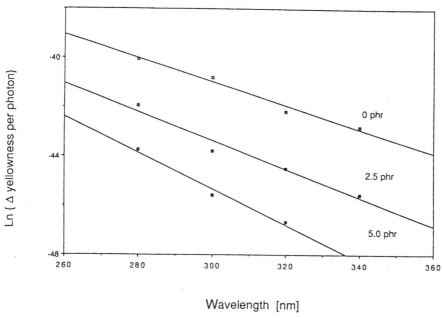

Figure 4   Monochromatic action spectra for light-induced yellowing in rigid PVC formulations containing different levels of titanium dioxide (0, 2.5, and 5 phr). (From Ref. 6.)

lem. Under these considerations we have studied the accelerated degradation of PVC introducing the chromophores into PVC with the irradiation of monochromatic radiation.

PVC films were irradiated with radiation from a low-pressure mercury lamp. By this irradiation, conjugated double bonds and carbonyl group formations in PVC were confirmed by UV and Fourier-transform infrared (FTIR) spectra. The preirradiated samples were irradiated with monochromatic radiation from the OLS at nine wavelengths from 275 to 500 nm. Photodegradation of these samples was estimated by UV-visible, FTIR spectroscopy, and gel permeation chromatography (GPC) measurements. The typical molecular weight distribution curves are shown in Fig. 5. At the longer wavelength irradiation, a decrease in molecular weight in the preirradiated PVC was confirmed.

Changes in the number-average molecular weight (n) represented by the relative molecular weight change (n/n_0) are plotted against the irradiation wavelength in Fig. 6.

In PVC samples without preirradiation, this value is 1 (showing no molecular weight change observed) with the irradiation by the wavelengths longer than 400 nm. The degradation of PVC is accelerated with preirradiation at all wavelengths, and the threshold wavelength of degradation shifts to the wavelength that is longer than 400 nm. This acceleration is dependent on the time of preirradiation. Accelerated photodegradation of preirradiated PVC is also confirmed by UV-visible and FTIR spectroscopic measurements. The results obtained from these measurements reasonably coincided with those from molecular weight measurement.

2. Polymethyl Methacrylate

a. *Effect of Irradiation Wavelength*

Polymethyl methacrylate (PMMA) films were irradiated with monochromatic radiation from the OLS in air and in a vacuum at 23°C. The effect of irradiation wavelength on the quantum yields of main-chain scission (Φ_{cs}) was studied (9,10). Values of Φ_{cs} were calculated from the viscosity-average molecular weights of PMMA before and after photoirradiations and by using the absorption spectrum of unirradiated PMMA film. The results are shown in Table 1.

Although Φ_{cs} values are almost the same by irradiating with 300 nm radiation, the Φ_{cs} values obtained in vacuum at 260 and 280 nm are about half those in air. From these results, we proposed that two main-chain scission processes are included in photodegradation of PMMA. A main-chain scission process,

Table 1 Quantum Yields of Main-chain Scission (Φ_{cs}) at Specified Irradiation Wavelengths in Air and In Vacuo

$\Phi_{cs} \times 10^4$	Wavelength (nm)		
	260	280	300
In air	2.07	2.34	4.31
In vacuo	0.84	1.06	4.21

Source: Ref. 10.

initiating from the side-chain scission, plays a role in the photolysis at shorter wavelengths (260 and 280 nm). This idea was confirmed by electron spin resonance (ESR) measurements of photoirradiated PMMA at 260, 280, and 300 nm in vacuum at 77 K and 23°C (11). By irradiating at 260 and 280 nm, the radicals pro-

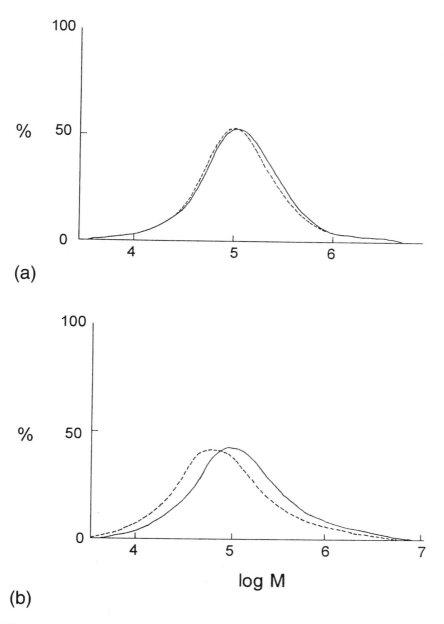

Figure 5   Molecular weight distribution curve of (a) photoirradiated PVC and (b) PVC preirradiated for 10 h: solid line, unirradiated; dashed line, irradiated at 450 nm. (From Ref. 8.)

duced from the side-chain scission, such as $\cdot CH_3$, $\cdot CHO$, and $\cdot COOCH_3$ were detected, adding to the radical resulting from the homolytic main-chain scission. Only the radical from the homolytic main-chain scission was produced by irradiation at 300 nm.

When irradiating in air, oxygen attack to an on-chain radical formed by side-chain scission may occur according to the Scheme 1:

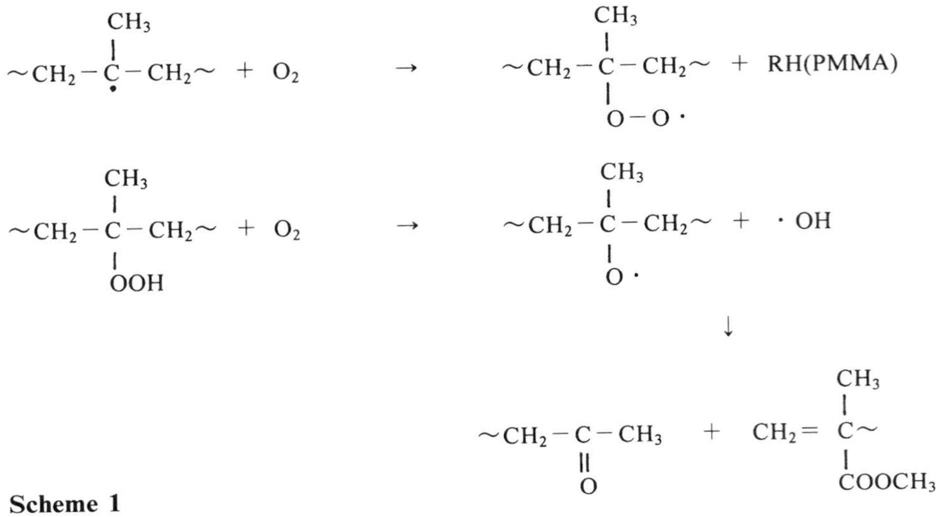

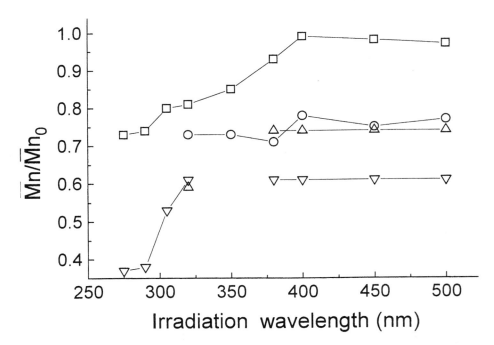

Scheme 1

Figure 6 Changes in the number-average molecular weight of photoirradiated PVC: Preirradiation time: □, 0 h; ○, 1 h; △, 3 h; ▽, 10 h. (From Ref. 8.)

Side-chain scission followed by main-chain scission of PMMA participates when the irradiations are carried out at 260 and 280 nm.

The threshold wavelength for the main-chain scission of PMMA was close to 320 nm.

b.   Effect of Photon Intensity

The effect of photon intensity (per unit time and unit square) on photodegradation of PMMA is a factor that should be explored for predicting the useful life of polymer material under terrestrial sunlight or other artificial light. PMMA films were irradiated changing the photon intensity at the same wavelength. Total photon fluence was adjusted to a constant value. Changes in viscosity molecular weight ($\bar{M}_v$) are used as a measure of degradation (9). A typical result is shown in Fig. 7 for the irradiation wavelength at 300 nm.

A linear relation between $\bar{M}_v$ and photon intensity was observed. This means that, when compared, at the same wavelength and total photon fluence, the decrease of molecular weight tends to be greater for the lower photon intensity. Similar trends were also found for irradiation at 260 and 280 nm.

c.   Accelerated Degradation by Additives

The threshold wavelength of main-chain scission of PMMA is close to 320 nm. Benzophenone (BP) is an excellent photosensitizer, having its λ_{max} at a wavelength longer than 320 nm. The threshold wavelength may shift to wavelengths longer than

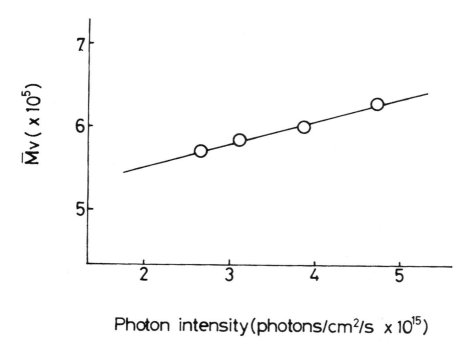

Figure 7   Effect of incident photon intensity on changes in viscosity-average molecular weight ($\bar{M}_v$): irradiation wavelength: 300 nm; total photon intensity, 3.83×10^{19} photons/cm^2. (From Ref. 9.)

320 nm, and an accelerated degradation of PMMA in the presence of BP can be expected. PMMA films containing BP in low concentrations (0.1, 0.3, 0.5, and 1.0 mol%) were irradiated with monochromatic radiation from the OLS (12). By the addition of BP to PMMA, the threshold wavelength of PMMA's main-chain scission shifted to 380 nm, as expected. The main-chain scission of PMMA was accelerated at any wavelength of irradiation; this acceleration was concentration dependent. Moreover, photo–cross-linking of PMMA was observed in this system, although PMMA is known to be a degradation-type polymer for photoirradiation. Gel formation in PMMA–BP is shown in Fig. 8.

The authors proposed the reaction scheme for accelerated degradation and cross-linking reaction of PMMA in the presence of BP.

The same authors also studied wavelength sensitivity in photodegradation of PMMA, in the presence of β-carotene, by optical absorption and GPC measurements (13). Main-chain scission without gel formation was observed. Accelerated main-chain scission of PMMA by β-carotene was confirmed by molecular weight changes in PMMA, as shown in Fig. 9. The threshold wavelength for main-chain scission shifted to the longer wavelength (400 nm) compared with that of PMMA with no additive (320 nm).

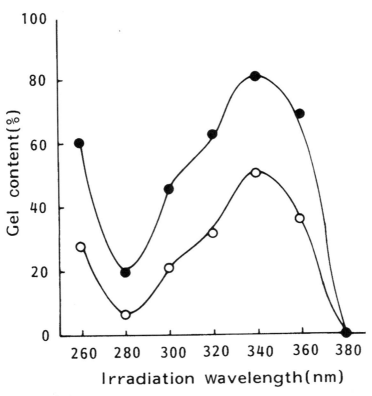

Figure 8   Changes in the gel content of BP–PMMA films photoirradiated with various wavelengths: concentration of BP, (○), 0.1 mol%; (●), 0.3 mol%. (From Ref. 12.)

3.   Polycarbonate

Irradiation of polycarbonate (PC) with polychromatic radiation leads to extensive yellowing in the polymer. Two different processes responsible for yellowing are suggested. One is the photo-Fries rearrangement, and the other is a photooxidative degradation process. Photo-Fries rearrangement of PC produces phenyl salicylate and dihydroxybenzophenone-type compounds. Photooxidative degradation produces an oxidation product, such as a carbonyl group. We have previously investigated the photodegradation of PC that was irradiated with a medium-pressure mercury lamp (14). Changes in chemical structure and in molecular weight were followed by UV-visible and FTIR spectra and by viscosity measurements. Reaction intermediates produced by photoirradiation were followed by ESR measurements. We proposed the reaction mechanism for both processes. Andrady et al. reported the action spectrum for yellowing of PC by monochromatic radiation using the OLS (15). The dependence of yellowing in the wavelength region 280–340 nm is presented by plotting Ln (yellowness index) as a function of irradiation wavelength. The plot is linear and is described by the following equation.

$$\mathrm{Ln}\, y = -24.2 - 0.082\lambda \qquad r = 0.99$$

We reported the wavelength sensitivity of the quantum yield of main-chain scission calculated from intrinsic viscosity as well as the estimated efficiency of the photo-Fries rearrangement (16), assuming the absorption band at 320 nm is related to the rearrangement products. The results are shown in Fig. 10.

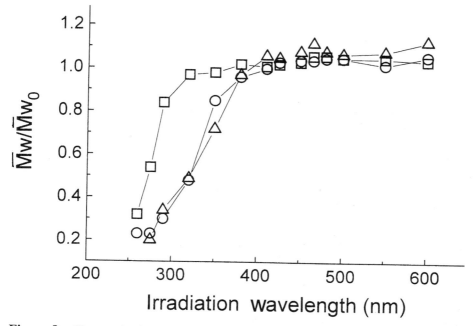

Figure 9   Changes in the molecular weight in photoirradiated PMMA and PMMA containing β-carotene: (□), PMMA; (○), PMMA containing 0.5 mol% of β-carotene; (△), PMMA containing 1.0 mol% of β-carotene. (From Ref. 13.)

Under our experimental conditions, changes in molecular weight and in UV spectra were not found in photoirradiation with radiation longer than 320 nm. Fukuda and Osawa also studied the photodegradation of PC using IR absorbance at 1687 cm^{-1} as a measure of photo-Fries rearrangement (17). Only the shorter-wavelength irradiation showed the product formation. No product formation was observed by irradiation with radiation wavelengths longer than 300 nm. Absorbance in the UV at 320 nm also showed wavelength dependence in the range of $\lambda = 260$–320 nm. These results coincide with our observations. All these studies were carried out under similar conditions using the same irradiation source (OLS), but on different grades of bisphenol A polycarbonate. Recently, Andrady compared these three data, as obtained by different authors, by the plots of Ln (damage/photon) versus irradiation wavelength (18). The results are compared in Table 2. The data show the dependence of action spectra on the type of photoreaction studied. He concluded from these data that the high degree of linearity observed was significant and showed that the effectiveness of photons bringing about photodamage was logarithmically related to photon energy.

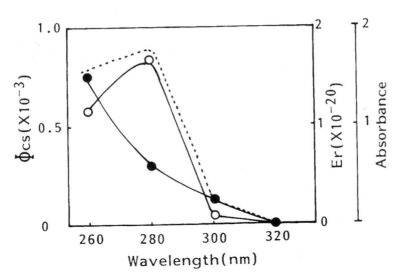

Figure 10 Quantum yield of main-chain scission (Φ_{cs}) (●) and efficiency of photo-Fries rearrangement (*Er*) (○) versus irradiation wavelength: The dashed line shows the absorption spectrum of unirradiated PC film.

Table 2 Comparison of Data from Three Different Experiments Represented as Ln(Damage per Photon) Versus Irradiation Wavelength

Ref.	Property	Gradient	Value r
Andrady et al. (15)	Yellowness index	−0.082	0.99
Torikai et al. (16)	Quantum yield (scission)	−0.044	0.99
Osawa et al. (17)	Absorbance (320 nm)	−0.059	0.88

Source: Ref. 18.

4. Polystyrene

Polystyrene (PS) undergoes yellowing by photoirradiation. Yellowing of PS by photoirradiation is due to the formation of conjugated double bonds or other types of chromophoric groups. Action spectra estimated by the changes in optical density at $\lambda = 310$ nm and by the number of main-chain scissions of melt-pressed PS films were reported (19). The changes in optical density at 310 nm per incident photon were a monotonically decreasing function of the wavelength in the range of $\lambda = 260$–320 nm. The threshold wavelength was close to 320 nm. Main-chain scission of PS was estimated by GPC measurement, and was observed by irradiation with 260, 280, and 300 nm. The most efficient irradiation wavelength for main-chain scission was 280 nm.

a. Polystyrene-Containing Flame Retardants

For PS containing 2 phr of flame retardant, either decabromodiphenylether (DBDE) or tetrabromobisphenol A (TBA), the maximum efficiency of main-chain scission was observed at 300 nm (19). The presence of these flame retardants, accelerates the main-chain scission of matrix PS, and the threshold wavelength shifted to a wavelength longer than 320 nm. On the basis of these experiments, we compared the effect of nine flame retardants, containing bromine in the molecule, on the efficiency of photodegradation of matrix PS (20). The melt-pressed films were photoirradiated with monochromatic radiation using the OLS. The concentration of each additive in PS was 1 phr. The photodegradation was estimated by UV-visible, FTIR, and GPC measurements. Flame retardants used as additives and their abbreviated names are summarized in Table 3.

The efficiency of photodegradation was compared at the irradiation wavelength of 280 nm, which was the most effective irradiation wavelength for photodegradation of PS for each flame retardant except one. Molecular weight distribution curves of photoirradiated PS and PS containing typical flame retardants are shown in Fig. 11.

Although the photodegradation of PS estimated by these three methods showed some irregularities in the order of efficiency of flame retardants, the most

Table 3   Flame Retardants and Their Abbreviated Names

Flame retardant	Abbreviated name
2,2-Bis[4-(2,3-dibromopropxy)-3,5-dibromophenyl]propane	TBA-BP
TBA-epoxy oligomer (uncapped)	TBA-EPO-U
TBA-epoxy oligomer (capped)	TBA-EPO-C
1,2,5,6,9,10-Hexabromocyclododecane	HBCD
Tris(tribromoneopentyl)phosphate	TBNPP
1,2-Bis(pentabromophenyl)ethane	PBPE
1,2-Bis(tetrabromophthalimido)ethane	ETBP
Tetrabromobisphenol-A	TBA
Decabromodiphenylether	DBDE

Source: Ref. 20.

efficient additive for the acceleration of the degradation of PS was TBA. On the contrary, ETBP or TBNPP can protect against the photodegradation of PS. The photostability of PS containing ETBP or TBNPP was excellent in these experiments.

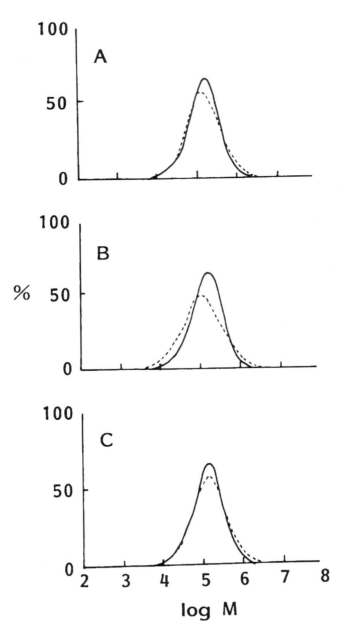

Figure 11   Molecular weight distribution curve of photoirradiated (A) PS; (B) PS–TBA; and (C) PS–ETBP: solid line, unirradiated; dashed line, irradiated at 280 nm to total photon fluence, 4.75×10^{19} photons/cm^2. (From Ref. 20.)

The order of efficiencies of flame retardants for the degradation of PS at a wavelength of 280 nm estimated by main-chain scission of PS is as follows:

$$\text{ETBP} < \text{TBNPP} < \text{none} < \text{TBA} - \text{EPO-C} < \text{PBPE} < \text{TBA-EPO-U} < \text{DBDE}$$
$$< \text{HBCD} < \text{TBA-BP} < \text{TBA}$$

The threshold wavelength of photodegradation of PS lies between 300 and 320 nm for each flame retardant, although it shifts to 360 nm for DBDE.

From the foregoing observations, the efficiency of photodegradation of PS that contains a flame retardant was thought to be affected by the chemical structure or other factors of the additives. The authors continued to study the effect of flame retardants on the photodegradation of matrix PS (21). Four kinds of flame retardants, containing bromine in the molecule and having the same frame structure, were compared. The flame retardants used were TBA, TBA-BP, TBA-EPO-U, and TBA-EPO-C. The effect of additive concentration on the efficiency of photodegradation of PS was also investigated. Irradiation of monochromatic radiation was carried out using the OLS. The degradation of matrix PS was followed by UV-visible and FTIR spectra for the chemical change produced. Changes in the molecular weight and its distribution were estimated by GPC. In all additives tested, the photodegradation of PS was accelerated, as in the former studies. The most effective wavelength for the main-chain scission of PS was 280 nm for the samples containing 1 phr of flame retardants, whereas this wavelength shifted to 300 nm for the samples containing 2 phr of flame retardant. A probable reaction scheme of the accelerated degradation in the presence of flame retardant was proposed on the basis of the experimental results. Differences in the efficiency of degradation may be interpreted in terms of chemical structure of the additive and compatibility of the additive with PS.

5. Polyolefins

Pure polyolefins contain only C–C and C–H bonds, and the bond dissociation energy of these bonds are reported to be 348 kJ/mol and 413 kJ/mol, respectively. Accordingly, the absorption spectra of polyolefins should not exceed 200 nm. Generally, however, they have an extended spectra to about 300 nm, which is responsible for the internal abnormal bonds, such as conjugated double bonds or many kinds of impurities. In spite of several studies reported on photooxidation of polyolefins, the details of the reaction mechanism of photodegradation remains unsolved. The authors studied the elementary processes of photodegradation of polypropylene (PP) by ESR measurements at 77 K and from elevated temperatures to ambient temperature (22). Alkyl radical formation is the first step in the photodegradation of polyolefins. The alkyl radicals formed at 77 K changes to an allyl-polyenyl-type radical by warming. In the presence of oxygen, the peroxy radical was detected by ESR measurement. The radical formation needs the photon absorption by some chromophores that are present in polyolefins as impurities or abnormal bonds. Primary initiation species in PE and PP are believed to be mainly ketones and hydroperoxides. Detailed information on photodegradation of polyolefins is given in a recent review (23).

a. Polypropylene

The photoirradiation of PP with monochromatic radiation using the OLS may lead
to the degradation of PP. For photoirradiated PP, the increase in the absorbance
at 310 nm is too low to measure in the wavelength range from 260 to 360 nm. Even
at 260 nm, only a minimal increase was observed. This situation is the same for
the measurements of GPC. Changes in the molecular weight of PP without additives
were not observed under our experimental condition (total photon fluence, 9.0×10^{19}
photons/cm^2) (18). For PP containing either DBDE or TBA as an additive at a 2%
level, main-chain scission of PP took place. The results are given in Table 4.

The rate of main-chain scission of PP containing DBDE or TBA was accelerated at specified irradiation wavelengths. The maximum efficiency of main-chain
scission was 280 nm for DBDE and 260 nm for TBA, respectively. In PP containing
flame retardant, carbonyl group formation was favored compared with that of PP
without additives. The threshold wavelengths of photodegradation are 320 and
360 nm for TBA and DBDE, respectively. This result is also supported by the
changes in the absorbance at 310 nm.

b. Polyethylene

Two kinds of PE—linear low-density PE (LLDPE) and high-density PE
(HDPE)—samples were photoirradiated with monochromatic radiation, and their
wavelength sensitivity was studied by spectral and molecular weight change analysis
(24). On photoirradiation of PE and additive (DBDE and TBA) containing PE, each
sample develops an increase in the intensity of absorbance in the region of
260–350 nm of the UV spectrum. The increase in the absorbance at 310 nm
(ΔOD_{310}) was chosen as a measure of degradation of PE, as in PS and PP. Although
we have not assigned the origin of this absorption band, one possibility is a conjugated double bond or ketonic-type compound. The ΔOD_{310} per exposed photon
is plotted against irradiation wavelength in Figs. 12 and 13 for LDPE and HDPE,
respectively.

The ΔOD_{310} is zero for the samples without additive, except when being
irradiated at 260 nm. For additive-containing PE, ΔOD_{310} increased at specified
wavelengths, as shown in Figs. 12 and 13. This result shows the acceleration of

Table 4 Changes in the Number-Average Molecular Weight of Photoirradiated PP,
PP–TBA, and PP–DBDE Samples

Irradiation wavelength (nm)	Total fluence (photons/cm^2) ($\times 10^{-18}$)	$\bar{M}_n (\times 10^{-4})$		
		PP	PP–TBA	PP–DBDE
	0	3.4	3.3	3.4
260	60	3.8	3.2	2.1
280	90	3.8	2.6	2.3
300	90	3.5	3.3	2.5
320	90	3.6	3.4	2.8
340	90	3.5	2.9	3.3
360	90	3.3	3.4	3.8

Source: Ref. 19.

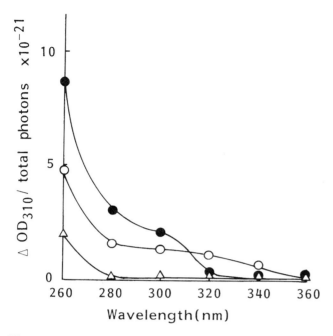

Figure 12 Changes in the optical density at 310 nm at various irradiation wavelengths: △, LDPE; ○, LDPE–DBDE (1.0 phr); ●, LDPE-TBA (1.0 phr). (From Ref. 24.)

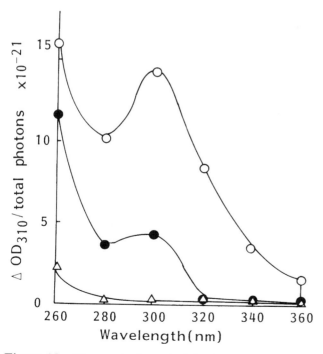

Figure 13 Changes in the optical density at 310 nm at various irradiation wavelengths: △, HDPE; ○, HDPE–DBDE (1.0 phr); ●, HDPE–TBA (1.0 phr). (From Ref. 24.)

the degradation of PE by flame retardant similar to that of PS and PP. A yellow coloration was formed when irradiations were carried out with wavelengths less than 320 nm for all additive-containing HDPE samples. No coloration was observed for the samples without additive. The deepest coloration was observed on irradiating the additive-containing HDPE samples at 300 nm. Little coloration was found for LDPE containing additive, except for the samples containing TBA as an additive when irradiated at 300 nm. The increase in ΔOD_{310} value may be responsible for this yellowing.

Carbonyl group formation was followed by the changes in carbonyl index (A_{1720}/A_{1720-0}). PE without additive shows the value of 1 under our experimental condition (total photon fluence, 8.0×10^{19} photons/cm^2). PE samples containing flame retardant show the increase in the absorption at 1720 cm^{-1}, and the most effective wavelength was 300 nm, similar to that in UV-spectral analysis.

Changes in the number-average molecular weight before and after photoirradiation were not found in all samples. We examined the changes in the molecular weight component using the molecular weight distribution (MWD) curve. Three major molecular weights for each PE sample were chosen for this treatment. Changes in the molecular weight component before and after photoirradiation were plotted against the irradiation wavelength. The following results were obtained from this analysis. Main-chain scission was favored at the irradiation wavelength of 300 nm for HDPE–DBDE and HDPE–TPA samples. The most effective irradiation wavelengths for cross-linking are 300 and 280 nm for LDPE–DBDE and LDPE–TBA samples, respectively. It is difficult to clarify the role of flame retardant in photodegradation of polyolefins, although the accelerated photodegradation of polyolefins by these additives was confirmed.

6. Polyphenylene Ether

Polyphenylene ether (PPE) is sensitive to photoirradiation, which is known to cause its photoyellowing. Takeda et al. studied the yellowing process by UV and FTIR spectroscopic measurements (25). Photodegradation of PPE took place with irradiation by wavelengths shorter than 300 nm. They also found an insoluble fraction in photoirradiated PPE, and suggested that the cross-linking products were responsible for the yellowing.

7. Polysulfone

Kuroda et al. reported that the threshold wavelength of photoyellowing of polysulfone (PSF) was close to 340 nm (26). They found that the action spectrum of photoyellowing did not coincide with the absorption spectrum of PSF. The action spectrum is similar to the fluorescense and phosphorescense spectra. Quenching efficiency for phosphorescense was similar to the stabilizing efficiency for photoyellowing in three kinds of triplet quenchers, as shown in Table 5. They concluded that the photoyellowing reaction took place mainly by triplet excited state.

8. Polymer Blends and Copolymer

a. *Polycarbonate–Polymethyl Methacrylate Blend*

On photoirradiation of PC–PMMA blend with polychromatic radiation, the photo-Fries rearrangement reaction of PC was not inhibited by blending with PMMA,

Table 5 Quenching Efficiency for Phosphorescence and Stabilizing Efficiency for Photocoloration of PSF[a]

Triplet quencher[b]	Er (kJ/mol)	Quenching efficiency[c] (%)	Stabilizing efficiency[d] (%)
Biphenyl	277	61	49
m-Terphenyl	272	63	61
o-Terphenyl	260	55	46

[a] Er of PSF; 272 kJ/mol.
[b] Concentration in PSF film; 5 mol%.
[c] Excitation at 320 nm.
[d] Irradiation wavelength; 320-nm—irradiation dose; 0.5×10^{19} photons/cm^2.
Source: Ref. 26.

whereas the photodegradation of PMMA was reduced in the presence of PC. A model for photodegradation mechanism of this blend was proposed by Fukuda and Osawa (27). To confirm this model, they studied the wavelength sensitivity of photodegradation of each component (17). The degree of photodegradation in PC was followed by the changes in the absorbance at 1687 cm^{-1} that was assigned to the product formed by photo-Fries rearrangement. Changes in this band were observed only during irradiations with wavelengths at 260 and 280 nm. No significant change in infrared spectrum of PMMA irradiated with monochromatic radiation was observed other than these two wavelengths. UV-spectral changes in PC and PMMA were measured, and the changes in absorbance of PC at 320 nm and of PMMA at 280 nm were plotted against the irradiation wavelength. The threshold wavelength for photodegradation of each polymer was estimated to be 300–320 nm for PC and about 255 nm for PMMA. The results indicate that the transmissivity of each component plays an important role in the photodegradation of PC–PMMA blends and the photodegradation of PMMA is inhibited by the screening of PC film.

b. Ethylene–Carbon Monoxide Copolymer

Ethylene–carbon monoxide copolymer (E–CO; CO, 2.4%) is a photodegradable plastic material, because the carbonyl group has the absorption band in the UV region. These polymer films were irradiated with monochromatic radiation and changes in the absorbance at 1719 cm^{-1} ($\nu_{c=0}$ was measured (28). Figure 14 shows the changes in optical density at 1719 cm^{-1} (OD$_{1719}$) versus irradiation wavelength. This value did not change with the irradiation wavelength between 340 and 500 nm. The OD$_{1719}$ decreases with the decrease of irradiation wavelength range of 260–320 nm. This result indicates that threshold wavelength of photodegradation of E–CO copolymer lies close to 320 nm.

B. Natural Polymers

A small study was reported on the wavelength sensitivity of natural polymers compared with that of synthetic polymers. Recently, studies on lignocellulose; chitosan, and collagen were reported.

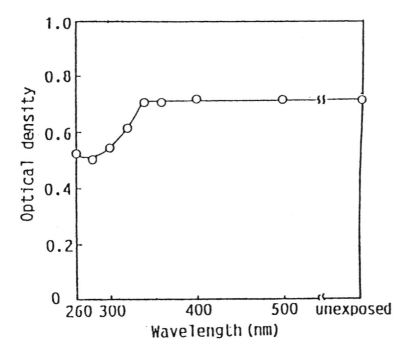

Figure 14 Optical density at 1719 cm^{-1} versus wavelengths of irradiated light for E–CO. (From Ref. 28.)

1. Lignocellulose

Lignin contains several chromophores and efficiently absorbs UV radiation and visible light. Both photoyellowing and bleaching were observed with UV irradiation. Photochemistry of the yellowing and photobleaching in lignin-containing pulp has been widely studied (29), although the mechanism and kinetics of these reactions are not fully elucidated. Andrady et al. studied the wavelength dependence of light-induced yellowing and bleaching dependence on the paper made from mechanical pulp by monochromatic radiation (30). They studied the samples exposed to monochromatic radiation in two modes: 1) constant exposure time at each wavelength; and 2) constant number of photons at each wavelength. Both photobleaching and photoyellowing were observed as wavelength-dependent phenomena in the wavelength range of 260–600 nm. Transition from photoyellowing to photobleaching was observed between 340 and 400 nm. The increase in yellowness (ΔYI) per available photon decreased logarithmically in the region of 260–340 nm, as shown in Fig. 15.

Action spectra obtained from these two types of exposure show the linear relation with a similar gradient (Table 6). The results also suggest that the change in yellowness index is a linear function of light intensity at a given wavelength.

2. Chitosan

Chitosan is derived from chitin, which is the second most abundant biopolymer (after cellulose), by deacetylation. Chitosan is a generic product that is a copolymer

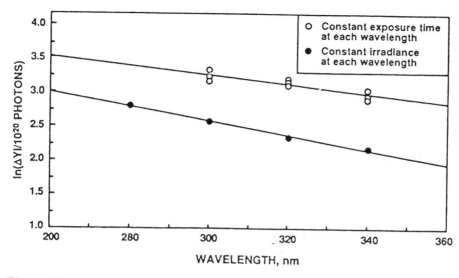

Figure 15 Log yellowing efficiency (ΔYI/10^{20} photons) as a function of wavelength for set U samples from both experiments. (From Ref. 30.)

Table 6 Intercept and Gradient of Action Spectra for Photoyellowing Under Constant Photon and Constant-Duration Exposure Conditions

Sample	Intercept	Gradient	r^2
	Constant-photon exposure (1×10^{20} photons)		
1	6.23	0.010	0.99
2	4.58	0.005	0.88
3	5.48	0.007	0.96
4	5.26	0.007	0.94
Average	5.76	0.011	0.99
	Constant-duration exposure (10 h)		
Average	—	0.011	0.94–0.99

Source Ref. 30.

consisting of (β-(1,4)-2-acetamino-D-glucose) units and (β-(1,4)-2-amino-D-glucose) units, with the latter usually exceeding 80%. Chitosans are thus described in terms of the degree of deacetylation and the average molecular weight. Chitosans form clear, tough, and water-insoluble films and have been used as a wet-strength additive in paper for packaging applications, nonwoven fabric, fibers, and other uses. The photodegradation of chitosan has not been reported. However, the chemical structure of chitosan is similar to that of cellulose, and the photodegradation of cellulose has been extensively studied. The authors investigated the wavelength sensitivity of photodegradation of chitosan on exposure to monochromatic radiation (31).

Solvent-casted chitosan films were irradiated with monochromatic radiation using the OLS at eight different wavelengths. UV-visible, FTIR spectroscopic, and viscosity measurements were made after irradiations at selected wavelengths. Changes in absorbance at $\lambda = 310$ nm per incident photon are given in Fig. 16.

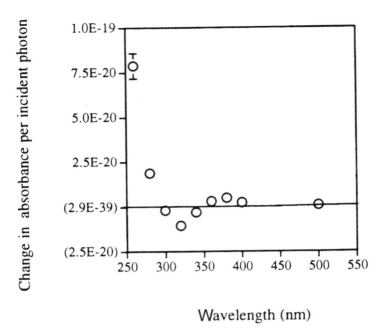

Figure 16   Changes in absorbance of chitosan films at 310 nm per photon incident on sample surface as a function of the wavelength of irradiation. (From Ref. 31.)

The photodamage estimated by the increase in the absorbance at $\lambda = 310$ nm decreases sharply with the irradiation wavelength. For several wavelength regions in UVA, the absorbance of irradiated sample is lower than that of the control sample. Both the increase and the decrease in the chromophore responsible for the absorbance at 310 nm were observed as wavelength dependent. They also studied the effect of intensity on the photodegradation of chitosan in the wavelength region 260–500 nm. As shown in Fig. 17, only the samples irradiated at 260 and 280 nm showed the intensity dependence of photodegradation. In these samples, the span of intensity levels studied was too narrow to draw any conclusions about the intensity effect.

The data obtained by the exposure at 280 nm suggest the linear relation between the decrease in the absorbance at 310 nm and the intensity of monochromatic radiation, indicating that some radical process contributes in photodegradation of chitosan. Changes in viscosity in dilute acetic acid solution were measured. Wavelengths shorter than 340 nm showed measurable decrease in viscosity by photoirradiation, indicating the main-chain scission of chitosan during irradiation. An insoluble gel fraction was not detected; hence, the cross-linking may not be included in this experimental condition. The photochemical changes in polymer structure can be followed by FTIR spectra. An increase in the carbonyl and amino groups in the polymer after irradiation was suggested by the FTIR spectroscopic analysis. From the data obtained, the following reaction schemes (Schemes 2 and 3) were proposed for the photodegradation of chitosan, although further work is needed to clarify a complete mechanism.

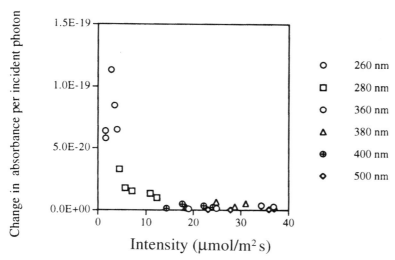

Figure 17 Intensity dependence of the efficiency of photodegradation as measured by the changes in absorbance at 310 nm of chitosan, obtained at different irradiating wavelengths. (From Ref. 31.)

3. Collagen

Collagen is one of the popular proteins consisting of various kinds of amino acids, mainly glucine, alanine, proline, and phenylalanine. Other kinds of amino acids were also included in collagen. Collagen is the constituent of human skin, bone, and nail. The photoaging of collagen by UVA or UVB radiation is well known. It is said that the insoluble fraction in collagen increases with the increase in age because of photocross-linking reactions in collagen. The photodegradation of collagen may result from denaturation of protein or photocross-linking between polymer molecules. Stratospheric ozone depletion induces the enhancement of UVB radiation reaching earth's surface; this radiation will cause skin cancer or other diseases by photoinduced damage in proteins. Under these considerations we have studied the wavelength sensitivity of photodegradation of collagen to clarify the reaction mechanism chemically (32). We also investigated the effect of vitamin E on photodegradation of collagen to find an effective quencher against photoinduced damage to collagen. Collagen type I (from calf skin) was dissolved in dilute acetic acid, and solvent-casted films were irradiated with polychromatic radiation from a medium-pressure mercury lamp and monochromatic radiation using the OLS. From the results of UV spectrum of photoirradiated collagen, the absorbance at $\lambda_{max} = 280$ and 235 nm increased in their intensities with the irradiation time. These increases may be attributed to the formation of tyrosine from phenylalanine for 280-nm band and the decrease of peptide linkage by photoirradiation for 235-nm one. Formation of tyrosine is said to be initial step for the photoaging of animal skin.

$$
\underset{\text{Phenylalanine}}{\overset{\overset{\text{NH}_2}{|}}{\text{Ph}-\text{CH}_2-\text{CHCOOH}}} \; \underset{\longleftarrow}{\overset{\longrightarrow}{}} \; \underset{\text{Tyrosine}}{\text{HO}-\overset{\overset{\text{NH}_2}{|}}{\text{Ph}-\text{CH}_2-\text{CHCOOH}}}
$$

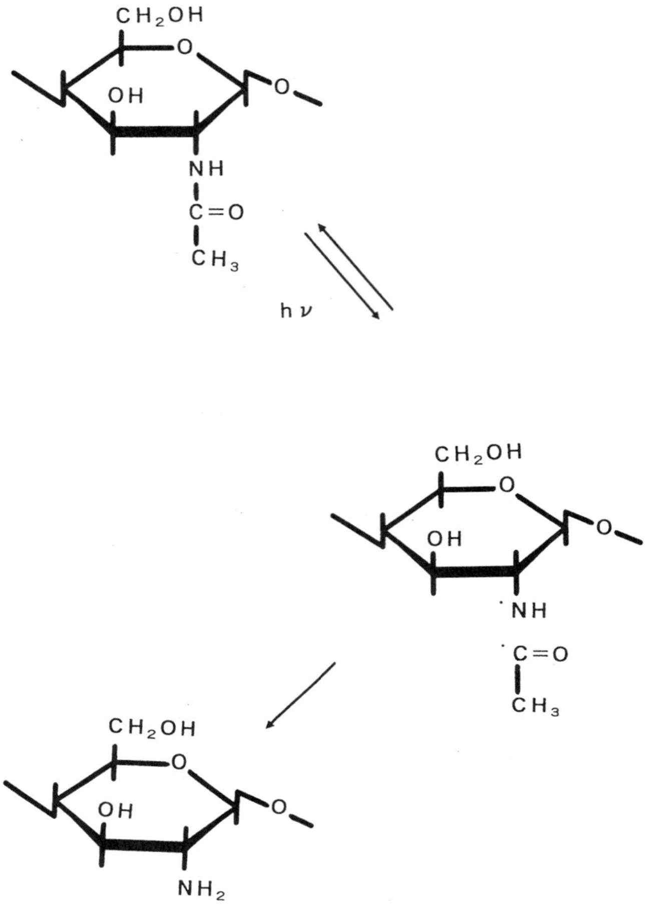

Scheme 2

The FTIR spectra of photoirradiated collagen were analyzed. The amide A (3350 cm^{-1}), amide B (3100 cm^{-1}), amide I (1650 cm^{-1}), and amide II (1550 cm^{-1}) bands are related to the peptide linkage of collagen. These bands decreased in their intensities by photoirradiation, showing the degradation of peptide linkage. Amide II and amide A bands shift to the longer wavelength with the irradiation time. Because these bands relate to the N–H bonds, degradation of the hydrogen bond is suggested. This means the triple-stranded helix of collagen may be loosened by photoirradiation.

The ESR spectra of photoirradiated collagen and collagen containing vitamin E as an additive were measured. The relative amount of radicals produced from collagen was plotted against the irradiation time. As shown in Fig. 18, the radical formation in the collagen that contained vitamin E (0.03 mol/kg) was efficiently suppressed. This indicates that vitamin E is an efficient protecting agent against photodegradation of collagen.

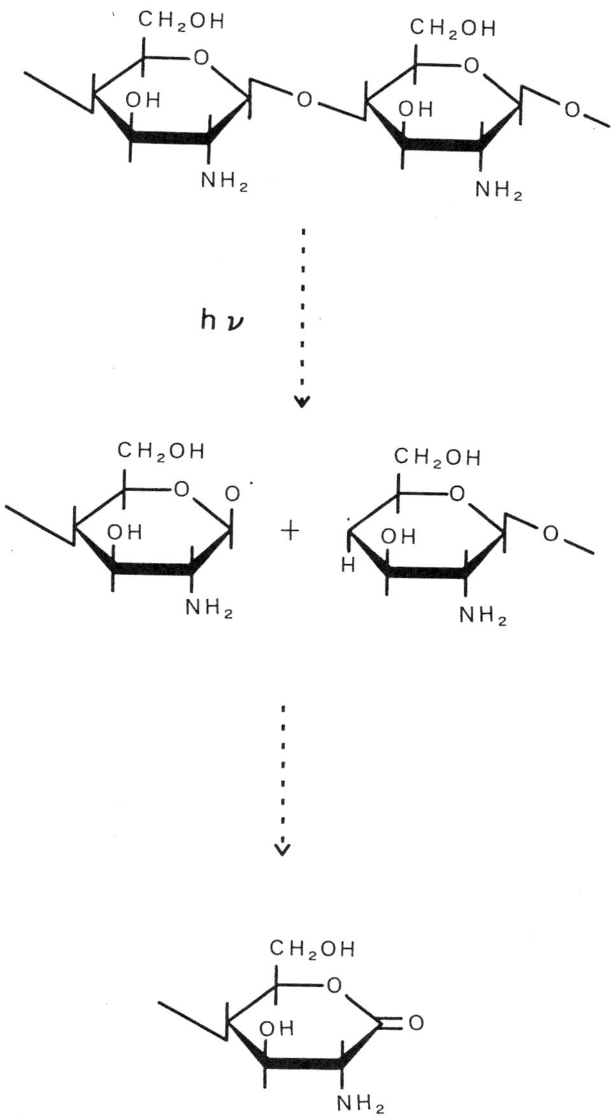

Scheme 3

On the basis of these experimental results, the wavelength sensitivity of photodegradation of collagen has been studied with monochromatic radiation by UV-visible spectral and FTIR measurements. By analyzing the UV spectra of photoirradiated collagen and collagen containing vitamin E, the increase in absorbance at 280 nm (ΔOD_{280}) showed specified behavior against the wavelengths of irradiation. The increase in this absorbance is attributed to the increase in the tyrosine fraction in collagen. On addition of vitamin E to collagen, the formation of tyrosine is suppressed in the wavelength region of 275–305 and 350–380 nm, as shown in Fig. 19.

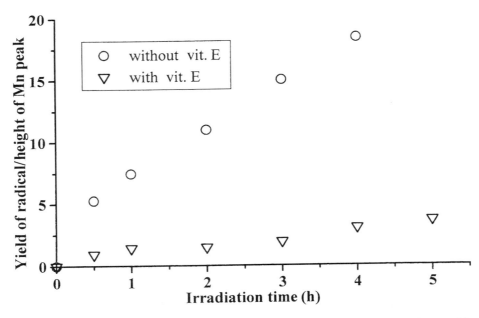

Figure 18 Relative yield of radicals formed in photoirradiated collagen films, with or without vitamin E, versus irradiation time.

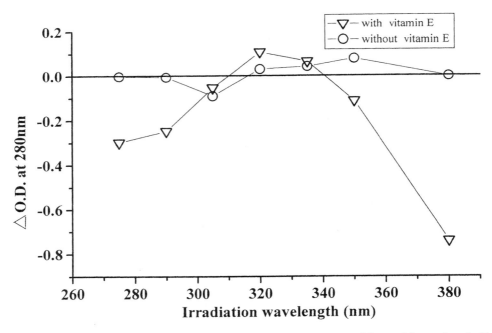

Figure 19 Changes in absorbance of photoirradiated collagen, with or without vitamin E, as a function of irradiation wavelength.

Changes in the optical densities of amide I and amide II bands in FTIR spectra were plotted against the irradiation wavelength, the effect of vitamin E on these bands was analyzed, and the suppression of the degradation of peptide linkage was confirmed. The photodegradation of collagen is wavelength dependent as is the suppression of degradation by vitamin E. The reaction scheme of photodegradation is very complex; however, these experimental results offer basic information on photoaging in animal skin.

IV. STUDIES OF WAVELENGTH SENSITIVITY USING A CUT-ON FILTER

The author has reviewed in the foregoing the studies using the OLS as the monochromatic radiation source. The wavelength sensitivity of polymers has been extensively studied by the use of cut-on filter technique. Studies on PVC (33), PC (34), PS (35), PP (36,37), PE (38,39), polyamide (40), sulfur-vulcanized rubber (41), natural rubber (42), paper (43) and wool (44) were reported. The details of the studies will be given in Chapter 16.

V. CONCLUDING REMARKS

The wavelength sensitivity of polymeric photodegradation has been reviewed. The monochromatic radiation source (OLS) and the experimental technique for monochromatic irradiations were also reviewed. In modern industry, polymeric materials are playing an important role and the amounts of the materials are increasing in various industrial fields. Reliable polymers that have stability under various environmental conditions are required. For this purpose, it is important to know the action spectra of polymers, which will offer crucial information on their photoinduced damage. Action spectra are also valid to design photodegradable polymers. There are a limited amount data on wavelength sensitivity of polymers. As data on their wavelength sensitivity are now accumulating, further development in this field is expected.

ACKNOWLEDGMENTS

The author would like to recognize her co-workers whose names appear in the cited references for their contributions to this review. The suggestions made by Dr. Anthony L. Andrady (Research Triangle Institute, North Carolina) on our studies are also acknowledged. The author wishes to thank Prof. Masakatsu Watanabe and Mr. Mamoru Kubota of the National Institue for Basic Biology for their advice and help in carrying out the irradiation studies.

REFERENCES

1. MJ Molina, FS Rowland. Stratospheric sink for chlorofluoromethanes: chlorine atom catalyzed destruction of ozone. Nature 249:810, 1974.
2. Stratospheric Ozone. EPA's Safety Assessment of Substituents for Ozone Depleting Chemicals. U.S. General Accounting Office (GAO), GAO/RCED 89-49, February 1989.
3. JF Rabek. Polymer Photodegradation. London: Chapman & Hall, 1995, pp 67–352.

4. M Watanabe. The Okazaki Large Spectrograph and its application to action spectroscopy, Proceedings of Ninth International Congress on Photobiology. Philadelphia, 1985, pp 37–44.

5. M Watanabe, M Furuya, Y Miyoshi, Y Inoue, I Iwahashi, K Matsumoto. Design and performance of the Okazaki Large Spectrograph for photobiological research. Photochem Photobiol 36:491–498, 1982.

6. AL Andrady, A Torikai, K Fueki. Photodegradation of rigid PVC formulations I. Wavelength sensitivity to light-induced yellowing by monochromatic light. J Appl Polym Sci 37:935–946, 1989.

7. AL Andrady, K Fueki, A Torikai. Photodegradation of rigid PVC formulations III. Wavelength sensitivity of photo-yellowing reaction in processed PVC formulations. J Appl Polym Sci 39:763–766, 1990.

8. A Torikai and H Hasegawa. Accelerated degradation and gel formation. Polym Degradn Stab 63:441–445, 1999.

9. A Torikai, M Ohno, K Fueki. Photodegradation of poly(methylmethacrylate) by monochromatic light: quantum yield, effect of wavelengths, and intensity of light. J Appl Polym Sci 41:1023–1032, 1990.

10. T Mitsuoka, A Torikai, K Fueki. Wavelength sensitivity of the photodegradation of polymethylmethacrylate. J Appl Polym Sci 47:1027–1032, 1993.

11. A Torikai, T Mitsuoka. Electron spin resonance studies of poly(methyl methacrylate) irradiated with monochromatic light. J Appl Polym Sci 55:1703–1706, 1995.

12. A Torikai, T Hattori, T Eguchi. Wavelength effect on the photoinduced reaction of polymethylmethacrylate. J Polym Sci A Polym Chem 33:1867–1871, 1995.

13. A Torikai, H Hasegawa. Wavelength effect on the accelerated photodegradation of polymethylmethacrylate. Polym Degrad Stab 61:361–364, 1998.

14. A Torikai, T Murata, K Fueki. Photo-induced reactions of polycarbonate studied by ESR, viscosity and optical absorption measurements. Polym Photochem 4:255–269, 1984.

15. AL Andrady, K Fueki, A Torikai. Spectral sensitivity of polycarbonate to light-induced yellowing. J Appl Polym Sci 42:2105–2107, 1991.

16. A Torikai, T Mitsuoka, K Fueki. Wavelength sensitivity of the photoinduced reaction in polycarbonate. J Polym Sci A Polym Chem 31:2785–2788, 1993.

17. Y Fukuda, Z Osawa. Wavelength effect on the photo-degradation of polycarbonate and poly(methylmethacrylate)—confirmation of the photo-degradation mechanism. Polym Degrad Stabil 34:75–84, 1991.

18. AL Andrady. Wavelength sensitivity in polymer photodegradation. Adv Polym Sci 128:49–94, 1996.

19. A Torikai, H Kato, K Tueki, Y Suzuki, F Okisaki, M Nagata. Photodegradation of polymer materials containing flame-cut agents. J Appl Polym Sci. 50:2185–2190, 1993.

20. A Torikai, T Kobatake, F Okisaki, H Shuyama. Photodegradatin of polystyrene containing flame-retardants: wavelength sensitivity and efficiency of degration. Polym Degrad Stabil 50:262–267, 1995.

21. A Torikai, T Kobatake, F Okisaki. Photodegradation of polystyrene containing flame-retardant: effect of chemical structure of the additives on the efficiency of degradation. J Appl Polym Sci 67:1293–1300, 1998.

22. A Torikai, K Suzuki, K Fueki. Photodegradation of polypropylene and polypropylene containing pyrene. Polym Photochem 3:379–390, 1983.

23. JF Rabek. Polymer photodegradation. London: Chapman & Hall, 1995, pp 69–92.

24. A Torikai, K Chigita, F Okisaki, M Nagata. Photooxidative degradation of polyethylene containing flame retardants by monochromatic light. J Appl Polym Sci 58:685–690, 1995.

25. T Asai, H Inoue, K Takeda. Photodegradation and the lifetime of materials in photo-irradiated polyphenylene ether by monochromatic radiation. Preprints, 7th Meeting of Materials Life Society, Japan. Tokyo, 1996, pp 26–27.
26. S Kuroda, N Sasaki, Z Osawa. Wavelength dependence of photodegradation of polysulfone and its excited states. Preprints, 8th Meeting of Materials Life Society, Japan, Tokyo, 1997, p 2627.
27. Z Osawa, Y Fukuda. Photo-degradation of blends of polycarbonate and poly(methyl methacrylate). Polym Degrad Stabil 32:285–297, 1991.
28. H Aoki, M Niregi, T Saito, S Akahori, S Kobayashi, Z Osawa, and Z Li. Photodegradation behavior of photodegradable ethylene–carbon monoxide copolymer. Polym Preprints Jpn 41:426, 1992.
29. JF Rabek. Polymer Photodegradation. London: Chapman & Hall, 1995, pp. 346–349.
30. AL Andrady, Ye Song, VR Parthasarathy, K Fueki, A Torikai. Photoyellowing of mechanical pulp. Part I: examining the wavelength sensitivity of light induced yellowing using monochromatic radiation. TAPPI J 74:162–168, 1991.
31. AL Andrady, A Torikai, T Kobatake. Spectral sensitivity of chitosan photodegradation. J Appl Polym Sci 62:1465–1471, 1996.
32. A Torikai, H Shibata. Effect of ultraviolet radiation on photodegradation of collagen. J Appl Polym Sci 73:1259–1265, 1999.
33. KG Martin, RI Tilley. Br Polym J 3:36, 1971.
34. AL Andrady, ND Searle, LFE Crewdson. Polym Degrad Stabil 35:235, 1992.
35. AL Andrady. Accelerated Environmental Exposure, Laboratory Testing and Recyclability Study of Photo/Biodegradable Plastics. Final Report to USEPA under Contract 68-01-4544 Task 11–60, January.
36. AL Andrady, JE Pegram, N Searle. (unpublished data).
37. Z Zhenfeng, Hu Xingzhou, Luo Zubo. Wavelength sensitivity of photodegradation of polypropylene. Polym Degrad Stabil 51:93–97, 1966.
38. P Trubiroha. International Conference on the Advances in Stabilization and Controlled Degradation of Polymers. 1:236, 1969.
39. Hu Xingzhou. Wavelength sensitivity of photodegradation of polyethylene. Polym Degrad Stabil (in press).
40. LD Johnson, WC Tincher, HC Bach. J Appl Polym Sci 13:1825, 1989.
41. S Yano. Rubber Chem Technol 54:1, 1981.
42. MJL Morand. Rubber Chem Technol 45:481, 1972.
43. AL Andrady, ND Searle. TAPPI J 78:131, 1995.
44. FG Lenox, MG King, IH Leaver, GC Ramsay, WE Savige. Appl Polym Symp 18:353, 1991.

16

Activation Spectra of Polymers and Their Application to Stabilization and Stability Testing

NORMA D. SEARLE

Plastics and Chemicals, Deerfield Beach, Florida

I. INTRODUCTION

Degradation of polymeric materials by the environment is due to a complex inter-action of the effects of all weather factors, including solar radiation, moisture, heat, atmospheric pollutants, and others. However, the radiation of the sun, particularly the ultraviolet (UV) portion, is mainly responsible for limiting the lifetime of materials exposed to the environment.

The importance of solar UV in deterioration of polymeric materials is because it has enough energy to initiate degradation processes by breaking organic chemical bonds. The latter is a critical primary step in the changes that occur in appearance and mechanical properties when these materials are exposed to the environment. Other weather factors contribute to degradation through their effect on the second-ary reactions, those that follow bond breakage.

The actinic effects of UV and visible radiation on a material are wavelength dependent because of the spectral selectivity of both absorption of the incident radiation and the quantum efficiencies of degradation. Therefore, the rates of degra-dation of a material exposed to equal total radiant energies from different types of light sources are determined by the spectral properties of the incident radiation. The mechanism and type of degradation of many materials also depend on the inci-dent wavelengths (1–9). Thus, the spectral emission of the radiation source is a criti-cal factor in its effect on a material.

The wavelengths responsible for degradation of a material by a specific radiation source are identified by its activation spectrum (10,11). The latter is a graphic representation of the relative amount of damage caused by individual spectral regions of the source to which the material is exposed. No adjustment is made for differences in flux densities of the wavelengths emitted by the source.

The activation spectrum provides information on the following: 1) the extent to which an artificial test source simulates the natural one in terms of the wavelengths responsible for the degradation; 2) the actinic spectral region for timing exposures; 3) the effective dosage; 4) the type of protective UV absorber required; 5) the screening effectiveness of the UV absorber; 6) the effect of thickness on the wavelength sensitivity; and 7) the dependence of type of degradation on incident wavelengths. Thus, it is an important tool in the design of meaningful stability tests and in the development of light-stable materials.

The factors that determine the activation spectrum are reviewed and the spectrographic and sharp-cut filter techniques used to obtain activation spectra are described in detail in this chapter. Its application to stabilization and stability testing is discussed and illustrated with examples. Comparison is made with action spectra that give the wavelength sensitivity of a material, independently of the spectral emission properties of the radiation source.

II. THE ACTIVATION SPECTRUM: DETERMINING FACTORS

The activation spectrum of a material is defined by 1) the spectral absorption properties of the material; 2) the emission properties of the light source; 3) the quantum efficiencies of the degradation processes initiated by the absorbed wavelengths; and 4) the type of degradation measured.

A. Spectral Absorption Properties of the Material

The spectral absorption properties of a material determine the fraction of incident radiation at each wavelength that a material is capable of absorbing. Only light absorbed can have any effect on the material. The mere incidence of light on the surface cannot cause any damage if it is not absorbed. Therefore, the relation between the absorption properties of the material and the emission properties of the light source is fundamental to the interaction of the radiation with the material and, thus, to the activation spectrum.

The spectral absorption properties are determined by the chemical structure of the material. The wide variation in UV absorption properties of polymers are shown in Fig. 1 (12,13), which compares the UV absorption spectra of various types of polymers in the form of 2-mil films. The data is presented in terms of absorbance, which is a linear function of the thickness of the sample and concentration of absorbing species. The emission spectrum of terrestrial solar radiation is included to show its relation to the absorption spectra.

Pure forms of aliphatic-type polymers (e.g., polyethylene, polypropylene, poly(vinyl chloride), and others, are not capable of absorbing terrestrial solar radiation. The main absorption band of these polymers usually peaks below 220 nm and the long wavelength absorption tail barely extends beyond the cut-on wavelength of sunlight on the earth's surface. The effect of weather on these polymers is due to the UV-absorbing impurities and thermal oxidation products introduced during polymerization and processing. Hydroperoxides and carbonyl groups

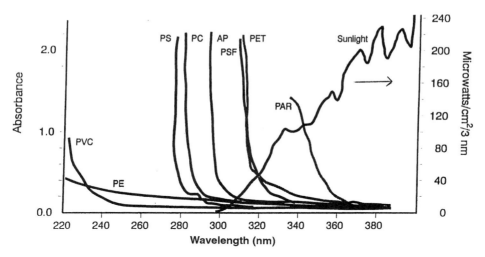

Figure 1 Ultraviolet absorption spectra of 0.05-mm polymer films and spectral irradiance of July noon sunlight (direct beam) at 41° north latitude: AP, aromatic polyester; PAR, polyarylate; PC, polycarbonate; PE, polyethylene; PET, poly(ethylene terephthalate); PS, polystyrene; PSF, polysulfone; PVC, poly(vinyl chloride). (From Ref. 13.)

attached to the polymer chain play a very substantial role in the weathering of these materials.

Because of their low concentration and small absorption coefficients, these impurities are not detected in the UV absorption spectra of the polymers. However, the severe degradation initiated by the small amount of radiation absorbed by the impurities can be readily measured and is used to define the activation spectrum. Depending on the variety and types of impurities present, these polymers can be degraded by the full range of solar UV. Often, the spectral region responsible for degradation shifts to a longer wavelength as the exposure progresses, because of formation of impurities with longer wavelength UV-absorbing properties.

In contrast to the aliphatic-type polymers, the long wavelength edge of the main absorption band of many of the aromatic-type polymers is in the 300- to 400-nm region. Thus, their structural components are capable of absorbing solar radiation, and because the wavelengths of solar radiation absorbed vary with the type of structural component, their activation spectra differ more than those of the aliphatic types. For example, a thin film of polyarylate is a strong absorber of wavelengths as long as 380 nm, whereas polystyrene and polycarbonate absorb very little solar radiation, particularly in the form of thin films. The effect of outdoor exposure on the latter two polymers is mainly due to the presence of UV-absorbing impurities similar to those in the aliphatic-type polymers.

B. Emission Properties of the Light Source

Each type of light source has a unique spectral power distribution, characterized by the wavelengths emitted and their relative intensities. Figures 2–4 show the emission properties of several types of light sources used for accelerated laboratory tests, in

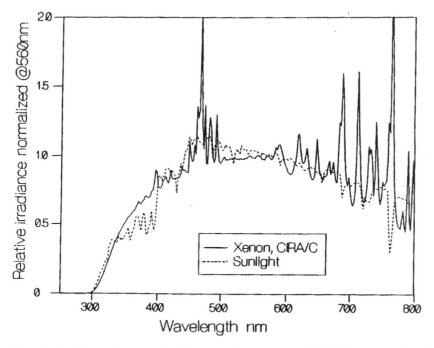

Figure 2 Spectral energy distributions of xenon arc with CIRA inner and soda lime outer glass filters and Miami daylight measured at noon at 26° south exposure.

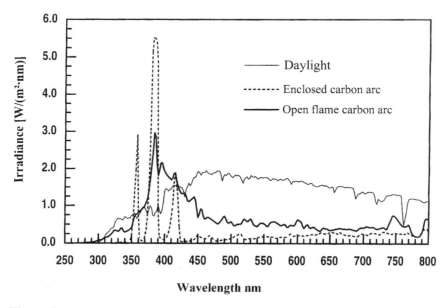

Figure 3 Spectral energy distributions of enclosed carbon arc, open-flame carbon arc with Corex D filters, and Miami peak daylight measured at noon at 26° south during the spring equinox. (Courtesy of Atlas Electric Devices.)

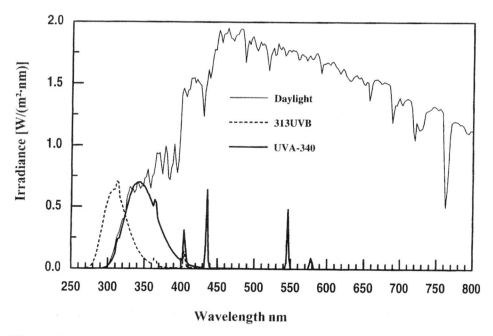

Figure 4   Spectral energy distributions of fluorescent UVB-313 and fluorescent UVA-340 lamps and Miami peak daylight, measured at noon at 26° south during the spring equinox. (Courtesy of Atlas Electric Devices.)

comparison with terrestrial solar radiation. Figure 2 shows that the xenon arc, when properly filtered, closely simulates the full spectrum of solar radiation. Although the UV portion of the latter is mainly responsible for the effects of the weather on both colored and colorless materials, the visible radiation absorbed by colored materials can cause deterioration of the polymeric components as well as color changes. The visible radiation absorbed also increases the temperature of the material, which accelerates the chemical reactions initiated by UV radiation.

Neither of the two types of carbon arcs shown in Fig. 3 provide a good representation of solar radiation in the UV or visible regions. The filtered open-flame carbon arc (also referred to as the Sunshine carbon arc) was an improvement over the enclosed carbon arc, the first type used for exposure tests. It gives a better match to solar radiation in the 300- to 350-nm region. However, they both deviate significantly from solar radiation in the long wavelength UV and visible regions. Depending on the type of filter used, the open-flame carbon arc emits some radiation below the solar cut-on.

The emissions of the two fluorescent UV lamps shown in Fig. 4 have very little similarity to daylight, except for the good match in the 300- to 350-nm region by the UVA-340 lamp. The UVB fluorescent lamp has a significant amount of energy below the solar cut-on, and both lamps lack energy in the long wavelength UV and visible regions compared with solar radiation.

For most materials, activation spectra based on exposure to the carbon arc or fluorescent UV sources can be expected to differ from their activation spectra based on solar radiation. Examples are given in Section V of the dependence of the activation spectrum on the light source for two types of polymers.

C. Quantum Efficiencies of Degradation by Absorbed Wavelengths

Although absorption of radiation is a prerequisite to photodegradation, the relation between the absorption properties of the material and spectral emission properties of the light source does not, in itself, define the activation spectrum. The latter also depends on the quantum efficiencies of degradation by the absorbed wavelengths. Only a small portion of the radiation absorbed by polymeric materials causes photochemical changes. Much of the radiation absorbed is eliminated in harmless ways. The number of moles of material degraded per mole of photons absorbed is the *quantum efficiency*. It is dependent on the energies of the photons absorbed in relation to the bond strengths of the material and the photochemical reactions that follow bond breakage.

That the activation spectrum is not defined solely by the spectral energy absorbed is illustrated in Fig. 5 (14). The activation spectrum of a film of polycarbonate, based on exposure to the borosilicate–glass-filtered xenon arc, is compared with the spectral energy absorbed by the polycarbonate film from the light source. The energy absorbed was calculated by multiplying, at 1-nm intervals, the irradiance of the source by the fraction of incident radiation the sample is capable of absorbing at that wavelength. The technique for determining the activation spectrum is described in Section III.B.

The largest amount of energy absorbed by the polycarbonate film is in the spectral region between 325 and 345 nm. Although the energy absorbed at each of the wavelengths in this region is about tenfold more than the energy absorbed

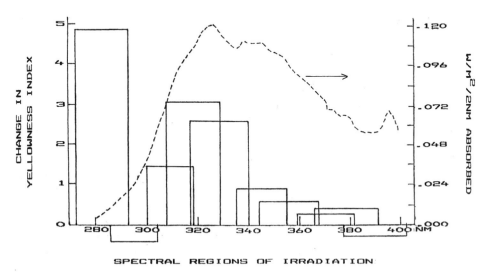

Figure 5 Activation spectrum of a 0.70-mm film of bisphenol-A polycarbonate exposed to borosilicate–glass-filtered xenon arc radiation, compared with spectral energy absorbed from the light source. (From Ref. 14.)

at wavelengths shorter than 300 nm, the activation spectrum shows that the latter causes the most intense yellowing.

The higher quantum efficiencies of the shorter wavelengths, based on yellowing of polycarbonate, are due to the greater potential for bond breakage by the higher energy photons associated with these wavelengths and different photochemistry than the degradation promoted by the longer wavelengths (4,5,15–17). Only the short wavelength photons are absorbed by the main structural components and have sufficient energy to cause bond breakage of these structures. The breakage is followed by photo-Fries rearrangement and formation of highly yellow-colored degradation species. The longer wavelength photons, which can be absorbed only by the impurities, initiate photooxidation processes, resulting in weaker yellow-colored species.

Although the shorter wavelength, higher energy photons can produce more severe degradation, the effect of different energy photons cannot be predicted quantitatively. Photodegradation involves complex reaction processes that are still not fully understood for most materials. Therefore, the activation spectrum must be determined experimentally.

D.   Type of Degradation Measured

The activation spectrum of some materials shifts with type of degradation measured because of the dependence of mechanism and type of degradation on wavelength in these materials. For example, loss in impact strength in acrylonitrile–butadiene–styrene (ABS) and carbonyl formation in polyethylene is caused by longer wavelengths than those that cause yellowing of these polymers (11). In many materials, tensile strength is affected mainly by short wavelengths, whereas longer wavelengths are primarily responsible for changes in elongation.

The effect of wavelength on the type of degradation results from differences in the extent to which the wavelengths are absorbed, and thus, to differences in their depth of penetration into the material. In TiO_2 and ZnO pigmented systems and aromatic-type polymers, because the shorter wavelengths are very strongly absorbed, they are completely absorbed very close to the surface. Thus, degradation by these wavelengths is confined to the surface layers. The longer, more weakly absorbed wavelengths, are able to penetrate deeper and cause bulk effects.

III.   ACTIVATION SPECTRUM TECHNIQUES

The two main techniques used for determining the effects of individual spectral regions of a polychromatic radiation source to obtain activation spectra are the spectrographic (18–22) and sharp-cut filter techniques (11,14,23–28). They are described in detail in this chapter. In several studies, interference filters were used as an alternative to the spectrograph (29,30).

The spectrographic technique determines the effects of narrower spectral regions than the sharp-cut filter technique, but it is generally only applicable to measurement of optical changes because of the small size of the specimen area irradiated by each spectral band, and it is limited to light sources that can be focused on the entrance slit of the spectrograph. In contrast, the filter technique is applicable to mechanical as well as to optical property changes and to any type of exposure used for samples. It is also more representative of the polychromatic exposure during use, since degradation caused by a specific spectral region is determined during simultaneous exposure to longer wavelengths than the spectral band of interest. On exposure to polychromatic radiation, a variety of photochemically initiated processes, including antagonistic reactions by short versus long wavelengths, such as yellowing versus bleaching, may occur simultaneously.

A. Spectrographic Technique

In the spectrographic technique, polychromatic light is spectrally dispersed by either a prism or grating into individual spectral bands. Each band is incident on a separate portion of the surface of a single specimen that is exposed in the focal plane of the spectrograph. The quartz prism spectrograph used by Hirt and co-workers for activation spectra by this technique was previously illustrated and described (18–20). The light source, focused by a lens system on a 0.5-mm–wide entrance slit, is separated into its spectral components by two quartz prisms. Two concave mirrors collimate and reflect the spectrally dispersed beam onto the exit window of the spectrograph. Spectral bands ranging from 280 to 410 nm are incident on a 1-cm–wide section of the sample, which is about the size of a microscope slide. It is placed in the focal plane at the exit window. The dispersion ranges from 10 nm/mm at the short wavelength end to 30 nm/mm at the long wavelength end. The effective speed of the spectrograph is $f/2$.

Wavelength positioning on the sample is calibrated by substituting a photographic plate for the sample at the exit window and exposing it to the spectrally dispersed emission lines of a low-pressure mercury arc focused on the entrance slit. The positions of the mercury emission lines on the photographic plate are determined by densitometry of the plate.

The sample is irradiated in the focal plane at the exit window until the optical changes produced are sufficient to provide a well-defined activation spectrum. Optical changes in the sample at a specific wavelength in the UV, visible, or infrared (IR) regions are determined using the appropriate spectrophotometer set at a fixed wavelength. A specially designed motorized adapter was used to move the sample at a constant rate of speed across a 0.5-mm slit inserted in the sample beam to obtain absorption measurements in 0.5-mm–wide sections of the sample (19,20). Change in absorption is related to wavelengths of incident radiation by means of the calibration procedure and is plotted as a function of the wavelength of irradiation to produce the activation spectrum.

For most of the activation spectra obtained by this technique, the radiation source was a 900-W high-pressure xenon arc. A borosilicate glass filter was placed between the lens system and entrance slit to reduce the short wavelength emission to simulate the cut-on of sunlight on the earth's surface. Activation spectra of some of the materials were also obtained by exposure to sunlight. A heliostat was used to follow the sun and send a steady beam of sunlight to an optical system that concentrated and focused it onto the entrance slit of the spectrograph (18,31).

B. Sharp Cut-On Filter Technique

In the sharp cut-on filter technique, a series of specially designed cut-on filters are interposed between the light source and replicate specimens of the polymeric material of interest. Spectral bands are defined by the difference in transmission of pairs of adjacent filters in the series. Nearly equal spectral band areas for each of the pairs were obtained by grinding the filters to the required thickness.

Figure 6 (14) shows the UV spectral transmission curves of some of the filters in the set of 15 used to define ten spectral bands between 272 and 402 nm. The cut-on of each filter is the wavelength at 5% transmission. Each progressively shorter wavelength filter transmits increasingly more of the short wavelength emission and a larger portion of the total UV than the longer wavelength filter. All filters transmit nearly all of the visible and infrared radiation, except for the reduction of about 10% at all wavelengths caused by specular reflection at the air–glass interface.

Filter 1 is used to determine the effect of exposure to the full radiation of the source except for the portion that is specularly reflected equally by all wavelengths. Filters 2–9 define the six spectral regions of irradiation shown in Fig. 7 (14). Each spectral band represents the incremental portion of the UV transmitted by the shorter wavelength filter of the pair. The peak is the maximum difference in transmission of the pair of filters. Convolution of the source radiation with each of the spectral bands may cause the band to shift somewhat, depending on the spectral irradiance of the source (32,33). The spectral range of the band at delta 20% transmission, approximately one-half the peak height, identifies the spectral region responsible for the difference in degradation in the two specimens exposed behind a pair of filters.

The specimens are mounted about 3-mm behind each of the filters to allow circulation of air between the filter and specimen. The specimens are backed with nonreflective black paper to prevent unfiltered scattered light from irradiating the back side. In the studies reported here, the filters were 2 × 2 in. If the measure-

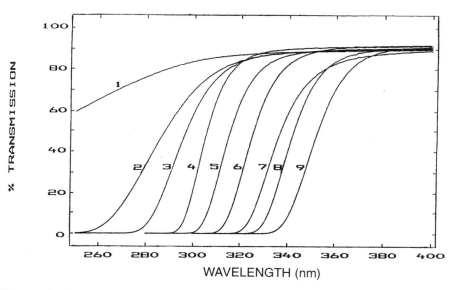

Figure 6 UV spectral transmission curves of nine of the sharp cut-on filters used for activation spectra. (From Ref. 14.)

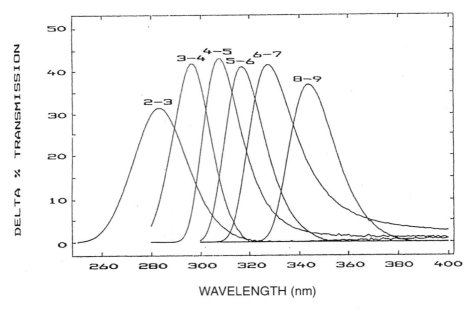

Figure 7   Spectral bands defined by the difference in transmittance between pairs of filters shown in Fig. 6. (From Ref. 14.)

ment technique allows the use of smaller samples, multiple specimens can be exposed behind each filter.

The filtered specimens are exposed to the light source until the difference in degradation in most of the pairs of specimens is measurable and statistically significant. A small adjustment is made, if necessary, to normalize the measured differences in degradation so that they represent the effects of equal spectral band areas. Adjustments greater than 10% usually do not give valid data because degradation is rarely a linear function of irradiance or radiant exposure. Because the filter transmission characteristics required to limit the adjustment to less than 10% are usually not provided by off-the-shelf filters, each is custom-ground to the appropriate thickness.

The changes in optical or mechanical properties caused by each of the spectral regions are plotted in bar graph form to produce the activation spectrum. The height of the bar represents the change in property and the width represents the spectral region responsible for the degradation. The width of most of the spectral bands is about 20 nm. As an alternative to a plot in bar graph form, the spectral region causing the change can be identified by the wavelength at the midpoint of the spectral band. The activation spectrum plotted by connecting the points has the same form as that obtained by the spectrographic technique.

IV.   ACTIVATION SPECTRA BY THE SPECTROGRAPHIC AND FILTER TECHNIQUES

Data obtained by both techniques are compared in Table 1 (11), which lists the spectral regions of the filtered xenon arc primarily responsible for the change in

Table 1 Activation Spectra Maxima Based on Filtered Xenon Arc Radiation

Polymer	Measured change	Filter technique		Spectrographic technique	
		Mil	Max. change (nm)	Mil	Max change (nm)
Polycarbonate	UV at 300 nm		—	4.5	295; 310–350
	UV at 340 nm	28	<300; 310–340	4.5	295; 310–340
	Yellowness	28	<300; 310–340	4.5	295; 310–340
Polystyrene	Yellowness	125	300–330	125	319
Polysulfone	UV at 330 nm	1	310	1	305
	Yellowness	1	320	1	310–320
	$C=O \cdot OH$		—	1	330
Polyethylene	Yellowness		—	4	310
	$C=O$		—	4	340
ABS	Yellowness	100	340–360	10	330
	Bleaching	100	>380	10	380–400
	Impact strength	100	350–380[a]		—
	Impact strength		>380[b]		—
Polypropylene	UV		—	15	295, 330, 370
	$C=O$	10	320	15	340–380
	Tensile strength	60	320–350[a]		—
	Tensile strength	60	360–380[b]		—
PVC	Yellowness	40	300–320	2	308–325[c]
Polyarylate	Yellowness	3	350		—
	Yellowness	60	385		—

[a] Short exposures.
[b] Extended exposures.
[c] Range for different PVC film.
Source: Ref. 11.

measured property for each of the materials. Good agreement was obtained between the two techniques, although the materials may not have been identical, because they were manufactured at different times and, in many cases, the thicknesses were different.

The activation spectra of polycarbonate by both techniques shows that increase in UV absorption and yellowing are caused by two distinctly separate spectral regions, 295–300 and 310–340 nm. The information, which is not shown, is that the shorter wavelength spectral region has a much more severe effect. The yellowing of polystyrene by the filter technique is mainly due to the 300- to 330-nm region of the filtered xenon arc, whereas the spectrographic technique locates the peak of the activation spectrum at 319 nm.

Both techniques show that yellowing of polysulfone is caused by longer wavelengths than those responsible for increase in UV absorption, and that bleaching in ABS is caused by longer wavelengths than those that cause yellowing. The filter technique shows that loss in impact strength of ABS is due to wavelengths shorter than those responsible for bleaching (34). The spectrographic technique shows that carbonyl formation in polyethylene is caused by longer wavelengths than

those responsible for yellowing. The shift in wavelength sensitivity with type of
degradation demonstrates the dependence of the latter on incident wavelengths.
The explanation for the shift in activation spectrum peak of polyarylate from
350 to 385 nm with increase in thickness from 3 to 60 mil (0.075–1.5 mm) will
be given in Section VI.B.

It is important to emphasize that the activation spectra given in the table are
specific to the filtered xenon arc and can not be extrapolated to other types of light
sources. However, similar data were obtained using terrestrial solar radiation
for many of the materials listed. Polycarbonate is unique in this respect because
it is very sensitive to wavelengths shorter than 300 nm that are present in the
Pyrex-glass-filtered xenon arc emission, but are absent in terrestrial solar radiation.
A radiation source with very different emission characteristics than either terrestrial
solar radiation or the filtered xenon arc can be expected to produce very different
activation spectra for most materials.

V. DEPENDENCE OF ACTIVATION SPECTRUM ON LIGHT SOURCE

The spectral energy distribution is a critical factor in the effect of a light source on
materials and is one of the major factors that determines the activation spectrum
of a material. The dependence of the activation spectrum on the light source is illustrated in this section by two materials, polycarbonate and polyarylate, exposed
to various light sources.

A. Polycarbonate Activation Spectra

The spectral sensitivity of a 30-mil (0.75-mm) film of unstabilized bisphenol-A
polycarbonate to four types of UV radiation are given in Figs. 5 and 8–10. The

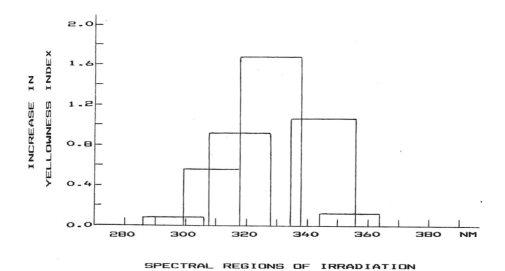

Figure 8 Activation spectrum of a 0.70-nm film of bisphenol-A polycarbonate exposed to
solar radiation (direct plus diffuse) at 26° south in Miami. (From Ref. 14.)

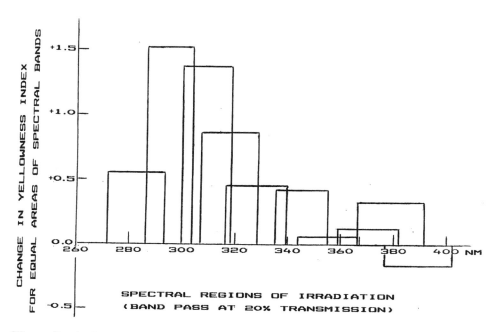

Figure 9 Activation spectrum of a 0.70-mm film of bisphenol-A polycarbonate exposed to the Corex D (potash lithia) glass-filtered open-flame carbon arc. (From Ref. 11.)

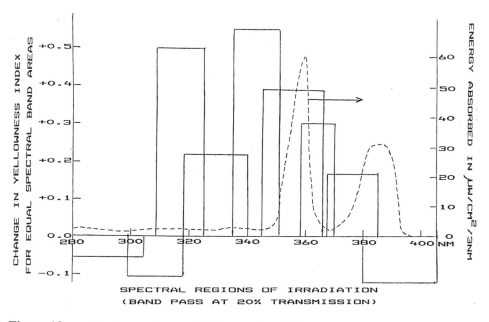

Figure 10 Activation spectrum of a 0.70-mm film of bisphenol-A polycarbonate exposed to enclosed carbon arc radiation compared with spectral energy absorbed from the light source. (From Ref. 11.)

activation spectra were based on yellowing and were obtained by the sharp-cut filter technique. Figure 8 (14) shows that yellowing by terrestrial solar radiation is mainly due to wavelengths between 310 and 350 nm and that the most severe yellowing during the 3-month exposure in Miami was caused by radiation in the 320- to 335-nm spectral region.

Figure 5 shows the activation spectrum of the same material exposed to the borosilicate-filtered xenon arc. The data given in Table 1 for the yellowing of polycarbonate was based on Fig. 5. It lists two spectral regions mainly responsible for yellowing: radiation shorter than 300 nm and that between 310 and 340 nm. The small amount of energy in the 300-nm region present in borosilicate-filtered xenon arc radiation and not in solar radiation has a particularly significant effect on polycarbonate for two reasons: 1) the onset of strong absorption by the structural components is in this region, and 2) these wavelengths cause photochemical processes that lead to strongly yellow-colored degradation products. The small difference in radiation in the 300-nm region would have very little effect on the activation spectra of polymers in which the long wavelength edge of the absorption band of the structural components is either at longer or shorter wavelengths.

Figure 9 (11) shows the activation spectrum of the polycarbonate film exposed to the open-flame carbon arc, filtered with a potash lithia glass filter (Corex D). It differs from the activation spectrum based on exposure to either solar radiation or the borosilicate-filtered xenon arc. The strongest yellowing by this source is caused by wavelengths in the 290- to 330-nm region. Wavelengths longer than 340 nm are more effective in yellowing polycarbonate than these wavelengths in the xenon arc or solar radiation because of the strong irradiance of the carbon arc in this region. Wavelengths longer than 380 nm cause some bleaching of the yellow species produced during processing.

The activation spectrum in Fig. 10 (11), based on exposure to enclosed carbon arc radiation, has a very different profile than those described in the foregoing. Yellowing is due to wavelengths in the 310- to 320-nm region, as well as to wavelengths longer than 340 nm. Although the source has negligible intensity below 320 nm, the photon energies are higher, and polycarbonate absorbs more radiation in this region than above 340 nm. In the latter region, because of the high intensity of the source, the large number of photons absorbed also causes significant yellowing, in spite of the weak absorption of polycarbonate. In contrast, yellowing by xenon arc and solar radiation in this region is negligible. The only similarity between this and the activation spectrum based on exposure to filtered xenon arc radiation is the distinct separation between the effects of the short and long wavelengths. However, the two actinic spectral regions differ in the two types of exposures. In addition, the activation spectrum based on enclosed carbon arc radiation shows that the 280- to 310-nm region reverses yellowing caused by longer wavelengths. The enclosed carbon arc is more effective than the open-flame carbon arc in bleaching thermally produced yellow-colored species by wavelengths longer than 380 nm.

B. Polyarylate Activation Spectra

Figures 11–13 (35) compare activation spectra based on yellowing of a thin film of polyarylate exposed to three types of light sources; solar radiation, the

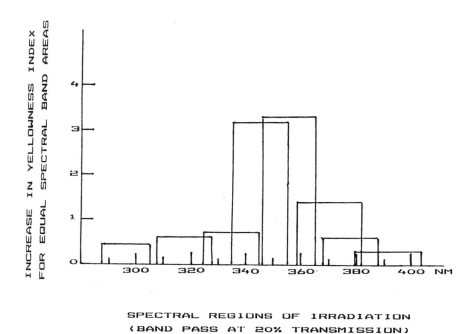

SPECTRAL REGIONS OF IRRADIATION
(BAND PASS AT 20% TRANSMISSION)

Figure 11 Activation spectrum of a 0.075-mm film of polyarylate exposed to solar radiation. (From Ref. 35.)

SPECTRAL REGIONS OF IRRADIATION
(BAND PASS AT 20% TRANSMISSION)

Figure 12 Activation spectrum of a 0.075-mm film of polyarylate exposed to borosilicate–glass-filtered xenon arc radiation. (From Ref. 35.)

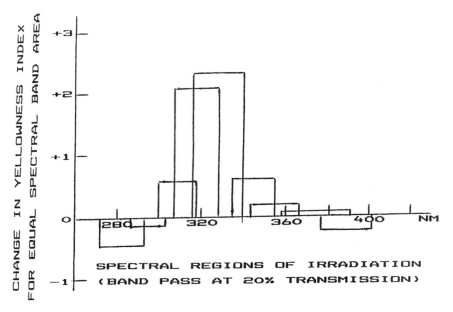

Figure 13 Activation spectrum of a 0.05-mm film of polyarylate exposed to fluorescent UVB lamps. (From Ref. 35.)

borosilicate-filtered xenon arc, and fluorescent UVB lamps. All were obtained by the sharp cut-on filter technique.

The activation spectrum in Fig. 11 shows that solar radiation in the spectral region of 335–365 nm causes most of the yellowing. Wavelengths shorter than 330 nm and longer than 380 nm have very little effect in yellowing the 3-mil (0.075-mm) film of polyarylate. Figure 12 shows that the wavelengths of the borosilicate-filtered xenon arc that initiate yellowing of the polyarylate film are the same as those in sunlight. Wavelengths shorter than 330 nm and longer than 380 nm have very little effect, whereas wavelengths between 335 and 365 nm cause most of the yellowing.

In contrast to the effect of these two light sources, the activation spectrum in Fig. 13 shows that yellowing of the polyarylate film by the fluorescent UVB lamps is primarily due to wavelengths between 310 and 340 nm. Obviously, this type of radiation is not representative of solar radiation in terms of the wavelengths that cause yellowing of polyarylate. Therefore, it cannot be reliably used for testing the effectiveness of an UV absorber in protecting polyarylate against solar radiation. One of the consequences of the fact that the damaging wavelengths vary with type of light source is that the effectiveness of an additive in protecting the material will depend on the source used for the evaluation.

VI. DEPENDENCE OF ACTIVATION SPECTRUM ON ABSORPTION PROPERTIES

The degradation of a material by incident radiation from a specific light source depends on the ability of the material to absorb the radiation and on the fate of the absorbed radiation. The latter largely depends on the relation between the energies of the photons absorbed and the bond energies. Because of differences among materials in their absorption properties and bond energies, the spectral region of the light source responsible for the degradation will vary with the type of material and its formulation. If the radiation emitted by the light source is absorbed by the structural components of a polymer, the activation spectrum will also depend on the thickness of the specimen.

A. Type of Polymeric Material

Differences in absorption properties of various polymeric materials were illustrated in Fig. 1 and were described in Section II.B. It was pointed out that solar radiation is absorbed only by the impurities in aliphatic-type polymers, but by both the impurities and structural components in many aromatic-type polymers. Because the UV-absorbing impurities are similar in many of the polymers, the activation spectra of the aliphatic-type polymers differ less than those of the aromatic types.

Differences among aromatic-type polymers in their spectral sensitivities to the same radiation source is illustrated by comparison of the activation spectrum of polyarylate in Fig. 12 with that of polysulfone in Fig. 14 (11). On exposure of thin films of these materials to the borosilicate-filtered xenon arc, the spectral regions primarily responsible for causing photochemical yellowing are 335–365 nm for polyarylate and 280–335 nm for polysulfone. The large difference in their activation

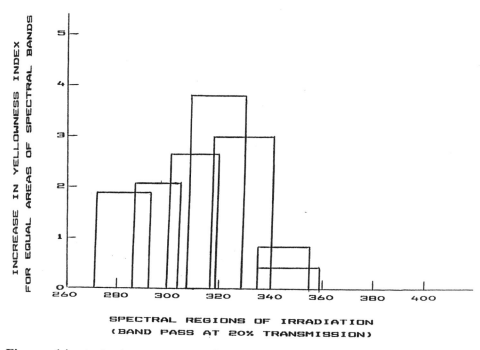

Figure 14 Activation spectrum of a 0.05-mm film of polysulfone exposed to borosilicate–glass-filtered xenon arc radiation. (From Ref. 11.)

spectra is not unexpected considering the differences in the wavelengths absorbed by their structural components and the spectral power distribution of the radiation source (see Fig. 1).

The ability of polyarylate to absorb longer wavelength UV radiation than polysulfone can readily account for its being degraded by longer wavelengths than polysulfone when they are exposed to solar radiation or to the filtered xenon arc. However, the insensitivity of polyarylate to short wavelengths, in spite of the fact that they are absorbed more strongly by polyarylate than by polysulfone, is due to protection of the polymer against these wavelengths by UV-absorbing photodegradation products (36,37). A polymeric *ortho*-dihydroxybenzophenone-type structure, typical of one of the types of UV absorbers, forms in a photo-Fries, rearrangement of the main structural component of polyarylate. Because it is concentrated in the surface layers, it is very effective in screening out radiation that it is capable of absorbing.

B. Thickness of Aromatic-Type Polymers

The spectral sensitivities of polymeric materials vary not only with type of polymer, but also with thickness. The data in Table 1 showed a shift from 350 to 385 nm in the activation spectrum peak of polyarylate exposed to the borosilicate–glass-filtered xenon arc when the thickness was increased from 3 to 60 mil (0.075–1.5 nm). This can be explained by the shift in the long wavelength edge of the main absorption band with thickness, which is illustrated in Fig. 15 (13). The latter compares the UV absorption curves of two thicknesses of a styrene cross-linked phthalic–maleic-type aromatic polyester, a 0.1-mm film and a 2.5-mm plaque.

Similar to the effect on the spectral absorption band of an absorbing component in solution when its concentration is increased, increase in thickness of a polymer causes the absorption band to broaden and the peak height to increase. Figure 15 shows only the long wavelength edge of the band. The long wavelength tail of the absorption band of the 2.5-mm plaque extends about 50 nm beyond that of the thinner specimen. Thus, the thicker specimen absorbs significantly more long wavelength radiation than the film. The latter is almost completely transparent to wavelengths longer than 320 nm, whereas the 2.5-mm specimen absorbs wavelengths as long as 360 nm and significantly more radiation than the thin film between 320 and 360 nm.

The effect of thickness on the activation spectrum of the polyester is shown in Fig. 16 (13). Activation spectra of a 0.1-mm film and a 3.1-mm plaque exposed to borosilicate–glass-filtered xenon arc radiation were obtained by the spectrographic technique. Although yellowing of the thin film is due almost entirely to radiation in the 290- to 320-nm spectral region, yellowing of the thicker specimen is due to wavelengths as long as 370 nm. The peak of the activation spectrum shifts from 300 nm for the thin film to 325 nm for the thick specimen. The reason the latter does not appear to be as sensitive as the thin film to the shorter wavelengths is the difference in exposure times for the two specimens.

The shift in harmful wavelengths to the longer wavelength spectral region with increase in thickness is due to a combination of factors: 1) the main absorption band of the polymer, which broadens with thickness, extends into the spectral region of the

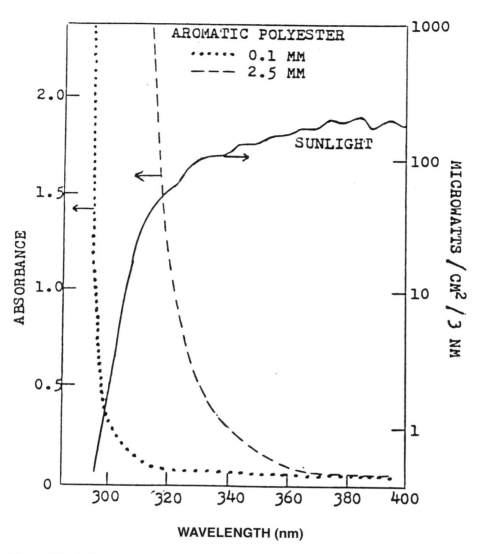

Figure 15 Effect of thickness on UV absorption of aromatic polyester and UV emission of solar radiation. (From Ref. 13.)

light source, 2) the increase of source intensity with wavelength coincides with the long-wavelength broadening of the absorption band, and 3) because of the logarithmic nature of the absorption process, thickness has a greater effect on the amount of light absorbed in the long wavelength region where polymer absorption is weak. The shift is very much dependent on the spectral energy distribution of the source. Thus, on exposure to fluorescent UVB lamps, the shift should be minor because this source lacks the long wavelength emission that the thicker specimen is capable of absorbing. The shift based on exposure to fluorescent UVA lamps should be somewhat more representative of the data in Fig. 16. However, for polymers, such as

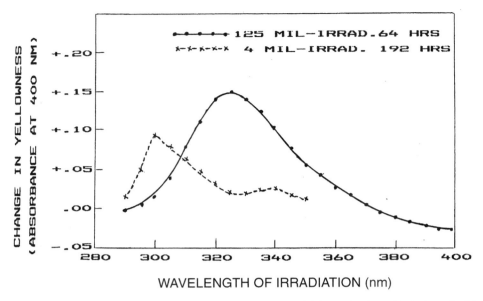

Figure 16 Activation spectra of 4-mil (0.1-mm)–thick and 125-mil (3.1-mm)–thick aromatic polyester exposed to borosilicate–glass-filtered xenon arc radiation. (From Ref. 13.)

polyarylate, aromatic polyurethanes, and other materials that are sensitive to wavelengths longer than 350 nm, the shift with thickness on exposure to fluorescent UVA lamps should also be smaller than the shifts on exposure to solar or filtered xenon arc radiation.

The effect of thickness on the wavelength sensitivity to solar radiation applies only to polymers in which the main absorption band extends into the spectral region of the light source. For aliphatic-type polymers, thickness has a negligible effect on absorption properties in this spectral region; hence, on the wavelength sensitivity to solar radiation.

VII. ACTIVATION VERSUS ACTION SPECTRA

Wavelength sensitivity or spectral sensitivity of a material are generic terms that can be used interchangeably to describe either an activation spectrum or an action spectrum. Both of the latter represent the relative sensitivities of a material to incident radiation of different wavelengths. They differ in that the activation spectrum is specific to both the material and the source of radiation, whereas the action spectrum does not depend on the spectral emission properties of the source to which the material is exposed. It is source-independent and is specific only to the material. The experimental techniques as well as the spectral sensitivity profiles differ.

The term *activation spectrum* was introduced in 1959 (31) to describe data obtained by the spectrographic technique developed to determine the relative spectral effects of a specific radiation source on a material. It was used to determine the spectral absorption properties required of an UV absorber designed to protect a polymeric material against the damaging effects of sunlight. Other applications are listed in Table 2 and described in Section IX. In this technique, spectral bands of near-monochromatic radiation are isolated from a continuum radiation source by a spectrograph. Pure monochromatic radiation is not obtainable in this way. Only a narrow band of wavelengths with a finite band width can be isolated. The sharp cut-on filter technique was a later development for determining the effect of different spectral regions of a specific radiation source based on exposure to polychromatic radiation.

Action spectra techniques were initially used for determining wavelength sensitivities of biological processes. These included the erythemal response (38–40) and pigmentation (41) of human skin, inactivation of DNA (42), inactivation (43) and UV-enhanced cell reactivation (44) of viruses, photosynthesis (45,46), and others. The experimental technique differs from that for activation spectra in that action spectra require quantitative measurements of incident radiation to make adjustments for the intensity variations of the radiation source with wavelength. The measured incident intensities in each spectral band are used either to adjust the length of exposure to each spectral band or to normalize the measured property changes so that they are based on either 1) equal radiant exposure to all spectral bands or 2) the radiant exposure by each spectral band required to produce the same predetermined change by all spectral bands. However, normalization is valid only if the reciprocity law applies (i.e., if property change is a linear function of intensity and exposure).

In many of the studies, incident radiation is measured in watts per square meter (W/m^2) and radiant exposure is reported in joules per square meter (J/m^2). In photobiology, the latter is often referred to as "fluence" (46,47) as an alternative to "dose" to distinguish energy incident on the surface from energy absorbed. Action spectra are also based on the number of photons of incident energy (42,47), a requirement for determining spectral quantum yields of degradation. In some of the early investigations of erythemal action spectra (48,49), the radiation consisted of line spectra from low-pressure arc sources (mostly mercury) with appropriate filters. Later studies used broadened lines plus continuum from high-pressure mercury–xenon or xenon arc sources, along with interference filters or a monochromator to isolate spectral bands.

Table 2 Applications of Activation Spectra

A. Stabilization of materials
    1. Light screening requirements
    2. Screening effectiveness of UV absorbers
    3. Screening mechanism of stabilizer
    4. Shift in wavelength sensitivity with type of degradation
B. Stability testing
    1. Selection of appropriate weathering device
    2. Antagonistic wavelength effects
    3. Effects of thickness on stabilizer evaluation
    4. Actinic wavelengths for timing exposures
    5. Effective radiant exposure data for predicting lifetimes
C. Appropriateness of use environment for photodegradable materials

---

Source: Ref. 11.

Source-independent wavelength sensitivities (i.e., action spectra) have been reported in recent years for various types of synthetic and natural polymers. Studies have included polyamides (50), polyarylate (50), polycarbonate (50–54), polyethylene (50,55), polybutadienes and other elastomers (56), poly(methyl methacrylate) (51,57), polypropylene (50,58), polystyrene (58,59), poly(vinyl chloride) (60), chitosan (61), and mechanical pulp (62,63).

Most of these studies were carried out by Torikai and co-workers using the Okazaki large grating spectrograph in which near-monochromatic radiation between 250 and 1000 nm from a 30-kW xenon short-arc lamp is spectrally dispersed by a double-blazed plane grating and projected onto a 10-m–long focal curve (64). Replicate specimens are exposed to selected spectral bands and incident radiation is measured in terms of number of photons. Exposures are generally for the same length of time to all spectral bands, with normalization of the measured degradation to equal number of photons for all spectral bands. In a few of the studies, exposures were adjusted to provide the same number of photons for all spectral bands. The latter type of exposure is preferable because it eliminates the need to normalize the degradation to a common amount of radiation.

For the action spectra by Trubiroha (50), referred to as spectral sensitivities, a smaller grating spectrograph was used to disperse the UV radiation from a 900-W high-pressure short-arc xenon lamp onto a single specimen mounted in the focal plane. The incident radiation in each spectral band was measured in terms of radiant energy, and the degradation caused by each of the bands was normalized to equal radiant exposure of 1 MJ/m^2.

A single-grating monochromator was used in one of the studies (63) and interference filters in another (56) to irradiate replicate specimens with selected spectral bands of near-monochromatic radiation from a xenon arc. In the former, action spectra were obtained based both on 1) the reciprocal of the exposure to each spectral band required to produce the same specified change by all bands, and 2) the calculated change produced by equal amounts of incident energy following exposure to all spectral bands for the same length of time. For the action spectra obtained with interference filters, the radiant exposure for all spectral bands was 5 mW/cm^2. Some of the action spectra were represented in terms of the calculated change per absorbed photon.

Because both types of wavelength sensitivities—activation and action spectra—have been obtained using near-monochromatic radiation, they cannot be differentiated from each other on the basis of type of exposure. Activation spectra have been erroneously differentiated from action spectra on the basis of exposure to polychromatic radiation for the former and near-monochromatic radiation for the latter (32,33). However, it is only the more recent polymer activation spectra that were obtained by the sharp cut-on filter technique using polychromatic radiation.

The action spectrum often mirrors the absorption spectrum of the photosensitive species (46). Therefore, for materials that contain known absorbing species with well-defined spectral absorption properties, action spectra can sometimes be used to identify the chromophore in a multicomponent system responsible for the photochemical changes (63). However, the success of this application of the action spectrum depends on knowledge of the absorbing species and decreases rapidly as the number of reacting chromophores increases. It is not generally applicable to polymeric materials because degradation is often due to a complex mixture of UV absorbing impurities that are unidentified.

The two types of wavelength sensitivity spectra bear no resemblance to one another. Because the contour of the activation spectrum is determined by the spectral power distribution of the light source as well as by the spectral response of the material, it exhibits peaks in spectral regions where the intensity of the light, its absorption by the material, and the quantum efficiency of degradation combine to produce maximum damage. Action spectra of many polymers are smooth curves showing increase in damage with decreasing wavelength. Because short wavelength photons are more energetic than long wavelength ones, they can be expected to cause more damage. Therefore, the damage per photon increases and the amount of energy required to produce equal damage decreases with decreasing wavelength. The difference in the two types of wavelength sensitivity spectra are illustrated in the next two sections with activation and action spectra obtained on identical samples of poly(vinyl chloride) and polycarbonate exposed to filtered xenon arc radiation.

A. Activation Versus Action Spectra of Poly(vinyl Chloride)

Wavelength sensitivities to yellowing of an extruded rigid poly(vinyl chloride) formulation, with and without rutile titanium dioxide (TiO_2), were determined by both techniques (24,60). The formulation is typical of those used for building siding and window profiles, except that the TiO_2 concentration of 2.5 parts per hundred (phr) is less than in the latter applications. Sample thickness was 1 mm.

1. Activation Spectra of PVC

The activation spectra of PVC (24) shown in Fig. 17 were obtained by the sharp-cut filter technique. The filtered samples were exposed in the borosilicate–glass-filtered xenon arc accelerated weathering apparatus. The photochemical yellowing by this radiation in both the unpigmented and pigmented PVC was largely due to wavelengths between 300 and 320 nm. Because pure PVC does not absorb the radiation emitted by the light source, the spectral sensitivity to yellowing is attributed to the photosensitive UV absorbing impurities. The data agree, within experimental accuracy, with the activation spectrum by the spectrographic technique reported by Hirt and Searle (65) for a 0.05-mm film of a commercial PVC homopolymer exposed to the same type of radiation. Yellowing was caused mainly by radiation between 310 and 325 nm. This study also showed that 5 phr of TiO_2 had no effect on the activation spectrum of the material.

2. Action Spectra of PVC

Action spectra of the PVC formulations (60) were obtained using the Okazaki Large Spectrograph equipped with a 30-kW xenon short-arc lamp. Specimens were placed at six spectral positions identified by the midpoint of the spectral band as 280, 300, 320, 340, 400, and 500 nm. All specimens were exposed for the same length of time. Based on the photon flux measured at each position, the measured yellowness index in each of the specimens was normalized to the change in yellowness index per incident photon. The data are plotted in Fig. 18 in terms of the natural log of the change as a function of wavelength of irradiation.

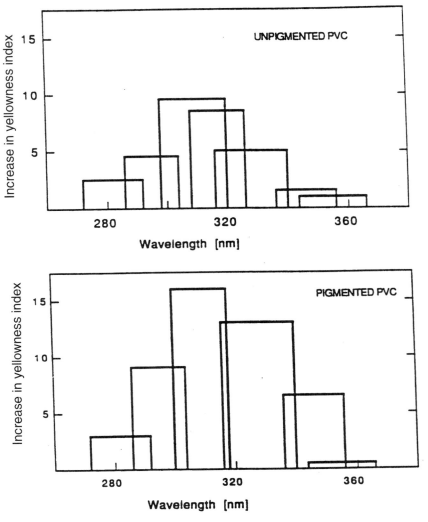

Figure 17 Activation spectra of rigid poly(vinyl chloride) with and without 2.5 phr TiO$_2$ exposed to borosilicate–glass-filtered xenon arc radiation. (From Ref. 24.)

The action spectra in Fig. 18 show that the shorter the wavelength the more severe the degradation per incident photon, typical of action spectra of most polymers. Yellowing was caused by wavelengths of 280, 300, 320, and 340 nm, whereas wavelengths of 400 and 500 nm reduced the yellow color in the unexposed samples (i.e., caused bleaching). The presence of rutile titanium dioxide protected the polymer against yellowing, but had very little effect on the action spectrum, judging from the slopes of the plotted data. That the action spectrum does not give any evidence of the spectral absorption of the photosensitive species may be partly due to an insufficient number of spectral bands used in the experiment.

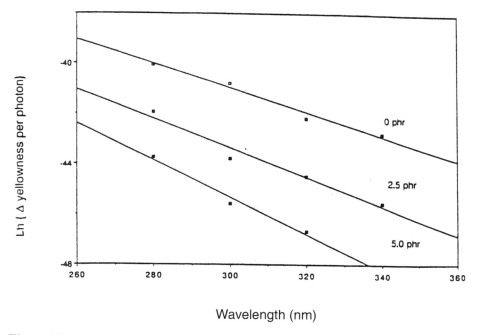

Ln (Δ yellowness per photon)

Wavelength (nm)

Figure 18 Action spectra of rigid poly(vinyl chloride) formulations with 0, 2.5 and 5.0 phr TiO_2. (From Ref. 60.)

B. Activation Versus Action Spectra of Polycarbonate

Activation and action spectra were obtained of unstabilized, clear bisphenol-A polycarbonate films based on formation of yellow-colored species and increase in UV absorption. Borosilicate glass-filtered xenon arc radiation was used for all the exposures.

1. Activation Spectra of Polycarbonate

The sharp cut-on filter (14) and spectrographic (22) techniques produced similar activation spectrum of polycarbonate. Figure 5 in Section II.C shows the activation spectrum of a 0.70-mm film of the polycarbonate obtained by the sharp cut-on filter technique. It identifies two spectral regions in borosilicate–glass-filtered xenon arc radiation responsible for most of the yellowing. These are wavelengths shorter than 300 nm and wavelengths between about 310 and 340 nm. The distinct separation of the effects of the two spectral regions support studies ascribing the yellowing in the two regions to different photosensitive species and different mechanisms of degradation (4,5,15–17).

Wavelengths shorter than 300 nm, strongly absorbed by the polycarbonate chromophores, cause photo-Fries rearrangement of the structural components, followed by formation of highly yellow-colored degradation products. Wavelengths between 310 and 340 nm, weakly absorbed by the impurities and defects, initiate mainly photooxidation processes. These also lead to yellow-colored degradation products. The effect of the two spectral regions is separated by a spectral band centered at about 300 nm that reduces the yellowing formed under longer wave-length radiation. The activation spectrum also shows that wavelengths longer than 380 nm bleach the yellow-colored species present in unexposed material.

Figure 19 (22) shows the activation spectra of a 0.1-mm film of polycarbonate obtained by the spectrographic technique. It represents the wavelength sensitivities based on increase in absorbance at two wavelengths in the UV and yellowing, which is measured at 400 nm. The light source was a borosilicate–glass-filtered 1000-W high-presure xenon arc. These activation spectra also show two well-defined spectral regions responsible for the optical changes. The most severe degradation is caused by wavelengths shorter than 300 nm. The longer wavelength actinic region ranges from about 310 to 350 nm, with wavelengths near 330 nm having maximum effect. The similarity of the activation spectra by the two techniques indicates the absence of any synergistic or antagonistic effects when the material is exposed to polychromatic radiation in the sharp cut-on filter technique.

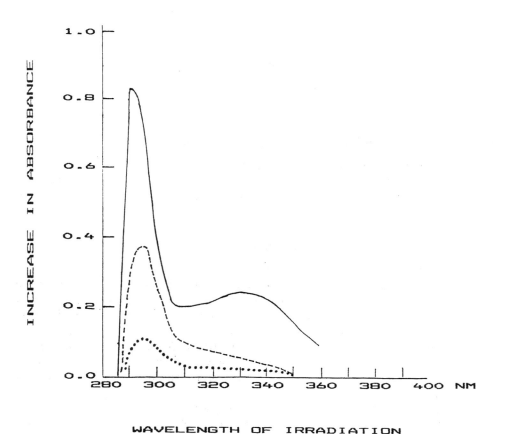

Figure 19 Activation spectra of 0.1-mm film of bisphenol-A polycarbonate by the spectrographic technique based on increase in absorbance at 300 nm (solid line), 350 nm (dashed line), and 400 nm (dotted line). Light source: borosilicate–glass-filtered 1000-W high-pressure xenon arc. (From Ref. 22.)

2. Action Spectra of Polycarbonate

In contrast with the activation spectra that show that two well-defined spectral regions of the filtered xenon arc are responsible for yellowing and increase in UV absorption of the polycarbonate film, the action spectrum (50–54) of a 0.70-mm film of the same material in Fig. 20 (50) is a smooth curve showing a sharp increase in sensitivity with decreasing wavelength. The data is plotted in terms of the radiant exposure required to decrease the transmittance at 360 nm by 10% as a function of wavelength of irradiation.

The effect of temperature was determined by comparing the spectral sensitivity at 10° and 50°C. An increase in temperature causes an increase in the rate of reaction following breakage of bonds by the absorbed radiation. The effect of a 40°C increase in temperature shifts the 10°C action spectrum about 10 nm to a longer wavelength. However, an increase in temperature does not change the action spectrum (i.e., the relative effect of different wavelengths).

C. Conversion from Activation to Action Spectrum

The experimental data for an action spectrum based on exposure to all spectral bands of the radiation source for the same length of time is an activation spectrum. Normalization converts it to an action spectrum. It was pointed out previously that

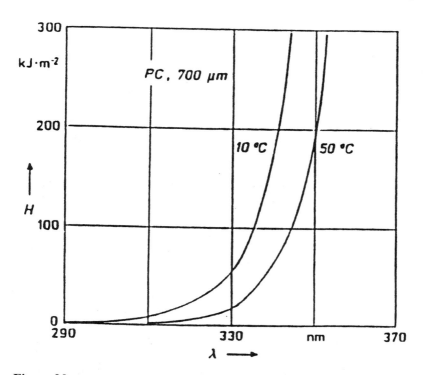

Figure 20 Action spectra of a 0.70-mm film of bisphenol-A polycarbonate based on amount of radiant energy (kJ/m^2) required for 10% decrease in transmittance at 360 nm. (From Ref. 50.)

this technique for producing an action spectrum is valid only if degradation is a linear function of intensity. Assuming the latter condition is met, activation spectra obtained by the spectrographic and interference filter techniques can be converted to action spectra. However, the conversion is not as straightforward for activation spectra obtained by the sharp cut-on filter technique. In addition to validation of the reciprocity law, lack of synergistic or antagonistic effects of exposure to all wavelengths longer than the cut-on of the filter must be established.

With the assumption that both the foregoing conditions were met, Allan et al. (32) calculated source-independent wavelength sensitivity spectra of polystyrene from data obtained by the sharp cut-on filter technique. A mathematical model was developed to predict weatherability of transparent polymers containing any chosen UV absorber. It requires information on the combined spectral absorption properties of the UV absorbing species in the polymer itself and by any UV absorbing additives. The model is primarily applicable to the actinic radiation absorbed by the main structural components of the polymer because the sensitizing impurities in the polymer and their UV absorption properties are generally unknown.

The model was tested by comparing predicted and experimentally measured yellowing rates of stabilized polystyrene. The data showed that the model underestimated the absolute degradation of all formulations. This can probably be attributed to the inability to include information on the UV absorbing impurities. Thus, the model would not be applicable to aliphatic-type polymers exposed to solar radiation unless the absorbing impurities have been characterized.

D. Conversion from Action to Activation Spectrum

In theory, the action spectrum can be converted to an activation spectrum by convoluting it against the spectral power distribution of the light source of interest. However, conversion of an action spectrum to an activation spectrum in this way is obviously not applicable to polycarbonate, nor would it be applicable to other materials for which the degradation mechanism changes with actrinic wavelength. The mathematical conversion from action to activation spectrum for materials in which all wavelengths promote the same mechanism and type of degradation is based on the assumption that the property change measured is a linear function of intensity and radiant exposure. One of the few polymeric materials for which this assumption has been valid is poly(vinyl chloride). When the assumption is not valid, any error caused by nonreciprocity is introduced twice if the action spectrum had been based on normalized data.

Determination of source-dependent wavelength sensitivity by the direct approach i.e., by experimentally determined activation spectra produces more valid data. Conversion from an action to an activation spectrum is recommended only when it is not possible to obtain the activation spectrum directly using the radiation source of interest. For example, the spectral effects of reduction of the stratospheric ozone layer on materials may not be amenable to direct determination unless the increase in solar UVB radiation on the earth's surface can be simulated by appropriately filtering an artificial source.

VIII. ACTIVATION SPECTRUM VERSUS QUANTUM EFFICIENCY SPECTRUM

The quantum efficiency spectrum or spectral quantum yield of photoprocesses is source-independent wavelength sensitivity and has a profile similar to that of the action spectrum of the material. It represents the fraction of photolytically effective photons at each wavelength and is obtained from the ratio of the number of moles of photoproducts formed divided by the number of moles of photons absorbed at that wavelength. Spectral quantum yield data is widely used for biological applications to convert total incident radiant exposure to effective dosage used in predicting lifetimes (see Sec. IX.B.5).

Because information on the absorption properties of the photosensitive species is required for determination of spectral quantum efficiencies, the latter can be determined only for the structual components of polymeric materials. The UV absorption properties of the defects and impurities that initiate degradation in both aliphatic and aromatic type polymers are usually unknown and are not measurable because of their low concentrations and weak absorption coefficients. However, both activation and action spectra of polymeric materials provide data for determination of effective dosage. Both are the product of the spectral absorption properties of the photosensitive components and the quantum efficiencies of the radiation absorbed. The activation spectrum includes the spectral power distribution of the radiation source.

IX. APPLICATIONS OF ACTIVATION SPECTRA

Applications of activation spectra to polymeric materials are listed in Table 2 and are described in the following sections. They include support of stabilization and stability testing of materials as well as prediction of lifetimes. Activation spectra also provide information on the potential degradation of photodegradable materials by the light in the use environments.

A. Stabilization of Materials

Information on the spectral response of a polymeric material to the source of radiation to which it will be exposed under use conditions is important in development of stabilizer systems to prolong its life. Activation spectra identify the light-screening requirements of a material as well as the screening effectiveness of an UV absorber. They are also useful in confirming additive protection by screening and indicating the occurrence of different mechanisms of degradation by the shift in spectral response with type of degradation measured.

1. Light-Screening Requirements

The activation spectrum determines the wavelengths in the exposure source that are harmful to a specific material and thus the type of UV absorber needed for optimum screening protection. For example, activation spectra based on solar radiation have shown that polycarbonate requires protection only against wavelengths shorter than 360 nm, whereas even a thin film of polyarylate requires protection against wavelengths as long as 380 nm. The absorption properties of the additive should closely match the activation spectrum obtained using the appropriate light source.

Activation spectra are also useful in determining whether incorporation of a UV absorber is a practical approach to stabilization. For example, the activation spectrum of a 1.5-mm (60-mil)–thick sample of polyarylate showed that significant yellowing was caused by solar simulated radiation between 380 and 390 nm (see Table 1). Thus, an UV absorber would offer very little protection to this form of polyarylate against solar radiation. Any additive capable of screening the harmful wavelengths would itself impart some yellow color to the material. The availability of this information provided by the activation spectrum of the material could have avoided time-consuming and costly investigations in an attempt to stabilize the material with UV absorbers (B. Dickinson, personal communication, 1987).

2. Screening Effectiveness of UV Absorbers

The effectiveness of an UV absorber in protecting a material depends on its ability to compete with the polymer in absorbing the actinic wavelengths. The relative effectiveness of different types of UV absorbers can be estimated based on the match of their absorption properties to the activation spectrum of the material. This assumes that other factors, such as dispersion, compatibility, and migration are similar.

The adequacy of the UV absorber in screening the harmful wavelengths can be evaluated by comparing the activation spectrum of the stabilized with that of the stabilized material. The comparison identifies spectral regions that may need further screening for more adequate protection. An example is given in Fig. 21 (13), which compares the activation spectra of the 3.1-mm–thick sample of the aromatic polyester with and without an ortho-dihydroxybenzophenone-type UV absorber.

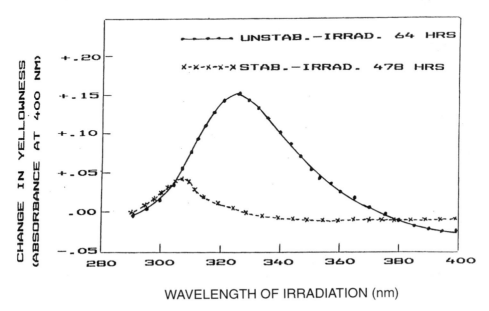

Figure 21 Effect of UV absorber on activation spectrum of a 3.1-mm plaque of aromatic polyester exposed to borosilicate–glass-filtered xenon arc radiation. (From Ref. 13.)

It shows that although the additive protects the polymer against most of the wavelengths that cause yellowing, it does not completely protect it against the shortest actinic wavelengths. Yellowing on prolonged exposure of the stabilized sample is due to wavelengths in the 290- to 325-nm region, with the strongest yellowing caused by the spectral region centered at 305 nm. Complete screening protection would require an additive with a higher absorption coefficient for these shorter wavelengths to more effectively compete with the strong absorption of the polymer in this region.

3. Screening Mechanism of Stabilizers

Comparison of the activation spectra of the stabilized and unstabilized polyester in Fig. 21 leaves little doubt that screening the harmful wavelengths is the prime means of stabilization by this type of additive. This does not rule out the possibility that it may also protect as a free radical scavenger or hydroperoxide decomposer by interfering with the mechanism of any degradation caused by absorption of light that is not completely screened out.

4. Dependence of Wavelength Sensitivity on Type of Degradation

The shift in wavelength sensitivity with the type of degradation measured indicates that different mechanisms may be responsible for the various types of degradation. The activation spectra listed in Table 1 show that yellowing, loss in impact strength, and bleaching of ABS are each caused by a different spectral region of borosilicate–glass-filtered xenon arc radiation. Yellowing and carbonyl formation in both polyethylene and polysulfone are also caused by different wavelengths. Thus, information is provided on the potential for improved stabilization with a combination of additives, each interfering with a different degradation mechanism.

B. Stability Testing

In development of new materials and formulations, laboratory accelerated light stability testing is carried out to compare these materials with those of known performance. It is expected that this data will relate to relative performance under use conditions and will be useful in prediction of lifetimes. Simulation of environmental conditions, particularly the type of radiation, is essential to providing meaningful test results. Activation spectra provide information useful in selecting the appropriate artificial test source and of demonstrating the importance of simulating the full spectrum of actinic wavelengths. The validity of the test data also depends on the form in which materials are tested and on use of actinic wavelengths for timing exposures. Information in support of the former and data for the latter are obtained from activation spectra, which also have application for predicting lifetimes.

1. Selection of Accelerated Test Source

Selection of the appropriate laboratory accelerated exposure device is critical to relating the results to natural exposures and to predicting the effects of long-term exposure under use conditions. The extent of the match of the harmful wavelengths in the artificial test source with those in the natural source is determined by comparison of activation spectra of the material obtained under each of the exposure conditions. Differences in harmful wavelengths can change the mechanism and type of degradation and produce invalid data on the effectiveness of an UV absorber or other type of stabilizer in protecting the material against degradation by the natural source. Laboratory accelerated test data is only applicable to prediction of lifetimes under use conditions if the mechanism and type of degradation are the same under both types of exposure.

2.   Antagonistic Effects of Spectral Regions

In many polymers the photochemical effects of different spectral regions are antagonistic to each other. For example, scission of chemical bonds, resulting in reduction in molecular weight and tensile strength of polyolefins, is caused by exposure to short wavelength UV radiation, whereas cross-linking, resulting in increased molecular weight and changes in elongation, is promoted by longer wavelengths. The opposing effects of different spectral regions is also manifested by the photobleaching phenomenon in which wavelengths between 350 and 450 nm reverse the yellowing caused by shorter wavelengths.

The wavelength dependence of yellowing versus bleaching in the 3.1-mm sample of the aromatic polyester is shown in Fig. 22 (13). The activation spectrum represents the spectral sensitivity to the borosilicate–glass-filtered xenon arc of a specimen that had been photochemically yellowed by a short wavelength UV source. The change in yellow color is based on measurement of absorption at 400 nm. The smooth curve was obtained by connecting data points, each representing the wavelength at the midpoint of the narrow spectral band to which the sample was exposed in the spectrograph. The straight line drawn across the figure at the zero value on the ordinate scale represents the yellow color of the photodegraded sample before it was placed in the spectrograph.

The activation spectrum shows that the yellow color is diminished by wavelengths longer than 350 nm. The 405-nm spectral band is most effective in reversing this color. Longer wavelengths are less effective because they are not as strongly absorbed by the yellow-colored species. Wavelengths shorter than 405 nm are more strongly absorbed, but because of the simultaneous formation of new yellow-colored species, bleaching is less pronounced in the region between 350 and 405 nm. Below 350 nm, further yellowing is the predominant effect. It increases with decreasing wavelength to a maximum at about 325 nm. Weaker yellowing by wavelengths shorter than 325 nm is due to the sharp decrease in intensity of the light source in this region.

The opposing effects of different spectral regions on the same material shown by the activation spectrum demonstrates the importance of closely simulating the natural source over the full range of actinic wavelengths by the laboratory accelerated test source. Because the net effect of the radiation source depends on the relative intensities of the short versus long wavelengths, simulation of the long wavelength UV and visible radiation can be just as important as simulation of the short wavelength UV radiation.

3.   Form of Test Samples

Activation spectra of different thicknesses of an aromatic-type polymeric material demonstrate the importance of testing these materials in the form in which they will be used in practice, whether testing is for the purpose of evaluating the performance of stabilizers, ranking the stability of materials, or predicting lifetimes. The shift

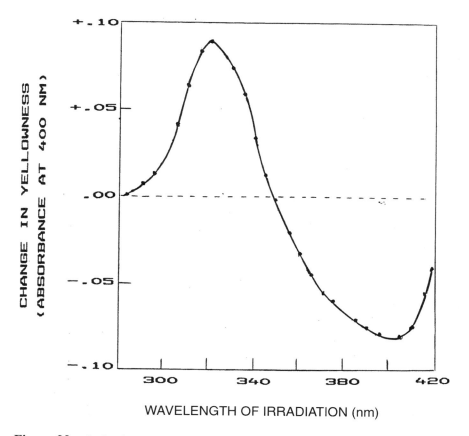

WAVELENGTH OF IRRADIATION (nm)

Figure 22 Activation spectrum of photochemically yellowed 3.1-mm plaque of aromatic polyester exposed to borosilicate–glass-filtered xenon arc radiation. (From Ref. 13.)

of activation spectra of aromatic-type polymers to longer wavelength with increase in thickness when they are exposed to solar radiation was discussed in Section VI.B. The screening requirements and the type of UV absorber used for the thicker forms of the polymers differ from those for thin films. An absorber that screens only short wavelengths would not be suitable for the thicker specimens, and evaluation of its effectiveness will vary, depending on the thickness of the specimen in which it is tested.

Because of the shift in damaging wavelengths with thickness, the mechanism and type of photodegradation can change with thickness. If this occurs, the effectiveness and type of stabilizer used to inhibit degradation by interfering with the mechanism will also depend on thickness. Owing to differences among materials in the effect of thickness on the absorption of light, change in thickness can also change the stability ranking of materials. For these reasons and others, prediction of lifetimes requires the testing of materials, particularly aromatic-type polymers, at the same thickness at which they will be used in practice.

4. "Timing" of Radiant Exposures

Correlations between laboratory accelerated and outdoor weathering tests as well as between natural weathering tests under different site and seasonal conditions are reported (66–72) to be improved when radiant exposures are "timed" in terms of total UV radiation, rather than calendar time or total solar radiation. It was suggested that "timing" of exposures, based on only the portion of the UV radiation shown by the activation spectrum of the material to be responsible for the degradation, is necessary to further improve correlations. Because all wavelengths in this region are not equally effective, a further refinement would be to time exposures based on "effective" irradiance by weighting the incident actinic radiation by the activation spectrum. It takes into account the fact that all wavelengths in the actinic region do not have an equal effect on a material.

5. Predicting Lifetimes

The destructive effect of weather on most materials is primarily due to the incident actinic radiation responsible for initiating the degradation. Assuming that other weather factors are reasonably well simulated, short-term laboratory accelerated tests can be used to predict service life of materials under field conditions from an experimentally determined damage function that relates effective dosage to material damage. The damage model may be a linear, power law, or exponential function, varying with type of material and its formulation, and should be determined over the range of temperature and humidity conditions to which the material is expected to be exposed.

Martin et al. (73,74) have suggested this approach to predicting lifetimes of polymeric materials based on its successful application in the biological field. In the latter, total incident radiation is converted to effective dosage (D_{Eff}) (i.e., the fraction effective in causing damage) by integrating over both the range of photolytically effective wavelengths and the duration of exposure the product of the following measured quantities:

1. $E_0(\lambda)$, the spectral irradiance of the source to which the material is exposed
2. $[1 - e^{-A(\lambda)}]$, the spectral absorption of the photosensitive components
3. $[\phi(\lambda)]$, the spectral quantum efficiencies of the absorbed radiation

Integration of the product of 1, 2, and 3 over the actinic wavelengths gives the effective irradiance (E_{Eff}) according to Eq. (1), and integration of effective irradiance over the total time of exposure (t) at a constant irradiance level gives the effective dosage according to Eq. (2). The validity is based on the assumption that the spectral quantum yield does not depend on the radiant flux or duration of exposure.

$$E_{\mathrm{Eff}} = \int_{\lambda_1}^{\lambda_2} E_o(\lambda)[1 - e^{-A(\lambda)}]\phi(\lambda)\,\mathrm{d}\lambda \tag{1}$$

$$D_{\mathrm{Eff}} = \int_0^t E_{\mathrm{Eff}}(t)\,\mathrm{d}t \tag{2}$$

Both quantities 2 and 3 require information on the absorption properties of the photosensitive species. Because these have not been identified in most polymeric materials, the approach used to determine total effective dosage for biological substances is not viable for polymeric materials. However, the activation spectrum is a direct experimental determination of the quantum efficiencies of degradation initiated by absorption of the spectral bands of radiation from the light source. It contains all the elements of items 1, 2, and 3, and eliminates the requirement for information on the absorption properties of the photosensitive species. Martin et al. (73) define the activation spectrum by Eq. (3) which shows that the factors responsible for the change in property of the material, Δ, by each incident spectral band of the light source $E_0(\lambda)$ are the spectral absorption of the photosensitive components and the spectral quantum efficiency of the absorbed radiation.

$$\Delta(\lambda) = E_o(\lambda)\left[1 - e^{-A(\lambda)}\right]\phi(\lambda) \tag{3}$$

Note that Eq. (3) differs from determination of E_{Eff} by Eq. (1) only in that the latter is integrated over the spectral region.

Thus, the activation spectrum can be used to determine the effective portion E_{Eff} of the total incident radiation by weighting the latter by the activation spectrum according to Eq. (4). The irradiance at the wavelength of the activation spectrum peak is normalized to 1.0 and the irradiances at the other wavelengths are multiplied by the fractional sensitivities. The sum is the effective irradiance.

$$E_{\text{Eff}} = \sum_{\lambda_t}^{\lambda_2} E_o(\lambda)\left[\frac{\Delta(\lambda)}{\Delta(\lambda_{\text{max}})}\right] \tag{4}$$

$\Delta(\lambda) = $ property change of material at wavelength
$\Delta(\lambda_{\text{max}}) = $ property change of material at wavelength of activation spectrum peak.

The effective dosage, D_{Eff}, is the ratio of the effective irradiance, E_{Eff}, divided by the total irradiance, E_{Total}, multiplied by the total dosage, D_{Total}, as per Eq. (5).

$$D_{\text{Eff}} = \frac{E_{\text{Eff}}}{E_{\text{Total}}} \times D_{\text{Total}} \tag{5}$$

For reliable prediction of lifetimes, the activation spectrum and the damage model relating effective dosage to material damage are determined using a source that closely simulates the source to which the materials will be exposed in the use environment.

C. Uses of Photodegradable Materials

Photodegradable materials are designed to rapidly disintegrate outdoors when exposed to daylight, but to maintain their desirable mechanical properties during use under indoor illumination, including window–glass-filtered daylight. Information on the wavelength sensitivity to indoor illumination is important in assessing the potential for degradation during use.

Activation spectra of two types of enhanced photodegradable polyethylenes, based on measurement of tensile elongation as the criterion of degradation, showed that these polymers can be expected to lose some of their tensile extensibility and strength when exposed to window–glass-filtered daylight (25). One of the polyethylenes was degraded by wavelengths as long as 338 nm at 60°C and 348 nm at 77°C. Thus, depending on the temperature under use conditions, protection against wavelengths as long as 350 nm may be required during indoor use of the material. The other polyethylene was sensitive to wavelengths as long as 328 and 338 nm, respectively, at the two temperatures.

REFERENCES

1. A Rivaton, D Sallet, J Lemaire. The photochemistry of bisphenol—A polycarbonate reconsidered. Polym Photochem 3:463–481, 1983.
2. J Lemaire, J-L Gardette, A Rivaton, A Roger. Dual photochemistries in aliphatic polyamides, bisphenol-A a polycarbonate and aromatic polyurethanes—a short review. Polym Degrad Stabil 15:1–13, 1986.
3. J-L Gardette, J Lemaire. Wavelength effects on the discoloration and oxidation of poly(vinyl chloride). Polym Degrad Stabil 25:293–306, 1989.
4. A Factor, ML Chu. The role of oxygen in the photo-ageing of bisphenol—A polycarbonate. Polym Degrad Stabil 2:203–223, 1980.
5. HS Munro, RS Allaker. Wavelength dependence of the surface photo-oxidation of bisphenol-A a polycarbonate. Polym Degrad Stabil 11:349–358, 1985.
6. A Rivaton, D Sallet, J Lemaire. The photo-chemistry of bisphenol—A polycarbonate reconsidered: Part 2—FTIR analysis of the solid-state photo-chemistry in "dry" conditions. Polym Degrad Stabil 14:1–22, 1986.
7. HS Munro, DT Clark. The surface photo-oxidation of bisphenol-A a polysulfone films as studied by ESCA. Polym Degrad Stabil 11:211–224, 1985.
8. MH Tabankia, J-L Gardette. Photo-oxidation of block copolymer (ether–ester) thermoplastic elastomers: Part 2—origins of the photo-yellowing. Polym Degrad Stabil 19:113–123, 1987.
9. ND Searle. Simulating solar UV and visible radiation to reproduce effects of weathering. Surface Coatings Australia, July 1997, pp 28–30.
10. ND Searle. The activation spectrum and its application to stabilization and weatherability tests. Preprint, SPE 43rd ANTEC, Washington, DC, 1985, pp 248–251.
11. ND Searle. Spectral factors in photodegradation: activation spectra using the spectrographic and sharp cut filter techniques. In: D Kockott, ed. International Symposium on Natural and Accelerated Weathering of Organic Materials. Essen, Germany, Sept 28–29, 1987. Lochem, The Netherlands: Atlas SFTS BV, 1988, Chap. B.
12. ND Searle. Weathering. In: Mark–Bikales–Overberger–Menges Encyclopedia of Polymer Science and Engineering, 2nd ed. Vol. 17. New York: John Wiley & Sons, 1989, pp 796–827.
13. ND Searle. Wavelength sensitivity of polymers. In: AV Patsis, ed. International Conference on Advances in the Stabilization and Controlled Degradation of Polymers. Vol. 1. Lucerne, Switzerland, May 1986, Vol. 1. Lancaster, PA: Technomic Publishing, 1989, pp 62–74.
14. AL Andrady, ND Searle, and LFE Crewdson. Wavelength sensitivity of unstabilized and UV stabilized polycarbonate to solar simulated radiation. Polym Degrad Stabil 35:235–247, 1992.
15. A Factor. Mechanisms of thermal and photodegradation of bisphenol A polycarbonate. Adv Chem Ser 249:59–76, 1996.
16. A Rivaton. Recent advances in bisphenol-A polycarbonate photodegradation. Polym Degrad Stabil 49:163–179, 1995.
17. A Rivaton. The bisphenol-A polycarbonate dual photochemistry. Angew Makromol Chem 216:147–153, 1994.
18. RC Hirt, NZ Searle, RG Schmitt. Ultraviolet degradation of plastics and the use of protective ultraviolet absorbers. SPE Trans 1(1):21–25, 1961.

19. NZ Searle, RC Hirt. Ultraviolet spectroscopic techniques for studying the photodegradation of high polymers. Preprint, Divs Polym Chem Org Coatings Plastics. Am Chem Soc 22(2):56–61, 1962.

20. RC Hirt, NZ Searle. Wavelength sensitivity and activation spectra of polymers. Preprint SPE RETEC, Dec 1964, pp 286–302.

21. LD Johnson, SC Tincher, HC Bach. Photodegradative wavelength dependence of thermally resistant organic polymers. J Appl Polym Sci 13:1825–1832, 1969.

22. PA Mullen, NZ Searle. The ultraviolet activation spectrum of polycarbonate. J Appl Polym Sci 14:765–776, 1970.

23. KG Martin, RI Tilley. Influence of radiation wavelength on photo-oxidation of unstabilized PVC. Br Polym J 3:36–40, 1971.

24. AL Andrady, ND Searle. Photodegradation of rigid PVC formulations. II spectral sensitivity to light-induced yellowing by polychromatic light. J Appl Polym Sci 37:2789–2802, 1989.

25. AL Andrady, JE Pegram, ND Searle. Wavelength sensitivity of enhanced photo-degradable polyethylenes ECO and LDPE/MX. J Appl Polym Sci 62:1457–1463, 1996.

26. JP Barren, JE Pickett, RJ Oliver. Effect of exposure conditions on the photodegradation of BPA polycarbonate/ABS blends. Proceedings of the Society of Plastics Engineers Color and Appearance Division Regional Technical Conference, St. Louis, MO, 1996.

27. Z Zhenfeng, H Xingzhou, and L Zubo. Wavelength sensitivity of photooxidation of polypropylene. Polym Degrad Stabil 51:93–97, 1996.

28. H Xingzhou. Wavelength sensitivity of photooxidation of polyethylene. Polym Degrad Stabil 55:131–134, 1997.

29. WC Warner, EE Gruber. Light aging of polyblend film under interference filters. Ind Eng Chem Prod Res Dev 5(3):219–221, 1966.

30. VG Kampf, K Sommer, E Zirngiebl. Bestimmung des Aktivierungsspektrums der Photodegradation von Polymeren. Farbe Lack, 95 Jahrgang 12:883–886, 1989.

31. RC Hirt, RG Schmitt, WL Dutton. Solarization studies on polyester resins using a heliostat–spectrometer. Solar Energy III:19–22, April 1959.

32. DS Allan, NL Maecker, DB Priddy. Modeling photodegradation in transparent polymers. Macromolecules 27:7621–7629, 1994.

33. AL Andrady. Wavelength sensitivity in polymer photodegradation. Adv Polym Sci 128:47–94, 1997.

34. ND Searle, NL Maecker, LFE Crewdson. Wavelength sensitivity of acrylonitrile–butadiene–styrene. J Polym Sci A Polym Chem 27:1341–1357, 1989.

35. ND Searle. Effect of light source emission on durability testing. In: WD Ketola, D Grossman, eds. Accelerated and Outdoor Durability Testing of Organic Materials. ASTM STP 1202, Philadelphia. American Society for Testing and Materials, 1993 pp 52–67.

36. VV Korshak, SV Vinogradova, SA Siling, SR Rafikov, ZY Fomina, and VV Rode. Synthesis and properties of self-protecting polyarylates. J Polym Sci A-1 7:157–172 1969.

37. SM Cohen, RH Young, AH Markhart. Transparent ultraviolet barrier coatings. J Polym Sci A-1 9:3263–3299, 1971.

38. DJ Cripps, CA Ramsay. Ultraviolet action spectrum with a prism-grating monochromator. Br J Dermatol 82:584–592, 1970.

39. PM Farr, BL Diffey. The erythemal response of human skin to ultraviolet radiation. Br J Dermatol 113:65–76, 1985.

40. CIE Research Note. A reference action spectrum for ultraviolet induced erythema in human skin. CIE J 6(1):17–22, 1987.

41. MM Cahn. Photosensitivity. J Soc Cosmet Chem 17:81–91, 1966.

42. JG Peak, CS Foote, MJ Peak. Protection by Dabco against inactivation of transforming DNA by near-ultraviolet light: action spectra and implications for involvement of singlet oxygen. Photochem Photobiol 34:45–49, 1981.

43. RM Detsch et al. Wavelength dependence of herpes simplex virus inactivation by UV radiation. Photochem Photobiol 32:173–176, 1980.

44. TP Coohill, LC James, SP Moore. The wavelength dependence of ultraviolet enhanced reactivation in a mammalian cell virus system. Photochem Photobiol 27:725–730, 1978.

45. NI Bishop. Comparison of the action spectra and quantum requirements for photosynthesis and photoreduction of *Scenedesmus*. Photochem Photobiol 6:621–628, 1967.

46. P Halldal. Automatic recording of action spectra of photobiological processes, spectrophotometric analyses, fluorescence measurements and recording of the first derivative of the absorption curve in one simple unit. Photochem Photobiol 10:23–34, 1969.

47. J Jagger, T Fossum, S Mc Caul. Ultraviolet irradiation of suspensions of micro-organisms: possible errors involved in the estimation of average fluence per cell. Photochem Photobiol 21:379–382, 1975.

48. WW Coblentz, R Stair. Data on the spectral erythemic reaction of the untanned human skin to ultraviolet radiation. Bur Stand J Res 12:13, 1934.

49. DF Robertson. Solar ultraviolet radiation in relation to sunburn and skin cancer. Med J Aust 2:1123–1132, 1968.

50. P Trubiroha. The spectral sensitivity of polymers in the spectral range of solar radiation. In: AV Patsis, ed. International Conference on Advances in the Stabilization and Controlled Degradation of Polymers. Vol. 1. Lucerne, Switzerland, May 1986, Lancaster, PA: Technomic Publishing, 1989, pp 236–241.

51. Y Fukuda, Z Osawa. Wavelength effect on the photo-degradation of polycarbonate and poly(methyl methacrylate)—Confirmation of the photo-degradation mechanism of PC/PMMA blends. Polym Degrad Stabil 34:75–84, 1991.

52. A Torikai, T Mitsuoka, K Fueki. Wavelength sensitivity of photoinduced reaction in polycarbonate. J Polym Sci A Polym Chem 31:2785–2788, 1993.

53. AL Andrady, K Fueki, A Torikai. Spectral sensitivity of polycarbonate to light-induced yellowing. J Appl Polym Sci 42:2105–2107, 1991.

54. JJ Laski, MH Chipalkatti. Discoloration of polycarbonate and other transparent polymers under combined thermal and ultraviolet conditions: a phenomenological study. Preprint ANTEC '95, May 7–11, 53rd Annual Technical Conf. SPE, Boston, MA, Technical Papers XLI, 3:3260–3264, 1995.

55. A Torikai, K Chigita, F Okisaki, M Nagata. Photooxidative degradation of polyethylene containing flame retardants by monochromatic light. J Appl Polym Sci 58:685–690, 1995.

56. JL Morand. Oxidation of photoexcited elastomers, rubber chemistry and technology. 45:481–518, 1972.

57. A Torikai, T Hattori, T Eguchi. Wavelength effect on the photoinduced reaction of poly(methyl methacrylate). J Polym Sci A Polym Chem 33:1867–1871, 1995.

58. A Torikai, H Kato, K Fueki, Y Suzuki, F Okisaki, and M Nagata. Photodegradation of polymer materials containing flame-cut agents. J Appl Polym Sci 50:2185–2190, 1993.

59. A Torikai, T Kobatake, F Okisaki, H Shuyama. Photodegradation of polystyrene containing flame-retardants: wavelength sensitivity and efficiency of degradation. Polym Degrad Stabil 50:261–267, 1995.

60. AL Andrady, A Torikai, K Fueki. Photodegradation of rigid PVC formulations. I. Wavelength sensitivity to light-induced yellowing by monochromatic light. J Appl Polym Sci 37:935–946, 1989.

61. AL Andrady, A Torikai, T Kobatake. Spectral sensitivity of chitosan photodegradation. J Appl Polym Sci 62:1465–1471, 1996.
62. AL Andrady, Y Song, VR Parthasarathy, K Fueki, A Torikai. Photoyellowing of mechanical pulps. I. Wavelength sensitivity to light–induced-yellowing by monochromatic radiation. TAPPI J 74(8):162–168, 1991.
63. I Forsskahl, H Tylli. Action spectra in the UV and visible region of light-induced changes of various refiner pulps. In: C Heitner, JC Scaiano, eds. Photochemistry of Lignocellulosic Materials. ACS Symp Ser 531:45–59, 1993.
64. A Torikai. Wavelength sensitivity of photodegradation of polymeric materials. Trends Photochem. Photobiol 3:101–115, 1994.
65. RC Hirt, NZ Searle. Energy characteristics of outdoor and indoor exposure sources and their relation to the weatherability of Plastics. In: MR Kamal, ed. Weatherability of Plastic Materials. Appl Polym Symp 4:61–83, 1967.
66. GA Zerlaut. Solar ultraviolet radiation: aspects of importance to the weathering of materials. In: WD Ketola, D Grossman, eds. Accelerated and Outdoor Durability Testing of Organic Materials. American Society for Testing and Materials STP 1202: Conshohocken, PA 1994 pp 3–26.
67. GA Zerlaut, MW Rupp, TE Anderson. Ultraviolet radiation as a timing technique for outdoor weathering of materials. Paper 850378, Proceedings SAE International Congress. Detroit, Feb. 1985.
68. RW Singleton, RK Kunkel, BS Sprague. Factors influencing the evaluation of actinic degradation of fibers. Textile Res J 35:228–237, 1965.
69. RW Singleton, PAC Cook. Factors influencing the evaluation of the actinic degradation of fibers part II: refinement of techniques for measuring degradation in weathering. Textile Res J 39:43–49, 1969.
70. GA Zerlaut, ML Ellinger. Precision spectral ultraviolet measurements and accelerated weathering. J Oil Color Chem Assoc 64:387–397, 1981.
71. GA Zerlaut. Accelerated weathering and precision spectral ultraviolet measurements. In: Permanence of Organic Coatings. ASTM STP 781. Conshohocken, PA: American Society for Testing and Materials, 1982 pp 10–34.
72. B Zahradnik, A Juriaanse. The influence of UV spectral distribution on the weathering of polyolefins. Preprint, ANTEC '84, 42nd Annual Technical Conference SPE, 1984, pp 397–400.
73. JW Martin, JA Lechner, RN Varner. Quantitative characterization of photodegradation effects of polymeric materials exposed in weathering environments. In: WD Ketola, D Grossman, eds. Accelerated and Outdoor Durability Testing of Organic Materials. ASTM STP 1202. Conshococken, PA: American Society for Testing and Materials, 1994 pp 27–51.
74. JW Martin, SC Saunders, FL Floyd, JP Wineburg. Methodologies for predicting the service lives of coating systems. Blue Bell, PA: Federation of Societies on Coatings Technology, 1996 pp 1–34.

17

Accelerated Weathering Test Design and Data Analysis

RICHARD M. FISCHER and WARREN D. KETOLA
3M Corporation, St. Paul, Minnesota

I. INTRODUCTION

Accelerated aging tests of polymers are conducted to determine how a material or product will perform in its end-use environment. When the end-use environment is outdoor weathering, the accelerated test may need to account for several environmental factors. Most often the environmental stresses that are simulated are light, heat, and moisture. Biological agents, pollutant gases, and airborne particulates can also play a major role in the service life of polymeric materials.

Accelerated weathering tests are most often conducted to address three related issues: The first of these is *comparative testing* for research and development programs. The product developer wants to know which is the best polymer system, colorant, stabilizer package, and such. The testing objective is to pick the best product construction to scale-up and release for sale. Simplistically, the question being asked is, Is material A better than material B?

The second area is *specification testing*, which is a product requirement usually set by the buyer. The question being asked of the accelerated test protocol is, Should I purchase material A? Closely related to this is *quality control testing* that ensures that the product to be sold meets the seller's requirements.

The third testing area tries to establish a *service life* for a potential product or, How long will material A last? This information is especially useful in product engineering (preventing over- or underdesigning of product durability) and in setting warranty periods.

During the 1970s and early 1980s, researchers throughout industry had serious concerns about accelerated test correlation with natural weathering (1–4).

Weathering specification tests often led to buyer–seller disputes, in that the seller would run the specification test and pass the product. The buyer would test the product to the same specification in his laboratory and fail the product. If service life predictions were attempted, the models were often overly simplistic, could not be repeated, and were of very limited applicability. These acceleration factor–service life models were often misused and had the potential of introducing products with a high probability of a field failure.

Two factors played a major role in the pessimism associated with the results obtained from accelerated weathering tests. The first and most important was "poor correlation to outdoor results." The second, not so obvious, factor was variability or lack of reproducibility in both laboratory and outdoor results.

The purpose of this chapter is to discuss a nonparametric approach to the following:

1. Address variability in both accelerated and exterior weathering tests.
2. Quantify and improve accelerated test correlation with natural weathering
3. Develop specification test protocols based on performance comparisons with control materials.
4. Establish service life estimates based on accelerated weathering test results.

II. VARIABILITY IN ACCELERATED WEATHERING TESTS

A. Sources of Variability

Variability in test results from accelerated weathering has been documented in several instances in the technical literature. This variability is not specific to a particular device or equipment manufacturer, nor does a particular device type stand out as being either better or worse than the rest. For example, Blakey reported that enclosed carbon arc devices testing identical specimens (a reference or control sample) demonstrated a wide range of gloss loss rates for TiO_2-pigmented coatings (5). The Working Group on Test Methods for Paints of the Association of Automobile Industries conducted a round-robin evaluation of automotive finishes in both air-cooled and water-cooled xenon arc devices (6). A wide spread in gloss values was reported for replicate specimens tested in different laboratories. In this same study, total color shift data was also highly variable between laboratories for certain paint types.
The components of variability come from all aspects of laboratory accelerated testing. The test material (7) and exposure apparatus design and operation (8) have been identified as major contributors. Other variability components are test duration, property measurement type and instrumentation, and the operator.

B. Absolute Precision

Because of these weathering-testing difficulties, the American Society for Testing and Materials (ASTM) subcommittee G03 03 (Simulated and Controlled Environmental Testing), which is part of main committee G-3 (Weathering and Durability), embarked on a round-robin evaluation of its standard practices. These include ASTM G 23 for carbon arc devices, G 26 for xenon arc devices, and G 53 for fluorescent UV–condensation devices (9–11). These standard practices are the parent documents for the operation of the most common laboratory accelerated testing devices and are referenced by numerous product standards and specifications.

The round-robins (12), completed in 1992, demonstrated the same variability between laboratories (and test devices) reported in earlier studies. Ten different colored polyvinyl chloride (PVC) tapes were exposed in several participating laboratories following specific test protocols. Gloss measurements were made at predetermined exposure intervals. The ten different PVC tapes used in these exposures were chosen specifically because of their demonstrated durability differences during most exposure tests. Table 1 shows the differences between each of the ten tapes and the differences in gloss between different devices running the same test cycle after a 6-week exposure period. The results in Table 1 are for fluorescent UV–condensation exposures, but the differences are typical of those found for the carbon arc and xenon arc exposures.

The data reported in Table 2 is the typical ASTM index of precision known as the difference "2"-standard deviation limit for reproducibility ($d2s_R$). For the exposures evaluated in the round-robin, the $d2s_R$ value means that a pair of samples tested in different devices running the same test cycle must differ in gloss value by more than $d2s_R$ to be considered different in gloss durability performance (with 95% confidence). For fluorescent UV exposures conducted according to ASTM G 53, the average $d2s_R$ for the ten tapes was 29.3. This means that, on the average, the PVC tapes tested in laboratory A must differ by more than 29.3 gloss units from

Table 1 The Mean 60° Gloss Values (Corrected to a Single Gloss Meter) for the Ten PVC Tapes from Each Laboratory After 6 Weeks Fluorescent UV Exposure Conducted According to G 53 Round-Robin

Sample	Lab I	Lab III	Lab IV	Lab V	Lab VI	Lab VII	Lab VIII
1	16	34	22	20	14	18	10
2	15	38	20	19	12	17	11
3	68	77	73	70	65	71	65
4	69	81	68	76	67	73	62
5	45	66	53	52	40	49	41
6	21	46	26	32	21	26	25
7	28	45	32	31	27	30	32
8	53	70	59	58	48	51	49
9	33	55	43	34	31	33	29
10	51	69	57	57	44	50	45

Table 2 Maximum Estimates of Reproducibility for the Standard Practices (for all Ten PVC Tapes)

Exposure device	ASTM standard practice	Mean $d2s_R$	Range $d2s_R$
Carbon arc	G 23	21.6	13.8–31.6
Xenon arc	G 26	44.0	36.4–55.5
Fluorescent UV	G 53	29.3	18.0–41.9

the same tapes tested in laboratory B to be considered different in performance. The large $d2s_R$ ranges shown in Table 2 indicate that the between-laboratory variability for some of the tapes was extremely high. The PVC tapes showing the worst reproducibility were those that lost the most gloss during exposure. These particular specimens were best able to demonstrate the differences in exposure severity between the laboratories. The data in Table 2 indicated that exposures according to all three of these basic standard practices had high variability between laboratories.

Round-robin evaluations (13) by the Fade and Weathering Committee—Transportation Division of the Industrial Fabrics Association International (IFAI) provided almost identical data analysis conclusions as the ASTM round-robins. Twenty-four different textiles were tested in nine laboratories following SAE J1885 (14). Total color shift (Cielab ΔE) was the measure of durability. The ΔE difference 2-sigma limit ranged from 2.89 to 7.58 units. As in the ASTM studies the level of variability depended on material type. In general, materials that experience more change during the exposure test also experience more variability. When a specimen is tested in two different laboratories to the same specification, large differences in gloss or color-shift values must often be obtained before real performance differences can be established.

It is a difficult task to precisely control stress levels in accelerated test devices in different laboratories. Small variations in temperature, light intensity, and wet periods can affect the degradation rate of polymeric materials, which over lengthy exposure periods can lead to poor absolute precision. This same situation, to an even greater degree, affects outdoor-weathering results. The absolute failure rates can vary significantly from year to year even at a specific location because of differing yearly temperatures, sunlight hours, and wet periods.

Results from the round-robin studies in ASTM subcommittee G03.03 show that absolute performance comparisons for a single material are not practical between laboratories. The same studies demonstrate that material performance rankings for a series of materials are reproducible. This requires the use of nonparametric statistics to compare test results between laboratories.

C. Rank Precision

Although the actual gloss values for a particular specimen (i.e., sample 2) from each laboratory is highly variable after the 6 weeks of exposure, inspection of the gloss data in Table 1 indicates that the performance ranking of the ten specimens is quite similar between laboratories. This can be more clearly seen by converting the gloss values to ranks, in which the lowest gloss is assigned a value of 1 and the highest gloss is assigned a value of 10. The results of this ranking are shown in Table 3. A quantitative measure of the degree of association between the gloss performance rankings for each laboratory is obtained by calculating Spearman's rank correlation coefficient, r_S [eq. (1)].

A description of methods of rank correlation can be found in Conover's book on nonparametric statistics (15). The r_S for each laboratory (bottom row of Table 3) is calculated by comparing the rank order from each laboratory with the average rank order for all the other laboratories. Spearman's rank correlation coefficient is analogous to Pearson's linear correlation coefficient, in that r_S ranges from 1.0 (perfect positive correlation; i.e., ranks are identical to zero (no correlation

Table 3 Rank Values for 60° Gloss Data in Table 1

Sample	Lab I	Lab III	Lab IV	Lab V	Lab VI	Lab VII	Lab VIII
1	2	1	2	2	2	2	1
2	1	2	1	1	1	1	2
3	9	9	10	9	9	9	10
4	10	10	9	10	10	10	9
5	6	6	6	6	6	6	6
6	3	4	3	4	3	3	3
7	4	3	4	3	4	4	5
8	8	8	8	8	8	8	8
9	5	5	5	5	5	5	4
10	7	7	7	7	7	7	7
r_S	1.0	0.98	0.99	0.99	1.0	1.0	0.96

or random ranks) to -1.0 (perfect negative correlation or ranks are reversed).

$$r_s = \frac{\sum_{i=1}^{n} R(X_i)R(Y_i) - n[(n+1)/2]^2}{\left[\sum_{i=1}^{n} R(X_i)^2 - n[(n+1)/2]^2\right]^{1/2} \left[\sum_{i=1}^{n} R(Y_i)^2 - n[(n+1)/2]^2\right]^{1/2}} \qquad (1)$$

where

$n =$ number of pairs of values
$R(X_i)$, $R(Y_i) =$ rank values for sample series X and Y

No absolute requirement for r_S can be established, although the weathering community has rather arbitrarily set 0.9 or higher as a desirable target value. Values of r_S well above 0.9 were observed for all the accelerated exposure tests when comparing ranks between laboratories (Table 4). Although the rate of degradation (test severity) varies greatly between laboratories (see Table 1), performance rankings for a series of materials remains quite reproducible. The test severity in a particular laboratory (within limits) determines the exposure period required to obtain significant performance differences between materials.

Again, the same characteristics of excellent rank correlation between laboratories held for the IFAI textile round-robins. The rank correlation (r_S) between laboratories in the IFAI studies was also quite high ranging from 0.92 to 0.99.

III. CORRELATION BETWEEN ACCELERATED AND OUTDOOR EXPOSURE RESULTS

A. Correlation Approaches

"Correlation" has been one of the most misused or abused terms in the weathering testing of polymers. Its use has ranged from observing the same failure modes in both the accelerated and outdoor exposure to an acceleration factor that suppos-

Table 4   Comparison of Rank Reproducibility for Accelerated Exposures (for all Ten PVC Tapes)

Exposure device	ASTM standard practice	Mean r_S	Range r_S
Carbon arc	G 23	0.95	0.92–0.97
Xenon arc	G 26	0.95	0.88–1.0
Fluorescent UV	G 53	0.99	0.96–1.0

edly relates the time in the accelerated test to time outdoors. As was demonstrated in the ASTM round-robins, correlation coefficients can also be used to establish the strength of the relations between accelerated and outdoor weathering tests.

B. Pearson or Spearman Correlation

The most widely used measure of the relationship between variables is the Pearson product moment correlation coefficient (r) which is described in any introductory statistics text (16). Pearson's r can also range from -1 to $+1$. The value of r, either positive or negative, indicates the strength of the relationship between accelerated and outdoor exposure results. Again an r value close to $+1$ is obviously desirable for weathering tests. Pearson's r has been used to describe accelerated "test goodness" when comparing color shift or gloss retention for materials weathered both in accelerated tests and outdoors (17–19). The strength of using Pearson's r is that if the relation between the results is linear, the square of $r(r^2)$ indicates the proportion of the variance in the accelerated result that is associated with the variance of the outdoor result. For example, if r is 0.9, r^2 ($\times$ 100) is 81%. This would mean that 81% of the result being measured in the accelerated test is also being measured by the outdoor test.

One of the biggest misuses of Pearson's r is with clustered data. An example of this is shown in Table 5. In the second and third columns of this table the total color shift (ΔE) is listed for a series of 29 automotive graphic films (attached to automotive painted steel panels with a pressure-sensitive adhesive). These specimens were weathered outdoors at Arizona 45° for 2 years and in a water-cooled xenon arc according to SAE 1960 (20). The graphic films were made up from a range of polymer types used in a variety of multilayer constructions. If just the Pearson's r value of 0.94 is reported, the "correlation" between the accelerated and outdoor exposure might be assumed to be quite good. As in any data analysis, a simple xy plot of the color shift values for the two exposures can add significant insight. Inspection of the scatter plot (Fig. 1) shows several outliers (graphics that had large color change values) in conjunction with the cluster of relatively durable films are providing a misleading linear fit with a high r value.

Nonparametric or Spearman rank correlation approaches have also been used to determine the relation between accelerated and outdoor weathering tests (18,21–24). As was demonstrated in the ASTM round-robin studies, the gloss, color shift, or other measured property value is converted to an ordinal rank value, (e.g., the color shift data used in Fig. 1 is assigned a rank in columns 4 and 5 in Table

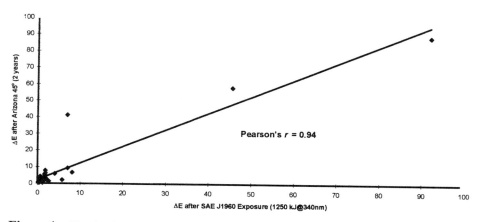

Figure 1 Total color shift scatter plot for 29 automotive graphic films after 2 years Arizona 45° and 1250 kJ of SAE J1960 exposure in a water-cooled xenon arc (Model Ci 65; Atlas Electric Devices, Chicago, IL).

5: 1, most color shift; 29, least). These rank values are plotted in Fig. 2 in a fashion similar to that done with the ΔE values in Fig. 1.

Spearman's rank correlation coefficient (r_S) is a measure of the tendency for "reversals" between the two tests. It is now possible to see that the Arizona rank performance order for color shift is not well predicted by the SAE J1960 test at this particular level of exposure. In making the judgement, "Is material A better than material B?" the chance of an error is quite high for this data set.

One drawback that deserves mention is that considerable information is lost when using Spearman's r_S instead of Pearson's r. The r^2 value no longer has particular meaning. There is also no measure of linearity between the response variables for the two exposures. Spearman's r_S can also give misleadingly high correlation results when several performance ties (ranks are the same) exist for similar specimens in both the accelerated and outdoor exposure. In addition, no specific value can be associated with either Pearson's or Spearman's correlation coefficients. A value of 0.8 is not twice as good as 0.4. The difference between 0.2 and 0.3 is not the same as the difference between 0.8 and 0.9. It is obvious that an accelerated test with a 0.9 correlation coefficient with outdoor exposure is superior to an accelerated test with 0.5, but little else can be inferred by the higher value.

C. Test Error Analysis

Estimating the chance of an accelerated test error or reversal is the most important part of the risk analysis involved in the use of accelerated tests for the development of exterior durable materials. A measure of the chance of a test error should have more intrinsic value or practical meaning than Spearman's r_S or Pearson's r. In the psychological or health sciences, r_S is typically used to look at very low levels of association between ranks to establish possible cause-and-effect relationships. Use of high r_S (0.9+) in weathering test development provides little information on the chance of test error.

Table 5　Color Shift and Rank Data for 29 Graphic Films Exposed at Arizona 45° for 2 Years and in SAE J1960 for 1250 kJ at 340 nm (Rank 1 = greatest ΔE)

Graphic specimen no.	ΔE AZ 45°	ΔE SAE J1960	Rank ΔE AZ 45°	Rank ΔE SAE J1960
1	0.35	0.35	29	27
2	1.61	0.84	19	21
3	2.50	1.97	14	11
4	4.14	0.61	10	22
5	3.89	1.54	11	15
6	5.42	1.59	9	14
7	7.84	1.78	5	13
8	1.41	2.58	22	8
9	41.28	6.92	3	5
10	0.71	0.16	26	29
11	0.93	0.43	25	26
12	9.20	7.06	4	4
13	0.46	1.19	28	17
14	0.99	0.50	24	23
15	0.58	1.11	27	19
16	6.75	8.14	6	3
17	3.09	0.29	12	28
18	58.11	45.62	2	2
19	6.19	1.87	7	12
20	88.75	92.20	1	1
21	1.49	1.15	21	18
22	1.37	0.49	23	24
23	2.16	2.20	16	9
24	2.39	2.19	15	10
25	1.50	1.48	20	16
26	5.87	4.05	8	7
27	2.14	5.81	17	6
28	1.85	1.02	18	20
29	2.69	0.47	13	25
	r of AZ45 to SAE 1960→	0.94	r_S of AZ45 to SAE J1960→	0.67

An approach that can be used to obtain an estimate of test error is to reduce a weathering study to its most simple element: Is material A more durable than material B? To be more specific, if any sample pair is picked at random from an accelerated weathering study, what are the chances that the result is reversed from actual service or exterior exposure. Evaluating weathering studies by determining the number of reversals out of a total number of pair comparison possibilities permits an estimate of the chance of a test error or, more simplistically, making a mistake in the durability comparison of samples A and B.

Random sample pair analysis can be extended to a series of materials exposed in both an accelerated test and an exterior environment, such as the data in Table 5. The approach is to determine the numbers of pair errors or reversals in the accelerated test using the exterior exposure as the correct result and ratio it to the total

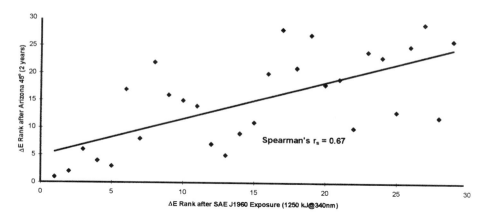

Figure 2 Ranks of total color shift data pairs (from Fig. 1) for 29 automotive graphic films after 2 years Arizona 45° and 1250 kJ of SAE J1960.

number of pair comparison possibilities. This can be accomplished by using a modified form of the equation used to calculate another nonparametric correlation coefficient, Kendall's tau (14). This modification is shown as Eq. (2).

$$PE_\% = \frac{100(N_D + 0.5N_T)}{n(n-1)/2} \qquad (2)$$

$PE_\%$ = number percent of pair errors in the accelerated test
N_D = number of pair reversals (discordant pairs)
N_T = number of pair ties (identical performance)
n = number of samples in the test series

The term, $n(n-1)/2$, is equal to the total number of possible pair combinations for "n" samples or, in this case, test materials. N_D is the number of sample pair reversals observed in the accelerated test compared with the exterior or "natural" exposure. N_T is the number of sample pairs with identical rank performance in the accelerated test. As with Spearman's r_S, a relatively simple computer program or spreadsheet macro can be written to calculate $PE_\%$ for any n samples.

A $PE_\%$ of 50% is analogous to r_S equal to zero when all the samples are identical (or indistinguishable) at the beginning of the exposure test. This means there is a 50 : 50 or random chance of choosing the more durable specimen from any random sample pair. If the specimens were ranked exactly in reverse order ($r_S = -1$), $PE_\%$ would be 100% because for every sample pair, the results from the accelerated test would be the opposite of that obtained outdoors. It then follows that a $PE_\%$ of 0% ($r_S = 1$) represents the perfect test (no reversals).

The $PE_\%$ for the exposure data in Table 5 can be calculated on either the raw color shift data or the ranked data and will provide the same result. In this case $PE_\%$ for SAE J1960 predicting Arizona 45° color-shift results is 25%. This means that for any "random" sample pair selected from the SAE J1960 results, the chances of the color-shift performance comparison being reversed from the actual outdoor results is 25%. In this instance the chance of test error is unacceptably high (one out of four random pair comparisons will be reversed).

In most instances, the chance of a reversal is directly related to the absolute separation in durability performance for the specimen pairs. Table 6 summarizes the results from further analysis of the "pair error" possibilities for the data in Table 5. This analysis demonstrates that the smaller the absolute difference in ΔE for the sample pair (from SAE J1960) being compared, the greater the chance of a reversal.

If the difference in ΔE between the selected random pair is between 0 and 1, the chance of a reversal from Arizona 45° results is 45.5%. For example, specimen 2 has a ΔE of 1.61, and specimen 8 has a ΔE of 1.41. The absolute ΔE difference is 0.2, which means there is a 45.5% chance actual ΔE ranking of this SAE J1960 result is reversed from the Arizona 45° result. The percentage pair error distribution of Table 6 is quite typical for accelerated tests. As the true performance difference between specimen pairs approaches zero, the chance of selecting the more durable material should be 50 : 50. As the difference is ΔE becomes greater, the chance of the true durability ranking of the selected pair being reversed becomes smaller until the difference is greater than 7 and no more reversals occur. For example, specimens 4 and 16 with an absolute DE difference of 7.53 indicates there is no chance of an error in this test. This simply means that the greater the difference in performance that is observed between any pair of specimens in the accelerated test, the greater the chance is that the result is "correct."

IV. DEVELOPING ACCELERATED TESTS WITH IMPROVED CORRELATION TO OUTDOOR EXPOSURES

A. Experimental

In the following examples, the accelerated weathering device used was an open-flame carbon arc (Model XW-WR; Atlas Electric Devices, Chicago, IL). The test cycle was 102 min of light (63°C black panel temperature) followed by 18 min of light and water spray (102–18 cycle). Test specimens were film tape graphics of two basic constructions. Type I is a 50 μm–thermoplastic film, usually pigmented, with a 20 to 25 μm pressure-sensitive adhesive underlayer. Type II is identical with type I except the thermoplastic film is coated with an ink and a protective topcoat

Table 6 Pair Error Distribution as a Function of the Absolute ΔE Difference Between the Specimens Being Compared from the SAE J1960 Exposure Results (Compare with AZ 45°)

Absolute ΔE between sample pairs	Total no. of pair possibilities	No. of errors	% pair errors
0–1	145	66	45.5
1–2	86	21	24.4
2–3	20	4	20
3–4	16	4	25
4–5	17	3	17.6
5–6	29	3	10.3
6–7	27	1	3.7
>7	66	0	0

making a four-layer, final construction. All specimens were applied to painted steel panels cut to sizes appropriate for specimen holders of each device. Gloss measurements were made with a Glossgard II–60° gloss meter (Pacific Scientific, Silver Springs, MD). Color measurements were made with a Spectrosensor II–colorimeter (Applied Color Systems, Inc., Princeton, NJ).

B. Correlation Profiles

When the exterior exposure rank for a series of materials has been established, a plot of r_s versus time in the accelerated test can be generated (Fig. 3). In this example, seven type I graphic films were exposed in the open-flame carbon arc and the results compared with those from a 12-month Florida 5°, backed exposure. Ranks were determined based on the amount of 60° gloss loss. The r_S and corresponding $PE_\%$ values (comparing the results from the accelerated test and those from the Florida exposure) were calculated after several exposure times in the carbon arc. At the beginning of the test, when gloss loss is zero for all the samples, r_S is zero ($PE_\%$ is 50%) because performance levels cannot be distinguished. As the test proceeds, performance differences become significant between the test specimens, and r_s climbs to a maximum value at 1000 h ($r_s$max or $PE_\%$min). Durability performance judgements should be made at this *point of maximum correlation*. The $r_s$max region (0.81) is identical with the $PE_\%$min region (16.6%) in Fig. 3. The 16.6% pair error value indicates a 1 : 6 chance of any randomly selected sample pair being reversed. As the test continues, r_S begins to fall ($PE_\%$ rises) and would eventually reach zero when all of the specimens have achieved a maximum level of gloss loss, so the relative performance differences are very small, and rank assignments are less reliable.

An example of rank correlation plots based on color shift for type II graphic films in the open-flame carbon arc test is shown in Fig. 4. The rank correlation

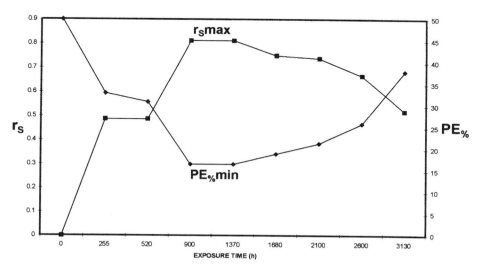

Figure 3 Correlation profile of open-flame carbon arc exposure (102–18 cycle) to 12-month Florida 5° backed exposure (seven type 1 graphic films—60° gloss retention).

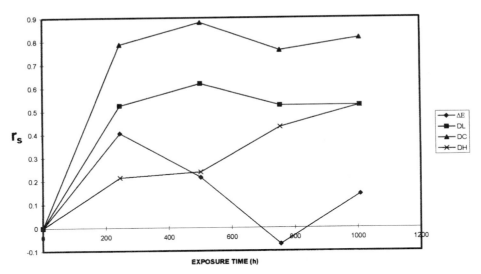

Figure 4 Spearman rank correlation profile of open-flame carbon arc (102–18 cycle) to 24-month Arizona 45° backed exposure (color shift of eight type II graphic films).

coefficient is calculated relative to 24 months of Arizona exposure. Not only can r_s be determined for total color shift (ΔE) but also for the color components; DL, DC, and DH. Higher r_S values are obtained for the color components than for total color shift. This is because there is greater spread in the data for the components (therefore, easier ranking) than for ΔE. ΔE values ranged from 2 to 4 while DL, DC, and DH ranged from -3 to 3.

C. Maximum Correlation Intervals

The point at which "maximum correlation" occurs depends on the weathering performance properties of the test materials and the severity of the specific accelerated exposure test. Therefore, for each new test, the precise exposure time that maximum correlation is achieved is unknown. For most tests, this time can be estimated by following the standard deviation of the measured performance property for the series of test specimens being evaluated. This is demonstrated in Fig. 5 where the standard deviation of gloss loss and Spearman rank correlation coefficient for the seven graphic films is plotted versus exposure time in the open-flame carbon arc. Standard deviation reaches a maximum value at the same exposure time that $r_s$max and PE$_\%$min occur (1000 h). Maximum rank correlation is achieved when the specimen performance differences are the greatest, which is reflected by the standard deviation of the measured performance property. It is at this point that ranks are more accurately assigned.

When developing an improved accelerated test, exposure conditions can be modified until r_S reaches an acceptable level for a particular series of specimens. High rank correlation for a specific material series does not ensure high correlation for other material types. The tracking of maximum r_s for materials and exposure

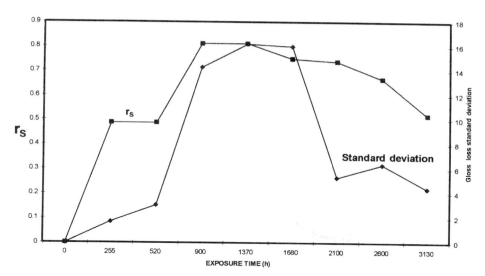

Figure 5 Standard deviation of 60° gloss loss and Spearman rank correlation during open-flame carbon arc exposure (102–18 cycle) for the seven type I graphic films referenced in Fig. 3.

tests over a period of time develops an understanding of the strengths and weaknesses of each test for particular failure modes and materials types.

D. Test Error Analysis of Traditional and Improved Correlation Weathering Tests

Significant improvements in accelerated test accuracy can be achieved by using nonparametric statistics as a tool for accelerated test development. Traditional industry tests used for many years had $PE_\%$ values of 25–50% for gloss loss. New accelerated tests developed since 1980 can achieve $PE_\%$ values of 10% or less. An example of this is shown in Fig. 6, where the $PE_\%$ value for 20 type II graphic films in open-flame carbon arc exposure (102–18 cycle) reaches only 40%. This is only slightly better than random guessing as to the real gloss retention durability of these films. The "improved correlation test" (25) reaches a $PE_\%$ of 8%, which means the chance of a reversal is one out of 12 random sample pair selections or a fivefold reduction in reversals.

Table 7 shows the percentage pair error distribution for the accelerated cycles in Fig. 6 after approximately 3600 h of exposure, compared with Florida 5° backed for 36 months. The improved correlation cycle has a typical distribution in that most of the errors occur when absolute gloss differences between specimen pairs are small (>10). The errors that occurred at gloss differences greater than 10 units were caused by two specimens. When large errors such as this occur, it can usually be attributed to an accelerated test weakness. Either the accelerated test has an unnatural stress for these materials, or it is missing a stress that is present outdoors. A careful analysis of materials such as these defines test weaknesses and can assist in future test devel-

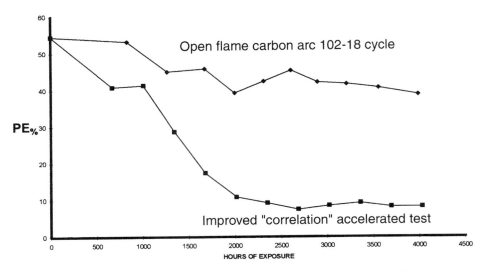

Figure 6 Percentage pair errors for 20 type II graphic films in open-flame carbon arc exposure and the improved correlation cycle compared with Florida 5° backed exposure (36 months).

Table 7 Pair Error Distribution as a Function of the Absolute 60° Gloss Difference for the 20 Type II Graphic Films in Open-Flame Carbon Arc Exposure (3565 h) and the Improved Correlation Cycle (3696 h) Compared with Florida 5° Backed Exposure (36 Months)

Absolute 60° gloss difference between sample pairs	Open-flame carbon arc exposure (3565 h) % pair errors	Improved correlation cycle (3696 h) % pair errors
0–10	34	24.4
11–20	62.9	4.2
21–30	70.6	9.5
31–40	23.5	5.3
41–50	5.3	0
51–60	34.8	0
60–70	42.9	0
>70	0	0

opment or modification. The open-flame carbon arc cycle has high-error levels at even large gloss level differences indicating severe problems in simulating natural stresses for a variety of the test materials.

E. General Correlation Observations

The pair error distribution for the open-flame carbon arc exposures is indicative of a test that has made significant reversals owing to unnatural stresses. In this particular instance, poor correlation was caused by the unnaturally short wavelength (nonsolar) light emitted by the Corex 7058 filtered open-flame carbon arc source. Gerlock (26) and Bauer et al. (27) have shown that short wavelength light in open-flame carbon arcs and xenon arc devices causes unnatural photochemistry with certain two-part polyurethane coatings. Use of "short-wavelength" light sources has been a popular approach in achieving rapid accelerated tests, but often the results from these tests do not relate well to outdoor exposures.

By definition, an accelerated test requires that the environmental stresses applied to the test specimens be more severe than those of nature. This distortion does alter reaction kinetics, can lead to changes in failure mechanisms, and ultimately cause weathering reversals. Because of these problems, it may be appropriate to limit an accelerated weathering study to individual polymer classes (intrafamily, instead of interfamily testing) and component changes within those classes (i.e., stabilizer or color differences).

When developing accelerated tests, there is always a compromise drawn between test speed and test accuracy. If one simulates outdoor conditions exactly, the acceleration factor is 1, but the test correlates precisely with outdoors results. In a practical sense, acceleration factors of roughly 3–6 may be the best that can be achieved with a test that does a reasonable job ranking the durabilities of a broad range of materials (e.g., between polymer and product construction types). As acceleration factors approach 10 or more, the test becomes much more specific and will be of value for a more-limited set of materials. This does not mean that "fast" tests are not useful. If good correlation can be established for a specific polymer type, test speed is certainly advantageous. But, do not assume that the good correlation established for one set of materials in a weathering test will extend to other polymer systems.

V. VARIABILITY EFFECT ON SPECIFICATION AND QUALITY CONTROL TESTS

One common method of setting a weathering specification is to require a specific maximum allowable level of change after a specific exposure duration. An excerpted quote from an actual specification (circa 1980) for high-gloss graphics states "After 24 months exposure at Florida 5°, the panels should not exhibit more than 15% loss of original gloss." Figure 7 shows the 60° gloss retention for a high-gloss, type II graphic exposure for two different 24-month periods at Florida 5°. The material weathered from 1987 to 1989 would pass the specification, whereas the identical material weathered between 1985 and 1987 would be rejected. This is caused by the typical yearly climatic variations that occur in outdoor exposures.

Another quote from a similar specification for high-gloss graphics states "After 1600 hours exposure in an (open flame carbon arc), the panels should not exhibit more than 15% loss of original gloss." Figure 8 shows the 60° gloss retention for the same high-gloss graphic exposed in 12 different open-flame carbon arcs, all running the same 102–18 exposure cycle. Obviously, one could choose a test device that would pass this material, or one that would reject it. This inherent variability in both natural and accelerated weathering results requires a new approach in designing specification tests.

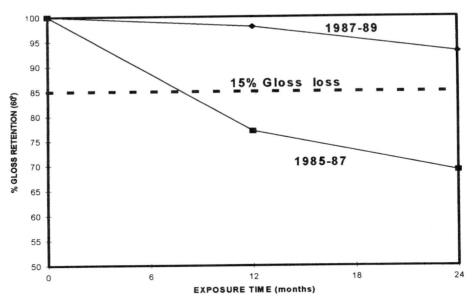

Figure 7   Percentage 60° gloss retention for a single lot of a type II high-gloss graphic film during different 24-month–exposure periods in Florida at 5°.

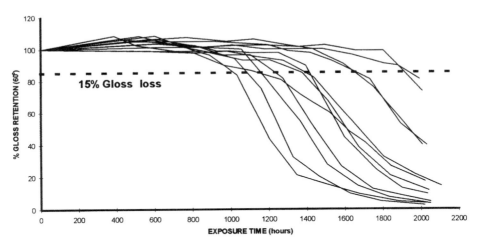

Figure 8   Gloss loss profiles for 12 identical type II graphic films exposed in 12 different open-flame carbon arc devices (102–18 cycle).

VI. SPECIFICATION TESTING BY COMPARISON WITH A CONTROL MATERIAL

A. Experimental Requirements

Although the absolute reproducibility between laboratories was poor for both gloss and color shift in the ASTM and IFIA round-robins (12,13), the rank performance for a series of specimens was quite reproducible. This meant that the durability ranking for a series of materials that was obtained in one laboratory would be very similar to the ranking obtained in another laboratory. This reproducibility in ranking can be used to address the variability in specification tests by comparing the weathering performance of the test material with that of a control material.

After selecting an appropriate accelerated test, the first step for establishing a weathering specification for a material is the selection of an appropriate control material. The primary requirement for the control material is that its weathering durability is well established in the specified exposure test. The control material and the candidate specimens should be similar polymer types and construction so failure modes are comparable. To simplify durability comparisons made between candidate specimens and the control material, the performance level of the control should be at a *minimum* acceptable level in actual service. Statistically significant differences in performance between the control material and the candidate specimens can be made using *analysis of variance*. Replicates for each specimen (control and test specimens) must be exposed during the same time period outdoors, or simultaneously in the same accelerated device. This eliminates the effect of poor reproducibility between outdoor exposure periods or among accelerated devices.

An example of using ranks for setting a specification is illustrated in Fig. 9. The performance data is 60° gloss loss after 2000 h of exposure in an open-flame carbon arc (102–18 cycle). The test specimens are 51 different type I and II graphic films. The control material lost 35 gloss units during the exposure period, which was deemed the minimum acceptable performance level for this particular accelerated test.

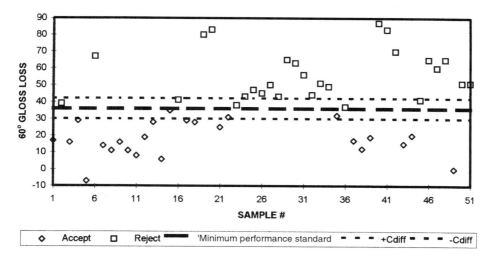

Figure 9 Open-flame carbon arc specification testing (2000 h, 102–18 cycle) of 51 type I and II graphic films: gloss loss compared with a control material (minimum performance standard).

B. Critical Differences

Dunnett (28) developed an equation for establishing a critical performance difference (C_{diff}) between a control specimen and the test specimen [Eq. (3)].

$$C_{\text{diff}} = d_k \sqrt{\frac{2(ms_w)}{n}} \tag{3}$$

where

d_k = table critical values of the Dunnett test for two-tailed comparisons of a
treatment mean with a control (for k means test specimens plus control)
and within degrees of freedom.

n = number of replicates in each sample or specimen type.

ms_w = mean square value (within) from the analysis of variance table

The dashed lines on either side of the control material's gloss level (see Fig. 9) is
the critical difference (C_{diff}). If the gloss loss of a test specimen falls outside these
limits, it is statistically different from the control material (with 95% confidence).
Statistical software packages can make Dunnett's calculation of a critical difference
much less tedious. A similar critical difference can be calculated using *analysis
of variance* (ANOVA) in Minitab 7.1 (29). The Minitab critical range is determined
by the following general equation.

$$CR_M = 2t_{df_w} \left(\frac{\text{SD}_{pooled}}{\sqrt{n_i}} \right) \tag{4}$$

t_{dfw} = critical value of t (statistical tables) for a two-tailed test at *within* degrees
of freedom.

$\text{SD}_{\text{pooled}}$ = pooled standard deviation from individual groups or, in this case,
test specimen types.

n_i = number of replicates in each group or specimen type.

The replicate 60° gloss values for the control material and the test materials are
entered in individual, adjacent columns, and the ANOVA operation is performed.
The Minitab output does not list a critical level but, instead, graphically displays
specimen performance differences.

C. Accept or Reject Plots

Figure 9 demonstrates that by simple inspection the acceptance or rejection of a
candidate material is now a trivial matter. If the gloss loss for the test specimen
is less than or equal to the control material (minimum performance standard), it
is accepted. If the test specimen has more gloss loss than the upper critical difference
limit of the minimum performance standard, it is rejected. The upper critical difference limit is the accept/reject line when using this pass/fail criteria.

D. Specification Test Goodness

After accounting for variability, a specification test is still only as good as the
predictive capability of the accelerated test. For the specification test example shown
in Fig. 9, the number of test reversals can be determined by direct comparison with

actual exterior exposure results for identical specimens. The same 51 graphic films were evaluated after 24 months of Florida 5° backed exposure (Fig. 10). The same control material lost 25 gloss units during that time period. When the rankings of the individual test specimens and the control material (more or less gloss than the control material) from the open-flame carbon arc and the Florida exposures were compared; there were 19 instances when the carbon arc exposure produced a result opposite that of the Florida exposure (Fig. 11). These 19 pair reversals out of the 51 total pair comparisons mean that there is a 37% chance of a test error.

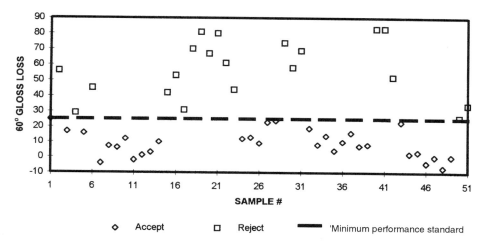

Figure 10 Florida specification testing (5° backed for 24 months) of 51 type I and II graphic films: gloss loss compared with a control material (minimum performance standard).

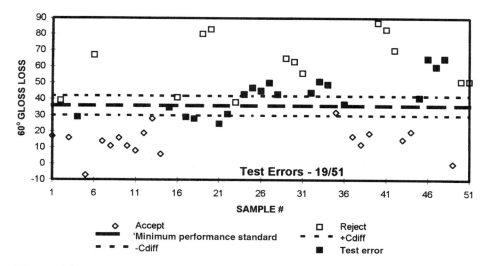

Figure 11 Open-flame carbon arc specification testing from Fig. 9 with test errors (compared with Florida exposure results, see Fig. 10) designated.

With a slightly greater than a one out of three chance of a reversal, this particular accelerated test is not of much use to the materials person who is depending on it to qualify acceptable materials for exterior usage. (i.e., should I purchase Material A?).

Figure 12 shows the test results for the same 51 graphic films after exposure to SAE J2020 (29) for 1000 h. In this particular accelerated test there were 24 reversals out of 51 comparisons or 47.1% test errors. This test is no better than a random guess in predicting the durability of these graphic films relative to the minimum performance standard.

Analogous to the open-flame carbon arc, the most likely cause for this poor test accuracy is the short-wavelength light emitted by FS-40 fluorescent lamp that is not present in sunlight.

A common belief is specification testing is that if the test is made severe enough it may indeed fail some good materials, but poor or less durable materials will be screened out. The results shown in Fig. 12 dispute this claim. Several comparatively durable graphic films fail, but several less durable materials are passed. Severe tests do not necessarily reduce the odds of a failure occurring under actual service conditions.

Figure 13 shows the test data for the same 51 graphic films exposed in the "improved correlation" accelerated test, cited previously (25). This time only six reversals occurred compared with the Florida results.

VII. SERVICE LIFE ESTIMATES FROM ACCELERATED WEATHERING

A. Linear Models

Since the inception of accelerated weathering testing the most common approach to estimating service life has been to determine an "acceleration factor" that relates

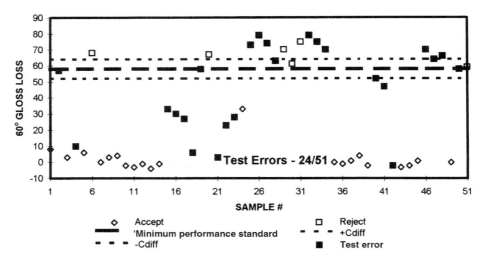

Figure 12 SAE J2020 specification testing (1000 h) of 51 type I and II graphic films: gloss loss compared with a control material (minimum performance standard).

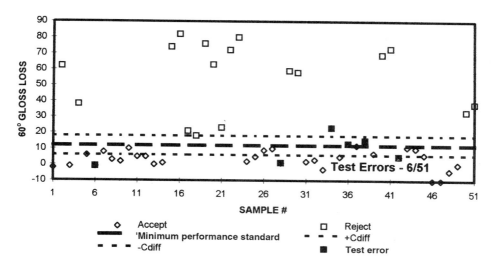

Figure 13 Proprietary accelerated specification testing (2000 h) of 51 type I and II graphic films: gloss loss compared with a control material (minimum performance standard).

the accelerated test to outdoor weathering. This typically is done by testing a material both outdoors and in an accelerated test. Assuming similar failure modes, the time to reach a predetermined property change is determined in both tests. These times are assumed to have a zero-intercept linear relation and are used to calculate an acceleration factor k_a [Eq. (5)].

$$k_a = \frac{t_{actual}}{t_{accel}}$$ (5)

Most of the time these acceleration factors are determined from experiments using only one material and a single data point from both exposures. This means that k_a does not account for sample, property measurement, exposure tests, climatic, and seasonal variations. Once k_a has been established for a particular material, it is often erroneously used for other materials that may have significantly different acceleration factors. Fedor reports acceleration factors ranging from 2 to 35 times faster than outdoor exposure (depending primarily on polymer type) for a study recently conducted with fluorescent UV condensation devices (31). The misuse of these types of acceleration factors has led to numerous field failures and has been the primary contributor to the general distrust of results from accelerated weathering that has developed over the years.

Wernstahl and Carlsson (32) calculate an [equivalent light dose factor] (ELDF) that is similar to k_a, except it is based on the relative light doses (outdoors or accelerated) required to achieve a specific level of deterioration, instead of time. Using light dose is often touted as being superior to time in equating exposure environments, but they point out that exposure temperature has a significant influence on their calculated ELDFs.

B. Nonlinear Approaches

Simms (33) has described the problems associated with linear acceleration factors and has developed a methodology for establishing a technically valid "accelerated shift factor" (ASF). Material dependent ASFs were determined through many replicate specimens and exposures (both accelerated and real-time). Even the calculation of valid accelerated shift factors is of little use with new or modified materials exposed in a single accelerated test, which is often the case in polymer product development.

It is readily apparent that simple linear models will not be adequate in predicting actual service life from accelerated test results. Several approaches have been proposed including time series analysis (34) and time–temperature–dose rate superposition (35). Brown (36) has discussed the use of various empirical models to predict service performance from accelerated test results. Hamid (37–39) has developed several empirical models to predict the property retention characteristics of polyolefins exposed in accelerated and severe exterior environments.

Jorgensen (40) identified key stress factors leading to the degradation of polymer solar reflectors and developed a model to predict performance loss in a variety of climates. Bauer (41) has described a comprehensive approach in producing time-to-failure distributions for the delamination of automotive top coats. The methodology first identifies basic failure mechanisms that are used to develop time-to-failure models. The models are merged with the variability associated with key parameters in manufacturing processes and in the exposure environment to develop a life distribution estimate.

C. Reliability-Based Methodologies

Both Jorgensen's and Bauer's proposed methodologies are closely related to reliability-based protocols described in detail by Martin et al. (42). As in Bauer's approach, every polymer or system must be treated separately. Acceleration factors are not assumed, but are determined for every new material through carefully designed accelerated tests. The tests are constructed so the material's failure response is established relative to the important environmental stresses (light, heat water, or other). Numerous replicate specimens are exposed in the accelerated tests so that life distribution information can be extracted and translated to the service life predictions (43). In addition, the outdoor environmental stresses must be well understood and quantified. Again, the most difficult task is the development of appropriate models to relate the accelerated data to the service environment.

D. Life Estimates by Comparison with Control Materials

In a far less sophisticated approach, comparative testing and rank correlation can be used to obtain service life estimates. First of all, as with all service life methods, the accelerated test must produce failure modes that correspond to actual service. In addition, the rank correlation between results from the accelerated test and those from actual service should be very high. Finally, failure should be well defined and quantified. There should be a certain acceptable minimum level of some material performance property, with failure defined as the moment the performance property falls below that level.

To estimate service life, ideally, several weathering control materials should be selected that have a sequential range of failure times. The identification of acceptable control materials is the most difficult task. If the actual service life of these control materials has been well defined, the service life of the candidate specimen can be bracketed by two of the control materials in the exposure test. If "good rank correlation" has been established between the accelerated test and actual service life, it is reasonable to expect that the actual service life of the candidate specimens will fall within the same performance range (compared with the control materials). The accelerated results depicted in Figs. 11 and 12 have many significant reversals; hence, the chance of making a life estimate error are great. Even with accelerated tests that more accurately rank materials relative to outdoors, there is always the risk of an error in the service life estimate because of reversals. The use of multiple controls may reduce the chance of these types of errors.

VIII. CONCLUSIONS

Rank correlation has proved to be useful in dealing with the ramifications of variability in weathering test results. This approach is successful because differing stress levels influence failure rates to a far greater degree than performance rankings. Once significant performance ranks are established for a series of materials on exposure, the performance tends to remain reasonably constant. This is why less severe environments rank well with severe environments, as long as the stresses are qualitatively the same.

Rank correlation (nonparametrics) is a useful tool for evaluating the predictive ability of an accelerated exposure. This quantification of "test goodness" has allowed the development of improved accelerated tests, at least, in comparing the relative durability of materials, which provides a more reliable answer to, Is material A better than Material B? In addition, if appropriate control materials or minimum performance standards can be selected, highly reproducible specification weathering tests can be developed: Should I buy material A? If the control materials have known service lives, these tests can be used to estimate service life times for new or modified products: How long will material A last?

Pair error analysis is a useful tool in estimating the probability of a test error. It can also be used to identify material types that do not "correlate" well in specific tests. The chance of a test error, and eventually a field failure occurring, is significantly reduced as knowledge accumulates for accelerated test strengths and weaknesses.

Natural outdoor weathering will never be replaced by accelerated testing. Not all stresses that are present in an outdoor exposure can be simulated in a laboratory accelerated test. Biological, pollution, and climatological affects on polymers still require a comprehensive evaluation of durability at a variety of outdoor exposure sites.

ASTM Committee G-3 has approached the problem of test variability on several fronts. First, new precision and bias statements have been approved for G 23, G 26, and G 53, and the performance-based standards that will soon replace these older documents. These statements will include a strong warning about the poor reproducibility between laboratories and stipulate that:

[P]erformance requirements for materials tested according to (these practices) be specified in terms of comparison (ranked) with a standard reference material. This reference material must be exposed simultaneously with the test material in the same device. The specific reference material used must be agreed upon by the concerned parties.

This means that "absolute specifications" such as setting a specific property level after a specific exposure period are no longer recommended.

In addition, G141-96 (44), a standard *Guide for Addressing Variability in Exposure Testing of Nonmetallic Materials* covers information on the sources of variability and strategies for its reduction in exposure testing.

REFERENCES

1. H Mark. Encyclopedia of Polymer Science and Technology. Vol. 4. 1970, pp 779–795.
2. ML Ellinger. JOCCA 62(4):136–141, 1979.
3. R Dreger. Machine Design 45:29–35, 1973.
4. W Papenroth. Defazet 6(282):102–108, 1974.
5. R Blakey. Prog Org Coat 13:279–296, 1985.
6. Association of Automobile Industries. J Coat Technol 58(734):57–65, 1986.
7. R Kinmonth, R Saxon, R King. Polym Eng Sci 10(5):309–313, 1970.
8. R Fischer, W Ketola, W Murray. Prog Org Coat 19:165–179, 1991.
9. ASTM G 23-93 Standard Practice for Operating Light (Carbon Arc Type) and Water Exposure Apparatus for Nonmetallic Materials. Philadelphia: American Society for Testing and Materials.
10. ASTM G 26-93 Standard Practice for Operating Light (Xenon Arc Type) and Water Exposure Apparatus for Nonmetallic Materials. Philadelphia: American Society for Testing and Materials.
11. ASTM G 53-93 Standard Practice for Operating Fluorescent UV and Condensation Exposure Apparatus for Nonmetallic Materials. Philadelphia: American Society for Testing and Materials.
12. R Fischer. Results of round robin studies of light- and water-exposure standard practices. In: W Ketola, D Grossman, eds. Accelerated and outdoor durability testing of organic materials, ASTM STP 1202. Philadelphia: American Society for Testing and Materials, 1993, pp 112–132.
13. RM Fischer, WD Ketola. Impact of recent research on the development and modification of ASTM weathering standards. In: RJ Herling, ed., Durability Testing of Non-Metallic Materials, ASTM STP 1294. Philadelphia: American Society for Testing and Materials, 1996, pp 7–23.
14. SAE Materials Standard, SAE J1885, Accelerated Exposure of Automotive Interior Materials using a Controlled Irradiance Water-Cooled Xenon Arc Apparatus. Warrendale, PA. Society of Automotive Engineers.
15. E Conover. Practical Nonparametric Statistics, 2nd ed., New York: John Wiley & Sons, 1980.
16. LL Havlicek, RD Crane. Practical Statistics for the Physical Sciences. Washington DC: American Chemical Society, 1988.
17. WJ Putman, M McGreer. Comparison of outdoor behind glass weathering exposure standard practices in Arizona and Florida. In: RJ Herling, ed. Durability Testing on Nonmetallic Materials, ASTM STP 1294. Philadelphia: American Society for Testing and Materials, 1996, pp 136–156.
18. L Crewdson. Correlation of Outdoor and Laboratory Accelerated Weathering Tests at Currently Used and Higher Irradiation Levels—Part II, Sunspots. Vol 23. 46, 4th Quarter 1993. Chicago: Atlas Electric Devices.

19. WJ Putman, M McGreer, D Pekara. Parametric control of a Fresnel reflecting concentrator outdoor accelerated weathering device. In: RJ Herling, ed. Durability Testing of Nonmetallic Materials, ASTM STP 1294. Philadelphia: American Society for Testing and Materials, 1996, pp 40–55.
20. SAE Materials Standard, SAE J1960, Accelerated Exposure of Automotive Exterior Materials using a Controlled Irradiance Water-Cooled Xenon Arc Apparatus. Warrendale, PA: Society of Automotive Engineers.
21. GW Grossman. J Coat Technol 49(633):45–54, 1977.
22. R Fischer. SAE Technical Paper Series, 841022, Warrendale, PA: SAE International, 15096–0001, 1984.
23. P Pagan. Polym Paint Color J 1/6(41/5) Sept 17, 1986.
24. U Biskup. Eur Coat J 6:452–457, 1995.
25. Unpublished information. Accelerated weathering test developed during the mid 1980's in 3M Weathering Resource Center.
26. J Gerlock. Polym Degrad Stabil 26:241–254, 1989.
27. D Bauer, M Paputa Peck, R Carter. J Coat Technol 59(755):123–129, 1987.
28. CW Dunnett. J Am Stat Assoc 50:1096–1121, 1955.
29. Minitab 7.1. State College, PA: Minitab Inc.
30. SAE Materials Standard, SAE J2020, Accelerated Exposure of Automotive Exterior Materials using a Fluorescent UV and Condensation Apparatus. Warrendale, PA: Society of Automotive Engineers.
31. G Fedor, P Brennan. Correlation between natural weathering and fluorescent UV exposures. In: RJ Herling, ed. Durability Testing of Non-Metallic Materials, ASTM STP 1294. Philadelphia: American Society for Testing and Materials, 1996, pp 91–105.
32. KM Wernstahl, B Carlsson. J Coat Technol 69(865):65–75, 1997.
33. JA Simms. J Coat Technol 59(748):45–53, 1987.
34. TK Rehfeldt. Prog Org Coat 15:261–268, 1987.
35. KT Gillen, RL Clough. Polym Degrad Stabil 24:137–168, 1989.
36. RP Brown. Polym Test 14:403–414, 1995.
37. SH Hamid, MB Amin. J Appl Polym Sci 55:1385–1394, 1995.
38. MB Amin, SH Hamid, F Rahman. J Appl Polym Sci 56:279–284, 1995.
39. SH Hamid, WH Prichard. J Appl Polym Sci 43:651–678, 1991.
40. GJ Jorgensen, H Kim, TJ Wendelin. Durability studies of solar reflector materials exposed to environmental stresses. In: RJ Herling, ed. Durability testing of Non-Metallic Materials, ASTM STP 1294. Philadelphia: American Society for Testing and Materials, 1996, pp 121–135.
41. DR Bauer. J Coat Technol 69(864):85–96, 1997.
42. JW Martin, SC Saunders, FL Floyd, JP Wineburg. Methodologies for predicting the service lives of coatings systems. In: Federation Series on Coatings Technology. Blue Bell, PA: Federation of Societies for Coatings Technology, June 1996.
43. P Schutyser, DY Perera. Double liaison—Phys Chem Econ Peintures Adhes 479–480:24–28, 1996.
44. ASTM G 141-96 Standard Guide for Addressing Variability in Exposure Testing on Nonmetallic Materials. Philadelphia: American Society for Testing and Materials.

18

Fundamental and Technical Aspects of the Photooxidation of Polymers

JEAN-LUC GARDETTE

Université Blaise Pascal (Clermont-Ferrand), Aubière, France

I. INTRODUCTION

One major problem associated with the applications of polymers is their instability to weathering (1–9). The formation of oxidation products appears as the major cause of this instability. Understanding the mechanisms by which these products are formed has been the subject of intensive research for the last 30 years. In addition to the academic interest in understanding the nature and the mechanistic behavior of these species in polymers, developing and adapting new stabilizers on a rational basis, and predicting the lifetime service of polymers are most valuable for practical applications of polymeric materials.

The fate of solid synthetic polymers that are aged in natural weathering conditions is not easily predicted from accelerated test methods (10–12). The major part of the industrial activity concerning the long-term durability of polymers is still based on empirical methods developed in the 1950s: the aging of the polymer is characterized by the changes in the values of in-use properties of the material. Many different methods exist, from the simple evaluation of the color changes, to the determination of more complex physical properties (13). The aging is provoked to the sample by application of all the stresses that the polymer could receive when exposed to natural environment: light (with an accurate simulation of the solar light distribution), heat, water, mechanical stresses, pollutants, and so on, including cycles of exposure to light and moisture. The weathering devices are then designed to simulate exactly the stresses that the polymeric samples could experience. In these conditions, no important acceleration of the aging can be expected. Consequently, the experiments are often carried out at "high" temperatures (60–80°C), in that way, increasing the rate at which aging occurs.

The prediction of the long-term durability on the basis of the empirical methods present some negative aspects that result, as much from the definition of the aging parameters as from the test methods carried out to determine experimentally the consequences of the weathering.

It appeared in the 1970s that complementary concepts were needed to study the complex phenomena involved in the oxidation degradation of polymers. Because no experimental simplification could be made, the development of these new concepts required novel experimental procedures and analytical techniques. Because most of the degradation results from chemical changes in the polymer, the reliability of the observed phenomena has to be controled at the molecular level through recognition of the chemical reactions of the macromolecules (14,15). The experimental conditions of the laboratory experiments can, therefore, be selected to observe the same "mechanism" in accelerated conditions and in weathering. The acceleration of the photoaging is due to higher light intensities (limited to the range of monophotonic excitation) and to higher temperatures (limited to the invariance of the activation energy). The chemical changes of the polymeric matrix are described by a simplified scheme that involves several reactions sequences, including the following:

The formation of the intermediate photoproducts
The conversion pathways
The routes leading to the formation of the final photoproducts that accumulate
   in the matrix

Several rules have to be strictly followed to ensure the relevance of the observed phenomena:

1. The chemical changes have to be studied on solid polymers to take into account any perturbation that would result from heterogeneities of the solid state.
2. Chemical evolution should only be considered at very small levels that do not exceed 1%. Above this extent, it is often observed that a complete loss of the mechanical properties of the polymer occurs.
3. Polychromatic light has to be used, and any wavelength shorter than 295 ± 5 nm has to be strictly filtered. Medium-pressure mercury arcs, which were avoided as light sources in the simulation units because they emit discrete narrow lines, can be used (16).
4. Any control of oxidation by diffusion of oxygen into the polymer must be recognized. To avoid oxygen starvation effects, irradiation of very thin films (a few tens of microns) may be required.
5. The temperature at the surface of the exposed samples is an important parameter of the photoaging (fairly high apparent activation energies of 5–15 kcal mol^{-1} can be determined). This means that the control of this parameter has to be seriously considered.
6. A correct balance between photo- and thermoaging has to be respected. Therefore, acceleration must result from an increase of both the light intensity and the temperature. Increasing the temperature to 60–70°C, but

working at moderate-light intensity can give important discrepancies, as observed, for example, in the polyamides (14).

7. The migration of additives must not interfere with the photochemical consumption of these additives.

8. It is necessary to identify the chemical transformations that accounts for the physical detriment of the polymer. Usually the most important route involves an oxidative mechanism, the products of which are formed in concentrations high enough to be observed by vibrational spectroscopy.

On the basis of the oxidation kinetic curves determined in both of the conditions of aging, it is, thereby possible to derive from the time scale of the laboratory experiments a true time scale for the chemical evolution throughout environmental weathering. Because the variations of physical properties of the material in outdoors conditions are primarily controlled by the chemical changes in the polymeric matrix, it becomes possible to predict the fate of the macroscopic system from the acceleration of the ongoing chemistry.

This chapter focusses on the analytical methods that permit analyzing the chemical changes occurring in solid polymeric materials submitted to photo-oxidative degradation under artificially accelerated conditions. Results obtained in the common polymers are chosen to detail the experimental procedures. The photooxidation mechanisms that can be established on the basis of identification of the various photoproducts resulting from aging are discussed. Moreover, particular attention is given to the microscopic aspects dealing with the distribution of the products within the aged polymeric material.

II. ANALYSIS OF THE CHEMICAL CHANGES

Vibrational spectroscopy is a very useful tool for characterization of the chemical and physical nature of polymers (17–19).

A. Characterization of Modifications of the IR Spectra

Photooxidation of polymers produces a complex mixture of different products. The precise identification and quantification of the oxidation products is essential to understand the mechanism of degradation. Probably the most informative analytical technique is infrared (IR) spectrometry.

Chemical changes resulting from aging involve the formation of functional groups specific to a particular type of polymer at rates that are strongly dependent on the chemical structure of the polymer. The formation of these groups generally leads to dramatic modifications of the infrared spectrum of the polymer. For example, Fig. 1 shows the infrared spectrum in transmission mode of a polypropylene film at different irradiation times (20). The sample was irradiated in a medium-accelerated artificial weathering device in the presence of air.

The main changes occur in the IR regions that are characteristic of the carbonylated and hydroxylated compounds, which correspond, respectively, to the 1900–1500 and 3800–3100 cm^{-1} ranges. The carbonyl band shows an absorption maximum at 1713 cm^{-1} and two shoulders at 1735 and 1780 cm^{-1}. The formation of hydroxylated photoproducts corresponds to the increase of a broad band with a maximum near 3400 cm^{-1} owing to hydrogen-bonded hydroxylated functions.

Plotting the variations of absorbance at 1713 cm^{-1} as a function of the irradiation time permits characterization of the oxidation kinetics of the polymer. Figure 2 shows the curve that is obtained for the polypropylene sample.

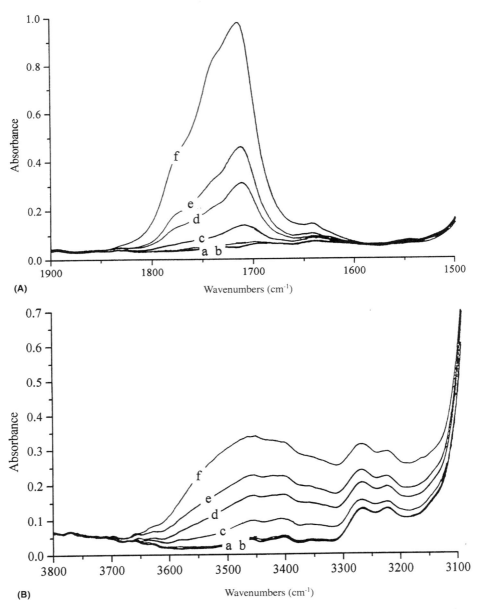

Figure 1 Modifications in the infrared spectrum of a polypropylene sample (thickness 100 μm) irradiated in a medium-accelerated artificial weathering unit SEPAP 12-24 for (b) 25 h; (c) 50 h; (d) 75 h; (e) 90 h; (f) 150 h; (a) initial spectrum before irradiation: (A): carbonyl region; (B): hydroxyl region. (From Ref. 20.)

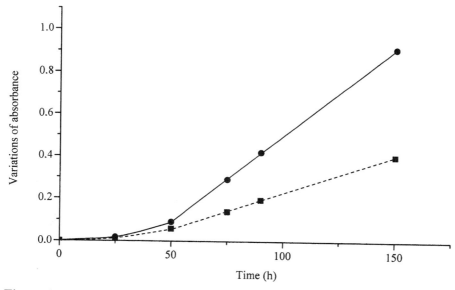

Figure 2 Kinetic curves of oxidation of a polypropylene sample (determined from the spectra of Fig. 1): (——), $\Delta D_{1713 cm^{-1}}$; (— —), $\Delta D_{3400 cm^{-1}}$.

The curve presents two different phases: the first, during which no formation of carbonyl product can be detected by IR spectroscopic analysis of the photooxidized sample. This phase, which corresponds to the induction period of the oxidation, results partly from the consumption of the residual processing antioxidants used to transform the polymer. In the second phase, the oxidation starts and the absorbance increases at a constant rate. The increase of the 1713 cm^{-1} absorbance of a 100-μm–thick polypropylene film has to be limited below 0.6–0.7 to avoid the loss of the mechanical properties of the film during irradiation.

Irradiation of a polystyrene film (thickness 90 μm) under the same standard conditions also leads to noticeable modifications of the IR spectrum (21). Figure 3 shows these modifications in the range 1900–1500 and 3800-3100 cm^{-1}.

Subtraction of the initial spectrum before irradiation from the spectra recorded after irradiation allows one to observe the global shape of the broad carbonyl and hydroxyl absorption bands that are formed.

Several maxima, or shoulders, are observed in the carbonyl band: 1690, 1698, 1732, and 1785 cm^{-1}. These bands result from the convolution of several bands the maxima of which can be determined by complementary analysis. A weak band is formed at 1515 cm^{-1}, and an absorption band, with a maximum close to 1605 cm^{-1}, can be noticed, even if the observation is hampered because of the important initial absorption band of polystyrene at 1600 cm^{-1} that interferes in the substraction of the spectra. The broad hydroxyl band that is formed is centered near 3450 cm^{-1}. A narrow band, with a high intensity, is observed in the region, with an absorption maxima at 3540 cm^{-1}, and a weak band at 3250 cm^{-1}.

Figure 4 shows the kinetic curve obtained by plotting the variations of absorbance at 1732 cm^{-1}, maximum of the carbonyl band, as a function of the irradiation time.

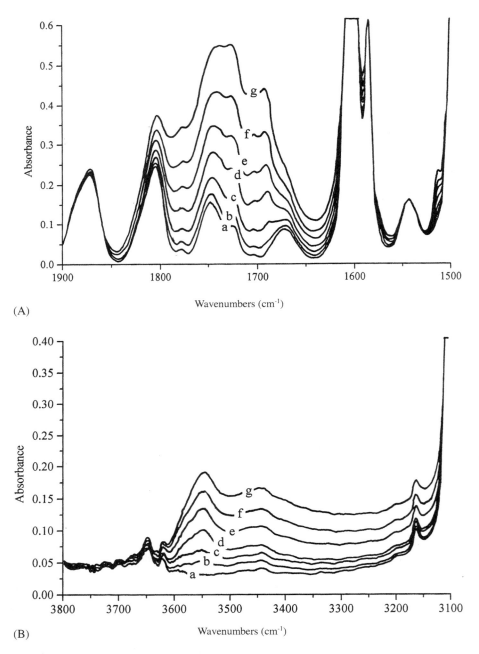

Figure 3 Modifications in the infrared spectrum of a polystyrene sample (thickness 90 μm) irradiated in a medium-accelerated artificial weathering unit SEPAP 12-24 for (b): 85 h; (c) 150 h; (d) 180 h; (e) 220 h; (f) 250 h; (g) 300 h; (a) initial spectrum before irradiation: (A), carbonyl region, (B): hydroxyl region. (From Ref. 21.)

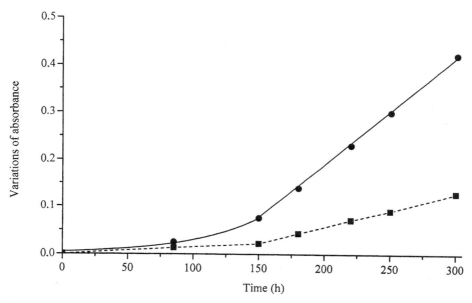

Figure 4 Kinetic curves of oxidation of a polypropylene sample (determined from the spectra of Fig. 3): (solid line), $\Delta D_{1732 cm^{-1}}$; (dashed line), $\Delta D_{3450 cm^{-1}}$.

Irradiation of poly(vinylethylether) thin films (thickness about 3 μm) on KBr windows, causes the following changes of the IR spectra of the samples (23): in the region 3800–3000 cm^{-1}: a broad band, with an absorption maximum at 3400 cm^{-1} appears and increases. In the region of the spectrum corresponding to the C–H-stretching vibrations, photooxidation leads to the decrease of the initial bands at 2974, 2941, 2931, and 2869 cm^{-1}, assigned on the basis of the literature data to the methyl and methylene groups (22). In the carbonyl region, a band with an absorption maximum at 1735 cm^{-1} develops and a shoulder at about 1725 cm^{-1} is observed for longer irradiation times (Fig. 5).

B. Identification of the Oxidation Photoproducts

The figures just presented show that the absorption bands that result from the formation of oxidation products often overlap to form complex absorption bands. The assignment of the various absorption maxima to well-identified products is then rather difficult. IR cannot differentiate between hydrogen-bonded alcohol and hydroperoxide OH groups, and it is only partially successful in the resolution of the mix of carbonyl species formed.

Several authors attempted the determination of the different carbonyl-containing functional groups in polypropylene by mathematical methods using an extinction coefficient evaluated for the various groups (24), nor by neutralization of acid groups followed by IR measurement of the ketones (25). Combined methods (chemical and spectrophotometric) were developed for determination of some carbonyl groups in polyethylene in the presence of others (26).

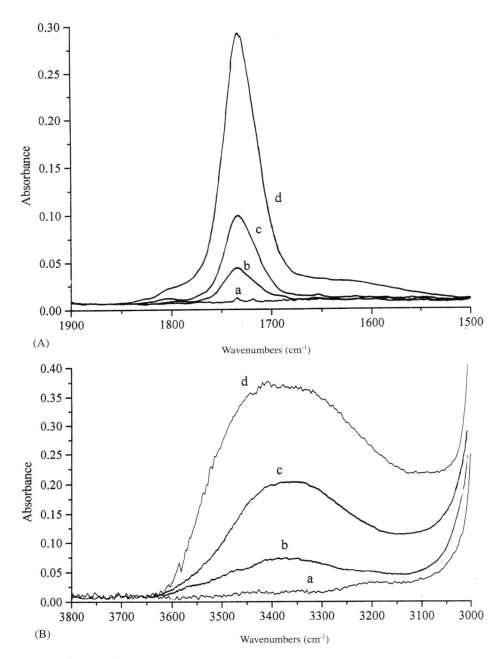

Figure 5 Modifications of the infrared spectrum of a poly(vinylmethylether) film (thickness 3 μm on KBr window) irradiated in a medium-accelerated artificial weathering unit SEPAP 12-24 for (b) 15 h; (c) 30 h; (d) 50 h; (a) initial spectrum before irradiation: (A), carbonyl region; (B): hydroxyl region. (From Ref. 23.)

The resolution of IR spectrometry can be extended by employing derivation reactions. Those methods consist of treating the oxidized polymer samples with reactive gas that can specifically convert the oxidation products that have been formed. The result expected from derivatization reactions is a simplification in the interpretation of the IR spectra resulting from the disappearance of specific bands and the appearance of the absorption bands of the derivated products.

The derivatization methods for the identification of oxidation products were applied to polyolefins by Carlsson et al. (27). The gases used and the groups quantified included diazomethane to convert acids and peracids to their respective methylester, sulfur tetrafluoride (SF_4) to convert acids to acyl fluorides, nitric oxide to convert alcohols and hydroperoxides to nitrites and nitrates, respectively, and phosgene to convert alcohol and hydroperoxides to chloroformates.

SF_4 derivatization is very informative because there is an important shift between the carboxyl band (C=O-stretching vibration) of the acyl fluoride and that of the corresponding acid. The fluorination of organic carbonyl compounds has been reviewed in detail (28,29). The carbonyl group of an acid is converted to a trifluoromethyl group. The reaction proceeds in two steps: formation of the acyl fluoride and replacement of the carbonyl oxygen.

$$R\,COOH + SF_4 \rightarrow R\,COF + HF + SOF_2$$
$$R\,COF + SF_4 \rightarrow R\,CF_3 + SOF_2$$

The first step occurs readily at or below room temperature, whereas the second step requires elevated temperatures. The literature indicates that acid anhydrides and carboxylic acids can also react with SF_4, but more rigorous conditions are required (30). In polymers no reaction or ketones was found under moderate conditions.

For oxidized polyethylene, SF_4 treatment leads to the generation of a C=O absorption at 1848 $^{-1}$ (27). This band was attributed to chain-end acyl fluorides formed by the reaction:

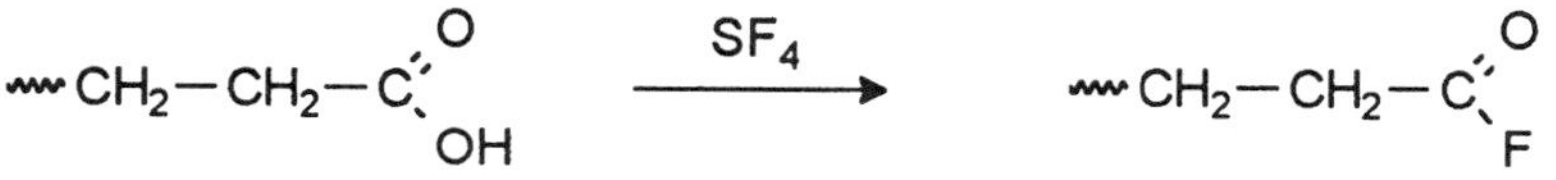

One of the advantages of SF_4 derivatization is that, depending on the structure of the acyl fluoride, noticeable shifts can be observed. The reactions of SF_4 with various carboxylic acids introduced into several polymeric matrices were studied (31). Several factors of influence were observed: the nature (aromatic or aliphatic) of the acid, the substitutive degree of the carbon atom in the α-position relative to the carboxylic group, and the nature of the polymer matrix. It was determined that the frequency of the C=O absorption band of the acyl fluoride band was lying from 1815 to 1853 cm^{-1} depending on the structure of the acid.

SF_4 treatment, carried out on a photooxidized sample of polypropylene, induces the formation of an absorption band with a maximum at 1841 cm^{-1} and a decrease of the absorbance at 1713 cm^{-1} (32). Comparisons with the model compounds (31) permits the maximum at 1841 cm^{-1} to be attributed to a methyl α-substituted acyl fluoride formed following the reaction:

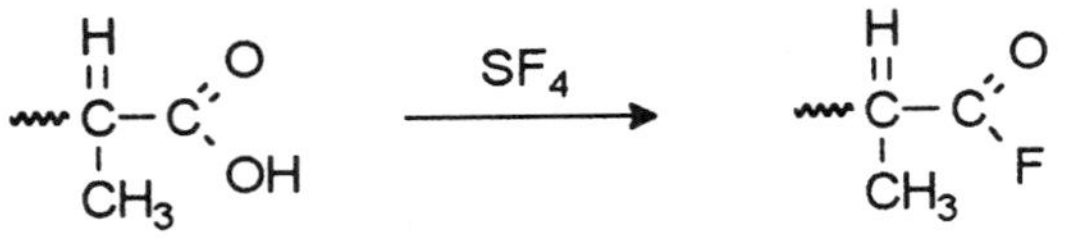

In the case of polybutyleneterephthalate (PBT), SF$_4$ treatments permitted identifying two different types of carboxylic acids formed by photochemical oxidation of the material (33,34). Chemical reaction of photooxidized PBT with SF$_4$ leads to the development of an intense band at 1815 cm^{-1} and a weaker one at 1841 cm^{-1}. These bands can, respectively, be attributed to the formation of an aromatic and an aliphatic acyl fluorides with the structures:

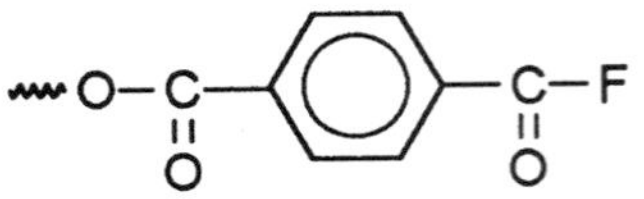

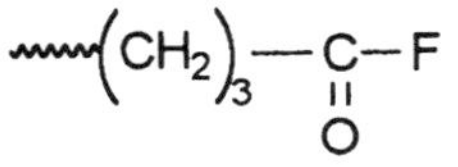

These results have confirmed that photooxidation of PBT leads to the formation of aromatic and aliphatic acids by direct scission of the C–O bond for the aromatic acid and by oxidation of the carbon atom α to the ester group for the aliphatic one (35).

Treatments with SF$_4$, carried out on photooxidized polystyrene films, leads to a decrease of the carbonyl absorption, whereas new absorption bands appear at 1813 and 1841 cm^{-1} (36). The comparison with model compounds treated by SF$_4$ in polystyrene films allows identification of the carboxylic acids to benzoic acid (giving the acyl fluoride at 1813 cm^{-1}) and an acid with the structure HO–C(=O)–CH$_2$ (with the acyl fluoride at 1841 cm^{-1}).

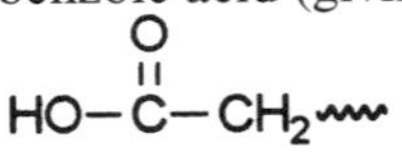

Treatments by SF$_4$ of photooxidized poly(vinylethylether) films narrows the carbonyl band at 1725 cm^{-1} and provokes the appearance of a band as 1845 cm^{-1} (23). The difference between the spectra after and before reacting the photooxidized sample with SF$_4$ shows the disappearance of a band at 1712 cm^{-1}, resulting from the conversion of a carboxylic acid, following the reaction:

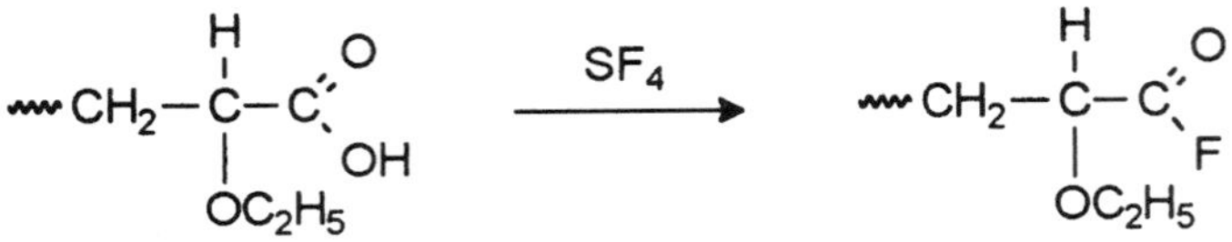

In addition to the generation of acyl fluoride absorption, SF$_4$ reactions provoke a complete loss of all the OH absorptions, which considerably simplifies the IR spectrum of the oxidized polymers. Moreover, this reaction leads to a narrowing of the envelope of the carbonyl band, for the absorptions, resulting from the hydrogen-bonded complexes that are formed with carboxylic acids and other adjacent hydroxylated products, are eliminated (27).

Derivatization reactions based on treatment of oxidized polymers by ammonia are very complementary to the treatments by SF$_4$. When carboxylic acids are treated with ammonia, or amines, salts are obtained (30):

$$R\,COOH + NH_3 \rightarrow R\,COO^- NH_4^+$$

When a salt is made of a carboxylic acid, the C=O and C–O bonds of the acid are replaced by two C $\dot{-}$ bonds. The antisymmetric CO_2 stretch band is usually seen at 1650–1540 cm^{-1} (22).

Acylation of amines by an ester gives amides following the reaction:

$$R\,COOR' + NH_3 \rightarrow R\,CONH_2 + R'OH$$

The C=O stretch band of unsubstituted amines absorbs about 1680–1640 cm^{-1}.

Treatments by NH_3 can be successfully applied to the determination of functional groups formed by oxidation of polypropylene (32). A notable decrease of the C=O band is observed, with the development of a broad band at 1560 cm^{-1}, resulting from the formation of carboxylates. A weak band, with an absorption maximum at 1670 cm^{-1}, is also observed. This band results from the formation of unsubstituted amide, and it can be associated with the decrease of the absorbance at 1735 cm^{-1}.

Reaction of photooxidized polystyrene films with ammonia leads to a decrease of the carbonyl bands between 1680 and 1800 cm^{-1} and to an increase of absorbance between 1680 and 1500 cm^{-1}, with maxima at 1668, 1553, and 1592 cm^{-1} (36). The absorption maximum at 1553 cm^{-1} corresponds to the C=O band of the carboxylate ion obtained by neutralization of benzoic acid by ammonia. The maximum at 1668 cm^{-1} that appears after the reaction with ammonia can be assigned to the C=O-stretching vibration of an amide group. The formation of this absorption band parallels the decrease of the bands at 1785 and 1732 cm^{-1}. The maximum at 1785 cm^{-1} can be assigned to benzoic anhydride, and the maximum at 1732 cm^{-1} to ester groups (or δ-lactones) formed by oxidation of the polymer.

Reaction of an oxidized poly(vinylethylether) sample with NH_3 leads to an important decrease of the carbonyl band at 1735 cm^{-1}. The consequence is the formation of two bands at 1670 and 1570 cm^{-1}. These maxima, which result from the formation of amide and carboxylate groups, show the presence of ester–formates (1735/1725 cm^{-1}) and carboxylic acid (1712 cm^{-1}) in the photooxidized samples (23).

Derivatization reaction of oxidized polyolefins with NO are particularly informative. The IR spectra of the nitrates produced from NO reaction of hydroperoxides present different absorption bands that can be used to discriminate primary, secondary, and tertiary hydroperoxides (37). Clear differences are also observed between the spectra of tertiary, secondary and primary nitrites formed from the corresponding alcohols (38). The potential of the derivatization by nitric oxide to identify the hydroxylated products formed by oxidation of polyolefins has been studied in details by Carlsson et al. (39). For comparison with the oxidized polymer, the authors recorded the spectra of model nitrates and nitrites with primary, secondary, and tertiary alkyl substituents and calculated their extinction coefficients. On the basis of these results, a quantification of the hydroxylated products, obtained by γ-oxidation of polypropylene and polyethylene (40), was obtained. Derivatization by nitric oxide reaction has been applied successfully to the identification of the oxidation products obtained by γ-oxidation of ethylenevinyl alcohol copolymers (41).

Nitric oxide is an efficient radical scavenger and an inhibitor of free radical reactions. A mechanism proposed to explain the reaction of hydroperoxides with nitric oxide is a hydrogen abstraction (27) following the reactions:

$$-\overset{|}{\underset{|}{C}}-OOH + NO \longrightarrow -\overset{|}{\underset{|}{C}}-OO^{\bullet} + HNO$$

$$-\overset{|}{\underset{|}{C}}-OO^{\bullet} + NO \longrightarrow \left[-\overset{|}{\underset{|}{C}}-OONO\right] \longrightarrow -\overset{|}{\underset{|}{C}}-ONO_2$$

A similar scheme could explain the formation of nitrites from alcohols (27).

The NO reaction has a good potential for identification of the hydroxylated photoproducts formed by oxidation of polyethylene and polypropylene. However, experiments carried out with aromatic polymers (polystyrene, MDI-based polyurethanes, polybutylene terephthalate) and with polyamides and unsaturated polymers, such as polybutadiene and polyisoprene, show that this method cannot be applied to these polymers because nitric oxide reacts with the unoxidized matrix to give nitrocompounds that present IR characteristics very similar to those of the expected nitrates and nitrites (36).

Hydroperoxides are readily decomposed by reduction by sulfur dioxide. The decomposition of hydroperoxides in oxidized polymers leads to the decrease of the absorbance of the OH-stretching vibration. Assuming an extinction coefficient of about 75 M^{-1} cm^{-1} for associated hydroperoxides (as determined for concentrated solutions of *tert*-butyl hydroperoxide), the treatment by sulfur dioxide coupled with IR analysis permits measuring the concentration of hydroperoxides in oxidized polymer. This method has been applied with success to various polymer such as poly(ether-urethane)s (42) and poly(ether-ester)s (43).

The iodometric method (44–48) consists of the reduction of hydroperoxides by sodium iodide in acidic medium according to the reaction:

$$R_3COOH + 3I^- + 2H^+ \rightleftharpoons R_3COH + H_2O + I_3^-$$

The concentration of I_3^- ions formed by this reaction can be measured by their UV absorption at 350 nm. The extinction coefficient determined for a model hydroperoxide (*tert*-butyl hydroperoxide) is 25,000 M^{-1} cm^{-1}.

Extensive studies with many different types of polymers oxidized under various conditions (photo-, thermo-, or γ-initiated) have shown that this method gives reliable quantitative information on –OOH species (24,49–53), if the hydroperoxides are thermally stable up to the temperature of refluxing solution (NaI in isopropanol) (42).

Hydroperoxides can be titrated following a method proposed for polyvinyl chloride (54) and polyethylene (55) based on the reduction by ferrous ions. This method has been applied to γ-oxidized polyolefins (56). Preswelling of the polymer is needed to ensure the penetration of the reagents into the polymer. As photooxidation leads to a partial cross-linking of polymers, which results in the formation of unsoluble fractions, the penetration into the oxidized zones is often hindered for highly oxidized polymers (42). This method, however, appears to be very reliable for polymers that can be totally dissolved before analysis (57,58). This titration method has considerable sensitivity and very low concentrations of hydroperoxides can be quantitatively measured.

C. Analysis of the Volatile Products

The photooxidative degradation of polymers can lead to significant formation of low molecular weight photoproducts, including water, carbon monoxide, and carbon dioxide. As a consequence of their low molecular weight, most of these photoproducts can migrate out of the polymeric sample. If this happens, analyzing the solid sample does not permit monitoring the formation of these photoproducts. The analysis of the solid sample then has to be completed by an analysis of the gas phase to take into account all of the photoproducts formed.

The formation of low molecular weight photoproducts is more likely to occur in polymers that have short, branched reactive groups. Quite important concentrations of volatile photoproducts can then be formed, so that the identification of these products is needed to understand the mechanisms by which these polymers degrade (59).

For polymers that are mostly linear, a precise identification of volatile products is not strictly required, because these products usually represent only a small fraction of the oxidation products. However, identifying the volatile products can help establish the oxidation mechanism.

Various methods exist for characterizing the low molecular weight products formed along with oxidation of solid polymers. Extraction of the products trapped in the solid matrix can be realized with an appropriate solvent, and the extract then analyzed by a chromatographic method (36). Extraction can be obtained by trapping the volatile products and analyze them by IR spectrometry with a gas cell (32), or by gas chromatography (24). More recently, a method based on mass spectrometry analysis of the volatiles evolved during irradiation was developed (59,60).

Collection of the volatiles evolved during irradiation of polypropylene shows that the major photooxidation products with low molecular weights are carbon monooxide and acetone (24,32,60). Several other products are detected at lower concentrations. These products include methane, carbon dioxide, water, methanol, acetic acid, and formic acid.

The photooxidation of polystyrene generates several oxidation products (61–66) that are formed with multiple chain scissions. Benzoic acid, benzophenone, benzaldehyde, benzoïc anhydride, and dibenzoylmethane have been identified by high-performance liquid chromatography (HPLC) analysis of the methanol extract of photooxidized samples (36).

III. CHARACTERIZATION OF HETEROGENEOUS OXIDATION

Depending on the conditions in which aging is carried out, heterogeneous effects can be produced, and then any attempt to correlate results obtained in different experimental conditions as, for example, accelerated artificial aging versus outdoor natural weathering, must take into account that heterogeneities can be induced by the accelerated test. The correlation that could be established would be meaningless, especially when the testing procedures are based on the monitoring of physical or mechanical changes.

The most widespread cause of heterogeneous degradation at the macroscopic level results from oxygen diffusion-limited effects (67–73). These effects are likely to be observed in conditions of accelerated aging, if the rate of oxygen consumption exceeds the rate of oxygen permeation; oxidation occurs in the surface layers, whereas the core remains practically unoxidized.

The importance of these effects depends on several parameters:

1. Intrinsic parameters, related to the material geometry (e.g., sample thickness), coupled with the oxygen consumption rate, which depends on the reactivity of the polymer, the nature of the additives, and the oxygen permeability of the material.
2. External parameters that result from the conditions in which the aging is carried out, and include both the light intensity and the sample temperature, as well as the oxygen pressure during the weathering (74).

The combination of these parameters leads to the definition of an optimal thickness that is required to ensure that no limitation of the oxidation rate by oxygen diffusion occurs.

Increasing the light intensity and the temperature to produce an accelerated degradation may lead to a heterogeneous oxidation if oxygen is consumed more rapidly than it can be resupplied by diffusion processes. Oxygen starvation can be avoided by reducing the thickness of the samples and, for example, in polyamides, it may be necessary to exposed films as thin as 40 μm (75). Another possibility consists of exposing pigmented systems, which limits the photochemical reactions to the very superficial layers and generally limits oxygen starvation effects.

The migration of additives is also an important factor that has to be considered. If the additives are nonmigrative, their influence is the same in both natural and accelerated aging. If the additives can migrate, their rate of diffusion must be fast enough to compensate for their consumption in the reactive layers. If this is not so, their concentration in the front layers can be replenished by submitting the samples to intermediate periods of obscurity.

Infrared spectroscopy offers the great advantage that it permits monitoring the effects of heterogeneous oxidation. Two very simple methods are generally used:

1. The intensity of an oxidation band can be measured as a function of the thickness of the film (76). If the absorbance varies linearly with the thickness (for reasonably oxidized samples), the oxidation may be expected to be homogeneous in the film. If oxidation is heterogeneous, a deviation from the linearity is observed above a definite thickness.
2. The second method is based on successive analysis of slices microtomed parallel to the irradiated edges (77,78). Profiles of oxidation can be directly rebuilt by plotting the oxidation of the different slices versus the depth at which the measurement is made.

The first method that consists in analyzing different thicknesses does not allow an oxygen-diffusion effect to be distinguished from a limitation of the oxidation that would result from the attenuation of the UV light as it passes through the polymer. The attenuation of light results from the light absorption either by an initial chromophor or by UV-absorbing photoproducts. The second method is not so easily applied to the analysis of thin films, for evident practical reasons. This method has to be reserved for monitoring the oxidation of relatively thick films (a few millimeters thickness) that can be microtomed. However, frequently, the surface of the oxidized samples becomes very brittle and collecting the microtome shavings that correspond to the outer layers is much more difficult, especially when very thin samples are needed (e.g., thickness about 20 μm).

Another approach that permits monitoring the IR spectra at the surface of the sample is attenuated total reflection (ATR) spectroscopy. This technique is easily applied to both thick and thin films. By varying the reflecting crystal and the angle of incidence, one may measure the depth dependence of oxidation in the range of 0.1 μm to a few micrometers (79). However, this technique can be successfully applied only if it is possible to apply an intimate contact between the polymer sample and the ATR crystal. This is not always possible with degraded materials because the outer oxidized surface roughens and often becomes brittle and deformed. Poor quantitative data are obtained most of the time for very oxidized samples. Roughened surfaces also result in reduced spectral quality. Moreover, with thin films, the surface of the sample is often not large enough to fit the surface of the ATR crystal.

Experimental techniques that are better adapted to the analysis of aged polymeric materials have been recently developed. These techniques are capable of profiling spatial variations in the concentration of oxidation photoproducts across the small distances of interest (10–20 μm):

1. The first method is based on the use of micro-Fourier transform IR (FTIR) spectroscopy and consists of analyzing a microtomed shaving of the photooxidized sample obtained in a plane perpendicular to the axis of irradiation (80,81). The IR spectra of narrow areas (about 10 μm in width) are successively monitored by moving the sample along the axis of irradiation, and the profile of concentration of the photooxidation products is then plotted by analyzing the different spectra obtained.

2. The second method is based on analysis by photoacoustic-FTIR spectroscopy (PAS-FTIR) and permits analyzing the surface of the material without any preparation of the sample (82). The thickness that is analyzed depends on the nature of the sample and on the characteristics of the optical bench of the spectrometer, and varies with the frequency. The analyzed layer can range from a few micrometers to several tens of micrometers. This technique is very well adapted to solid samples that are often difficult to prepare for analysis by classic IR. The measurements concern the surface of the sample, but efforts are made to adapt this technique for mapping spatial variations along the axis of the profile (83).

The experimental procedure followed to monitor the depth dependence of the oxidation by micro-FTIR spectroscopy consists first in slicing an oxidized polymer sample with a microtome in a plane perpendicular to the irradiation axis (in thin polymer films, the sample is first embedded in an inclusion resin). A thickness of about 50 μm is chosen. The slices that are obtained are then placed under the objective of an IR microscope and analyzed by transmission of the light through a small area delimited by an image-masking aperture. The sample is moved along an axis that corresponds to the irradiation axis by successive steps and the IR spectra of each zone are successively recorded (Fig. 6).

This procedure permits measuring afterward the absorbance corresponding to a well-defined oxidation photoproduct. The variations of the absorbance of the analyzed zone are plotted as a function of the distance from the edge of the sample. By moving the sample from the exposed front layers to the unexposed rear layers, the oxidation profile can then be determined. The main limitation in terms of spatial resolution is imposed by the diffraction of the IR light through the small aperture. High-performances microscopes that are equipped with a double-masking aperture permit the physical limit of the diffraction (approximately 10 μm) to be approached.

Photoacoustic detection coupled with IR spectroscopy permits monitoring the IR spectra of the surface of solid samples. This technique requires virtually no sample preparation and is relatively independent of the surface morphology of the sample. The photoacoustic effect is produced when an intensity-modulated light impinges on a sample (83–87). The sample under study is placed in a sealed PAS cell which also contains an inert gas and a sensitive microphone. The IR spectrum is obtained by measuring the heat generated from the sample owing to absorption processes. The heat is transferred by thermal diffusion to the surrounding gas, which causes a small boundary layer of gas to expand and contract, resulting in a pressure wave within the cell. This pressure wave is detected by the microphone, and the electric signal is Fourier transformed and recorded in the form of an IR spectrum.

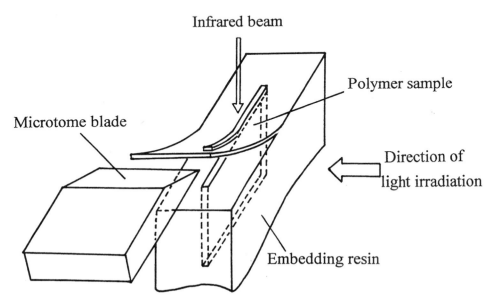

Figure 6   Schematic representation of the analysis method for preparation of thin microtome cross-sections.

Materials are classified depending on the relation between their absorption coefficient and thermal diffusion length. Theoretical predictions for thermally thick and optically transparent samples show that the magnitude of the photoacoustic signal generated from the surface of the samples depends on the modulation frequency and varies as $\omega^{-3/2}$. According to the theory, the thermal diffusion length μ is related to the modulation frequency ω through the equation:

$$\mu = \left(\frac{k}{\rho C \omega}\right)^{1/2}$$

where k is the thermal conductivity, ρ is the density, C is the specific heat, ω is the modulation frequency ($\omega = 2\pi V$) with V the mirror velocity (cm s^{-1}) and the wavenumber (cm^{-1}).

Heterogeneous oxidation resulting from oxygen diffusion-limited effects are observed in polymers having a weak permeability to oxygen (88) (PVC, PVDC, and so on), or in those polymers that are very sensitive to photochemical degradation (71) (elastomers, ABS, and such). Pronounced oxidation profiles are also monitored when the oxidation is limited by the penetration of light in the polymer. Such effects have been reported for several aromatic polymers [PET, PBT (33), PS (89), PEN (90), aromatic PA (91)].

Figure 7 shows the oxidation profile of an ABS sample photooxidized by exposure in a medium accelerated artificial weathering device. The dissymmetry between the oxidation within the front and the rear layers results from the attenuation of light passing through the sample.

IV. PHOTOOXIDATION MECHANISMS

The wavelengths of the radiation from the sun that reaches the surface of the earth extend from the infrared into the ultraviolet with a cutoff at close to 300 nm depending on the atmospheric conditions. The quantum energies associated with wavelengths in the region 300–400 nm are sufficient to break chemical bonds in most polymers. The presence of chromophoric groups is required to absorb the incident radiation. If one considers the chemical structure of various aliphatic, saturated polymers, no absorption of sunlight spectrum is expected. Yet most polymers are susceptible to photodegradation to some extent. Conversely, the photosensitivity of many aromatic polymers, such as polycarbonate or polybutylene terephthalate, is attributable to their strong inherent UV absorption above 300 nm.

The discussion on the initiation of photodegradation of "nonabsorbing" polymers has been the subject of many papers. The true nature of the chromophores responsible for photoinitiation of the degradation is not unambiguously established, and probably never will be. The reason the transparent polymers can absorb light is the presence of chromophoric impurities. The chemical nature and the relative importance of these chromophores essentially depends on the thermal history of the polymer: conditions of polymerization, processing, and storage. The chromophoric additives, such as pigments, dyes, or stabilizers, plasticizers and such, will not be described here. Another chromophor, which has been actively studied, is the polymer–oxygen charge transfer complex. The formation of charge transfer complexes with oxygen was first reported for polypropylene (94). On absorption of UV

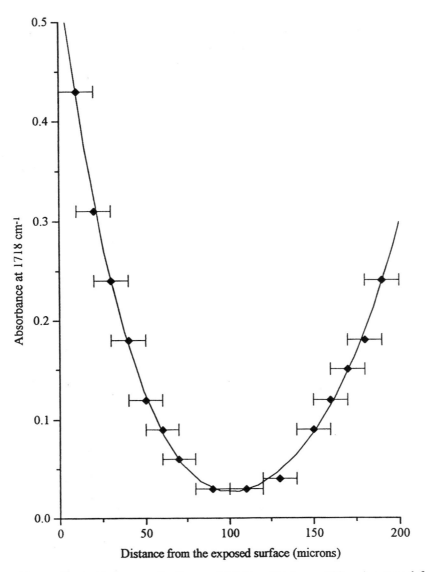

Figure 7 Oxidation profile for an ABS film (thickness 200 μm) exposed for 100 h in a SEPAP 12-24 unit (sample temperature 60°C).

light, these species generate hydroperoxides or their derived radicals. The efficiency of this mechanism, however is, believed to be low.

A. Polypropylene

The potential chromophores, together with their relative importance in initiation, have been discussed by Carlsson and Wiles (92,93). These authors have concluded that, of the many chromophoric impurities, hydroperoxides were the most susceptible to satisfy the kinetic parameters of photooxidation during the early stages. The introduction of these species during the polymerization or the processing of the polymer and their further decomposition can be described by the well-established Bolland–Gee autooxidation mechanism, as shown in Scheme 1:

$$Polymer \xrightarrow{\Delta} P^{\bullet}$$

Propagation

$$P^{\bullet} + O_2 \rightarrow PO_2^{\bullet}$$
$$PO_2^{\bullet} + PH \rightarrow POOH + P^{\bullet}$$

Termination

$$2\,PO_2^{\bullet} \rightarrow inert\ products$$
$$PO_2^{\bullet} + P^{\bullet} \rightarrow POOP$$
$$P^{\bullet} + P^{\bullet} \rightarrow P\!-\!P$$

Branching

$$POOH \xrightarrow{\Delta} PO^{\bullet} + PO^{\bullet}$$

Scheme 1 Autooxidation mechanism showing formation and decomposition of hydroperoxides during the processing.

Hydroperoxides absorb weakly up to 360 nm ($\epsilon^{360\,nm} = 1\ M^{-1}\ cm^{-1}$). The light quanta produced by solar irradiation are energetically sufficient to cleave PO–OH bonds, which have a dissociation energy of about 45 kcal mol^{-1}. The photolysis of hydroperoxides produces very reactive radicals that can propagate the oxidation.

Most of the products generated by photooxidation of polypropylene have been identified and quantified (24,25,32,95–98). These identifications permit the proposal of a simplified mechanism that accounts for the main routes of the photoinitiated oxidation of polypropylene.

If we consider that most of the primary oxidation products (hydroperoxides) in polypropylene are tertiary, with secondary representing about 10%, the probability of a direct oxidation of methylene or methyl groups to form α-methylated acids CH(CH$_3$) $-$ COOH appears rather low. Carboxylic structures could be formed from the acyl groups resulting from Norrish I processes of the intermediate ketone groups that are obtained by oxidation of the tertiary carbon atom:

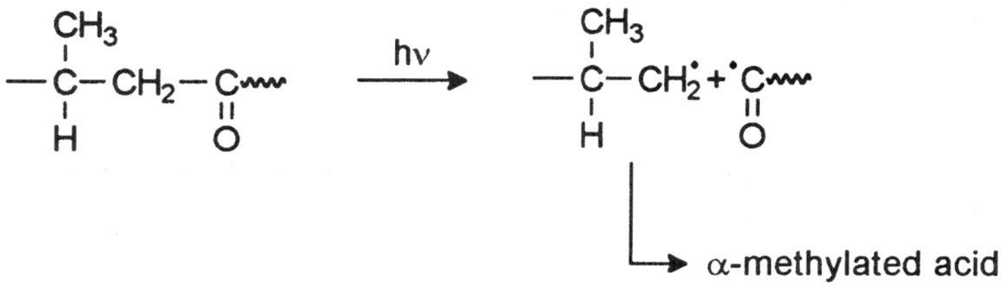

However, if one considers the kinetic curves corresponding to the formation of ketones and acids in photooxidized polypropylene, it appears that a supplementry route for acid formation has to be proposed. Acyl groups are indeed obtained along with the ketones by β-scission of alkoxy radicals formed by oxidation of the tertiary carbon:

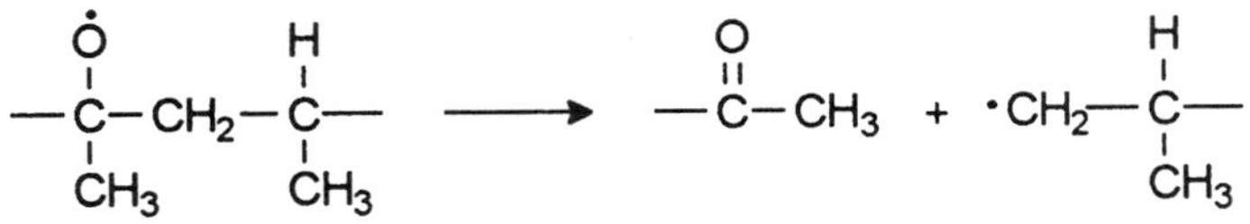

$$
\begin{array}{c}
\overset{\overset{\displaystyle \cdot}{O}}{\underset{\underset{\displaystyle CH_3}{|}}{-C}}-CH_2-\overset{\overset{\displaystyle H}{|}}{\underset{\underset{\displaystyle CH_3}{|}}{C}}- \quad\longrightarrow\quad -\overset{\overset{\displaystyle O}{\|}}{C}-CH_3 \;+\; {}^{\cdot}CH_2-\overset{\overset{\displaystyle H}{|}}{\underset{\underset{\displaystyle CH_3}{|}}{C}}-
\end{array}
$$

The primary radical so formed can be oxidized into a primary hydroperoxide, then into an aldehyde (95), and finally into an α-methylated carboxylic acid:

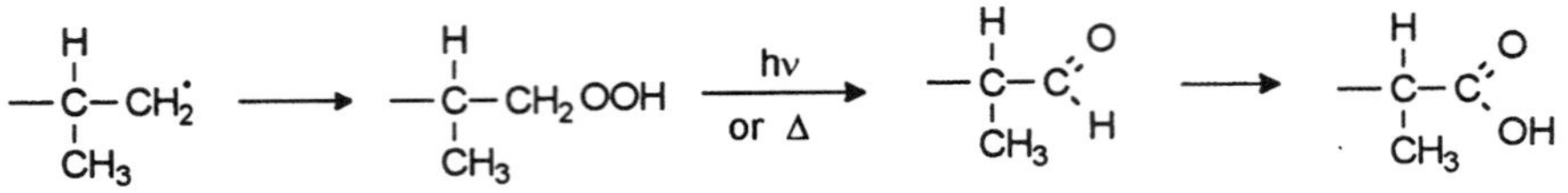

$$
-\overset{\overset{H}{|}}{\underset{\underset{CH_3}{|}}{C}}-CH_2^{\cdot} \;\longrightarrow\; -\overset{\overset{H}{|}}{\underset{\underset{CH_3}{|}}{C}}-CH_2OOH \;\xrightarrow[\text{or }\Delta]{h\nu}\; -\overset{\overset{H}{|}}{\underset{\underset{CH_3}{|}}{C}}-C\overset{\displaystyle O}{\underset{\displaystyle H}{\diagup}} \;\longrightarrow\; -\overset{\overset{H}{|}}{\underset{\underset{CH_3}{|}}{C}}-C\overset{\displaystyle O}{\underset{\displaystyle OH}{\diagup}}
$$

Oxidation of the primary radical competes with an isomerization that gives a more stable tertiary radical. Oxidation of the tertiary radical leads to the formation of acetone, the formation of which is clearly shown by analysis of the volatile products (24,32):

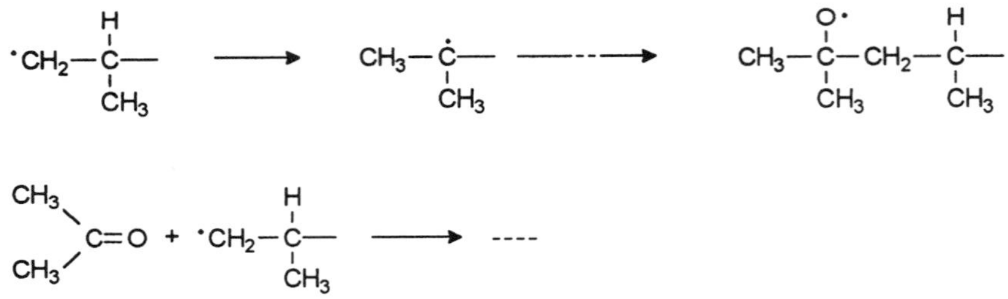

$$
{}^{\cdot}CH_2-\overset{\overset{H}{|}}{\underset{\underset{CH_3}{|}}{C}}- \;\longrightarrow\; CH_3-\overset{\overset{\cdot}{}}{\underset{\underset{CH_3}{|}}{C}}- \;\dashrightarrow\; CH_3-\overset{\overset{O^{\cdot}}{|}}{\underset{\underset{CH_3}{|}}{C}}-CH_2-\overset{\overset{H}{|}}{\underset{\underset{CH_3}{|}}{C}}-
$$

$$
\underset{CH_3}{\overset{CH_3}{\diagdown\diagup}}C{=}O \;+\; {}^{\cdot}CH_2-\overset{\overset{H}{|}}{\underset{\underset{CH_3}{|}}{C}}- \;\longrightarrow\; \text{----}
$$

This reaction regenerates the primary radical ${}^{\cdot}CH_2{-}\overset{\overset{H}{|}}{\underset{\underset{CH_3}{|}}{C}}{-}$ that can be involved in a further reaction producing acetone and acids.

The β-scission of the alkoxy radical may also involve the homolysis of the side bond, giving a methyl radical and a chain-end ketone:

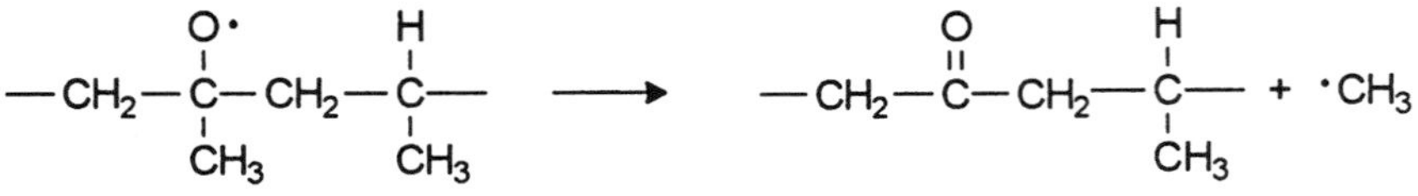

$$
-CH_2-\overset{\overset{O^{\cdot}}{|}}{\underset{\underset{CH_3}{|}}{C}}-CH_2-\overset{\overset{H}{|}}{\underset{\underset{CH_3}{|}}{C}}- \;\longrightarrow\; -CH_2-\overset{\overset{O}{\|}}{C}-CH_2-\overset{\overset{H}{|}}{\underset{\underset{CH_3}{|}}{C}}- \;+\; {}^{\cdot}CH_3
$$

Oxidation of the methyl radical would account for the formation of methanol, detected by analysis of the gas and liquid phases (32).

The primary radical responsible for the formation of acids and acetone could be formed in a Norrish type I photolysis of the chain-end ketones or the macroketones:

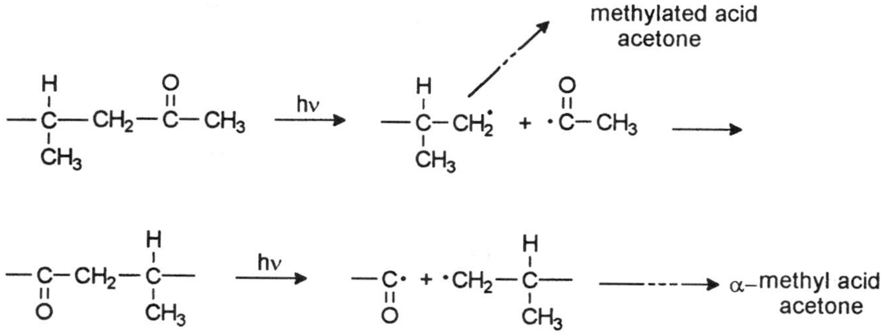

In addition to the fact that this behavior is not supported by the kinetic data, these routes should not be considered important because the stoichiometries in photooxidation and thermooxidation are very similar. The photolysis of the chain-end ketones is, however, the source of production of acetic acid.

Norris type II reactions of ketonic groups that would lead to vinylidenes are not likely to occur significantly because no information of an unsaturated group is observed.

B. Polystyrene

Polystyrene does not absorb the shorter wavelengths of the sun spectrum. It shows strong absorption at 250–280 nm owing to the phenyl group. The light-induced degradation of commercial polystyrene is attributed to the scission of the weak O–O bonds of the peroxide groups that result from the processing. Abstraction of the tertiary-bonded hydrogen occurs preferentially, leading to the formation of a benzyl-type radical.

Tertiary polystyryl radicals, $-CH_2\dot{C}(Ph)CH_2-$, have been identified by ESR measurements (99). The lifetime of polystyryl radicals at room temperature in vacuum is rather long (100). Secondary radicals, $-CH(Ph)-\dot{C}HCH(Ph)-$, could be formed, but no experimental observation supports this hypothesis (101). Macroradicals can react with oxygen molecules producing polymer peroxy radicals, $-CH_2C(Ph)(OO^\bullet)CH_2-$, which have also been identified by ESR spectroscopy.

Abstraction of hydrogen on the macromolecular chain by the peroxy radical so formed leads to hydroperoxydes $-CH_2C(Ph)(OOH)CH_2-$.

The decomposition of hydroperoxyl groups either by photolysis or by thermolysis leads to the formation of an alkoxy macroradical, $-CH_2C(Ph)(O^\bullet)CH_2-$, that may react in one of several ways:

1. By abstraction of a hydrogen atom to the polymeric backbone, hydroxyl groups $-CH_2C(Ph)(OH)CH_2-$ are formed.
2. β-Scission of the macroalkoxy radical may occur. Two types of scission are possible: (1) scission of the C–Ph bond and (2) scission of the C–CH$_2$ bond. Scission 1 leads to a chain ketone $-CH_2C(O)CH_2-$. Scission 1 would also generate a benzyl radical that is the precursor of benzene (102). Scission

2 leads to an acetophenone-type structure $-CH_2C(O)Ph$. The formation of such ketonic end groups in photooxidized polystyrene has been reported (103). Scission 2 also leads to a macroradical $^\bullet CH_2CH(Ph)-CH_2-$. This radical may react by isomerization to yield a more stable tertiary radical $CH\overset{\bullet}{C}_3(Ph)CH_2-$.

By oxidation of the tertiary radical $CH_3\overset{\bullet}{C}(Ph)CH_2-$, an alkoxy radical $CH_3C(Ph)(O^\bullet)CH_2-$ is obtained. The reactivity of this radical is very similar to the reactivity of the radical $-CH_2C(Ph)(O^\bullet)CH_2-$. The only difference is that the β-scission of this radical gives end-chain ketones $CH_3C(O)CH_2-$, acetophenone groups $PhC(O)CH_2-$, and acetophenone.

Photochemical oxidation of the end-chain ketone $CH_3C(O)CH_2-$ involving Norrish type I reaction leads to a carboxylic acid $HOOCCH_2-$ that can be obtained also by photochemical oxidation of the chain ketone $-CH_2C(O)CH_2-$ and also leads to acetic and formic acids.

Acetophenone end groups $PhC(O)CH_2-$ formed by β-scission of macroalkoxy radical $-C(Ph)(O^\bullet)CH_2-$ react photochemically following two different routes: scission of the $C-CH_2$ bond gives a radical $PhC(O)$ that can abstract a hydrogen atom or add to oxygen. The first reaction leads to benzaldehyde. Benzaldehyde is oxidized in benzoic acid. Oxidation of the radical $Ph\overset{\bullet}{C}(O)$ gives the same reaction. The scission of the $C-Ph$ bond leads to radicals $-CH_2\overset{\bullet}{C}(O)$. Oxidation of this radical may be the precursor of the carboxylic acid $-CH_2-COOH$.

The photooxidation of polystyrene causes a discoloration of the polymer owing to increased absorption in the visible region. The formation of the chromophoric groups responsible for the discoloration has been interpreted, by analog to the photooxidation of benzene, as due to a ring-opening reaction of the phenyl groups (104,105). This reaction results in the formation of mucodialdehyde groups:

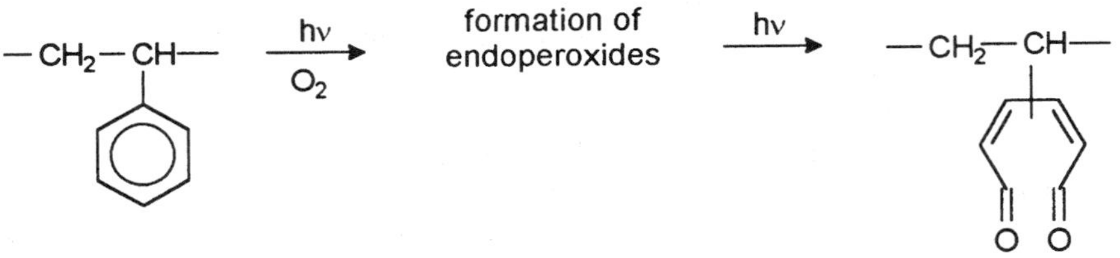

Other degradation products (e.g., quinone, diketones, conjugated double bonds) also have been proposed for the photooxidative yellowing of polystyrene.

Recent results have shown that the photoyellowing of polystyrene could be associated with the formation of thermally unstable photoproducts (102,106). It was observed that the thermolysis of a photooxidized sample led to a dramatic increase of absorbance in the visible spectra. Moreover, it was reported that the modifications of the spectrum induced by the thermolysis were totally reversible if the sample was submitted to a further irradiation. This reversible effect could be repeated several times without any modification of the IR spectrum of the samples. On the basis of these experiments, it was proposed that the products responsible for the yellowing of photooxidized polystyrene could be identified as benzalacetophenone structures formed by hydrogen abstraction in the α-position of an oxidized carbon atom (107):

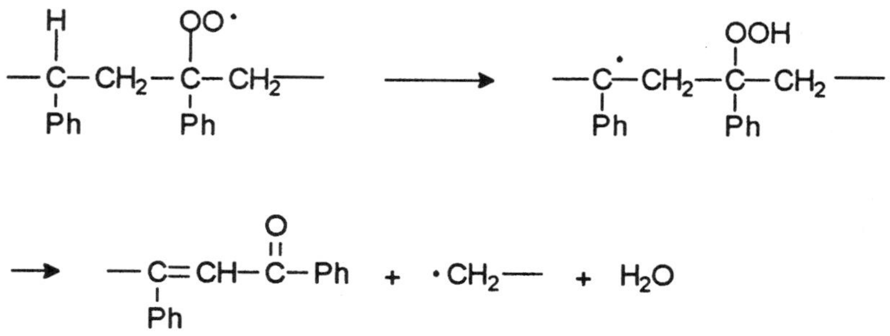

The photothermal equilibrium between the Z and E forms of the benzalacetophenone groups would be responsible for the reversible effects observed. The formation of α, β-unsaturated ketones has also been proposed:

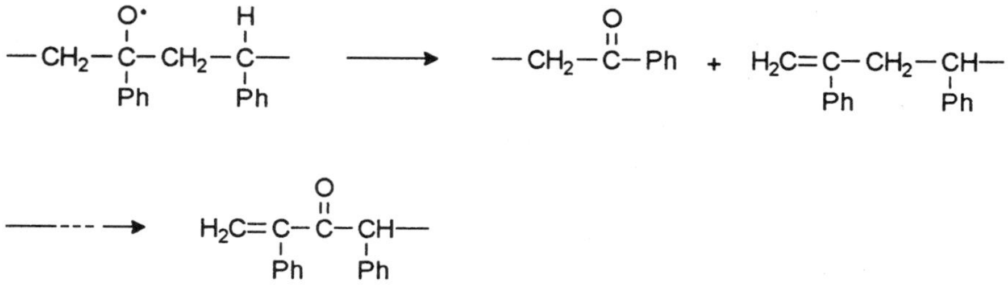

These ketonic groups can lead to photothermal equilibrium between the ketonic and enolic forms.

V. CONCLUSIONS

The extensive research carried out in the past 30 years in the field of polymer degradation has considerably increased our knowledge of the mechanisms by which polymers degrade with exposure to natural or artificial aging. The analysis of the chemical changes permit recognizing the main degradation routes involved in the deterioration of the properties of the material and ensure the relevance of the laboratory experiments. On the basis of these analyses, it generally appears possible to convert the data from the laboratory to the environment.

REFERENCES

1. B Ranby, JF Rabek. Photodegradation, Photooxidation and Photostabilization of Polymers. London: Wiley, 1975, pp 493–500.
2. N Grassie, ed. Developments in Polymer degradation. Vols 1–7. London: Elsevier Applied Science, 1978–1987.
3. JF Mckellar, NS Allen. Photochemistry of Man-Made Polymers. London: Elsevier Applied Science 1979.
4. NS Allen, ed. Developments in Polymer Photochemistry. Vols 1–3. London: Elsevier Applied Science 1980–1983.

5. W Schnabel. Polymer Degradation Munich: Hansen, 1981, pp 95–127.
6. N Grassie, G Scott. Polymer Degradation and Stabilization, Cambridge: Cambridge University Press, 1985.
7. KS Minsker, SV Kolesov, GE Zaikov. Degradation and Stabilization of Vinylchloride Based Polymers. Oxford: Pergamon Press, 1988.
8. SH Hamid, MB Amin, AG Maadhah, eds. Handbook of Polymer Degradation. New York: Marcel Dekker, 1992.
9. JF Rabek. Photodegradation of polymers. Berlin: Springer-Verlag, 1996.
10. RP Brown. Survey of status of test methods for accelerated durability testing. Polym Test 10:3–30, 1991.
11. RP Brown. Predictive techniques and models for durability tests. Polym Test 14:403–414, 1995.
12. JL Scott. Real time evaluation as a test for evaluating materials. J Vinyl Technol 16:116–123, 1994.
13. J Wypich. Weathering Handbook. Toronto: Chemtec Publishing, 1990, pp 222–261.
14. J Lemaire, R Arnaud, JL Gardette, JM Ginhac, Ly Tang, E Fanton. Vieillissement des polymères: empirisme ou science? Caoutchoucs Plast 593:147–150, 1979.
15. J Lemaire, JL Gardette, J Lacoste, P Delprat, D Vaillant. Mechanisms of photooxidation of polyolefins: prediction of lifetime in weathering conditions. In: RL Clough, NC Billingham, KT Gillen, eds. Polymer Durability. Washington: ACS, 1996, pp 577–598.
16. A Rivaton, JL Gardette, J Lemaire. Photoviellissement: évaluation des sources lumineuses. Caoutchoucs Plast 651:81–85, 1985.
17. HW Siesler, K Holland-Moritz. Infrared and Raman Spectroscopy of Polymers. New York: Marcel Dekker, 1980.
18. PC Painter, MM Coleman, JL Koenig. The Theory of Vibrational Spectroscopy and Its Applications to Polymeric Systems. New York: Wiley Interscience, 1982.
19. DI Bower, WF Maddams. The Vibrational Spectroscopy of Polymers. New York: Cambridge University Press, 1989.
20. P Delprat. Etude du Mécanisme de photooxydation de copolymères polypropylène/polyéthylène. PhD dissertation l'Université Blaise Pascal, Clermont-Ferrand (France), 1993.
21. B Mailhot. Etude des mécanismes de photooxydation du polystyrène du poly-acrylonitrile et du poly(styrène-co-acrylonitrile). PhD dissertation de l'Université Blaise Pascal, Clermont-Ferrand (France), 1993.
22. D Liu-Vien, WB Colthup, WG Fateley, JE Grassely. The Handbook of Infrared and Raman Characteristic Frequencies of Organic Molecules. San Diego: Academic Press, 1991.
23. F Posada. Photooxydation de polyuréthanes à base de copolymères d'oléfines fluorées et d'éthers allyliques et vinyliques. PhD dissertation l'Université Blaise Pascal, Clermont-Ferrand (France), 1997.
24. DJ Carlsson, DM Wiles. The photodegradation of polypropylene films. II. Photolysis of ketonic oxidation products. Macromolecules 2:587–597, 1969.
25. JH Adams. Analysis of the nonvolatile oxidation products of polypropylene. III. Photodegradation. J Poly Sci A1 8:1279–1288, 1970.
26. ZS Fodor, M Iring, F Tüdös, T Kelen. Determination of carbonyl containing functional groups in oxidized polyethylene. J Polym Sci Polym Chem Ed 22:2539–2550, 1984.
27. DJ Carlsson, R Brousseau, Can Zhang, DM Wiles. Identification of products from polyolefin oxidation by derivatization reactions. ACS Symp Ser 364:376–389, 1988.
28. WR Hasek, WC Smith, VA Engelhardt. The chemistry of sulfur tetrafluoride. II. The fluorination of organic carbonyl compounds. J Chem Soc 82:543–551, 1960.
29. WC Smith. The chemistry of sulfur tetrafluoride. Angew Chem Int Ed 1:467–475, 1962.

30. J March. Advanced Organic Chemistry: Reactions, Mechanisms and Structure, 2nd ed. Tokyo: McGraw-Hill Kogakusha, 1977.

31. C Wilhelm, JL Gardette. Infrared analysis of carboxylic acids formed in polymer photooxidation. J Appl Polym Sci 51:1411–1420, 1994.

32. P Delprat, X Duteurtre, JL Gardette. Photooxidation of unstabilized and HALS-stabilized polyphasic ethylene–propylene polymers. Polym Degrad Stabil 50:1–12, 1995.

33. A Casu, JL Gardette. Photolysis and photooxidation of poly(butylene terephthalate)–fiber glass systems. Polymer 36:4005–4009, 1995.

34. A Rivaton. Photochemistry of poly(butyleneterephthalate): 2-identification of the IR-absorbing photooxidation products. Polym Degrad Stabil 41:297–310, 1993.

35. MH Tabankia, JL Gardette. Photochemical degradation of polybutyleneterephthalate: Part 1–Photooxidation and photolysis at long wavelengths. Polym Degrad Stabil 14:351–365, 1986.

36. B Mailhot, JL Gardette. Polystyrene photooxidation. 1. Identification of the IR-absorbing photoproducts formed at short and long wavelengths. Macromolecules 25:4119–4126, 1992.

37. J Reid Shelton, RF Kopczewski. Nitric oxide induced free-radical reactions. J Org Chem 32:2908–2910, 1967.

38. P Tarte. Rotational isomerisation as a general property of alkyl nitrites. J Chem Phys 20:1570–1575, 1952.

39. DJ Carlsson, R Brousseau, Can Zhang, DM Wiles. Polyolefin oxidation: quantification of alcohol and hydroperoxide products by nitric oxide reactions. Polym Degrad Stabil 17:303–318, 1987.

40. J Lacoste, DJ Carlsson, S Falicki, DM Wiles. Polyethylene hydroperoxide decomposition products. Polym Degrad Stabil 34:309–323, 1991.

41. DJ Carlsson, S Chmela, DM Wiles. The oxidative degradation of ethylene/vinyl alcohol copolymers. Polym Degrad Stabil 31:255–267, 1991.

42. JL Gardette, J Lemaire. Advantages and limits of hydroperoxide titration methods in solid polymers. Polym Photochem 7:409–416, 1986.

43. MH Tabankia, JL Philippart, JL Gardette. Photooxidation of block copoly-(ether–ester) thermoplastic elastomers. Polym Degrad Stabil 12:348–362, 1985.

44. CD Wagner, RM Smith. ED Peters. Evaluation of the ferrous–titanous method. Anal Chem 19:982–984, 1947.

45. VR Kokatnur, M. Jelling. Iodometric determination of peroxygen in organic compounds. J Am Chem Soc 63:1432–1433, 1941.

46. RD Mair, AJ Graupner. Determination of organic peroxides by iodine liberations procedures. Anal Chem 36:194–204, 1964.

47. DK Banerjee, CC Budke. Spectrophotometric determination of organic peroxides in unsaturates. Anal Chem 36:2367–2380, 1964.

48. R Hiatt, WMJ Strachan. Effect of structure on the thermal stability of hydroperoxides. J Org Chem 28:1893–1894, 1963.

49. E Niki, C Decker, FR Mayo. Ageing and degradation of polyolefins. 1. Peroxide-initiated oxidation of atactic polypropylene. J Polym Sci Polym Chem Ed 11:2813–2845, 1973.

50. DJ Carlsson, J Lacoste. Hydroperoxides measurement in oxidized polyolefins. Polym Degrad Stabil 32:377–386, 1991.

51. JM Ginhac, JL Gardette, R Arnaud, J Lemaire. Influence of hydroperoxides on the photothermal oxidation of polyethylene. Makromol Chem 182:1017–1025, 1981.

52. JL Gardette, J Lemaire. Oxydation photothermique d'élastomères de polyuréthannes thermoplastiques. 1. Propriétés des hydroperoxydes formés. Makromol Chem 182:2723–2736, 1981.

53. J Lemaire, R Arnaud, JL Gardette. The role of hydroperoxides in photooxidation of polyolefins, polyamides and polyurethane elastomers. Pure Appl Chem 55:1603–1614, 1983.

54. G Zeppenfeld. Die angewendung der Eisenrhodanidmethode zur quantitativen bestimmung des Peroxidgehaltes in strahlenoxydiertem Polyvinylchlorid. Makromol Chem 90:169–176, 1966.

55. MU Amin, G Scott, LMK Tillikeratne. Mechanism of the photoinitiation process in polyethylene. Eur Polym J 11:85–89, 1975.

56. J Petruj, J Marchal. Mechanism of ketone formation in the thermooxidation and radiolytic oxidation of low density polyethylene. J Radiat Phys Chem 16:27–36, 1980.

57. Ly Tang, D Sallet, J Lemaire. Photochemistry of polyundecanamides. 1. Mechanism of photooxidation at short and long wavelengths. Macromolecules 15:1432–1437, 1982.

58. P Gauvin, JL Philippart, J Lemaire, D Sallet. Photooxidation de polyéther-block-amides. Makromol Chem 186:1167–1180, 1985.

59. JL Philippart, F Posada, JL Gardette. Photooxidation of cured fluorinated polymers. III. Quantitative study of the photooxidation by mass spectrometry. Polym Degrad Stabil 53:33–37, 1996.

60. JL Philippart, F Posada, JL Gardette. Mass spectroscopy analysis of volatile photoproducts in photooxidation of polypropylene. Polym Degrad Stabil 49:285–290, 1995.

61. N Grassie, NA Weir. The photooxidation of polymers. III. Photooxidation of polystyrene. J Appl Polym Sci 9:987–998, 1965.

62. JA Lawrence, NA Weir. Photodecomposition of polystyrene on long-wave ultraviolet irradiation: a possible mechanism of initiation of photooxidation. J Polym Sci Polym Chem Ed 11:105–118, 1973.

63. C Crouzet, J Marchal. Oxidative degradation of polystyrene. J Appl Polym Sci 35:151–160, 1979.

64. PC Lucas, RS Porter. Carbonyl functional groups in photo-oxidized polystyrene. Polym Degrad Stabil 13:287–295, 1985.

65. G Geuskens, G-L Vinh. Photooxidation of polymers. VII. A reinvestigation of the photooxidation of polystyrene based on a model compound study. Eur Polym J 18:307–311, 1982.

66. NA Weir, A Ceccarelli. Photodecomposition of polystyrene hydroperoxide. II. Reactions in the solid state. Polym Degrad Stabil 41:93–101, 1993.

67. AV Cunliffe, A Davis. Photooxidation of thick polymer samples. Part II: The influence of oxygen diffusion on the natural and artificial weathering of polyolefins. Polym Degrad Stabil 4:17–37, 1982.

68. M Rappon. Depth profiling of polymer photooxidation. Eur Polym J 22:319–322, 1986.

69. X Jouan, C Adam, D Fromageot, JL Gardette, J Lemaire. Microscopic determinations of photoproducts profiles in photooxidized matrices. Polym Degrad Stabil 25:247–265, 1989.

70. KT Gillen, RL Clough. Quantitative confirmation of simple theoretical models for diffusion. Limited oxidation. In: RL Clough, SW Shalaby, eds. Radiation Effects on Polymers. Washington DC: American Chemical Society, 1991, pp 457–472.

71. X Jouan, JL Gardette. Photooxidation of ABS at long wavelengths. J Polym Sci A Polym Chem 29:685–696, 1991.

72. NS Allen, SJ Palmer, GP Marshall, JL Gardette. Environmental oxidation processes in yellow gaz pipe: implications for electrowelding. Polym Degrad Stabil 56:265–274, 1997.

73. KT Gillen, RL Clough. Techniques for monitoring heterogeneous oxidation of polymers. In: NP Cheremisinoff, ed. Handbook of Polymer Science and Technology. New York: Marcel Dekker, 1989, pp 167–202.

74. JL Gardette, S Gaumet, JL Philippart. Influence of the experimental conditions on the photooxidation of poly(vinylchloride). J Appl Polym Sci 48:1885–1895, 1993.
75. T Ly, D Sallet, J Lemaire. Photochemistry of polyundecanamides. I. Mechanisms of photooxidation at long and short wavelengths. Macromolecules 15:1432–1437, 1982.
76. JL Gardette, JL Philippart. Perturbation of the photochemistry of poly(vinylchloride) by a conventional aromatic plasticizer. J Photochem Photobiol A43:221–231, 1988.
77. GE Schoolenberg, P Vink. Ultra-violet degradation of polypropylene. 1. Degradation profile and thickness of the embrittled surface layer. Polymer 32:432–437, 1991.
78. NS Allen, GP Marshall, C Vasiliou, LM Moore, JL Kotecha, JL Gardette. Oxidation processes in blue water pipe. Polym Degrad Stabil 20:315–324, 1988.
79. DJ Carlsson, DM Wiles. Photooxidation of polypropylene films. IV. Surface changes studied by attenuated total reflection spectroscopy. Macromolecules 4:174–179, 1970.
80. X Jouan, JL Gardette. Development of a micro (FTIR) spectrometric method for characterization of heterogeneities in polymer films. Polymer 28:329–331, 1987.
81. JL Gardette. Micro (FTIR) spectroscopic profiling of aged polymer films. Spectrosc Eur 5:28–32, 1993.
82. P Delprat, JL Gardette. Analysis of the photooxidation of polymer materials by photoacoustic Fourier transform infra-red spectroscopy. Polymer 34:933–937, 1993.
83. PA Dolby, R McIntyre. Depth profiling of multilayer polymer systems using FTIR phase analysis photoacoustic spectroscopy. Polymer 32:586–589, 1991.
84. MW Urban. Photoacoustic Fourier transform infrared spectroscopy: a new method for characterization of coatings. J Coat Technol 59:29–33, 1987.
85. RO Carter III, MC Paputa Peck, DR Bauer. The characterization of polymer surfaces by photoacoustic Fourier transform infrared spectroscopy. Polym Degrad Stabil 23:121–134, 1989.
86. B Jasse. Fourier-transform infrared photoacoustic spectroscopy of synthetic polymers. J Macromol Sci Chem A 26:43–67, 1989.
87. R Dittmar, RA Palmer, RO Carter III. Fourier transform photoacoustic spectroscopy of polymers. Appl Spectr Rev 29:171–231, 1994.
88. JL Gardette, J Lemaire. Prediction of the long-term outdoor weathering of poly(vinylchloride). J Vinyl Technol 15:113–117, 1993.
89. B Mailhot, JL Gardette. Polystyrene photooxidation. 2. A pseudo wavelengths effect. Macromolecules 25:4127–4133, 1992.
90. J Scheirs, JL Gardette. Photooxidation and photolysis of poly(ethylenenaphthalate). Polym Degrad Stabil 56:339–350, 1997.
91. R Arnaud, E Fanton, JL Gardette. Photochemical behaviour of semi-aromatic polyamides. Polym Degrad Stabil 45:361–369, 1994.
92. DJ Carlsson, DM Wiles. The photooxidative degradation of polypropylene. Part 1. Photooxidation and photoinitiation processes. J Macromol Sci Rev Macromol Chem C 14:65–106, 1976.
93. DL Carlsson, A Garton, DM Wiles. Initiation of polypropylene photooxidation. 2. Potential processes and their relevance to stability. Macromolecules 9:695–701, 1976.
94. K Tsuji, T Seiki. Observation of absorption spectra due to charge transfer complexes of polymers with oxygen and its possible contribution to radical formation in polymers by ultraviolet irradiation. Polym Lett 8:817–821, 1970.
95. JH Adams. Analysis of the non volatile oxidation products of polypropylene. 1. Thermal oxidation. J Polym Sci A-1 8:1077–1090, 1970.
96. DJ Carlsson, DM Wiles. The photodegradation of polypropylene films. III. Photolysis of polypropylene hydroperoxides. Macromolecules 2:597–606, 1969.
97. J Lacoste, D Vaillant, DJ Carlsson. Gamma-, Photo- and thermally-initiated oxidation of isotactic polypropylene. J Polym Sci A 31:715–722, 1993.

98. D Vaillant, J Lacoste, G Dauphin. The oxidation mechanism of polypropylene: contribution of ^{13}C-NMR spectroscopy. Polym Degrad Stabil 45:355–360, 1994.
99. RF Cozzens, WB Moniz, RB Fox. ESR study of the photolysis of polystyrene and poly(α-methylstyrene). J Chem Phys 48:581–585, 1968.
100. A Torikai, T Takenchi, K Fueki. Photodegradation of polystyrene and polystyrene containing benzophenones. Polym Photochem 3:307–320, 1983.
101. NA Weir. Photo and photooxidative reactions of polystyrene and of the ring substituted polystyrene. In: N Grassie, ed. Developments in Polymer Degradation. Vol 4. London: Applied Science, 1982, pp 143–184.
102. Z Khalil, S Michaille, J Lemaire. Photooxydation du polystyrène à grandes longueurs d'onde. Makromol Chem 188:1743–1756, 1987.
103. G Geuskens, D Baeyens-Volent, G Delaunois, Q Lu-Vinh, W Piret, C David. Photooxidation of polymers. 1. A quantitative study of the chemical reactions resulting from irradiation of polystyrene at 253.7 nm in the presence of oxygen. Eur Polym J 14:291–297, 1978.
104. JF Rabek, B Ranby. Studies on the photooxidation mechanism of polymers. 1. Photolysis and photooxidation of polystyrene. J Polym Sci Polym Chem Ed 12:273–294, 1974.
105. PC Lucas, RS Porter. Hydroperoxides from phenyl ring reactions of photooxidized polystyrene. Polym Degrad Stabil 22:175–184, 1988.
106. X Jouan, JL Gardette. Photo-oxidation of ABS: Part 2—origin of the photodiscoloration on irradiation at long wavelengths. Polym Degrad Stabil 36:91–96, 1992.
107. LA Wall, DW Brown. γ-Irradiation of poly(methyl methacrylate) and polystyrene. J Phys Chem 61:129–136, 1956.

19

Lifetime Prediction of Plastics

S. HALIM HAMID and IKRAM HUSSAIN
King Fahd University of Petroleum & Minerals, Dhahran, Saudi Arabia

I. INTRODUCTION

The deterioration of a polymeric material depends on how and to what extent it interacts with its surroundings. Outdoor uses of plastic products include buildings and construction, agriculture and horticulture, automobiles and airplanes, solar heating equipment, and packaging. It has been estimated that roughly half the annual tonnage of polymer is employed outdoors (1), where performance is often limited by weathering. Degradation of plastics during outdoor exposure is influenced to varying degrees by all natural meteorological phenomena. Heat, radiation (ultraviolet and infrared), rain, humidity, atmospheric contaminants, thermal cycling, and oxygen content of air, all contribute to the degradation of plastics subjected to outdoor exposure. None of these factors are constant in any one location, and weather conditions vary widely with location. The useful lifetime of plastics needs to be predicted for planning their maintenance and replacement. To attain maximum accuracy in predicting the useful lifetime of an outdoor exposed plastic, all components of the anticipated exposure environment must be considered. This is best accomplished by conducting exposure trials in that environment (2).

It is now generally recognized that stabilization against degradation is necessary if the useful lifetime of a polymer is to be extended sufficiently to meet design requirements for long-term applications. The stabilization of the polymer is still undergoing transition, from an art to a science, as the mechanism of degradation becomes more fully understood. A scientific approach to stabilization can be approached only when there is an understanding of the reactions that lead to degradation (3).

Weather-induced degradation involves the simultaneous action of sunlight, oxygen, temperature, and harmful atmospheric emissions; oxidative deterioration of thermoplastic polymers during processing involves the simultaneous action of heat, mechanical forces, and oxygen (4). The utilization of plastics in outdoor applications has grown rapidly with the overall development of industries, and new commercial and industrial requirements for more inexpensive and durable materials. One of the major problems faced by the plastic products used outdoors is weather-induced degradation. The extent of plastic degradation can be evaluated in terms of measuring certain properties considering 1) prolonged exposure to the harsh weather conditions, and 2) various representative locations in a geographic location.

The ability to predict critical properties of weather-induced degraded plastics is of great importance and usefulness. It helps in determining the useful lifetime (durability) of the plastic products for better planning in maintenance and replacement. Prediction of the lifetime of polymers, or the time to failure of specified properties during long-term service, is of great commercial importance. Thus, the highest permissible service temperature or the recommended time of service under specified conditions can be estimated to assess the scope of application of polymeric materials, in particular, engineering thermoplastics (5,6).

This chapter describes the main factors affecting weathering degradation, advances being made in the lifetime prediction methods for outdoor exposed plastics, and some case studies describing the lifetime prediction of plastics weathered in a near equatorial region.

II. ROLE OF WEATHER PARAMETERS IN WEATHERING

The primary weather factors responsible for degradation reactions in outdoor exposures of polymers are solar radiation, atmospheric oxygen, ambient temperature, humidity, wind, air pollutants, and contaminants. These factors act synergistically; therefore, the combined effect is different from those obtained by exposure to individual components. The following discussion is devoted to the important weather parameters mainly responsible for degradation reactions. This will help in understanding the degradation phenomena and their effect on lifetime determination of polymers.

A. Solar Radiation

Most polymeric materials lose their properties mainly owing to photooxidative attack. This implies a combined action of oxygen and sunlight on their chemical structure. The three aspects of sunlight—sunshine duration, total solar radiation, and ultraviolet (UV) radiation—can be used to quantify sunlight as an agent of deterioration. Many commercial polymers are degraded outdoors, primarily because of photodegradation, and efforts have been made to relate their resistance to the number of sunshine hours they can withstand before they lose their useful properties. The search for a quantitative relation between sunshine hours and global solar radiation has been a subject of considerable interest (7,8).

Total solar radiation reaching earth's surface is only about one-half of two-thirds its value at the outer limits of the atmosphere. The most deleterious of the environmental factors in weathering phenomena is the UV section of radiation, which is responsible for the weather-induced degradation of plastics (9,10). The sun sends a continuous spectrum of energy radiation to the earth, whereby various types of radiation are usually differentiated on the basis of the wavelengths. The physical nature of these radiations is the same, and they vary only in wavelength and, therefore, their photon energy. The wavelength has an inverse relation with quantum energy (11). The extraterrestrial radiation of the sun is largely constant in its spectral composition. Its intensity depends on the angle of incidence. The changes as a result of the tilt of the earth's axis and the rotation of the earth in a daily and annual cycle are linked with the geographic latitude of the place of measurement (12).

The atmosphere modifies the solar spectrum unevenly over the spectral range. The effect is more pronounced in the infrared (IR) and UV regions than it is in the visible range. The radiation wavelength ranges reaching the earth surface are from about 290 to 1400 nm (13–15). The portion of the sun's spectrum between 290 and 400 nm is important to polymers because it includes the highest-energy region and is the only portion that can cause direct harm to unmodified polymers (16). The visible region of the solar spectrum, 400–800 nm, does not usually cause direct harm to polymers, but can do so by interaction with sensitizing substances in the polymers. The IR portion, less than 800 nm, is generally considered harmless in a photochemical sense, but it may have a role in the thermal oxidative degradation of some polymers.

The UV portion of the radiation is mainly responsible for the degradation of polymers; the energy at visible and high wavelengths is too low to damage the chemical structure of a polymer. Because of its chemical structure every polymer is susceptible to photochemical degradation at a particular wavelength. The absorption of UV radiation and its concomitant degradative effects vary for each individual polymer. Stability is strongly dependent on the specific chemical and molecular structure of the polymer, variations in structure lead to differences in the absorption ranges of individual polymers. In theory, pure polyolefins, such as polyethylene, do not contain functional groups that would be capable of absorbing UV radiation. However, polyolefins are known to absorb, and be degraded by, UV radiation. Sources of UV-absorbing chromophores are

> Polymer structure
> Feedstock or process impurities
> Residual catalyst
> Thermal-processing degradation products
> Antioxidants and transformation products
> Colorants
> Fillers
> Other additives (flame retardant, or other)

Ultraviolet energy is available from sunlight to break many of the chemical bonds in organic compounds. Because most polymers should not absorb at wavelength greater than 300 nm, it has been assumed that the aforementioned sources of UV-absorbing chromophores present in polymers are responsible for weathering (17). It has been discussed that it is UVB radiation with a shorter wavelength that is mainly responsible for photodamaging, discoloration, and loss of mechanical properties. The severity of the effect of UVB levels on the lifetime of polymer depends on both the geographic location of exposure and the susceptibility of the particular material to UVB radiation (18,19).

B. Atmospheric Oxygen

Residual double bonds in some molecules, such as polyethylene are especially susceptible to attack by atmospheric oxygen, although most polymers react very slowly with oxygen. However, oxidation is greatly promoted by elevated temperatures and UV radiation, and the reactions of polymers with oxygen under these conditions can be very complex (20). Oxygen is not usually considered as an experimental variable in the study of polymer weatherability because the oxygen concentration in the environment is substantially constant. Moreover, most weathering phenomena occur at the surface of the plastic, which is in equilibrium with air (21).

C. Ambient Temperature

Under extreme outdoor exposure conditions, a plastic sample may reach about 77°C. The temperature of an object in sunlight is usually significantly higher than that of the surrounding air. The difference between the air and surface temperature depends on factors such as radiation intensity, wind speed, object shape, and the nature of the material, in particular its surface finish, color, heat capacity, and thermal conductivity (22). The weathering process depends on temperature along with other factors, and the maximum temperature attained by the polymers is not sufficient to promote bond cleavage of any structures likely to be found in commercial plastics. The role of heat in the outdoor degradation of plastics is in accelerating processes otherwise induced, such as hydrolysis, secondary photochemical reactions, or the oxidation of trace contaminants (23). It is a well-experienced fact that deterioration rates of polymers in the tropical zones far exceed those in temperature zones. Simulated weathering experiments have shown that the oxidation rates of polyethylene exposed to 300-nm radiation increase fourfold from 10° to 50°C (24).

D. Humidity

The significant role of water in weather-induced degradation of plastics lies in the combination of its unique physical properties with its chemical reactivity (25). Water can have at least three kinds of effects that are important for the degradation of polymers. One is chemical, hydrolysis of labile bonds, such as those of polyesters or polyamides; a second is physical, destroying the bond between a polymer and a filler, such as glass fiber or pigment and resulting in chalking or fiber bloom. Rain can wash away water-soluble degradation products from the exposed surface, and moisture swells, softens, and plasticizes selected plastics. A third is photochemical, involving the generation of hydroxyl radicals or other reactive species that can promote a host of free radical reactions (26).

Accurate climatological data are necessary to determine the causes of degradation and for the development of new stabilized plastic products with longer lifetimes. The standard instruments for measuring the important weather parameters should be located in the vicinity of the weathering trials test site to have realistic profile of meteorological data needed for correlation and modeling.

III. LIFETIME PREDICTION METHODS USED FOR EXPOSED PLASTICS

Polymers exposed in an outdoor environment need to be assessed for their durability or lifetime. Prediction of the lifetime of polymer materials in outdoor conditions is still a difficult problem. Researchers have tried to predict the lifetime of polymers aged outside by accelerated weathering tests with two types of methods. These methods include simulation and mechanistic approaches. The simulation approach is normally based on the relevance of some observed phenomenon that is deduced from the physical and chemical changes under artificially accelerated and natural exposures. However, in the mechanistic approach, the observed phenomenon is controlled at a molecular level in the polymer matrix. In a mechanistic approach the chemical nature and spatial distribution of intermediate and final groups formed in polymer chain or branches in the exposed samples are determined using spectrometric or microspectrometric techniques. In this approach when a common mechanism has been observed between kinetics and artificial aging kinetics, it leads to determining the required acceleration factor.

A. Recent Advances in Lifetime Prediction Methods

A reliable lifetime prediction method is of importance in terms of assuring the durability of stabilized plastics products in use today and to develop new plastic formulations for expanding use in day-to-day life and engineering applications. Currently, progress has been made in the understanding of degradation and stabilization mechanisms, and also, some advances have been made in lifetime prediction methods. A brief review of the important methods used for durability assessment in lifetime prediction of polymers is given to promote an understanding of the field. Prediction of lifetimes of weather-induced degradation of stabilized polyethylene using artificially accelerated weathering and natural exposures in a tropical climate is discussed in detail. Some case studies of stabilized polyethylene (PE) films and polyvinyl chloride (PVC) pipes exposed in natural environment of Dhahran, Saudi Arabia are presented. Tidjani (31) in his Fourier transform infrared (FTIR)-based studies showed that the lack of correlation may be due to the differences in degradation mechanisms in accelerated and natural outdoor testing.

Outdoor aging is too slow to be useful in the development of stabilized formulations or for quality control of plastic products. This has led to the development of accelerated weathering tests. Unfortunately, not all accelerated weathering devices accurately predict the relative durability of specific stabilized plastic products. Often it is desirable to perform accelerated weathering tests with the highest level of correlation and acceleration possible.

The commonly used accelerated weathering devices are weatherometer, Xenotester, Suntester, QUV, UVCON, SEPAP, and EMMAQUA. It has been reported that most of the accelerated weathering devices show poor correlation between the stability measured with the devices and those of outdoors (27–29).

However, it can be said that in selecting the accelerated weathering method two factors are of utmost importance. These are *correlation* and *acceleration* of the accelerated test results with those of real-time outdoor test results. Acceleration is a factor that indicates how rapidly the weathering test can be conducted in a weathering device compared with outdoor weathering.

One of the studies shows that the acceleration factor for photooxidation of polyethylene depends on the degradation parameters measured. For oxygen uptake, the acceleration factor was 2.5, and for carbonyl and end-unsaturation formation rates the acceleration factors were larger (30,31). The differences between accelerated and outdoor weathering may be due to a change in mechanisms, leading to a decrease of temperature or an increase of the oxygen pressure during accelerated weathering. These factors may lead to initiation of radicals through a charge-transfer complex (CTC) and, therefore, a better correlation with outdoor weathering. In accelerated weathering most of the oxygen is consumed through propagation reactions whereas in outdoor weathering most of the oxygen is consumed by an air initiation reaction by a CTC of oxygen. Basically, the environmental factors, such as light intensity, spectral distribution, and temperature, have an influence on the degradation rates during exposures of polymers. These factors may lead to differences in oxygen uptake, changes in IR spectra, and loss of mechanical properties.

It is possible to achieve a good relation between an accelerated test and an outdoor weathering trial if the degradation-determining factors are controlled and accelerated in the same way to have same reaction prevailing throughout the testing period.

For PE photodegradation, the basic mechanisms are very well understood. However, there is a discussion going on about the relative importance of the different reactions in the PE matrix. In PE the radical initiating capabilities of added ketones and hydroperoxide formed by thermal oxidation are low (32,33). The suggestion was made that hydroperoxides in PE do not initiate photooxidation, because they decompose without forming radicals. But Gugumus (34,35) proposed an alternative mechanism of radical formation. He proposed that the hydroperoxide is the main source of new radical formation. This is formed through CTC of the polymer with oxygen. Fractionation of the CTC leads to the formation of a *trans*-vinylene group and hydrogen peroxide. Hydrogen peroxide can initiate oxidation by thermal or photochemical decomposition. Other CTCs are also possible, which may lead to unsaturation, or cross-links, instead of *trans*-vinylene groups (36).

The conversion of oxygen into other products depends on the kind of exposure. A higher oxygen uptake is necessary to bring about the same drop of the elongation at break and an increase of carbonyl and end-saturation absorption in the IR spectra for outdoor versus accelerated weathering.

It was suggested by Gijsman et al. (37) that in accelerated weathering most of the oxygen is consumed through propagation reaction and gives all expected products. In outdoor weathering, most of the oxygen is consumed by an initiation reaction caused by a CTC of oxygen. The higher conversion of oxygen by a CTC during outdoor weathering may be due to higher stability of these complexes at low temperatures. Therefore, a decrease of temperature or an increase in oxygen pressure during accelerated weathering will result in more initiation of radical formation through a CTC and, as a result, a better correlation with outdoor weathering is possible.

The lifetime prediction of elastomers exposed to accelerated thermal conditions was studied by Gillen et al. (38). For elastomers aged in the air environment, the long-term degradation usually involves oxygen (39). Attempts have been made to accelerate these reactions by exposing the materials to elevated temperatures, but a diffusion-limited oxidation occurs (40–42). Accelerated thermal aging studies are extrapolated to use temperature conditions by using Arrhenius lifetime prediction methods. Therefore, according to the Arrhenius method, the temperature dependence of the rate of an individual chemical reaction is proportional to $\exp(-E_a/RT)$, where E_a is the activation energy, R is the ideal gas constant, and T is the absolute temperature. In general, polymers can be described by a series of chemical reactions, each assumed to have Arrhenius behavior. Kinetic analysis of these reactions results in a steady-state rate expression, where E_a represents the effective activation energy for the number of reactions in the degradation. If we assume that the number of reactions remains unchanged throughout the temperature range, a linear relationship will exist between the logarithm of the time to a certain amount of property change and $1/T$. The value of E_a can be obtained from the slope of the line. If the relative number of degradation reactions change with change in T, the effective E_a would be changed, as this change will result in a curved Arrehenius plot.

There are, however, certain limitations to the Arrhenius method. For example, it is valid for ultimate tensile elongation, but does apply to ultimate tensile strength. Another problem with the Arrhenius approach is that the assumption that E_a derived under the accelerated conditions remains constant at lower temperatures and that these E_a values can be used to extrapolate the accelerated results to low-temperatures use conditions and thereby allow the predictions for longer duration. To confirm this assumption, several years of exposures at low temperatures will be needed until the elastomer will fail; thus, it becomes practically impossible. An alternative approach to monitor oxygen consumption rates at various temperatures was adopted that verifies that the same E_a is suitable for use in the low-temperature extrapolation region. This method may be used for exposed samples having oxidation processes predominate the reaction mechanisms in the degradation (39). Elastomeric polymers are usually used at close to room temperature and are meant for long-term usage. Therefore, lifetime assessment is of importance for such applications. A common approach involves accelerating the chemical reaction underlying the degradation by aging at several elevated temperatures and monitoring the degradation through change in ultimate tensile strength and elongation at break. The accelerated high-temperature results are extrapolated to use temperature based on the Arrhenius method (43).

PVC is a very important commodity polymer. Trabilco has tried to obtain lifetime estimates (44). The PVC cladding was exposed in natural weather of Wellington, New Zealand. Samples were also exposed in an artificially accelerated weathering device. Changes in impact strength, ductile–brittle transition temperature (DBTT), yellowness index, chalking, and tensile strength were monitored for the exposed cladding samples. The DBTT shows a definite trend, as it has larger increases for both the naturally and artificially exposed samples.

Plasticized and nonplasticized PVC films are used as coatings. Exposures of PVC formulations lead to degradation by thermal (45–47) or photochemical (48–50) processes, both of which can be oxidative or nonoxidative. In each of the foregoing cases dehydrochlorination occurs, and propagation occurs in a stepwise (zip-elimination) manner. The dehydrochlorination leads to formation of long, conjugated polymer sequences (–CHCH–)n in the backbone, which gives a red-brown color to PVC.

Hollande and Laurent have studied color changes in plasticized PVC films by exposing these in an accelerated weathering device at 50°C (51). UV spectroscopy was used to follow degradation by measuring optical density (OD) for absorption bands between 250 and 500 nm. These bands are characteristic of polyene sequences. The plasticized PVC films developed a darker color because the plasticizer also takes part in the discoloring of PVC. The OD increases with aging time for plasticized PVC formulations.

Ethylene–propylene rubber (EPR) is used for electrical insulation purpose. A new approach was used by Guegen et al. (52) to predict the lifetime of EPR. It consists of building a graph (temperature–dose rate) in which four domains are distinguished. The concept of irradiation temperature synergism in the radio-chemical aging is used in this study. This method needs both thermal and radiochemical aging experiments. It is then possible to define three (temperature–dose rate) boundaries separating four domains, each one representing a given life-time prediction model. These tentative boundaries for EPR rubber have synergistic interactions between thermolytic and radiolytic initiation of oxidative degradation process. For natural weathering conditions there are no kinetic data available, therefore, a lifetime prediction has not been determined.

Substituted polyacetylenes are used as membranes for oxygen-enrichment process. These membranes encounter air at high temperatures. As a result, they lose their mechanical properties owing to oxidation of the polymer. Gonzalez et al., who used a kinetic model for the thermooxidative degradation process (53), have done lifetime predictions of poly(2-hexene) films exposed to high temperatures in an oven at 140°C. The weight loss obtained through thermogravimetry has been kinetically modeled. This model represents two opposing effects, including an increasing weight process caused by polymer oxidation, and a decreasing weight process caused by volatilization of the oxidation products. This mechanism was proposed earlier (54). Once the weight loss is quantified, an approximate equation for the lifetime of polymers can be proposed as a function of temperature. The weight loss varies linearly with chain scission, which indicates that weight loss corresponds to a value of molecular weight (MW) at any time and temperature. Consequently, a critical MW corresponds to a critical weight loss value. For this case critical weight loss of 2% has been assumed. Given the critical weight loss, a lifetime can be predicted at any temperature using the proposed equation as a function of temperature.

Polypropylene (PP) is a semicrystalline polymer that is used in films, tapes, and injection-molded articles for a variety of commercial and industrial applications. The oxidative degradation of PP is a diffusion-controlled process. The extent of degradation will decrease with the increase in the degree of crystallinity and mol-ecular orientation, following the effects of these parameters on oxygen diffusion (55,56). However, much of the oxidation occurs through radical chain reaction involving low-MW radicals produced by photolysis or thermal cleavage. The crystallinity and the molecular orientation also determine the mobility of radicals (57) and, therefore, control the rate of termination through recombination or dis-proportionation. Photooxidative degradation of isotactic PP molded samples was conducted by exposing them in an accelerating UV device (58). The extent of degradation was monitored by gel permeation chromatograph (GPC), FTIR, and scanning electron-microscopy (SEM). They suggested that crystallinity was the main factor for controlling the kinetics of photodegradation. The trend observed for the effect of structure on the extent of chemical degradation, however, is not reflected in the mechanical behavior. The results of melting temperature of the recrystallized modules were consistent with those by GPC and FTIR in the analysis-of-depth profile. This indicates that they can be used as a guide to the extent of photodegradation of PP.

In another study on stabilized PP films (59), polypropylene films were stabilized with low and high molecular weight hindered amine light stabilizers (HALS) and were exposed to moderately accelerated conditions in SEPAP 12.24 unit, ultra-accelerated photoaging in SEPAP 50.24, and natural aging. The FTIR spectroscopy was performed on the aged films and the maximum of the carbonyl absorption appeared at 1735 and 1713 cm^{-1} (representative of acid groups) and absorption at 3077, 1640, and 910 cm^{-1} also appeared. These absorptions may be assigned to vinyl group buildup that indicates photooxidation of ethylene segments of the ethylene–propylene copolymer (EPR). Evolution of optical density relative to irradiation time for moderate and ultra-accelerated conditions has been obtained. Similar variations were also observed for stabilized films exposed in outdoor weathering.

Photodegradable PE films, stabilized with HALS, were also exposed in moderate accelerated conditions, ultra-accelerated photoaging, and natural aging (59). The conclusions of this study include the following: 1) evolution mechanism throughout weathering and accelerated aging allows control of relevancy of phenomenon, 2) simulation techniques showed irrelevant phenomena in accelerated conditions, 3) the aging matrix should not be considered as a homogeneous (photo) reaction at a molecular scale, and 4) the data collected at the initial time in nonaccelerated conditions be extrapolated to the real lifetime of polymers using conventional kinetics. Therefore, experimental acceleration appears as a necessity in such studies. It was pointed out that deviations are generally due to either bimolecular reactions between intermediate species, the concentrations of which in real-time conditions are too low to interact, or to diffusional processes of reactants that are slower under accelerated conditions.

Another approach for determining lifetime is proposed using degradation profiles of thick high-density polyethylene (HDPE) samples exposed in an artifically accelerated apparatus and in outdoor weathering condition (60). The main objective of this study was to find the critical degradation profiles that result in the failure of the HDPE samples stabilized with a phenolic, antioxidant-type process stabilizer. Samples were aged in Xenotester 1200, ATLAS weatherometer, and the outdoor weather of Florida and Delft, The Netherlands. Degradation profiles were obtained by microtoming the surface and measuring the mechanical strength using microfil tensile testing system (MTTS), carbonyl and vinyl index applying FTIR spectrometry at 1712 cm^{-1} and 909 cm^{-1}, respectively, and density measurements. The results showed that exposed samples have limited degradation depth, which is the maximum depth at which the degradation profile differs from the profile of an unexposed sample. However, sample exposed at Delft show an increase in degradation depth. The degradation depth is important, as it may determine the lifetime to failure (61,62). When different exposure conditions are compared, the only variations seems to be the depth at which changes took place and the slope at which the degree of degradation decreased from the surface toward the middle. Generally, the degree of degradation increases, whereas the slope decreases while going from Xenotester to weatherometer, to Florida outdoor exposure.

The need to accelerate the weathering conditions has led to the development of several weathering devices (Table 1). These include the weatherometer, Xenotester, Suntester, QUV, UVCON, and SEPAP. Gijsman et al. (63) discusses comparison of UV degradation of PE in accelerated testing and sunlight. It has been reported that the stability measured with these devices shows poor correlation with those measured in outdoor weather (27,64,65). This lack of correlation may be attributable to the different degradation mechanism prevailing in accelerated and outdoor testing (66). The study shows the following main conclusions: 1) a comparison of the rate of degradation in accelerated and summer outdoor weather showed that the acceleration factor depends on the degradation parameter measured; 2) the acceleration factor for oxygen uptake was 2.5, and was larger for carbonyl and end-unsaturation formation rates; 3) the differences in accelerated and outdoor weathering are probably due to a change in the mechanism leading to oxygen uptake. They suggested that, in accelerated weathering, most of the oxygen is consumed through the propagation reactions, whereas in outdoor exposures, most of the oxygen is consumed by an initiation reaction caused by a CTC of oxygen and the polymer. A decrease of temperature or an increase in oxygen pressure during accelerated weathering will lead to more initiation through a CTC and thus a better correlation with outdoor weathering.

Lifetime prediction of HALS-stabilized LDPE and PP was attempted (67). Two models, based on physical parameters, were used for estimation of HALS stabilization efficiency in polyolefins. That these models are empirical relationed, has been suggested (68), and theoretical model for additive loss proposed (69,70). Moison's model (68) has been tested for HALS-stabilized LDPE and PP with considerations that an efficient light stabilizer should be highly soluble in the polymer, whereas its diffusion rate has to be minimal. Earlier attempts to use this model failed because the additive was lost by dissolution in a contacting liquid and not by evaporation from the surface into air (71). But Sampers (72) considered the rate of loss of UV-absorber from LDPE and showed a very good fit of the experimental data and gave good predicted values from the model. Defining a critical stabilizer concentration has successfully used in the other model for additive loss (69,70). The critical concentration of HALS is different for PP and LDPE. This difference could be related to different photooxidation initiation mechanisms of HALS-stabilized

Table 1 Accelerated Weathering Devices Used for Plastics Evaluation

Method	Type
ATLAS CXWA carbon arc weatherometer	Carbon arc
Q-Panel QUV (340 or 313A bulb)	Fluorescent result
ATLAS SUNTEST CPS benchtop xenon	Xenon arc
ATLAS Ci 65 xenon arc weatherometer	Xenon arc
Suntester, Hanau	Xenon lamp
SEPAP 12.24 University of Clermont-Ferrand	Moderately accelerating
SEPAP 50.24, University of Clermont-Ferrand	Ultra-accelerating
EMMAQUA, DSET, Fresnel reflector	Solar concentrator

polyethylene and polypropylenes as proposed by Gugumus (73). The calculated lifetime predictions from the model agreed well with the data obtained from different accelerated weathering machines. It was also pointed out that the physical parameters could affect the shape of dependence of HALS stabilization performance on its initial concentration in polymers.

Cross-linked polyethylene is generally used for the manufacture of cable insulation and pipes for hot-water transport. The thermal oxidation of β-ray–cross-linked polyethylene was studied in the temperature range of 90–180°C (74). The duration of induction period (DIP) was obtained using phenol depletion, ultimate elongation, density, carbonyl development, color change, and weight loss determinations. The Arrhenius law was applied to different DIP values. Previously the induction period was linked to the residual stabilizer concentration (75–77). Phenol depletion is the only parameter, that allowed defining a critical concentration. This method gives a lowest lifetime value, which gives a safety margin in the prediction.

IV. LIFETIME PREDICTION OF POLYMER IN NEAR EQUATORIAL REGION

A. Mathematical Modeling of Weather-Induced Degradation

The weathering of plastics is dependent on almost all parameters of the environment (78–81). The weather is variable from time to time and from place to place; hence, the outdoor trial results obtained in different seasons, years, or locations have been inadequate to fully ascertain the degradation pattern. Hamid and Prichard (82) conducted an extensive study using a mathematical-modeling technique, that describes the weather-induced degradation of linear low-density PE (LLDPE) material. The main objectives of this study are as follows: to develop mathematical models that can describe the weathering trial data to predict the lifetime and help in understanding the complex weathering degradation phenomenon occuring in the material as a result of natural weathering (83). LLDPE film products were exposed to natural outdoor weather of Dhahran, Saudi Arabia for a period of 1 year. Exposures were carried out on wooden racks fabricated according to the ASTM D 1435. The racks were adjusted to expose the plastic samples facing south at an angle of 45° from the horizontal (84). Samples were withdrawn on a monthly basis.

The degradation was monitored in terms of changes in mechanical properties, carbonyl groups, and percentage crystallinity of the exposed LLDPE films. The weathering site of Dhahran (26.32°N, 50.13°E) is situated just north of the Tropic of Cancer on the eastern coastal plain of Saudi Arabia. The environment is very much desert-like. The mean monthly average temperature during July and August is about 37°C, with a maximum temperature reaching about 50°C. During the cool season, the mean monthly temperature is about 20°C lower than the modest months. The maximum monthly mean solar radiation is near 480 langleys. Annual precipitation totals are low, typically close to 80 mm.

Three different models were developed to represent the independent variables of 1) tensile strength (TS); 2) carbonyl growth (CA); and 3) percentage crystallinity (CY) in terms of the following dependent variables: average monthly UV radiation

(UV), cumulative monthly UV radiation, cumulative total solar radiation (CR), average total solar radiation (RD), and average monthly temperature (AT).

A step-wise regression method was used to eliminate the insignificant variables (85,86) because the potential outliers of the meteorological parameters can misrepresent total degradation behavior (87,88). The computational analysis has been carried out using a statistical analysis system (SAS) software package.

Model I $TS = 220.52 - 0.58\,AT - 0.84\,CU - 2.12\,UV$

Model II $CA = 0.22 - 0.125\,CU + 0.144\,UV - 0.004\,RD + 0.005\,CR$

Model III $CY = 43.31 - 0.29\,AT + 0.08\,CU + 0.00008\,CR$

The mechanical property was more dependent on the UV portion of the total solar radiation, carbonyl development was synergistically affected by UV and total solar radiation, and the percentage crystallinity was affected by UV, total solar radiation, and temperature. Humidity and other weather parameters played a less significant role in the weather-induced degradation of LLDPE properties.

B. Lifetime Prediction from Accelerated and Natural Weathering Trials

Accelerated weathering of plastics can be achieved by continuous exposure to light, elevated temperatures, and humidity. The intensity of radiation by exposure to high-energy wavelengths can accelerate the degradation process (89). The data obtained from artificially accelerated weathering trial can be correlated with natural weathering trials using mathematical models and correlations (90). There are inconsistencies in the reproducibility of weathering trial results for the material exposed at different times in a certain location. Correlation of such data with those of artificially accelerated weathering trials does not prove to be accurate. One of the major reasons is the consideration given to the geographic climatological data in the test criteria adopted in setting the artifically accelerated weathering apparatus and also the variation in natural weather conditions with time. Furthermore, the variability of climatic conditions between various locations and at different times in the same location makes it difficult to extrapolate natural weathering results from one location or time to another (91).

A weathering study was carried out (92) on HALS-stabilized LDPE film samples exposed at five different locations in Saudi Arabia. The details of the meteorological data for these sites are given in Table 2.

The main objective of this study was to develop correlation between natural and artificial weathering, which can predict lifetime in a short time span for near-equatorial region. Natural weathering trials on five exposure sites were carried out on aluminum racks designed to the ASTM D 1435 (93), for a duration of 2 years. Artificially accelerated weathering trials were carried out in ATLAS Ci 65 xenon arc weatherometer for 5000 h according to the ASTM D 2565 (94).

The weatherometer was set on automatic irradiance control mode, with an irradiance level of 0.35 W/m^2 at 340 nm. The degradation rate was monitored by measuring carbonyl absorbance in the region 1700–1750 cm^{-1}. The growth in carbonyl absorbance for the samples exposed at five different sites is presented in Fig. 1. During another exposure trial, chain scission and cross-linking took place simultaneously (95). The rate of cross-linking was higher at the initial stages than

Table 2 Meteorological Data of Exposure Sites

Exposure site	Zone	Latitude	Max. temp. in last 10 y (°C)	Max. monthly mean solar radiation (langleys)	Max. monthly relative humidity (%)
Dhahran	Coastal eastern	26.32°	49.5	480	90
Riyadh	Central	25.51°	48.0	575	77
Jeddah	Coastal western	21.29°	48.0	520	86
Tabuk	Northern	28.23°	44.9	475	68
Baha	Southern	18.13°	38.6	555	60

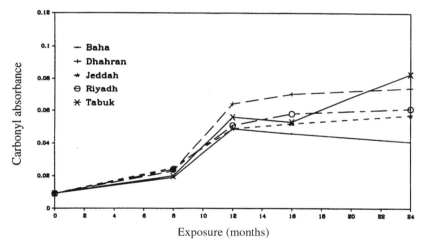

Figure 1 Changes in carbonyl absorbance of polyethylene samples as a function of exposure time.

the chain scission. However, chain scission dominated during the later stage (96). There was lower carbonyl growth during the initial 8 months of exposures. After this period there was a step change in carbonyl growth. The slower initial carbonyl development may be due to the active role of stabilizer played by HALS during early stages of natural weathering.

Thermal analysis was carried out using DSC. The results are presented in Figs. 2 and 3 showing the changes in percentage crystallinity and crystalline-melt temperature (T_m), respectively. A consistent behavior is exhibited in terms of T_m for samples exposed at different location. However, an increasing trend in percentage crystallinity can be observed for all locations. Crystallinity increase has also been reported (97), and it has been indicated that imperfect crystalline region of LDPE are believed to degrade because of cross-linking, whereas chain scission predominates in the amorphous matrix (98). This leads to secondary crystallization in an amorphous phase, which is later inhibited by the decreasing mobility of chains owing to branching and cross-linking.

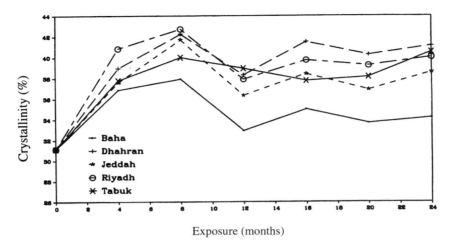

Figure 2   Changes in percentage crystallinity of PE samples as a function of exposure time.

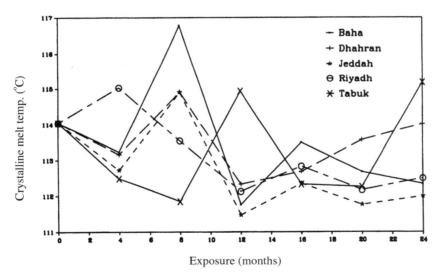

Figure 3   Changes in crystalline melting temperature of PE samples as a function of exposure time.

The increase in crystallinity is generally attributed to the degradation of ultra strong chains, which is caused by chain scission owing to radiation-induced reactions (99). The changes in molecular weight were determined by GPC analysis. A drop of 30% of the initial value of peak molecular weight (M_p) was observed for samples exposed for 2 years. This drop in M_p was a direct consequence of chain scission reactions that took place.

Mechanical properties, the most important performance characteristics, were determined and the changes in percentage elongation at break are presented in Fig.

4. A consistent initial behavior and then a continuous drop in property were observed. The consistent behavior may be due to the dominance of the cross-linking reactions over the chain scission during the early stages (100).

The changes in mechanical properties were also monitored for samples exposed in an artificially accelerated weathering device, as shown in Fig. 5. A perfect correlation between natural and artificially accelerated weathering is not practically possible because not all the possible weathering parameters can be considered in

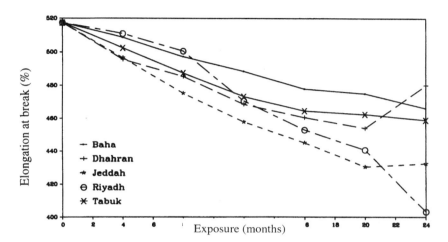

Figure 4   Changes in elongation at break of polyethylene as a function of exposure time.

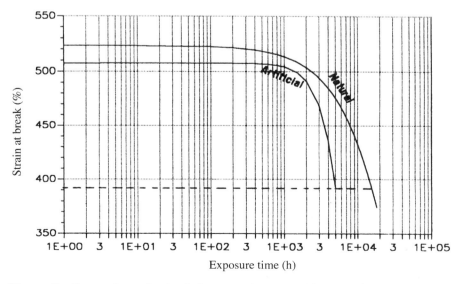

Figure 5   Comparison of polyethylene samples exposed in natural and artificially accelerated environment.

an experiment design. Moreover, different materials respond differently to the accelerating parameters of the weathering environment (101).

The following models were tried to correlate the data:

1. Straight line $Y = a + bx$
2. Parabolic $Y = a + bx + cx^2$
3. Exponential $Y = ax^b$
4. Cubic $Y = a + bx + cx^2 + dx^3$

The criteria for selecting a model were the value of r^2 (coefficient of determination). The best model is that in which has r^2 approaching 1 (102). SAS was used for all computational work.

The result showed that following parabolic model had the maximum r^2 value.

$$YN = 523.241 - 10075 \times 10^{-6} XN + 1/0 \times 10^{-7} XN^2$$

$$r^2 = 0.994$$

$$YA = 507.258 + 19956 \times 10^{-7} XA - 5.0 \times 10^{-6} \times A^2$$

$$r^2 = 0.878$$

where

YN = percentage elongation at break values for natural weathering
YA = percentage elongation at break values for artificial weathering
XN = natural exposure time
YA = artificial exposure time

Figure 5 shows a change in mechanical properties for both natural and artificial exposure trials. The *acceleration factor* determined for the artificially accelerated weathering is about three times that of natural weathering trials (5,000 h of artificial weathering = 14,000 h of natural weathering). From this study it is shown that drop in mechanical properties can be considered as a quick means to predict a polymer's lifetime from artificially accelerated and natural weathering trials.

C.   Lifetime Prediction of Greenhouse Film Exposed to Outdoor Weather

The most deleterious of the environmental factors in the weathering phenomenon is the UV portion of radiation, which is responsible for the weather-induced degradation of plastics (103,104). Early failures of greenhouse films in tropical climates cause heavy losses in terms of production, plastic materials, and operational interruption; maintenance replacements costs are also incurred for repairing new greenhouse films.

A 200-μm–thick greenhouse film based on LDPE and stabilized with HALS was studied for its performance in actual greenhouse environment (105). The objective of this study was to correlate the natural weathering trial data with the weather conditions and exposure duration to ascertain the lifetime of LDPE films exposed in a desert-like climate. Greenhouse films were exposed for 3 years in actual greenhouses at Dhahran. The variation in total solar radiation with time at exposure site of Dhahran is shown in Fig. 6. Samples were withdrawn on a bimonthly basis.

Characterization of the weathering film samples was performed using the critical performance properties of tensile strength and percentage elongation at break.

The durability of HALS- and Ni quencher-stabilized LDPE greenhouse films exposed on aluminum racks and also on actual greenhouses has been compared (106). This study showed that films mounted in the actual greenhouse degraded more in comparison with films mounted on the aluminum racks. The spectral emissions with wavelengths between 7 and 14 μm are important part of energy losses from soil and the plants inside a greenhouse. The partial prevention of this dissipation of the thermal energy from greenhouse film during cool night hours influences the degradation reactions in plastic films. Pesticides, based on sulfur and halogen compounds, can accelerate the degradation reactions. The chemical changes in terms of carbonyl absorbance are presented in Fig. 7. The higher values of carbonyl absorbance for exposed film on greenhouse versus exposure racks clearly indicate the severity of photooxidation reactions. The mechanical property results show good correlation with FTIR results. Figure 8 shows the stress-at-break values for the samples exposed on racks and greenhouses. The film usually breaks at a point of contact with galvanized iron or the greenhouse frame.

A software package of SAS was used to estimate the unknown coefficients in the regression model (107). The following empirical models, given in Table 3, are available from the literature (108,109). The mean maximum and minimum deviations obtained for these models are presented in Table 3. Some of the relation, which gives good results for Dhahran, is listed in Table 4.

The investigation showed that model 2 gives lowest percentage mean deviation and is relatively simpler. A comparison of experimental data and the predicted values for tensile strength as a function of exposure time is shown in Fig. 9.

$$T_s = C_0 + C_1(T_m) + C_2(T_m)^2$$

Where, T_s = tensile strength (MPa), T_m is time in months, $C_0 = 23.88$ (tensile

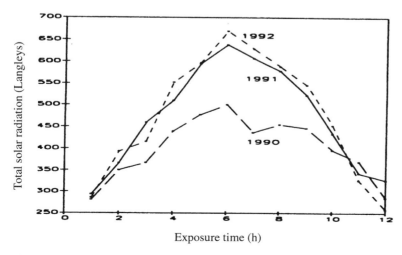

Figure 6 Variation in total solar radiation with time at Dhahran.

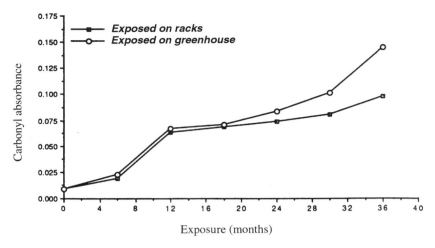

Figure 7   Carbonyl absorbance versus exposure time for greenhouse film exposed on model greenhouse and aluminum racks.

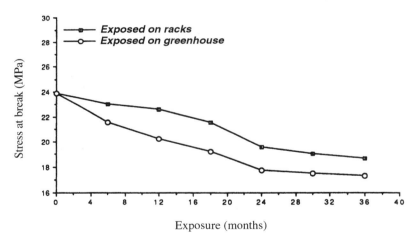

Figure 8   Stress at break versus exposure time for greenhouse film exposed on model greenhouse and aluminum racks.

Table 3   Results of Regression Analysis Using Existing Empirical Models

Relation	Mean deviation (%)	Max. deviation (%)	Ref.
$y = A \exp[B(t - c)]$	SAS failed to converge	—	7
$y = K - A \exp(-Bt^2)$	14.6	28.0	8
$y = C + A \exp(-Bt^n)$	14.6	28.0	9

Table 4 Results of Regression Analysis Used for Tensile Strength for Dhahran Site

Sample no.	Relation	Mean deviation (%)	Max. deviation (%)
1	$T_s = C_0 + C_1(T_M) + C_2(T_M)C_3$	2.5	8.1
2	$T_s = C_0 + C_1(T_M) + C_2(T_M)^2$	2.5	8.7
3	$T_s = C_0 + C_1(T_M)^{1/2} + C_2(T_M)^2$	2.8	9.3
4	$T_s = C_0 \exp(-C_1(T_M)^2 + C_2(T_M))$	41.7	92.9

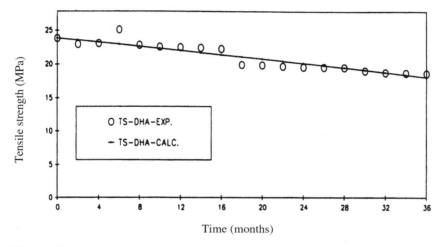

Figure 9 Experimental data and predicted values of tensile strength as a function of exposure time at Dhahran.

strength of virgin samples at $T_m = 0$), C_1, C_2 are constants. Also $C_1 = -0.149$, $C_2 = -0.000329$. The coefficient of determination r^2 was 0.86.

The same form of model was also used to predict the percentage elongation at break values. The following model was obtained for the percentage elongation at break as an independent value using the SAS package.

$$PE = 517.5 - 4.438(T_m) + 0.0576(T_m)^2$$

The mean and maximum deviations were 1.2 and 5.1%, respectively. The coefficient of determination was 0.89, which is within limits for engineering purpose (105).

D. Durability of PVC Pipe in the Natural Environment

Polyvinyl chloride pipes used in Saudi Arabia are exposed to high UV dosage—170–190 Kly/year, temperatures of 13–50°C, relative humidity of 31–85%, dust, and other environmental factors during transportation, storage, and usage. During outdoor weathering, complex reactions occur in the presence of oxygen.

Degradation may start from chromophores in the polymer chain, but their precise nature is questionable (107). UV radiation excites the polymer chain, causing radicals to form that further react with other chains of PVC, causing dehydrochlorination, bond cleavage, and cross-linking (108). Dehydrochlorination results in the formation of colored polyene sequences, which can be seen on the pipe surface as color changes. The radicals may also react with oxygen to start chain cleavage or cross-linking and form hydroperoxide and carbonyl groups.

A study was conducted (109) on white TiO_2-based pipes, containing lead stabilizers, by exposing them for 2 years in the natural weather of Dhahran, Saudi Arabia and Florida. The objective of this study was to assess the service lifetime of PVC pipes weathered in natural weather of Dhahran and Florida.

PVC pipes are formulated with a range of stabilizers to combat degradation effects caused by UV radiation and heat (110). The mean protection against UV attack is provided by titanium dioxide (TiO_2) incorporated at 1–5% in pipe compounds. The pigment TiO_2 strongly absorbs UV radiation at wavelengths below 720 nm (110). The energy associated with these wavelengths is 284–412 kJ/mol, which is enough to break the C–H bond (410 kJ/mol) and form an alkyl radical or break the C–Cl bond (325 kJ/mol) to form polenyl radicals. The incorporation of TiO_2 inhibits the formation of radicals and subsequent photooxidation. However, complete stoppage of the formation of radicals is not possible because radicals are also formed as existing polyenes, carbonyl groups, chain ends, and other impurities.

Degradation was monitored in terms of carbonyl group growth, changes in glass transition temperature (T_g), and mechanical properties. Figure 10 presents the changes in T_g relative to exposure time in Dhahran and Florida. A downward trend of T_g is observed for both exposure sites. However, the degree of drop in T_g is more for samples exposed at Dhahran. This may be attributable to higher UV radiation doses, temperatures, and humidity factors present in Dhahran.

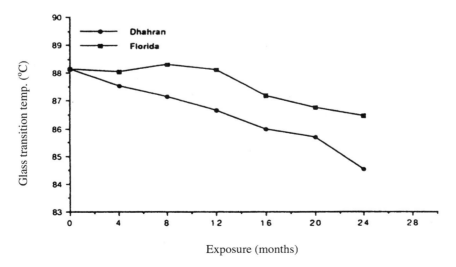

Figure 10 Changes in glass transition temperature relative to exposure time for the exposed PVC samples.

The drop in T_g also indicates the dehydrochlorination and oxidative degradation reactions occurred in the PVC. The dehydrochlorination reaction causes a change in color from white to light brown. Prolonged exposure allows the photooxidation and cross-linking reaction to dominate (111–113).

An overall increasing trend of carbonyl absorbance is observed for both sides as shown in Fig. 11. The extent of development of carbonyl groups is more at Dhahran. This carbonyl transmits electronic excitation to the vinyl chloride units, followed by the formation of radicals, which degrade the polymer.

Changes in mechanical properties are an authentic indicator of in-service performance of pipes. Figures 12 and 13 show the changes in stress at break and percentage elongation at break with time.

The stress at break was lowered by 43% in Dhahran as compared with 26% in Florida within 24 months. Similarly, percentage elongation at break dropped by 66% for samples exposed in Dhahran and 49% for samples exposed in Florida. The service lifetime determined in terms of loss in percentage elongation at break is 47% in the initial 12 months. However, it took 16 months of outdoor exposure in Florida to drop to the same level of mechanical properties. This was also pointed out in other studies; 6 weeks are enough to create microcracks on the surface of PVC pipes, which results in the loss of mechanical, physical, and chemical properties, owing to the high UV dosages and temperatures that are encountered in the region (114,115). Therefore, a much higher level of TiO_2 and UV-stabilization are needed to enhance the PVC pipe lifetime for use in tropical climates.

V. CONCLUSIONS

Prediction of the lifetime of polymeric materials exposed in an outdoor environment can be achieved, but not with full accuracy. Basically two types of methods, which

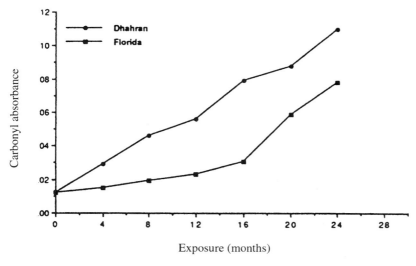

Figure 11   Changes in carbonyl absorbance relative to exposure time for the exposed PVC samples.

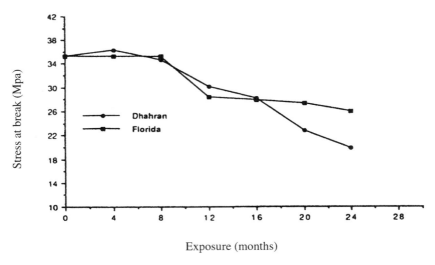

Figure 12   Changes in stress at break relative to exposure time for the exposed PVC samples.

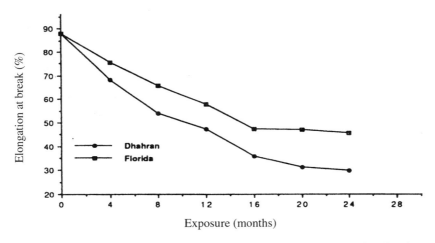

Figure 13   Changes in percentage elongation relative to exposure time for the exposed PVC samples.

are based on simulation and mechanistic approaches, have been used by different researchers. The simulation approach is normally based on the relevance of some observed phenomenon that is deduced from the physical and chemical changes in an artificially accelerated and a natural exposure. In mechanistic approach the chemical nature and spatial distribution of intermediate and final groups formed in the polymer chain or branches in the exposed samples are determined using spectrometric techniques. Given the weathering kinetics and artificial aging kinetics the needed acceleration factor can be calculated.

Outdoor aging is too slow to be useful in development of stabilized formulations; therefore, accelerated aging is very much required to conduct weathering trials. The acceleration factors obtained for the photooxidation of polymers will depend on the degradation parameters measured. Some of the important parameters are oxygen uptake, carbonyl, end-unsaturation, and mechanical properties. It has been observed that the experimental design of artificially accelerated weathering is important because very high acceleration leads to a change in the mechanism that leads to oxygen uptake. In accelerated weathering, most oxygen is consumed through a propagation reaction, whereas in outdoor weathering, most of the oxygen is consumed by air initiation reaction caused by a charge transfer complex (CTC) of oxygen. It was suggested that a decrease of temperature or an increase of oxygen pressures during accelerated weathering might lead to more initiation through a CTC and then a better correlation with outdoor weathering.

The theoretical model proposed for additive loss caused by exposure of polyolefins (69,70) and the empirical model (68) showed a very good fit of the experimental data with the predicted values. By defining a critical concentration, the lifetime of the stabilized polyethylene formulation can be calculated from the proposed model for the accelerated weathering trials. The physical parameters can influence the dependence of HALS stabilization performance on its initial concentration.

In high-temperature applications of polymeric materials, a common approach is to accelerate the chemical reaction underlying the aging at elevated temperatures and monitoring the degradation through change in ultimate tensile strength and elongation at break. The accelerated high-temperature results are related to use temperature, based on the Arrhenius method.

The lifetime of a polymer can also be determined using a kinetic model for thermooxidative processes. The weight loss obtained through thermogravimetry can be kinetically modeled. The weight ion may vary linearly with chain scission, which corresponds to a value of molecular weight at any time and temperature. Based on a critical weight loss of 2% the lifetime can be predicted at any temperature.

The aging polymer should not be considered as a homogeneous (photo)reactor at a molecular scale. The data collected at the initial time under nonaccelerated conditions cannot be extrapolated to the real lifetime of polymers using conventional kinetics. For such cases, experimental acceleration is a necessity. The deviations in the artificially accelerated and natural weathering trial results are generally due to either bimolecular reactions between intermediate species, the concentrations of such in real-time conditions are too low to interact, or to diffusion processes of reactants that are slower in accelerated conditions.

Given the lifetime evaluation of plastic study conducted in the near-equatorial region, the following conclusions have been drawn:

A weathering trial in natural weather is extremely necessary to obtain real-time performance data of stabilized plastics products. For lifetime evaluation of new plastic formulations natural weathering trials should be conducted in the weather in which they will be used.

Mathematical modeling to represent independent variables of percentage elongation at break, tensile strength, carbonyl growth, and percentage crystallinity in terms of dependent variables, including average monthly UV radiation, cumulative total solar radiation, average total solar radiation, and average monthly temperature, can be developed using regression techniques. Models developed for mechanical properties were more dependent on the UV portion of total solar radiation; carbonyl development was synergistically affected by UV and total solar radiation. The percentage crystallinity was affected by cumulative monthly UV radiation, total solar radiation, and average monthly temperature. Other weather parameters play a less significant role in the weather-induced degradation of LLDPE.

The results of lifetime studies conducted in LDPE films show that data obtained from artificially accelerated weathering trials can be correlated with the natural weathering trial data using parabolic models for tensile strength and percentage elongation at break. The acceleration factor for accelerated weathering is about three times that of natural weathering trials (i.e., 5,000 h of accelerated weathering is equivalent to 14,000 h of natural weathering).

Studies conducted on HALS-stabilized LDPE greenhouse film weathered on real greenhouses led to development of two models based for tensile strength and percentage elongation at break as a function of exposure time in months. The correlations were good enough to serve the engineering calculations. Furthermore, the greenhouse film exposed on real greenhouse versus films exposed on aluminum racks suffered more. The reasons attributed for higher degradation on the real greenhouse includes partial prevention of this dissipation of thermal energy—spectral emissions with wavelengths between 7–14 μm—from plants and soil, and from film during night hours can influence the degradative reactions. Moreover, adverse effects of sulfur- and halogen-based pesticides on the HALS stabilizers can accelerate the degradation reactions in the films.

PVC pipes stabilized with lead stabilizer showed faster degradation and lower lifetime for pipes exposed at Dhahran, Saudi Arabia compared with Florida. A service lifetime of about 12 months for Dhahran and 16 months for Florida was obtained through natural weathering trials. A downward trend of T_g, development of carbonyl groups, and a drop in mechanical properties are indicative of the dehydrochlorination and oxidation reactions. Dehydrochlorination reactions caused a change in color from white to light brown owing to formation of polyene sequences. Longer exposure will allow photooxidation and cross-linking to dominate in the PVC matrix. The pigment titanium dioxide strongly absorbs below 420 nm, thus it inhibits radical formation and subsequent photooxidation. Pipes designed for a longer lifetime for near-equator region will need a higher level of titanium dioxide, together with other UV stabilizers.

ACKNOWLEDGMENTS

The authors wish to acknowledge the support of the Center for Refining and Petrochemicals of King Fahd University of Petroleum & Minerals in conducting the research work for the foregoing studies. The authors express their thanks to A. Maroof for typing this work, Alim Rizvi for assistance in the experimental parts of the research work, and J. H. Khan for technical support in the various research studies.

REFERENCES

1. A Davis, D Sims. Weathering of Polymers. London: Applied Science, 1983, p. 184.
2. FS Qureshi, SH Hamid, AG Maadhah, MB Amin. In: Proceedings of the Conference of Plastic Pipes VII, Bath, UK, 20–22 Sept. 1988.
3. WL Hawkins. Polymer Degradation and Stabilization. Berlin: Springer-Verlag, 1984.
4. W Schnabel. Polymer Degradation, Principles and Practical Applications. Berlin: Hanser, 1981.
5. AH Fraser. High Temperature Resistant Polymers. New York: Interscience, 1968.
6. JH Flynn. Thermochim Acta 134:115, 1988.
7. E Kay, A Davis, GL Palmer. Recommended procedures for the effective study of the natural weathering behavior of plastics. In: Weathering of Plastics and Rubbers. PRI Symp C2.1–C2.12, 1976.
8. AV Cunliffe, A Davis. Polym Degrad Stabil 4:17–37, 1982.
9. FH Winslow, W Matreyek, AM Trozzolo. Polym Preprint, ACS Div Polym Chem 10:1271–1280, 1969.
10. MR Kamal. Polym Eng Sci 10:108–121, 1970.
11. B Ranby, JF Rabek. Photodegradation, Photooxidation and Photostabilization of Polymers. London: John Wiley & Sons, 1975.
12. KL Coulson. Solar Terrestrial Radiation. New York: Academic Press, 1975.
13. R Ghani. Processing, heat and light stabilization of polyolefins. Presented at a seminar at KFUPM/RI, Dhahran, Saudi Arabia, October 4, 1987.
14. AG Maadhah, SH Hamid, FS Qureshi. Role of photodegradable polymers in packing and packaging materials. Proceedings of Conference on Utilization of Plastic Materials in Packing and Packaging Industry in the Kingdom. Yanbu, Saudi Arabia, December 1988.
15. AL Andrady, MB Amin, AG Maadhah, SH Hamid, K Fueki, A Torikai, Material damage. In: United Nations, Environmental Program Environmental Effects Panel Report, Chapter 6: 55–60, 1989.
16. PP Klemchuk. Antioxidants. Reprint from Ulman's Encyclopedia of Industrial Chemistry. 1985, pp A3:91–111.
17. AM Trozzolo. In: WL Hawkins, ed. Photooxidation of Polyolefins, Polymer Stabilization. New York: Wiley, 1972.
18. AL Anrady, MB Amin, SH Hamid, X Hu, A Torikai. Effects of increased solar UV-radiation materials. In: Environmental Effects of Ozone Depletion: Assessment. UNEP Report, 1994, pp 101–110.
19. AL Andrady, K Fueki, A Torikai. J Appl Polym Sci 39:763–766, 1989.
20. NS Allen. Polym Degrad Stabil 2:155–161 1980.
21. FS Qureshi, S Hamid, MB Amin, AG Maadhah. Polym Plast Technol Eng 28: 663–760, 1989.
22. MA Abdelrehman, SAM Said, AN Shuaib. Solar Energy 40:219–225, 1988.
23. SH Hamid, MB Amin, AG Maadhah, JH Khan. Weathering degradation of greenhouse plastic films. Proceedings, First Saudi Symposium on Energy, Utilization and Conservation, Jeddah, Saudi Arabia, March 4–7, 1990.
24. FH Winslow, W Matreyek, AM Trozzolo. Soc Plast Eng 18:766–772, 1972.
25. AL Andrady. Plastics in marine environment. In: Proceedings Symposium Degradable Plastics. Washington, DC: SPI, 1987, pp 22–25.
26. MR Kamal, R Saxon. Appl Polym Sym 4:1–28, 1967.
27. F Gugumus. In: G. Scott, ed. Developments in Polymer Stabilization. London: Applied Science, 1987, pp 239–289.
28. T Laus. Plast Rubbers Mater Appl 2:77, 1977.

29. DJ Kingsnorth, DGM Wood. In: The Weathering of Plastics and Rubber. PRI Symposium, 1976, p E2.

30. GN Foster. in: J. Pospisil, PP Klemchuk, eds. Oxidation Inhibition in Organic Materials. Boca Raton, FL: CRC, 1990, pp 314–331.

31. A Tidjani, R Arnaud. Polym Degrad Stabil 39:285, 1983.

32. R Arnaud, J Moison, Y Lemaire. Macromolecules 17:332, 1984.

33. F Gugumus. Makromol Chem Macromol Symp 25:1, 1989.

34. F Gugumus. Makromol Chem Makromol Symp 27:129, 1989.

35. F Gugumus. Polym Degrad Stabil 34:205, 1991.

36. P Gijsman, J Hennekens, D Tummers. Polym Degrad Stabil 39:225, 1993.

37. P Gijsman, J Hennekens, K Jenssen. Comparison of UV Degradation of PE in accelerated test and sunlight. Polymer Durability. ACS Symp SER 249:621–636, 1993.

38. KT Gillen, RL Clough, J Wise. Prediction of elastomer lifetimes from accelerated thermal-aging experiments. Polymer Durability. ACS Symp Ser 249:557–575, 1993.

39. N Grassie, G Scott. Polymer Degradation and Stabilization. London: Cambridge University Press 1985, p. 86.

40. G Amerongeu. Rubber Chem Technol 37:1065–1152, 1964.

41. KT Gillen, RL Clough. In: N, Cheremisinoff, ed. Handbook of Polymer Science and Technology, New York: Marcel Dekker, 1989, pp 167–202.

42. KT Gillen, RL Clough Polymer 33:4358–4366, 1992.

43. TW Dakin. AIEE Trans 67:113–118, 1948.

44. N Trebilco. Corrosion Austr 20(3):6–9, 1994.

45. J Verdu. Viellissement des plastiques. AFNOR Tech, Paris 1984.

46. S Gaumet, JL Gardette. Polym Degrad Stabil 33:17, 1991.

47. J Lemaire, R Arnaud, JL Gardette. Polym Degrad Stabil 33:277, 1991.

48. JL Gardette, J Lemaire. Polym Degrad Stabil 25:293, 1991.

49. F Castillo, G Martinez, R Sastre, J Millan, V Bellanger, BD Gupta, J Verdu. Polym Degrad Stabil 27:1, 1990.

50. T Gancheva, P Genova, A Marinova. Angew Makromol Chem 158/159:107, 1988.

51. S Hallande, JL Laurent. Polym Degrad Stabil 55:141–145, 1997.

52. V Guegen, L Audouin, B Pinel, J Verdu. Polym Degrad Stabil 46:113–122, 1994.

53. VJR Gonzalez, JA Gonzalez-Marcos, FA Delgado, C Gonzalez-Oritz, JI Gutierrez-Oritz. Chem Eng Sci 51:1113–1120, 1996.

54. T Masuda, T Higashimura. Adv Polym Sci 81:121–165, 1987.

55. N Martakis, M Niaovnakis, D Pissinissis. J Appl Polym Sci 51:313, 1994.

56. R Baumhardt-Neto, MA De Paoli. Polym Degrad Stabil 40:59, 1993.

57. NY Rapopart, SI Berulava, AL Kovarskii, IN Musayelyan, YA Yershov, VB Miller. Polym Sci USSR, A17, 2901, 1975 [Trans Vysokonol Soyed A17, 2521, 1975].

58. MS Robelo JR White. Polym Degrad Stabil 56:55–73, 1997.

59. J Lemaire, J Gardette, J Lacoste, P Delprat, D Vaillant. Mechanism of photooxidation of polyolefins: prediction of lifetime in weathering conditions. In: Polymer Durability. ACS Sump Ser 249:577–598, 1993.

60. JCM de Bruijn. Degradation profiles of thick HDPE samples after outdoor and artificial weathering. In: Polymer Durability. ACS Symp Ser 249:599–620, 1993.

61. GE Schoolenberg. PhD dissertation, Delft University of Technology, Delft, 1988.

62. L Rolland. PhD dissertation, Illinois Institute of Technology, Chicago, IL, 1983.

63. P Gijsman, J Hennekens, K Janssen. Comparison of UV degradation of PE in accelerated test and sunlight. In: Polymer Durability. ACS Symp Ser 249:621–636, 1993.

64. JAJM Vincent, JMA Jansen, JJH Nijsten. Proceedings of the International Conference on Advances in the Stabilization and Controlled Degradation of Polymers. Lucene, Switzerland, 1982.

65. T Laus. Plast Rubbers Mater Appl 2:77, 1977.

66. A Tidjani, D Anicet, R Arnaud, J Appl Polym Sci 47:211–216, 1993.
67. J Malik, DQ Tuan, E Spirk. Polym Degrad Stabil 47:1–8, 1995.
68. JY Moison. In: J Lomya, ed. Polymer Permeability. London: Elsevier, 1985, p 199.
69. NC Billingham, PD Calvert. In: G Scott ed. Developments in Polymer Stabilization, 3rd ed. London: Applied Science Publishers, 1980, p 139.
70. PD Calvert, NCT Billingham. Appl Polym Sci 24:357, 1979.
71. J Malik, A Hrivik and D Alexyoua. Polym Degrad Stabil 35:125, 1992.
72. JTEH Sampers, Physical loss of UV stabilize from LDPE films. In: 34th IUPAC, International Symposium Macromolecules, Prague, 13–18 July, 1993, p 40, 167.
73. F Gugumus. Polym Degrad Stabil 40:167, 1993.
74. L Audouin, V Langlois, J Verdu. Angew Makromol Chem 232:1–12, 128, 1995.
75. E Kramer, J Koppelmann, J Dobrowsky. J Therm Anal 35:443, 1989.
76. JB Howard. Proc. 31st SPE, Montreal, 1973, p 408.
77. GN Foster. In: A Patsis, ed., Advances in the Stabilization and Controlled Degradation of Polymers. Vol. 1. Basel: Technomic Publishing, 1989.
78. SH Hamid, WH Prichard, Polym Plast Technol Eng 27:303–334, 1988.
79. SH Hamid, AG Maadhah, FS Qureshi, MB Amin. Arab J Sci Eng 13:503–531, 1988.
80. SH Hamid, FS Qureshi, MB Amin, AG Maadhah. Polym Plast Technol Eng 28:475–492 1989.
81. FS Qureshi, SH Hamid, AG Maadhah, MB Amin. Prog Rubber Plast Technol 5:1–14, 1989.
82. SH Hamid, WH Prichard. J Appl Polym Sci 43:561–678, 1981.
83. BW Lindgren, GW McElrath. Introduction to Probability and Statistics. London: Macmillan, 1971.
84. A Davis, D Sims. Weathering of Polymers. London: Applied Science, 1983, p 661.
85. DR Cox, EJ Shell, Int Stat Rev. 23:51–59, 1974.
86. RJ Freund, RC Littell. SAS for Linear Models, A Guide to the ANOVA and GLM Procedure. Cary, NC: SAS Institute, 1981.
87. RJ Beckman, RD Cook. Technometrics 25:119–149, 1983.
88. CL Mallows. Technometrics 15:661–675, 1973.
89. MR Kamal, B Huang. In: SH Hamid, MB Amin, AG Maadhah, eds. Handbook of Polymer Degradation. New York: Marcel Dekker, 1992, pp 127–168.
90. JL Scott. In: AV Patsis, ed. Proceedings International Conference on Advances in the Stabilization and Controlled Degradation of Polymers. Lancaster, UK: Technomic Publishing, 1989, pp 75–85.
91. CA Brighton. In: SH Pinner, ed. Weathering of Plastic. New York: Gordon & Breach Science Publishers, 1966.
92. SH Hamid, MB Amin. J Appl Polym Sci 55:1385–1394, 1995.
93. ASTM Standard Method, ASTM D 1435-85. Outdoor Weathering of Plastics. v 8.01.
94. ASTM Standard Method, ASTM D 2565-79. Operating Xenon Arc-type Light-Exposure Apparatus with and without Water for Exposure of Plastics. v. 8.02.
95. F Severini, R Gallo, S Ipsale, N De Fenti. Polym Degrad Stabil 14:341, 1986.
96. F Gugumus. Stabilization of polyolefins. Presented at the European Symposium of Polymeric Material, Lyon, France, September 14–18, 1987.
97. DR Gee, TP Melin. Polymers 11:178, 1976.
98. N Khraishi, A Al-Rabaidi. Polym Degrad Stabil 32:105, 1991.
99. A Shinde, R Salovey. J Polym Sci Polym Phys Ed 23:1681, 1985.
100. MC Sebaa, J Pouyet. J Appl Polym Sci 45:1049, 1992.
101. LFE Crewdson. First International Symposium on Weatherability (ISW). Tokyo, May 12–13, 1992.
102. DC Montgomery, EA Peck. Introduction to Linear Regression Analysis. New York: Wiley, 1982.

103. FH Winslow, W Matreyek, Paper presented at ACS Div Polym Chem meeting in Chicago, 1964, pp 552–557.
104. K Furneuax, J Ledbury, A Davis. Polym Degrad Stabil 3:431, 1981.
105. MB Amin, SH Hamid, F Rahman. J Appl Polym Sci 47:279–284, 1994.
106. JH Khan, SH Hamid. Polym Degrad Stabil 48:137–143, 1995.
107. SAS Institute Inc. SAS User's Guide, Version 6, 4th ed., v.2. Cary, NC: SAS, 1990.
108. WH Daiger, WH Madison. J Paint Technol 39:399, 1967.
109. I Hussain, SH Hamid, JH Khan. J Vinyl Add Technol 1(3):137–141, 1995.
110. MR Kamal. Polym Eng Sci 10:119, 1970.
111. WH Starnes. ACS Symp Ser 151:197, 1981.
112. JF Rabek, G Canback, J Luky, B Ranby. J Polym Sci 14:1447, 1976.
113. EB Rabinvitch, JW Summers, WE Northcott. J Vinyl Technol 15:214, 1993.
114. F Mahmud, I Hussain, AG Maadhah, MB Amin. Arab J Sci Eng 13, p 210, 1988.
115. F Mahmud, I Hussain. RAPRA Q Rev J 7:4, 1991.

20

Kinetic Modeling of Low-Temperature Oxidation of Hydrocarbon Polymers

LUDMILA AUDOUIN, LAURENCE ACHIMSKY, and JACQUES VERDU
École Nationale Supérieure d'Arts et Métiers (ENSAM), Paris, France

I. INTRODUCTION

Since the pioneering work of RAPRA research workers 50 years ago (1,2), there is consensus that the low-temperature oxidation of saturated hydrocarbon polymers, such as polyethylene (PE) or polypropylene (PP), results from a radical chain process with a two step propagation:

(II) $P^{\bullet} + O_2 \quad \rightarrow \quad PO_2^{\bullet}$ (k_2)

(III) $PO_2^{\bullet} + PH \quad \rightarrow \quad PO_2H + P^{\bullet}$ (k_3)

There is a wide variety of mechanistic schemes differing by the nature of initiation, termination, branching, and side reactions (3–6). The present chapter will focus on the following cases:

1. Only low-temperature ($T<150°C$) oxidations are studied. In the conditions under study, the thermolysis or the photolysis of the polymer is negligible. Radicals are generated only from thermolabile groups. These latter are essentially hydroperoxides, POOH. The reaction generates its own initiator (Fig. 1) so that the process can be called a "closed-loop" mechanism.

2. The oxygen is in excess. Because reaction II is very fast [$k_2 \sim 10^8 \, \mathrm{Lmol^{-1} s^{-1}}$ (6)], all the $P^{\bullet}$ radicals are readily transformed into $PO_2^{\bullet}$, so that terminations involving $P^{\bullet}$ radicals can be neglected and, in principle, only one type of termination ($PO_2^{\bullet} + PO_2^{\bullet}$) is to be taken into

727

consideration. This hypothesis is appropriated to the study of thin films, powders, or superficial layers of bulk samples at oxygen pressures higher than a critical value P_c corresponding to a critical concentration C_c of O_2 in the polymer (Fig. 2) (7).

It is noteworthy that P_c is not necessarily lower than the atmospheric pressure $(2.10^4$ Pa) (8,9). Surprisingly, despite their high practical interest, these schemes have rarely been as well investigated, for the unimolecular as for bimolecular POOH decomposition (10–12). Some of their important properties were discovered only very recently. These schemes, in their simplest version presented here, cannot be considered the definitive solution for kinetic modeling of polymer oxidation, but rather, as a good point of departure for further investigations. Their main interest is to express in an unambiguous, mathematical form, validity criteria for kinetic models or experimental approaches and to allow a quantitative description of properties, such as the spatial heterogeneity of oxidation, which is now experimentally well established for polypropylene. The aim of this chapter is to review our recent studies on these topics.

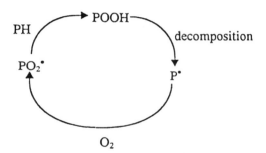

Figure 1 Schematic of a closed-loop oxidation mechanism.

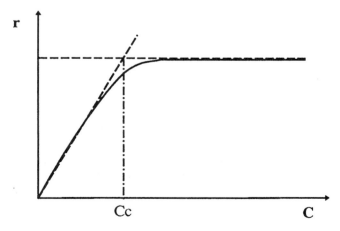

Figure 2 Shape of the change of oxidation rate with oxygen concentration and determination of C_c. (From Ref. 7.)

II. CLOSED-LOOP MECHANISTIC SCHEMES

A. Justification

Our interest focuses on schemes of which the general form is:

(I)	$\lambda POOH$	$\rightarrow$	$\alpha P^{\bullet}$	k_i
(II)	$P^{\bullet} + O_2$	$\rightarrow$	$PO_2^{\bullet}$	k_a
(III)	$PO_2^{\bullet} + PH$	$\rightarrow$	$PO_2H + P^{\bullet}$	k_p
(IV)	$PO_2^{\bullet} + PO_2^{\bullet}$	$\rightarrow$	inact. prod.	k_t

There are a priori arguments against such simple schemes. The most important ones are linked to the experimentally well-documented hypercomplexity and heterogeneity of the oxidation process.

The hypercomplexity is, for instance, attested to by the great diversity of stable oxidation products, as illustrated by the shape of the carbonyl band for PP thermal oxidation (Fig. 3).

Figure 3 calls for two important remarks:

1. The absorption results from the overlapping of many elementary bands: five or six are reported by Adams (13), but computer-assisted deconvolution can easily detect more than ten species (14).
2. This diversity is established at very low conversions: the spectra corresponding to different exposure times are almost intimate (15). This behavior is usually interpreted in terms of heterogeneous oxidation (see Sec. IV). The observations are consistent with one of the foregoing schemes

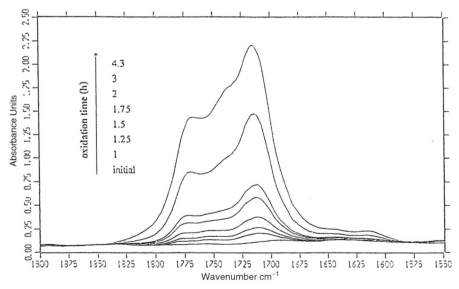

Figure 3 IR band of carbonyls in PP thermal oxidation at 150°C. Increasing absorbances correspond to increasing exposure times.

only if the written steps correspond in fact to a set of elementary processes to which they are kinetically equivalent. Let us consider, for instance, POOH decomposition. It can be written:

$$
\begin{array}{llll}
\text{POOH} & \rightarrow & \text{PO}^{\bullet} + \text{OH}^{\bullet} & (k_1) \\
\text{OH}^{\bullet} + \text{PH} & \rightarrow & \text{H}_2\text{O} + \text{P}^{\bullet} & (k_2) \\
\text{PO}^{\bullet} + \text{PH} & \rightarrow & \text{POH} + \text{P}^{\bullet} & (k_3) \\
\text{PO}^{\bullet} & \rightarrow & \text{P} = \text{O} + \text{P}^{\bullet}\ (\beta = \text{scission}) & (k_4)
\end{array}
$$

In a first approach, radicals $P^{\bullet}$ resulting from β-scissions will not be distinguished from radicals $P^{\bullet}$ resulting from hydrogen abstraction. $OH^{\bullet}$ and $P^{\bullet}$ radicals are very reactive, they cannot be observed by electron spin resonance (ESR) at ambient temperatures. The foregoing set of reactions is thus kinetically equivalent to:

$$
\text{(I)}\quad \text{POOH} \quad \rightarrow \quad 2\text{P}^{\bullet} \quad (k_i)
$$

provided that $k_i = k_1$: the formation of $P^{\bullet}$ radicals in initiation processes is governed by POOP decomposition.

The same reasoning could be applied to termination processes. The mechanisms of these are not fully elucidated, at least for PP (16), and their discussion would be beyond the scope of this chapter. It will be assumed in the following that a virtual step (IV) of bimolecular terminating combination of $PO_2^{\bullet}$ radicals is kinetically equivalent to the ensemble of real termination processes, keeping in mind that the validity of this hypothesis remains to be demonstrated.

Finally one sees that the experimentally found diversity of oxidation products is not necessarily inconsistent with a simple kinetic scheme.

The heterogeneity of the oxidation in solid state has now been well established by various means (16–19). Concerning kinetics, the key problem can be summarized by the following question: Is it or is it not allowable to use conventional kinetic modeling when local concentrations of reactive species differ noticeably from the average (macroscopic) concentration? Section IV is devoted to this very important question. Our approach is based on two remarks:

1. The case for which oxidation is heterogeneous at a given scale, but homogeneous on a smaller scale, cannot be set out a priori. In this case, the classic kinetic modeling would be acceptable. What we need, thus, is a quantitative approach to heterogeneity.

2. If this quantitative study of heterogeneity would give ambiguous results, there is still the possibility to check classic kinetic models. Their failure would not be a proof of the nonvalidity of the approach, because the chosen model is not necessarily pertinent. In contrast, their success to predict the polymer behavior would be the best proof of the validity of the approach, at least if it is not due to the use of ad hoc adjustable parameters.

B.   Common Features of Closed-Loop Models in the Stationary State

Let us recall the general mechanistic scheme:

$$\lambda POOH \quad\rightarrow\quad \alpha P^{\bullet} + \beta PO_2^{\bullet} \qquad\qquad (k_i)$$
$$P^{\bullet} + O_2 \quad\rightarrow\quad PO_2^{\bullet} \qquad\qquad (k_a)$$
$$PO_2^{\bullet} + PH \quad\rightarrow\quad PO_2H + P^{\bullet} \qquad\qquad (k_p)$$
$$PO_2^{\bullet} + PO_2^{\bullet} \quad\rightarrow\quad \text{inactive product} + O_2 \qquad\qquad (k_t)$$

One can distinguish when $\lambda = 1$ (unimolecular with $\alpha = 2$, and $\beta = 0$) and $\lambda = 2$ (bimolecular, with $\alpha = 1$ and $\beta = 1$) hydroperoxide decomposition. Cases where $\lambda > 2$ (hydroperoxide "hundles") can also be envisaged (20). These systems are characterized by the existence of a stationary state in which the rates of POOH formation and destruction are equal. Then,

$$\left(\frac{d[POOH]}{dt}\right) = k_p[PH][PO_2^{\bullet}] - \lambda k_i[POOH]^{\lambda} = 0, \qquad \text{that is,}$$

$$[PO_2^{\bullet}]_{\infty} = \frac{\lambda k_i}{k_p[PH]}[POOH]_{\infty}^{\lambda} \qquad\qquad (1)$$

In the stationary state, the radical concentration tends to become constant so that:

$$\frac{d[Rad]}{dt} = (\alpha + \beta)k_i[POOH]^{\lambda} - 2k_p[PO_2^{\bullet}]^2 \qquad \text{that is,}$$

$$[PO_2^{\bullet}]_{\infty} = \left[\frac{(\alpha + \beta)k_i}{2k_t}[POOH]_{\infty}^{\lambda}\right]^{1/2} \qquad\qquad (2)$$

It becomes

$$[POOH]_{\infty}^{\lambda} = \left(\frac{\alpha + \beta}{2k_t}\right)^{\lambda/2}\frac{k_p[PH]}{\lambda k_i^{1/2}} \qquad \text{that is,}$$

$$[POOH]_{\infty} = \left(\frac{\alpha + \beta}{2k_t}\right)^{1/\lambda}\left(\frac{k_p[PH]}{\lambda}\right)^2\frac{1}{k_i^{1/\lambda}} \qquad \text{and} \qquad (3)$$

$$[PO_2^{\bullet}]_{\infty} = \left(\frac{\alpha + \beta}{\lambda}\right)\frac{k_p[PH]}{2k_t} \qquad\qquad (4)$$

The initiation and termination rates are equal to

$$r_{i\infty} = r_{t\infty} = (\alpha + \beta)k_i[POOH]_{\infty}^{\lambda} = \left(\frac{\alpha + \beta}{\lambda}\right)^2\left(\frac{k_p(PH)}{\sqrt{2k_t}}\right)^2 \qquad\qquad (5)$$

$$[P^{\bullet}]_{\infty} = \frac{1}{k_a[O_2]}\left(\alpha k_i[POOH]_{\infty}^{\lambda} + k_p[PH][PO_2^{\bullet}]_{\infty}\right) = \frac{\alpha + \beta}{\lambda k_a[O_2]}\frac{k_p^2[PH]^2}{2k_t}\left(\frac{\alpha}{\lambda} + 1\right)$$
$$\qquad\qquad (6)$$

The oxygen consumption rate is given by

$$r_\infty = \frac{d[O_2]}{dt} = -k_a[O_2][P^\bullet]_\infty + k_t[PO_2^\bullet]_\infty^2 = -\left(k_p\frac{[PH]^2}{k_t}\left(\frac{\alpha+\beta}{2\lambda}\right)\left(\frac{\alpha+2\lambda-\beta}{2\lambda}\right)\right) \quad (7)$$

where the minus sign denotes consumption. These results call for the following comments.

1. In the stationary state the kinetic chain length KCL is

$$KCL_\infty = \frac{k_p[PH]\left[PO_2^\bullet\right]_\infty}{r_{i\infty}} = \frac{\lambda}{\alpha+\beta} \quad (8)$$

At the beginning of exposure, the reaction is slightly autoaccelerated (see later). It can be considered in a first approximation that the initiation rate is constant, and that the hypothesis of stationary state is valid. So that the initial kinetic chain length is

$$KCL_0 = \frac{k_p[PH]}{\sqrt{2r_{i0}k_t}} \quad (9)$$

$r_{i0} = (\alpha + \beta)k_i [POOH]_0^\lambda$ is extremely low so that $KCL_0 \gg KCL_\infty$. It can be thus seen that the concept of the kinetic chain length is not very useful here, because KCL varies continuously (decreases) during the autoaccelerated period if oxidation to tend toward $\lambda/(\alpha+\beta)$ (i.e., to 1/2 in the unimolecular POOH decomposition, and to 1 in the bimolecular POOH decomposition). It is usual, to simplify calculations, to consider long kinetic chain length values. It is clear that this assumption is not valid here: the generation of O_2 from termination reactions cannot be neglected relatively to the O_2 addition to radicals $P^\bullet$. Incidentally, it can be observed that the hypothesis that $[P^\bullet]_\infty = 0$ is absurd.

2. All the stationary radical concentrations and stationary rates essentially depend on the term $k_p[PH]/\sqrt{2k_t}$. Classically, this term was used to represent the "intrinsic oxidability" of the polymer, essentially linked to the structure of the monomer unit. An important review on its values in polymers and model compounds has been published (21). We see that in the case of closed-loop models, this characteristic would, rather, be linked to the square of this term: $k_p^2(PH)^2/2k_t$.

3. The stationary-state characteristics do not depend on the initiation rate r_i (except for POOH concentration). This is a very important feature that must allow one to identify a closed-loop mechanism (see following).

4. The stationary-state rates are expected to obey the Arrhenius law, with an activation energy.

$$E_\infty = 2E_p - E_t$$

where E_p and E_t are the respective activation energies of propagation (hydrogen abstraction) and termination.

C.   General Approach to the Behavior Characteristics of Closed-Loop Systems During the Induction Period

We will try, at least in a first approach, to describe the whole oxidation kinetics, including the induction period and the stationary state, by a single function. In other words, it is assumed that there is no discontinuity at the end of the induction period, despite the suggestive shape of the kinetic curves at low temperature (22). In our approach the apparent discontinuity is linked to the chosen scale for the measured quantity q and to the sensitivity limit of the measurement method. A change appears only when q becomes higher than the sensitivity threshold. This is confirmed in PP thermal oxidation by the use of a very sensitive chemiluminescence method (23).

There are many possible definitions of the induction period. For us, it is the time to reach stationary state. The best way to determine induction time t_i consists of using oxygen absorption $[C = f(t)]$ or the buildup of any species formed in initiation or termination steps, for instance carbonyls $[X = f(t)]$ curves, or integral curves of POOH build-up $\int_0^t y\,dt = f(t)$ or chemiluminescence intensity $\int_0^t I\,dt = f(t)$ curves. All these curves have the shape of Fig. 4.

In reality, the curves generally display a more or less autoretardant character in their final part. This is because

1. The substrate consumption is generally not taken into account in the models.
2. The existence of an important volatilization process (24,25), which makes it difficult to quantitatively analyze the process at high conversions.

In the most disfavorable cases, these processes become important just after the end of the induction period, and the determination of stationary-state characteristics becomes difficult or impossible. This is especially true when POOH buildup curves are used. Oxygen absorption (20,25,26) or carbonyl buildup curves (27,28), at least

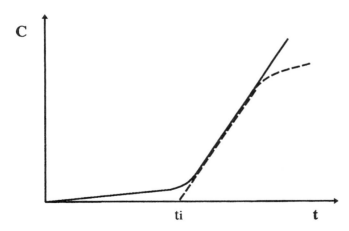

Figure 4   Shape of oxidation kinetic curves, a graphical determination of the induction time t_i: solid line, theoretical curve; dashed line, experimental curve.

at low temperature, (T<150°C) display in contrast a linear section sufficiently large to allow determination of stationary-state characteristics, with reasonable accuracy.

To predict the polymer's behavior during the induction period, we have to resolve the following set of differential equations derived from the foregoing mechanistic scheme:

$$\frac{d[P^\bullet]}{dt} = \alpha k_i [POOH]^\lambda - k_a [O_2][P^\bullet] + k_p [PH][PO_2^\bullet] \tag{10}$$

$$\frac{d[PO_2^\bullet]}{dt} = k_a [O_2][P^\bullet] - k_p [PH][PO_2^\bullet] - 2k_t [PO_2^\bullet]^2 + \beta k_i [POOH]^\lambda \tag{11}$$

$$\frac{d[POOH]}{dt} = k_p [PH][PO_2^\bullet] - \lambda k_i [POOH]^\lambda \tag{12}$$

In principle, the hypothesis of stationary state for radical concentrations is not permitted when the overall radical concentration must increase during the whole induction period. However, it will be shown in the following that in certain cases, the hypothesis of the stationary state leads to analytical expressions that can be considered as good approximations of the general solution.

To simplify the resolution of these systems, it is convenient to use reduced variables such as

$$x = \frac{[PO_2^\bullet]}{[PO_2^\bullet]_\infty}; \qquad y = \frac{[POOH]}{[POOH]_\infty}; \qquad \text{and } z = \frac{[P^\bullet]}{[P^\bullet]_\infty}$$

so that we have the following boundary conditions (Table 1).

The boundary condition for hydroperoxide concentration needs some comments: First, $[POOH]_0$ cannot be zero, otherwise oxidation would not begin. But in most of the practical cases $[POOH]_0$ will be very low beyond the sensitivity limit of most of the titration methods. In other words, $[POOH]_0$ (or y_0) will appear as an adjustable parameter in the kinetic model.

Second, $[POOH]_0$ does not necessarily correspond to the real initial POOH concentration, but rather, to the hypothetical POOH initial concentration kinetically equivalent to the radical-producing species initially present in the sample. The problem of the nature and effect of species responsible for initial steps of oxidation has been the subject of a large number of studies, especially in photooxidation (29–31). One can cite, for instance, initiator residues (i.e., peroxides or transition metal salts), atmospheric contaminants (especially polynuclear aromatics), oxidation products formed during processing, polymer–oxygen charge transfer complexes, structural defects formed during polymerization or others. Their contribution to the initiation of oxidation chains is expected to be constant or decreasing with time (if they

Table 1 Boundary Conditions for the Reduced Variables

Variable	$t = 0$	$t \to \infty$
x	0	1
y	y_0	1
z	0	1

are consumed) whereas the contribution of POOH groups increases in an autoaccelerated way. Thus, after a more or less long time, the contribution of nonhydroperoxidic groups becomes negligible and it can be reasonably considered that the polymer behaves as a pure closed-loop system (Fig. 5).

Depending on the mathematical structure of the differential Eqs. (10)–(12), one can envisage two extreme cases:

1. The autoaccelerated character of oxidation is very sensitive to initial conditions. In this case, the choice of [POOH]$_0$ values will be very important. Let us notice that such a dependence is favorable to the buildup of spatial heterogeneities because the autoacceleration must emphasize eventual spatial fluctuations of the concentration of the initiating species that was initially present.
2. The autoaccelerated character is insensitive to initial conditions. In this case, the behavior must be almost independent of the nature and concentration of the initially present species.

D. The Unimolecular Closed-Loop Scheme

Let us now consider the example in which initiation is due to unimolecular POOH decomposition:

$$
\begin{array}{llll}
\text{POOH} & \rightarrow & 2\text{P}^{\bullet} & (k_u) \\
\text{P}^{\bullet} + \text{O}_2 & \rightarrow & \text{PO}_2^{\bullet} & (k_a) \\
\text{PO}_2^{\bullet} + \text{PH} & \rightarrow & \text{PO}_2\text{H} + \text{P}^{\bullet} & (k_p) \\
\text{PO}_2^{\bullet} + \text{PO}_2^{\bullet} & \rightarrow & \text{inactive product} + \text{O}_2 & (k_t)
\end{array}
$$

The corresponding system of differential equations has been resolved by both analytical (32) and numerical (33) methods.

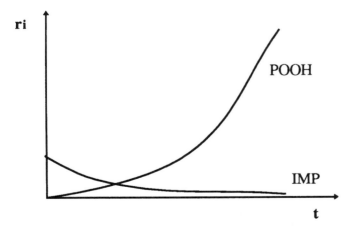

Figure 5 Schematic of time variations in the contribution to initiation of hydroperoxides (POOH) and other radical-producing species.

The analytical methods needs the hypothesis of the stationary state for radical concentrations. It leads to the following expression for $y_0 \ll 1$ (the general case has been reported in Ref. 33).

$$[PO_2^{\bullet}]_{\infty} = \frac{k_p[PH]}{k_t} ; \qquad [PO_2^{\bullet}] = [PO_2^{\bullet}]_{\infty}\left(1 - \exp-\frac{k_u t}{2}\right) \qquad (13)$$

$$[PO_2H]_{\infty} = \frac{k_p^2[PH]^2}{k_u k_t} \qquad [PO_2H] = [PO_2H]_{\infty}\left(1 - \exp-\frac{k_u t}{2}\right)^2 \qquad (14)$$

$$[P^{\bullet}]_{\infty} = \frac{3k_p^2[PH]^2}{k_t . k_a[O_2]} \qquad (15)$$

The rate of product formation from initiation or termination is given by

$$\frac{dX}{dt} = \psi' k_u[POOH] + \psi'' k_t[PO_2^{\bullet}]^2 = \psi k_u[POOH]$$

Since the rates of initiation and termination are equal, ψ' and ψ'' are linked to the yield of corresponding processes.

By integration one obtains

$$X = \psi \frac{k_p^2[PH]^2}{k_t}\left(t - \frac{3}{k_u} + \frac{4}{k_u}\exp\left(-\frac{k_u t}{2}\right) - \frac{1}{k_u}\exp(-k_u t)\right] \qquad (16)$$

The rate of oxygen absorption is given by

$$\frac{dC}{dt} = -k_a[O_2][P^{\bullet}] + k_t[PO_2^{\bullet}]^2$$

$$\frac{dC}{dt} = -2k_u[POOH] - k_p[PH][PO_2^{\bullet}] + k_t[PO_2^{\bullet}]^2 \qquad (17)$$

$$C = \frac{k_p^2}{k_t}\left(2t - \frac{5}{k_u} + \frac{6}{k_u}\exp\left(-\frac{k_u t}{2}\right) - \frac{1}{k_u}\exp(-k_u t)\right]$$

Thus, the induction time for POOH or for initiation/termination products is $t_{ix} = 3/k_u$ and for oxygen absorption and presumably for weight gain linked to oxygen absorption is $t_{ic} = 2.5/k_u$.

The numerical method (no hypothesis of stationary state) leads to results of the same order (33). For a large range (many orders of magnitude) of variation of k_u, $k_p[PH]$, and $k_a[O_2]$, and for $y_0 \leq 10^{-2}$, one obtains for POOH, initiation and termination products. $2/k_u \leq t_{ix} \leq 5/k_u$, and for the oxygen absorption $1.8/k_u \leq t_{ic} \leq 4/k_u$, with generally $t_{ix} - t_{ic} \sim 0.5/k_u$ as found in the analytical resolution.

Indeed, the shorter the time, the higher the error from the hypothesis of stationary state in the analytical model. But, from a practical point of view, this error is not important because it significantly modifies the calculated values only in the range where these latter are very low, out of the sensitivity range of the most widely used characterization methods. An important characteristic of this scheme, put in evidence by numerical resolution, is that termination and initiation kinetics are undistinguishable by kinetic analysis, even at low conversions. The corresponding induction times differ by less than 1% in the great majority of cases. Thus, it appears difficult, for instance, to identify the origin of chemiluminescence (POOH decomposition or $PO_2^\bullet$ combination ?) from only kinetic considerations.

The numerical model needs the knowledge of four parameters: y_0, k_u, $k_p[PH]$, and $k_a[O_2]$ k_u can be determined from independent experiments of hydroperoxide decomposition in an inert atmosphere (34,35), although the results of these latter can carry some ambiguities relative to the reaction order or the existence of two types (slow and fast decomposing) hydroperoxides. The other parameters are very difficult or practically impossible (low y_0 values) to determine by independent methods. They can, in principle, be identified by trial and error procedures, searching the parameter set giving the best fitting of experimental curves, but the model is not very sensitive to variations of $k_p[PH]$ and $k_a[O_2]$ beyond certain values.

Furthermore, calculations are time-consuming, especially when the parameters differ by several orders of magnitude. For all these reasons, the analytical method has to be preferred to the numerical one. The reason why it gives good results, despite that it is based on a wrong hypothesis, can be found in the mathematical structure of differential equations: three variables (in the reduced form x, y, and z, see foregoing) are linked in such a way that they cannot differ one from another by one order of magnitude. In the usual ageing conditions, the three time constants contained by the equations are always in the following order:

$$k_u^{-1} >> \left[k_p[PH]\right]^{-1} >> (k_2[O_2])^{-1}$$

Thus, the POOH decomposition (time constant k_u^{-1}), which is the slowest process, is expected to govern the whole kinetics. This is valid, indeed, in only the conditions under study (high oxygen concentrations and low temperatures). It is obvious that at low O_2 concentrations, $(k_a[O_2])^{-1}$ can become higher than $(k_p[PH])^{-1}$ and even than k_u^{-1}. In the same way, the activation energy of $k_p[PH]$ is lower than this of k_u, such that there is a critical (elevated) temperature above which $(k_p[PH])^{-1} > k_u^{-1}$.

From a practical point of view, the analytical model (for low initial POOH concentrations or, more generally, for low concentrations of initiating species) contains only two variables: k_u and $(k_p[PH])^2/2k_t$. They can be easily determined from oxygen absorption kinetic curves, k_u being obtained from the induction time t_u: $k_u = 2.5/t_u$; and $k_p^2[PH]^2/k_t$ being obtained from the slope of the linear areas (stationary state) of the curves. A third parameter ψ must be added for the species formed in initiation or termination reactions. ψ, which is the yield of the species under consideration (carbonyl, hydroxyl, chain scission, chemiluminescence, on other), is expected to be temperature-dependent in the general case. For instance, for carbonyl, ψ depends on the competition between hydrogen abstraction and β-scission from alkoxy radicals. Oxygen absorption must thus be preferred to the other methods because its modeling needs only two parameters against three for the other analytical methods, (four for chemiluminescence, for which the quantum yield of emission and its eventual temperature variation must also be taken into account). Mass changes (36) and hydroperoxide buildup (37,38) can also be used, but stationary-state characteristics can be difficult to determine with these methods, strongly influenced by side reactions occurring at moderate to high conversions. The density increase, slightly complicated by morphological changes, can be used in the same manner as oxygen absorption (39). The dielectric constant is expected to behave more or less the same as density. Let us notice that each initiation event gives one water molecule. Thus, an accurate titration of H_2O evolved during oxidation would allow the determination of the initiation rate and would be complementary to oxygen absorption measurement, rather linked to propagation and termination.

E.   The Biomolecular Closed-Loop Scheme

The mechanistic scheme is now

$$
\begin{aligned}
2POOH & \rightarrow & PO_2^{\bullet} + P^{\bullet} & \quad (k_b) \\
P^{\bullet} + O_2 & \rightarrow & PO_2^{\bullet} & \quad (k_a) \\
PO_2^{\bullet} + PH & \rightarrow & PO_2H + P^{\bullet} & \quad (k_p) \\
PO_2^{\bullet} + PO_2^{\bullet} & \rightarrow & \text{inactive product} + O_2 & \quad (k_t)
\end{aligned}
$$

The corresponding system of differential equations has been resolved—to our knowledge—by the analytical method using only the hypothesis of stationary state for radical concentrations. This resolution leads to the following:

$$
Y + \frac{Y_\infty}{(Y_\infty - Y_0)/Y_0 \exp(-Kt) + 1} \qquad Y = [POOH] \tag{18}
$$

with

$$
Y_\infty = \frac{k_p[PH]}{2(k_b k_t)^{1/2}} \quad \text{and} \quad K = k_p[PH] \left(\frac{k_b}{k_t}\right)^{1/2}
$$

The concentration of initiation–termination products is given by

$$
\frac{dX}{dt} = \psi k_b[POOH]^2
$$

and by integration we obtain

$$
X = \frac{Y_\infty^2}{K} \left\{ \ln \frac{b + \exp Kt}{b + 1} + b \left(\frac{1}{b + \exp k_t} - \frac{1}{b + 1} \right) \right\} \qquad \text{with} \qquad b = \frac{Y_\infty - Y_0}{Y_0} \tag{19}
$$

The curve $X = f(t)$ tends toward an asymptotic straight-line equation.

$$
X = \frac{Y_\infty^2}{K} \left\{ Kt - \ln(b + 1) - \frac{b}{b + 1} \right\} \tag{20}
$$

The induction time for X is thus

$$
t_{ix} = \frac{1}{K} \left(\ln(b + 1) + \frac{b}{b + 1} \right) \tag{21}
$$

For low values of the initial hydroperoxide concentration, for instance, for $b > 100$, it can be written with a good approximation.

$$t_{ix} \sim \frac{1 + \ln b}{K}$$

Let us recall that

$$b = \frac{[POOH]_\infty - [POOH]_0}{[POOH]_0}$$

so that

$$t_{ix} \sim \frac{1 - \ln y_0}{K}$$

where $y_0 = [POOH]_0/[POOH]_\infty$ is the reduced hydroperoxide concentration at $t = 0$ as previously defined.

The rate of oxygen absorption is given by

$$r = \frac{d[O_2]}{dt} = k_2[O_2][P^\bullet] + k_t[PO_2^\bullet]^2 = -K[POOH] - \left(\frac{k_b}{k_t}\right)^{1/2} \frac{d[POOH]}{dt} \qquad (22)$$

where

$$K = k_p[PH]\left(\frac{k_b}{k_t}\right)^{1/2}$$

The oxygen uptake is thus

$$C = [POOH]_\infty \left[\ln\frac{b + \exp Kt}{b+1} + \left(\frac{k_b}{k_t}\right)^{1/2} \left(\frac{1}{1 + b\exp(-Kt)} - y_0\right) \right] \qquad (23)$$

For long oxidation times ($t \gg K^{-1}$) this expression simplifies as follows:

$$C = [POOH]_\infty \left[Kt - \ln(1 + b) + \left(\frac{k_b}{k_t}\right)^{1/2}(1 - y_0) \right] \qquad (24)$$

Since $b \gg 1$ and $(k_b/k_t)^{1/2} \ll 1$, it becomes

$$C \sim [POOH]_\infty(Kt - \ln b)$$

so that $t_{ic} \sim \dfrac{\ln b}{K}$ \qquad (25)

We can see that the induction time for oxygen absorption is lower than the induction time for initiation/termination products.

$$\Delta t_i = t_{ix} - t_{ic} = \frac{1}{K} \qquad (26)$$

The relative difference $\Delta t_i/t_{ic} = 1/\ln b$ is presumably observable for low values of b

(i.e., high initial hydroperoxide concentrations) but not significant for very low $[POOH]_0$ values.

The stationary state rate of oxygen absorption is given by:

$$r_\infty = K[POOH]_\infty = \frac{k_p^2[PH]^2}{2k_t} \tag{27}$$

against $r_\infty = 2k_p^2[PH]^2/k_t$ for the unimolecular scheme.

If one supposes, as previously reported (40), that bimolecular POOH decomposition is favored at low temperatures, whereas the unimolecular one is favored at high temperatures, then the Arrhenius plot of the stationary state rate would have the shape of Fig. 6.

The numerical resolution of the bimolecular scheme is in progress, but here the investigation would be limited to a short discussion on time constants. Using the previously defined reduced variables, one obtains the following set of differential equations:

$$\frac{dx}{dt} = k_p[PH]\left(\frac{y^2}{2} + \frac{3}{2}z - x - x^2\right) \tag{28}$$

$$\frac{dy}{dt} = k_p[PH]\left(\frac{k_b}{k_t}\right)^{1/2}(x - y^2) \tag{29}$$

$$\frac{dz}{dt} = k_a[O_2]\left(\frac{y^2}{4} - \frac{3}{4}z + \frac{x}{2}\right) \tag{30}$$

The variables are interrelated by the following inequalities:

$$x > y^2; \qquad y^2 > \frac{3}{4}z; \qquad \text{and} \qquad \frac{9}{8}z > x^2$$

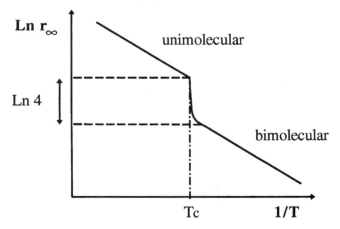

Figure 6 Presumed shape of the Arrhenius plot for the stationary state rate of oxygen absorption in the theoretical case of transition between bimolecular and unimolecular kinetic regimens.

These inequalities indicate that the species corresponding to x, y, and z must grow in time scales of the same order of magnitude. The time constants contained in the equations are

$$\tau_i = \left(k_p[PH]\left(\frac{k_b}{k_t}\right)^{1/2}\right)^{-1} \qquad \tau_p = \left(k_p[PH]\right)^{-1} \qquad \text{and} \qquad \tau_0 = (k_a[O_2])^{-1}$$

for respectively PO_2H, $PO_2^{\bullet}$, and $P^{\bullet}$ growth. In the aging conditions under consideration, $k_b \ll k_t$ so that $\tau_i \gg \tau_p \gg \tau_0$. Thus, because the slowest process governs the whole kinetics, one expects that the time constant characteristic of the autoacceleration is τ_i. In other words, the induction time must be sharply related to τ_i (as found in the analytical resolution where t_i is proportional to K^{-1} and $K^{-1} = \tau_i$).

F. Comparison of Unimolecular and Bimolecular Schemes

The common features of both schemes have been reported in the first paragraph of this chapter. The specific features are the following:

1. Influence of Initial POOH Concentration ([POOH]$_0$, y_0)

For given exposure conditions, the relations between the induction time and the initial hydroperoxide concentration are schematized in Fig. 7. At low POOH concentrations, the induction time tends to become independent of [POOH]$_0$ for the unimolecular process, whereas it increases continuously when [POOH]$_0$ decreases for the biomolecular process.

This property is interesting because it opens the way to experimental checking (see later). In certain domains, it could orient the research. One sees, for instance that in photochemical oxidations initiated by the unimolecular POOH decomposition (as generally postulated for POOH photolysis), the determination of the species responsible for initial photochemical acts is practically useless,

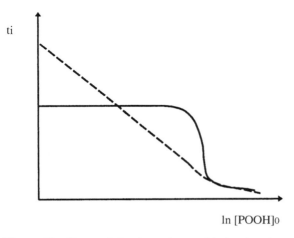

Figure 7 Shape of the dependence of induction time with the initial POOH concentration for the unimolecular (solid line) and bimolecular (dashed line) schemes.

and no significant result can be expected from tentative stabilization by polymer purification. In contrast, these efforts make sense in biomolecular initiation.

2. Role of the Initiation Rate Constant on the Duration of the Induction Period

In the unimolecular process, t_i depends on only the initiation rate constant k_u, whereas in the bimolecular process t_i depends on a composite rate constant K proportional to the square root of the initiation rate constant k_b. In photooxidation, this property, in principle, must allow one to identify the initiation mechanism from experiments made at different light intensities I. As a matter of fact, it can be assumed that k_u and k_b are proportional to I so that one expects that:

$$t_i \quad \alpha \quad I^{-1} \qquad \text{in the unimolecular case}$$

$$t_i \quad \alpha \quad I^{-1/2} \qquad \text{in the bimolecular case}$$

For thermal oxidation, attention must be focused on activation energies

For the unimolecular process: $E_i = E_u$
For the bimolecular process: $E_i = E_p + \frac{1}{2}E_b - \frac{1}{2}E_t$

where E_u, E_b, E_p, and E_t are the activation energies of k_u, k_b, k_p, and k_t. It is noteworthy that $E_p - \frac{1}{2}E_t = \frac{1}{2}E_s$, where E_s is the activation energy of the stationary state. Furthermore, $E_u > E_b$.

E_u is expected to be about 150 kJ mol^{-1} for pure hydroperoxides, but its value can be lowered by the presence of catalytic impurities, for instance Ti^{3+} from initiator residues.

3. Differences Between Induction Times Determined by Different Methods

The induction time for oxygen absorption t_i is always shorter than the induction time for initiation/termination products buildup t_{ix}:

$$t_{ix} - t_{iO_2} \sim \frac{0.5}{k_i} \qquad \text{(unimolecular scheme)}$$

$$t_{ix} - t_{iO_2} = \frac{1}{K} \qquad \text{(biomolecular scheme)}$$

That the induction time is higher for carbonyl buildup than for weight gain (linked to oxygen absorption) has been experimentally checked for PP (39), but the results cannot be used to distinguish between both mechanisms owing to data scatter.

4. Sharpness of the Transition Between the Autoaccelerated and the Stationary-State Regimens

The difference in the shape of kinetic curves is illustrated in Fig. 8. The transition is noticeably sharper for bimolecular than for unimolecular POOH decomposition.

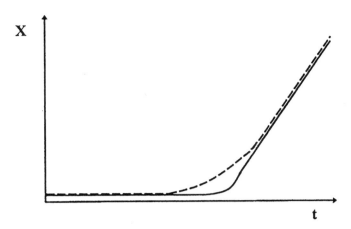

Figure 8 Shape of kinetic curves for bimolecular (solid line) and unimolecular (dashed line) processes having close induction times and the same stationary-state rate.

III. EXPERIMENTAL CHECKING OF THE MODELS

A. Introduction

The discussion will be essentially limited to the thermal oxidation of PP in the 80–150°C temperature range. Notice that there is a change of kinetic regimen at about 80°C; for instance, the activation energy of induction time is lower at less than 80°C than it is at a higher temperature (41), whereas there is no change in Arrhenius parameters for the stationary-state characteristics. In the frame of closed-loop schemes, this behavior indicates a change of the initiation mechanism, all the other reaction steps being unaffected.

One possible explanation (41) is linked to the presence of small quantities of catalytic residues responsible for a dual initiation mechanism, for instance in the unimolecular mechanism:

Uncatalyzed reaction	POOH	$\rightarrow$ 2P$^\bullet$	(k_u)
Catalyzed reaction	POOH + M$^+$	$\rightarrow$ γP$^\bullet$	(k_c)

The overall initiation rate is thus:

$$r_i = k_i[\text{POOH}] = (k_u + \gamma k_c[M^+])[\text{POOH}]$$

The induction time is given by:

$$t_i = \frac{2.5}{k_i} = \frac{2.5}{k_u + \gamma k_c[M^+]}$$

The activation energy of k_c is lower than that of the activation energy of k_u so that the Arrhenius plot of t_i is expected to have the shape of Fig. 9.

The transition is expected to occur at $t_c = \frac{E_u - E_c}{R \ln 2 k_{uo}/\gamma k_{co}[M^+]}$

where k_{u0} and k_{co} are the respective preexponential factors of k_u and k_c.

Here, only if $T > T_c$ ($\sim 80°C$) will be investigated because it corresponds to the domain at which most of the experimental data are available.

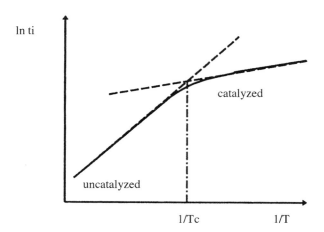

Figure 9 Schematic of the Arrhenius plot of induction time for catalyzed POOH decomposition.

We will try to answer these two successive questions:

1. Does the PP oxidation display characteristics of a closed-loop scheme?
2. In case of a positive answer to the first question, is it possible to identify the initiation mechanism?

B. Characteristics of the Steady State

In the chosen approaches, the steady state appears as an asymptotic straight-line, but this is because the substrate consumption and the occurrence of secondary reactions are not taken into account. Methods, such as oxygen absorption (25,26) or carbonyl titration (27,28), reveal however, the existence of linear sections of kinetics curves that are sufficiently long to be interpreted in terms of a stationary state.

The main characteristic of a stationary state, whatever the order of POOH decomposition, is that its rate is independent of the initiation rate constant k_i. If, thus we have the possibility to modify k_i, all the other parameters being constant, we expect to obtain kinetic curves having the shape of Fig. 10.

A first experimental way of checking consists of comparing samples differing by the concentration of catalytic impurities. For instance, Celina et al. compared the chemiluminescence emission of individual grains of a reactor powder (18). There are strong differences from grain to grain in the induction times, whereas the maximum rates differ considerably less. The differences in the induction times are presumably due to differences in the concentration of catalytic residues. Gijsman et al. (26) compared oxygen absorption kinetics of various PP samples differing by the concentration of voluntarily incorporated Ti^{3+} salts. Here, also, it appears that the catalyst plays an important role on the induction period, but no significant role on the stationary state rate (26).

Another possible way of checking the closed-loop nature of the oxidation mechanism consists of comparing photooxidation experiments differing only by the light intensity (which controls only the initiation rate). It is expected, in principle, that the stationary-state rate is independent of the light intensity. The closed-loop

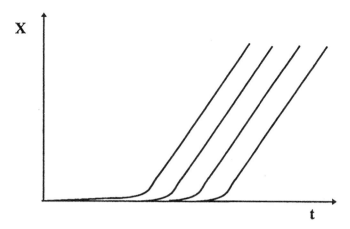

Figure 10 Expected shape of oxidation kinetic curves for samples differing by only the initiation rate constant k_i.

scheme, however, is valid only if ketones do not play an important role in initiation (through Norrish I process), which remains to be establish.

Anyhow, that catalysts do not modify the stationary-state rate can be considered a strong proof in favor of a closed-loop mechanism.

C. Characteristics of the Induction Period

Literature values of t_i or values graphically determined from literature curves have been compiled (21,42), and completed a few years later (39), to give the Arrhenius plot of Fig. 11.

Despite the great diversity of polymer sources (worldwide from 1950 to 1995), most of the points are close to a single straight-line of activation energy of about $100 \, kJ \, mol^{-1}$. This seems to be consistent with the hypothesis that POOH decomposition is unimolecular and that most of the samples were relatively clean (low initial concentrations of POOH, low concentrations of catalytic residues). As a matter of fact, in bimolecular initiation, this behavior would result from a coincidence or from an unexplained constancy of $[POOH]_0$.

The value of the apparent activation energy, higher than $100 \, kJ \, mol^{-1}$ (39), is also rather in favor of the unimolecular scheme, but not decisive.

D. Discussion

The foregoing results are globally in favor of the unimolecular mechanism, but they cannot be considered as rigorous proofs. Unfortunately, literature data do not allow one to suppress completely the ambiguity. Analytical studies, especially IR spectra in the stretching OH region (43,44) reveal that POOH groups are hydrogen-bonded, even at low conversions. This is in principle a situation favorable to bimolecular decomposition. However, H bonds are considerably weaker than O–O bonds so that they are expected to dissociate at moderately high temperatures (44,45), thus, allowing POOH groups to react separately. Therefore, the occurrence of POOH

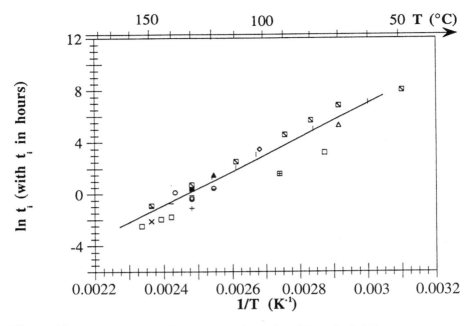

Figure 11   Arrhenius plot of literature values of t_i. (From Ref. 39.)

unimolecular decomposition is not necessarily contradictory to the existence of POOH clusters.

Kinetic studies of the decomposition of the polymer hydroperoxides in an inert atmosphere are expected to give a clear answer to the question of initiation molecularity, but here also, a doubt remains. As a matter of fact, Chien and Jabloner (44) and, more recently, Gijsman et al. (22), by chemical titration obtained POOH decay curves interpretable in terms of superposition of two first-order processes, corresponding to the coexistence of fast- and slow-decomposing hydroperoxides. Billingham et al. (46) used chemiluminescence and obtained decay curves well fitted, in the whole conversion range, by the second-order model. However, here, the mechanism of light emission remains unknown, and the intensity is not necessarily proportional to the POOH decomposition rate, especially if chemiluminescence results from the $PO_2^{\bullet}$ bimolecular combination, as frequently proposed.

A possible way of investigation this domain consists of studying the transition between the two modes of initiation. The first example is relative to the transition between uncatalyzed and catalyzed POOH decomposition. The redox catalysis of POOH thermolysis has been widely studied (6,26,37). It is always described, as a first-order (relatively to POOH) process. At high catalyst concentration, we are thus in unimolecular scheme. At a low catalyst concentration, in contrast, most of the POOH decomposition events are uncatalyzed and the mechanism is unknown. There are two possibilities:

1. The uncatalyzed initiation mechanism is unimolecular: the stationary-state rate r_∞ takes the same value at low and high catalyst concentration.

2. The uncatalyzed initiation mechanism is bimolecular: r_∞ must be $k_p^2[PH]^2/2k_t$ at low catalyst concentration and $2k_p^2[PH]^2/k_t$ at high catalyst concentration.

Experimental results have been published (26) for Ti^{3+} catalyzed PP thermal oxidation. It appears that the stationary-state rate is almost independent—within experimental scatter—of the Ti^{3+} concentration, which seems to be clearly in favor of the hypothesis of unimolecular POOH decomposition.

At this stage of our investigation, it seems that the unimolecular scheme must be preferred to the bimolecular one, although certain points, such as the sharpness of autoacceleration at the end of the induction period, remain to be elucidated. Indeed, it is possible to envisage the combination of both processes. Here, one can imagine that there is a critical hydroperoxide concentration corresponding approximately to

$$[POOH]_c \sim \frac{k_u}{k_b}$$

such as the unimolecular decomposition predominates in initial steps until $[POOH] = [POOH]_c$, and then the bimolecular decomposition becomes responsible for most of the initiation steps. Then, the induction period duration would be essentially scaled by the unimolecular rate constant k_u, but the transition at the end of the induction period and the stationary-state characteristics would be governed by the bimolecular process.

From a practical point of view, one can be sure that there is a pair of k_u and k_b values able to give a good fit of experimental curves. There is no peculiar difficulty in resolving numerically the system of differential equations corresponding to this scheme, but the inverse problem (i.e., the identification of k_u and k_b values) from experimental results, appears to be very difficult.

IV. KINETIC MODELING AND SPATIAL HETEROGENEITY OF OXIDATION

A. Introduction

Large-scale heterogeneities can be observed in the oxidation of industrial parts. They can be due to processing (inhomogeneous thermal and shear fields; skin-core structure, owing to fast cooling; insufficient additive dispersion, and so on), or to oxidation thickness gradients, resulting from O_2 diffusion control of the kinetics, or from screen effects in photooxidation. These heterogeneities will not be considered here, but it will be kept in mind that they can affect the conclusions of kinetic analyses. For instance molecular weight measurements made on samples of thicknesses more than about 100 μm would be questionable. The object of this chapter is the study of the effect of small-scale heterogeneity, as evidenced by optical microscopy associated with imaging techniques. There are three main causes of this heterogeneity:

1. Morphology: crystalline domains are impermeable to oxygen so that oxidation occurs only in the amorphous phase (47–49). The interspherulitic amorphous phase is characterized by a higher content of irregular structures and a higher molecular mobility than the

intraspherulitic (interlamellar) phase, hence, it could be more reactive than this latter (50,51).

2. Low molecular mobility could favor local propagation around initiation centers (51–55). The formation of colored spots in thermally aged PP samples can be attributed to this characteristic (56). Monte Carlo simulations of the oxidation spreading have been proposed (57).

3. Intramolecular propagation is important in PP (16). It can justify the existence of hydrogen-bonded hydroperoxide clusters, even at low conversions (43).

The oxidation heterogeneity at the scale accessible by optical microscopy and at the morphological scale is unquestionable (58). But what is important from the practical point of view is the small-scale distribution of oxidation rates. Is it or is it not homogeneous? The answer is not trivial because the main cause of observable heterogeneity is presumably the crystalline morphology. One can imagine situations in which, despite the existence of experimental evidences of heterogeneity, kinetic modeling is valid because oxidation would be homogeneous at the small scale. Is it possible to quantify the degree of heterogeneity? To characterize this latter by some failure in kinetic modeling? To take it into account in the build-up of new kinetic models? The aim of this section is to explore some ways of investigating these domains.

B. Quantitative Approaches of Heterogeneity from Molecular Weight Distribution Studies

Let us consider a volume unit of the polymer divided in N equal microdomains, having a volume v sufficiently small to be homogeneous. If N_i is the number of the microdomains of type i for which the oxidation rate at a given time is r_i, one can define a distribution function: $N_i = f(r_i)$ of oxidation rates at the time t.

In a semicrystalline polymer, there is a nonreactive fraction Nx_c, where x_c is the crystallinity ratio. What is important in practice is the rate distribution within the amorphous phase. Some possible cases are shown schematically in Fig. 12.

In Fig. 12a; low polydispersity, oxidation is homogeneous in the amorphous phase. The local rate r is linked to the overall (macroscopic) rate $\bar{r}$ by:

$$r = \bar{r}/(1 - x_c)$$

In Fig. 12b, high polydispersity, there is no simple relation between the local rate r_i and the overall rate $\bar{r}$. Classic kinetic modeling is, in principle, not valid. In Fig. 12c, the rate distribution is binodal, which could correspond to the interor intraspherulitic amorphous phase duality. Many subcases can be imagined. The one presented in the Fig. 12c is when oxidation is homogeneous in each phase, but with distinct rates. In this case, classic kinetic modeling, in principle, can be used. One expects a two-step kinetic behavior (Fig. 13).

Generally, experimental curves reveal a one-step behavior, even when very sensitive methods are used (18,22,23). Nevertheless, chemiluminescence imaging seems to indicate that interspherulitic domains are considerably brighter than intraspherulitic ones (19,56,59). Is it an artifact linked to some screen effect of crystallites? Or is oxidation of intraspherulitic domains so slow that we generally

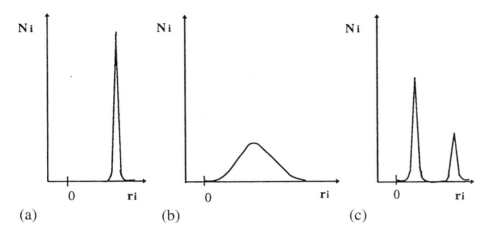

Figure 12 Some typical cases of oxidation rate distribution: (a) low polydispersity; (b) high polydispersity; (c) binodal rate distribution.

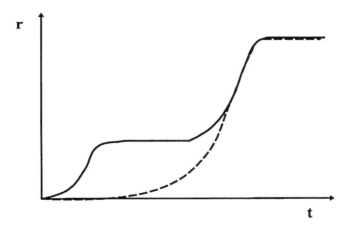

Figure 13 Expected shape of the time change of oxidation rate for a binodal rate distribution of type (c). The generally observed behavior is schematized by the dashed line.

observe only the first stage corresponding to the oxidation of the interspherulitic domains?

Molecular weight distribution (MWD) determinations, in principle, must allow us to answer these questions. As a matter of fact, the distributions schematized in Fig. 12 must lead to the changes of MWD schematized in Fig. 14.

Before a discussion of these examples, let us recall some possible causes of a misinterpretation of MWD data (generally obtained by steric exclusion chromatography at $T > 150°C$):

1. Measurements are to be made on thin films or on microtome sections parallel to the exposed surface of bulk samples. If the sample thickness

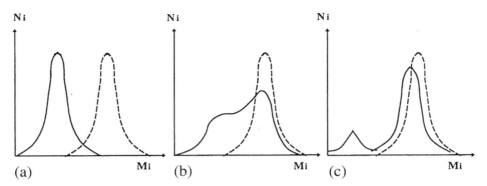

Figure 14 Schematic molecular weight distribution before (dashed line) and after (solid line) degradation for (a), (b), and (c) of Fig. 12.

    is higher than typically 50 μm, oxidation, gradients caused by oxygen diffusion control or screen effect can contribute to an increase of the polydispersity.

2.   The number-average molar mass $\bar{M}_n$ is sometimes difficult to determine accurately because there is practically a continuum in the oligomer range (0–2000 g mol^{-1}), whereas calculations need a truncature of the distribution ($\bar{M}_n$ is overestimated, whereas the $\bar{M}_w$ value is more accurate).

3.   At the measurement temperature, the hydroperoxide and peroxide lifetime is considerably shorter than the measurement time so that secondary thermolysis can affect the quantitative data.

These problems have been discussed recently (60), and it was concluded that polydispersity index values can be used to discuss the homogeneity on heterogeneity of the reaction. In industrial polyolefins, such as PE or PP, the initial polydispersities are generally relatively high: $\bar{M}_w/\bar{M}_n > 3$. Here, an homogeneous, random chain scission must lead to a polydispersity decrease until an asymptotic value of 2 (61). Indeed, any heterogeneity is expected to increase the polydispersity, as illustrated in Fig. 14. Furthermore, when oxidation would occur principally in the interspherulitic zone, as suggested by chemiluminescence imaging (59), we would expect a binodal distribution with a slow variation of the main component as schematized in Fig. 14c.

Experimental MWD data obtained in PP photooxidation (62), similar to that of thermooxidation (63), can be summarized as follows:

1.   No residual peak corresponding to an eventual unreactive phase is observed.

2.   No new peak at low M_w; no shoulder in the main peak is observed.

3.   The polydispersity index decreases significantly.

Thus, PP low-temperature oxidation seems to belong to example (a) of Fig. 14 (i.e., to homogeneous oxidation). It is noteworthy that using steric exclusion chromatography having a relative sensitivity of 10% in terms of the molar mass, we are able to detect a number of chain scission events n (min i) such as:

$$n(\text{min } i) \sim \frac{0.1}{\overline{M}_{no}}$$

For an industrial sample having $\overline{M}_{no}$ of approximately 2.10^5 g mol^{-1}, this corresponds to n (min i) of about 5.10^{-7} mol g^{-1}. Common analytical methods, such as IR spectrophotometry, are not so sensitive. It can be effectively observed that $\overline{M}_n$ begins to decrease during the induction period for oxygen absorption (62,63).

That the polydispersity decreases and, thus, the oxidation behaves as a homogeneous random-chain scission process despite the existence of crystallites, indicates that each chain probably possesses many amorphous and crystalline segments as illustrated by Fig. 15. Oxidative chain scissions can occur only on the ap or f chain portions. If their number per chain is sufficiently high, the system can behave as a random chain scission, whereas it is not, in reality, a true random chain scission. This problem would merit a detailed investigation, but here, it will be only noted that the results disagree with the hypothesis of a marked heterogeneity.

C. Kinetic Approaches to Heterogeneity

There are basically two ways to consider heterogeneity: the first one is to start from the concept of spatial heterogeneity, whereas the second one is to convert spatial heterogeneity into temporal heterogeneity and to characterize this latter by a distribution of induction times (or any other aging characteristic time). The chosen approach can, in fact, depend on the cause of heterogeneity. At least three distinct cases have to be considered:

1. Heterogeneous Distribution of Initially Present Radical Sources (POOH)

Let us consider a partition of the sample in equal volume elements the size of which is such that each volume element initially contains at least one radical source (i.e., mathematically speaking one POOH group). In the chosen conditions, each radical species involved in the chain process: ($P^\bullet$, $PO_2^\bullet$, $PO^\bullet$, $OH^\bullet$) is characterized by its diffusivity D and its lifetime τ, from which one can determine an average migration length $d \sim (D\tau)^{\frac{1}{2}}$. Let us consider the highest pathlength d_M presumably associated with the most mobile species $OH^\bullet$. If one compares d_M with the size of volume elements l (which is a decreasing function of the initial POOH concentration), one can distinguish between two cases:

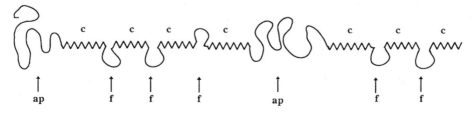

Figure 15   Schematic of the structure of a given chain: ap, chain portion in the amorphous phase; c, chain portion belonging to the crystalline phase; f, chain fold.

$d_M \geq l$:   here, the classic kinetic modeling is, in principle, valid.

$d_M \ll l$:   here, oxidation is expected to propagate locally around initiation sites, by growth of oxidized domains, the classic kinetic modeling is not valid. This situation has been studied by Celina et al. (59), and a new kinetic model has recently been proposed (64,65).

The case where $d_M \geq l$ can, in fact, be divided in two subcases:

1.   If initiation is unimolecular and $[POOH]_0$ is not too high, there must be a tendency toward homogenization because oxidation kinetics are almost independent of $[POOH]_0$.
2.   If, on the contrary, initiation is bimolecular, the oxidation rate of a given domain will be an increasing function of the local initial POOH concentration, and autoacceleration will amplify the initial differences so that oxidation will appear heterogeneous. Although this problem would need a more detailed investigation, it seems reasonable to suppose that if oxidation displays the characteristic features of an homogeneous reaction, this means that initiation is probably unimolecular. In bimolecular initiation, homogeneity would probably result from peculiar circumstances.

2.   Heterogeneous Distribution of Redox Catalysts

Metal salt particles are more or less homogeneously distributed in the sample volume. They accelerate local oxidation so that they potential sources of heterogeneity. In the presence of catalysts, the POOH decomposition is expected to be unimolecular, and the catalyst is expected to affect essentially the initiation rate constant k_u (see foregoing). According to the unimolecular scheme characteristics, local variations of the catalyst concentration are thus expected to lead to local variations of the induction time ($t_i \sim 3/k_i$). If we are able to find a quantity varying nonmonotonically with the aging time, then the characteristic time associated with this variation will depend on the degree of heterogeneity of the reaction, as illustrated by the example of stabilized polymers (Fig. 16).

If oxidation is homogeneous, all the volume elements are synchronous, they display the same induction time, so that the transition at the end of the induction period is sharp. If oxidation is heterogeneous, there is a spreading of induction times and the transition appears more progressive than in the preceding case. Here, the width of the second derivative peak can be considered as a measure of the degree of heterogeneity.

If homogeneity is judged by the sharpness of the transition at the end of the induction period, for instance, from the value of $\Delta t_i / t_i$, Δt_i being the second derivative peak width, no doubt, the oxidation of stabilized polyolefins is generally highly homogeneous. Have stabilizers a homogenizing effect? There are published data, for instance on thermal oxidative of cross-linked polyethylene, where it clearly appears that antioxidant diffusion cannot play an homogenizing effect in the time scale of the transition (66) because it is too slow.

Unfortunately this method cannot be easily used in all the cases of unstabilized polymers because sometimes the maximum of the second derivative lies in a region where the conversion is too low to permit accurate determinations; however, gravimetric curves can be used (see following).

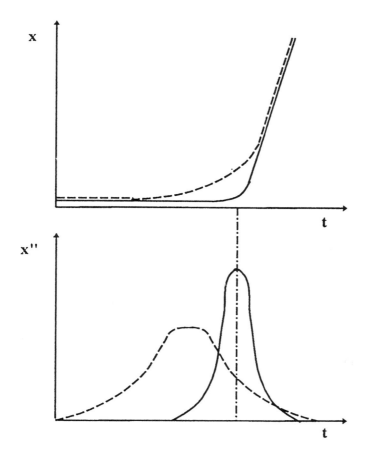

Figure 16 Shape of oxidation kinetic curves (top) and their second derivative (bottom) for the thermooxidative aging of stabilized polyolefins: solid line, experimental data and theoretical curve for homogeneous degradation; dashed line, heterogeneous degradation.

3. Heterogeneous Distribution of Molecular Mobilities

Below a certain critical temperature, propagation and termination, which involve segmental motions, must become diffusion controlled. If there are local variations of the molecular mobility, linked for instance to the proximity of the crystalline phase, then one expects variations of the ratio $k_p^2[PH]^2/k_t$. In the frame of the bimolecular scheme, they would have an influence on the induction time, and the transition would be broadened, whereas in the frame of the unimolecular scheme, they would affect only the value of the stationary state rate, but not the shape of oxidation curves.

D. Kinetic Investigations on Gravimetric Curves

Mass variations during thermooxidation (24,36,39,67) as well as during photooxidation (68) of polyolefins generally have the shape of Fig. 17. The curves display an induction period, a positive maximum, followed by a mass loss of which

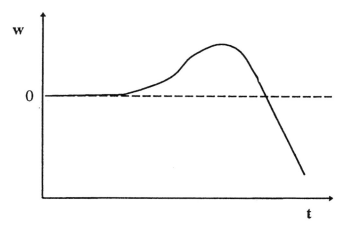

Figure 17 Shape of an isothermal gravimetric curve for low-temperature oxidation of hydrocarbon polymers.

the rate tends, more or less rapidly, toward an asymptotic value. This is the existence of a steady state during which the loss of molecular fragments, resulting from chain scissions, predominate largely over the mass gain from oxygen incorporation into the polymer.

At least in a first approach, closed-loop schemes can be used to model this process, assuming that the formation of volatile molecules occurs during the initiation or termination steps (both being kinetically equivalent), and that mass loss is not diffusion controlled (thin films).

In the unimolecular case, the initiation step can be written

$$POOH \quad \rightarrow \quad 2P^{\bullet} + H_2O + \beta V \qquad (k_u)$$

H_2O, coming from $OH^{\bullet}$ radicals, is distinguished from the other volatile molecules, which are represented by an average species V of molar mass M_v, formed with a yield β.

The rate of mass variation is then:

$$\frac{dW}{dt} = \frac{32}{\rho}\frac{d[O_2]}{dt} - \frac{18}{\rho}\frac{d[H_2O]}{dt} - \frac{\beta M_v}{\rho}\frac{d[V]}{dt}$$

where dW/dt is the rate of mass variation per polymer mass unit and ρ the polymer density. By using the previously established equations it becomes

$$\frac{dW}{dt} = \frac{k_p^2[PH]^2}{\rho k_t}\left[(46 - \beta M_v) - (60 - 2\beta M_v)\exp\left(-\frac{k_u t}{2}\right) + (14 - \beta M_v)\exp(-k_u t)\right]$$

The stationary state rate is

$$\left(\frac{dW}{dt}\right)_\infty = \frac{k_p^2[PH]^2}{\rho k_t}(46 - \beta M_v)$$

It is negative (mass loss), so that $\beta M_v > 46$ g mol^{-1}.
    The maximum is given by

$$\left(\frac{dW}{dt}\right)_M = 0 \quad \rightarrow \quad t_M = \frac{2}{k_u}\ln\frac{\beta M_v - 14}{\beta M_v - 46}$$

For PP, it can be observed that $t_M \sim t_i \sim 3/k_u$, so that

$$\ln\frac{\beta M_v - 14}{\beta M_v - 46} \sim \frac{3}{2} \qquad \beta M_v \sim 55 \text{ g mol}^{-1}$$

In the bimolecular case, the initiation step can be written:

$$POOH + POOH \rightarrow P^\bullet + PO_2^\bullet + H_2O + \beta M_v \qquad (k_b)$$

By using the same reasoning as before, one obtains:

$$\frac{dW}{dt} = \frac{[POOH]_\infty}{\rho}\frac{1}{[(1 + b\exp(-Kt)]^2}\left[K\left(23 - \frac{\beta M_v}{2}\right) + 32Kb\left(1 + \sqrt{\frac{k_b}{k_t}}\right)\exp(-Kt)\right]$$

where

$$[POOH]_\infty = \frac{k_p^2[PH]^2}{2(k_b k_t)^{1/2}}; \qquad K = \left(k_p[PH]\sqrt{\frac{k_b}{k_t}}\right) \qquad \text{and}$$

$$b = \frac{[POOH]_\infty - [POOH]_0}{[POOH]_0}$$

the condition for a predominating mass loss in a stationary state is the same as in the unimolecular case:

$$\beta M_v > 46 \text{ g mol}^{-1}$$

The time of maximum mass gain t_M is given by (provided that $\sqrt{(k_b/k_t)} \ll 1$):

$$t_M \approx \frac{1}{K}\ln\frac{64\,b}{\beta M_v - 46}$$

For low initial POOH concentrations ($b \gg 1$), it can be recalled that the induction time is

$$t_i \sim \frac{1}{K}(\ln b + 1)$$

Since, experimentally, it has been found that $t_M \sim t_i$, it becomes

$$\ln\left(\frac{64}{\beta M_v - 46}\right) \sim 1 \qquad \text{so that} \qquad \beta M_v \sim 70 \,\mathrm{g\,mol^{-1}}$$

The value of βM_v is higher than in the unimolecular case, but of the same order.

Relatively fast mass loss has been observed at relatively low temperature; for instance, in the photooxidation at 40°C (68), or in thermal oxidation at 50°C (39) for PP. This indicates that volatile products have a relatively low molar mass; thus, β is not far from unity. Thus, we have to imagine very efficient mechanisms of volatile formation to explain the predominance of mass loss, for instance sequential back-biting reactions (68). A possible alternative could be, for instance, acetone release from β scission (69) (Fig. 18). The efficiency of such a hypothetical process depends on the efficiency of the 1–2 hydrogen migration. This latter is thermodynamically favored because tertiary radicals are considerably more stable than primary ones, but indeed, the oxidation of primary radicals could be competitive.

The shape of gravimetric curves calculated from the models is illustrated in Fig. 19. One can see that bimolecular curves are closer to experimental ones than are unimolecular curves.

To simulate heterogeneity, one can start from the hypothesis that a given quantity X, for instance, the initiation rate constant k_i or the initial POOH concentration y_0 is nonuniformly distributed in the sample volume. The distribution function could, for instance, be gaussian:

$$Y(X) = \frac{1}{\sigma\sqrt{2\pi}} \exp\left[-\frac{(X - \bar{X})^2}{2\sigma^2}\right]$$

where σ is the standard deviation characterizing the degree of heterogeneity and is the average value of $\bar{X}$ (Fig. 20).

The distribution is arbitrarily truncated at k_0 and k_n.

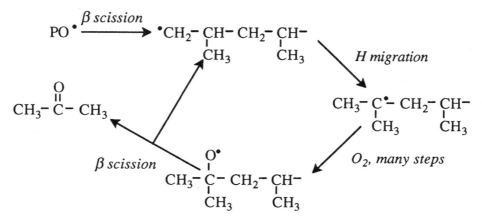

Figure 18 Hypothetical mechanism of acetone formation from alkoxy β-scission. (From Ref. 69.)

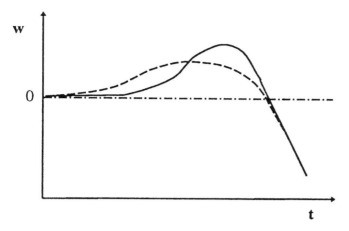

Figure 19 Shape of gravimetric curves for unimolecular (dashed line) and bimolecular (solid line) models. The dashed straight line indicates zero level of weight loss.

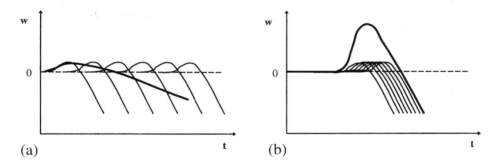

Figure 20 Shape of the gaussian distribution for the kinetic variable X.

If the volume is divided in N elements, the number of elements in which $k_i = k_j$ can be written: $k_j = k_0 + jh$

$$n(k_j) = \frac{N}{\sigma\sqrt{2M}} \int_{k_0+(j-1)h}^{h_0+jh} \frac{\exp\left[-\left(x - \bar{k}_i\right)^2\right]}{2\sigma^2} \, dx$$

where h is an arbitrarily small interval of k_i and $\bar{k}_i$ is the average value of k_i. For a volume element in which the rate constant is k_j, the mass change is

$$m_\infty(t) = m_0 + M(t, k_j) \qquad \text{where } M(t, k_j) \text{ is the mass increase at time } t.$$

Thus, for the whole sample

$$M(t) = \sum_{j=0}^{n} m_j(t) = m_0 N + M(t, k_0) + M(t, k_0 + h) + \dots$$

where $m_0 N$ is the initial mass, so

$$\Delta M_t = M(t) - M(0) = \sum_{j=0}^{n} n(k_j) \, M(t, k_j)$$

and

$$W + \frac{M(t) - M(0)}{M(0)} = \frac{1}{m_0 N} \sum_{j=0}^{n} n(k_j) - M(t, k_j)$$

that is,

$$W = \frac{1}{m_0 \sigma \sqrt{2H}} \sum_{j=0}^{n} \left[\int_{k_0 + (j-1)(6\sigma/n)}^{n_0 + j(6\sigma/n)} \exp\left(-\frac{(x - \bar{k}_i)^2}{2\sigma^2}\right) dx \right] M\left(t, k_0 + \frac{6\sigma j}{n}\right)$$

taking $k_0 = \bar{k}_j - 3\sigma$. In this case, the integral is 99.73% of the whole gaussian integral. An example of computation for the bimolecular model is shown in Fig. 21.

The same approach has been used assuming a gaussian distribution of $[POOH]_0$ (k_i being constant) for the bimolecular scheme (Fig. 22).

In both simulations, it appears that experimental data are in agreement with the hypothesis of a low degree of heterogeneity, for the model (bimolecular) giving the sharper discontinuity in the gravimetric curve. It seems that in the frame of the chosen models, the results are clearly in favor of the hypothesis of homogeneous degradation.

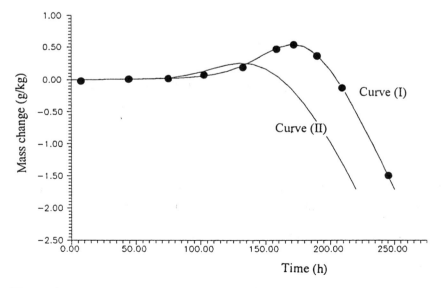

Figure 21 Kinetic modeling of mass changes in PP thermal oxidation at 90°C: points, experimental results; curve (I), bimolecular model with a gaussian distribution of k_i using $\sigma = 0.001$; curve (II), bimolecular model with a gaussian distribution of k_i using $\sigma = 0.017$.

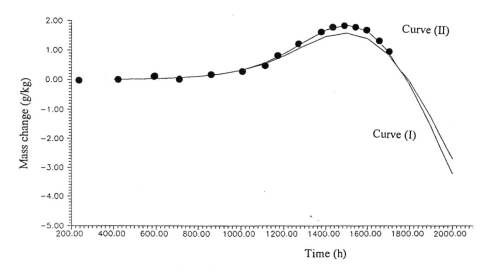

Figure 22 Kinetic modeling of mass changes in PP thermal oxidation at 60°C: points, experimental results; curve (I), model with a gaussian distribution of y_0 using $\sigma = 2.10^{-6}$; curve (II), model with a gaussian distribution of y_0 using $\sigma_0 = 2.10^{-4}$.

V. CONCLUSIONS

Oxidation of semicrystalline hydrocarbon polymers in a solid state is heterogeneous. Even though classic kinetic modeling is based on the hypothesis that hydroperoxides are the unique source of radicals, closed-loop schemes display good predictive properties. The models, for instance, predict the following:

1. The existence of an induction period, followed by a stationary state during which the decomposition of POOH equilibrates their formation.
2. The stationary-state rate is independent of the initiation rate constant (linked, for instance, to the catalyst concentration). This is a general property of closed-loop models.
3. The apparent activation energy decreases with the conversion.
4. The induction period is almost independent of the initial state (this is a property of unimolecular schemes).
5. The induction time is of the order of the POOH lifetime determined independently (also a property of unimolecular schemes).
6. The induction time is shorter for oxygen absorption than for carbonyl buildup.

Most of the kinetic data are in favor of a unimolecular POOH, rather than a bimolecular one, but some aspects, for instance, the sharpness of the transition between induction period and stationary state, remain somewhat ambiguous and need supplementary research.

To justify the use of classic kinetic modeling, a quantitative approach to the problem of heterogeneity has been tried. Two relatively strong proofs of homogeneity have been found:

1. Polydispersity does not increase during oxidation.
2. The sharpness of the transition at the end of the induction period is close to the theoretical sharpness of a homogeneous process.

From the practical point of view, closed-loop models have the advantage of being simple. The unimolecular model contains only two parameters: $k_p^2 PH^2/k_t$ and k_u, whereas the bimolecular model contains three parameters $k_p[PH]/\sqrt{k_t}$, k_b, and $[POOH]_0$. It remains to be explained in detail why such simple models are able to represent so complex a reality. The problem of termination mechanisms in PP is, for instance, especially interesting because it is suspected that primary radicals coming from β-scissions play a key role in terminations. This aspect is not taken into account in the models under study. A study of the extension of these models to stabilized systems is in progress at our laboratory.

REFERENCES

1. JLB Bolland, G Gee. Kinetic studies in the chemistry of rubber and related materials II. The kinetics of oxidation of unconjugated olefins. Trans Faraday Soc 42:236–243, 1946.
2. JLB Bolland, G Gee. Kinetic studies in the chemistry of rubber and related materials III. Thermochemistry and mechanisms of olefin oxidation. Trans Faraday Soc 42:244–252, 1946.
3. WL Hawkins. In: The Thermal Oxidation of Polyolefins. Mechanisms of Degradation and Stabilization. New York: John Wiley & Sons, 1972, pp 77–94.
4. L Reich, SS Stivala. Oxidation of simple hydrocarbons in the absence of inhibitors and accelerators. In: Autoxidation of Hydrocarbons and Polymers, Kinetics and Mechanisms. New York: Marcel Dekker, 1969, pp 31–127.
5. VS Pudov, MB Neiman. The thermo-oxidative degradation of isotactic polypropylene. In: AS Kuz'minskii, ed. The Ageing and Stabilisation of Polymers. New York: Elsevier Publishing, 1971, pp 1–27.
6. Y Kamiya, E Niki. Oxidative degradative. In: HHG Jellinek, ed. Aspects of Degradation and Stabilisation of Polymers. New York: Elsevier Scientific, 1978, pp 79–147.
7. G Papet, L Audouin-Jirackova, J Verdu. Diffusion controlled radiochemical oxidation of low density polyethylene—II. Kinetic modelling. Radiat Phys Chem 33:329–338, 1989.
8. C Anton-Prinet, J Dubois, G Mur, M Gay, L Audouin, J Verdu. Photoageing of rigid PVC II. Degradation thickness profiles. Polym Degrad Stabil 60:275–282, 1998.
9. C Anton-Prinet, G Mur, M Gay, L Audouin, J Verdu. Photoageing of rigid PVC III. Influence of exposure conditions on the thickness distribution of photoproducts. Polym Degrad Stabil 60:283, 1998.
10. SS Stivala, L Reich, PG Kelleher. Kinetics of thermal oxidation of isotactic polypropylene by infrared spectroscopy. Makromol Chem 59:28–42, 1962.
11. SG Kiryushkin, YA Shlyapnokov. Diffusion-controlled polymer oxidation. Polym Degrad Stabil 23:185–192, 1989.
12. CR Boss, JCW Chien. Oxygen diffusion limitation in autoxidation of polypropylene. J Polym Sci A-1 4:1543–1551, 1966.
13. JH Adams. Analysis of the non volatile oxidation products of polypropylene. I. Thermal oxidation. J Polym Sci A-1 8: 1077–1090, 1970.
14. DJ Carlsson, DM Wiles. The photooxidation of polypropylene films III. Photolysis of polypropylene hydroperoxides. Macromolecules 2:597–606, 1969.
15. GA George, M Celina, AM Vassallo, PA Cole-Clarke. Real-time analysis of the thermal oxidation of polyolefins by FT-IR emission. Polym Degrad Stabil 48:199–210, 1995.

16. NC Billingham. Localization of oxidation in polypropylene. Makromol Chem Makromol Symp 28:145–163, 1989.

17. M Celina, GA George. A heterogeneous model for the thermal oxidation of solid polypropylene from chemiluminescence analysis. Polym Degrad Stabil 40:323–335, 1993.

18. M Celina, GA George, NC Billingham. Physical spreading of oxidation in solid polypropylene as studied by chemiluminescence. Polym Degrad Stabil 42:335–344, 1993.

19. DJ Lacey, V Dudler. Chemiluminescence from polypropylene. Part 1: Imaging thermal oxidation of unstabilized film. Polym Degrad Stabil 51:101–108, 1996.

20. G Geuskens, F Debie, MS Kamamba, G Nedelkos. New aspects of the photooxidation of polyolefins. Polym Photochem 5:313–331, 1984.

21. FR Mayo. Relative reactivities in oxidation of polypropylene and polypropylene models. Macromolecules 11:942–946, 1978.

22. P Gijsman, J Hennekens, J Vincent. The mechanism of the low temperature oxidation of polypropylene. Polym Degrad Stabil 42:95–105, 1993.

23. GA George, GT Egglestone, SZ Riddel. Chemiluminescence studies of the degradation and stabilization of polymers. Polym Eng Sci 23:412–418, 1983.

24. L Matisova-Rychla, J Rychly, J Verdu, LA Audouin, K Csmorova. Chemiluminescence and thermogravimetric study of thermal oxidation of polypropylene. Polym Degrad Stabil 49:51–55, 1995.

25. HC Beachel, DL Beck. Thermal oxidation of deutereted polypropylenes. J Polym Sci A 3:457–468, 1968.

26. P Gijsman, J Hennekens J Vincent. The influence of temperature and catalyst residues on the degradation of unstabilized polypropylene. Polym Degrad Stabil 39:272–277, 1993.

27. JCW Chien, CR Boss. Polymer reactions V. Kinetics of autoxidation of polypropylene. J Polym Sci A-1 5:3091–3101, 1967.

28. F Tüdös, M Iring, T Kelen. Oxidation of polyolefins. In: A Patsis, ed. Proceedings of the International Conference on Advances in Stabilization and Controlled Degradation of Polymers. Luzern 1985, pp, 86–98.

29. DJ Carlsson, DM Wiles. The photooxidative degradation of polypropylene. Part I. Photooxidation and photoinitiation process. J Macromol Sci Rev Macromol Chem C 14:65–106, 1976.

30. G Geuskens, F Debie, MS Kamamba, G Nedelkos. New aspects of photooxidation of polyolefins. Polym Photochem 5:313–331, 1984.

31. G Scott. Development in the photooxidation and photostabilization of polymers. Polym Degrad Stabil 10:97–125, 1985.

32. L Audouin, V Gueguen, A Tcharkhtchi, J Verdu. "Close loop" mechanistic scheme for hydrocarbon polymer oxidation. J Polym Sci Polym Chem 33:921–927, 1995.

33. S Verdu, J Verdu. A new kinetic model for PP thermal oxidation at moderate temperature. Macromolecules 30:2262, 1997.

34. L Matisova-Rychla, Z Fedor, J Rychly. Decomposition of peroxides of oxidized polypropylene studies by the chemiluminescence method. Polym Degrad Stabil 3:371–382, 1981.

35. P Gijsman, J Hennekens, J Vincent. The role of peroxides in the low temperature oxidation of polypropylene. In: Proceedings of 11th International Conference on Advances in the Stabilization and Controlled Degradation of Polymers. Luzern, Switzerland, 1989.

36. J Rychly, L Matisova-Rychla, K Csmorova, L Achimsky, L Audouin, A Tcharkhtchi, J Verdu. Kinetics of mass change in oxidation of polypropylene. Polym Degrad Stabil 58:269–279, 1997.

37. DM Brown, A Fish. The extension to long-chain alkanes and to high temperatures of the hydroperoxide chain mechanism of autoxidation. Proc R Soc A 308:547–568, 1969.

38. V Stannet, RB Mesrobian. Discuss Faraday Soc 14:9, 1953.

39. L Achimsky, L Audouin, J Verdu. Kinetic study of the thermal oxidation of polypropylene. Polym Degrad Stabil (in press).

40. F Gugumus. Re-examination of the role of hydroperoxides in polyethylene and polypropylene: chemical and physical aspects of hydroperoxides in polyethylene. Polym Degrad Stabil 49:29–50, 1995.

41. L Achimsky, L Audouin, J Verdu. On a transition at 80°C in polypropylene oxidation kinetics. Polym Degrad Stabil (in press).

42. JCW Chien, TDS Wang. Autoxidation of polyolefins. Absolute rate constants and effect of morphology. Macromolecules 8:920:928, 1975.

43. JCW Chien, EJ Vandenberg, H Jabloner. Polymer reactions III. Structure of polypropylene hydroperoxide. J Polym Sci A-1 6:393–402, 1968.

44. JCW Chien, H Jabloner. Polymer reactions IV. Thermal decomposition of polypropylene hydroperoxides. J Polym Sci A-1 6:402, 1968.

45. R Hiatt. Hydroperoxides. In: D Swern, Ed. Organic peroxides. New York: Wiley-Interscience, 1971, pp 45–46.

46. NC Billingham, ETH Then, JP Gijsman. Chemiluminescence from peroxides in polypropylene. Part I. Relation of luminescence to peroxide content. Polym Degrad Stabil 34:263–277, 1991.

47. B Knight, PD Calvert, NC Billingham. Localization of oxidation in polypropylene. Polymer 26:1713–1718, 19.

48. Z Osawa, T Saito, Y Kimura. Thermal oxidation of fractionated polypropylene in solution. J Appl Polym Sci 22:563–569, 1978.

49. HP Frank, H Lehner. Thermooxidative and photooxidative aging of polypropylene; separation of heptane soluble and insoluble fractions. In: DL Allara, L Hawkins, ed. Stabilization and Degradation of Polymers. Adv Chem Ser 169. Washington DC: ACS, 1978.

50. NC Billingham, GA George. Chemiluminescence from oxidation of polypropylene: some comments on a kinetic approach. J Polym Sci B Polym Phys 28:257–265, 1990.

51. VB Miller, MB Neiman, YA Shlyapnikov. Vysokomol Soed 1:1690–1703, 1959.

52. M Kryszewski, M Mucha. The influence of morphology of semicrystalline polymers on their thermooxidative stability. In: Proceedings of 11th International Conference on Advances in the Stabilization and Controlled Degradation of Polymers. Luzern, Switzerland, 1989.

53. M Mucha, K Kryszewski. The effect of morphology on the thermal stability of isotactic polypropylene in air. Colloid Polym Sci 258:743–752, 1980.

54. TA Bogaevskaya, BA Gromov, VB Miller, TV Monakhova, YB Shlyapnikov. Effect of the supermolecular structure of polypropylene on its oxidation kinetics. Vysokomol Soed A 14:1552–1556, 1972.

55. AL Buchachenko. Specificity of the solid phase oxidation of polyolefins. J Polym Sci Symp 57:299–310, 1976.

56. P Richters. Initiation process in the oxidation of polypropylene. Macromolecules 3:262–265, 1970.

57. N Rapoport. Stress effects on polymer oxidation. Semin ENSAM, 1994.

58. M Inoue. Spherulitic crystallization and cracking during heat ageing of polypropylene. J Polym Sci 55:443–450, 1961.

59. M Celina, GA George, DJ Lacey, NC Billingham. Chemiluminescence imaging of the oxidation of polypropylene. Polym Degrad Stabil 47:353–356, 1995.

60. J Verdu. On the autoaccelerated character of the branched oxidation of polyolefins. Macromol Symp 115:165–181, 1997.

61. W Schnabel. Degradation by high energy radiation. In: HHG Jellinek, ed. Aspects of Degradation and Stabilization of Polymers. New York: Elsevier Scientific, 1978, pp 149–189.

62. S Girois, L Audouin, J Verdu, P Delprat, G Marot. Molecular weight changes during the photooxidation of isotactic polypropylene. Polym Degrad Stabil 51:125–132, 1996.

63. L Achimsky. Etude cinétique de la thermooxydation du polypropylène. Dissertation ENSAM, Paris, 1996.

64. M Celina, GA George. Heterogeneous and homogeneous kinetic analyses of the thermal oxidation of polypropylene. Polym Degrad Stabil 50:89–99, 1995.

65. M Celina, GA George, C Lerf, G Cash, D Weddel. A spreading model for the oxidation of polypropylene. Macromol Symp 115:69–93, 1997.

66. V Langlois, L Audouin, P Courtois, J Verdu. Thermooxidative ageing of crosslinked polyethylene: stabilizer consumption and lifetime prediction. Polym Degrad Stabil 40:399–409, 1993.

67. N Guarrotxena, L Audouin, J Verdu. Oxidation kinetic of poly(4-methylpentene). In: Proceedings of 18th International Conference on Advances in the Stabilization and Controlled Degradation of Polymers, Luzern, Switzerland, 1996.

68. S Girois, L Audouin, P Delprat, J Verdu. Weight loss mechanism in the photooxidation of polypropylene. Polym Degrad Stabil 51:133–134, 1996.

69. P Delprat, X Duteurtre, JL Gardette. Photooxidation of unstabilized and HALS-stabilized polyphasic ethylene–propylene polymers. Polym Degrad Stabil 50:1–12, 1995.

Index